The Human Genome

" The Complete Nucleic Acid Sequence for Human DNA "

Edited by Paul F. Kisak

Contents

Chapter 1

Human genome

For a non-technical introduction to the topic, see Introduction to genetics.

The **human genome** is the complete set of nucleic acid sequence for humans (*Homo sapiens*), encoded as DNA within the 23 chromosome pairs in cell nuclei and in a small DNA molecule found within individual mitochondria. Human genomes include both protein-coding DNA genes and noncoding DNA. Haploid human genomes, which are contained in germ cells (the egg and sperm gamete cells created in the meiosis phase of sexual reproduction before fertilization creates a zygote) consist of three billion DNA base pairs, while diploid genomes (found in somatic cells) have twice the DNA content. While there are significant differences among the genomes of human individuals (on the order of 0.1%),[1] these are considerably smaller than the differences between humans and their closest living relatives, the chimpanzees (approximately 4%[2]) and bonobos.

The Human Genome Project produced the first complete sequences of individual human genomes, with the first draft sequence and initial analysis being published on February 12, 2001.[3] The human genome was the first of all vertebrates to be completely sequenced. As of 2012, thousands of human genomes have been completely sequenced, and many more have been mapped at lower levels of resolution. The resulting data are used worldwide in biomedical science, anthropology, forensics and other branches of science. There is a widely held expectation that genomic studies will lead to advances in the diagnosis and treatment of diseases, and to new insights in many fields of biology, including human evolution.

Although the sequence of the human genome has been (almost) completely determined by DNA sequencing, it is not yet fully understood. Most (though probably not all) genes have been identified by a combination of high throughput experimental and bioinformatics approaches, yet much work still needs to be done to further elucidate the biological functions of their protein and RNA products. Recent results suggest that most of the vast quantities of noncoding DNA within the genome have associated biochemical activities, including regulation of gene expression, organization of chromosome architecture, and signals controlling epigenetic inheritance.

There are an estimated 20,000-25,000 human protein-coding genes. The estimate of the number of human genes has been repeatedly revised down from initial predictions of 100,000 or more as genome sequence quality and gene finding methods have improved, and could continue to drop further.[4][5] Protein-coding sequences account for only a very small fraction of the genome (approximately 1.5%), and the rest is associated with non-coding RNA molecules, regulatory DNA sequences, LINEs, SINEs, introns, and sequences for which as yet no function has been elucidated.[6]

1.1 Molecular organization and gene content

The total length of the human genome is over 3 billion base pairs. The genome is organized into 22 paired chromosomes, plus the X chromosome (one in males, two in females) and, in males only, one Y chromosome. These are all large linear DNA molecules contained within the cell nucleus. The genome also includes the mitochondrial DNA, a comparatively small circular molecule present in each mitochondrion. Basic information about these molecules and their gene content, based on a reference genome that does not represent the sequence of any specific individual, are provided in the following table. (Data source: Ensembl genome browser release 68, July 2012)

Table 1 (above) summarizes the physical organization and gene content of the human reference genome, with links to the original analysis, as published in the Ensembl database at the European Bioinformatics Institute (EBI) and Wellcome Trust Sanger Institute. Chromosome lengths were estimated by multiplying the number of base pairs by 0.34 nanometers, the distance between base pairs in the DNA double helix. The number of proteins is based on the number of initial precursor mRNA transcripts, and does not include products of alternative pre-mRNA splicing, or modifications to protein structure that occur after translation.

The number of variations is a summary of unique DNA sequence changes that have been identified within the sequences analyzed by Ensembl as of July, 2012; that number is expected to increase as further personal genomes are sequenced and examined. In addition to the gene content shown in this table, a large number of non-expressed functional sequences have been identified throughout the human genome (see below). Links open windows to the reference chromosome sequence in the EBI genome browser. The table also describes prevalence of genes encoding structural RNAs in the genome.

MiRNA, or MicroRNA, functions as a post-transcriptional regulator of gene expression. Ribosomal RNA, or rRNA, makes up the RNA portion of the ribosome and is critical in the synthesis of proteins. Small nuclear RNA, or snRNA, is found in the nucleus of the cell. Its primary function is in the processing of pre-mRNA molecules and also in the regulation of transcription factors. SnoRNA, or Small nucleolar RNA, primarily functions in guiding chemical modifications to other RNA molecules.

1.1.1   Completeness of the human genome sequence

Although the human genome has been completely sequenced for all practical purposes, there are still hundreds of gaps in the sequence. A recent study noted more than 160 euchromatic gaps of which 50 gaps were closed.[7] However, there are still numerous gaps in the heterochromatic parts of the genome which is much harder to sequence due to numerous repeats and other intractable sequence features.

1.2   Coding vs. noncoding DNA

The content of the human genome is commonly divided into coding and noncoding DNA sequences. Coding DNA is defined as those sequences that can be transcribed into mRNA and translated into proteins during the human life cycle; these sequences occupy only a small fraction of the genome (<2%). Noncoding DNA is made up of all of those sequences (ca. 98% of the genome) that are not used to encode proteins.

Some noncoding DNA contains genes for RNA molecules with important biological functions (noncoding RNA, for example ribosomal RNA and transfer RNA). The exploration of the function and evolutionary origin of noncoding DNA is an important goal of contemporary genome research, including the ENCODE (Encyclopedia of DNA Elements) project, which aims to survey the entire human genome, using a variety of experimental tools whose results are indicative of molecular activity.

Because non-coding DNA greatly outnumbers coding DNA, the concept of the sequenced genome has become a more focused analytical concept than the classical concept of the DNA-coding gene.[8][9]

1.3   Mutation Rate of Human Genome

Mutation rate of human genome is a very important factor in calculating evolutionary time points. Researchers calculated the number of genetic variations between human and apes. Dividing that number by age of fossil of most recent common ancestor of humans and ape, researchers calculated the mutation rate. Recent studies using next generation sequencing technologies concluded a slow mutation rate which doesn't add up with human migration pattern time points and suggesting a new evolutionary time scale.[10] 100,000 year old human fossil found in Israel threw more questions on human migration time points.[10]

1.4 Coding sequences (protein-coding genes)

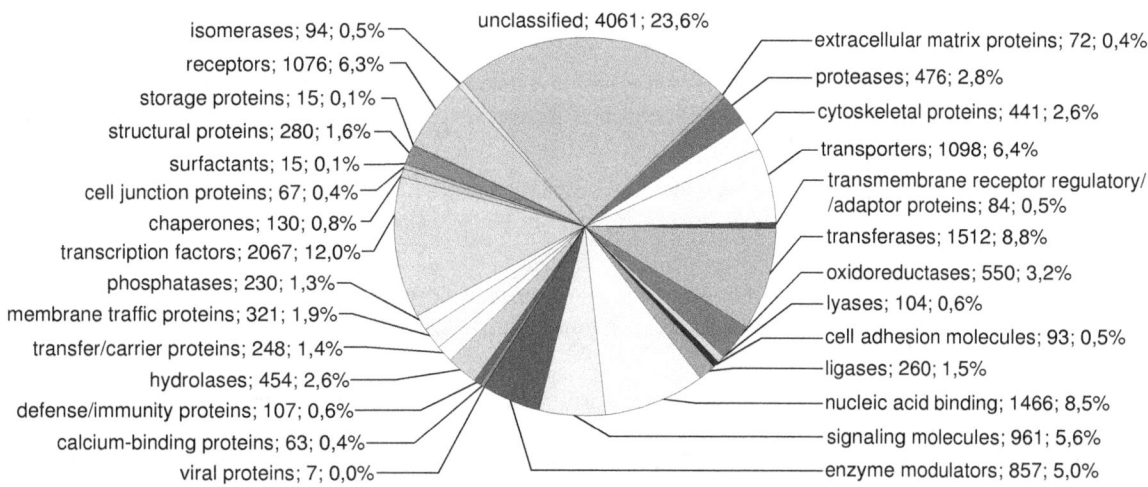

Human genes categorized by function of the transcribed proteins, given both as number of encoding genes and percentage of all genes.[11]

Protein-coding sequences represent the most widely studied and best understood component of the human genome. These sequences ultimately lead to the production of all human proteins, although several biological processes (e.g. DNA rearrangements and alternative pre-mRNA splicing) can lead to the production of many more unique proteins than the number of protein-coding genes.

The complete modular protein-coding capacity of the genome is contained within the exome, and consists of DNA sequences encoded by exons that can be translated into proteins. Because of its biological importance, and the fact that it constitutes less than 2% of the genome, sequencing of the exome was the first major milepost of the Human Genome Project.

Number of protein-coding genes. About 20,000 human proteins have been annotated in databases such as Uniprot.[12] Historically, estimates for the number of protein genes have varied widely, ranging up to 2,000,000 in the late 1960s,[13] but several researchers pointed out in the early 1970s that the estimated mutational load from deleterious mutations placed an upper limit of approximately 40,000 for the total number of functional loci (this includes protein-coding and functional non-coding genes).[14]

The number of human protein-coding genes is not significantly larger than that of many less complex organisms, such as the roundworm and the fruit fly. This difference may result from the extensive use of alternative pre-mRNA splicing in humans, which provides the ability to build a very large number of modular proteins through the selective incorporation of exons.

Protein-coding capacity per chromosome. Protein-coding genes are distributed unevenly across the chromosomes, ranging from a few dozen to more than 2000, with an especially high gene density within chromosomes 19, 11, and 1 (Table 1). Each chromosome contains various gene-rich and gene-poor regions, which may be correlated with chromosome bands and GC-content . The significance of these nonrandom patterns of gene density is not well understood.[15]

Size of protein-coding genes. The size of protein-coding genes within the human genome shows enormous variability (Table 2). For example, the gene for histone H1a (HIST1HIA) is relatively small and simple, lacking introns and encoding mRNA sequences of 781 nt and a 215 amino acid protein (648 nt open reading frame). Dystrophin (DMD) is the largest protein-coding gene in the human reference genome, spanning a total of 2.2 MB, while Titin (TTN) has the longest coding sequence (80,780 bp), the largest number of exons (364), and the longest single exon (17,106 bp). Over the whole genome, the median size of an exon is 122 bp (mean = 145 bp), the median number of exons is 7 (mean = 8.8), and the median coding sequence encodes 367 amino acids (mean = 447 amino acids; Table 21 in[6]).

Table 2. Examples of human protein-coding genes. Chrom, chromosome. Alt splicing, alternative pre-mRNA splicing. (Data source: Ensembl genome browser release 68, July 2012)

1.5   Noncoding DNA (ncDNA)

Main article: Noncoding DNA

Noncoding DNA is defined as all of the DNA sequences within a genome that are not found within protein-coding exons, and so are never represented within the amino acid sequence of expressed proteins. By this definition, more than 98% of the human genomes is composed of ncDNA.

Numerous classes of noncoding DNA have been identified, including genes for noncoding RNA (e.g. tRNA and rRNA), pseudogenes, introns, untranslated regions of mRNA, regulatory DNA sequences, repetitive DNA sequences, and sequences related to mobile genetic elements.

Numerous sequences that are included within genes are also defined as noncoding DNA. These include genes for noncoding RNA (e.g. tRNA, rRNA), and untranslated components of protein-coding genes (e.g. introns, and 5' and 3' untranslated regions of mRNA).

Protein-coding sequences (specifically, coding exons) constitute less than 1.5% of the human genome.[6] In addition, about 26% of the human genome is introns.[16] Aside from genes (exons and introns) and known regulatory sequences (8–20%), the human genome contains regions of noncoding DNA. The exact amount of noncoding DNA that plays a role in cell physiology has been hotly debated. Recent analysis by the ENCODE project indicates that 80% of the entire human genome is either transcribed, binds to regulatory proteins, or is associated with some other biochemical activity.[5]

It however remains controversial whether all of this biochemical activity contributes to cell physiology, or whether a substantial portion of this is the result transcriptional and biochemical noise, which must be actively filtered out by the organism.[17] Excluding protein-coding sequences, introns, and regulatory regions, much of the non-coding DNA is composed of: Many DNA sequences that do not play a role in gene expression have important biological functions. Comparative genomics studies indicate that about 5% of the genome contains sequences of noncoding DNA that are highly conserved, sometimes on time-scales representing hundreds of millions of years, implying that these noncoding regions are under strong evolutionary pressure and positive selection.[18]

Many of these sequences regulate the structure of chromosomes by limiting the regions of heterochromatin formation and regulating structural features of the chromosomes, such as the telomeres and centromeres. Other noncoding regions serve as origins of DNA replication. Finally several regions are transcribed into functional noncoding RNA that regulate the expression of protein-coding genes (for example[19]), mRNA translation and stability (see miRNA), chromatin structure (including histone modifications, for example[20]), DNA methylation (for example[21]), DNA recombination (for example[22]), and cross-regulate other noncoding RNAs (for example[23]). It is also likely that many transcribed noncoding regions do not serve any role and that this transcription is the product of non-specific RNA Polymerase activity.[17]

1.5.1   Pseudogenes

Main article: Pseudogenes

Pseudogenes are inactive copies of protein-coding genes, often generated by gene duplication, that have become nonfunctional through the accumulation of inactivating mutations. **Table 1** shows that the number of pseudogenes in the human genome is on the order of 13,000,[24] and in some chromosomes is nearly the same as the number of functional protein-coding genes. Gene duplication is a major mechanism through which new genetic material is generated during molecular evolution.

For example, the olfactory receptor gene family is one of the best-documented examples of pseudogenes in the human genome. More than 60 percent of the genes in this family are non-functional pseudogenes in humans. By comparison, only 20 percent of genes in the mouse olfactory receptor gene family are pseudogenes. Research suggests that this is a species-specific characteristic, as the most closely related primates all have proportionally fewer pseudogenes. This genetic discovery helps to explain the less acute sense of smell in humans relative to other mammals.[25]

1.5.2 Genes for noncoding RNA (ncRNA)

Main article: Noncoding RNA

Noncoding RNA molecules play many essential roles in cells, especially in the many reactions of protein synthesis and RNA processing. ncRNAs include tRNA, ribosomal RNA, microRNA, snRNA and other non-coding RNA genes including about 60,000 long non coding RNAs (lncRNAs).[5][26][27][28] It should be noted that while the number of reported lncRNA genes continues to rise and the exact number in the human genome is yet to be defined, many of them are argued to be non-functional.[29]

Many ncRNAs are critical elements in gene regulation and expression. Noncoding RNA also contributes to epigenetics, transcription, RNA splicing, and the translational machinery. The role of RNA in genetic regulation and disease offers a new potential level of unexplored genomic complexity.[30]

1.5.3 Introns and untranslated regions of mRNA

In addition to the ncRNA molecules that are encoded by discrete genes, the initial transcripts of protein coding genes usually contain extensive noncoding sequences, in the form of introns, 5'-untranslated regions (5'-UTR), and 3'-untranslated regions (3'-UTR). Within most protein-coding genes of the human genome, the length of intron sequences is 10- to 100-times the length of exon sequences (Table 2).

1.5.4 Regulatory DNA sequences

The human genome has many different regulatory sequences which are crucial to controlling gene expression. Conservative estimates indicate that these sequences make up 8% of the genome,[31] however extrapolations from the ENCODE project give that 20[32]–40%[33] of the genome is gene regulatory sequence. Some types of non-coding DNA are genetic "switches" that do not encode proteins, but do regulate when and where genes are expressed (called enhancers).[34]

Regulatory sequences have been known since the late 1960s.[35] The first identification of regulatory sequences in the human genome relied on recombinant DNA technology.[36] Later with the advent of genomic sequencing, the identification of these sequences could be inferred by evolutionary conservation. The evolutionary branch between the primates and mouse, for example, occurred 70–90 million years ago.[37] So computer comparisons of gene sequences that identify conserved non-coding sequences will be an indication of their importance in duties such as gene regulation.[38]

Other genomes have been sequenced with the same intention of aiding conservation-guided methods, for exampled the pufferfish genome.[39] However, regulatory sequences disappear and re-evolve during evolution at a high rate.[40][41][42]

As of 2012, the efforts have shifted toward finding interactions between DNA and regulatory proteins by the technique ChIP-Seq, or gaps where the DNA is not packaged by histones (DNase hypersensitive sites), both of which tell where there are active regulatory sequences in the investigated cell type.[31]

1.5.5 Repetitive DNA sequences

Repetitive DNA sequences comprise approximately 50% of the human genome.[43]

About 8% of the human genome consists of tandem DNA arrays or tandem repeats, low complexity repeat sequences that have multiple adjacent copies (e.g. "CAGCAGCAG..."). The tandem sequences may be of variable lengths, from two nucleotides to tens of nucleotides. These sequences are highly variable, even among closely related individuals, and so are used for genealogical DNA testing and forensic DNA analysis.

Repeated sequences of fewer than ten nucleotides (e.g. the dinucleotide repeat $(AC)_n$) are termed microsatellite sequences. Among the microsatellite sequences, trinucleotide repeats are of particular importance, as sometimes occur within coding regions of genes for proteins and may lead to genetic disorders. For example, Huntington's disease results from an expansion of the trinucleotide repeat $(CAG)_n$ within the *Huntingtin* gene on human chromosome 4. Telomeres (the ends of linear chromosomes) end with a microsatellite hexanucleotide repeat of the sequence $(TTAGGG)_n$.

Tandem repeats of longer sequences (arrays of repeated sequences 10–60 nucleotides long) are termed minisatellites.

1.5.6   Mobile genetic elements (transposons) and their relics

Transposable genetic elements, DNA sequences that can replicate and insert copies of themselves at other locations within a host genome, are an abundant component in the human genome. The most abundant transposon lineage, *Alu*, has about 50,000 active copies,[44] and can be inserted into intragenic and intergenic regions.[45] One other lineage, LINE-1, has about 100 active copies per genome (the number varies between people).[46] Together with non-functional relics of old transposons, they account for over half of total human DNA.[47] Sometimes called "jumping genes", transposons have played a major role in sculpting the human genome. Some of these sequences represent endogenous retroviruses, DNA copies of viral sequences that have become permanently integrated into the genome and are now passed on to succeeding generations.

Mobile elements within the human genome can be classified into LTR retrotransposons (8.3% of total genome), SINEs (13.1% of total genome) including Alu elements, LINEs (20.4% of total genome), SVAs and Class II DNA transposons (2.9% of total genome).

1.6   Genomic variation in humans

Main articles: Human genetic variation and Human genetic clustering

1.6.1   Human Reference Genome

With the exception of identical twins, all humans show significant variation in genomic DNA sequences. The Human Reference Genome (HRG) is used as a standard sequence reference.

There are several important points concerning the Human Reference Genome--

- The HRG is a haploid sequence. Each chromosome is represented once.

- The HRG is a composite sequence, and does not correspond to any actual human individual.

- The HRG is periodically updated to correct errors and ambiguities.

- The HRG in no way represents an "ideal" or "perfect" human individual. It is simply a standardized representation or model that is used for comparative purposes.

1.6.2   Measuring human genetic variation

Most studies of human genetic variation have focused on single-nucleotide polymorphisms (SNPs), which are substitutions in individual bases along a chromosome. Most analyses estimate that SNPs occur 1 in 1000 base pairs, on average, in the euchromatic human genome, although they do not occur at a uniform density. Thus follows the popular statement that "we are all, regardless of race, genetically 99.9% the same",[48] although this would be somewhat qualified by most geneticists. For example, a much larger fraction of the genome is now thought to be involved in copy number variation.[49] A large-scale collaborative effort to catalog SNP variations in the human genome is being undertaken by the International HapMap Project.

The genomic loci and length of certain types of small repetitive sequences are highly variable from person to person, which is the basis of DNA fingerprinting and DNA paternity testing technologies. The heterochromatic portions of the human genome, which total several hundred million base pairs, are also thought to be quite variable within the human population (they are so repetitive and so long that they cannot be accurately sequenced with current technology). These regions contain few genes, and it is unclear whether any significant phenotypic effect results from typical variation in repeats or heterochromatin.

Most gross genomic mutations in gamete germ cells probably result in inviable embryos; however, a number of human diseases are related to large-scale genomic abnormalities. Down syndrome, Turner Syndrome, and a number of other diseases result from nondisjunction of entire chromosomes. Cancer cells frequently have aneuploidy of chromosomes and chromosome arms, although a cause and effect relationship between aneuploidy and cancer has not been established.

Mapping human genomic variation

Whereas a genome sequence lists the order of every DNA base in a genome, a genome map identifies the landmarks. A genome map is less detailed than a genome sequence and aids in navigating around the genome.[50][51]

An example of a variation map is the HapMap being developed by the International HapMap Project. The HapMap is a haplotype map of the human genome, "which will describe the common patterns of human DNA sequence variation."[52] It catalogs the patterns of small-scale variations in the genome that involve single DNA letters, or bases.

Researchers published the first sequence-based map of large-scale structural variation across the human genome in the journal *Nature* in May 2008.[53][54] Large-scale structural variations are differences in the genome among people that range from a few thousand to a few million DNA bases; some are gains or losses of stretches of genome sequence and others appear as re-arrangements of stretches of sequence. These variations include differences in the number of copies individuals have of a particular gene, deletions, translocations and inversions.

1.6.3 SNP Frequency across the Human Genome

Single-nucleotide polymorphisms (SNPs) do not occur homogeneously across the human genome. In fact, there is enormous diversity in SNP frequency between genes, reflecting different selective pressures on each gene as well as different mutation and recombination rates across the genome. However, studies on SNPs are biased towards coding regions, the data generated from them are unlikely to reflect the overall distribution of SNPs throughout the genome. Therefore, the SNP consortium protocol was designed to identify SNPs with no bias towards coding regions and the 100000 TSC SNPs generally reflect sequence diversity across the human chromosomes.[55]

Changes in **non-coding sequence** and synonymous changes in **coding sequence** are generally more common than non-synonymous changes, reflecting greater selective pressure reducing diversity at positions dictating amino acid identity. Transitional changes are more common than transversions, with CpG dinucleotides showing the highest mutation rate, presumably due to deamination.

1.6.4 Personal genomes

See also: Personal genomics

A personal genome sequence is a (nearly) complete sequence of the chemical base pairs that make up the DNA of a single person. Because medical treatments have different effects on different people due to genetic variations such as single-nucleotide polymorphisms (SNPs), the analysis of personal genomes may lead to personalized medical treatment based on individual genotypes.

The first personal genome sequence to be determined was that of Craig Venter in 2007. Personal genomes had not been sequenced in the public Human Genome Project to protect the identity of volunteers who provided DNA samples. That sequence was derived from the DNA of several volunteers from a diverse population.[56] However, early in the Venter-led Celera Genomics genome sequencing effort the decision was made to switch from sequencing a composite sample to using DNA from a single individual, later revealed to have been Venter himself. Thus the Celera human genome sequence released in 2000 was largely that of one man. Subsequent replacement of the early composite-derived data and determination of the diploid sequence, representing both sets of chromosomes, rather than a haploid sequence originally reported, allowed the release of the first personal genome.[57] In April 2008, that of James Watson was also completed.

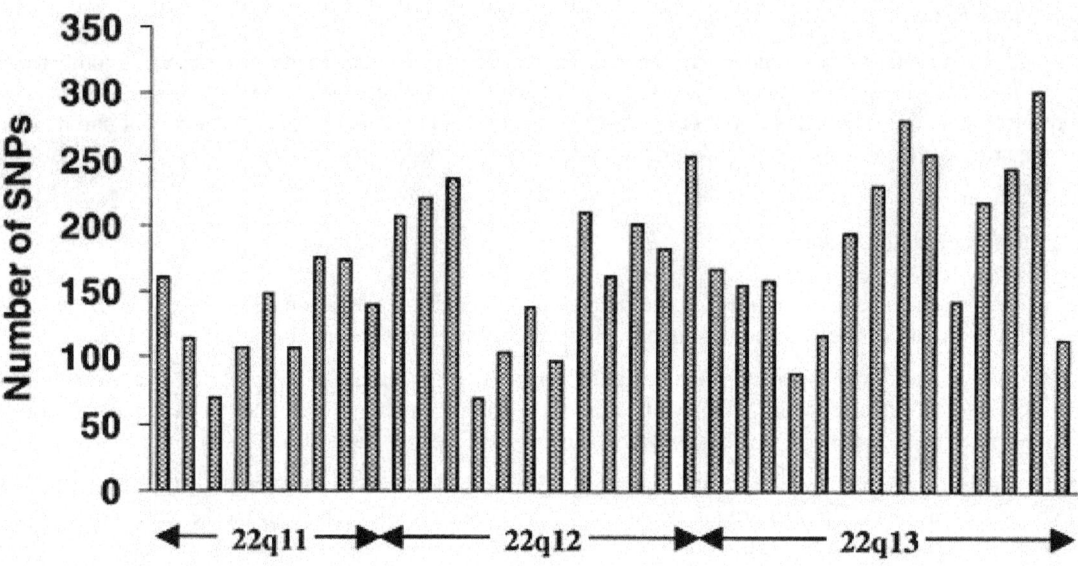

TSC SNP distribution along the long arm of chromosome 22 (taken from the TSC website at http://snp.cshl.org/). Each column represents a 1 Mb interval; the approximate cytogenetic position is given on the x-axis. Clear peaks and troughs of SNP density can be seen, possibly reflecting different rates of mutation, recombination and selection.

Since then hundreds of personal genome sequences have been released,[58] including those of Desmond Tutu,[59][60] and of a Paleo-Eskimo.[61] In November 2013, a Spanish family made their personal genomics data obtained by direct-to-consumer genetic testing with 23andMe publicly available under a Creative Commons public domain license. This is believed to be the first such public genomics dataset for a whole family.[62]

The sequencing of individual genomes further unveiled levels of genetic complexity that had not been appreciated before. Personal genomics helped reveal the significant level of diversity in the human genome attributed not only to SNPs but structural variations as well. However, the application of such knowledge to the treatment of disease and in the medical field is only in its very beginnings.[63] Exome sequencing has become increasingly popular as a tool to aid in diagnosis of genetic disease because the exome contributes only 1% of the genomic sequence but accounts for roughly 85% of mutations that contribute significantly to disease.[64]

1.7   Human genetic disorders

For more details on this topic, see Genetic disorder.

Most aspects of human biology involve both genetic (inherited) and non-genetic (environmental) factors. Some inherited variation influences aspects of our biology that are not medical in nature (height, eye color, ability to taste or smell certain compounds, etc.). Moreover, some genetic disorders only cause disease in combination with the appropriate environmental factors (such as diet). With these caveats, genetic disorders may be described as clinically defined diseases caused by genomic DNA sequence variation. In the most straightforward cases, the disorder can be associated with variation in a single gene. For example, cystic fibrosis is caused by mutations in the CFTR gene, and is the most common recessive disorder in caucasian populations with over 1,300 different mutations known.[65]

Disease-causing mutations in specific genes are usually severe in terms of gene function, and are fortunately rare, thus genetic disorders are similarly individually rare. However, since there are many genes that can vary to cause genetic disorders, in aggregate they constitute a significant component of known medical conditions, especially in pediatric medicine. Molecularly characterized genetic disorders are those for which the underlying causal gene has been identified, currently there are approximately 2,200 such disorders annotated in the OMIM database.[65]

Studies of genetic disorders are often performed by means of family-based studies. In some instances population based approaches are employed, particularly in the case of so-called founder populations such as those in Finland, French-Canada, Utah, Sardinia, etc. Diagnosis and treatment of genetic disorders are usually performed by a geneticist-physician trained in clinical/medical genetics. The results of the Human Genome Project are likely to provide increased availability of genetic testing for gene-related disorders, and eventually improved treatment. Parents can be screened for hereditary conditions and counselled on the consequences, the probability it will be inherited, and how to avoid or ameliorate it in their offspring.

As noted above, there are many different kinds of DNA sequence variation, ranging from complete extra or missing chromosomes down to single nucleotide changes. It is generally presumed that much naturally occurring genetic variation in human populations is phenotypically neutral, i.e. has little or no detectable effect on the physiology of the individual (although there may be fractional differences in fitness defined over evolutionary time frames). Genetic disorders can be caused by any or all known types of sequence variation. To molecularly characterize a new genetic disorder, it is necessary to establish a causal link between a particular genomic sequence variant and the clinical disease under investigation. Such studies constitute the realm of human molecular genetics.

With the advent of the Human Genome and International HapMap Project, it has become feasible to explore subtle genetic influences on many common disease conditions such as diabetes, asthma, migraine, schizophrenia, etc. Although some causal links have been made between genomic sequence variants in particular genes and some of these diseases, often with much publicity in the general media, these are usually not considered to be genetic disorders *per se* as their causes are complex, involving many different genetic and environmental factors. Thus there may be disagreement in particular cases whether a specific medical condition should be termed a genetic disorder. The categorized table below provides the prevalence as well as the genes or chromosomes associated with some human genetic disorders.

1.8   Evolution

See also: Human evolution and Chimpanzee Genome Project

Comparative genomics studies of mammalian genomes suggest that approximately 5% of the human genome has been conserved by evolution since the divergence of extant lineages approximately 200 million years ago, containing the vast majority of genes.[67][68] The published chimpanzee genome differs from that of the human genome by 1.23% in direct sequence comparisons.[69] Around 20% of this figure is accounted for by variation within each species, leaving only ~1.06% consistent sequence divergence between humans and chimps at shared genes.[70] This nucleotide by nucleotide difference is dwarfed, however, by the portion of each genome that is not shared, including around 6% of functional genes that are unique to either humans or chimps.[71]

In other words, the considerable observable differences between humans and chimps may be due as much or more to genome level variation in the number, function and expression of genes rather than DNA sequence changes in shared genes. Indeed, even within humans, there has been found to be a previously unappreciated amount of copy number variation (CNV) which can make up as much as 5 – 15% of the human genome. In other words, between humans, there could be +/- 500,000,000 base pairs of DNA, some being active genes, others inactivated, or active at different levels. The full significance of this finding remains to be seen. On average, a typical human protein-coding gene differs from its chimpanzee ortholog by only two amino acid substitutions; nearly one third of human genes have exactly the same protein translation as their chimpanzee orthologs. A major difference between the two genomes is human chromosome 2, which is equivalent to a fusion product of chimpanzee chromosomes 12 and 13[72] (later renamed to chromosomes 2A and 2B, respectively).

Humans have undergone an extraordinary loss of olfactory receptor genes during our recent evolution, which explains our relatively crude sense of smell compared to most other mammals. Evolutionary evidence suggests that the emergence of color vision in humans and several other primate species has diminished the need for the sense of smell.[73]

1.9   Mitochondrial DNA

The human mitochondrial DNA is of tremendous interest to geneticists, since it undoubtedly plays a role in mitochondrial disease. It also sheds light on human evolution; for example, analysis of variation in the human mitochondrial genome has led to the postulation of a recent common ancestor for all humans on the maternal line of descent (see Mitochondrial Eve).

Due to the lack of a system for checking for copying errors, mitochondrial DNA (mtDNA) has a more rapid rate of variation than nuclear DNA. This 20-fold increase in the mutation rate allows mtDNA to be used for more accurate tracing of maternal ancestry. Studies of mtDNA in populations have allowed ancient migration paths to be traced, such as the migration of Native Americans from Siberia or Polynesians from southeastern Asia. It has also been used to show that there is no trace of Neanderthal DNA in the European gene mixture inherited through purely maternal lineage.[74] Due to the restrictive all or none manner of mtDNA inheritance, this result (no trace of Neanderthal mtDNA) would be likely unless there were a large percentage of Neanderthal ancestry, or there was strong positive selection for that mtDNA (for example, going back 5 generations, only 1 of your 32 ancestors contributed to your mtDNA, so if one of these 32 was pure Neanderthal you would expect that ~3% of your autosomal DNA would be of Neanderthal origin, yet you would have a ~97% chance to have no trace of Neanderthal mtDNA).

1.10   Epigenome

See also: Epigenetics

Epigenetics describes a variety of features of the human genome that transcend its primary DNA sequence, such as chromatin packaging, histone modifications and DNA methylation, and which are important in regulating gene expression, genome replication and other cellular processes. Epigenetic markers strengthen and weaken transcription of certain genes but do not affect the actual sequence of DNA nucleotides. DNA methylation is a major form of epigenetic control over gene expression and one of the most highly studied topics in epigenetics. During development, the human DNA methylation profile experiences dramatic changes. In early germ line cells, the genome has very low methylation levels. These low levels generally describe active genes. As development progresses, parental imprinting tags lead to increased methylation activity.[75][76]

Epigenetic patterns can be identified between tissues within an individual as well as between individuals themselves. Identical genes that have differences only in their epigenetic state are called **epialleles**. Epialleles can be placed into three categories: those directly determined by an individual's genotype, those influenced by genotype, and those entirely independent of genotype. The epigenome is also influenced significantly by environmental factors. Diet, toxins, and hormones impact the epigenetic state. Studies in dietary manipulation have demonstrated that methyl-deficient diets are associated with hypomethylation of the epigenome. Such studies establish epigenetics as an important interface between the environment and the genome.[77]

1.11   See also

- Genetics

- Genomics

- Genographic Project

- Genomic organization

- INTS7 (gene)

- Noncoding DNA

- Whole genome sequencing

- Universal Declaration on the Human Genome and Human Rights

1.12 References

[1] Abecasis GR, Auton A, Brooks LD, DePristo MA, Durbin RM, Handsaker RE, Kang HM, Marth GT, McVean GA (Nov 2012). "An integrated map of genetic variation from 1,092 human genomes". *Nature* **491** (7422): 56–65. doi:10.1038/nature11632. PMID 23128226.

[2] Varki A, Altheide TK (Dec 2005). "Comparing the human and chimpanzee genomes: searching for needles in a haystack". *Genome Research* **15** (12): 1746–58. doi:10.1101/gr.3737405. PMID 16339373.

[3] International Human Genome Sequencing Consortium Publishes Sequence and Analysis of the Human Genome

[4] International Human Genome Sequencing Consortium (Oct 2004). "Finishing the euchromatic sequence of the human genome". *Nature* **431** (7011): 931–45. Bibcode:2004Natur.431..931H. doi:10.1038/nature03001. PMID 15496913.

[5] Pennisi E (Sep 2012). "Genomics. ENCODE project writes eulogy for junk DNA". *Science* **337** (6099): 1159, 1161. doi:10.1126/science.337.6099.1159. PMID 22955811.

[6] International Human Genome Sequencing Consortium (Feb 2001). "Initial sequencing and analysis of the human genome". *Nature* **409** (6822): 860–921. doi:10.1038/35057062. PMID 11237011.

[7] Chaisson MJ, Huddleston J, Dennis MY, Sudmant PH, Malig M, Hormozdiari F, Antonacci F, Surti U, Sandstrom R, Boitano M, Landolin JM, Stamatoyannopoulos JA, Hunkapiller MW, Korlach J, Eichler EE (Jan 2015). "Resolving the complexity of the human genome using single-molecule sequencing". *Nature* **517** (7536): 608–11. doi:10.1038/nature13907. PMID 25383537.

[8] Ken Waters (2007-03-07). "Molecular Genetics". Stanford Encyclopedia of Philosophy. Retrieved 2013-07-18.

[9] Lisa Gannett (2008-10-26). "The Human Genome Project". Stanford Encyclopedia of Philosophy. Retrieved 2013-07-18.

[10] Callaway E (2012). "Studies slow the human DNA clock". *Nature* **489** (7416): 343–4. doi:10.1038/489343a. PMID 22996522.

[11] PANTHER Pie Chart at the PANTHER Classification System homepage. Retrieved May 25, 2011

[12] List of human proteins in the Uniprot Human reference proteome; accessed 28 Jan 2015

[13] Kauffman SA (Mar 1969). "Metabolic stability and epigenesis in randomly constructed genetic nets". *Journal of Theoretical Biology* (Elsevier) **22** (3): 437–67. doi:10.1016/0022-5193(69)90015-0. PMID 5803332.

[14] Ohno, S. (1972). "An argument for the genetic simplicity of man and other mammals". *Journal of Human Evolution* **1** (6): 651–662. doi:10.1016/0047-2484(72)90011-5.

[15] M. Huang, H. Zhu, B. Shen, G. Gao, "A non-random gait through the human genome", *3rd International Conference on Bioinformatics and Biomedical Engineering* (UCBBE, 2009), 1–3

[16] Gregory TR (Sep 2005). "Synergy between sequence and size in large-scale genomics". *Nature Reviews. Genetics* **6** (9): 699–708. doi:10.1038/nrg1674. PMID 16151375.

[17] Palazzo AF, Akef A (Jun 2012). "Nuclear export as a key arbiter of "mRNA identity" in eukaryotes". *Biochimica Et Biophysica Acta* **1819** (6): 566–77. doi:10.1016/j.bbagrm.2011.12.012. PMID 22248619.

[18] Ludwig MZ (Dec 2002). "Functional evolution of noncoding DNA". *Current Opinion in Genetics & Development* **12** (6): 634–9. doi:10.1016/S0959-437X(02)00355-6. PMID 12433575.

[19] Martens JA, Laprade L, Winston F (Jun 2004). "Intergenic transcription is required to repress the Saccharomyces cerevisiae SER3 gene". *Nature* **429** (6991): 571–4. Bibcode:2004Natur.429..571M. doi:10.1038/nature02538. PMID 15175754.

[20] Tsai MC, Manor O, Wan Y, Mosammaparast N, Wang JK, Lan F, Shi Y, Segal E, Chang HY (Aug 2010). "Long noncoding RNA as modular scaffold of histone modification complexes". *Science* **329** (5992): 689–93. Bibcode:2010Sci...329..689T. doi:10.1126/science.1192002. PMC 2967777. PMID 20616235.

[21] Bartolomei MS, Zemel S, Tilghman SM (May 1991). "Parental imprinting of the mouse H19 gene". *Nature* **351** (6322): 153–5. Bibcode:1991Natur.351..153B. doi:10.1038/351153a0. PMID 1709450.

[22] Kobayashi T, Ganley AR (Sep 2005). "Recombination regulation by transcription-induced cohesin dissociation in rDNA repeats". *Science* **309** (5740): 1581–4. Bibcode:2005Sci...309.1581K. doi:10.1126/science.1116102. PMID 16141077.

[23] Salmena L, Poliseno L, Tay Y, Kats L, Pandolfi PP (Aug 2011). "A ceRNA hypothesis: the Rosetta Stone of a hidden RNA language?". *Cell* **146** (3): 353–8. doi:10.1016/j.cell.2011.07.014. PMC 3235919. PMID 21802130.

[24] Pei B, Sisu C, Frankish A, Howald C, Habegger L, Mu XJ, Harte R, Balasubramanian S, Tanzer A, Diekhans M, Reymond A, Hubbard TJ, Harrow J, Gerstein MB (2012). "The GENCODE pseudogene resource". *Genome Biology* **13** (9): R51. doi:10.1186/gb-2012-13-9-r51. PMC 3491395. PMID 22951037.

[25] Gilad Y, Man O, Pääbo S, Lancet D (Mar 2003). "Human specific loss of olfactory receptor genes". *Proceedings of the National Academy of Sciences of the United States of America* **100** (6): 3324–7. Bibcode:2003PNAS..100.3324G. doi:10.1073/pnas.0535 697100.PMC152291. PMID12612342.

[26] Iyer MK, Niknafs YS, Malik R, Singhal U, Sahu A, Hosono Y, Barrette TR, Prensner JR, Evans JR, Zhao S, Poliakov A, Cao X, Dhanasekaran SM, Wu YM, Robinson DR, Beer DG, Feng FY, Iyer HK, Chinnaiyan AM (Mar 2015). "The landscape of long noncoding RNAs in the human transcriptome". *Nature Genetics* **47** (3): 199–208. doi:10.1038/ng.3192. PMID 25599403.

[27] Eddy SR (Dec 2001). "Non-coding RNA genes and the modern RNA world". *Nature Reviews. Genetics* (Nature Publishing Group) **2** (12): 919–29. doi:10.1038/35103511. PMID 11733745.

[28] Managadze D, Lobkovsky AE, Wolf YI, Shabalina SA, Rogozin IB, Koonin EV (2013). "The vast, conserved mammalian lincRNome". *PLoS Computational Biology* **9** (2): e1002917. doi:10.1371/journal.pcbi.1002917. PMID 23468607.

[29] Palazzo AF, Lee ES (2015). "Non-coding RNA: what is functional and what is junk?". *Frontiers in Genetics* **6**: 2. doi:10.3389/f PMID 25674102. gene.2015.00002.

[30] Mattick JS, Makunin IV (Apr 2006). "Non-coding RNA". *Human Molecular Genetics.* 15 Spec No 1: R17–29. doi:10.1093/. PMID 16651366.

[31] Bernstein BE, Birney E, Dunham I, Green ED, Gunter C, Snyder M (Sep 2012). "An integrated encyclopedia of DNA elements in the human genome". *Nature* **489** (7414): 57–74. doi:10.1038/nature11247. PMC 3439153. PMID 22955616.

[32] Birney E (5 September 2012). "ENCODE: My own thoughts". *Ewan's Blog: Bioinformatician at large.*

[33] Stamatoyannopoulos JA (Sep 2012). "What does our genome encode?". *Genome Research* **22**(9): 1602–11. doi:10.1101/gr.14 PMC 3431477. PMID 22955972.

[34] Carroll SB, Gompel N, Prudhomme B (May 2008). "Regulating Evolution". *Scientific American*: 60–67.

[35] Miller JH, Ippen K, Scaife JG, Beckwith JR (1968). "The promoter-operator region of the lac operon of Escherichia coli". *J. Mol. Biol.* **38** (3): 413–20. doi:10.1016/0022-2836(68)90395-1. PMID 4887877.

[36] Wright S, Rosenthal A, Flavell R, Grosveld F (1984). "DNA sequences required for regulated expression of beta-globin genes in murine erythroleukemia cells". *Cell* **38** (1): 265–73. doi:10.1016/0092-8674(84)90548-8. PMID 6088069.

[37] Nei M, Xu P, Glazko G (Feb 2001). "Estimation of divergence times from multiprotein sequences for a few mammalian species and several distantly related organisms". *Proceedings of the National Academy of Sciences of the United States of America* **98** (5): 2497–502. Bibcode:2001PNAS...98.2497N. doi:10.1073/pnas.051611498. PMC 30166. PMID 11226267.

[38] Loots GG, Locksley RM, Blankespoor CM, Wang ZE, Miller W, Rubin EM, Frazer KA (Apr 2000). "Identification of a coordinate regulator of interleukins 4, 13, and 5 by cross-species sequence comparisons". *Science* **288** (5463): 136–40. Bibcode:2000Sci...288..136L. doi:10.1126/science.288.5463.136. PMID 10753117. Summary

[39] Meunier M. "Genoscope and Whitehead announce a high sequence coverage of the Tetraodon nigroviridis genome". Genoscope. Archived from the original on 16 October 2006. Retrieved 2006-09-12.

[40] Romero IG, Ruvinsky I, Gilad Y (Jul 2012). "Comparative studies of gene expression and the evolution of gene regulation". *Nature Reviews. Genetics* **13** (7): 505–16. doi:10.1038/nrg3229. PMID 22705669.

[41] Schmidt D, Wilson MD, Ballester B, Schwalie PC, Brown GD, Marshall A, Kutter C, Watt S, Martinez-Jimenez CP, Mackay S, Talianidis I, Flicek P, Odom DT (May 2010). "Five-vertebrate ChIP-seq reveals the evolutionary dynamics of transcription factor binding". *Science* **328** (5981): 1036–40. doi:10.1126/science.1186176. PMC 3008766. PMID 20378774.

[42] Wilson MD, Barbosa-Morais NL, Schmidt D, Conboy CM, Vanes L, Tybulewicz VL, Fisher EM, Tavaré S, Odom DT (Oct 2008). "Species-specific transcription in mice carrying human chromosome 21". *Science* **322** (5900): 434–8. doi:10.1126/science.1160930.PMC3717767. PMID 18787134.

[43] Treangen TJ, Salzberg SL (Jan 2012). "Repetitive DNA and next-generation sequencing: computational challenges and solutions". *Nature Reviews. Genetics* **13** (1): 36–46. doi:10.1038/nrg3117. PMC 3324860. PMID 22124482.

[44] Bennett EA, Keller H, Mills RE, Schmidt S, Moran JV, Weichenrieder O, Devine SE (Dec 2008). "Active Alu retrotransposons in the human genome". *Genome Research* **18** (12): 1875–83. doi:10.1101/gr.081737.108. PMC 2593586. PMID 18836035.

[45] Liang KH, Yeh CT. "A gene expression restriction network mediated by sense and antisense Alu sequences located on protein-coding messenger RNAs". *BMC Genomics* **14**: 325. doi:10.1186/1471-2164-14-325. PMC 3655826. PMID 23663499.

[46] Brouha B, Schustak J, Badge RM, Lutz-Prigge S, Farley AH, Moran JV, Kazazian HH (Apr 2003). "Hot L1s account for the bulk of retrotransposition in the human population". *Proceedings of the National Academy of Sciences of the United States of America* **100** (9): 5280–5. doi:10.1073/pnas.0831042100. PMC 154336. PMID 12682288.

[47] Barton NH, Briggs DE, Eisen JA, Goldstein DB, Patel NH (2007). *Evolution.* Cold Spring Harbor, NY: Cold Spring Harbor Laboratory Press. ISBN 0-87969-684-2.

[48] from Bill Clinton's 2000 State of the Union address

[49] Nature. "Global variation in copy number in the human genome : Article : Nature". Nature. Retrieved 2009-08-09.

[50] "What's a Genome?". Genomenewsnetwork.org. 2003-01-15. Retrieved 2009-05-31.

[51] NCBI_user_services (2004-03-29). "Mapping Factsheet". Ncbi.nlm.nih.gov. Retrieved 2009-05-31.

[52] "About the Project". HapMap. Retrieved 2009-05-31.

[53] "2008 Release: Researchers Produce First Sequence Map of Large-Scale Structural Variation in the Human Genome". genome.gov. Retrieved 2009-05-31.

[54] Kidd JM, Cooper GM, Donahue WF, Hayden HS, Sampas N, Graves T, et al. (May 2008). "Mapping and sequencing of structural variation from eight human genomes". *Nature* **453** (7191): 56–64. doi:10.1038/nature06862. PMC 2424287. PMID 18451855.

[55] http://www.ncbi.nlm.nih.gov/pubmed/11005795

[56] "Human Genome Project Completion: Frequently Asked Questions". genome.gov. Retrieved 2009-05-31.

[57] Singer, Emily (September 4, 2007). "Technology Review". *Technology review.* Retrieved May 25, 2010.

[58] "Complete Genomics Adds 29 High-Coverage, Complete Human Genome Sequencing Datasets to Its Public Genomic Repository".

[59] Ian Sample (17 February 2010). "Desmond Tutu's genome sequenced as part of genetic diversity study". *The Guardian.*

[60] Schuster SC, Miller W, Ratan A, Tomsho LP, Giardine B, Kasson LR, Harris RS, Petersen DC, Zhao F, Qi J, Alkan C, Kidd JM, Sun Y, Drautz DI, Bouffard P, Muzny DM, Reid JG, Nazareth LV, Wang Q, Burhans R, Riemer C, Wittekindt NE, Moorjani P, Tindall EA, Danko CG, Teo WS, Buboltz AM, Zhang Z, Ma Q, Oosthuysen A, Steenkamp AW, Oostuisen H, Venter P, Gajewski J, Zhang Y, Pugh BF, Makova KD, Nekrutenko A, Mardis ER, Patterson N, Pringle TH, Chiaromonte F, Mullikin JC, Eichler EE, Hardison RC, Gibbs RA, Harkins TT, Hayes VM (2010). "Complete Khoisan and Bantu genomes from southern Africa". *Nature* **463** (7283): 943–7. doi:10.1038/nature08795. PMC 3890430. PMID 20164927.

[61] Rasmussen M, Li Y, Lindgreen S, Pedersen JS, Albrechtsen A, Moltke I, et al. (Feb 2010). "Ancient human genome sequence of an extinct Palaeo-Eskimo". *Nature* **463** (7282): 757–62. Bibcode:2010Natur.463..757R. doi:10.1038/nature08835. PMC 3951495. PMID 20148029.

[62] Corpas M, Cariaso M, Coletta A, Weiss D, Harrison AP, Moran F, Yang H (November 12, 2013). "A Complete Public Domain Family Genomics Dataset". *BioRxiv.* doi:10.1101/000216. Retrieved November 15, 2013.

[63] Gonzaga-Jauregui C, Lupski JR, Gibbs RA (2012). "Human genome sequencing in health and disease". *Annual Review of Medicine* **63**: 35–61. doi:10.1146/annurev-med-051010-162644. PMID 22248320.

[64] Choi M, Scholl UI, Ji W, Liu T, Tikhonova IR, Zumbo P, Nayir A, Bakkaloğlu A, Ozen S, Sanjad S, Nelson-Williams C, Farhi A, Mane S, Lifton RP (Nov 2009). "Genetic diagnosis by whole exome capture and massively parallel DNA sequencing". *Proceedings of the National Academy of Sciences of the United States of America* **106** (45): 19096–101. Bibcode:2009PNAS..10619096C. doi:10.1073/pnas.0910672106. PMC 2768590. PMID 19861545.

[65] Online Mendelian Inheritance in Man (OMIM)

[66] "Sickle-cell anaemia – Report by the Secretariat" (pdf). *Fifty-ninth World Health Assembly*. World Health Organization. 24 April 2006.

[67] Waterston RH, Lindblad-Toh K, Birney E, Rogers J, Abril JF, Agarwal P, Agarwala R, Ainscough R, Alexandersson M, et al. (Dec 2002). "Initial sequencing and comparative analysis of the mouse genome". *Nature***420**(6915): 520–62. Bibcode:2002Natur .420..520W.doi:10.1038/nature01262. PMID 12466850. the proportion of small (50–100 bp) segments in the mammalian genome that isunder (purifying) selection can be estimated to be about 5%.
This proportion is much higher than can be explained by protein-coding sequences alone, implying that the genome contains many additional features (such as untranslated regions, regulatoryelements, non-protein-coding genes, and chromosomal structural elements) under selection for biological function.

[68] Birney E, Stamatoyannopoulos JA, Dutta A, Guigó R, Gingeras TR, Margulies EH, et al. (Jun 2007). "Identification and analysis of functional elements in 1% of the human genome by the ENCODE pilot project". *Nature* **447** (7146): 799–816. doi:10.1038/nature05874. PMC 2212820. PMID 17571346.

[69] The Chimpanzee Sequencing and Analysis Consortium (Sep 2005). "Initial sequence of the chimpanzee genome and comparison with the human genome". *Nature* **437** (7055): 69–87. Bibcode:2005Natur.437...69.. doi:10.1038/nature04072. PMID 16136131. We calculate the genome-wide nucleotide divergence between human and chimpanzee to be 1.23%, confirming recent results from more limited studies.

[70] The Chimpanzee Sequencing and Analysis Consortium (Sep 2005). "Initial sequence of the chimpanzee genome and comparison with the human genome". *Nature* **437** (7055): 69–87. Bibcode:2005Natur.437...69.. doi:10.1038/nature04072. PMID 16136131. we estimate that polymorphism accounts for 14–22% of the observed divergence rate and thus that the fixed divergence is ~1.06% or less

[71] Demuth JP, De Bie T, Stajich JE, Cristianini N, Hahn MW (2006). "The evolution of mammalian gene families". *PLoS One* **1** (1): e85. Bibcode:2006PLoSO...1...85D. doi:10.1371/journal.pone.0000085. PMC 1762380. PMID 17183716. Our results imply that humans and chimpanzees differ by at least 6% (1,418 of 22,000 genes) in their complement of genes, which stands in stark contrast to the oft-cited 1.5% difference between orthologous nucleotide sequences

[72] The Chimpanzee Sequencing and Analysis Consortium (Sep 2005). "Initial sequence of the chimpanzee genome and comparison with the human genome". *Nature* **437** (7055): 69–87. Bibcode:2005Natur.437...69.. doi:10.1038/nature04072. PMID 16136131. Human chromosome 2 resulted from a fusion of two ancestral chromosomes that remained separate in the chimpanzee lineage
Olson MV, Varki A (Jan 2003). "Sequencing the chimpanzee genome: insights into human evolution and disease". *Nature Reviews. Genetics* **4** (1): 20–8. doi:10.1038/nrg981. PMID 12509750. Large-scale sequencing of the chimpanzee genome is now imminent.

[73] Gilad Y, Wiebe V, Przeworski M, Lancet D, Pääbo S (Jan 2004). "Loss of olfactory receptor genes coincides with the acquisition of full trichromatic vision in primates". *PLoS Biology* **2** (1): E5. doi:10.1371/journal.pbio.0020005. PMC 314465. PMID 14737185. Our findings suggest that the deterioration of the olfactory repertoire occurred concomitant with the acquisition of full trichromatic color vision in primates.

[74] Sykes, Bryan (2003-10-09). "Mitochondrial DNA and human history". The Human Genome. Retrieved 2006-09-19.

[75] Misteli T (Feb 2007). "Beyond the sequence: cellular organization of genome function". *Cell***128**(4): 787–800. doi:10.1016/j. PMID 17320514.

[76] Bernstein BE, Meissner A, Lander ES (Feb 2007). "The mammalian epigenome". *Cell***128**(4): 669–81. doi:10.1016/j.cell.2007 PMID 17320505.

[77] Scheen AJ, Junien C (May–Jun 2012). "[Epigenetics, interface between environment and genes: role in complex diseases]". *Revue Médicale De Liège* **67** (5-6): 250–7. PMID 22891475.

1.13 External links

- The National Human Genome Research Institute

- Ensembl **The Ensembl Genome Browser Project**

- National Library of Medicine human genome viewer

- UCSC Genome Browser.

- Human Genome Project.

- The National Office of Public Health Genomics

- New findings challenge established views about human genome

- INMEGEN: Complete genetic map of some mexican native groups

- Missing bits of DNA may define humans

Chapter 2

Human Genome Project

The **Human Genome Project** (**HGP**) was an international scientific research project with the goal of determining the sequence of chemical base pairs which make up human DNA, and of identifying and mapping all of the genes of the human genome from both a physical and functional standpoint.[1] It remains the world's largest collaborative biological project.[2] The project was proposed and funded by the US government; planning started in 1984, got underway in 1990, and was declared complete in 2003. A parallel project was conducted outside of government by the Celera Corporation, or Celera Genomics, which was formally launched in 1998. Most of the government-sponsored sequencing was performed in twenty universities and research centers in the United States, the United Kingdom, Japan, France, Germany, and China.[3]

The Human Genome Project originally aimed to map the nucleotides contained in a human haploid reference genome (more than three billion). The "genome" of any given individual is unique; mapping "the human genome" involves sequencing multiple variations of each gene.[4]

2.1 Project

2.1.1 History

In May, 1985 Robert Sinsheimer organized a workshop to discuss sequencing the human genome,[5] but for a number of reasons the NIH was uninterested in pursuing the proposal. The following March, the Santa Fe Workshop was organized by Charles DeLisi and David Smith of the Department of Energy's Office of Health and Environmental Research (OHER).[6] At the same time Renato Dulbecco proposed whole genome sequencing in an essay in Science.[7] James Watson followed two months later with a workshop held at the Cold Spring Harbor Laboratory.

The fact that the Santa Fe workshop was motivated and supported by a Federal Agency opened a path, albeit a difficult and tortuous one (Cook-Deegan),[8] for converting the idea into public policy. In a memo to the Assistant Secretary for Energy Research (Alvin Trivelpiece), Charles DeLisi, who was then Director of OHER, outlined a broad plan for the project.[9] This started a long and complex chain of events which led to approved reprogramming of funds that enabled OHER to launch the Project in 1986, and to recommend the first line item for the HGP, which was in President Regan's 1988 budget submission (Cook-Deegan),[10] and ultimately approved by the Congress. Of particular importance in Congressional approval was the advocacy of Senator Peter Domenici, whom DeLisi had befriended.[11] Domenici chaired the Senate Committee on Energy and Natural Resources, as well as the Budget Committee, both of which were key in the DOE budget process. Congress added a comparable amount to the NIH budget, thereby beginning official funding by both agencies.

Dr. Alvin Trivelpiece sought and obtained the approval of DeLisi's proposal by Deputy Secretary William Flynn Martin. This chart[12] was used in the spring of 1986 by Trivelpiece, then Director of the Office of Energy Research in the Department of Energy, to brief Martin and Under Secretary Joseph Salgado regarding his intention to reprogram $4 million to initiate the project with the approval of Secretary Herrington. This reprogramming was followed by a line item budget of $16 million in the Reagan Administration's 1987 budget submission to Congress.[13] It subsequently passed

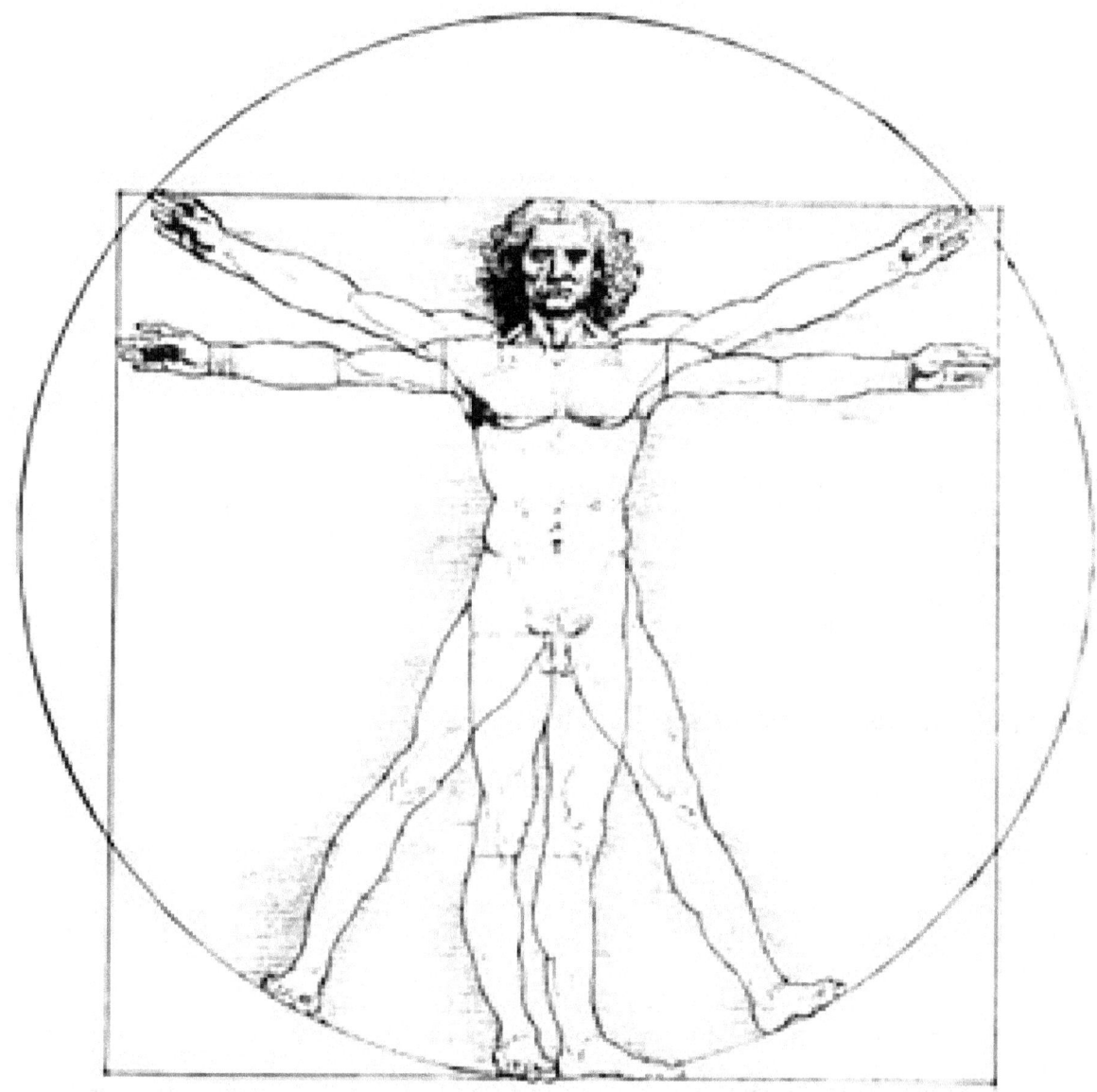

Logo

both Houses. The Project was planned for 15 years.[14]

Candidate technologies were already being considered for the proposed undertaking at least as early as 1985.[15]

In 1990, the two major funding agencies, DOE and NIH, developed a memorandum of understanding in order to co-ordinate plans and set the clock for the initiation of the Project to 1990.[16] At that time, David Galas was Director of the renamed "Office of Biological and Environmental Research" in the U.S. Department of Energy's Office of Science and James Watson headed the NIH Genome Program. In 1993, Aristides Patrinos succeeded Galas and Francis Collins succeeded James Watson, assuming the role of overall Project Head as Director of the U.S. National Institutes of Health (NIH) National Center for Human Genome Research (which would later become the National Human Genome Research Institute). A working draft of the genome was announced in 2000 and the papers describing it were published in February 2001. A more complete draft was published in 2003, and genome "finishing" work continued for more than a decade.

The $3-billion project was formally founded in 1990 by the US Department of Energy and the National Institutes of Health, and was expected to take 15 years.[17] In addition to the United States, the international consortium comprised geneticists in the United Kingdom, France, Australia, China and myriad other spontaneous relationships.[18]

Due to widespread international cooperation and advances in the field of genomics (especially in sequence analysis), as well as major advances in computing technology, a 'rough draft' of the genome was finished in 2000 (announced jointly by U.S. President Bill Clinton and the British Prime Minister Tony Blair on June 26, 2000).[19] This first available rough draft assembly of the genome was completed by the Genome Bioinformatics Group at the University of California, Santa Cruz, primarily led by then graduate student Jim Kent. Ongoing sequencing led to the announcement of the essentially complete genome on April 14, 2003, two years earlier than planned.[20][21] In May 2006, another milestone was passed on the way to completion of the project, when the sequence of the last chromosome was published in *Nature*.[22]

2.1.2   State of completion

The project did not aim to sequence all the DNA found in human cells. It sequenced only "euchromatic" regions of the genome, which make up about 90% of the genome. The other regions, called "heterochromatic" are found in centromeres and telomeres, and were not sequenced under the project.[23]

The Human Genome Project was declared complete in April 2003. An initial rough draft of the human genome was available in June 2000 and by February 2001 a working draft had been completed and published followed by the final sequencing mapping of the human genome on April 14, 2003. Although this was reported to be 99% of the euchromatic human genome with 99.99% accuracy a major quality assessment of the human genome sequence was published on May 27, 2004 indicating over 92% of sampling exceeded 99.99% accuracy which was within the intended goal.[24] Further analyses and papers on the HGP continue to occur.[25]

2.2   Applications and proposed benefits

The sequencing of the human genome holds benefits for many fields, from molecular medicine to human evolution. The Human Genome Project, through its sequencing of the DNA, can help us understand diseases including: genotyping of specific viruses to direct appropriate treatment; identification of mutations linked to different forms of cancer; the design of medication and more accurate prediction of their effects; advancement in forensic applied sciences; biofuels and other energy applications; agriculture, animal husbandry, bioprocessing; risk assessment; bioarcheology, anthropology and evolution. Another proposed benefit is the commercial development of genomics research related to DNA based products, a multibillion-dollar industry.

The sequence of the DNA is stored in databases available to anyone on the Internet. The U.S. National Center for Biotechnology Information (and sister organizations in Europe and Japan) house the gene sequence in a database known as GenBank, along with sequences of known and hypothetical genes and proteins. Other organizations, such as the UCSC Genome Browser at the University of California, Santa Cruz,[26] and Ensembl[27] present additional data and annotation and powerful tools for visualizing and searching it. Computer programs have been developed to analyze the data, because the data itself is difficult to interpret without such programs. Generally speaking, advances in genome sequencing technology have followed Moore's Law, a concept from computer science which states that integrated circuits can increase in complexity at an exponential rate.[28] This means that the speeds at which whole genomes can be sequenced can increase at a similar rate, as was seen during the development of the above-mentioned Human Genome Project.

2.3   Techniques and analysis

The process of identifying the boundaries between genes and other features in a raw DNA sequence is called genome annotation and is the domain of bioinformatics. While expert biologists make the best annotators, their work proceeds slowly, and computer programs are increasingly used to meet the high-throughput demands of genome sequencing projects. Beginning in 2008, a new technology known as RNA-seq was introduced that allowed scientists to directly sequence the messenger RNA in cells. This replaced previous methods of annotation, which relied on inherent properties of the DNA sequence, with direct measurement, which was much more accurate. Today, annotation of the human genome and other genomes relies primarily on deep sequencing of the transcripts in every human tissue using RNA-seq. These experiments have revealed that over 90% of human genes contain at least one and usually several alternative splice variants, in which the exons are combined in different ways to produce 2 or more gene products from the same locus.

The genome published by the HGP does not represent the sequence of every individual's genome. It is the combined mosaic of a small number of anonymous donors, all of European origin. The HGP genome is a scaffold for future work in identifying differences among individuals. Subsequent projects sequenced the genomes of multiple distinct ethnic groups, though as of today there is still only one "reference genome."

2.3.1 Findings

Key findings of the draft (2001) and complete (2004) genome sequences include:

1. There are approximately 20,500[29] genes in human beings, the same range as in mice.

2. The human genome has significantly more segmental duplications (nearly identical, repeated sections of DNA) than had been previously suspected.[30][31]

3. At the time when the draft sequence was published fewer than 7% of protein families appeared to be vertebrate specific.[32]

2.3.2 Accomplishment

The Human Genome Project was started in 1990 with the goal of sequencing and identifying all three billion chemical units in the human genetic instruction set, finding the genetic roots of disease and then developing treatments. It is considered a Mega Project because the human genome has approximately 3.3 billion base-pairs. With the sequence in hand, the next step was to identify the genetic variants that increase the risk for common diseases like cancer and diabetes.[16][33]

It was far too expensive at that time to think of sequencing patients' whole genomes. So the National Institutes of Health embraced the idea for a "shortcut", which was to look just at sites on the genome where many people have a variant DNA unit. The theory behind the shortcut was that, since the major diseases are common, so too would be the genetic variants that caused them. Natural selection keeps the human genome free of variants that damage health before children are grown, the theory held, but fails against variants that strike later in life, allowing them to become quite common. (In 2002 the National Institutes of Health started a $138 million project called the HapMap to catalog the common variants in European, East Asian and African genomes.)[34]

The genome was broken into smaller pieces; approximately 150,000 base pairs in length.[33] These pieces were then ligated into a type of vector known as "bacterial artificial chromosomes", or BACs, which are derived from bacterial chromosomes which have been genetically engineered. The vectors containing the genes can be inserted into bacteria where they are copied by the bacterial DNA replication machinery. Each of these pieces was then sequenced separately as a small "shotgun" project and then assembled. The larger, 150,000 base pairs go together to create chromosomes. This is known as the "hierarchical shotgun" approach, because the genome is first broken into relatively large chunks, which are then mapped to chromosomes before being selected for sequencing.[35][36]

Funding came from the US government through the National Institutes of Health in the United States, and a UK charity organization, the Wellcome Trust, as well as numerous other groups from around the world. The funding supported a number of large sequencing centers including those at Whitehead Institute, the Sanger Centre, Washington University in St. Louis, and Baylor College of Medicine.[17][37]

The United Nations Educational, Scientific and Cultural Organization (UNESCO) served as an important channel for the involvement of developing countries in the Human Genome Project.[38]

2.4 Public versus private approaches

In 1998, a similar, privately funded quest was launched by the American researcher Craig Venter, and his firm Celera Genomics. Venter was a scientist at the NIH during the early 1990s when the project was initiated. The $300,000,000

Celera effort was intended to proceed at a faster pace and at a fraction of the cost of the roughly $3 billion publicly funded project. The Celera approach was able to proceed at a much more rapid rate, and at a lower cost than the public project because it relied upon data made available by the publicly funded project.[39]

Celera used a technique called whole genome shotgun sequencing, employing pairwise end sequencing,[40] which had been used to sequence bacterial genomes of up to six million base pairs in length, but not for anything nearly as large as the three billion base pair human genome.

Celera initially announced that it would seek patent protection on "only 200–300" genes, but later amended this to seeking "intellectual property protection" on "fully-characterized important structures" amounting to 100–300 targets. The firm eventually filed preliminary ("place-holder") patent applications on 6,500 whole or partial genes. Celera also promised to publish their findings in accordance with the terms of the 1996 "Bermuda Statement", by releasing new data annually (the HGP released its new data daily), although, unlike the publicly funded project, they would not permit free redistribution or scientific use of the data. The publicly funded competitors were compelled to release the first draft of the human genome before Celera for this reason. On July 7, 2000, the UCSC Genome Bioinformatics Group released a first working draft on the web. The scientific community downloaded about 500 GB of information from the UCSC genome server in the first 24 hours of free and unrestricted access.[41]

In March 2000, President Clinton announced that the genome sequence could not be patented, and should be made freely available to all researchers. The statement sent Celera's stock plummeting and dragged down the biotechnology-heavy Nasdaq. The biotechnology sector lost about $50 billion in market capitalization in two days.

Although the working draft was announced in June 2000, it was not until February 2001 that Celera and the HGP scientists published details of their drafts. Special issues of *Nature* (which published the publicly funded project's scientific paper)[42] and *Science* (which published Celera's paper[43]) described the methods used to produce the draft sequence and offered analysis of the sequence. These drafts covered about 83% of the genome (90% of the euchromatic regions with 150,000 gaps and the order and orientation of many segments not yet established). In February 2001, at the time of the joint publications, press releases announced that the project had been completed by both groups. Improved drafts were announced in 2003 and 2005, filling in to approximately 92% of the sequence currently.

2.5   Genome donors

In the IHGSC international public-sector Human Genome Project (HGP), researchers collected blood (female) or sperm (male) samples from a large number of donors. Only a few of many collected samples were processed as DNA resources. Thus the donor identities were protected so neither donors nor scientists could know whose DNA was sequenced. DNA clones from many different libraries were used in the overall project, with most of those libraries being created by Dr. Pieter J. de Jong's lab. Much of the sequence (>70%) of the reference genome produced by the public HGP came from a single anonymous male donor from Buffalo, New York (code name RP11).[44][45]

HGP scientists used white blood cells from the blood of two male and two female donors (randomly selected from 20 of each) – each donor yielding a separate DNA library. One of these libraries (RP11) was used considerably more than others, due to quality considerations. One minor technical issue is that male samples contain just over half as much DNA from the sex chromosomes (one X chromosome and one Y chromosome) compared to female samples (which contain two X chromosomes). The other 22 chromosomes (the autosomes) are the same for both sexes.

Although the main sequencing phase of the HGP has been completed, studies of DNA variation continue in the International HapMap Project, whose goal is to identify patterns of single-nucleotide polymorphism (SNP) groups (called haplotypes, or "haps"). The DNA samples for the HapMap came from a total of 270 individuals: Yoruba people in Ibadan, Nigeria; Japanese people in Tokyo; Han Chinese in Beijing; and the French Centre d'Etude du Polymorphisme Humain (CEPH) resource, which consisted of residents of the United States having ancestry from Western and Northern Europe.

In the Celera Genomics private-sector project, DNA from five different individuals were used for sequencing. The lead scientist of Celera Genomics at that time, Craig Venter, later acknowledged (in a public letter to the journal *Science*) that his DNA was one of 21 samples in the pool, five of which were selected for use.[46][47]

In 2007, a team led by Jonathan Rothberg published James Watson's entire genome, unveiling the six-billion-nucleotide genome of a single individual for the first time.[48]

2.6 Developments

The work on interpretation and analysis of genome data is still in its initial stages. It is anticipated that detailed knowledge of the human genome will provide new avenues for advances in medicine and biotechnology. Clear practical results of the project emerged even before the work was finished. For example, a number of companies, such as Myriad Genetics, started offering easy ways to administer genetic tests that can show predisposition to a variety of illnesses, including breast cancer, hemostasis disorders, cystic fibrosis, liver diseases and many others. Also, the etiologies for cancers, Alzheimer's disease and other areas of clinical interest are considered likely to benefit from genome information and possibly may lead in the long term to significant advances in their management.[34][49]

There are also many tangible benefits for biologists. For example, a researcher investigating a certain form of cancer may have narrowed down his/her search to a particular gene. By visiting the human genome database on the World Wide Web, this researcher can examine what other scientists have written about this gene, including (potentially) the three-dimensional structure of its product, its function(s), its evolutionary relationships to other human genes, or to genes in mice or yeast or fruit flies, possible detrimental mutations, interactions with other genes, body tissues in which this gene is activated, and diseases associated with this gene or other datatypes. Further, deeper understanding of the disease processes at the level of molecular biology may determine new therapeutic procedures. Given the established importance of DNA in molecular biology and its central role in determining the fundamental operation of cellular processes, it is likely that expanded knowledge in this area will facilitate medical advances in numerous areas of clinical interest that may not have been possible without them.[50]

The analysis of similarities between DNA sequences from different organisms is also opening new avenues in the study of evolution. In many cases, evolutionary questions can now be framed in terms of molecular biology; indeed, many major evolutionary milestones (the emergence of the ribosome and organelles, the development of embryos with body plans, the vertebrate immune system) can be related to the molecular level. Many questions about the similarities and differences between humans and our closest relatives (the primates, and indeed the other mammals) are expected to be illuminated by the data in this project.[34][51]

The project inspired and paved the way for genomic work in other fields, such as agriculture. For example, by studying the genetic composition of Tritium aestivum, the world's most commonly used bread wheat, great insight has been gained into the ways that domestication has impacted the evolution of the plant.[52] Which loci are most susceptible to manipulation, and how does this play out in evolutionary terms? Genetic sequencing has allowed these questions to be addressed for the first time, as specific loci can be compared in wild and domesticated strains of the plant. This will allow for advances in genetic modification in the future which could yield healthier, more disease-resistant wheat crops.

2.7 Ethical, legal and social issues

At the onset of the Human Genome Project several ethical, legal, and social concerns were raised in regards to how increased knowledge of the human genome could be used to discriminate against people. One of the main concerns of most individuals was the fear that both employers and health insurance companies would refuse to hire individuals or refuse to provide insurance to people because of a health concern indicated by someone's genes.[53] In 1996 the United States passed the Health Insurance Portability and Accountability Act (HIPAA) which protects against the unauthorized and non-consensual release of individually identifiable health information to any entity not actively engaged in the provision of healthcare services to a patient.[54]

Along with identifying all of the approximately 20,000–25,000 genes in the human genome, the Human Genome Project also sought to address the ethical, legal, and social issues that were created by the onset of the project. For that the Ethical, Legal, and Social Implications (ELSI) program was founded in 1990. Five percent of the annual budget was allocated to address the ELSI arising from the project.[17][55]

Whilst the project may offer significant benefits to medicine and scientific research, some authors have emphasised the need to address the potential social consequences of mapping the human genome. "Molecularising disease and their possible cure will have a profound impact on what patients expect from medical help and the new generation of doctors' perception of illness."[56]

2.8   See also

- 1000 Genomes Project

- Chimpanzee Genome Project

- ENCODE

- EuroPhysiome

- Genome Compiler

- HUGO Gene Nomenclature Committee

- Human Brain Project

- Human Connectome Project

- Human Cytome Project

- Human Microbiome Project

- Human proteome project

- Human Variome Project

- Neanderthal Genome Project

- Sanger Institute

- The Genographic Project

- List of biological databases

2.9   References

[1] Robert Krulwich (2001-04-17). *Cracking the Code of Life* (Television Show). PBS.

[2] "Economic Impact of the Human Genome Project – Battelle" (PDF). Retrieved 1 August 2013.

[3] "Human Genome Project Completion: Frequently Asked Questions". *genome.gov*.

[4] Harmon, Katherine (2010-06-28). "Genome Sequencing for the Rest of Us". Scientific American. Retrieved 2010-08-13.

[5] Sinsheimer, Robert (1989). "The Santa Cruz Workshop, May 1985". *Genomics* **5**: 954. doi:10.1016/0888-7543(89)90142-0.

[6] DeLisi, Charles (October 2008). "Conferences That Changed the World". *Nature*: 455. doi:10.1038/455876a.

[7] Dulbecco, Renato (1986). "Turning Point in Cancer Research, Sequencing the Human Genome". *Science* **231** (4742): 1055–1056. doi:10.1126/science.3945817. PMID 3945817.

[8] Gene Wars, Op Cit.

[9] "Search". *georgetown.edu*.

[10] Gene Wars, Op.Cit.p 102

[11] "President Clinton Awards the Presidential Citizens Medals". *nara.gov*.

[12] http://www.wpainc.com/Archive/Trivelpiece/HGP%20Presenation.jpg

[13] DeLisi, Charles (2008). "Meetings that changed the world: Santa Fe 1986: Human genome baby-steps". *Nature* **455** (7215): 876. Bibcode:2008Natur.455..876D. doi:10.1038/455876a.

[14] DeLisi, Charles (1988). "The Human Genome Project". *American Scientist* **76**: 488. Bibcode:1988AmSci..76..488D.

[15] DeLisi, Charles (2001). "Genomes: 15 Years Later A Perspective by Charles Deli, HEP Pioneer". *Human Genome News* **11**: 3–4. Retrieved 2005-02-03.

[16] "About the Human Genome Project: What is the Human Genome Project". The Human Genome Management Information System (HGMIS). 2011-07-18. Retrieved 2011-09-02.

[17] Human Genome Information Archive. "About the Human Genome Project". U.S. Department of Energy & Human Genome Project program. Retrieved 1 August 2013.

[18] Collins F; Galas D (1993-10-01). "A New Five-Year Plan for the United States: Human Genome Program". National Human Genome Research Institute. Retrieved 1 August 2013.

[19] "White House Press Release". Retrieved 2006-07-22.

[20] Noble, Ivan (2003-04-14). "Human genome finally complete". *BBC News*. Retrieved 2006-07-22.

[21] Kolata, Gina (15 April 2013). "Human Genome, Then and Now". *The New York Times*. Retrieved 24 April 2014.

[22] "Guardian Unlimited IUK Latest I Human Genome Project finalised". *The Guardian* (London). Archived from the original on October 12, 2007. Retrieved 2006-07-22.

[23] "The Human Genome Project FAQ". *Genoscope*. Centre National de Séquençage. Retrieved 12 February 2015.

[24] Schmutz, Jeremy; Wheeler, Jeremy; Grimwood, Jane; Dickson, Mark; Yang, Joan; Caoile, Chenier; Bajorek, Eva; Black, Stacey; Chan, Yee Man; Denys, Mirian; Escobar, Julio; Flowers, Dave; Fotopulos, Dea; Garcia, Carmen; Gomez, Maria; Gonzales, Eidelyn; Haydu, Lauren; Lopez, Frederick; Ramirez, Lucia; Retterer, James; Rodriguez, Alex; Rogers, Stephanie; Salazar, Angelica; Tsai, Ming; Myers, Richard M. (2004). "Quality assessment of the human genome sequence". *Nature* **429** (6990): 365–368. Bibcode:2004Natur.429..365S. doi:10.1038/nature02390. PMID 15164052.

[25] "Landmark Human Genome Project Papers". *ornl.gov*.

[26] "An Overview of the Human Genome Project".

[27] "Ensembl Genome Browser". *ensembl.org*.

[28] Mardis, E. (2008). "The impact of next-generation sequencing technology on genetics". *Trends in Genetics* **24** (3): 133. doi:10.1016/j.tig.2007.12.007. PMID 18262675.

[29] "An Overview of the Human Genome Project". *genome.gov*.

[30] International Human Genome Sequencing Consortium (IHGSC) (2004). "Finishing the euchromatic sequence of the human genome". *Nature* **431** (7011): 931–945. Bibcode:2004Natur.431..931H. doi:10.1038/nature03001. PMID 15496913.

[31] Spencer, Geoff (20 December 2004). "International Human Genome Sequencing Consortium Describes Finished Human Genome Sequence". *NIH Nes Release*. National Institutes of Health.

[32] Bryant, J. A (2007). *Design and information in biology: From molecules to systems*. p. 108. ISBN 9781853128530. ...brought to light about 1200 protein families. Only 94 protein families, or 7%, appear to be vertebrate specific

[33] Wellcome Trust Sanger Institute. "The Human Genome Project: a new reality". Wellcome Trust Sanger Institute, Genome Research Limited. Retrieved 1 August 2013.

[34] Naidoo N; Pawitan Y; Soong R; Cooper DN; Ku CS (2011). "Human genetics and genomics a decade after the release of the draft sequence of the human genome". *Hum Genomics* **5** (6): 577–622. doi:10.1186/1479-7364-5-6-577. PMC 3525251. PMID 22155605.

[35] "Celera: A Unique Approach to Genome Sequencing". *ocf.berkeley.edu*. Biocomputing. 2006. Retrieved 1 August 2013.

[36] Davidson College (2002). "Sequencing Whole Genomes: Hierarchical Shotgun Sequencing v. Shotgun Sequencing". *bio.davids* Department of Biology, Davidson College. Retrieved 1 August 2013.

[37] Human Genome Project Information Archive (2013). "U.S. & International HGP Research Sites". U.S. Department of Energy & Human Genome Project. Retrieved 1 August 2013.

[38] Vizzini, Casimiro (March 19, 2015). "The Human Variome Project: Global Coordination in Data Sharing". *Science & Diplomacy* **4** (1).

[39] Venter, J. C.; Adams, M. D.; Myers, E. W.; Li, P. W.; Mural, R. J.; Sutton, G. G.; Smith, H. O.; Yandell, M.; Evans, C. A. (2001-02-16). "The sequence of the human genome". *Science (New York, N.Y.)* **291** (5507): 1304–1351. doi:10.1126/science.1058040. ISSN 0036-8075. PMID 11181995.

[40] Roach JC; Boysen C; Wang K; Hood L (1995). "Pairwise end sequencing: a unified approach to genomic mapping and sequencing". *Genomics* **26** (2): 345–353. doi:10.1016/0888-7543(95)80219-C. PMID 7601461.

[41] Center for Biomolecular Science & Engineering. "The Human Genome Project Race". Center for Biomolecular Science and Engineering. Retrieved 1 August 2013.

[42] International Human Genome Sequencing Consortium (2001). "Initial sequencing and analysis of the human genome" (PDF). *Nature* **409** (6822): 860–921. doi:10.1038/35057062. PMID 11237011.

[43] Venter, JC; et al. (2001). "The sequence of the human genome"(PDF).*Science***291**(5507): 1304–1351. Bibcode:2001Sci...29 doi:10.1126/science.1058040. PMID 11181995.

[44] Osoegawa, Kazutoyo; Mammoser, AG; Wu, C; Frengen, E; Zeng, C; Catanese, JJ; De Jong, PJ (2001). "A Bacterial Artificial Chromosome Library for Sequencing the Complete Human Genome". *Genome Research* **11** (3): 483–96. doi:10.1101/gr.169601. PMC 311044. PMID 11230172.

[45] Tuzun, E; et al. (2005). "Fine-scale structural variation of the human genome". *Nature Genetics***37**(7): 727–737. doi:10.1038/n PMID 15895083.

[46] Kennedy D (2002). "Not wicked, perhaps, but tacky". *Science* **297** (5585): 1237. doi:10.1126/science.297.5585.1237. PMID 12193755.

[47] Venter D (2003). "A Part of the Human Genome Sequence". *Science* **299** (5610): 1183–4. doi:10.1126/science.299.5610.1183. PMID 12595674.

[48] Wadman, Meredith (2008-04-16). "James Watson's genome sequenced at high speed". *Nature News* **452** (7189): 788–788. doi:10.1038/452788b.

[49] Gonzaga-Jauregui C; Lupski JR; Gibbs RA (2012). "Human genome sequencing in health and disease". *Annu Rev Med* **63** (1): 35–61. doi:10.1146/annurev-med-051010-162644. PMID 22248320.

[50] Snyder M, Du J; Gerstein M (2012). "Personal genome sequencing: current approaches and challenges". *Genes Dev* **24** (5): 423–431. doi:10.1101/gad.1864110. PMID 20194435.

[51] Lander ES (2011). "Initial impact of the sequencing of the human genome". *Nature***479**(7333): 187–197. Bibcode:2011Natur doi:10.1038/nature09792. PMID 21307931.

[52] Peng, J; Sun, E; Nevo, D (2011). "Domestication Evolution, Genetics And Genomics In Wheat". *Molecular Breeding* **28** (3): 281–301. doi:10.1007/s11032-011-9608-4.

[53] Greely, Henry (1992). *The Code of Codes: Scientific and Social Issues in the Human Genome Project*. Cambridge, Massachusetts: Harvard University Press. pp. 264–65. ISBN 0-674-13646-2.

[54] US Department of Health and Human Services. "Understanding Health Information Privacy".

[55] Genetics Home Reference (2013). "What were some of the ethical, legal, and social implications addressed by the Human Genome Project?". *ghr.nlm.nih.gov*. Retrieved 1 August 2013.

[56] Rheinberger, H.J. (2000). *Living and Working with the New Medical Technologies*. Cambridge: Cambridge University Press. p. 20.

2.10 Further reading

- McElheny, Victor K. (2010). *Drawing the Map of Life: Inside the Human Genome Project*. Basic Books. ISBN 978-0-465-03260-0. 361 pages. Examines the intellectual origins, history, and motivations of the project to map the human genome; draws on interviews with key figures.

- Collins, F. (2006). *The Language of God: A Scientist Presents Evidence for Belief*. Free Press. ISBN 0-7432-8639-1. OCLC 65978711.

- Venter, J. Craig (October 18, 2007). *A Life Decoded: My Genome: My Life*. New York, New York: Viking Adult. ISBN 0-670-06358-4. OCLC 165048736.

- *Gene Wars*. New York and London: W W Norton and Company. 1994.

2.11 External links

- Human Genome Project official information page

- National Human Genome Research Institute (NHGRI). NHGRI led the National Institutes of Health's contribution to the International Human Genome Project. This project, which had as its primary goal the sequencing of the three thousand million base pairs that make up human genome, was successfully completed in April 2003.

- Human Genome News. Published from 1989 to 2002 by the US Department of Energy, this newsletter was a major communications method for coordination of the Human Genome Project. Complete online archives are available.

- The HGP information pages Department of Energy's portal to the international Human Genome Project, Microbial Genome Program, and Genomics:GTL systems biology for energy and environment

- yourgenome.org: The Sanger Institute public information pages has general and detailed primers on DNA, genes and genomes, the Human Genome Project and science spotlights.

- Ensembl project, an automated annotation system and browser for the human genome

- UCSC genome browser, This site contains the reference sequence and working draft assemblies for a large collection of genomes. It also provides a portal to the ENCODE project.

- Nature magazine's human genome gateway, including the HGP's paper on the draft genome sequence

- Wellcome Trust Human Genome website A free resource allowing you to explore the human genome, your health and your future.

- Learning about the Human Genome. Part 1: Challenge to Science Educators. ERIC Digest.

- Learning about the Human Genome. Part 2: Resources for Science Educators. ERIC Digest.

- *Patenting Life* by Merrill Goozner

- Prepared Statement of Craig Venter of Celera Venter discusses Celera's progress in deciphering the human genome sequence and its relationship to healthcare and to the federally funded Human Genome Project.

- Cracking the Code of Life Companion website to 2-hour NOVA program documenting the race to decode the genome, including the entire program hosted in 16 parts in either QuickTime or RealPlayer format.

- Lone Dog L (1999). "Whose genes are they? The Human Genome Diversity Project". *J Health Soc Policy* **10** (4): 51–66. doi:10.1300/J045v10n04_04. PMID 10538186.

- Bioethics Research Library Numerous original documents at Georgetown University.

- Works by Human Genome Project at Project Gutenberg

- Project Gutenberg hosts e-texts for Human Genome Project, titled *Human Genome Project, Chromosome Number #* (# denotes 01-22, X and Y). This information is raw sequence, released in November 2002; access to entry pages with download links is available through https://www.gutenberg.org/etext/3501 for Chromosome 1 sequentially to https://www.gutenberg.org/etext/3524 for the Y Chromosome. Note that this sequence might not be considered definitive due to ongoing revisions and refinements. In addition to the chromosome files, there is a supplementary information file dated March 2004 which contains additional sequence information.

- Works by or about Human Genome Project at Internet Archive

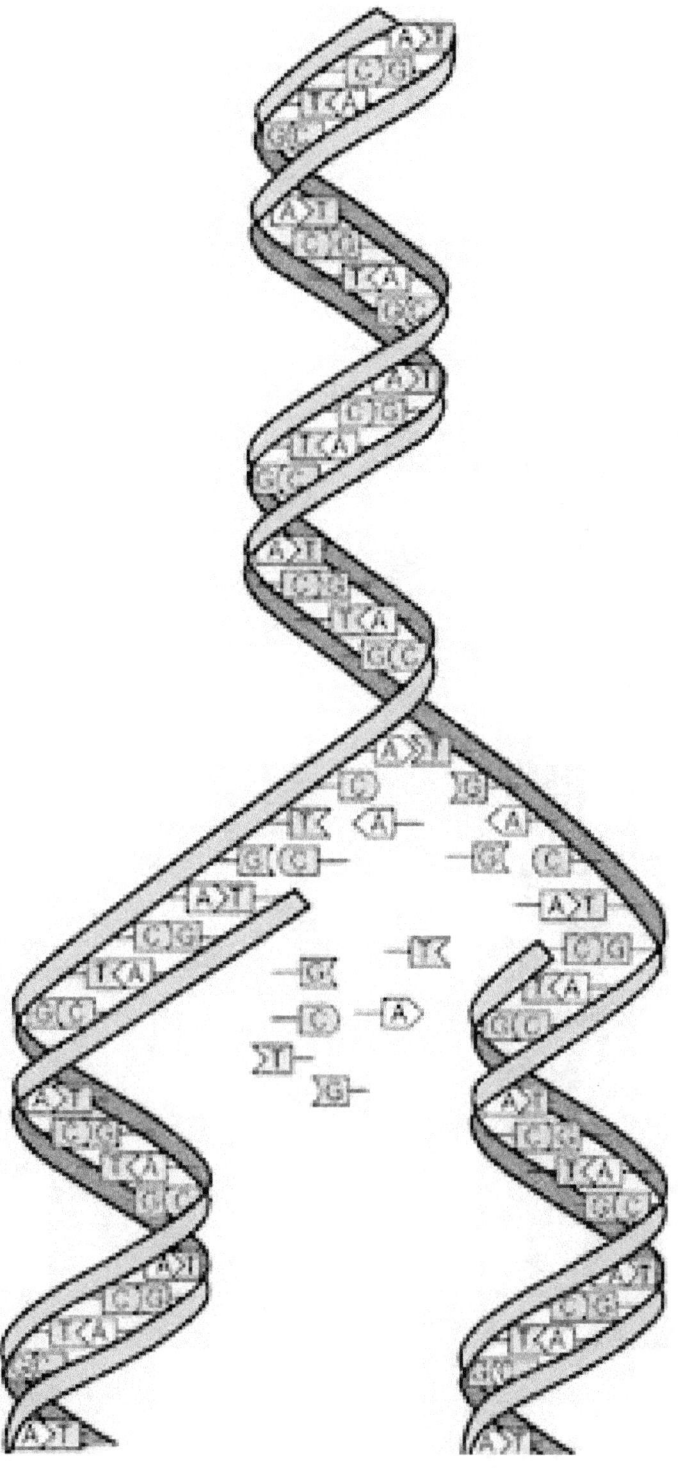

DNA replication

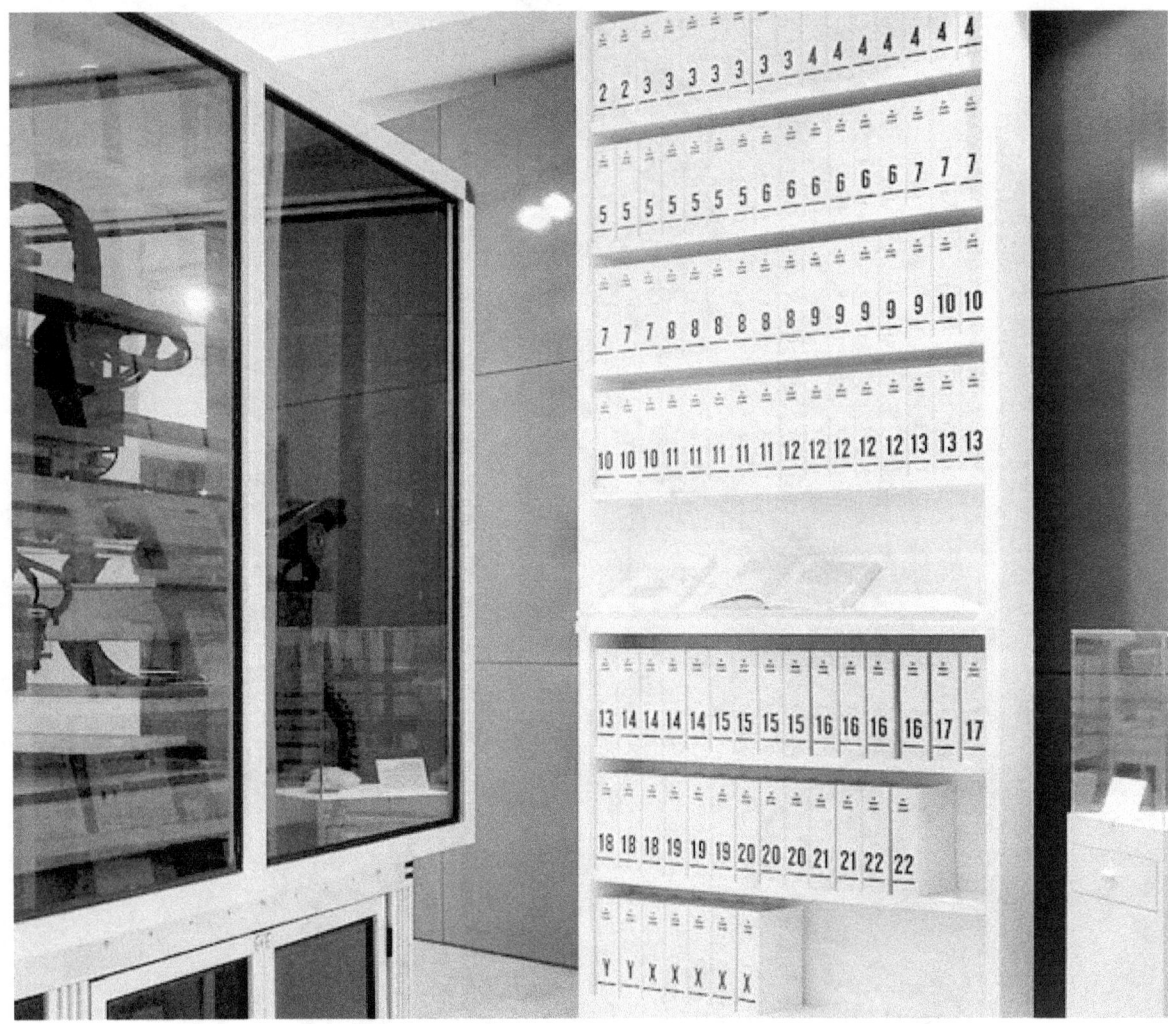

The first printout of the human genome to be presented as a series of books, displayed at the Wellcome Collection, London

Chapter 3

Genome

For a non-technical introduction to the topic, see Introduction to genetics. For other uses, see Genome (disambiguation). In modern molecular biology and genetics, the **genome** is the genetic material of an organism. It consists of DNA (or

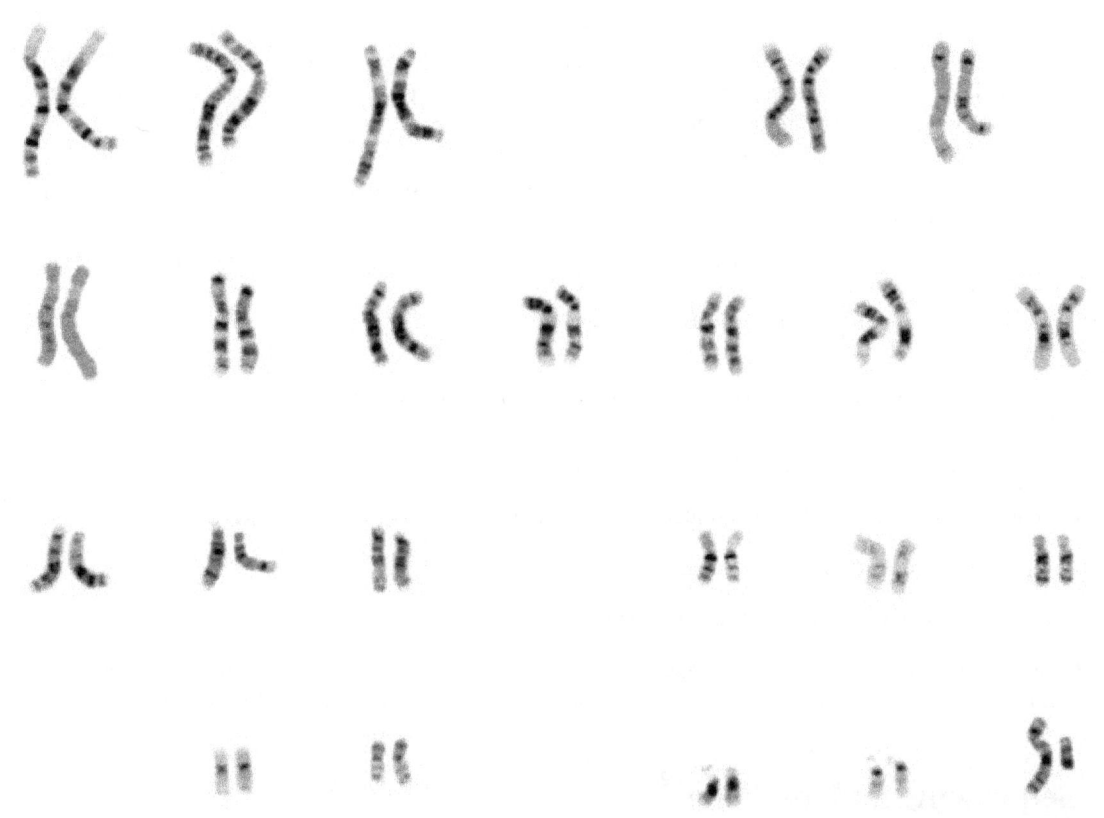

An image of the 46 chromosomes making up the diploid genome of a human male. (The mitochondrial chromosome is not shown.)

RNA in RNA viruses). The genome includes both the genes and the non-coding sequences of the DNA/RNA.[1]

3.1   Origin of term

The term was created in 1920 by Hans Winkler,[2] professor of botany at the University of Hamburg, Germany. The Oxford Dictionary suggests the name to be a blend of the words *gene* and *chromosome*.[3] However, see omics for a more thorough discussion. A few related *-ome* words already existed—such as *biome*, *rhizome*, forming a vocabulary into which *genome* fits systematically.[4]

3.2   Overview

Some organisms have multiple copies of chromosomes: diploid, triploid, tetraploid and so on. In classical genetics, in a sexually reproducing organism (typically eukarya) the gamete has half the number of chromosomes of the somatic cell and the genome is a full set of chromosomes in a diploid cell. The halving of the genetic material in gametes is accomplished by the segregation of homologous chromosomes during meiosis.[5] In haploid organisms, including cells of bacteria, archaea, and in organelles including mitochondria and chloroplasts, or viruses, that similarly contain genes, the single or set of circular or linear chains of DNA (or RNA for some viruses), likewise constitute the genome. The term *genome* can be applied specifically to mean what is stored on a complete set of nuclear DNA (i.e., the "nuclear genome") but can also be applied to what is stored within organelles that contain their own DNA, as with the "mitochondrial genome" or the "chloroplast genome". Additionally, the genome can comprise non-chromosomal genetic elements such as viruses, plasmids, and transposable elements.[6]

When people say that the genome of a sexually reproducing species has been "sequenced", typically they are referring to a determination of the sequences of one set of autosomes and one of each type of sex chromosome, which together represent both of the possible sexes. Even in species that exist in only one sex, what is described as a "genome sequence" may be a composite read from the chromosomes of various individuals. Colloquially, the phrase "genetic makeup" is sometimes used to signify the genome of a particular individual or organism. The study of the global properties of genomes of related organisms is usually referred to as genomics, which distinguishes it from genetics which generally studies the properties of single genes or groups of genes.

Both the number of base pairs and the number of genes vary widely from one species to another, and there is only a rough correlation between the two (an observation known as the C-value paradox). At present, the highest known number of genes is around 60,000, for the protozoan causing trichomoniasis (see List of sequenced eukaryotic genomes), almost three times as many as in the human genome.

An analogy to the human genome stored on DNA is that of instructions stored in a book:

- The book (genome) would contain 23 chapters (chromosomes);

- Each chapter contains 48 to 250 million letters (A,C,G,T) without spaces;

- Hence, the book contains over 3.2 billion letters total;

- The book fits into a cell nucleus the size of a pinpoint;

- At least one copy of the book (all 23 chapters) is contained in most cells of our body. The only exception in humans is found in mature red blood cells which become enucleated during development and therefore lack a genome.

3.3   Sequencing and mapping

For more details on this topic, see Genome project.

In 1976, Walter Fiers at the University of Ghent (Belgium) was the first to establish the complete nucleotide sequence of a viral RNA-genome (Bacteriophage MS2). The next year Fred Sanger completed the first DNA-genome sequence: Phage Φ-X174, of 5386 base pairs.[7] The first complete genome sequences among all three domains of life were released within a short period during the mid-1990s: The first bacterial genome to be sequenced was that of Haemophilus influenzae, completed by a team at The Institute for Genomic Research in 1995. A few months later, the first eukaryotic genome was

```
CATGACGTCGCGGACAACCCAGAATTGTCTTGAGCGATGGTAAGATCTAACCTCACTGCCGGGGGAGGCTCATAC
CTGGGGCTTTACTGATGTCATACCGTCTTGCACGGGGATAGAATGACGGTGCCCGTGTCTGCTTGCCTCGAAGCA
ATTTTCTGAAAGTTACAGACTTCGATTAAAAAGATCGGACTGCGCGTGGGCCCGGAGAGACATGCGTGGTAGTCA
TTTTTCGACGTGTCAAGGACTCAAGGGAATAGTTTGGCGGGAGCGTTACAGCTTCAATTCCCAAAGGTCGCAAGA
CGATAAAATTCAACTACTGGTTTCGGCCTAATAGGTCACGTTTTATGTGAAATAGAGGGGAACCGGCTCCCAAAT
CCCTGGGTGTTCTATGATAAGTCCTGCTTTATAACACGGGCGGTTAGGTTAAATGACTCTTCTATCTTATGGTG
ATCCAAGCGCCCGCTAATTCTGTTCTGTTAATGTTCATACCAATACTCACATCACATTAGATCAAAGGATCCCCG
AGCCCAGTCGCAAGGGTCTGCTGCTGTTGTCGACGCCTCATGTTACTCCTGGAATCTACCTGCCCTCCCCTCACC
GGTTAAGGCGTGTGATCGACGATGCAGGTATACATCGGCTCGGACCTACAGTGGTCGATCGACTGGCTACTGGCT
TCGCGGTTCGGCGCGTAGTTGAGTGCGATAACCCAACCGGTGGCAAGTAGCAAGAAGACCTACCTGGGTCACCTT
AGACAACCTAACTAATAGTCTCTAACGGGGAATTACCTTTACCAGTCTCATGCCTCCAATATATCTGCACCGCTT
CAATGATATCGCCCACAGAAAGTAGGGTCTCAGGTATCGCATACGCCGCGCCCGGGTCCCAGCTACGCTCAGGAC
GACAGTAGAGAGCTATTGTGTAATTCAGGCTCAGCATTCATCGACCTTTCCTGTTGTGAATATTGTGCTAATGCA
TCTCGTCCGTAACGATCTGGGGGGCAAAACCGAATATCCGTATTCTCGTCCTACGGGTCCACAATGAGAAAGTCC
TGCGCGTGATCGTCAGTTAAGTTAAATTAATTCAGGCTACGGTAAACTTGTAGTGAGCTAAGAATCACGGGAATC
ACGGGTTCGCTACAGATGAACTGAATTTATACACGGACAACTCATCGCCCATTTGGGCGTGGGCACCGCAGATCA
AAAGTGGCAGATTAGGAGTGCTTGATCAGGTTAGCAGGTGGACTGTATCCAACAGCGCATCAAACTTCAATAAAT
CCAAAGCGTTGTAGTGGTCTAAGCACCCCTGAACAGTGGCGCCCATCGTTAGCGTAGTACAACCCTTCCCCCTTG
AGGTGCGACATGGGGCCAGTTAGCCTGCCCTATATCCCTTGCACACGTTCAATAAGAGGGGCTCTACAGCGCCGC
TTTTTAAATTAGGATGCCGACCCCATCATTGGTAACTGTATGTTCATAGATATTTCTTCAGGAGTAATAGCGACA
AGCTGACACGCAAGGGTCAACAATAATTTCTACTATCACCCCGCTGAACGACTGTCTTTGCAAGAACCAACTGGG
CTTAGATTCGCGTCCTAACGTAGTGAGGGCCGAGTCATATCATAGATCAGGCATGAGAAACCGACGTCGAGTCTA
CACACGAGTTGTAAACAACTTGATTGCTATACTGTAGCTACCGCAAGGATCTCCTACATCAAAGACTACTGGGCG
ATCTGGATCCGAGTCAGAAATACGAGTTAATGCAAATTTACGTAGACCGGTGAAAACACGTGCCATGGGTTGCGT
AGACCGTAGTCAGAAGTGTGGCGCGCTATTCGTACCGAACCGGTGGAGTATACAGAATTGCTCTTCTACGACGTA
AGGAGCTCGGTCCCCAATGCACGCCAAAAAAGGAATAAAGTATTCAAACTGCGCATGGTCCCTCCGCCGGTGGCA
CTATTATCCATCCGAACGTTGAACCTACTTCCTCGGCTTATGCTGTCCTCAACAGTATCGCTTATGAATCGCATG
CGGCTGTGGATCTTAACGGCCACATTCTTAATTCCGACCGATCACCGATCGCCTTTCCTCGCTGGTACAATGAGT
ACTAAGTTATCCAGATCAAGGTTTGAACGGACTCGTATGACATGTGTGACTGAACCCGGGAGGAAATGCAGAGAA
CTGTTTCAAGGCCTCTGCTTTGGTATCACTCAATATATTCAGACCAGACAAGTGGCAAAATTTCGTGCGCCTCTC
CTAGGTATTCACGCAACCGTCGTAACATGCACTAAGGATAACTAGCGCCAGGGGGGCATACTAGGTCCCGGAGCT
AAAGACTACCCTATGGATTCCTTGGAGCGGGGACAATGCAGACCGGTTACGACACAATTATCGGGATCGTCTAGA
GGTATTATTAGCAAGACAATAAAGGACATTGCACAGAGACTTATTAGAATTCAACAAACAGGATCATATCATGCG
GTGTTGGGTCGGGCAAGTCCCCGAAGCTCGGCCAAAAGATTCGCCATGGAACCGTCTGGTCCTGTTAGCGTGTAC
GCCTGCTCCTGTTCCGGGTACCATAGATAGACTGAGATTGCGTCAAAAAATTGCGGCGAAAATAGAGGGGCTCCT
TGTAGAAATACCAGACTGGGGAATTTAAGCGCTTTCCACTATCTGAGCGACTAAACATCAACAAATGCGTCTACT
CGAATCCGCAGTAGGCAATTACAACCTGGTTCAGATCACTGGTTAATCAGGGATGTCTTCATAAGATTATACTTG
CCCCGACGCGACAGCTCTTCAAGGGGCCGATTTTTGGACTTCAGATACGCTAGAATTTAAAGGGTCTCTTACACC
TGCTGCGGCCTGCAGGGACCCCTAGAACTTGCCGCCTACTTGTCTCAGTCTAATAACGCGCGAAGCCGTGGGGCA
CGTGACCTTAAGTCGCAGAGCGAGTGATGAATTTGGGACGCTAATATGGGTGAATAGAGACTTATATCATCAGGG
```

Part of DNA sequence - prototypification of complete genome of virus

completed, with sequences of the 16 chromosomes of budding yeast *Saccharomyces cerevisiae* published as the result of a European-led effort begun in the mid-1980s. The first genome sequence for an archaeon, *Methanococcus jannaschii*, was completed in 1996, again by The Institute for Genomic Research.

The development of new technologies has made it dramatically easier and cheaper to do sequencing, and the number of complete genome sequences is growing rapidly. The US National Institutes of Health maintains one of several comprehensive databases of genomic information.[8] Among the thousands of completed genome sequencing projects include those for rice, a mouse, the plant *Arabidopsis thaliana*, the puffer fish, and the bacteria E. coli. In December 2013, scientists first sequenced the entire *genome* of a Neanderthal, an extinct species of humans. The genome was extracted from the toe bone of a 130,000-year-old Neanderthal found in a Siberian cave.[9][10]

New sequencing technologies, such as massive parallel sequencing have also opened up the prospect of personal genome sequencing as a diagnostic tool, as pioneered by Manteia Predictive Medicine. A major step toward that goal was the completion in 2007 of the full genome of James D. Watson, one of the co-discoverers of the structure of DNA.[11]

Whereas a genome sequence lists the order of every DNA base in a genome, a genome map identifies the landmarks. A genome map is less detailed than a genome sequence and aids in navigating around the genome. The Human Genome

Project was organized to map and to sequence the human genome. A fundamental step in the project was the release of a detailed genomic map by Jean Weissenbach and his team at the Genoscope in Paris.[12][13]

3.4   Genome compositions

Genome composition is used to describe the make up of contents of a haploid genome, which should include **genome size**, proportions of **non-repetitive DNA** and **repetitive DNA** in details. By comparing the genome compositions between genomes, scientists can better understand the evolutionary history of a given genome.

When talking about genome composition, one should distinguish between prokaryotes and eukaryotes as the big differences on contents structure they have. In prokaryotes, most of the genome (85–90%) is non-repetitive DNA, which means coding DNA mainly forms it, while non-coding regions only take a small part.[14] On the contrary, eukaryotes have the feature of exon-intron organization of protein coding genes; the variation of repetitive DNA content in eukaryotes is also extremely high. In mammals and plants, the major part of the genome is composed of repetitive DNA.[15]

Most biological entities that are more complex than a virus sometimes or always carry additional genetic material besides that which resides in their chromosomes. In some contexts, such as sequencing the genome of a pathogenic microbe, "genome" is meant to include information stored on this auxiliary material, which is carried in plasmids. In such circumstances then, "genome" describes all of the genes and information on non-coding DNA that have the potential to be present.

In eukaryotes such as plants, protozoa and animals, however, "genome" carries the typical connotation of only information on chromosomal DNA. So although these organisms contain chloroplasts or mitochondria that have their own DNA, the genetic information contained by DNA within these organelles is not considered part of the genome. In fact, mitochondria are sometimes said to have their own genome often referred to as the "mitochondrial genome". The DNA found within the chloroplast may be referred to as the "plastome".

3.4.1   Genome size

Genome size is the total number of DNA base pairs in one copy of a haploid genome. The genome size is positively correlated with the morphological complexity among prokaryotes and lower eukaryotes; however, after mollusks and all the other higher eukaryotes above, this correlation is no longer effective.[15][17] This phenomenon also indicates the mighty influence coming from repetitive DNA act on the genomes.

Since genomes are very complex, one research strategy is to reduce the number of genes in a genome to the bare minimum and still have the organism in question survive. There is experimental work being done on minimal genomes for single cell organisms as well as minimal genomes for multi-cellular organisms (see Developmental biology). The work is both *in vivo* and *in silico*.[18][19]

Here is a table of some significant or representative genomes. See #See also for lists of sequenced genomes.

3.4.2   Proportion of non-repetitive DNA

The **proportion of non-repetitive DNA** is calculated by using the length of non-repetitive DNA divided by genome size. Protein-coding genes and RNA-coding genes are generally non-repetitive DNA.[60] A bigger genome does not mean more genes, and the proportion of non-repetitive DNA decreases along with increasing genome size in higher eukaryotes.[15]

It had been found that the proportion of non-repetitive DNA can vary a lot between species. Some *E. coli* as prokaryotes only have non-repetitive DNA, lower eukaryotes such as *C. elegans* and fruit fly, still possess more non-repetitive DNA than repetitive DNA.[15][61] Higher eukaryotes tend to have more repetitive DNA than non-repetitive ones. In some plants and amphibians, the proportion of non-repetitive DNA is no more than 20%, becoming a minority component.[15]

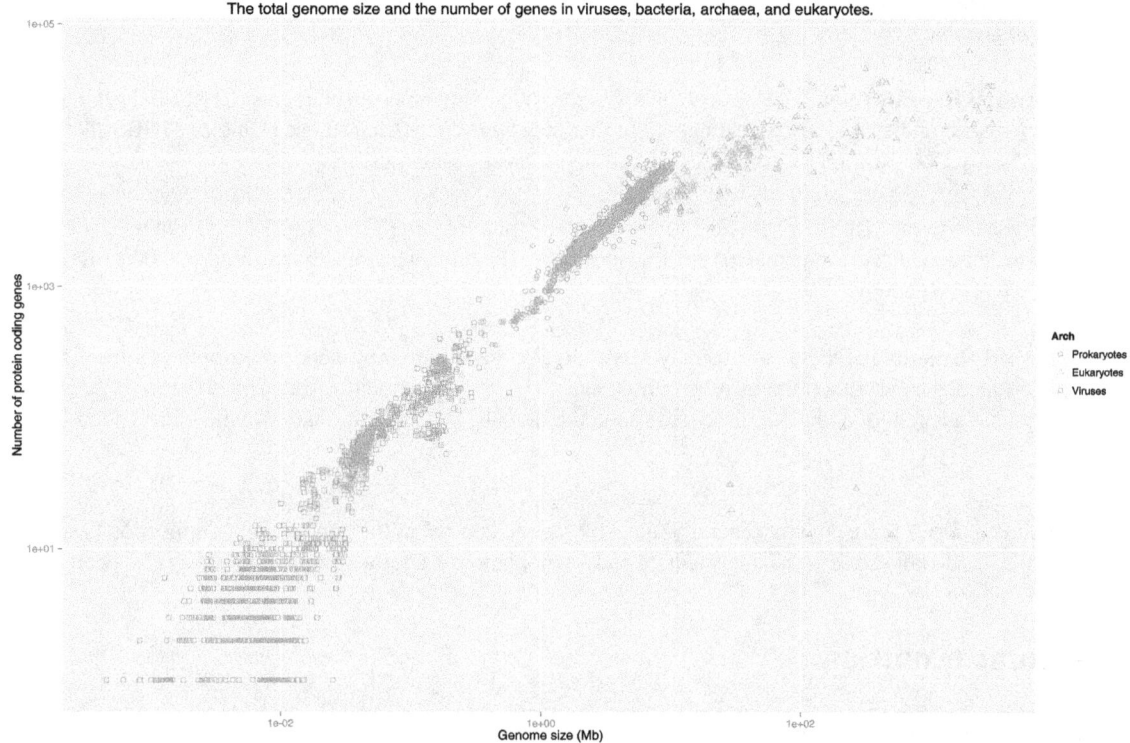

Log-log plot of the total number of annotated proteins in genomes submitted to GenBank as a function of genome size.[16]

3.4.3 Proportion of repetitive DNA

The **proportion of repetitive DNA** is calculated by using length of repetitive DNA divide by genome size. There are two categories of repetitive DNA in genome: tandem repeats and interspersed repeats.[62]

Tandem repeats

Tandem repeats are usually caused by slippage during replication, unequal crossing-over and gene conversion,[63] satellite DNA and microsatellites are forms of tandem repeats in the genome.[64] Although tandem repeats count for a significant proportion in genome, the largest proportion in mammalian is the other type, interspersed repeats.

Interspersed repeats

Interspersed repeats mainly come from transposable elements (TEs), but they also include some protein coding gene families and pseudogenes. Transposable elements are able to integrate into the genome at another site within the cell.[14][65] It is believed that TEs are an important driving force on genome evolution of higher eukaryotes.[66] TEs can be classified into two categories, Class 1 (retrotransposons) and Class 2 (DNA transposons).[65]

Retrotransposons   **Retrotransposons** can be transcribed into RNA, which are then duplicated at another site into the genome.[67] Retrotransposons can be divided into Long terminal repeats (LTRs) and Non-Long Terminal Repeats (Non-LTR).[66]

Long Terminal Repeats (LTRs)   similar to retroviruses, which have both gag and pol genes to make cDNA from RNA and proteins to insert into genome, but LTRs can only act within the cell as they lack the env gene in retroviruses.[65]

It has been reported that LTRs consist of the largest fraction in most plant genome and might account for the huge variation in genome size.[68]

Non-Long Terminal Repeats (Non-LTRs)   can be divided into long interspersed elements (LINEs), short interspersed elements (SINEs) and Penelope-like elements. In *Dictyostelium discoideum*, there is another DIRS-like elements belong to Non-LTRs. Non-LTRs are widely spread in eukaryotic genomes.[69]

Long interspersed elements (LINEs)   are able to encode two Open Reading Frames (ORFs) to generate transcriptase and endonuclease, which are essential in retrotransposition. The human genome has around 500,000 LINEs, taking around 17% of the genome.[70]

Short interspersed elements (SINEs)   are usually less than 500 base pairs and need to co-opt with the LINEs machinery to function as nonautonomous retrotransposons.[71] The Alu element is the most common SINEs found in primates, it has a length of about 350 base pairs and takes about 11% of the human genome with around 1,500,000 copies.[66]

DNA transposons   **DNA transposons** generally move by "cut and paste" in the genome, but duplication has also been observed. Class 2 TEs do not use RNA as intermediate and are popular in bacteria, in metazoan it has also been found.[66]

3.5   Genome evolution

Genomes are more than the sum of an organism's genes and have traits that may be measured and studied without reference to the details of any particular genes and their products. Researchers compare traits such as *chromosome number* (karyotype), genome size, gene order, codon usage bias, and GC-content to determine what mechanisms could have produced the great variety of genomes that exist today (for recent overviews, see Brown 2002; Saccone and Pesole 2003; Benfey and Protopapas 2004; Gibson and Muse 2004; Reese 2004; Gregory 2005).

Duplications play a major role in shaping the genome. Duplication may range from extension of short tandem repeats, to duplication of a cluster of genes, and all the way to duplication of entire chromosomes or even entire genomes. Such duplications are probably fundamental to the creation of genetic novelty.

Horizontal gene transfer is invoked to explain how there is often extreme similarity between small portions of the genomes of two organisms that are otherwise very distantly related. Horizontal gene transfer seems to be common among many microbes. Also, eukaryotic cells seem to have experienced a transfer of some genetic material from their chloroplast and mitochondrial genomes to their nuclear chromosomes.

3.6   See also

- Bacterial genome size

- Genome Browser

- Genome project

- Genome-wide association study

- Genomics

    - Genome Compiler

- List of sequenced eukaryotic genomes

- List of sequenced animal genomes

- List of sequenced archaeal genomes

- List of sequenced bacterial genomes

- List of sequenced fungi genomes

- List of sequenced plastomes

- List of sequenced protist genomes

- Metagenomics

- Microbiome

- Molecular epidemiology

- Molecular pathological epidemiology

- Molecular pathology

- Nucleic acid sequence

- Pan-genome

- Precision medicine

- Sequenceome

- Whole genome sequencing

- Genome topology

3.7 References

[1] Ridley, M. (2006). *Genome*. New York, NY: Harper Perennial. ISBN 0-06-019497-9

[2] Winkler, HL (1920). *Verbreitung und Ursache der Parthenogenesis im Pflanzen- und Tierreiche*. Jena: Verlag Fischer.

[3] "definition of Genome in Oxford dictionary". Retrieved 25 March 2014.

[4] Lederberg, Joshua; McCray, Alexa T. (2001). "'Ome Sweet 'Omics – A Genealogical Treasury of Words" (PDF). *The Scientist* **15** (7).

[5] Griffiths JF; Gelbart WM; Lewontin RC; Wessler SR; Suzuki DT; Miller JH (2005). *Introduction to Genetic Analysis*. New York: W.H. Freeman and Co. pp. 34–40, 473–476, 626–629. ISBN 0-7167-4939-4.

[6] Madigan M; Martinko J, eds. (2006). *Brock Biology of Microorganisms* (11th ed.). Prentice Hall. ISBN 0-13-144329-1.

[7]

[8] "Genome Home". 2010-12-08. Retrieved 27 January 2011.

[9] Zimmer, Carl (December 18, 2013). "Toe Fossil Provides Complete Neanderthal Genome". *New York Times*. Retrieved 18 December 2013.

[10] Prüfer, Kay; Racimo, Fernando; Patterson, Nick; Jay, Flora; Sankararaman, Sriram; Sawyer, Susanna; Heinze, Anja; Renaud, Gabriel; Sudmant, Peter H.; De Filippo, Cesare; Li, Heng; Mallick, Swapan; Dannemann, Michael; Fu, Qiaomei; Kircher, Martin; Kuhlwilm, Martin; Lachmann, Michael; Meyer, Matthias; Ongyerth, Matthias; Siebauer, Michael; Theunert, Christoph; Tandon, Arti; Moorjani, Priya; Pickrell, Joseph; Mullikin, James C.; Vohr, Samuel H.; Green, Richard E.; Hellmann, Ines; Johnson, Philip L. F.; et al. (December 18, 2013). "The complete genome sequence of a Neanderthal from the Altai Mountains". *Nature (journal)* **505** (7481): 43–49. Bibcode:2014Natur.505...43P. doi:10.1038/nature12886. Retrieved 18 December 2013.

[11] Wade, Nicholas (2007-05-31). "Genome of DNA Pioneer Is Deciphered". *The New York Times*. Retrieved 2 April 2010.

[12] "What's a Genome?". Genomenewsnetwork.org. 2003-01-15. Retrieved 27 January 2011.

[13] NCBI_user_services (2004-03-29). "Mapping Factsheet". Retrieved 27 January 2011.

[14] Koonin, Eugene V.; Wolf, Yuri I. (2010). "Constraints and plasticity in genome and molecular-phenome evolution". *Nature Reviews Genetics* **11** (7): 487–498. doi:10.1038/nrg2810. PMC 3273317. PMID 20548290.

[15] Lewin, Benjamin (2004). *Genes VIII* (8th ed.). Upper Saddle River, NJ: Pearson/Prentice Hall. ISBN 0-13-143981-2.

[16] Koonin, Eugene V. (2011-08-31). *The Logic of Chance: The Nature and Origin of Biological Evolution*. FT Press. ISBN 9780132542494.

[17] Gregory TR; Nicol JA; Tamm H; Kullman B; Kullman K; Leitch IJ; Murray BG; Kapraun DF; Greilhuber J; Bennett MD (3 January 2007). "Eukaryotic genome size databases". *Nucleic Acids Research* **35** (Database): D332–D338. doi:10.1093/nar/gkl828.

[18] Glass JI; Assad-Garcia N; Alperovich N; Yooseph S; Lewis MR; Maruf M; Hutchison CA 3rd; Smith HO; Venter JC (2006). "Essential genes of a minimal bacterium". *Proc Natl Acad Sci USA* **103** (2): 425–30. Bibcode:2006PNAS..103..425G. doi:10.1073/pnas.0510013103. PMC 1324956. PMID 16407165.

[19] Forster AC; Church GM (2006). "Towards synthesis of a minimal cell". *Mol Syst Biol.* **2** (1): 45. doi:10.1038/msb4100090. PMC 1681520. PMID 16924266.

[20] Mankertz P (2008). "Molecular Biology of Porcine Circoviruses". *Animal Viruses: Molecular Biology*. Caister Academic Press. ISBN 978-1-904455-22-6.

[21] Fiers W; Contreras, R.; Duerinck, F.; Haegeman, G.; Iserentant, D.; Merregaert, J.; Min Jou, W.; Molemans, F.; Raeymaekers, A.; Van Den Berghe, A.; Volckaert, G.; Ysebaert, M. (1976). "Complete nucleotide-sequence of bacteriophage MS2-RNA – primary and secondary structure of replicase gene". *Nature* **260** (5551): 500–507. Bibcode:1976Natur.260..500F. doi:10.1038/260500a0. PMID 1264203.

[22] Fiers, W.; Contreras, R.; Haegeman, G.; Rogiers, R.; Van De Voorde, A.; Van Heuverswyn, H.; Van Herreweghe, J.; Volckaert, G.; Ysebaert, M. (1978). "Complete nucleotide sequence of SV40 DNA". *Nature* **273** (5658): 113–120. Bibcode:1978Natur.2 doi:10.1038/273113a0. PMID205802.

[23] Sanger, F.; Air, G.M.; Barrell, B.G.; Brown, N.L.; Coulson, A.R.; Fiddes, J.C.; Hutchison, C.A.; Slocombe, P. M.; Smith, M. (1977). "Nucleotide sequence of bacteriophage phi X174 DNA". *Nature* **265** (5596): 687–695. Bibcode:1977Natur.265..687S. doi:10.1038/265687a0. PMID 870828.

[24] "Virology – Human Immunodeficiency Virus And Aids, Structure: The Genome And Proteins Of HIV". Pathmicro.med.sc.edu. 2010-07-01. Retrieved 27 January 2011.

[25] Thomason, Lynn; Court, Donald L.; Bubunenko, Mikail; Costantino, Nina; Wilson, Helen; Datta, Simanti; Oppenheim, Amos (2007). "Recombineering: genetic engineering in bacteria using homologous recombination". *Current Protocols in Molecular Biology*. Chapter 1: Unit 1.16. doi:10.1002/0471142727.mb0116s78. ISBN 0471142727. PMID 18265390.

[26] Court, D. L.; Oppenheim, A. B.; Adhya, S. L. (2007). "A new look at bacteriophage lambda genetic networks". *Journal of Bacteriology* **189** (2): 298–304. doi:10.1128/JB.01215-06. PMC 1797383. PMID 17085553.

[27] Sanger, F.; Coulson, A.R.; Hong, G.F.; Hill, D.F.; Petersen, G.B. (1982). "Nucleotide sequence of bacteriophage lambda DNA". *Journal of Molecular Biology* **162** (4): 729–73. doi:10.1016/0022-2836(82)90546-0. PMID 6221115.

[28] Legendre, M; Arslan, D; Abergel, C; Claverie, JM (2012). "Genomics of Megavirus and the elusive fourth domain of lifel journal". *Communicative & Integrative Biology* **5** (1): 102–106. doi:10.4161/cib.18624. PMC 3291303. PMID 22482024.

[29] Philippe, N.; Legendre, M.; Doutre, G.; Coute, Y.; Poirot, O.; Lescot, M.; Arslan, D.; Seltzer, V.; Bertaux, L.; Bruley, C.; Garin, J.; Claverie, J.-M.; Abergel, C. (2013). "Pandoraviruses: Amoeba Viruses with Genomes Up to 2.5 Mb Reaching That of Parasitic Eukaryotes". *Science* **341** (6143): 281–6. Bibcode:2013Sci...341..281P. doi:10.1126/science.1239181. PMID 23869018.

[30] Bennett, G. M.; Moran, N. A. (5 August 2013). "Small, Smaller, Smallest: The Origins and Evolution of Ancient Dual Symbioses in a Phloem-Feeding Insect". *Genome Biology and Evolution* **5** (9): 1675–1688. doi:10.1093/gbe/evt118. PMID 23918810.

[31] Shigenobu, S; Watanabe, H; Hattori, M; Sakaki, Y; Ishikawa, H (Sep 7, 2000). "Genome sequence of the endocellular bacterial symbiont of aphids Buchnera sp. APS". *Nature* **407** (6800): 81–6. doi:10.1038/35024074. PMID 10993077.

[32] Fleischmann R; Adams M; White O; Clayton R; Kirkness E; Kerlavage A; Bult C; Tomb J; Dougherty B; Merrick J; McKenney; Sutton; Fitzhugh; Fields; Gocyne; Scott; Shirley; Liu; Glodek; Kelley; Weidman; Phillips; Spriggs; Hedblom; Cotton; Utterback; Hanna; Nguyen; Saudek; et al. (1995). "Whole-genome random sequencing and assembly of Haemophilus influenzae Rd". *Science* **269** (5223): 496–512. Bibcode:1995Sci...269..496F. doi:10.1126/science.7542800. PMID 7542800.

[33] Frederick R. Blattner; Guy Plunkett III; et al. (1997). "The Complete Genome Sequence of Escherichia coli K-12". *Science* **277** (5331): 1453–1462. doi:10.1126/science.277.5331.1453. PMID 9278503.

[34] Challacombe, Jean F.; Eichorst, Stephanie A.; Hauser, Loren; Land, Miriam; Xie, Gary; Kuske, Cheryl R.; Steinke, Dirk (15 September 2011). Steinke, Dirk, ed. "Biological Consequences of Ancient Gene Acquisition and Duplication in the Large Genome of Candidatus Solibacter usitatus Ellin6076". *PLoS ONE* **6** (9): e24882. Bibcode:2011PLoSO...624882C. doi:10.1371/journal.pone.0024882. PMC 3174227. PMID 21949776.

[35] Rocap, G.; Larimer, F. W.; Lamerdin, J.; Malfatti, S.; Chain, P.; Ahlgren, N. A.; Arellano, A.; Coleman, M.; Hauser, L.; Hess, W. R.; Johnson, Z. I.; Land, M.; Lindell, D.; Post, A. F.; Regala, W.; Shah, M.; Shaw, S. L.; Steglich, C.; Sullivan, M. B.; Ting, C. S.; Tolonen, A.; Webb, E. A.; Zinser, E. R.; Chisholm, S. W. (2003). "Genome divergence in two Prochlorococcus ecotypes reflects oceanic niche differentiation". *Nature* **424** (6952): 1042–7. Bibcode:2003Natur.424.1042R. doi:10.1038/nature01947. PMID 12917642.

[36] Dufresne, A.; Salanoubat, M.; Partensky, F.; Artiguenave, F.; Axmann, I. M.; Barbe, V.; Duprat, S.; Galperin, M. Y.; Koonin, E. V.; Le Gall, F.; Makarova, K. S.; Ostrowski, M.; Oztas, S.; Robert, C.; Rogozin, I. B.; Scanlan, D. J.; De Marsac, N. T.; Weissenbach, J.; Wincker, P.; Wolf, Y. I.; Hess, W. R. (2003). "Genome sequence of the cyanobacterium Prochlorococcus marinus SS120, a nearly minimal oxyphototrophic genome". *Proceedings of the National Academy of Sciences* **100** (17): 10020–5. Bibcode:2003PNAS..10010020D. doi:10.1073/pnas.1733211100. PMC 187748. PMID 12917486.

[37] Meeks, J. C.; Elhai, J; Thiel, T; Potts, M; Larimer, F; Lamerdin, J; Predki, P; Atlas, R (2001). "An overview of the genome of Nostoc punctiforme, a multicellular, symbiotic cyanobacterium". *Photosynthesis Research* **70** (1): 85–106. doi:10.1023/A:10138 40025518.PMID16228364.

[38] Parfrey LW; Lahr DJG; Katz LA (2008). "The Dynamic Nature of Eukaryotic Genomes". *Molecular Biology and Evolution* **25** (4): 787–94. doi:10.1093/molbev/msn032. PMC 2933061. PMID 18258610.

[39] ScienceShot: Biggest Genome Ever, comments: "The measurement for Amoeba dubia and other protozoa which have been reported to have very large genomes were made in the 1960s using a rough biochemical approach which is now considered to be an unreliable method for accurate genome size determinations."

[40] Fleischmann A; Michael TP; Rivadavia F; Sousa A; Wang W; Temsch EM; Greilhuber J; Müller KF & Heubl G (2014). "Evolution of genome size and chromosome number in the carnivorous plant genus *Genlisea* (Lentibulariaceae), with a new estimate of the minimum genome size in angiosperms". *Annals of Botany* **114** (8): 1651–1663. doi:10.1093/aob/mcu189. PMID 25274549.

[41] Greilhuber J; Borsch T; Müller K; Worberg A; Porembski S & Barthlott W (2006). "Smallest angiosperm genomes found in Lentibulariaceae, with chromosomes of bacterial size". *Plant Biology* **8** (6): 770–777. doi:10.1055/s-2006-924101. PMID 17203433.

[42] Tuskan, GA; Difazio, S; Jansson, S; Bohlmann, J; Grigoriev, I; Hellsten, U; Putnam, N; Ralph, S; Rombauts, S; Salamov, A; Schein, J; Sterck, L; Aerts, A; Bhalerao, RR; Bhalerao, RP; Blaudez, D; Boerjan, W; Brun, A; Brunner, A; Busov, V; Campbell, M; Carlson, J; Chalot, M; Chapman, J; Chen, GL; Cooper, D; Coutinho, PM; Couturier, J; Covert, S; Cronk, Q; Cunningham, R; Davis, J; Degroeve, S; Déjardin, A; Depamphilis, C; Detter, J; Dirks, B; Dubchak, I; Duplessis, S; Ehlting, J; Ellis, B; Gendler, K; Goodstein, D; Gribskov, M; Grimwood, J; Groover, A; Gunter, L; Hamberger, B; Heinze, B; Helariutta, Y; Henrissat, B; Holligan, D; Holt, R; Huang, W; Islam-Faridi, N; Jones, S; Jones-Rhoades, M; Jorgensen, R; Joshi, C; Kangasjärvi, J; Karlsson, J; Kelleher, C; Kirkpatrick, R; Kirst, M; Kohler, A; Kalluri, U; Larimer, F; Leebens-Mack, J; Leplé, JC; Locascio, P; Lou, Y; Lucas, S; Martin, F; Montanini, B; Napoli, C; Nelson, DR; Nelson, C; Nieminen, K; Nilsson, O; Pereda, V; Peter, G; Philippe, R; Pilate, G; Poliakov, A; Razumovskaya, J; Richardson, P; Rinaldi, C; Ritland, K; Rouzé, P; Ryaboy, D; Schmutz, J; Schrader, J; Segerman, B; Shin, H; Siddiqui, A; Sterky, F; Terry, A; Tsai, CJ; Uberbacher, E; Unneberg, P; Vahala, J; Wall, K; Wessler, S; Yang, G; Yin, T; Douglas, C; Marra, M; Sandberg, G; Van de Peer, Y; Rokhsar, D (Sep 15, 2006). "The genome of black cottonwood, Populus trichocarpa (Torr. & Gray)". *Science* **313** (5793): 1596–604. Bibcode:2006Sci...313.1596T. doi:10.1126/science.1128691. PMID 16973872.

[43] PELLICER, JAUME; FAY, MICHAEL F.; LEITCH, ILIA J. (15 September 2010). "The largest eukaryotic genome of them all?". *Botanical Journal of the Linnean Society* **164** (1): 10–15. doi:10.1111/j.1095-8339.2010.01072.x.

[44] Lang D; Zimmer AD; Rensing SA; Reski R (October 2008). "Exploring plant biodiversity: the Physcomitrella genome and beyond". *Trends Plant Sci* **13** (10): 542–549. doi:10.1016/j.tplants.2008.07.002. PMID 18762443.

[45] "Saccharomyces Genome Database". Yeastgenome.org. Retrieved 27 January 2011.

[46] Galagan JE, Calvo SE, Cuomo C, Ma LJ, Wortman JR, Batzoglou S, Lee SI, Baştürkmen M, Spevak CC, Clutterbuck J, Kapitonov V, Jurka J, Scazzocchio C, Farman M, Butler J, Purcell S, Harris S, Braus GH, Draht O, Busch S, D'Enfert C, Bouchier C, Goldman GH, Bell-Pedersen D, Griffiths-Jones S, Doonan JH, Yu J, Vienken K, Pain A, Freitag M, Selker EU, Archer DB, Peñalva MA, Oakley BR, Momany M, Tanaka T, Kumagai T, Asai K, Machida M, Nierman WC, Denning DW, Caddick M, Hynes M, Paoletti M, Fischer R, Miller B, Dyer P, Sachs MS, Osmani SA, Birren BW (2005). "Sequencing of Aspergillus nidulans and comparative analysis with A. fumigatus and A. oryzae". *Nature* **438** (7071): 1105–15. Bibcode:2005Natur.438.1105G. doi:10.1038/nature04341. PMID 16372000.

[47] Leroy, S., S. Bouamer, S. Morand, and M. Fargette (2007). Genome size of plant-parasitic nematodes. Nematology 9: 449-450.

[48] Gregory TR (2005). "Animal Genome Size Database". http://www.genomesize.com. External link in |publisher= (help)

[49] The *C. elegans* Sequencing Consortium (1998). "Genome sequence of the nematode *C. elegans*: a platform for investigating biology". *Science* **282** (5396): 2012–2018. doi:10.1126/science.282.5396.2012. PMID 9851916.

[50] Adams MD; Celniker SE; Holt RA; et al. (2000). "The genome sequence of *Drosophila melanogaster*". *Science* **287** (5461): 2185–95. Bibcode:2000Sci...287.2185.. doi:10.1126/science.287.5461.2185. PMID 10731132. Retrieved 25 May 2007.

[51] Honeybee Genome Sequencing Consortium; Weinstock; Robinson; Gibbs; Weinstock; Weinstock; Robinson; Worley; Evans; Maleszka; Robertson; Weaver; Beye; Bork; Elsik; Evans; Hartfelder; Hunt; Robertson; Robinson; Maleszka; Weinstock; Worley; Zdobnov; Hartfelder; Amdam; Bitondi; Collins; Cristino; Evans (October 2006). "Insights into social insects from the genome of the honeybee Apis mellifera". *Nature* **443** (7114): 931–49. Bibcode:2006Natur.443..931T. doi:10.1038/nature05260. PMC 2048586. PMID 17073008.

[52] The International Silkworm Genome (2008). "The genome of a lepidopteran model insect, the silkworm Bombyx mori". *Insect Biochemistry and Molecular Biology* **38** (12): 1036–1045. doi:10.1016/j.ibmb.2008.11.004. PMID 19121390.

[53] Wurm Y; Wang, J.; Riba-Grognuz, O.; Corona, M.; Nygaard, S.; Hunt, B. G.; Ingram, K. K.; Falquet, L.; Nipitwattanaphon, M.; Gotzek, D.; Dijkstra, M. B.; Oettler, J.; Comtesse, F.; Shih, C.-J.; Wu, W.-J.; Yang, C.-C.; Thomas, J.; Beaudoing, E.; Pradervand, S.; Flegel, V.; Cook, E. D.; Fabbretti, R.; Stockinger, H.; Long, L.; Farmerie, W. G.; Oakey, J.; Boomsma, J. J.; Pamilo, P.; Yi, S. V.; et al. (2011). "The genome of the fire ant *Solenopsis invicta*". *PNAS* **108** (14): 5679–5684. Bibcode:2011PNAS..108.5679W. doi:10.1073/pnas.1009690108. PMC 3078418. PMID 21282665. Retrieved 1 February 2011.

[54] Church, DM; Goodstadt, L; Hillier, LW; Zody, MC; Goldstein, S; She, X; Bult, CJ; Agarwala, R; Cherry, JL; DiCuccio, M; Hlavina, W; Kapustin, Y; Meric, P; Maglott, D; Birtle, Z; Marques, AC; Graves, T; Zhou, S; Teague, B; Potamousis, K; Churas, C; Place, M; Herschleb, J; Runnheim, R; Forrest, D; Amos-Landgraf, J; Schwartz, DC; Cheng, Z; Lindblad-Toh, K; Eichler, EE; Ponting, CP; Mouse Genome Sequencing, Consortium (May 5, 2009). Roberts, Richard J, ed. "Lineage-specific biology revealed by a finished genome assembly of the mouse". *PLoS Biology* **7** (5): e1000112. doi:10.1371/journal.pbio.1000112. PMC 2680341. PMID 19468303.

[55] "Human Genome Project Information Site Has Been Updated". Ornl.gov. 2013-07-23. Retrieved 6 February 2014.

[56] Venter, J. C.; Adams, M.; Myers, E.; Li, P.; Mural, R.; Sutton, G.; Smith, H.; Yandell, M.; Evans, C.; Holt, R. A.; Gocayne, J. D.; Amanatides, P.; Ballew, R. M.; Huson, D. H.; Wortman, J. R.; Zhang, Q.; Kodira, C. D.; Zheng, X. H.; Chen, L.; Skupski, M.; Subramanian, G.; Thomas, P. D.; Zhang, J.; Gabor Miklos, G. L.; Nelson, C.; Broder, S.; Clark, A. G.; Nadeau, J.; McKusick, V. A.; Zinder, N. (2001). "The Sequence of the Human Genome". *Science* **291** (5507): 1304–1351. Bibcode:2001Sci...291.1304V. doi:10.1126/science.1058040. PMID 11181995.

[57] Crollius, HR; Jaillon, O; Dasilva, C; Ozouf-Costaz, C; Fizames, C; Fischer, C; Bouneau, L; Billault, A; Quetier, F; Saurin, W; Bernot, A; Weissenbach, J (2000). "Characterization and Repeat Analysis of the Compact Genome of the Freshwater Pufferfish Tetraodon nigroviridis". *Genome Research* **10** (7): 939–949. doi:10.1101/gr.10.7.939. PMC 310905. PMID 10899143.

[58] Olivier Jaillon; et al. (21 October 2004). "Genome duplication in the teleost fish Tetraodon nigroviridis reveals the early vertebrate proto-karyotype". *Nature* **431** (7011): 946–957. Bibcode:2004Natur.431..946J. doi:10.1038/nature03025. PMID 15496914.

[59] "Tetraodon Project Information". Retrieved 17 October 2012.

[60] Britten, RJ; Davidson, EH (June 1971). "Repetitive and non-repetitive DNA sequences and a speculation on the origins of evolutionary novelty". *The Quarterly review of biology* **46** (2): 111–38. doi:10.1086/406830. PMID 5160087.

[61] Naclerio, G; Cangiano, G; Coulson, A; Levitt, A; Ruvolo, V; La Volpe, A (1992-07-05). "Molecular and genomic organization of clusters of repetitive DNA sequences in Caenorhabditis elegans". *Journal of Molecular Biology* **226** (1): 159–68. doi:10.1016/0022-2836(92)90131-3. PMID 1619649.

[62] Stojanovic, edited by Nikola (2007). *Computational genomics : current methods*. Wymondham: Horizon Bioscience. ISBN 1-904933-30-0.

[63] Li, YC; Korol, AB; Fahima, T; Beiles, A; Nevo, E (December 2002). "Microsatellites: genomic distribution, putative functions and mutational mechanisms: a review". *Molecular ecology* **11** (12): 2453–65. doi:10.1046/j.1365-294X.2002.01643.x. PMID 12453231.

[64] Schlötterer, C (December 2000). "Microsatellite analysis indicates genetic differentiation of the neo-sex chromosomes in Drosophila americana americana". *Heredity* **85** (Pt 6): 610–6. doi:10.1046/j.1365-2540.2000.00797.x. PMID 11240628.

[65] Wessler, S. R. (13 November 2006). "Eukaryotic Transposable Elements and Genome Evolution Special Feature: Transposable elements and the evolution of eukaryotic genomes". *Proceedings of the National Academy of Sciences* **103** (47): 17600–17601. Bibcode:2006PNAS..10317600W. doi:10.1073/pnas.0607612103.

[66] Kazazian, H. H. (12 March 2004). "Mobile Elements: Drivers of Genome Evolution". *Science* **303** (5664): 1626–1632. Bibcode:2004Sci...303.1626K. doi:10.1126/science.1089670. PMID 15016989.

[67] Deininger PL; Moran JV; Batzer MA; Kazazian, HH Jr. (December 2003). "Mobile elements and mammalian genome evolution". *Current opinion in genetics & development* **13** (6): 651–8. doi:10.1016/j.gde.2003.10.013. PMID 14638329.

[68] Kidwell MG; Lisch DR (March 2000). "Transposable elements and host genome evolution". *Trends in ecology & evolution* **15** (3): 95–99. doi:10.1016/S0169-5347(99)01817-0. PMID 10675923.

[69] Richard G.-F., Kerrest A; Dujon B (3 December 2008). "Comparative Genomics and Molecular Dynamics of DNA Repeats in Eukaryotes". *Microbiology and Molecular Biology Reviews* **72** (4): 686–727. doi:10.1128/MMBR.00011-08. PMC 2593564. PMID 19052325.

[70] Cordaux R; Batzer MA (1 October 2009). "The impact of retrotransposons on human genome evolution". *Nature Reviews Genetics* **10** (10): 691–703. doi:10.1038/nrg2640. PMC 2884099. PMID 19763152.

[71] Han, Jeffrey S.; Boeke, Jef D. (1 August 2005). "LINE-1 retrotransposons: Modulators of quantity and quality of mammalian gene expression?". *BioEssays* **27** (8): 775–784. doi:10.1002/bies.20257. PMID 16015595.

3.8   Further reading

- Benfey, P.; Protopapas, A.D. (2004). *Essentials of Genomics*. Prentice Hall.

- Brown, Terence A. (2002). *Genomes 2*. Oxford: Bios Scientific Publishers. ISBN 978-1-85996-029-5.

- Gibson, Greg; Muse, Spencer V. (2004). *A Primer of Genome Science* (Second ed.). Sunderland, Mass: Sinauer Assoc. ISBN 0-87893-234-8.

- Gregory (2005). T. Ryan, ed. *The Evolution of the Genome*. Elsevier. ISBN 0-12-301463-8.

- Reece, Richard J. (2004). *Analysis of Genes and Genomes*. Chichester: John Wiley & Sons. ISBN 0-470-84379-9.

- Saccone, Cecilia; Pesole, Graziano (2003). *Handbook of Comparative Genomics*. Chichester: John Wiley & Sons. ISBN 0-471-39128-X.

- Werner, E. (2003). "In silico multicellular systems biology and minimal genomes". *Drug Discov Today* **8** (24): 1121–1127. doi:10.1016/S1359-6446(03)02918-0. PMID 14678738.

3.9   External links

- UCSC Genome Browser – view the genome and annotations for more than 80 organisms.

- genomecenter.howard.edu

- Build a DNA Molecule

- Some comparative genome sizes

- DNA Interactive: The History of DNA Science

- DNA From The Beginning

- All About The Human Genome Project—from Genome.gov

- Animal genome size database

- Plant genome size database

- GOLD:Genomes OnLine Database

- The Genome News Network

- NCBI Entrez Genome Project database

- NCBI Genome Primer

- GeneCards—an integrated database of human genes

- Visualization of nucleotide sequence - prototypification of complete genome of virus, sequence of 5418 nucleotides

- BBC News – Final genome 'chapter' published

- IMG (The Integrated Microbial Genomes system)—for genome analysis by the DOE-JGI

- GeKnome Technologies Next-Gen Sequencing Data Analysis—next-generation sequencing data analysis for Illumina and 454 Service from GeKnome Technologies.

Chapter 4

Reference genome

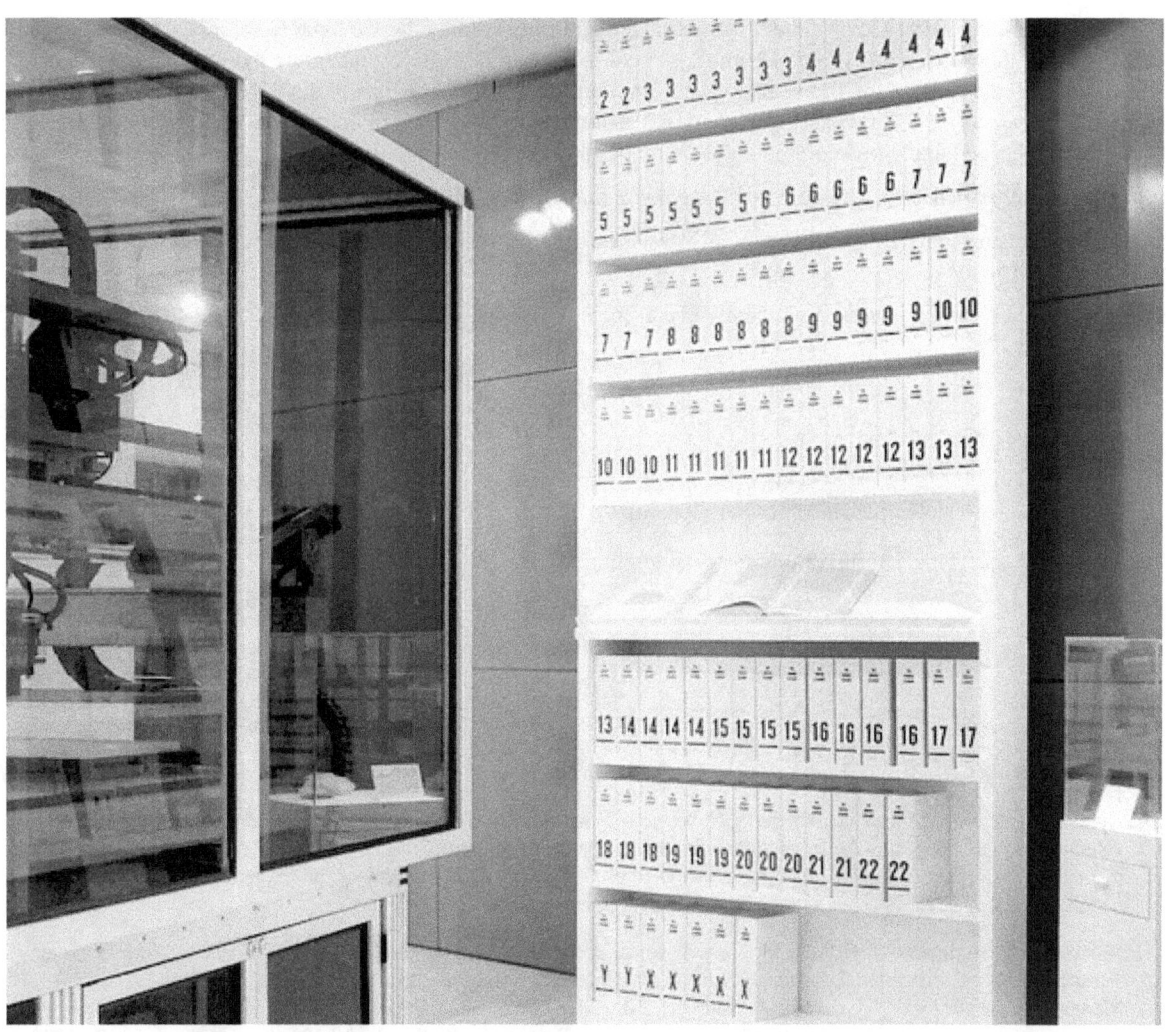

The first printout of the human reference genome presented as a series of books, displayed at the Wellcome Collection, London

A **reference genome** (also known as a **reference assembly**) is a digital nucleic acid sequence database, assembled by scientists as a representative example of a species' set of genes. As they are often assembled from the sequencing of

41

DNA from a number of donors, reference genomes do not accurately represent the set of genes of any single person. Instead a reference provides a haploid mosaic of different DNA sequences from each donor. For example *GRCh37*, the Genome Reference Consortium human genome (build 37) is derived from thirteen anonymous volunteers from Buffalo, New York.[1][2][3] The ABO blood group system differs among humans, but the human reference genome contains only an O allele (although the other alleles are annotated).<ref name2"Guide">Scherer, Stewart (2008). *A short guide to the human genome*. CSHL Press. p. 135. ISBN 0-87969-791-1.</ref>

As the cost of DNA sequencing falls, and new full genome sequencing technologies emerge, more genome sequences continue to be generated. Reference genomes are typically used as a guide on which new genomes are built, enabling them to be assembled much more quickly and cheaply than the initial Human Genome Project. Most individuals with their entire genome sequenced, such as James D. Watson, had their genome assembled in this manner.[4][5] For much of a genome, the reference provides a good approximation of the DNA of any single individual. But in regions with high allelic diversity, such as the major histocompatibility complex in humans and the major urinary proteins of mice, the reference genome may differ significantly from other individuals.[6][7][8] Comparison between the reference (build 36) and Watson's genome revealed 3.3 million single nucleotide polymorphism differences, while about 1.4 percent of his DNA could not be matched to the reference genome at all.[2][4] For regions where there is known to be large scale variation, sets of alternate loci are assembled alongside the reference locus.

The human and mouse reference genomes are maintained and improved by the Genome Reference Consortium (GRC), a group of fewer than 20 scientists from a number of genome research institutes, including the European Bioinformatics Institute, the National Center for Biotechnology Information, the Sanger Institute and McDonnell Genome Institute at Washington University in St. Louis. GRC continues to improve reference genomes by building new alignments that contain fewer gaps, and fixing misrepresentations in the sequence.

The human reference genome *GRCh38* was released on 24 December 2013.[9]

The previous human reference genome (*GRCh37*) was the nineteenth version. This build contained around 250 gaps, whereas the first version had ~150,000 gaps.[1]

Reference genomes can be accessed online at several locations, using dedicated browsers such as Ensembl or UCSC Genome Browser.[10]

4.1   Tutorials

- Reference Genome Sequencing: Conifer Genomics
- Selective Sequencing Through Combinatorial Pooling

4.2   References

[1] Editorial (October 2010). "E pluribus unum". *Nature Methods* **331** (5): 331. doi:10.1038/nmeth0510-331.

[2] Wade, Nicholas (May 31, 2007). "Genome of DNA Pioneer Is Deciphered". New York Times. Retrieved February 21, 2009.

[3] Donors were recruited by advertisement in *The Buffalo News*, on Sunday, March 23, 1997. The first ten male and ten female volunteers were invited to make an appointment with the project's genetic counselors and donate blood from which DNA was extracted. As a result of how the DNA samples were processed, about 80 percent of the reference genome came from eight people and one male, designated RP11, accounts for 66 percent of the total.

[4] Wheeler DA, Srinivasan M, Egholm M, Shen Y, Chen L, McGuire A, He W, Chen YJ, Makhijani V, Roth GT, Gomes X, Tartaro K, Niazi F, Turcotte CL, Irzyk GP, Lupski JR, Chinault C, Song XZ, Liu Y, Yuan Y, Nazareth L, Qin X, Muzny DM, Margulies M, Weinstock GM, Gibbs RA, Rothberg JM. (2008). "The complete genome of an individual by massively parallel DNA sequencing". *Nature* **452** (7189): 872–6. Bibcode:2008Natur.452..872W. doi:10.1038/nature06884. PMID 18421352.

[5] The exception to this is J. Craig Venter whose DNA was sequenced and assembled using shotgun sequencing methods.

[6] MHC Sequencing Consortium (1999). "Complete sequence and gene map of a human major histocompatibility complex". *Nature* **401** (6756): 921–923. Bibcode:1999Natur.401..921T. doi:10.1038/44853. PMID 10553908.

[7] Logan DW, Marton TF, Stowers L (2008). Vosshall, Leslie B., ed. "Species specificity in major urinary proteins by parallel evolution". *PLoS ONE* **3** (9): e3280. Bibcode:2008PLoSO...3.3280L. doi:10.1371/journal.pone.0003280. PMC 2533699. PMID 18815613.

[8] Hurst J, Beynon RJ, Roberts SC, Wyatt TD. (October 2007). *Urinary Lipocalins in Rodenta:is there a Generic Model?*. Chemical Signals in Vertebrates 11. Springer New York. ISBN 978-0-387-73944-1.

[9] New human genome assembly (GRCh38) released, NCBI news

[10] Flicek P, Aken BL, Beal K, et al. (January 2008). "Ensembl 2008". *Nucleic Acids Res.* **36** (Database issue): D707–14. doi:10.1093/nar/gkm988. PMC 2238821. PMID 18000006.

4.3 External links

- Genome Reference Consortium

Chapter 5

Whole genome sequencing

"Genome sequencing" redirects here. For the sequencing only of DNA, see DNA sequencing.
Whole genome sequencing (also known as **WGS**, **full genome sequencing**, **complete genome sequencing**, or **entire**

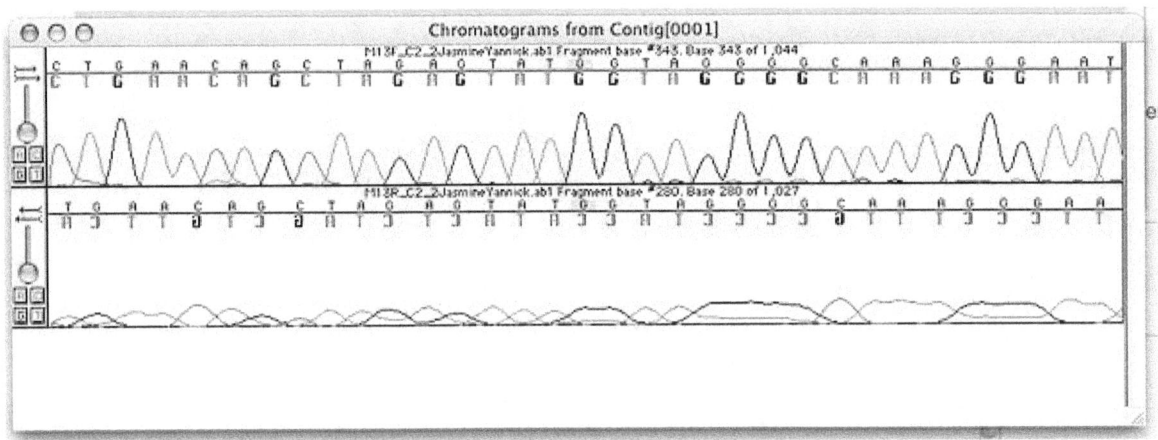

Electropherograms are commonly used to sequence portions of genomes.[1]

genome sequencing) is a laboratory process that determines the complete DNA sequence of an organism's genome at a single time. This entails sequencing all of an organism's chromosomal DNA as well as DNA contained in the mitochondria and, for plants, in the chloroplast.

Whole genome sequencing should not be confused with DNA profiling, which only determines the likelihood that genetic material came from a particular individual or group, and does not contain additional information on genetic relationships, origin or susceptibility to specific diseases.[2] Also unlike full genome sequencing, SNP genotyping covers less than 0.1% of the genome. Almost all truly complete genomes are of microbes; the term "full genome" is thus sometimes used loosely to mean "greater than 95%". The remainder of this article focuses on nearly complete human genomes.

High-throughput genome sequencing technologies have largely been used as a research tool and are currently being introduced in the clinics.[3][4][5] In the future of personalized medicine, whole genome sequence data will be an important tool to guide therapeutic intervention.[6] The tool of gene sequencing at SNP level is also used to pinpoint functional variants from association studies and improve the knowledge available to researchers interested in evolutionary biology, and hence may lay the foundation for predicting disease susceptibility and drug response.[7]

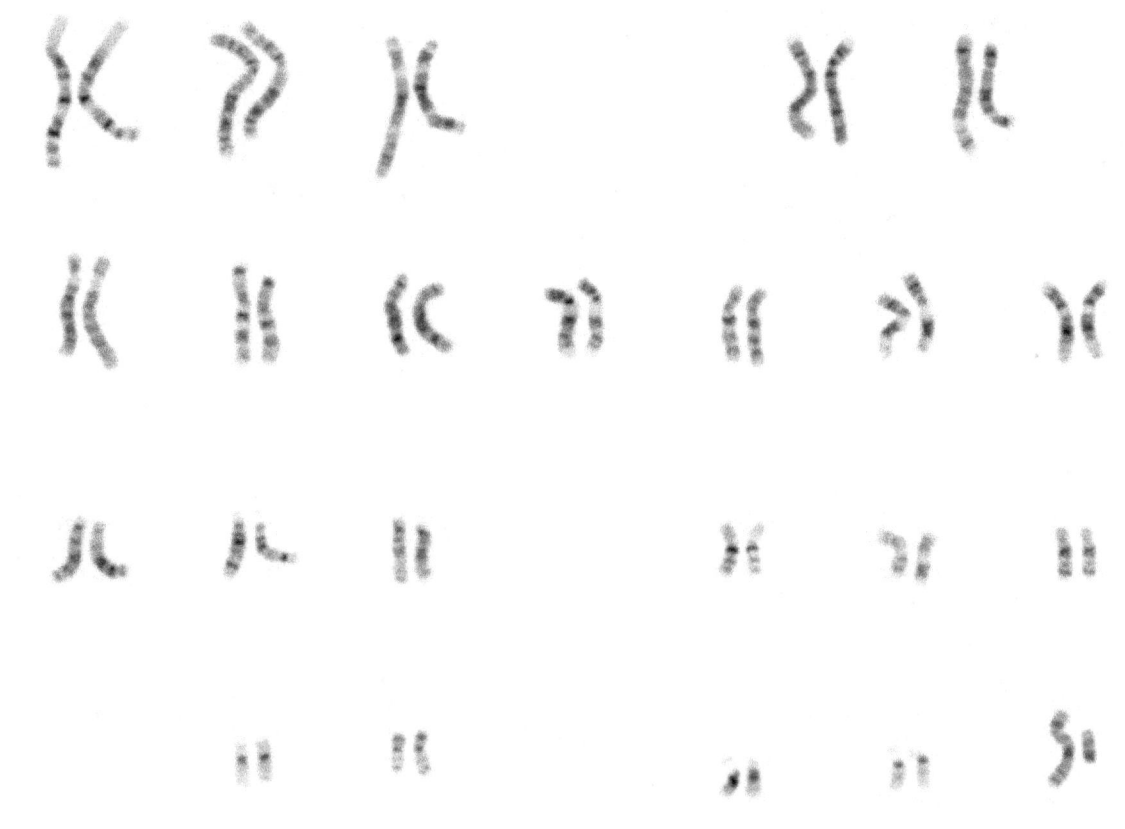

An image of the 46 chromosomes, making up the diploid genome of human male. (The mitochondrial chromosome is not shown.)

5.1 A brief history of whole genome sequencing

The shift from manual DNA sequencing methods such as Maxam-Gilbert sequencing and Sanger sequencing in the 1970s and 1980s to more rapid, automated sequencing methods in the 1990s played a crucial role in giving scientists the ability to sequence whole genomes.[9] *Haemophilus influenzae*, a commensal bacterium which resides in the human respiratory tract was the first organism to have its entire genome sequenced (**Figure 2.1**). The entire genome of this bacterium was published in 1995.[10] The genomes of *H. influenzae*, other Bacteria, and some Archaea were the first to be sequenced - largely due to their small genome size. *H. influenzae* has a genome of 1,830,140 base pairs of DNA.[10] In contrast, eukaryotes, both unicellular and multicellular such as *Amoeba dubia* and humans (*Homo sapiens*) respectively, have much larger genomes (see C-value paradox).[11] *Amoeba dubia* has a genome of 700 billion nucleotide pairs spread across thousands of chromosomes.[12] Humans contain fewer nucleotide pairs (about 3.2 billion in each germ cell - note the exact size of the human genome is still being revised) than *A. dubia* however their genome size far outweighs the genome size of individual bacteria.[13]

The first bacterial and archaeal genomes, including that of *H. influenzae*, were sequenced by Shotgun sequencing.[10] In 1996, the first eukaryotic genome - that of the yeast *Saccharomyces cerevisiae* was sequenced. *S. cerevisiae*, a model organism in biology has a genome of only around 12 million nucleotide pairs.[14] *S. cerevisiae* was the first *unicellular* eukaryote to have its whole genome sequenced. The first *multicellular* eukaryote, and animal, to have its whole genome sequenced was the nematode worm: *Caenorhabditis elegans* in 1998 (**Figure 2.2**).[15] Eukaryotic genomes are sequenced by several methods including Shotgun sequencing of short DNA fragments and sequencing of larger DNA clones from DNA libraries (see library (biology)) such as Bacterial artificial chromosomes (BACs) and Yeast artificial chromosomes

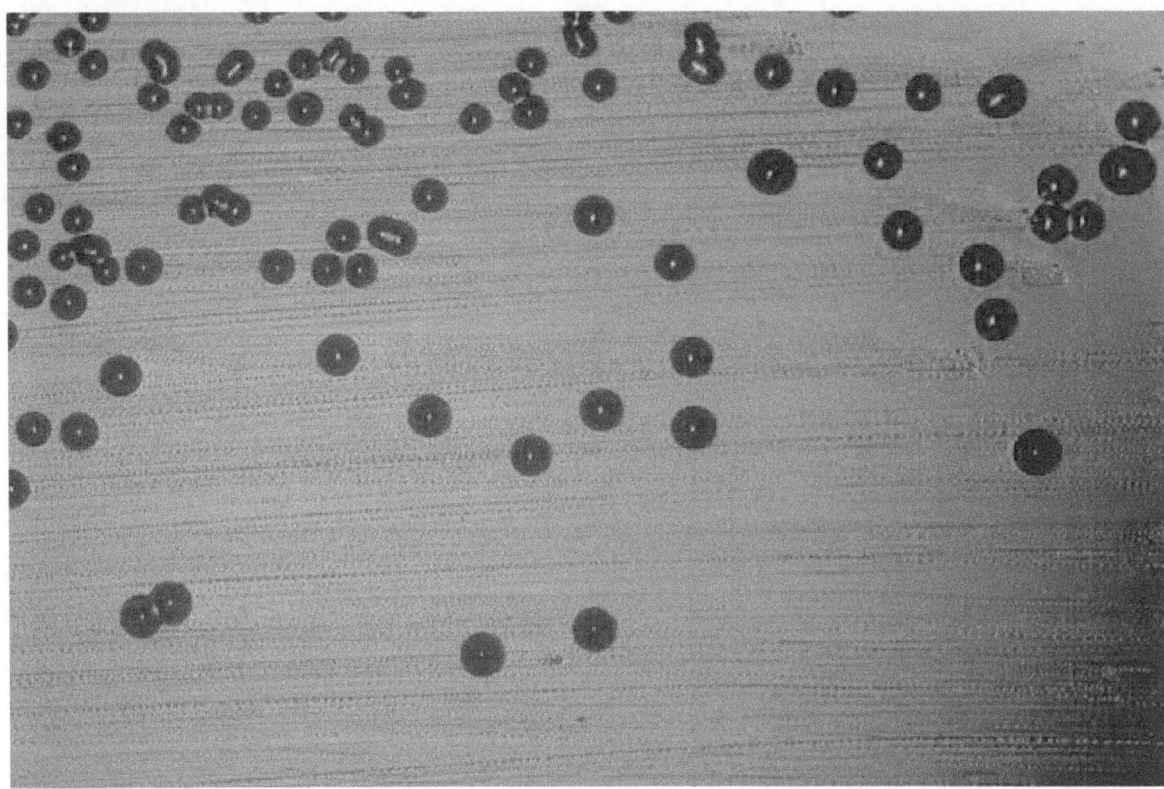

Figure 2.1 *The first whole genome to be sequenced was of the bacterium* Haemophilus influenzae.

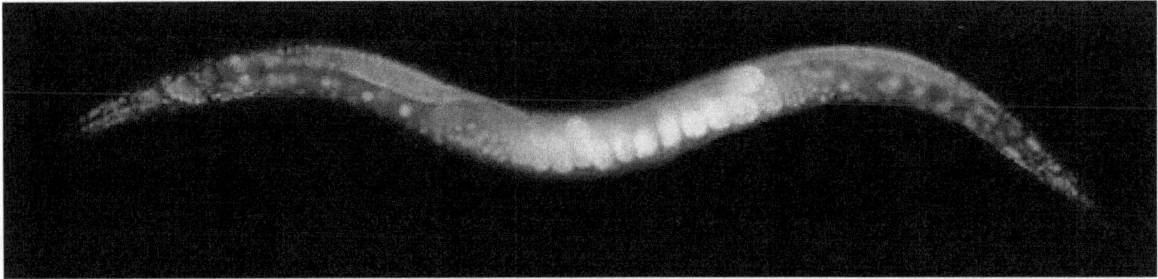

Figure 2.2 *The worm* Caenorhabditis elegans *was the first animal to have its whole genome sequenced.*

(YACs).[16]

In 1999, the entire DNA sequence of human chromosome 22, the shortest human autosome, was published.[17] By the year 2000, the second animal and second invertebrate (yet first insect) genome was sequenced - that of the fruit fly *Drosophila melanogaster* (**Figure 2.3**) - a popular choice of model organism in experimental research.[18] The first plant genome - that of the model organism *Arabidopsis thaliana* - was also fully sequenced by 2000 (**Figure 2.4**).[19] By 2001, a **draft** of the entire human genome sequence was published.[20] The genome of the laboratory mouse *Mus musculus* was completed in 2002 (**Figure 2.5**).[21]

In 2004, the Human Genome Project published the human genome.[22]

Currently, thousands of genomes have been sequenced.

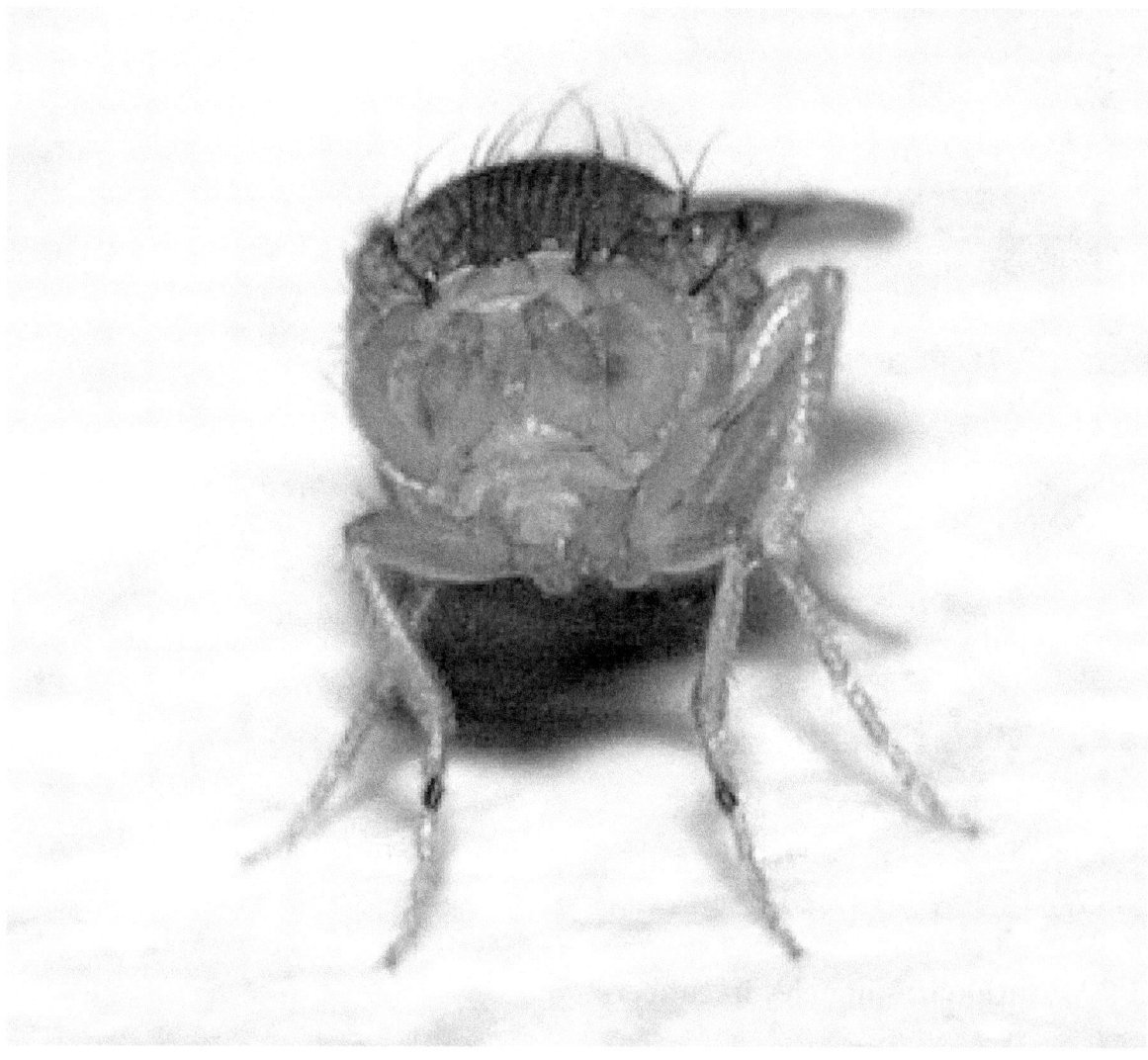

Figure 2.3 Drosophila melanogaster's *whole genome was sequenced in 2000.*

5.2 Cells used for sequencing

Almost any biological sample containing a full copy of the DNA—even a very small amount of DNA or ancient DNA—can provide the genetic material necessary for full genome sequencing. Such samples may include saliva, epithelial cells, bone marrow, hair (as long as the hair contains a hair follicle), seeds, plant leaves, or anything else that has DNA-containing cells.

The genome sequence of a single cell selected from a mixed population of cells can be determined using techniques of *single cell genome sequencing*. This has important advantages in environmental microbiology in cases where a single cell of a particular microorganism species can be isolated from a mixed population by microscopy on the basis of its morphological or other distinguishing characteristics. In such cases the normally necessary steps of isolation and growth of the organism in culture may be omitted, thus allowing the sequencing of a much greater spectrum of organism genomes.[23]

Single cell genome sequencing is being tested as a method of preimplantation genetic diagnosis, wherein a cell from the embryo created by in vitro fertilization is taken and analyzed before embryo transfer into the uterus.[24] After implantation, cell-free fetal DNA can be taken by simple venipuncture from the mother and used for whole genome sequencing of the fetus.[25]

Figure 2.4 Arabidopsis thaliana *was the first plant genome sequenced.*

5.3 Mutation frequencies in cancers

Whole genome sequencing has established the mutation frequency for whole human genomes. The mutation frequency in the whole genome between generations for humans (parent to child) is about 70 new mutations per generation.[26][27] An even lower level of variation was found comparing whole genome sequencing in blood cells for a pair of monozygotic (identical twins) 100-year-old centenarians.[28] Only 8 somatic differences were found, though somatic variation occurring in less than 20% of blood cells would be undetected.

In the specifically protein coding regions of the human genome, it is estimated that there are about 0.35 mutations that would change the protein sequence between parent/child generations (less than one mutated protein per generation).[29]

Cancers, however, have much higher mutation frequencies. The particular frequency depends on tissue type, whether there is a mis-match DNA repair deficiency, and exposure to DNA damaging agents such as UV-irradiation or components of tobacco smoke. Tuna and Amos have summarized the mutation frequencies per megabase (Mb),[30] as shown in the table (along with the indicated frequencies of mutations per genome).

The high mutation frequencies in cancers reflect the genome instability characteristic of cancers.

5.4 Early techniques

Sequencing of nearly an entire human genome was first accomplished in 2000 partly through the use of shotgun sequencing technology. While full genome shotgun sequencing for small (4000–7000 base pair) genomes was already in use

Figure 2.5 *The genome of the lab mouse* Mus musculus *was published in 2002.*

in 1979,[31] broader application benefited from pairwise end sequencing, known colloquially as *double-barrel shotgun sequencing*. As sequencing projects began to take on longer and more complicated genomes, multiple groups began to realize that useful information could be obtained by sequencing both ends of a fragment of DNA. Although sequencing both ends of the same fragment and keeping track of the paired data was more cumbersome than sequencing a single end of two distinct fragments, the knowledge that the two sequences were oriented in opposite directions and were about the length of a fragment apart from each other was valuable in reconstructing the sequence of the original target fragment.

The first published description of the use of paired ends was in 1990 as part of the sequencing of the human HPRT locus,[32] although the use of paired ends was limited to closing gaps after the application of a traditional shotgun sequencing approach. The first theoretical description of a pure pairwise end sequencing strategy, assuming fragments of constant length, was in 1991.[33] In 1995 Roach et al. introduced the innovation of using fragments of varying sizes,[34] and demonstrated that a pure pairwise end-sequencing strategy would be possible on large targets. The strategy was subsequently adopted by The Institute for Genomic Research (TIGR) to sequence the entire genome of the bacterium *Haemophilus influenzae* in 1995,[35] and then by Celera Genomics to sequence the entire fruit fly genome in 2000,[36] and subsequently the entire human genome. Applied Biosystems, now called Life Technologies, manufactured the automated capillary sequencers utilized by both Celera Genomics and The Human Genome Project.

It took 10 years and 50 scientists spanning the globe to sequence the genome of Elaeis guineensis *(oil palm). This genome was particularly difficult to sequence because it had many repeated sequences which are difficult to organise.*[8]

While capillary sequencing was the first approach to successfully sequence a nearly full human genome, it is still too expensive and takes too long for commercial purposes. Because of this, since 2005 capillary sequencing has been progressively displaced by newer technologies such as pyrosequencing, SMRT sequencing, and nanopore technology;[37] all of these new technologies nevertheless continue to employ the basic shotgun strategy, namely, parallelization and template generation via genome fragmentation.

Because the sequence data that is produced can be quite large (for example, there are approximately six billion base pairs in each human diploid genome), genomic data is stored electronically and requires a large amount of computing power and storage capacity. Full genome sequencing would have been nearly impossible before the advent of the microprocessor, computers, and the Information Age.

5.5   Current techniques

One possible way to accomplish the cost-effective high-throughput sequencing necessary to accomplish full genome sequencing is by using nanopore technology, which is a patented technology held by Harvard University and Oxford Nanopore Technologies and licensed to biotechnology companies.[38] To facilitate their full genome sequencing initiatives, Illumina licensed nanopore sequencing technology from Oxford Nanopore Technologies and Sequenom licensed the technology from Harvard University.[39][40]

Another possible way to accomplish cost-effective high-throughput sequencing is by utilizing fluorophore technology. Pacific Biosciences is currently using this approach in their SMRT (single molecule real time) DNA sequencing technology.

Complete Genomics has developed DNA Nanoball (DNB) technology that arranges DNA on self-assembling arrays.[42]

Pyrosequencing is a method of DNA sequencing based on the sequencing by synthesis principle.[43] The technique

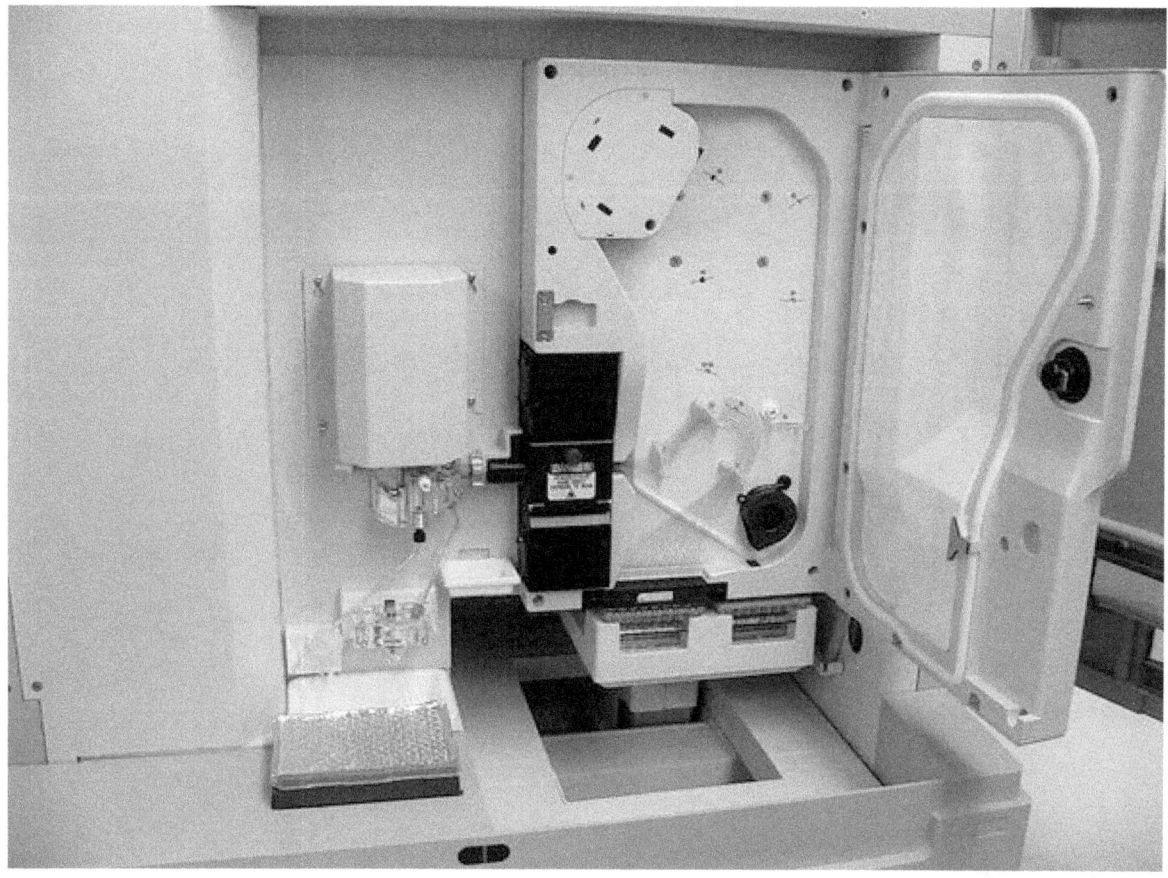

An ABI PRISM 3100 Genetic Analyzer. Such capillary sequencers automated the early efforts of sequencing genomes.

was developed by Pål Nyrén and his student Mostafa Ronaghi at the Royal Institute of Technology in Stockholm in 1996,[44][45][46] and is currently being used by 454 Life Sciences as a basis for a full genome sequencing platform.[47]

5.6 Commercialization

A number of public and private companies are competing to develop a full genome sequencing platform that is commercially robust for both research and clinical use,[48] including Illumina,[49] Knome,[50] Sequenom,[51] 454 Life Sciences,[52] Pacific Biosciences,[53] Complete Genomics,[54] Helicos Biosciences,[55] GE Global Research (General Electric), Affymetrix, IBM, Intelligent Bio-Systems,[56] Life Technologies and Oxford Nanopore Technologies.[57] These companies are heavily financed and backed by venture capitalists, hedge funds, and investment banks.[58][59]

5.6.1 Incentive

In October 2006, the X Prize Foundation, working in collaboration with the J. Craig Venter Science Foundation, established the Archon X Prize for Genomics,[60] intending to award US$10 million to "the first Team that can build a device and use it to sequence 100 human genomes within 10 days or less, with an accuracy of no more than one error in every 1,000,000 bases sequenced, with sequences accurately covering at least 98% of the genome, and at a recurring cost of no more than $1,000 per genome".[61] An error rate of 1 in 1,000,000 bases, out of a total of approximately six billion bases in the human diploid genome, would mean about 6,000 errors per genome. The error rates required for widespread clinical use, such as predictive medicine[62] is currently set by over 1,400 clinical single gene sequencing tests[63] (for

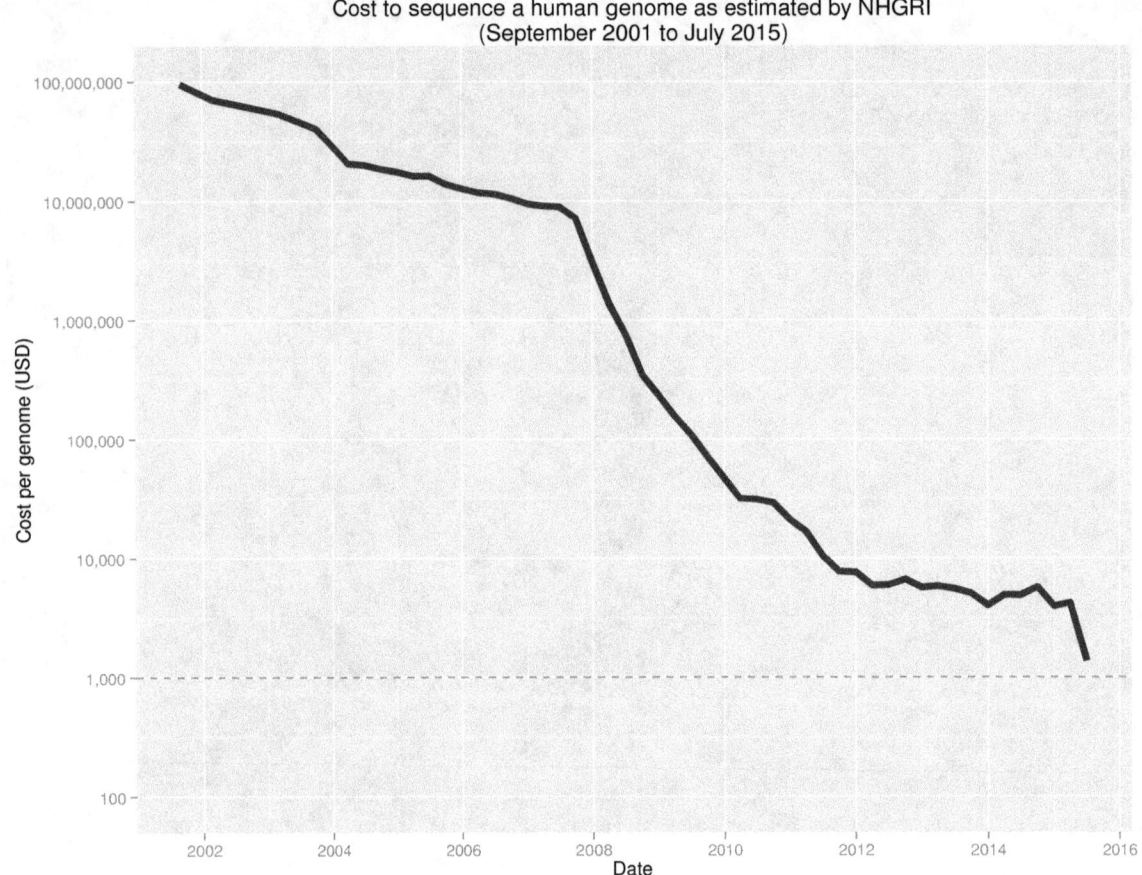

Total cost of sequencing a human genome as calculated by the NHGRI.

example, errors in BRCA1 gene for breast cancer risk analysis). As of August 2013, the Archon X Prize for Genomics has been cancelled.[64]

5.6.2   2007

In 2007, Applied Biosystems started selling a new type of sequencer called SOLiD System.[65] The technology allowed users to sequence 60 gigabases per run.[66]

5.6.3   2009

In March 2009, it was announced that Complete Genomics has signed a deal with the Broad Institute to sequence cancer patients' genomes and will be sequencing five full genomes to start.[67] In April 2009, Complete Genomics announced that it plans to sequence 1,000 full genomes between June 2009 and the end of the year and that they plan to be able to sequence one million full genomes *per year* by 2013.[68]

In June 2009, Illumina announced that they were launching their own Personal Full Genome Sequencing Service at a depth of 30× for $48,000 per genome.[69][70] Jay Flatley, Illumina's President and CEO, stated that "during the next five years, perhaps markedly sooner," the price point for full genome sequencing will fall from $48,000 to under $1,000.[71]

In August 2009, the founder of Helicos Biosciences, Stephen Quake, stated that using the company's Single Molecule

Sequencer he sequenced his own full genome for less than $50,000. He stated that he expects the cost to decrease to the $1,000 range within the next two to three years.[72]

In August 2009, Pacific Biosciences secured an additional $68 million in new financing, bringing their total capitalization to $188 million.[73] Pacific Biosciences said they are going to use this additional investment in order to prepare for the upcoming launch of their full genome sequencing service in 2010.[74] Complete Genomics followed by securing another $45 million in a fourth round venture funding during the same month.[75] Complete Genomics has also made the claim that it will sequence 10,000 full genomes by the end of 2010.[76]

GE Global Research is also part of this race to commercialize full genome sequencing as they have been working on creating a service that will deliver a full genome for $1,000 or less.[77][78]

In October 2009, IBM announced that they were also in the heated race to provide full genome sequencing for under $1,000, with their ultimate goal being able to provide their service for US$100 per genome.[79] IBM's full genome sequencing technology, which uses nanopores, is known as the "DNA Transistor".[80]

In November 2009, Complete Genomics published a peer-reviewed paper in *Science* demonstrating its ability to sequence a complete human genome for $1,700.[81] If true, this would mean the cost of full genome sequencing has come down exponentially within just a single year from around $100,000 to $50,000 and now to $1,700. This consumables cost was clearly detailed in the *Science* paper.[82] However, Complete Genomics has previously released statements that it was unable to follow through on. For example, the company stated it would officially launch and release its service during the "summer of 2009", provide a "$5,000" full genome sequencing service by the "summer of 2009", and "sequence 1,000 genomes between June 2009 and the end of 2009" – all of which, as of November 2009, have not yet occurred.[68][68][83][84] Complete Genomics launched its R&D human genome sequencing service in October 2008 and its commercial service in May 2010. The company sequenced 50 genomes in 2009. Since then, it has significantly increased the throughput of its genome sequencing factory and was able to sequence and analyze 300 genomes in Q3 2010.

Also in November 2009, Complete Genomics announced that it was beginning a large-scale human genome sequencing study of Huntington's disease (up to 100 genomes) with the Institute for Systems Biology.

5.6.4   2010

In March 2010, Researchers from the Medical College of Wisconsin announced the first successful use of Genome Wide sequencing to change the treatment of a patient.[85] This story was later retold in a Pulitzer prize winning article [86] and touted as a significant accomplishment in the journal *Nature*[87] and by the director of the NIH in presentations at congress.

In June 2010, Illumina lowered the cost of its individual sequencing service to $19,500 from $48,000.

5.6.5   2011

Knome provides full genome sequencing (98%) services for US$39,500 for consumers, or $29,500 for researchers (depending on their requirements).[88][89]

In May 2011, Illumina lowered its Full Genome Sequencing service to $5,000 per human genome, or $4,000 if ordering 50 or more.[90] Helicos Biosciences, Pacific Biosciences, Complete Genomics, Illumina, Sequenom, ION Torrent Systems, Halcyon Molecular, NABsys, IBM, and GE Global appear to all be going head to head in the race to commercialize full genome sequencing.[37][78]

5.6.6   2012

In January 2012, Life Technologies introduced a sequencer claimed to decode a human genome in one day for $1,000 although these claims have yet to be validated by customers on commercial devices.[91] A UK firm spun out from Oxford University has come up with a DNA sequencing machine (the MinION) the size of a USB memory stick which costs $900 and can sequence smaller genomes (but not full human genomes in the first version).[92] (While Oxford Nanopore stated in February that they would target having a sequencer in commercial early access by the end of 2012, this did not occur.)

In November 2012, *Gene by Gene, Ltd* started offering whole genome sequencing at an introductory price of $5,495 (with a minimum requirement of 3 samples per order). Currently the price is $6,995 and the minimum requirement has been removed.[93][94][95] However, more recent verification on the price for whole genome sequencing, as posted on the Web site of "Gene by Gene" (2015-09-07) shows that instead of decreasing, the price has significantly increased, to 9,995.00 USD for a basic whole genome sequence without analysis, or 10,395.00 USD for a whole genome sequence with alignment and variant calling.

A series of publications in 2012 showed the utility of SMRT sequencing from Pacific Biosciences in generating full genome sequences with de novo assembly.[96] Some of these papers reported automated pipelines that could be used for generating these whole-genome assemblies.[97][98] Other papers demonstrated how PacBio sequence data could be used to upgrade draft genomes to complete genomes.[99]

5.6.7   2014

A February 2014 news item stated that Illumina had announced "the long-awaited $1,000 human genome ... in the form of the company's HiSeq X 10 system."[100]

5.7   Disruption to DNA array market

Full genome sequencing provides information on a genome that is orders of magnitude larger than that provided by the previous leader in genotyping technology, DNA arrays. For humans, DNA arrays currently provide genotypic information on up to one million genetic variants,[101][102][103] while full genome sequencing will provide information on all six billion bases in the human genome, or 3,000 times more data. Because of this, full genome sequencing is considered a disruptive innovation to the DNA array markets as the accuracy of both range from 99.98% to 99.999% (in non-repetitive DNA regions) and their consumables cost of $5000 per 6 billion base pairs is competitive (for some applications) with DNA arrays ($500 per 1 million basepairs).[52] Agilent, another established DNA array manufacturer, is working on targeted (selective region) genome sequencing technologies.[104] It is thought that Affymetrix, the pioneer of array technology in the 1990s, has fallen behind due to significant corporate and stock turbulence and is currently not working on any known full genome sequencing approach.[105][106][107] It is unknown what will happen to the DNA array market once full genome sequencing becomes commercially widespread, especially as companies and laboratories providing this disruptive technology start to realize economies of scale. It is postulated, however, that this new technology may significantly diminish the total market size for arrays and any other sequencing technology once it becomes commonplace for individuals and newborns to have their full genomes sequenced.[108]

5.8   Sequencing versus analysis

In principle, full genome sequencing can provide raw data on all six billion nucleotides in an individual's DNA. However, it does not provide an analysis of what that information means or how it might be utilized in various clinical applications, such as in medicine to help prevent disease. Work toward that goal is continuously moving forward.

A 2015 study[109] done at Children's Mercy Hospital in Kansas City detailed the use of full genome sequencing including full analysis. The process took a record breaking 26 hours[110] and was done using Illumina HiSeq machines, the Edico Genome Dragen Processor, and several custom designed software packages. Most of this acceleration was achieved using the newly developed Dragen Processor which brought the analysis time down from 15 hours to 40 minutes.

5.9   Diagnostic use and societal impact

Further information: Personal genomics

Inexpensive, time-efficient full genome sequencing will be a major accomplishment not only for the field of genomics, but for the entire human civilization because, for the first time, individuals will be able to have their entire genome sequenced. Utilizing this information, it is speculated that health care professionals, such as physicians and genetic counselors, will eventually be able to use genomic information to predict what diseases a person may get in the future and attempt to either minimize the impact of that disease or avoid it altogether through the implementation of personalized, preventive medicine. Full genome sequencing will allow health care professionals to analyze the entire human genome of an individual and therefore detect all disease-related genetic variants, regardless of the genetic variant's prevalence or frequency. This will enable the rapidly emerging medical fields of predictive medicine and personalized medicine and will mark a significant leap forward for the clinical genetic revolution. Full genome sequencing is clearly of great importance for research into the basis of genetic disease and has shown significant benefit to a subset of individuals with rare disease in the clinical setting.[111][112][113][114] Illumina's CEO, Jay Flatley, stated in February 2009 that "A complete DNA read-out for every newborn will be technically feasible and affordable in less than five years, promising a revolution in healthcare" and that "by 2019 it will have become routine to map infants' genes when they are born".[115] This potential use of genome sequencing is highly controversial, as it runs counter to established ethical norms for predictive genetic testing of asymptomatic minors that have been well established in the fields of medical genetics and genetic counseling.[116][117][118][119] The traditional guidelines for genetic testing have been developed over the course of several decades since it first became possible to test for genetic markers associated with disease, prior to the advent of cost-effective, comprehensive genetic screening. It is established that norms, such as in the sciences and the field of genetics, are subject to change and evolve over time.[120][121] It is unknown whether traditional norms practiced in medical genetics today will be altered by new technological advancements such as full genome sequencing.

Currently available newborn screening for childhood diseases allows detection of rare disorders that can be prevented or better treated by early detection and intervention. Specific genetic tests are also available to determine an etiology when a child's symptoms appear to have a genetic basis. Full genome sequencing, in addition has the potential to reveal a large amount of information (such as carrier status for autosomal recessive disorders, genetic risk factors for complex adult-onset diseases, and other predictive medical and non-medical information) that is currently not completely understood, may not be clinically useful to the child during childhood, and may not necessarily be wanted by the individual upon reaching adulthood.[122] In addition to predicting disease risk in childhood, genetic testing may have other benefits (such as discovery of non-paternity) but may also have potential downsides (genetic discrimination, loss of anonymity, and psychological impacts).[123] Many publications regarding ethical guidelines for predictive genetic testing of asymptomatic minors may therefore have more to do with protecting minors and preserving the individual's privacy and autonomy to know or not to know their genetic information, than with the technology that makes the tests themselves possible.[124]

Due to recent cost reductions (see above) whole genome sequencing has become a realistic application in DNA diagnostics. In 2013, the 3Gb-TEST consortium obtained funding from the European Union to prepare the health care system for these innovations in DNA diagnostics.[125][126] Quality assessment schemes, Health technology assessment and guidelines have to be in place. The 3Gb-TEST consortium has identified the analysis and interpretation of sequence data as the most complicated step in the diagnostic process.[127] At the Consortium meeting in Athens in September 2014, the Consortium coined the word *genotranslation* for this crucial step. This step leads to a so-called *genoreport*. Guidelines are needed to determine the required content of these reports.

5.10 Ethical concerns

The majority of ethicists insist that the privacy of individuals undergoing genetic testing must be protected under all circumstances.[128] Data obtained from whole genome sequencing can not only reveal much information about the individual who is the source of DNA, but it can also reveal much probabilistic information about the DNA sequence of close genetic relatives.[129] Furthermore, the data obtained from whole genome sequencing can also reveal much useful predictive information about the relatives present and future health risks.[130] This raises important questions about what obligations, if any, are owed to the family members of the individuals who are undergoing genetic testing. In the Western/European society, tested individuals are usually encouraged to share important information on the genetic diagnosis with their close relatives since the importance of the genetic diagnosis for offspring and other close relatives is usually one of the reasons for seeking a genetic testing in the first place.[128] Nevertheless, Sijmons et al. (2011) also mention that a major ethical dilemma can develop when the patients refuse to share information on a diagnosis that is made for serious genetic disorder that is highly preventable and where there is a high risk to relatives carrying the same disease mutation.[129] Under such circumstances, the clinician may suspect that the relatives would rather know of the diagnosis and hence the clinician can face a conflict of interest with respect to patient-doctor confidentiality.[129]

Another major privacy concern is the scientific need to put information on patient's genotypes and phenotypes into the public scientific databases such as the locus specific databases.[129] Although only anonymous patient data are submitted to the locus specific databases, patients might still be identifiable by their relatives in the case of finding a rare disease or a rare missense mutation.[129]

5.11   People with public genome sequences

The first nearly complete human genomes sequenced were J. Craig Venter's (American at 7.5-fold average coverage) in 2007,[131][132][133] followed by James Watson's (American at 7.4-fold),[134][135][136] a Han Chinese (YH at 36-fold),[137] a Yoruban from Nigeria (at 30-fold),[138] a female leukemia patient (at 33 and 14-fold coverage for tumor and normal tissues),[139] and Seong-Jin Kim (Korean at 29-fold).[140] The first two persons with their full genome sequenced, James Watson and Craig Venter, two American scientists of European ancestry, were found to be genetically more closely related to and having more alleles in common with Korean scientist, Seong-Jin Kim (1,824,482 and 1,736,340, respectively) than with each other (1,715,851).[141] Steve Jobs was among the first 20 people to have their whole genome sequenced, reportedly for the cost of $100,000.[142] As of June 2012, there are 69 nearly complete human genomes publicly available.(reference - page not found)[143] Commercialization of full genome sequencing is in an early stage and growing rapidly.

5.12   See also

- Whole Exome Sequencing

- DNA sequencing

- DNA microarray

- DNA profiling

- Medical genetics

- Nucleic acid sequence

- Human Genome Project

- Personal Genome Project

- Genomics England

- List of sequenced eukaryotic genomes

- List of sequenced bacterial genomes

- List of sequenced archaeal genomes

- Predictive medicine

- Personalized medicine

5.13 References

[1] Alberts, Bruce; Johnson, Alexander; Lewis, Julian; Raff, Martin; Roberts, Keith; Walter, Peter (2008). "8". *Molecular biology of the cell* (5th ed.). New York: Garland Science. p. 550. ISBN 0-8153-4106-7.

[2] Kijk magazine, 01 January 2009

[3] Gilissen (Jul 2014). "Genome sequencing identifies major causes of severe intellectual disability". *Nature* **511** (7509): 344–7. doi:10.1038/nature13394. PMID 24896178.

[4] Nones, K; Waddell, N; Wayte, N; Patch, AM; Bailey, P; Newell, F; Holmes, O; Fink, JL; Quinn, MC; Tang, YH; Lampe, G; Quek, K; Loffler, KA; Manning, S; Idrisoglu, S; Miller, D; Xu, Q; Waddell, N; Wilson, PJ; Bruxner, TJ; Christ, AN; Harliwong, I; Nourse, C; Nourbakhsh, E; Anderson, M; Kazakoff, S; Leonard, C; Wood, S; Simpson, PT; Reid, LE; Krause, L; Hussey, DJ; Watson, DI; Lord, RV; Nancarrow, D; Phillips, WA; Gotley, D; Smithers, BM; Whiteman, DC; Hayward, NK; Campbell, PJ; Pearson, JV; Grimmond, SM; Barbour, AP (29 October 2014). "Genomic catastrophes frequently arise in esophageal adenocarcinoma and drive tumorigenesis". *Nature communications* **5**: 5224. doi:10.1038/ncomms6224. PMID 25351503.

[5] van El, CG; Cornel, MC; Borry, P; Hastings, RJ; Fellmann, F; Hodgson, SV; Howard, HC; Cambon-Thomsen, A; Knoppers, BM; Meijers-Heijboer, H; Scheffer, H; Tranebjaerg, L; Dondorp, W; de Wert, GM (June 2013). "Whole-genome sequencing in health care. Recommendations of the European Society of Human Genetics". *European journal of human genetics : EJHG*. 21 Suppl 1: S1–5. PMID 23819146.

[6] Mooney, Sean (Sep 2014). "Progress towards the integration of pharmacogenomics in practice". *Human Genetics*. doi:10.1007/s00439-014-1484-7. PMID 25238897.

[7] Fareed M., Afzal M (2013). "Single nucleotide polymorphism in genome-wide association of human population: A tool for broad spectrum service". *Egyptian Journal of Medical Human Genetics* **14**: 123–134. doi:10.1016/j.ejmhg.2012.08.001.

[8] Marx, Vivien (11 September 2013). "Next-generation sequencing: The genome jigsaw". *Nature* **501** (7466): 263–268. doi:10.1038/501261a.

[9] al.], Bruce Alberts ... [et (2008). *Molecular biology of the cell* (5th ed.). New York: Garland Science. p. 551. ISBN 0-8153-4106-7.

[10] Fleischmann, R.; Adams, M.; White, O; Clayton, R.; Kirkness, E.; Kerlavage, A.; Bult, C.; Tomb, J.; Dougherty, B.; Merrick, J.; al., e. (28 July 1995). "Whole-genome random sequencing and assembly of Haemophilus influenzae Rd". *Science* **269** (5223): 496–512. doi:10.1126/science.7542800. PMID 7542800.

[11] Eddy, Sean R. (November 2012). "The C-value paradox, junk DNA and ENCODE". *Current Biology* **22** (21): R898–R899. doi:10.1016/j.cub.2012.10.002. PMID 23137679.

[12] PELLICER, JAUME; FAY, MICHAEL F.; LEITCH, ILIA J. (15 September 2010). "The largest eukaryotic genome of them all?". *Botanical Journal of the Linnean Society* **164** (1): 10–15. doi:10.1111/j.1095-8339.2010.01072.x.

[13] Human Genome Sequencing Consortium, International (21 October 2004). "Finishing the euchromatic sequence of the human genome". *Nature* **431** (7011): 931–945. doi:10.1038/nature03001. PMID 15496913.

[14] Goffeau, A.; Barrell, B. G.; Bussey, H.; Davis, R. W.; Dujon, B.; Feldmann, H.; Galibert, F.; Hoheisel, J. D.; Jacq, C.; Johnston, M.; Louis, E. J.; Mewes, H. W.; Murakami, Y.; Philippsen, P.; Tettelin, H.; Oliver, S. G. (25 October 1996). "Life with 6000 Genes" (PDF). *Science* **274** (5287): 546–567. doi:10.1126/science.274.5287.546. PMID 8849441.

[15] The C. elegans Sequencing Consortium (11 December 1998). "Genome Sequence of the Nematode C. elegans: A Platform for Investigating Biology". *Science* **282** (5396): 2012–2018. doi:10.1126/science.282.5396.2012. PMID 9851916.

[16] al.], Bruce Alberts ... [et (2008). *Molecular biology of the cell* (5th ed.). New York: Garland Science. p. 552. ISBN 0-8153-4106-7.

[17] Dunham, I. "The DNA sequence of human chromosome 22". *nature.com*.

[18] Adams MD, Celniker SE, Holt RA; et al. (2000-03-24). "The Genome Sequence of Drosophila melanogaster". *Science* **287**: 2185–2195. doi:10.1126/science.287.5461.2185. PMID 10731132.

[19] "Analysis of the genome sequence of the flowering plant Arabidopsis thaliana.". *Nature***408**: 796–815. 2000-12-14. doi:10.1038/ PMID 11130711.

[20] Venter JC, Adams MD, Myers EW; et al. (2001-02-16). "The Sequence of the Human Genome". *Science* **291**: 1304–1351. doi:10.1126/science.1058040. PMID 11181995.

[21] Waterston RH, Lindblad-Toh K, Birney E; et al. (2002-10-31). "Initial sequencing and comparative analysis of the mouse genome". *Nature* **420**: 520–562. doi:10.1038/nature01262. PMID 12466850.

[22] "Finishing the euchromatic sequence of the human genome". *Nature* **431**: 931–945. 07.09.2004. doi:10.1038/nature03001. PMID 15496913. Check date values in: |date= (help)

[23] Braslavsky, Ido; et al. (2003). "Sequence information can be obtained from single DNA molecules". *Proc Natl Acad Sci USA* **100** (7): 3960–3984. doi:10.1073/pnas.0230489100. PMC 153030. PMID 12651960.

[24] Single-cell Sequencing Makes Strides in the Clinic with Cancer and PGD First Applications from Clinical Sequencing News. By Monica Heger. October 02, 2013

[25] Yurkiewicz, I. R.; Korf, B. R.; Lehmann, L. S. (2014). "Prenatal whole-genome sequencing--is the quest to know a fetus's future ethical?". *New England Journal of Medicine* **370** (3): 195–7. doi:10.1056/NEJMp1215536. PMID 24428465.

[26] Roach JC; Glusman G; Smit AF; et al. (April 2010). "Analysis of genetic inheritance in a family quartet by whole-genome sequencing". *Science* **328** (5978): 636–9. doi:10.1126/science.1186802. PMC 3037280. PMID 20220176.

[27] Campbell CD; Chong JX; Malig M; et al. (November 2012). "Estimating the human mutation rate using autozygosity in a founder population". *Nat. Genet.* **44** (11): 1277–81. doi:10.1038/ng.2418. PMC 3483378. PMID 23001126.

[28] Ye K; Beekman M; Lameijer EW; Zhang Y; Moed MH; van den Akker EB; Deelen J; Houwing-Duistermaat JJ; Kremer D; Anvar SY; Laros JF; Jones D; Raine K; Blackburne B; Potluri S; Long Q; Guryev V; van der Breggen R; Westendorp RG; 't Hoen PA; den Dunnen J; van Ommen GJ; Willemsen G; Pitts SJ; Cox DR; Ning Z; Boomsma DI; Slagboom PE (December 2013). "Aging as accelerated accumulation of somatic variants: whole-genome sequencing of centenarian and middle-aged monozygotic twin pairs". *Twin Res Hum Genet* **16** (6): 1026–32. doi:10.1017/thg.2013.73. PMID 24182360.

[29] Keightley PD (February 2012). "Rates and fitness consequences of new mutations in humans". *Genetics* **190** (2): 295–304. doi:10.1534/genetics.111.134668. PMC 3276617. PMID 22345605.

[30] Tuna M; Amos CI (November 2013). "Genomic sequencing in cancer". *Cancer Lett.* **340**(2): 161–70. doi:10.1016/j.canlet.201 PMID 23178448.

[31] Staden R (June 1979). "A strategy of DNA sequencing employing computer programs". *Nucleic Acids Res.* **6** (7): 2601–10. doi:10.1093/nar/6.7.2601. PMC 327874. PMID 461197.

[32] Edwards, A; Caskey, T (1991). "Closure strategies for random DNA sequencing". *Methods: A Companion to Methods in Enzymology* **3** (1): 41–47. doi:10.1016/S1046-2023(05)80162-8.

[33] Edwards A; Voss H; Rice P; Civitello A; Stegemann J; Schwager C; Zimmermann J; Erfle H; Caskey CT; Ansorge W (April 1990). "Automated DNA sequencing of the human HPRT locus". *Genomics* **6** (4): 593–608. doi:10.1016/0888-7543(90)90493-E. PMID 2341149.

[34] Roach JC; Boysen C; Wang K; Hood L (March 1995). "Pairwise end sequencing: a unified approach to genomic mapping and sequencing". *Genomics* **26** (2): 345–53. doi:10.1016/0888-7543(95)80219-C. PMID 7601461.

[35] Fleischmann RD; Adams MD; White O; Clayton RA; Kirkness EF; Kerlavage AR; Bult CJ; Tomb JF; Dougherty BA; Merrick JM; McKenney; Sutton; Fitzhugh; Fields; Gocyne; Scott; Shirley; Liu; Glodek; Kelley; Weidman; Phillips; Spriggs; Hedblom; Cotton; Utterback; Hanna; Nguyen; Saudek; et al. (July 1995). "Whole-genome random sequencing and assembly of Haemophilus influenzae Rd". *Science* **269** (5223): 496–512. Bibcode:1995Sci...269..496F. doi:10.1126/science.7542800. PMID 7542800.

[36] Adams, MD; et al. (2000). "The genome sequence of Drosophila melanogaster". *Science* **287**(5461): 2185–95. Bibcode:2000S doi:10.1126/science.287.5461.2185. PMID 10731132.

[37] Mukhopadhyay R (February 2009). "DNA sequencers: the next generation". *Anal. Chem.* **81**(5): 1736–40. doi:10.1021/ac8027. PMID 19193124.

[38] "Harvard University and Oxford Nanopore Technologies Announce Licence Agreement to Advance Nanopore DNA Sequencing and other Applications". Nanotechwire. August 5, 2008. Retrieved 2009-02-23.

[39] "Illumina and Oxford Nanopore Enter into Broad Commercialization Agreement". Reuters. January 12, 2009. Retrieved 2009-02-23.

[40]

[41] "Single Molecule Real Time (SMRT) DNA Sequencing". Pacific Biosciences. Retrieved 2009-02-23.

[42] "Complete Human Genome Sequencing Technology Overview" (PDF). Complete Genomics. 2009. Retrieved 2009-02-23.

[43] "Definition of pyrosequencing from the Nature Reviews Genetics Glossary". Retrieved 2008-10-28.

[44] Ronaghi M; Uhlén M; Nyrén P (July 1998). "A sequencing method based on real-time pyrophosphate". *Science* **281** (5375): 363, 365. doi:10.1126/science.281.5375.363. PMID 9705713.

[45] Ronaghi M; Karamohamed S; Pettersson B; Uhlén M; Nyrén P (November 1996). "Real-time DNA sequencing using detection of pyrophosphate release". *Anal. Biochem.* **242** (1): 84–9. doi:10.1006/abio.1996.0432. PMID 8923969.

[46] Nyrén P (2007). "The history of pyrosequencing". *Methods Mol. Biol.* **373**: 1–14. doi:10.1385/1-59745-377-3:1. ISBN 1-59745-377-3. PMID 17185753.

[47] http://files.shareholder.com/downloads/CRGN/0x0x53381/386c4aaa-f36e-4b7a-9ff0-c06e61fad31f/211559.pdf

[48] "Article : Race to Cut Whole Genome Sequencing Costs Genetic Engineering & Biotechnology News — Biotechnology from Bench to Business". Genengnews.com. Retrieved 2009-02-23.

[49] "Whole Genome Sequencing Costs Continue to Drop". Eyeondna.com. Retrieved 2009-02-23.

[50] Harmon, Katherine (2010-06-28). "Genome Sequencing for the Rest of Us". Scientific American. Retrieved 2010-08-13.

[51] San Diego/Orange County Technology News. "Sequenom to Develop Third-Generation Nanopore-Based Single Molecule Sequencing Technology". Freshnews.com. Retrieved 2009-02-24.

[52] "Article : Whole Genome Sequencing in 24 Hours Genetic Engineering & Biotechnology News — Biotechnology from Bench to Business". Genengnews.com. Retrieved 2009-02-23.

[53] "Pacific Bio lifts the veil on its high-speed genome-sequencing effort". VentureBeat. Retrieved 2009-02-23.

[54] "Bio-IT World". Bio-IT World. 2008-10-06. Retrieved 2009-02-23.

[55] "With New Machine, Helicos Brings Personal Genome Sequencing A Step Closer". Xconomy. 2008-04-22. Retrieved 2011-01-28.

[56] "Whole genome sequencing costs continue to fall: $300 million in 2003, $1 million 2007, $60,000 now, $5000 by year end". Nextbigfuture.com. 2008-03-25. Retrieved 2011-01-28.

[57] "Han Cao's nanofluidic chip could cut DNA sequencing costs dramatically". Technology Review.

[58] John Carroll (2008-07-14). "Pacific Biosciences gains $100M for sequencing tech". FierceBiotech. Retrieved 2009-02-23.

[59] Sibley, Lisa (2009-02-08). "Complete Genomics brings radical reduction in cost". *Silicon Valley / San Jose Business Journal* (Sanjose.bizjournals.com). Retrieved 2009-02-23.

[60] Carlson, Rob (2007-01-02). "A Few Thoughts on Rapid Genome Sequencing and The Archon Prize — synthesis". Synthesis.cc. Retrieved 2009-02-23.

[61] "PRIZE Overview: Archon X PRIZE for Genomics".

[62] Bentley DR (December 2006). "Whole-genome re-sequencing". *Curr. Opin. Genet. Dev.* **16**(6): 545–552. doi:10.1016/j.gde PMID 17055251.

[63] "GeneTests.org".

[64] Diamandis, Peter. "Outpaced by Innovation: Canceling an XPRIZE". *Huffington Post.*

[65] "SOLiD System — a next-gen DNA sequencing platform announced". Gizmag.com. 2007-10-27. Retrieved 2009-02-24.

[66] "The $1000 Genome: Coming Soon?". Dddmag.com. 2010-04-01. Retrieved 2011-01-28.

[67] "Complete Genomics, Broad Institute Forge Cancer Sequencing Collaboration". Bio-IT World. Retrieved 2011-01-28.

[68] Walsh, Fergus (2009-04-08). "Era of personalised medicine awaits". *BBC News*. Retrieved 2010-05-03.

[69] "Individual genome sequencing — Illumina, Inc.". Everygenome.com. Retrieved 2011-01-28.

[70] "Illumina launches personal genome sequencing service for $48,000 : Genetic Future". Scienceblogs.com. Retrieved 2011-01-28.

[71] "Illumina demos concept iPhone app for genetic data sharing". mobihealthnews. 2009-06-10. Retrieved 2011-01-28.

[72] Wade, Nicholas (2009-08-11). "Cost of Decoding a Genome Is Lowered". *The New York Times*. Retrieved 2010-05-03.

[73] Camille Ricketts (2009-08-13). "Pacific Biosciences takes $68M as genome sequencing becomes more competitive". VentureBeat. Retrieved 2011-01-28.

[74] "Pacific Biosciences Raises Additional $68 Million in Financing". FierceBiotech. 2009-08-12. Retrieved 2011-01-28.

[75] "Silicon Valley startup Complete Genomics promises low-cost DNA sequencing". *San Jose Mercury News*. Mercurynews.com. Retrieved 2011-01-28.

[76] "Silicon Valley Startup Complete Genomics Promises Low-Cost DNA Sequencing". Istockanalyst.com. 2009-08-24. Retrieved 2011-01-28.

[77] Jacquin Niles. "Explaining Sequencing | The Daily Scan". GenomeWeb. Retrieved 2011-01-28.

[78] "NHGRI Awards More than $50M for Low-Cost DNA Sequencing Tech Development". *Genome Web*. 2009.

[79] JOHN MARKOFF (October 5, 2009). "I.B.M. Joins Pursuit of $1,000 Personal Genome". *The Newyork Times*. Retrieved May 15, 2013.

[80] Shankland, Stephen (2009-10-06). "IBM Research jumps into genetic sequencing | Deep Tech". *CNET News*. News.cnet.com. Retrieved 2011-01-28.

[81]

[82] Drmanac R, Sparks AB, Callow MJ et al.: Human genome sequencing using unchained base reads on self-assembling DNA nanoarrays" *Science* 327(5961), 78-81 (2010)

[83] "Broad Institute to use Complete Genomics to sequence genomes of cancer patients : Genetic Future". Scienceblogs.com. Retrieved 2011-01-28.

[84] "Five Thousand Bucks for Your Genome". Technology Review. 2008-10-20. Retrieved 2009-02-23.

[85] http://journals.lww.com/geneticsinmedicine/Abstract/2011/03000/Making_a_definitive_diagnosis__Successful_clinical.15

[86] "One In A Billion: A boy's life, a medical mystery".

[87] "US clinics quietly embrace whole-genome sequencing".

[88] Herper, Matthew (2010-06-03). "Your Genome is Coming". Forbes. Retrieved 2010-08-13.

[89] Lauerman, John (2009-02-05). "Complete Genomics Drives Down Cost of Genome Sequence to $5,000". Bloomberg.com. Retrieved 2011-01-28.

[90] "Illumina Announces $5,000 Genome Pricing".

[91] http://www.lifetechnologies.com/us/en/home/about-us/news-gallery/press-releases/2012/life-techologies-itroduces-the-bech html.html

[92] https://www.nytimes.com/2012/02/18/health/oxford-nanopore-unveils-tiny-dna-sequencing-device.html

[93] "Products". dnadtc.com. Retrieved 28 November 2012.

[94] "Gene By Gene Launches DNA DTC". *The Wall Street Journal*. 29 November 2012. Retrieved 29 November 2012.

[95] Vorhaus, Dan (29 November 2012). "DNA DTC: The Return of Direct to Consumer Whole Genome Sequencing". genomic-slawreport.com. Retrieved 30 November 2012.

[96] "Finished bacterial genomes from shotgun sequence data" (PDF).

[97] Koren, Sergey (July 2012). "Hybrid error correction and de novo assembly of single-molecule sequencing reads". *NatureBiotechnology* **30** (7): 693–700. doi:10.1038/nbt.2280. PMID 22750884.

[98] A Klammer, Aaron (July 2012). "A hybrid approach for the automated finishing of bacterial genomes". *NatureBiotechnology* **30** (7). Retrieved 1 July 2012.

[99] "Mind the Gap:Upgrading Genomes with Pacific Biosciences RS Long-Read Sequencing Technology". *PLoS ONE* **7**: e47768. doi:10.1371/journal.pone.0047768.

[100] "Illumina Sequencer Enables $1,000 Genome". News: Genomics & Proteomics. *Gen. Eng. Biotechnol. News* (paper) **34** (4). 15 February 2014. p. 18.

[101] "Genomics Core". Gladstone.ucsf.edu. Retrieved 2009-02-23.

[102] Nishida N; Koike A; Tajima A; Ogasawara Y; Ishibashi Y; Uehara Y; Inoue I; Tokunaga K (2008). "Evaluating the performance of Affymetrix SNP Array 6.0 platform with 400 Japanese individuals". *BMC Genomics* **9** (1): 431. doi:10.1186/1471-2164-9-431. PMC 2566316. PMID 18803882.

[103] Petrone, Justin. "Illumina, DeCode Build 1M SNP Chip; Q2 Launch to Coincide with Release of Affy's 6.0 SNP Array | BioArray News | Arrays". GenomeWeb. Retrieved 2009-02-23.

[104] "Agilent Technologies Announces Licensing Agreement with Broad Institute to Develop Genome-Partitioning Kits to Streamline Next-Generation Sequencing".

[105] "Affymetrix stock slumps 30% on forecast". *Sacramento Business Journal* (Sacramento.bizjournals.com). 2008-07-25. Retrieved 2009-02-23.

[106] Bluis, John (2006-04-24). "Affymetrix Gets Chipped Again". Fool.com. Retrieved 2009-02-23.

[107] "The chips are down". *Nature* **444** (7117): 256–7. November 2006. Bibcode:2006Natur.444..256.. doi:10.1038/444256a. PMID 17108930.

[108] Coombs A (October 2008). "The sequencing shakeup". *Nat. Biotechnol.* **26** (10): 1109–12. doi:10.1038/nbt1008-1109. PMID 18846083.

[109] http://www.genomemedicine.com/content/7/1/100

[110] http://spectrum.ieee.org/tech-talk/biomedical/diagnostics/new-genetic-technologies-diagnose-critically-ill-infants-within-26

[111] Ng SB; Buckingham KJ; Lee C; et al. (January 2010). "Exome sequencing identifies the cause of a mendelian disorder". *Nat. Genet.* **42** (1): 30–5. doi:10.1038/ng.499. PMC 2847889. PMID 19915526.

[112] Hannibal MC; Buckingham KJ; Ng SB; et al. (July 2011). "Spectrum of MLL2 (ALR) mutations in 110 cases of Kabuki syndrome". *Am. J. Med. Genet. A* **155A** (7): 1511–6. doi:10.1002/ajmg.a.34074. PMC 3121928. PMID 21671394.

[113] Worthey EA; Mayer AN; Syverson GD; et al. (March 2011). "Making a definitive diagnosis: successful clinical application of whole exome sequencing in a child with intractable inflammatory bowel disease". *Genet. Med.* **13** (3): 255–62. doi:10.1097/GIM.0b013e3182088158. PMID 21173700.

[114] Goh V; Helbling D; Biank V; Jarzembowski J; Dimmock D (June 2011). "Next Generation Sequencing Facilitates The Diagnosis In A Child With Twinkle Mutations Causing Cholestatic Liver Failure". *J Pediatr Gastroenterol Nutr* **54** (2): 291–4. doi:10.1097/MPG.0b013e318227e53c. PMID 21681116.

[115] Henderson, Mark (2009-02-09). "Genetic mapping of babies by 2019 will transform preventive medicine". London: Times Online. Retrieved 2009-02-23.

[116] McCabe LL; McCabe ER (June 2001). "Postgenomic medicine. Presymptomatic testing for prediction and prevention". *Clin Perinatol* **28** (2): 425–34. doi:10.1016/S0095-5108(05)70094-4. PMID 11499063.

[117] Nelson RM; Botkjin JR; Kodish ED; et al. (June 2001). "Ethical issues with genetic testing in pediatrics". *Pediatrics* **107** (6): 1451–5. doi:10.1542/peds.107.6.1451. PMID 11389275.

[118] Borry P; Fryns JP; Schotsmans P; Dierickx K (February 2006). "Carrier testing in minors: a systematic review of guidelines and position papers". *Eur. J. Hum. Genet.* **14** (2): 133–8. doi:10.1038/sj.ejhg.5201509. PMID 16267502.

[119] Borry P; Stultiens L; Nys H; Cassiman JJ; Dierickx K (November 2006). "Presymptomatic and predictive genetic testing in minors: a systematic review of guidelines and position papers". *Clin. Genet.* **70** (5): 374–81. doi:10.1111/j.1399-0004.2006.00692.x. PMID 17026616.

[120] Mesoudi A; Danielson P (August 2008). "Ethics, evolution and culture". *Theory Biosci.* **127** (3): 229–40. doi:10.1007/s12064-008-0027-y. PMID 18357481.

[121] Ehrlich PR; Levin SA (June 2005). "The evolution of norms". *PLoS Biol.* **3** (6): e194. doi:10.1371/journal.pbio.0030194. PMC 1149491. PMID 15941355.

[122] Mayer AN; Dimmock DP; Arca MJ; et al. (March 2011). "A timely arrival for genomic medicine". *Genet. Med.* **13** (3): 195–6. doi:10.1097/GIM.0b013e3182095089. PMID 21169843.

[123] Ayday E; De Cristofaro E; Hubaux JP; Tsudik G (2015). "The Chills and Thrills of Whole Genome Sequencing". *ArXiv Repository.* arXiv:1306.1264. Bibcode:2015arXiv1306.1264.

[124] Borry, P.; Evers-Kiebooms, G.; Cornel, MC; Clarke, A; Dierickx, K; Public Professional Policy Committee (PPPC) of the European Society of Human Genetics (ESHG) (2009). "Genetic testing in asymptomatic minors Background considerations towards ESHG Recommendations". *Eur J Hum Genet* **17** (6): 711–9. doi:10.1038/ejhg.2009.25. PMC 2947094. PMID 19277061.

[125] "Introducing diagnostic applications of '3Gb-testing' in human genetics".

[126] "Beyond public health genomics: proposals from an international working group". *Eur J Public Health* **24**: 877–879. Aug 2014. doi:10.1093/eurpub/cku142. PMID 25168910.

[127] "RD-Connect News: 18 July 2014, Issue 7".

[128] Sijmons, R.H; Van Langen, I.M (2011). "A clinical perspective on ethical issues in genetic issues". *Accountability in Research: Policies and Quality Assurance* **18** (3): 148–162. doi:10.1080/08989621.2011.575033.

[129] Sijmons, R.H.; Van Langen, I.M (2011). "A clinical perspective on ethical issues in genetic testing". *Accountability in Research: Policies and Quality Assurance* **18** (3): 148–162. doi:10.1080/08989621.2011.575033.

[130] McGuire, Amy, L; Caulfield, Timothy (2008). "Science and Society: Research ethics and the challenge of whole-genome sequencing". *Nature Reviews: Genetics* **9** (2): 152–156. doi:10.1038/nrg2302.

[131] Wade, Nicholas (September 4, 2007). "In the Genome Race, the Sequel Is Personal". New York Times. Retrieved February 22, 2009.

[132] Nature. "Access : All about Craig: the first 'full' genome sequence". Nature. Retrieved 2009-02-24.

[133] Levy S; Sutton G; Ng PC; Feuk L; Halpern AL; Walenz BP; Axelrod N; Huang J; Kirkness EF; Denisov G; Lin Y; MacDonald JR; Pang AW; Shago M; Stockwell TB; Tsiamouri A; Bafna V; Bansal V; Kravitz SA; Busam DA; Beeson KY; McIntosh TC; Remington KA; Abril JF; Gill J; Borman J; Rogers YH; Frazier ME; Scherer SW; Strausberg RL; Venter JC (September 2007). "The diploid genome sequence of an individual human". *PLoS Biol.* **5** (10): e254. doi:10.1371/journal.pbio.0050254. PMC 1964779. PMID 17803354.

[134] Wade, Wade (June 1, 2007). "DNA pioneer Watson gets own genome map". International Herald Tribune. Retrieved February 22, 2009.

[135] Wade, Nicholas (May 31, 2007). "Genome of DNA Pioneer Is Deciphered". New York Times. Retrieved February 21, 2009.

[136] Wheeler DA; Srinivasan M; Egholm M; Shen Y; Chen L; McGuire A; He W; Chen YJ; Makhijani V; Roth GT; Gomes X; Tartaro K; Niazi F; Turcotte CL; Irzyk GP; Lupski JR; Chinault C; Song XZ; Liu Y; Yuan Y; Nazareth L; Qin X; Muzny DM; Margulies M; Weinstock GM; Gibbs RA; Rothberg JM (2008). "The complete genome of an individual by massively parallel DNA sequencing". *Nature* **452** (7189): 872–6. Bibcode:2008Natur.452..872W. doi:10.1038/nature06884. PMID 18421352.

[137] Wang J; Wang, Wei; Li, Ruiqiang; Li, Yingrui; Tian, Geng; Goodman, Laurie; Fan, Wei; Zhang, Junqing; Li, Jun; Zhang, Juanbin, Juanbin; Guo, Yiran, Yiran; Feng, Binxiao, Binxiao; Li, Heng, Heng; Lu, Yao, Yao; Fang, Xiaodong, Xiaodong; Liang, Huiqing, Huiqing; Du, Zhenglin, Zhenglin; Li, Dong, Dong; Zhao, Yiqing, Yiqing; Hu, Yujie, Yujie; Yang, Zhenzhen, Zhenzhen; Zheng, Hancheng, Hancheng; Hellmann, Ines, Ines; Inouye, Michael, Michael; Pool, John, John; Yi, Xin, Xin; Zhao, Jing, Jing; Duan, Jinjie, Jinjie; Zhou, Yan, Yan; et al. (2008). "The diploid genome sequence of an Asian individual". *Nature* **456** (7218): 60–65. Bibcode:2008Natur.456...60W. doi:10.1038/nature07484. PMC 2716080. PMID 18987735.

[138] Bentley DR; Balasubramanian S; et al. (2008). "Accurate whole human genome sequencing using reversible terminator chemistry". *Nature* **456** (7218): 53–9. Bibcode:2008Natur.456...53B. doi:10.1038/nature07517. PMC 2581791. PMID 18987734.

[139] Ley TJ; Mardis ER; Ding L; Fulton B; McLellan MD; Chen K; Dooling D; Dunford-Shore BH; McGrath S; Hickenbotham M; Cook L; Abbott R; Larson DE; Koboldt DC; Pohl C; Smith S; Hawkins A; Abbott S; Locke D; Hillier LW; Miner T; Fulton L; Magrini V; Wylie T; Glasscock J; Conyers J; Sander N; Shi X; Osborne JR; et al. (2008). "DNA sequencing of a cytogenetically normal acute myeloid leukaemia genome". *Nature* **456** (7218): 66–72. Bibcode:2008Natur.456...66L. doi:10.1038/nature07485. PMC 2603574. PMID 18987736.

[140] Ahn SM; Kim TH; Lee S; Kim D; Ghang H; Kim D; Kim BC; Kim SY; Kim WY; Kim C; Park D; Lee YS; Kim S; Reja R; Jho S; Kim CG; Cha JY; Kim KH; Lee B; Bhak J; Kim SJ (2009). "The first Korean genome sequence and analysis: Full genome sequencing for a socio-ethnic group". *Genome Research* **19** (9): 1622–9. doi:10.1101/gr.092197.109. PMC 2752128. PMID 19470904.

[141] Barbujani, Guido; Pigliucci, Massimo (2013). "Human races"(PDF).*Current Biology***23**(5): R185–R187. doi:10.1016/j.cub.2 ISSN 0960-9822. PMID 23473555. Retrieved 2 December 2013. *What does this imply for the existence of human races?* Basically, that people with similar genetic features can be found in distant places, and that each local population contains a vast array of genotypes. Among the first genomes completely typed were those of James Watson and Craig Venter, two U.S. geneticists of European origin; they share more alleles with Seong-Jin Kim, a Korean scientist (1,824,482 and 1,736,340, respectively) than with each other (1,715,851).

[142] Lohr, Steve (2011-10-20). "New Book Details Jobs's Fight Against Cancer". *The New York Times*.

[143] "Complete Human Genome Sequencing Datasets to its Public Genomic Repository".

5.14 External links

- Archon X Prize for Genomics

- James Watson's Personal Genome Sequence

- AAAS/Science: Genome Sequencing Poster

- Outsmart Your Genes: Book that discusses full genome sequencing and its impact upon health care and society

- Whole genome linkage analysis

Chapter 6

Introduction to genetics

This article is a non-technical introduction to the subject. For the main encyclopedia article, see Genetics.
Genetics glossary
DNA

A long molecule that looks like a twisted ladder. It is made of four types of simple units and the sequence of these units carries information, just as the sequence of letters carries information on a page.

Nucleotides

They form the rungs of the DNA ladder and are the repeating units in DNA. There are four types of nucleotides (A, T, G and C) and it is the sequence of these nucleotides that carries information.

Chromosome

A package for carrying DNA in the cells. They contain a single long piece of DNA that is wound up and bunched together into a compact structure. Different species of plants and animals have different numbers and sizes of chromosomes.

Gene

A segment of DNA. Genes are like sentences made of the "letters" of the nucleotide alphabet, between them genes direct the physical development and behavior of an organism. Genes are like a recipe or instruction book, providing information that an organism needs so it can build or do something - like making an eye or a leg, or repairing a wound.

Allele

The different forms of a given gene that an organism may possess. For example, in humans, one allele of the eye-color gene produces green eyes and another allele of the eye-color gene produces brown eyes.

Genome

The complete set of genes in a particular organism.

Genetic engineering

When people change an organism by adding new genes, or deleting genes from its genome.

Mutation

An event that changes the sequence of the DNA in a gene.

Genetics is the study of genes — what they are, what they do, and how they work. Genes are made up of molecules inside the nucleus of a cell that are strung together in such a way that the sequence carries information: that information determines how living organisms inherit phenotypic traits, (features) determined by the genes they received from their parents and thereby going back through the generations. For example, offspring produced by sexual reproduction usually look similar to each of their parents because they have inherited some of each of their parents' genes. Genetics identifies which features are inherited, and explains how these features pass from generation to generation. In addition to inheritance, genetics studies how genes are turned on and off to control what substances are made in a cell - gene expression; and how a cell divides - mitosis or meiosis.

Some phenotypic traits can be seen, such as eye color while others can only be detected, such as blood type or intelligence. Traits determined by genes can be modified by the animal's surroundings (environment): for example, the general design of a tiger's stripes is inherited, but the specific stripe pattern is determined by the tiger's surroundings. Another example is a person's height: it is determined by both genetics and nutrition.

Genes are made of DNA, which is divided into separate pieces called chromosomes. Humans have 46: 23 pairs, though this number varies between species, for example many primates have 24 pairs. Meiosis creates special cells, sperm in males and eggs in females, which only have 23 chromosomes. These two cells merge into one during the fertilization stage of sexual reproduction, creating a zygote in which a nucleic acid double helix divides, with each single helix occupying one of the daughter cells, resulting in half the normal number of genes. The zygote then divides into four daughter cells by which time genetic recombination has created a new embryo with 23 pairs of chromosomes, half from each parent. Mating and resultant mate choice result in sexual selection. In normal cell division (mitosis) is possible when the double helix separates, and a complement of each separated half is made, resulting in two identical double helices in one cell, with each occupying one of the two new daughter cells created when the cell divides.

Chromosomes all contain four nucleotides, abbreviated C (cytosine), G (guanine), A (adenine), or T (thymine), which line up in a particular sequence and make a long string. There are two strings of nucleotides coiled around one another in each chromosome: a double helix. C on one string is always opposite from G on the other string; A is always opposite T. There are about 3.2 billion nucleotide pairs on all the human chromosomes: this is the human genome. The order of the nucleotides carries genetic information, whose rules are defined by the genetic code, similar to how the order of letters on a page of text carries information. Three nucleotides in a row - a triplet - carry one unit of information: a codon.

The genetic code not only controls inheritance: it also controls gene expression, which occurs when a portion of the double helix is uncoiled, exposing a series of the nucleotides, which are within the interior of the DNA. This series of exposed triplets (codons) carries the information to allow machinery in the cell to "read" the codons on the exposed DNA, which results in the making of RNA molecules. RNA in turn makes either amino acids or microRNA, which are responsible for all of the structure and function of a living organism; i.e. they determine all the features of the cell and thus the entire individual. Closing the uncoiled segment turns off the gene.

Heritability means the information in a given gene is not always exactly the same in every individual in that species, so the same gene in different individuals does not give exactly the same instructions. Each unique form of a single gene is called an allele; different forms are collectively called polymorphisms. As an example, one allele for the gene for hair color and skin cell pigmentation could instruct the body to produce black pigment, producing black hair and pigmented skin; while a different allele of the same gene in a different individual could give garbled instructions that would result in a failure to produce any pigment, giving white hair and no pigmented skin: albinism. Mutations are random changes in genes creating new alleles, which in turn produce new traits, which could help, harm, or have no new effect on the individual's likelihood of survival; thus, mutations are the basis for evolution.

6.1 Inheritance in biology

6.1.1 Genes and inheritance

Genes are pieces of DNA that contain information for synthesis of ribonucleic acids (RNAs) or polypeptides. Genes are inherited as units, with two parents dividing out copies of their genes to their offspring. This process can be compared with mixing two hands of cards, shuffling them, and then dealing them out again. Humans have two copies of each of their genes, and make copies that are found in eggs or sperm—but they only include *one* copy of each type of gene. An egg and sperm join to form a complete set of genes. The eventually resulting offspring has the same number of genes as their parents, but for any gene one of their two copies comes from their father, and one from their mother.[1]

The effects of this mixing depend on the types (the alleles) of the gene. If the father has two copies of an allele for red hair, and the mother has two copies for brown hair, all their children get the two alleles that give different instructions, one for red hair and one for brown. The hair color of these children depends on how these alleles work together. If one allele dominates the instructions from another, it is called the *dominant* allele, and the allele that is overridden is called the *recessive* allele. In the case of a daughter with alleles for both red and brown hair, brown is dominant and she ends up with brown hair.[2]

Although the red color allele is still there in this brown-haired girl, it doesn't show. This is a difference between what you see on the surface (the traits of an organism, called its phenotype) and the genes within the organism (its genotype). In this example you can call the allele for brown "B" and the allele for red "b". (It is normal to write dominant alleles with capital letters and recessive ones with lower-case letters.) The brown hair daughter has the "brown hair phenotype" but her genotype is Bb, with one copy of the B allele, and one of the b allele.

Now imagine that this woman grows up and has children with a brown-haired man who also has a Bb genotype. Her eggs will be a mixture of two types, one sort containing the B allele, and one sort the b allele. Similarly, her partner will produce a mix of two types of sperm containing one or the other of these two alleles. When the transmitted genes are joined up in their offspring, these children have a chance of getting either brown or red hair, since they could get a genotype of BB = brown hair, Bb = brown hair or bb = red hair. In this generation, there is therefore a chance of the recessive allele showing itself in the phenotype of the children - some of them may have red hair like their grandfather.[2]

Many traits are inherited in a more complicated way than the example above. This can happen when there are several genes involved, each contributing a small part to the end result. Tall people tend to have tall children because their children get a package of many alleles that each contribute a bit to how much they grow. However, there are not clear groups of "short people" and "tall people", like there are groups of people with brown or red hair. This is because of the large number of genes involved; this makes the trait very variable and people are of many different heights.[3] Despite a common misconception, the green/blue eye traits are also inherited in this complex inheritance model.[4] Inheritance can also be complicated when the trait depends on interaction between genetics and environment. For example, malnutrition does not change traits like eye color, but can stunt growth.[5]

6.1.2 Inherited diseases

Some diseases are hereditary and run in families; others, such as infectious diseases, are caused by the environment. Other diseases come from a combination of genes and the environment.[6] Genetic disorders are diseases that are caused by a single allele of a gene and are inherited in families. These include Huntington's disease, Cystic fibrosis or Duchenne muscular dystrophy. Cystic fibrosis, for example, is caused by mutations in a single gene called *CFTR* and is inherited as a recessive trait.[7]

Other diseases are influenced by genetics, but the genes a person gets from their parents only change their risk of getting a disease. Most of these diseases are inherited in a complex way, with either multiple genes involved, or coming from both genes and the environment. As an example, the risk of breast cancer is 50 times higher in the families most at risk, compared to the families least at risk. This variation is probably due to a large number of alleles, each changing the risk a little bit.[8] Several of the genes have been identified, such as *BRCA1* and *BRCA2*, but not all of them. However, although some of the risk is genetic, the risk of this cancer is also increased by being overweight, drinking a lot of alcohol and not exercising.[9] A woman's risk of breast cancer therefore comes from a large number of alleles interacting with her environment, so it is very hard to predict.

6.2 How genes work

6.2.1 Genes make proteins

Main article: Genetic code

The function of genes is to provide the information needed to make molecules called proteins in cells.[1] Cells are the smallest independent parts of organisms: the human body contains about 100 trillion cells, while very small organisms like bacteria are just one single cell. A cell is like a miniature and very complex factory that can make all the parts needed to produce a copy of itself, which happens when cells divide. There is a simple division of labor in cells - genes give instructions and proteins carry out these instructions, tasks like building a new copy of a cell, or repairing damage.[10] Each type of protein is a specialist that only does one job, so if a cell needs to do something new, it must make a new protein to do this job. Similarly, if a cell needs to do something faster or slower than before, it makes more or less of the protein responsible. Genes tell cells what to do by telling them which proteins to make and in what amounts.

Proteins are made of a chain of 20 different types of amino acid molecules. This chain folds up into a compact shape, rather like an untidy ball of string. The shape of the protein is determined by the sequence of amino acids along its chain and it is this shape that, in turn, determines what the protein does.[10] For example, some proteins have parts of their surface that perfectly match the shape of another molecule, allowing the protein to bind to this molecule very tightly. Other proteins are enzymes, which are like tiny machines that alter other molecules.[11]

The information in DNA is held in the sequence of the repeating units along the DNA chain.[12] These units are four types of nucleotides (A,T,G and C) and the sequence of nucleotides stores information in an alphabet called the genetic code. When a gene is read by a cell the DNA sequence is copied into a very similar molecule called RNA (this process is called transcription). Transcription is controlled by other DNA sequences (such as promoters), which show a cell where genes are, and control how often they are copied. The RNA copy made from a gene is then fed through a structure called a ribosome, which translates the sequence of nucleotides in the RNA into the correct sequence of amino acids and joins these amino acids together to make a complete protein chain. The new protein then folds up into its active form. The process of moving information from the language of RNA into the language of amino acids is called translation.[13]

If the sequence of the nucleotides in a gene changes, the sequence of the amino acids in the protein it produces may also change - if part of a gene is deleted, the protein produced is shorter and may not work any more.[10] This is the reason why different alleles of a gene can have different effects in an organism. As an example, hair color depends on how much of a dark substance called melanin is put into the hair as it grows. If a person has a normal set of the genes involved in making melanin, they make all the proteins needed and they grow dark hair. However, if the alleles for a particular protein have different sequences and produce proteins that can't do their jobs, no melanin is produced and the person has white skin and hair (albinism).[14]

6.2.2 Genes are copied

Main article: DNA replication

Genes are copied each time a cell divides into two new cells. The process that copies DNA is called DNA replication.[12] It is through a similar process that a child inherits genes from its parents, when a copy from the mother is mixed with a copy from the father.

DNA can be copied very easily and accurately because each piece of DNA can direct the creation of a new copy of its information. This is because DNA is made of two strands that pair together like the two sides of a zipper. The nucleotides are in the center, like the teeth in the zipper, and pair up to hold the two strands together. Importantly, the four different sorts of nucleotides are different shapes, so for the strands to close up properly, an **A** nucleotide must go opposite a **T** nucleotide, and a **G** opposite a **C**. This exact pairing is called base pairing.[12]

When DNA is copied, the two strands of the old DNA are pulled apart by enzymes; then they pair up with new nucleotides and then close. This produces two new pieces of DNA, each containing one strand from the old DNA and one newly made strand. This process is not predictably perfect as proteins attach to a nucleotide while they are building and cause a change in the sequence of that gene. These changes in DNA sequence are called mutations.[15] Mutations produce new alleles of genes. Sometimes these changes stop the functioning of that gene or make it serve another advantageous function, such as the melanin genes discussed above. These mutations and their effects on the traits of organisms are one of the causes of evolution.[16]

6.3 Genes and evolution

Further information: Evolution, Introduction to evolution, and History of evolutionary thought

A population of organisms evolves when an inherited trait becomes more common or less common over time.[16] For instance, all the mice living on an island would be a single population of mice: some with white fur, some gray. If over generations, white mice became more frequent and gray mice less frequent, then the color of the fur in this population of mice would be evolving. In terms of genetics, this is called an increase in allele frequency.

Alleles become more or less common either by chance in a process called genetic drift, or by natural selection.[17] In natural selection, if an allele makes it more likely for an organism to survive and reproduce, then over time this allele becomes more common. But if an allele is harmful, natural selection makes it less common. In the above example, if the island were getting colder each year and snow became present for much of the time, then the allele for white fur would favor survival, since predators would be less likely to see them against the snow, and more likely to see the gray mice. Over time white mice would become more and more frequent, while gray mice less and less.

Mutations create new alleles. These alleles have new DNA sequences and can produce proteins with new properties.[18] So if an island was populated entirely by black mice, mutations could happen creating alleles for white fur. The combination of mutations creating new alleles at random, and natural selection picking out those that are useful, causes adaptation. This is when organisms change in ways that help them to survive and reproduce.

6.4 Genetic engineering

Main article: Genetic engineering

Since traits come from the genes in a cell, putting a new piece of DNA into a cell can produce a new trait. This is how genetic engineering works. For example, rice can be given genes from a maize and a soil bacteria so the rice produces beta-carotene, which the body converts to Vitamin A.[19] This can help children suffering from Vitamin A deficiency. Another gene being put into some crops comes from the bacterium *Bacillus thuringiensis*; the gene makes a protein that is an insecticide. The insecticide kills insects that eat the plants, but is harmless to people.[20] In these plants, the new genes are put into the plant before it is grown, so the genes are in every part of the plant, including its seeds.[21] The plant's offspring inherit the new genes, which has led to concern about the spread of new traits into wild plants.[22]

The kind of technology used in genetic engineering is also being developed to treat people with genetic disorders in an experimental medical technique called gene therapy.[23] However, here the new gene is put in after the person has grown up and become ill, so any new gene is not inherited by their children. Gene therapy works by trying to replace the allele that causes the disease with an allele that works properly.

6.5 See also

- Common misunderstandings of genetics

- Epigenetics

- Full genome sequencing

- History of genetics

- Genetics in simple English

- List of basic genetics topics

- Molecular genetics

- Predictive medicine

- Timeline of the history of genetics

6.6 References

[1] *University of Utah Genetics Learning Center animated tour of the basics of genetics*. Howstuffworks.com. Retrieved 2008-01-24.

[2] MELANOCORTIN 1 RECEPTOR, Accessed 27 November 2010

[3] Multifactorial Inheritance Health Library, Morgan Stanley Children's Hospital, Accessed 20 May 2008

[4] Eye color is more complex than two genes, Athro Limited, Accessed 27 November 2010

[5] "Low income kids' height doesn't measure up by age 1". University of Michigan Health System. Retrieved May 20, 2008.

[6] requently Asked Questions About Genetic Disorders NIH, Accessed 20 May 2008

[7] Cystic fibrosis Genetics Home Reference, NIH, Accessed 16 May 2008

[8] Peto J (June 2002). "Breast cancer susceptibility-A new look at an old model". *Cancer Cell* **1** (5): 411–2. doi:10.1016/S1535-6108(02)00079-X. ISSN 1535-6108. PMID 12124169.

[9] What Are the Risk Factors for Breast Cancer? American Cancer Society, Accessed 16 May 2008

[10] The Structures of Life National Institute of General Medical Sciences, Accessed 20 May 2008

[11] Enzymes HowStuffWorks, Accessed 20 May 2008

[12] What is DNA? Genetics Home Reference, Accessed 16 May 2008

[13] DNA-RNA-Protein Nobelprize.org, Accessed 20 May 2008

[14] What is Albinism? The National Organization for Albinism and Hypopigmentation, Accessed 20 May 2008

[15] Mutations The University of Utah, Genetic Science Learning Center, Accessed 20 May 2008

[16] Brain, Marshall. "How Evolution Works". *How Stuff Works: Evolution Library*. Howstuffworks.com. Retrieved 2008-01-24.

[17] Mechanisms: The Processes of Evolution Understanding Evolution, Accessed 20 May 2008

[18] Genetic Variation Understanding Evolution, Accessed 20 May 2008

[19] Staff Golden Rice Project Retrieved 5 November 2012

[20] Tifton, Georgia: A Peanut Pest Showdown USDA, accessed 16 May 2008

[21] Genetic engineering: Bacterial arsenal to combat chewing insects GMO Safety, Jul 2010

[22] Genetically engineered organisms public issues education Cornell University, Accessed 16 May 2008

[23] Staff (November 18, 2005). "Gene Therapy" (FAQ). *Human Genome Project Information*. Oak Ridge National Laboratory. Retrieved 2006-05-28.

6.7 External links

- Introduction to Genetics, University of Utah

- Introduction to Genes and Disease, NCBI open book

- Genetics glossary, A talking glossary of genetic terms.

- Animated guide to cloning

- Khan Academy on YouTube

- What Color Eyes Would Your Children Have? Genetics of human eye color: An interactive introduction

- Double Helix Game from the Nobel Prize website. Match CATG bases with each other, and other games

- Transcribe and translate a gene, University of Utah

- StarGenetics software simulates mating experiments between organisms that are genetically different across a range of traits

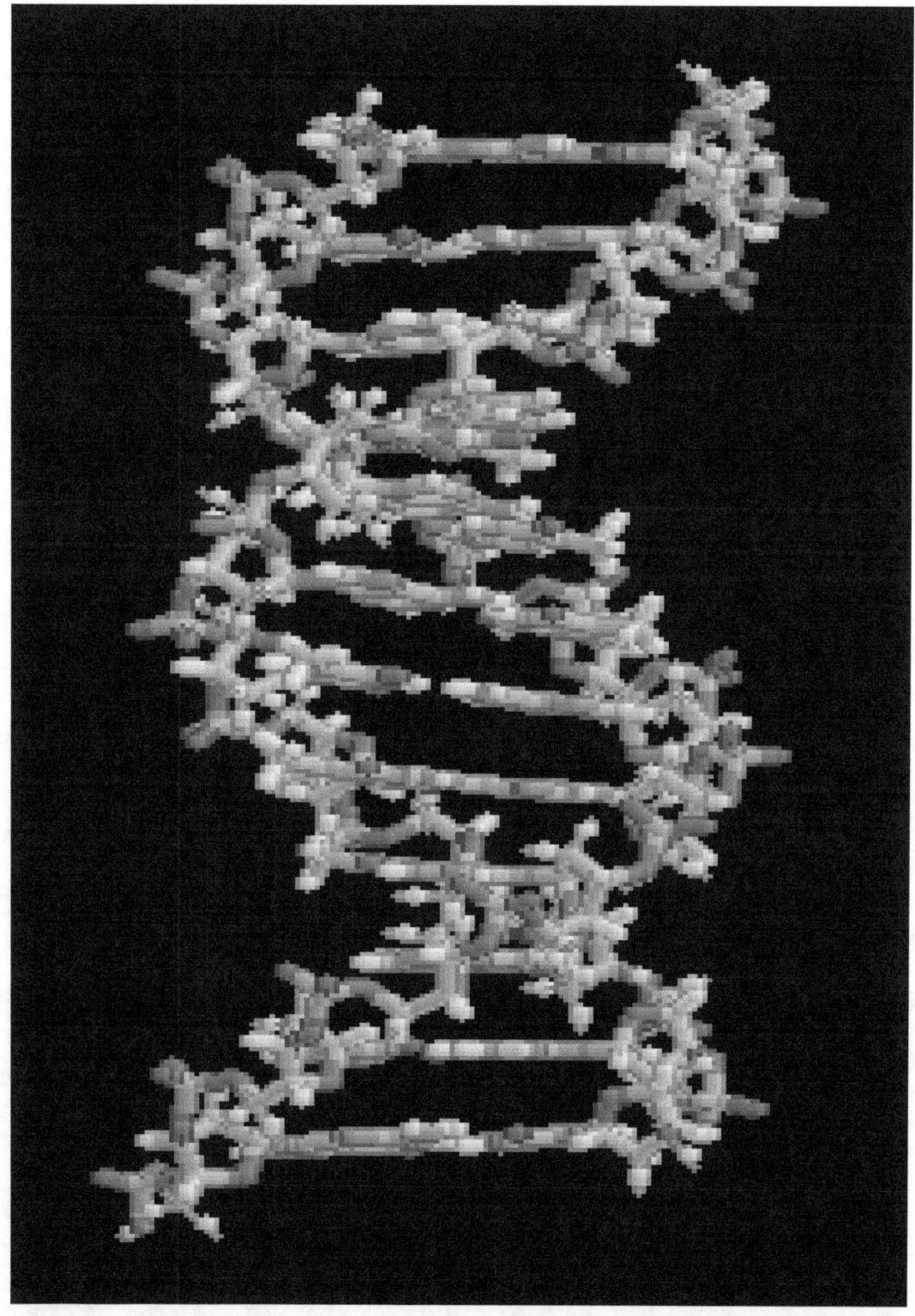

Red hair is a recessive trait.

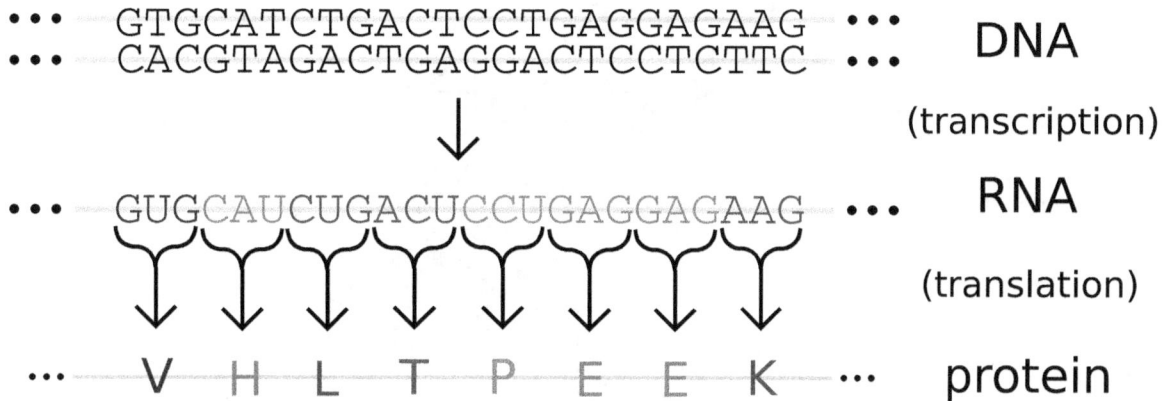

Genes are expressed by being transcribed into RNA, and this RNA then translated into protein.

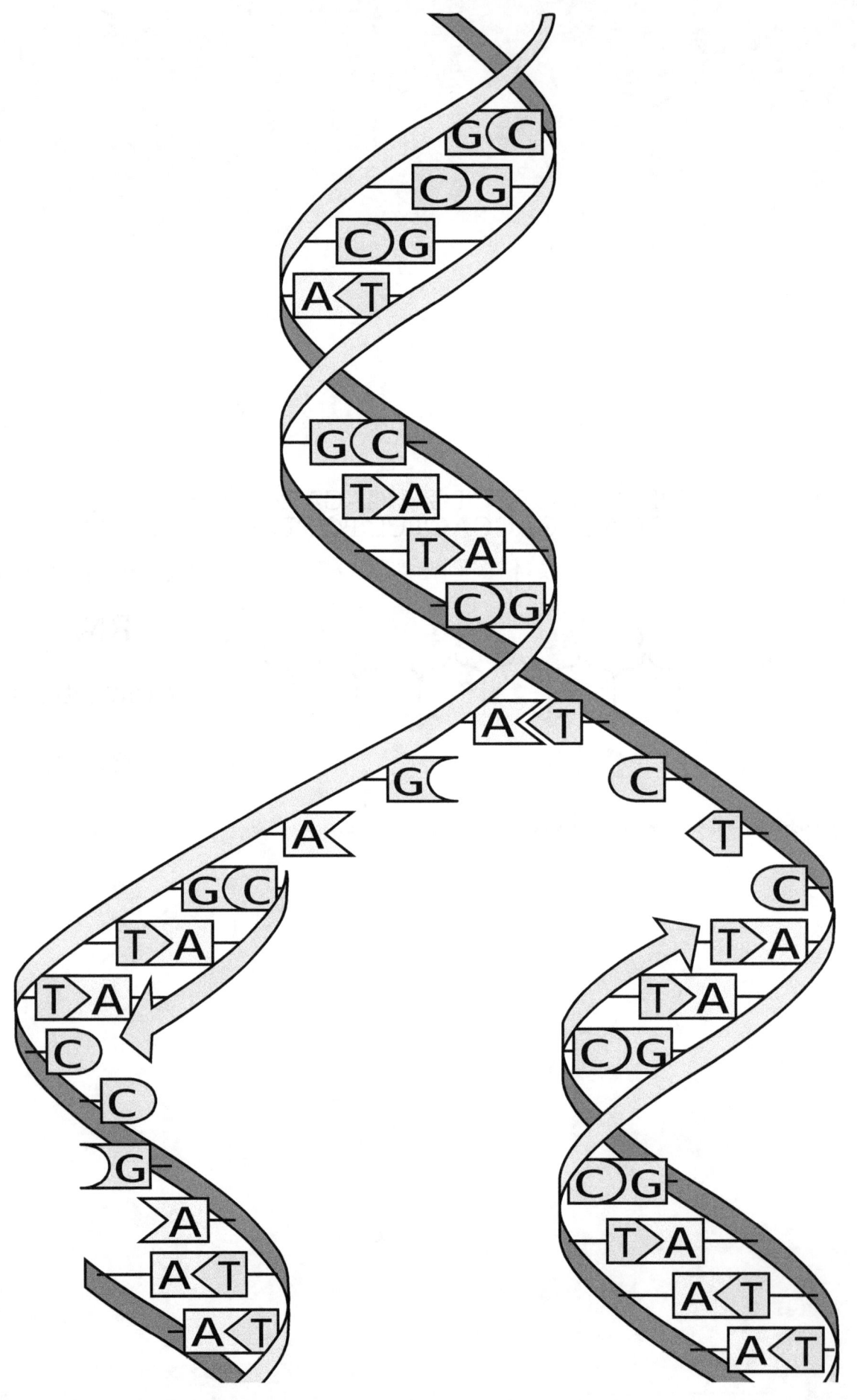

Mice with different coat colors.

Chapter 7

Personal genomics

Personal genomics is the branch of genomics concerned with the sequencing and analysis of the genome of an individual. The genotyping stage employs different techniques, including single-nucleotide polymorphism (SNP) analysis chips (typically 0.02% of the genome), or partial or full genome sequencing. Once the genotypes are known, the individual's genotype can be compared with the published literature to determine likelihood of trait expression and disease risk.

Automated sequencers have increased the speed and reduced the cost of sequencing, making it possible to offer genetic testing to consumers.

7.1   Use of personal genomics in predictive and precision medicine

Predictive medicine is the use of the information produced by personal genomics techniques when deciding what medical treatments are appropriate for a particular individual. Precision medicine is focused on "a new taxonomy of human disease based on molecular biology" [1]

Examples of the use of predictive and precision medicine include inherited medical genomics, cancer genomics and pharmacogenomics. In pharmacogenomics genetic information can be used to select the most appropriate drug to prescribe to a patient. The drug should be chosen to maximize the probability of obtaining the desired result in the patient and minimize the probability that the patient will experience side effects. Genetic information may allow physicians to tailor therapy to a given patient, in order to increase drug efficacy and minimize side effects. As of Oct 2012 there are 167 examples of drug gene pairs for which this information is currently useful in clinical practice and this number has been growing rapidly. [2]

Disease risk may be calculated based on genetic markers and genome-wide association studies for common medical conditions, which are multifactorial and include environmental components in the assessment. Diseases which are individually rare (less than 200,000 people affected in the USA) are nevertheless collectively common (affecting roughly 8-10% of the US population[3]). Over 2500 of these diseases (including a few more common ones) have predictive genetics of sufficiently high clinical impact that they are recommended as medical genetic tests available for single genes (and in whole genome sequencing) and growing at about 200 new genetic diseases per year. [4]

7.2   Cost of sequencing an individual's genome

The cost of sequencing a human genome is dropping rapidly, due to the continual development of new, faster, cheaper DNA sequencing technologies such as "next generation DNA sequencing".

The National Human Genome Research Institute, part of the U.S. National Institutes of Health, has set a target to be able to sequence a human-sized genome for US$100,000 by 2009 and US$1,000 by 2014. [5][6]

There are 6 billion base pairs in the diploid human genome. Statistical analysis reveals that a coverage of approximately ten

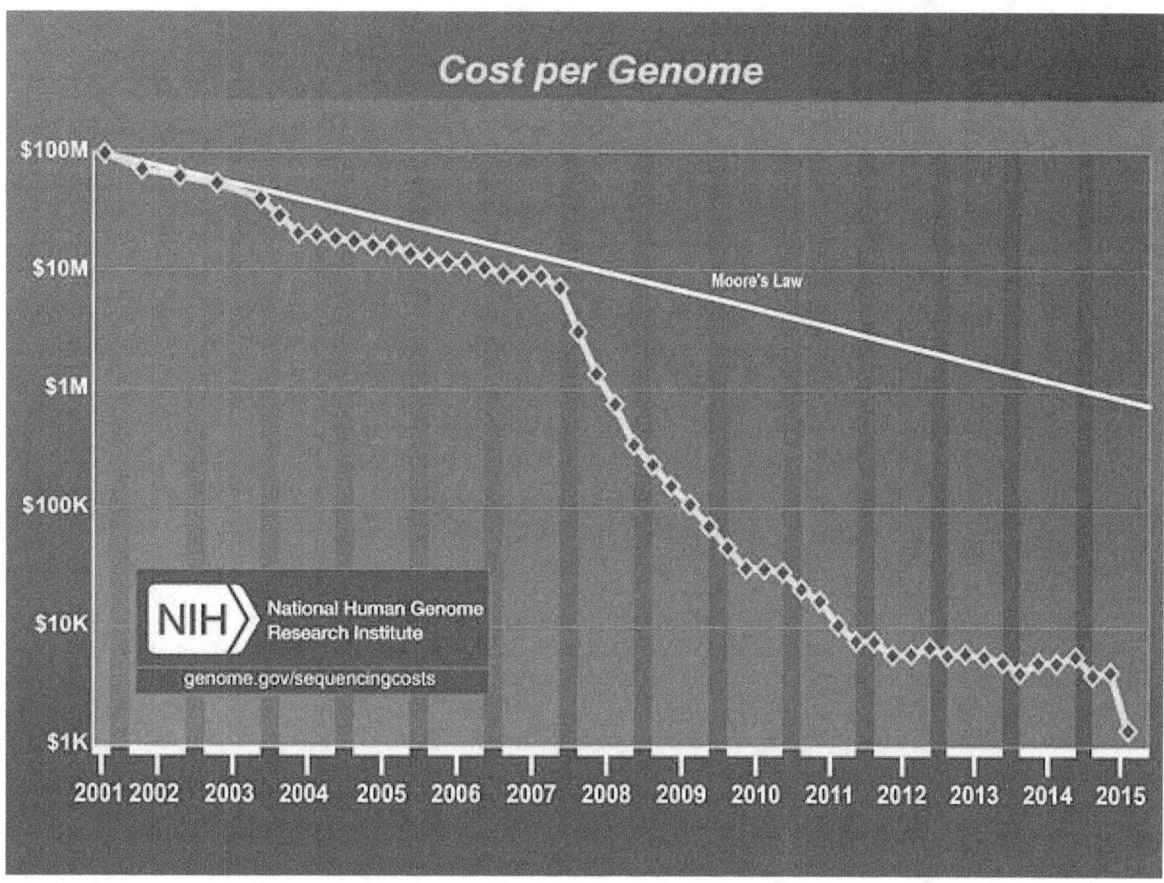

Typical cost of sequencing a human-sized genome, on a logarithmic scale. Note the drastic trend faster than Moore's law beginning in January 2008 as post-Sanger sequencing came online at sequencing centers.[5]

times is required to get coverage of both alleles in 90% human genome from 25 base pair reads with shotgun sequencing.[7] This means a total of 60 billion base pairs that must be sequenced. An Applied Biosystems SOLiD, Illumina or Helicos[8] sequencing machine can sequence 2 to 10 billion base pairs in each $8,000 to $18,000 run. The cost must also take into account personnel costs, data processing costs, legal, communications and other costs. One way to assess this is via commercial offerings. The first such whole diploid genome sequencing (6 billion bp, 3 billion from each parent) was from Knome and their price dropped from $350,000 in 2008 to $99,000 in 2009.[9][10] This inspects 3000-fold more bases of the genome than SNP chip-based genotyping, identifying both novel and known sequence variants, some relevant to personal health or ancestry.[11] In June 2009, Illumina announced the launch of its own Personal Full Genome Sequencing Service at a depth of 30X for $48,000 per genome.[12] In 2010, they cut the price to $19,500.[13]

In 2009, Complete Genomics of Mountain View announced that it would provide full genome sequencing for $5,000, from June 2009.[14] This will only be available to institutions, not individuals.[15] Prices are expected to drop further over the next few years through economies of scale and increased competition.[16][17] As of 2014, full exome sequencing was offered by Gentle for less than $2,000, including personal counseling along with the results. [18]

7.3 Projects and services already available

- Sequencing.com provides free, unlimited data storage for all genetic data[19] in high-security data centers[20] and software applications to analyze the data based on their patent-pending Real-Time Personalization™ technology.[21] The company invented an open API that translates genetic code into software code so that third-party developers without any training in genetics can integrate genetic code into their own apps.[22] The company also runs the not-for-profit Altruist Endeavor, which is an open data initiative consisting of a free, publically accessible online repository of anonymous genetic data in order to enable genetic research.[23]

- The Genographic Project is a project of the National Geographic Society and IBM to collect DNA samples to map historical human migration patterns. Launched in 2005, with 500,000 public participants as of December 2012, it helped to create the direct-to-consumer (DTC) genetic testing industry.

- The Personal Genome Project (PGP) is a long term, large cohort study based at Harvard Medical School which aims to sequence and publicize the complete genomes and medical records of 100,000 volunteers, in order to enable research into personal genomics and personalized medicine.

- SNPedia is a wiki that collects and shares information about the consequences of DNA variations, and through the associated program Promethease, anyone who has obtained DNA data about themselves (from any company) can get a free, independent report containing risk assessments and related information.

- deCODEme.com charged $1100 to carry out genotyping of approximately 1 million SNPs and provided risk estimates for 47 diseases as well as ancestry analyses. However, sales of genetic scans direct to consumer through deCODEme have now been discontinued.

- Navigenics began offering SNP-based genomic risk assessments as of April 2008. Navigenics is medically focused and emphasizes a clinician's and genetic counselor's role in interpreting results. Affymetrix Genome-Wide Human SNP Array 6.0, which genotypes 900,000 SNPs. Navigenics' service has now been discontinued.

- Pathway Genomics analyzes over 100 genetic markers to identify genetic risk for common health conditions such as melanoma, prostate cancer and rheumatoid arthritis.

- 23andMe launched in 2007, and sells mail order kits for SNP genotyping. The information is stored in a user profile and used to estimate the genetic risk of the consumer for over 240 diseases and conditions, as well as ancestry analysis. 23andMe utilizes a DNA array manufactured by Illumina.

- Illumina, Oxford Nanopore Technologies, Sequenom, Pacific Biosciences, Complete Genomics, and 454 Life Sciences are commercializing full genome sequencing but do not provide genetic analysis or counselling.[24][25][26][27][28]

- Gene by Gene provides Whole Genome Sequencing from $6995 to $7595, and other variants.

- Life Technologies

- Mapmygenome offers several SNP and NGS panels for personal genomics. See

7.4   Ethical issues

Genetic discrimination is discriminating on the basis of information obtained from an individual's genome. Genetic non-discrimination laws have been enacted in some US states[29] and at the federal level, by the Genetic Information Nondiscrimination Act (GINA). The GINA legislation prevents discrimination by health insurers and employers, but does not apply to life insurance or long-term care insurance. Given the ethical concerns about presymptomatic genetic testing of minors,[30][31][32][33] it is likely that personal genomics will first be applied to adults who can provide consent to undergo such testing, although genome sequencing is already proving valuable for children if any symptoms are present.[34]

Patients will need to be educated on interpreting their results and what they should be rationally taking from the experience. It is not only the average person who needs to be educated in the dimensions of their own genomic sequence but also professionals, including physicians and science journalists, who must be provided with the knowledge required to inform and educate their patients and the public [35] Examples of such efforts include the Personal Genetics Education Project (pgEd) and the Smithsonian collaboration with NHGRI[36]

7.5 Other issues

Full sequencing of the genome can identify polymorphisms that are so rare that no conclusions may be drawn about their impact, creating uncertainty in the analysis of individual genomes, particularly in the context of clinical care. Czech medical geneticist Eva Macháčková writes: "In some cases it is difficult to distinguish if the detected sequence variant is a causal mutation or a neutral (polymorphic) variation without any effect on phenotype. The interpretation of rare sequence variants of unknown significance detected in disease-causing genes becomes an increasingly important problem."[37]

There is a heavy debate as to how relevant the results of personal genome kits are and whether or not the ramifications of knowing one's predisposition to a disease is worth the potential psychological stress. There are also three potential problems associated with the validity of personal genome kits. The first issue is the test's validity. Handling errors of the sample increases the likelihood for errors which could affect the test results and interpretation. The second affects the clinical validity, which could affect the test's ability to detect or predict associated disorders. The third problem is the clinical utility of personal genome kits and associated risks, and the benefits of introducing them into clinical practices.[38]

Doctors are currently conducting tests for which some are not correctly trained to interpret the results. Many are unaware of how SNPs respond to one another. This results in presenting the client with potentially misleading and worrisome results which could strain the already overloaded health care system.[39] This may antagonize the individual to make uneducated decisions such as unhealthy lifestyle choices and family planning modifications. Moreover, negative results which may potentially be inaccurate, theoretically decrease the quality of life and mental health of the individual (such as increased depression and extensive anxiety).

There is also controversy regarding the concerns with companies testing individual DNA. There are issues such as "leaking" information, the right to privacy and what responsibility the company has to ensure this does not happen. Regulation rules are not clearly laid out. What is still not determined is who legally owns the genome information: the company or the individual whose genome has been read. There have been published examples of personal genome information being exploited.[40] Additional privacy concerns, related to, e.g., genetic discrimination, loss of anonymity, and psychological impacts, have been increasingly pointed out by the academic community [41] as well as government agencies.[42]

Conversely, sequencing one's genome would allow for more personalized medical treatments using pharmacogenomics; the use of genetic information to select appropriate drugs.[43] Treatments can be catered to the individual and the certain genetic predispositions they may have (such as personalized chemotherapy).

7.6 Popular culture

The 1997 science fiction film GATTACA presents a future society, where personal genomics is readily available to anyone, and explores its societal impact.

7.7 See also

- Human genome map

- Human Genome Project

- Single-nucleotide polymorphism

- Population genomics

- Full Genome Sequencing

- Bioinformatics

- Genomics

- Personalized medicine

- Systems biology

- Transcriptomics

- Omics

- Population groups in biomedicine

- Genomic counseling

7.8   References

[1] *Toward Precision Medicine: Building a Knowledge Network for Biomedical Research and a New Taxonomy of Disease.* The National Academies Press. 2011.

[2] "The Pharmacogenomics Knowledgebase".

[3] "NIH Office of Rare Disease Research".

[4] "Gene Tests".

[5] Wetterstrand, Kris (21 May 2012). "DNA Sequencing Costs: Data from the NHGRI Large-Scale Genome Sequencing Program". *Large-Scale Genome Sequencing Program.* National Human Genome Research Institute. Retrieved 24 May 2012.

[6] "Coming Soon: Your Personal DNA Map?". News.nationalgeographic.com. 28 October 2010. Retrieved 19 October 2011.

[7] "JDW-genome-supp-mat-march-proof.doc" (PDF). Retrieved 19 October 2011.

[8] "True Single Molecule Sequencing (tSMS): Helicos BioSciences". Helicosbio.com. Retrieved 19 October 2011.

[9] "Knome Lowers Price of Full Genome From $350,000 to $99,000". The Genetic Genealogist.

[10] Karow, Julia (19 May 2009). "Knome Adds Exome Sequencing, Starts Offering Services to Researchers". GenomeWeb. Retrieved 24 February 2010.

[11] Harmon, Katherine (28 June 2010). "Genome Sequencing for the Rest of Us". Scientific American. Retrieved 13 August 2010.

[12] "Individual genome sequencing – Illumina, Inc.". Everygenome.com. Retrieved 19 October 2011.

[13] "Illumina Cutting Personal Genome Sequencing Price by 60% | GPlus.com". Glgroup.com. 4 June 2010. Retrieved 19 October 2011.

[14] Karow, Julia. "Complete Genomics to Offer $5,000 Human Genome as a Service Business in Q2 2009 | In Sequence | Sequencing". GenomeWeb. Retrieved 19 October 2011.

[15] Lauerman, John (5 February 2009). "Complete Genomics Drives Down Cost of Genome Sequence to $5,000". Bloomberg. Retrieved 19 October 2011.

[16]

[17] "Illumina launches personal genome sequencing service for $48,000 : Genetic Future". Scienceblogs.com. Retrieved 19 October 2011.

[18] {http://www.healthcarejournallr.com/the-journal/contents-index/features/563-what-a-tangled-web-we-weave.html}

[19] https://sequencing.com/

[20] https://sequencing.com/researcher/

[21] https://sequencing.com/about-sequencingcom

[22] https://sequencing.com/developer/

[23] https://sequencing.com/altruist-endeavor

[24] January 2009+BW20090112 "Illumina and Oxford Nanopore Enter into Broad Commercialization Agreement" Check |url= value (help). Reuters. 12 January 2009. Retrieved 23 February 2009.

[25] http://www.genomeweb.com/sequenom-licenses-nanopore-technology-harvard-develop-third-generation-sequencer

[26] "Single Molecule Real Time (SMRT) DNA Sequencing". Pacific Biosciences. Retrieved 23 February 2009.

[27] "Complete Human Genome Sequencing Technology Overview" (PDF). Complete Genomics. 2009. Retrieved 23 February 2009.

[28] http://files.shareholder.com/downloads/CRGN/0x0x53381/386c4aaa-f36e-4b7a-9ff0-c06e61fad31f/211559.pdf

[29] "Genetics and Health Insurance State Anti-Discrimination Laws".

[30] McCabe LL; McCabe ER (June 2001). "Postgenomic medicine. Presymptomatic testing for prediction and prevention". *Clin Perinatol* **28** (2): 425–34. doi:10.1016/S0095-5108(05)70094-4. PMID 11499063.

[31] Nelson RM; Botkjin JR; Kodish ED; et al. (June 2001). "Ethical issues with genetic testing in pediatrics". *Pediatrics* **107** (6): 1451–5. doi:10.1542/peds.107.6.1451. PMID 11389275.

[32] Borry P; Fryns JP; Schotsmans P; Dierickx K (February 2006). "Carrier testing in minors: a systematic review of guidelines and position papers". *Eur. J. Hum. Genet.* **14** (2): 133–8. doi:10.1038/sj.ejhg.5201509. PMID 16267502.

[33] Borry P; Stultiens L; Nys H; Cassiman JJ; et al. (November 2006). "Presymptomatic and predictive genetic testing in minors: a systematic review of guidelines and position papers". *Clin. Genet.* **70** (5): 374–81. doi:10.1111/j.1399-0004.2006.00692.x. PMID 17026616.

[34] Mark Johnson & Kathleen Gallagher (27 Feb 2011). "One in a Billion. Nic Volker case may be the leading edge of a wave moving across genetic medicine". Milwaukee Journal Sentinel.

[35] Lunshof, Jeantine; Mardis, Elaine. [Retrieved from http://www.future-science-group.com/_img/pics/Mardis_-_Foreward.pdf "Navigenics – How it works"] Check |url= value (help) (PDF). Future Medicine Magazine. Retrieved 30 March 2012.

[36] "Genome: Unlocking Life's Code, Smithsonian Exhibit". Retrieved 7 Jun 2013.

[37] Macháčková, Eva (2003). "Disease-causing mutations versus neutral polymorphism: Use of bioinformatics and DNA diagnosis". *Cas Lek Cesk* (Czech Republic: Ceskoslovenska Lekarska Spolecnost) **142** (3): 150–153. PMID 12756842.

[38] Hunter DJ; Khoury MJ; Drazen JM (January 2008). "Letting the genome out of the bottle—will we get our wish?". *N. Engl. J. Med.* **358** (2): 105–7. doi:10.1056/NEJMp0708162. PMID 18184955.

[39] Lea DH; Skirton H; Read CY; Williams JK (March 2011). "Implications for educating the next generation of nurses on genetics and genomics in the 21st century". *J Nurs Scholarsh* **43** (1): 3–12. doi:10.1111/j.1547-5069.2010.01373.x. PMID 21342419.

[40] Gurwitz D; Bregman-Eschet Y (July 2009). "Personal genomics services: whose genomes?". *Eur. J. Hum. Genet.* **17** (7): 883–9. doi:10.1038/ejhg.2008.254. PMC 2986500. PMID 19259127.

[41] De Cristofaro, E. (2012). "Whole Genome Sequencing: Innovation Dream or Privacy Nightmare?". *ArXiv Repository*.

[42] Presidential Commission for the Study of Bioethical Issues (2012). "Privacy and Progress in Whole Genome Sequencing" (PDF).

[43] Blow N (October 2007). "Genomics: the personal side of genomics". *Nature* **449** (7162): 627–30. doi:10.1038/449627a. PMID 17914399.

7.9 Bibliography

- Dudley & Karczewski (2013). *Exploring Personal Genomics*. Oxford University Press. ISBN 978-0199644490.

- Sweet K; Michaelis R (May 2011). *The Busy Physician's Guide to Genetics, Genomics and Personalized Medicine* (1st ed.). Springer Scientific Press. ISBN 978-94-007-1147-1.

- Cadwalladr, Carole (8 June 2013). "What happened when I had my genome sequenced". *The Guardian*. Retrieved 10 July 2013.

7.10   External links

- Personalgenome.org

- Personalgenome.net

- Open Personal Genomics Consortium

- Genomics.org

- DNATest.org

- Personal Genomics Blog

- SNPedia

- Personal Genomics Institute (PGI)

- National Geographic Genographic Project

- Genomes to People (G2P)

Chapter 8

DNA

For a non-technical introduction to the topic, see Introduction to genetics. For other uses, see DNA (disambiguation).

Deoxyribonucleic acid (◀ᵈⁱ/diˌɒksiˌraɪbəˌnju:ˌkleɪ.ik ˈæsɪd/; **DNA**) is a molecule that carries most of the genetic instructions used in the development, functioning and reproduction of all known living organisms and many viruses. DNA is a nucleic acid; alongside proteins and carbohydrates, nucleic acids compose the three major macromolecules essential for all known forms of life. Most DNA molecules consist of two biopolymer strands coiled around each other to form a double helix. The two DNA strands are known as polynucleotides since they are composed of simpler units called nucleotides.[1] Each nucleotide is composed of a nitrogen-containing nucleobase—either cytosine (C), guanine (G), adenine (A), or thymine (T)—as well as a monosaccharide sugar called deoxyribose and a phosphate group. The nucleotides are joined to one another in a chain by covalent bonds between the sugar of one nucleotide and the phosphate of the next, resulting in an alternating sugar-phosphate backbone. According to base pairing rules (A with T, and C with G), hydrogen bonds bind the nitrogenous bases of the two separate polynucleotide strands to make double-stranded DNA. The total amount of related DNA base pairs on Earth is estimated at 5.0×10^{37}, and weighs 50 billion tonnes.[2] In comparison, the total mass of the biosphere has been estimated to be as much as 4 TtC (trillion tons of carbon).[3]

DNA stores biological information. The DNA backbone is resistant to cleavage, and both strands of the double-stranded structure store the same biological information. Biological information is replicated as the two strands are separated. A significant portion of DNA (more than 98% for humans) is non-coding, meaning that these sections do not serve as patterns for protein sequences.

The two strands of DNA run in opposite directions to each other and are therefore anti-parallel. Attached to each sugar is one of four types of nucleobases (informally, *bases*). It is the sequence of these four nucleobases along the backbone that encodes biological information. Under the genetic code, RNA strands are translated to specify the sequence of amino acids within proteins. These RNA strands are initially created using DNA strands as a template in a process called transcription.

Within cells, DNA is organized into long structures called chromosomes. During cell division these chromosomes are duplicated in the process of DNA replication, providing each cell its own complete set of chromosomes. Eukaryotic organisms (animals, plants, fungi, and protists) store most of their DNA inside the cell nucleus and some of their DNA in organelles, such as mitochondria or chloroplasts.[4] In contrast, prokaryotes (bacteria and archaea) store their DNA only in the cytoplasm. Within the chromosomes, chromatin proteins such as histones compact and organize DNA. These compact structures guide the interactions between DNA and other proteins, helping control which parts of the DNA are transcribed.

DNA was first isolated by Friedrich Miescher in 1869. Its molecular structure was identified by James Watson and Francis Crick in 1953, whose model-building efforts were guided by X-ray diffraction data acquired by Rosalind Franklin. DNA is used by researchers as a molecular tool to explore physical laws and theories, such as the ergodic theorem and the theory of elasticity. The unique material properties of DNA have made it an attractive molecule for material scientists and engineers interested in micro- and nano-fabrication. Among notable advances in this field are DNA origami and DNA-based hybrid materials.[5]

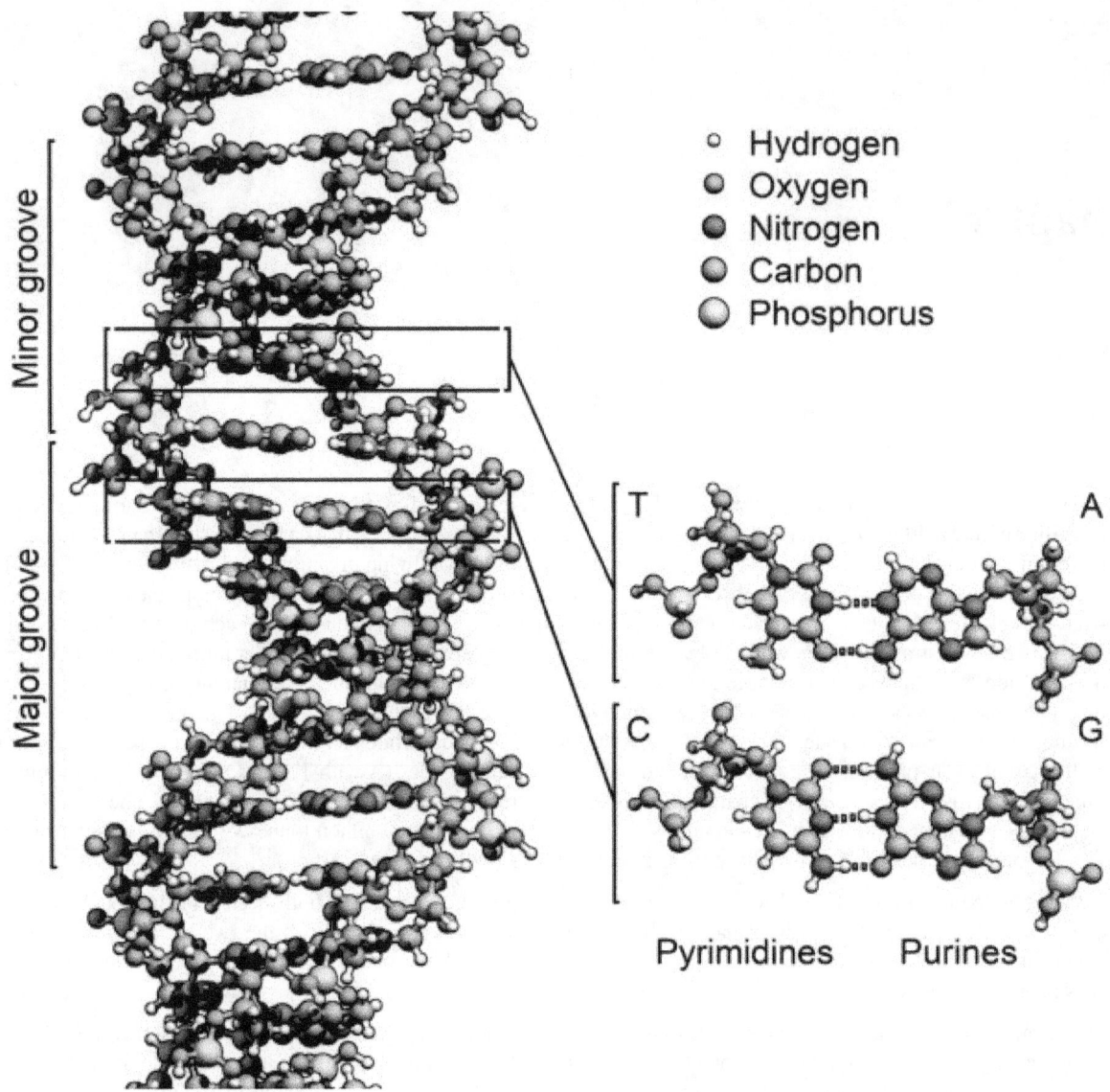

○ Hydrogen
◉ Oxygen
● Nitrogen
◕ Carbon
◯ Phosphorus

T   A

C   G

Pyrimidines   Purines

The structure of the DNA double helix. The atoms in the structure are colour-coded by element and the detailed structure of two base pairs are shown in the bottom right.

8.1 Properties

DNA is a long polymer made from repeating units called nucleotides.[6][7] DNA was first identified and isolated by Friedrich Miescher in 1869 at the University of Tübingen, a substance he called *nuclein*, and the double helix structure of DNA was first discovered in 1953 by Watson and Crick at the University of Cambridge, using experimental data collected by Rosalind Franklin and Maurice Wilkins. The structure of DNA is non-static,[8] all species comprises two helical chains each coiled round the same axis, and each with a pitch of 34 ångströms (3.4 nanometres) and a radius of 10 ångströms (1.0 nanometre).[9] According to another study, when measured in a particular solution, the DNA chain measured 22 to 26 ångströms wide (2.2 to 2.6 nanometres), and one nucleotide unit measured 3.3 Å (0.33 nm) long.[10] Although each individual repeating unit is very small, DNA polymers can be very large molecules containing millions of nucleotides. For instance, the DNA in the largest human chromosome, chromosome number 1, consists of approximately 220 million base pairs[11] and would be 85 mm long if straightened.

In living organisms DNA does not usually exist as a single molecule, but instead as a pair of molecules that are held tightly together.[12][13] These two long strands entwine like vines, in the shape of a double helix. The nucleotide repeats contain both the segment of the backbone of the molecule, which holds the chain together, and a nucleobase, which interacts with the other DNA strand in the helix. A nucleobase linked to a sugar is called a nucleoside and a base linked to a sugar and one or more phosphate groups is called a nucleotide. A polymer comprising multiple linked nucleotides (as in DNA) is called a polynucleotide.[14]

The backbone of the DNA strand is made from alternating phosphate and sugar residues.[15] The sugar in DNA is 2-deoxyribose, which is a pentose (five-carbon) sugar. The sugars are joined together by phosphate groups that form phosphodiester bonds between the third and fifth carbon atoms of adjacent sugar rings. These asymmetric bonds mean a strand of DNA has a direction. In a double helix the direction of the nucleotides in one strand is opposite to their direction in the other strand: the strands are *antiparallel*. The asymmetric ends of DNA strands are called the 5′ (*five prime*) and 3′ (*three prime*) ends, with the 5′ end having a terminal phosphate group and the 3′ end a terminal hydroxyl group. One major difference between DNA and RNA is the sugar, with the 2-deoxyribose in DNA being replaced by the alternative pentose sugar ribose in RNA.[13]

The DNA double helix is stabilized primarily by two forces: hydrogen bonds between nucleotides and base-stacking interactions among aromatic nucleobases.[17] In the aqueous environment of the cell, the conjugated π bonds of nucleotide bases align perpendicular to the axis of the DNA molecule, minimizing their interaction with the solvation shell and therefore, the Gibbs free energy. The four bases found in DNA are adenine (abbreviated A), cytosine (C), guanine (G) and thymine (T). These four bases are attached to the sugar/phosphate to form the complete nucleotide, as shown for adenosine monophosphate. Adenine pairs with thymine and guanine pairs with cytosine. It was represented by A-T base pairs and G-C base pairs.[18][19]

8.1.1 Nucleobase classification

The nucleobases are classified into two types: the purines, A and G, being fused five- and six-membered heterocyclic compounds, and the pyrimidines, the six-membered rings C and T.[13] A fifth pyrimidine nucleobase, uracil (U), usually takes the place of thymine in RNA and differs from thymine by lacking a methyl group on its ring. In addition to RNA and DNA a large number of artificial nucleic acid analogues have also been created to study the properties of nucleic acids, or for use in biotechnology.[20]

Uracil is not usually found in DNA, occurring only as a breakdown product of cytosine. However, in a number of bacteriophages – *Bacillus subtilis* bacteriophages PBS1 and PBS2 and *Yersinia* bacteriophage piR1-37 – thymine has been replaced by uracil.[21] Another phage - Staphylococcal phage S6 - has been identified with a genome where thymine has been replaced by uracil.[22]

Base J (beta-d-glucopyranosyloxymethyluracil), a modified form of uracil, is also found in a number of organisms: the flagellates *Diplonema* and *Euglena*, and all the kinetoplastid genera.[23] Biosynthesis of J occurs in two steps: in the first step a specific thymidine in DNA is converted into hydroxymethyldeoxyuridine; in the second HOMedU is glycosylated to form J.[24] Proteins that bind specifically to this base have been identified.[25][26][27] These proteins appear to be distant relatives of the Tet1 oncogene that is involved in the pathogenesis of acute myeloid leukemia.[28] J appears to act as a termination signal for RNA polymerase II.[29][30]

8.1.2 Grooves

Twin helical strands form the DNA backbone. Another double helix may be found tracing the spaces, or grooves, between the strands. These voids are adjacent to the base pairs and may provide a binding site. As the strands are not symmetrically located with respect to each other, the grooves are unequally sized. One groove, the major groove, is 22 Å wide and the other, the minor groove, is 12 Å wide.[31] The width of the major groove means that the edges of the bases are more accessible in the major groove than in the minor groove. As a result, proteins such as transcription factors that can bind to specific sequences in double-stranded DNA usually make contact with the sides of the bases exposed in the major groove.[32] This situation varies in unusual conformations of DNA within the cell (*see below*), but the major and minor grooves are always named to reflect the differences in size that would be seen if the DNA is twisted back into the ordinary B form.

8.1.3   Base pairing

Further information: Base pair

In a DNA double helix, each type of nucleobase on one strand bonds with just one type of nucleobase on the other strand. This is called complementary base pairing. Here, purines form hydrogen bonds to pyrimidines, with adenine bonding only to thymine in two hydrogen bonds, and cytosine bonding only to guanine in three hydrogen bonds. This arrangement of two nucleotides binding together across the double helix is called a base pair. As hydrogen bonds are not covalent, they can be broken and rejoined relatively easily. The two strands of DNA in a double helix can therefore be pulled apart like a zipper, either by a mechanical force or high temperature.[33] As a result of this complementarity, all the information in the double-stranded sequence of a DNA helix is duplicated on each strand, which is vital in DNA replication. Indeed, this reversible and specific interaction between complementary base pairs is critical for all the functions of DNA in living organisms.[7]

Top, a **GC** base pair with three hydrogen bonds. Bottom, an **AT** base pair with two hydrogen bonds. Non-covalent hydrogen bonds between the pairs are shown as dashed lines.

The two types of base pairs form different numbers of hydrogen bonds, AT forming two hydrogen bonds, and GC forming three hydrogen bonds (see figures, right). DNA with high GC-content is more stable than DNA with low GC-content.

As noted above, most DNA molecules are actually two polymer strands, bound together in a helical fashion by noncovalent bonds; this double stranded structure (**dsDNA**) is maintained largely by the intrastrand base stacking interactions, which are strongest for G,C stacks. The two strands can come apart – a process known as melting – to form two single-stranded DNA molecules (**ssDNA**) molecules. Melting occurs at high temperature, low salt and high pH (low pH also melts DNA, but since DNA is unstable due to acid depurination, low pH is rarely used).

The stability of the dsDNA form depends not only on the GC-content (% G,C basepairs) but also on sequence (since stacking is sequence specific) and also length (longer molecules are more stable). The stability can be measured in various ways; a common way is the "melting temperature", which is the temperature at which 50% of the ds molecules are converted to ss molecules; melting temperature is dependent on ionic strength and the concentration of DNA. As a result, it is both the percentage of GC base pairs and the overall length of a DNA double helix that determines the strength of the association between the two strands of DNA. Long DNA helices with a high GC-content have stronger-interacting strands, while short helices with high AT content have weaker-interacting strands.[34] In biology, parts of the DNA double helix that need to separate easily, such as the TATAAT Pribnow box in some promoters, tend to have a high AT content, making the strands easier to pull apart.[35]

In the laboratory, the strength of this interaction can be measured by finding the temperature necessary to break the hydrogen bonds, their melting temperature (also called *Tm* value). When all the base pairs in a DNA double helix melt, the strands separate and exist in solution as two entirely independent molecules. These single-stranded DNA molecules (*ssDNA*) have no single common shape, but some conformations are more stable than others.[36]

8.1.4   Sense and antisense

Further information: Sense (molecular biology)

A DNA sequence is called "sense" if its sequence is the same as that of a messenger RNA copy that is translated into protein.[37] The sequence on the opposite strand is called the "antisense" sequence. Both sense and antisense sequences can exist on different parts of the same strand of DNA (i.e. both strands can contain both sense and antisense sequences). In both prokaryotes and eukaryotes, antisense RNA sequences are produced, but the functions of these RNAs are not entirely clear.[38] One proposal is that antisense RNAs are involved in regulating gene expression through RNA-RNA base pairing.[39]

A few DNA sequences in prokaryotes and eukaryotes, and more in plasmids and viruses, blur the distinction between sense and antisense strands by having overlapping genes.[40] In these cases, some DNA sequences do double duty, encoding one protein when read along one strand, and a second protein when read in the opposite direction along the other strand.

In bacteria, this overlap may be involved in the regulation of gene transcription,[41] while in viruses, overlapping genes increase the amount of information that can be encoded within the small viral genome.[42]

8.1.5 Supercoiling

Further information: DNA supercoil

DNA can be twisted like a rope in a process called DNA supercoiling. With DNA in its "relaxed" state, a strand usually circles the axis of the double helix once every 10.4 base pairs, but if the DNA is twisted the strands become more tightly or more loosely wound.[43] If the DNA is twisted in the direction of the helix, this is positive supercoiling, and the bases are held more tightly together. If they are twisted in the opposite direction, this is negative supercoiling, and the bases come apart more easily. In nature, most DNA has slight negative supercoiling that is introduced by enzymes called topoisomerases.[44] These enzymes are also needed to relieve the twisting stresses introduced into DNA strands during processes such as transcription and DNA replication.[45]

8.1.6 Alternate DNA structures

Further information: Molecular Structure of Nucleic Acids: A Structure for Deoxyribose Nucleic Acid, Molecular models of DNA, and DNA structure

DNA exists in many possible conformations that include A-DNA, B-DNA, and Z-DNA forms, although, only B-DNA and Z-DNA have been directly observed in functional organisms.[15] The conformation that DNA adopts depends on the hydration level, DNA sequence, the amount and direction of supercoiling, chemical modifications of the bases, the type and concentration of metal ions, as well as the presence of polyamines in solution.[46]

The first published reports of A-DNA X-ray diffraction patterns—and also B-DNA—used analyses based on Patterson transforms that provided only a limited amount of structural information for oriented fibers of DNA.[47][48] An alternate analysis was then proposed by Wilkins *et al.*, in 1953, for the *in vivo* B-DNA X-ray diffraction/scattering patterns of highly hydrated DNA fibers in terms of squares of Bessel functions.[49] In the same journal, James Watson and Francis Crick presented their molecular modeling analysis of the DNA X-ray diffraction patterns to suggest that the structure was a double-helix.[9]

Although the "B-DNA form" is most common under the conditions found in cells,[50] it is not a well-defined conformation but a family of related DNA conformations[51] that occur at the high hydration levels present in living cells. Their corresponding X-ray diffraction and scattering patterns are characteristic of molecular paracrystals with a significant degree of disorder.[52][53]

Compared to B-DNA, the A-DNA form is a wider right-handed spiral, with a shallow, wide minor groove and a narrower, deeper major groove. The A form occurs under non-physiological conditions in partially dehydrated samples of DNA, while in the cell it may be produced in hybrid pairings of DNA and RNA strands, as well as in enzyme-DNA complexes.[54][55] Segments of DNA where the bases have been chemically modified by methylation may undergo a larger change in conformation and adopt the Z form. Here, the strands turn about the helical axis in a left-handed spiral, the opposite of the more common B form.[56] These unusual structures can be recognized by specific Z-DNA binding proteins and may be involved in the regulation of transcription.[57]

8.1.7 Alternative DNA chemistry

For a number of years exobiologists have proposed the existence of a shadow biosphere, a postulated microbial biosphere of Earth that uses radically different biochemical and molecular processes than currently known life. One of the proposals was the existence of lifeforms that use arsenic instead of phosphorus in DNA. A report in 2010 of the possibility in the bacterium GFAJ-1, was announced,[58][58][59] though the research was disputed,[59][60] and evidence suggests the bacterium actively prevents the incorporation of arsenic into the DNA backbone and other biomolecules.[61]

8.1.8   Quadruplex structures

Further information: G-quadruplex

At the ends of the linear chromosomes are specialized regions of DNA called telomeres. The main function of these regions is to allow the cell to replicate chromosome ends using the enzyme telomerase, as the enzymes that normally replicate DNA cannot copy the extreme 3′ ends of chromosomes.[62] These specialized chromosome caps also help protect the DNA ends, and stop the DNA repair systems in the cell from treating them as damage to be corrected.[63] In human cells, telomeres are usually lengths of single-stranded DNA containing several thousand repeats of a simple TTAGGG sequence.[64]

These guanine-rich sequences may stabilize chromosome ends by forming structures of stacked sets of four-base units, rather than the usual base pairs found in other DNA molecules. Here, four guanine bases form a flat plate and these flat four-base units then stack on top of each other, to form a stable G-quadruplex structure.[66] These structures are stabilized by hydrogen bonding between the edges of the bases and chelation of a metal ion in the centre of each four-base unit.[67] Other structures can also be formed, with the central set of four bases coming from either a single strand folded around the bases, or several different parallel strands, each contributing one base to the central structure.

In addition to these stacked structures, telomeres also form large loop structures called telomere loops, or T-loops. Here, the single-stranded DNA curls around in a long circle stabilized by telomere-binding proteins.[68] At the very end of the T-loop, the single-stranded telomere DNA is held onto a region of double-stranded DNA by the telomere strand disrupting the double-helical DNA and base pairing to one of the two strands. This triple-stranded structure is called a displacement loop or D-loop.[66]

Branched DNA can form networks containing multiple branches.

8.1.9   Branched DNA

Further information: Branched DNA and DNA nanotechnology

In DNA fraying occurs when non-complementary regions exist at the end of an otherwise complementary double-strand of DNA. However, branched DNA can occur if a third strand of DNA is introduced and contains adjoining regions able to hybridize with the frayed regions of the pre-existing double-strand. Although the simplest example of branched DNA involves only three strands of DNA, complexes involving additional strands and multiple branches are also possible.[69] Branched DNA can be used in nanotechnology to construct geometric shapes, see the section on uses in technology below.

8.2   Chemical modifications and altered DNA packaging

Structure of cytosine with and without the 5-methyl group. Deamination converts 5-methylcytosine into thymine.

8.2.1   Base modifications and DNA packaging

Further information: DNA methylation, Chromatin remodeling

The expression of genes is influenced by how the DNA is packaged in chromosomes, in a structure called chromatin. Base modifications can be involved in packaging, with regions that have low or no gene expression usually containing high levels of methylation of cytosine bases. DNA packaging and its influence on gene expression can also occur by covalent modifications of the histone protein core around which DNA is wrapped in the chromatin structure or else by remodeling carried out by chromatin remodeling complexes (see Chromatin remodeling). There is, further, crosstalk between DNA methylation and histone modification, so they can coordinately affect chromatin and gene expression.[70]

For one example, cytosine methylation, produces 5-methylcytosine, which is important for X-chromosome inactivation.[71] The average level of methylation varies between organisms – the worm *Caenorhabditis elegans* lacks cytosine methylation, while vertebrates have higher levels, with up to 1% of their DNA containing 5-methylcytosine.[72] Despite the importance of 5-methylcytosine, it can deaminate to leave a thymine base, so methylated cytosines are particularly prone to mutations.[73] Other base modifications include adenine methylation in bacteria, the presence of 5-hydroxymethylcytosine in the brain,[74] and the glycosylation of uracil to produce the "J-base" in kinetoplastids.[75][76]

8.2.2 Damage

Further information: DNA damage (naturally occurring), Mutation, DNA damage theory of aging

DNA can be damaged by many sorts of mutagens, which change the DNA sequence. Mutagens include oxidizing agents, alkylating agents and also high-energy electromagnetic radiation such as ultraviolet light and X-rays. The type of DNA damage produced depends on the type of mutagen. For example, UV light can damage DNA by producing thymine dimers, which are cross-links between pyrimidine bases.[78] On the other hand, oxidants such as free radicals or hydrogen peroxide produce multiple forms of damage, including base modifications, particularly of guanosine, and double-strand breaks.[79] A typical human cell contains about 150,000 bases that have suffered oxidative damage.[80] Of these oxidative lesions, the most dangerous are double-strand breaks, as these are difficult to repair and can produce point mutations, insertions and deletions from the DNA sequence, as well as chromosomal translocations.[81] These mutations can cause cancer. Because of inherent limitations in the DNA repair mechanisms, if humans lived long enough, they would all eventually develop cancer.[82][83] DNA damages that are naturally occurring, due to normal cellular processes that produce reactive oxygen species, the hydrolytic activities of cellular water, etc., also occur frequently. Although most of these damages are repaired, in any cell some DNA damage may remain despite the action of repair processes. These remaining DNA damages accumulate with age in mammalian postmitotic tissues. This accumulation appears to be an important underlying cause of aging.[84][85][86]

Many mutagens fit into the space between two adjacent base pairs, this is called *intercalation*. Most intercalators are aromatic and planar molecules; examples include ethidium bromide, acridines, daunomycin, and doxorubicin. For an intercalator to fit between base pairs, the bases must separate, distorting the DNA strands by unwinding of the double helix. This inhibits both transcription and DNA replication, causing toxicity and mutations.[87] As a result, DNA intercalators may be carcinogens, and in the case of thalidomide, a teratogen.[88] Others such as benzo[*a*]pyrene diol epoxide and aflatoxin form DNA adducts that induce errors in replication.[89] Nevertheless, due to their ability to inhibit DNA transcription and replication, other similar toxins are also used in chemotherapy to inhibit rapidly growing cancer cells.[90]

8.3 Biological functions

DNA usually occurs as linear chromosomes in eukaryotes, and circular chromosomes in prokaryotes. The set of chromosomes in a cell makes up its genome; the human genome has approximately 3 billion base pairs of DNA arranged into 46 chromosomes.[91] The information carried by DNA is held in the sequence of pieces of DNA called genes. Transmission of genetic information in genes is achieved via complementary base pairing. For example, in transcription, when a cell uses the information in a gene, the DNA sequence is copied into a complementary RNA sequence through the attraction between the DNA and the correct RNA nucleotides. Usually, this RNA copy is then used to make a matching protein sequence in a process called translation, which depends on the same interaction between RNA nucleotides. In alternative fashion, a cell may simply copy its genetic information in a process called DNA replication. The details of these functions are covered in other articles; here the focus is on the interactions between DNA and other molecules that mediate the function of the genome.

8.3.1 Genes and genomes

Further information: Cell nucleus, Chromatin, Chromosome, Gene, Noncoding DNA

Genomic DNA is tightly and orderly packed in the process called DNA condensation to fit the small available volumes of the cell. In eukaryotes, DNA is located in the cell nucleus, as well as small amounts in mitochondria and chloroplasts. In prokaryotes, the DNA is held within an irregularly shaped body in the cytoplasm called the nucleoid.[92] The genetic information in a genome is held within genes, and the complete set of this information in an organism is called its genotype. A gene is a unit of heredity and is a region of DNA that influences a particular characteristic in an organism. Genes contain an open reading frame that can be transcribed, as well as regulatory sequences such as promoters and enhancers, which control the transcription of the open reading frame.

In many species, only a small fraction of the total sequence of the genome encodes protein. For example, only about 1.5% of the human genome consists of protein-coding exons, with over 50% of human DNA consisting of non-coding repetitive sequences.[93] The reasons for the presence of so much noncoding DNA in eukaryotic genomes and the extraordinary differences in genome size, or *C-value*, among species represent a long-standing puzzle known as the "C-value enigma".[94] However, some DNA sequences that do not code protein may still encode functional non-coding RNA molecules, which are involved in the regulation of gene expression.[95]

Some noncoding DNA sequences play structural roles in chromosomes. Telomeres and centromeres typically contain few genes, but are important for the function and stability of chromosomes.[63][97] An abundant form of noncoding DNA in humans are pseudogenes, which are copies of genes that have been disabled by mutation.[98] These sequences are usually just molecular fossils, although they can occasionally serve as raw genetic material for the creation of new genes through the process of gene duplication and divergence.[99]

8.3.2   Transcription and translation

Further information: Genetic code, Transcription (genetics), Protein biosynthesis

A gene is a sequence of DNA that contains genetic information and can influence the phenotype of an organism. Within a gene, the sequence of bases along a DNA strand defines a messenger RNA sequence, which then defines one or more protein sequences. The relationship between the nucleotide sequences of genes and the amino-acid sequences of proteins is determined by the rules of translation, known collectively as the genetic code. The genetic code consists of three-letter 'words' called *codons* formed from a sequence of three nucleotides (e.g. ACT, CAG, TTT).

In transcription, the codons of a gene are copied into messenger RNA by RNA polymerase. This RNA copy is then decoded by a ribosome that reads the RNA sequence by base-pairing the messenger RNA to transfer RNA, which carries amino acids. Since there are 4 bases in 3-letter combinations, there are 64 possible codons (4^3 combinations). These encode the twenty standard amino acids, giving most amino acids more than one possible codon. There are also three 'stop' or 'nonsense' codons signifying the end of the coding region; these are the TAA, TGA, and TAG codons.

8.3.3   Replication

Further information: DNA replication

Cell division is essential for an organism to grow, but, when a cell divides, it must replicate the DNA in its genome so that the two daughter cells have the same genetic information as their parent. The double-stranded structure of DNA provides a simple mechanism for DNA replication. Here, the two strands are separated and then each strand's complementary DNA sequence is recreated by an enzyme called DNA polymerase. This enzyme makes the complementary strand by finding the correct base through complementary base pairing, and bonding it onto the original strand. As DNA polymerases can only extend a DNA strand in a 5′ to 3′ direction, different mechanisms are used to copy the antiparallel strands of the double helix.[100] In this way, the base on the old strand dictates which base appears on the new strand, and the cell ends up with a perfect copy of its DNA.

8.3.4   Extracellular nucleic acids

Naked extracellular DNA (eDNA), most of it released by cell death, is nearly ubiquitous in the environment. Its concentration in soil may be as high as 2 μg/L, and its concentration in natural aquatic environments may be as high at 88 μg/L.[101] Various possible functions have been proposed for eDNA: it may be involved in horizontal gene transfer;[102] it may provide nutrients;[103] and it may act as a buffer to recruit or titrate ions or antibiotics.[104] Extracellular DNA acts as a functional extracellular matrix component in the biofilms of a number of bacterial species. It may act as a recognition factor to regulate the attachment and dispersal of specific cell types in the biofilm;[105] it may contribute to biofilm formation;[106] and it may contribute to the biofilm's physical strength and resistance to biological stress.[107]

8.4 Interactions with proteins

All the functions of DNA depend on interactions with proteins. These protein interactions can be non-specific, or the protein can bind specifically to a single DNA sequence. Enzymes can also bind to DNA and of these, the polymerases that copy the DNA base sequence in transcription and DNA replication are particularly important.

8.4.1 DNA-binding proteins

Further information: DNA-binding protein
Interaction of DNA (shown in orange) with histones (shown in blue). These proteins' basic amino acids bind to the acidic phosphate groups on DNA.

Structural proteins that bind DNA are well-understood examples of non-specific DNA-protein interactions. Within chromosomes, DNA is held in complexes with structural proteins. These proteins organize the DNA into a compact structure called chromatin. In eukaryotes this structure involves DNA binding to a complex of small basic proteins called histones, while in prokaryotes multiple types of proteins are involved.[108][109] The histones form a disk-shaped complex called a nucleosome, which contains two complete turns of double-stranded DNA wrapped around its surface. These non-specific interactions are formed through basic residues in the histones making ionic bonds to the acidic sugar-phosphate backbone of the DNA, and are therefore largely independent of the base sequence.[110] Chemical modifications of these basic amino acid residues include methylation, phosphorylation and acetylation.[111] These chemical changes alter the strength of the interaction between the DNA and the histones, making the DNA more or less accessible to transcription factors and changing the rate of transcription.[112] Other non-specific DNA-binding proteins in chromatin include the high-mobility group proteins, which bind to bent or distorted DNA.[113] These proteins are important in bending arrays of nucleosomes and arranging them into the larger structures that make up chromosomes.[114]

A distinct group of DNA-binding proteins are the DNA-binding proteins that specifically bind single-stranded DNA. In humans, replication protein A is the best-understood member of this family and is used in processes where the double helix is separated, including DNA replication, recombination and DNA repair.[115] These binding proteins seem to stabilize single-stranded DNA and protect it from forming stem-loops or being degraded by nucleases.

In contrast, other proteins have evolved to bind to particular DNA sequences. The most intensively studied of these are the various transcription factors, which are proteins that regulate transcription. Each transcription factor binds to one particular set of DNA sequences and activates or inhibits the transcription of genes that have these sequences close to their promoters. The transcription factors do this in two ways. Firstly, they can bind the RNA polymerase responsible for transcription, either directly or through other mediator proteins; this locates the polymerase at the promoter and allows it to begin transcription.[117] Alternatively, transcription factors can bind enzymes that modify the histones at the promoter. This changes the accessibility of the DNA template to the polymerase.[118]

As these DNA targets can occur throughout an organism's genome, changes in the activity of one type of transcription factor can affect thousands of genes.[119] Consequently, these proteins are often the targets of the signal transduction processes that control responses to environmental changes or cellular differentiation and development. The specificity of these transcription factors' interactions with DNA come from the proteins making multiple contacts to the edges of the DNA bases, allowing them to "read" the DNA sequence. Most of these base-interactions are made in the major groove, where the bases are most accessible.[32]

8.4.2 DNA-modifying enzymes

Nucleases and ligases

Nucleases are enzymes that cut DNA strands by catalyzing the hydrolysis of the phosphodiester bonds. Nucleases that hydrolyse nucleotides from the ends of DNA strands are called exonucleases, while endonucleases cut within strands. The most frequently used nucleases in molecular biology are the restriction endonucleases, which cut DNA at specific sequences. For instance, the EcoRV enzyme shown to the left recognizes the 6-base sequence 5′-GATATC-3′ and makes a cut at the vertical line. In nature, these enzymes protect bacteria against phage infection by digesting the phage DNA when it enters the bacterial cell, acting as part of the restriction modification system.[121] In technology, these sequence-specific nucleases are used in molecular cloning and DNA fingerprinting.

Enzymes called DNA ligases can rejoin cut or broken DNA strands.[122] Ligases are particularly important in lagging strand DNA replication, as they join together the short segments of DNA produced at the replication fork into a complete copy of the DNA template. They are also used in DNA repair and genetic recombination.[122]

Topoisomerases and helicases

Topoisomerases are enzymes with both nuclease and ligase activity. These proteins change the amount of supercoiling in DNA. Some of these enzymes work by cutting the DNA helix and allowing one section to rotate, thereby reducing its level of supercoiling; the enzyme then seals the DNA break.[44] Other types of these enzymes are capable of cutting one DNA helix and then passing a second strand of DNA through this break, before rejoining the helix.[123] Topoisomerases are required for many processes involving DNA, such as DNA replication and transcription.[45]

Helicases are proteins that are a type of molecular motor. They use the chemical energy in nucleoside triphosphates, predominantly ATP, to break hydrogen bonds between bases and unwind the DNA double helix into single strands.[124] These enzymes are essential for most processes where enzymes need to access the DNA bases.

Polymerases

Polymerases are enzymes that synthesize polynucleotide chains from nucleoside triphosphates. The sequence of their products are created based on existing polynucleotide chains—which are called *templates*. These enzymes function by repeatedly adding a nucleotide to the 3′ hydroxyl group at the end of the growing polynucleotide chain. As a consequence, all polymerases work in a 5′ to 3′ direction.[125] In the active site of these enzymes, the incoming nucleoside triphosphate base-pairs to the template: this allows polymerases to accurately synthesize the complementary strand of their template. Polymerases are classified according to the type of template that they use.

In DNA replication, DNA-dependent DNA polymerases make copies of DNA polynucleotide chains. In order to preserve biological information, it is essential that the sequence of bases in each copy are precisely complementary to the sequence of bases in the template strand. Many DNA polymerases have a proofreading activity. Here, the polymerase recognizes the occasional mistakes in the synthesis reaction by the lack of base pairing between the mismatched nucleotides. If a mismatch is detected, a 3′ to 5′ exonuclease activity is activated and the incorrect base removed.[126] In most organisms, DNA polymerases function in a large complex called the replisome that contains multiple accessory subunits, such as the DNA clamp or helicases.[127]

RNA-dependent DNA polymerases are a specialized class of polymerases that copy the sequence of an RNA strand into DNA. They include reverse transcriptase, which is a viral enzyme involved in the infection of cells by retroviruses, and telomerase, which is required for the replication of telomeres.[62][128] Telomerase is an unusual polymerase because it contains its own RNA template as part of its structure.[63]

Transcription is carried out by a DNA-dependent RNA polymerase that copies the sequence of a DNA strand into RNA. To begin transcribing a gene, the RNA polymerase binds to a sequence of DNA called a promoter and separates the DNA strands. It then copies the gene sequence into a messenger RNA transcript until it reaches a region of DNA called the terminator, where it halts and detaches from the DNA. As with human DNA-dependent DNA polymerases, RNA polymerase II, the enzyme that transcribes most of the genes in the human genome, operates as part of a large protein complex with multiple regulatory and accessory subunits.[129]

8.5 Genetic recombination

Structure of the Holliday junction intermediate in genetic recombination. The four separate DNA strands are coloured red, blue, green and yellow.[130]

Further information: Genetic recombination

A DNA helix usually does not interact with other segments of DNA, and in human cells the different chromosomes even occupy separate areas in the nucleus called "chromosome territories".[131] This physical separation of different chromosomes is important for the ability of DNA to function as a stable repository for information, as one of the few times chromosomes interact is in chromosomal crossover which occurs during sexual reproduction, when genetic recombination occurs. Chromosomal crossover is when two DNA helices break, swap a section and then rejoin.

Recombination allows chromosomes to exchange genetic information and produces new combinations of genes, which increases the efficiency of natural selection and can be important in the rapid evolution of new proteins.[132] Genetic recombination can also be involved in DNA repair, particularly in the cell's response to double-strand breaks.[133]

The most common form of chromosomal crossover is homologous recombination, where the two chromosomes involved share very similar sequences. Non-homologous recombination can be damaging to cells, as it can produce chromosomal translocations and genetic abnormalities. The recombination reaction is catalyzed by enzymes known as recombinases, such as RAD51.[134] The first step in recombination is a double-stranded break caused by either an endonuclease or damage to the DNA.[135] A series of steps catalyzed in part by the recombinase then leads to joining of the two helices by at least one Holliday junction, in which a segment of a single strand in each helix is annealed to the complementary strand in the other helix. The Holliday junction is a tetrahedral junction structure that can be moved along the pair of chromosomes, swapping one strand for another. The recombination reaction is then halted by cleavage of the junction and re-ligation of the released DNA.[136]

8.6 Evolution

Further information: RNA world hypothesis

DNA contains the genetic information that allows all modern living things to function, grow and reproduce. However, it is unclear how long in the 4-billion-year history of life DNA has performed this function, as it has been proposed that the earliest forms of life may have used RNA as their genetic material.[137][138] RNA may have acted as the central part of early cell metabolism as it can both transmit genetic information and carry out catalysis as part of ribozymes.[139] This ancient RNA world where nucleic acid would have been used for both catalysis and genetics may have influenced the evolution of the current genetic code based on four nucleotide bases. This would occur, since the number of different bases in such an organism is a trade-off between a small number of bases increasing replication accuracy and a large number of bases increasing the catalytic efficiency of ribozymes.[140] However, there is no direct evidence of ancient genetic systems, as recovery of DNA from most fossils is impossible because DNA survives in the environment for less than one million years, and slowly degrades into short fragments in solution.[141] Claims for older DNA have been made, most notably a report of the isolation of a viable bacterium from a salt crystal 250 million years old,[142] but these claims are controversial.[143][144]

Building blocks of DNA (adenine, guanine and related organic molecules) may have been formed extraterrestrially in outer space.[145][146][147] Complex DNA and RNA organic compounds of life, including uracil, cytosine and thymine, have also been formed in the laboratory under conditions mimicking those found in outer space, using starting chemicals, such as pyrimidine, found in meteorites. Pyrimidine, like polycyclic aromatic hydrocarbons (PAHs), the most carbon-rich chemical found in the universe, may have been formed in red giants or in interstellar dust and gas clouds.[148]

8.7   Uses in technology

8.7.1   Genetic engineering

Further information: Molecular biology, nucleic acid methods and genetic engineering

Methods have been developed to purify DNA from organisms, such as phenol-chloroform extraction, and to manipulate it in the laboratory, such as restriction digests and the polymerase chain reaction. Modern biology and biochemistry make intensive use of these techniques in recombinant DNA technology. Recombinant DNA is a man-made DNA sequence that has been assembled from other DNA sequences. They can be transformed into organisms in the form of plasmids or in the appropriate format, by using a viral vector.[149] The genetically modified organisms produced can be used to produce products such as recombinant proteins, used in medical research,[150] or be grown in agriculture.[151][152]

8.7.2   DNA profiling

Further information: DNA profiling

Forensic scientists can use DNA in blood, semen, skin, saliva or hair found at a crime scene to identify a matching DNA of an individual, such as a perpetrator.This process is formally termed DNA profiling, but may also be called "genetic fingerprinting". In DNA profiling, the lengths of variable sections of repetitive DNA, such as short tandem repeats and minisatellites, are compared between people. This method is usually an extremely reliable technique for identifying a matching DNA.[153] However, identification can be complicated if the scene is contaminated with DNA from several people.[154] DNA profiling was developed in 1984 by British geneticist Sir Alec Jeffreys,[155] and first used in forensic science to convict Colin Pitchfork in the 1988 Enderby murders case.[156]

The development of forensic science, and the ability to now obtain genetic matching on minute samples of blood, skin, saliva or hair has led to a re-examination of a number of cases. Forensic scientists can analyze types of DNA: nuclear DNA or mitochondrial DNA; nuclear DNA can individualize evidence, while mitochondrial DNA, or mtDNA, can only classify the maternal inheritance of the sample taken in for analysis. Evidence can now be uncovered that was not scientifically possible at the time of the original examination. Combined with the removal of the double jeopardy law in some places, this can allow cases to be reopened where previous trials have failed to produce sufficient evidence to convince a jury. People charged with serious crimes may be required to provide a sample of DNA for matching purposes. The most obvious defence to DNA matches obtained forensically is to claim that cross-contamination of evidence has taken place. This has resulted in meticulous strict handling procedures with new cases of serious crime. DNA profiling is also used to identify victims of mass casualty incidents.[157] As well as positively identifying bodies or body parts in serious accidents, DNA profiling is being successfully used to identify individual victims in mass war graves – matching to family members.

DNA profiling is also used in DNA paternity testing in order to determine if someone is the biologicalparent or grandparent of a child with the probability of parentage is typically 99.99% when the alleged parent is biologically related to the child. Normal DNA sequencing methods happen after birth but there are new methods to test paternity while the mother is still pregnant.[158]

8.7.3   DNA enzymes or catalytic DNA

Further information: Deoxyribozyme

Deoxyribozymes, also called DNAzymes or catalytic DNA are first discovered in 1994.[159] They are mostly single stranded DNA sequences isolated from a large pool of random DNA sequences through a combinatorial approach called in vitro selection or SELEX. DNAzymes catalyze variety of chemical reactions including RNA/DNA cleavage, RNA/DNA ligation, amino acids phosphorylation/dephosphorylation, carbon-carbon bond formation, and etc. DNAzymes can enhance catalytic rate of chemical reactions up to 100,000,000,000-fold over the uncatalyzed reaction.[160] The most extensively studied class of DNAzymes are RNA-cleaving DNAzymes which have been used in detection of different metal ions and designing therapeutic agents. Several metal-specific DNAzymes have been reported including the GR-5 DNAzyme (lead-specific),[159] the CA1-3 DNAzymes (copper-specific),[161] the 39E DNAzyme (uranyl-specific) and the NaA43 DNAzyme (sodium-specific).[162] The NaA43 DNAzyme, which is reported to be more than 10,000-fold selective for sodium over other metal ions, was used to make a real-time sodium sensor in living cells.

8.7.4 Bioinformatics

Further information: Bioinformatics

Bioinformatics involves the development of techniques to store, data mine, search and manipulate biological data, including DNA nucleic acid sequence data. These have led to widely applied advances in computer science, especially string searching algorithms, machine learning and database theory.[163] String searching or matching algorithms, which find an occurrence of a sequence of letters inside a larger sequence of letters, were developed to search for specific sequences of nucleotides.[164] The DNA sequence may be aligned with other DNA sequences to identify homologous sequences and locate the specific mutations that make them distinct. These techniques, especially multiple sequence alignment, are used in studying phylogenetic relationships and protein function.[165] Data sets representing entire genomes' worth of DNA sequences, such as those produced by the Human Genome Project, are difficult to use without the annotations that identify the locations of genes and regulatory elements on each chromosome. Regions of DNA sequence that have the characteristic patterns associated with protein- or RNA-coding genes can be identified by gene finding algorithms, which allow researchers to predict the presence of particular gene products and their possible functions in an organism even before they have been isolated experimentally.[166] Entire genomes may also be compared, which can shed light on the evolutionary history of particular organism and permit the examination of complex evolutionary events.

8.7.5 DNA nanotechnology

Further information: DNA nanotechnology

DNA nanotechnology uses the unique molecular recognition properties of DNA and other nucleic acids to create self-assembling branched DNA complexes with useful properties.[167] DNA is thus used as a structural material rather than as a carrier of biological information. This has led to the creation of two-dimensional periodic lattices (both tile-based and using the "DNA origami" method) as well as three-dimensional structures in the shapes of polyhedra.[168] Nanomechanical devices and algorithmic self-assembly have also been demonstrated,[169] and these DNA structures have been used to template the arrangement of other molecules such as gold nanoparticles and streptavidin proteins.[170]

8.7.6 History and anthropology

Further information: Phylogenetics and Genetic genealogy

Because DNA collects mutations over time, which are then inherited, it contains historical information, and, by comparing DNA sequences, geneticists can infer the evolutionary history of organisms, their phylogeny.[171] This field of phylogenetics is a powerful tool in evolutionary biology. If DNA sequences within a species are compared, population geneticists can learn the history of particular populations. This can be used in studies ranging from ecological genetics to anthropology; For example, DNA evidence is being used to try to identify the Ten Lost Tribes of Israel.[172][173]

8.7.7 Information storage

Main article: DNA digital data storage

In a paper published in *Nature* in January 2013, scientists from the European Bioinformatics Institute and Agilent Technologies proposed a mechanism to use DNA's ability to code information as a means of digital data storage. The group was able to encode 739 kilobytes of data into DNA code, synthesize the actual DNA, then sequence the DNA and decode the information back to its original form, with a reported 100% accuracy. The encoded information consisted of text files and audio files. A prior experiment was published in August 2012. It was conducted by researchers at Harvard University, where the text of a 54,000-word book was encoded in DNA.[174][175]

8.8   History of DNA research

Further information: History of molecular biology

DNA was first isolated by the Swiss physician Friedrich Miescher who, in 1869, discovered a microscopic substance in the pus of discarded surgical bandages. As it resided in the nuclei of cells, he called it "nuclein".[176][177] In 1878, Albrecht Kossel isolated the non-protein component of "nuclein", nucleic acid, and later isolated its five primary nucleobases.[178][179] In 1919, Phoebus Levene identified the base, sugar and phosphate nucleotide unit.[180] Levene suggested that DNA consisted of a string of nucleotide units linked together through the phosphate groups. Levene thought the chain was short and the bases repeated in a fixed order. In 1937, William Astbury produced the first X-ray diffraction patterns that showed that DNA had a regular structure.[181]

In 1927, Nikolai Koltsov proposed that inherited traits would be inherited via a "giant hereditary molecule" made up of "two mirror strands that would replicate in a semi-conservative fashion using each strand as a template".[182][183] In 1928, Frederick Griffith in his experiment discovered that traits of the "smooth" form of *Pneumococcus* could be transferred to the "rough" form of the same bacteria by mixing killed "smooth" bacteria with the live "rough" form.[184][185] This system provided the first clear suggestion that DNA carries genetic information—the Avery–MacLeod–McCarty experiment—when Oswald Avery, along with coworkers Colin MacLeod and Maclyn McCarty, identified DNA as the transforming principle in 1943.[186] DNA's role in heredity was confirmed in 1952, when Alfred Hershey and Martha Chase in the Hershey–Chase experiment showed that DNA is the genetic material of the T2 phage.[187]

In 1953, James Watson and Francis Crick suggested what is now accepted as the first correct double-helix model of DNA structure in the journal *Nature*.[9] Their double-helix, molecular model of DNA was then based on a single X-ray diffraction image (labeled as "Photo 51")[188] taken by Rosalind Franklin and Raymond Gosling in May 1952, as well as the information that the DNA bases are paired—also obtained through private communications from Erwin Chargaff in the previous years.

Experimental evidence supporting the Watson and Crick model was published in a series of five articles in the same issue of *Nature*.[189] Of these, Franklin and Gosling's paper was the first publication of their own X-ray diffraction data and original analysis method that partially supported the Watson and Crick model;[48][190] this issue also contained an article on DNA structure by Maurice Wilkins and two of his colleagues, whose analysis and *in vivo* B-DNA X-ray patterns also supported the presence *in vivo* of the double-helical DNA configurations as proposed by Crick and Watson for their double-helix molecular model of DNA in the previous two pages of *Nature*.[49] In 1962, after Franklin's death, Watson, Crick, and Wilkins jointly received the Nobel Prize in Physiology or Medicine.[191] Nobel Prizes were awarded only to living recipients at the time. A debate continues about who should receive credit for the discovery.[192]

In an influential presentation in 1957, Crick laid out the central dogma of molecular biology, which foretold the relationship between DNA, RNA, and proteins, and articulated the "adaptor hypothesis".[193] Final confirmation of the replication mechanism that was implied by the double-helical structure followed in 1958 through the Meselson–Stahl experiment.[194] Further work by Crick and coworkers showed that the genetic code was based on non-overlapping triplets of bases, called codons, allowing Har Gobind Khorana, Robert W. Holley and Marshall Warren Nirenberg to decipher the genetic code.[195] These findings represent the birth of molecular biology.

8.9   See also

- Autosome

- Crystallography

- DNA-encoded chemical library

- DNA microarray

- DNA sequencing

- DNA, RNA and proteins: The three essential macromolecules of life

- Genetic disorder

- Haplotype

- Nucleic acid modeling

- Meiosis

- Nucleic acid double helix

- Nucleic acid notation

- Nucleic acid sequence

- Pangenesis

- Phosphoramidite

- Southern blot

- X-ray scattering techniques

- Xeno nucleic acid

- *Proteopedia DNA*

- RNA

- Deoxyribozyme

8.10   References

[1] Purcell, Adam. "DNA". *Basic Biology.*

[2] Nuwer, Rachel (18 July 2015). "Counting All the DNA on Earth". *The New York Times* (New York: The New York Times Company). ISSN 0362-4331. Retrieved 2015-07-18.

[3] "The Biosphere: Diversity of Life". *Aspen Global Change Institute.* Basalt, CO. Retrieved 2015-07-19.

[4] Russell, Peter (2001). *iGenetics.* New York: Benjamin Cummings. ISBN 0-8053-4553-1.

[5] Mashaghi A, Katan A (2013). "A physicist's view of DNA". *De Physicus* **24e** (3): 59–61. arXiv:1311.2545v1. Bibcode:2013arXiv1311.2545M.

[6] Saenger, Wolfram (1984). *Principles of Nucleic Acid Structure.* New York: Springer-Verlag. ISBN 0-387-90762-9.

[7] Alberts, Bruce; Johnson, Alexander; Lewis, Julian; Raff, Martin; Roberts, Keith; Walters, Peter (2002). *Molecular Biology of the Cell; Fourth Edition.* New York and London: Garland Science. ISBN 0-8153-3218-1. OCLC 145080076 48122761 57023651 69932405.

[8] Irobalieva, Rossitza N.; Fogg, Jonathan M.; Catanese Jr, Daniel J.; Sutthibutpong, Thana; Chen, Muyuan; Barker, Anna K.; Ludtke, Steven J.; Harris, Sarah A.; Schmid, Michael F. (2015-10-12). "Structural diversity of supercoiled DNA". *Nature Communications* **6**. doi:10.1038/ncomms9440. PMC 4608029. PMID 26455586.

[9] Watson JD, Crick FH (1953). "A Structure for Deoxyribose Nucleic Acid" (PDF). *Nature* **171** (4356): 737–738. Bibcode:1953Natur.171..737W. doi:10.1038/171737a0. PMID 13054692.

[10] Mandelkern M, Elias JG, Eden D, Crothers DM (1981). "The dimensions of DNA in solution". *J Mol Biol* **152** (1): 153–61. doi:10.1016/0022-2836(81)90099-1. PMID 7338906.

[11] Gregory SG, Barlow KF, McLay KE, Kaul R, Swarbreck D, Dunham A; et al. (2006). "The DNA sequence and biological annotation of human chromosome 1". *Nature* **441** (7091): 315–21. Bibcode:2006Natur.441..315G. doi:10.1038/nature04727. PMID 16710414.

[12] Watson JD, Crick FH (1953). "A Structure for Deoxyribose Nucleic Acid" (PDF). *Nature* **171** (4356): 737–738. Bibcode:1953Natur.171..737\ doi:10.1038/171737a0. PMID 13054692. Retrieved 4 May 2009.

[13] Berg J., Tymoczko J. and Stryer L. (2002) *Biochemistry*. W. H. Freeman and Company ISBN 0-7167-4955-6

[14] Abbreviations and Symbols for Nucleic Acids, Polynucleotides and their Constituents IUPAC-IUB Commission on Biochemical Nomenclature (CBN). Retrieved 3 January 2006.

[15] Ghosh A, Bansal M (2003). "A glossary of DNA structures from A to Z". *Acta Crystallogr D* **59** (4): 620–6. doi:10.1107/S0907444903003251. PMID 12657780.

[16] Created from PDB 1D65

[17] Yakovchuk P, Protozanova E, Frank-Kamenetskii MD (2006). "Base-stacking and base-pairing contributions into thermal stability of the DNA double helix". *Nucleic Acids Res.* **34** (2): 564–74. doi:10.1093/nar/gkj454. PMC 1360284. PMID 16449200.

[18] Burton E. Tropp - *"Molecular Biology"*- Jones and Barlett Learning, ISBN 978-0-7637-8663-2

[19] https://www.mun.ca - *Watson-Crick Structure of DNA - 1953*

[20] Verma S, Eckstein F (1998). "Modified oligonucleotides: synthesis and strategy for users". *Annu. Rev. Biochem.* **67**: 99–134. doi:10.1146/annurev.biochem.67.1.99. PMID 9759484.

[21] Kiljunen S, Hakala K, Pinta E, Huttunen S, Pluta P, Gador A, Lönnberg H, Skurnik M (2005). "Yersiniophage phiR1-37 is a tailed bacteriophage having a 270 kb DNA genome with thymidine replaced by deoxyuridine". *Microbiology* **151** (12): 4093–4102. doi:10.1099/mic.0.28265-0. PMID 16339954.

[22] Uchiyama J, Takemura-Uchiyama I, Sakaguchi Y, Gamoh K, Kato SI, Daibata M, Ujihara T, Misawa N, Matsuzaki S (2014) Intragenus generalized transduction in *Staphylococcus* spp. by a novel giant phage. ISME J. 2014 Mar 6. doi:10.1038/ismej.2014.29

[23] Simpson L (1998). "A base called J". *Proc Natl Acad Sci USA* **95** (5): 2037–2038. Bibcode:1998PNAS...95.2037S. doi:10.1073/pnas.95.5.203 PMC 33841. PMID 9482833.

[24] Borst P, Sabatini R (2008). "Base J: discovery, biosynthesis, and possible functions". *Annual review of microbiology* **62**: 235–51. doi:10.1146/annurev.micro.62.081307.162750. PMID 18729733.

[25] Cross M, Kieft R, Sabatini R, Wilm M, de Kort M, van der Marel GA, van Boom JH, van Leeuwen F, Borst P (1999). "The modified base J is the target for a novel DNA-binding protein in kinetoplastid protozoans". *The EMBO Journal* **18** (22): 6573–6581. doi:10.1093/emboj/18.22.6573. PMC 1171720. PMID 10562569.

[26] DiPaolo C, Kieft R, Cross M, Sabatini R (2005). "Regulation of trypanosome DNA glycosylation by a SWI2/SNF2-like protein". *Mol Cell* **17** (3): 441–451. doi:10.1016/j.molcel.2004.12.022. PMID 15694344.

[27] Vainio S, Genest PA, ter Riet B, van Luenen H, Borst P (2009). "Evidence that J-binding protein 2 is a thymidine hydroxylase catalyzing the first step in the biosynthesis of DNA base J". *Molecular and biochemical parasitology* **164** (2): 157–61. doi:10.1016/j.molbiopara.2008.12.001. PMID 19114062.

[28] Iyer LM, Tahiliani M, Rao A, Aravind L (2009). "Prediction of novel families of enzymes involved in oxidative and other complex modifications of bases in nucleic acids". *Cell Cycle* **8** (11): 1698–1710. doi:10.4161/cc.8.11.8580. PMC 2995806. PMID 19411852.

[29] van Luenen HG, Farris C, Jan S, Genest PA, Tripathi P, Velds A, Kerkhoven RM, Nieuwland M, Haydock A, Ramasamy G, Vainio S, Heidebrecht T, Perrakis A, Pagie L, van Steensel B, Myler PJ, Borst P (2012). "Leishmania". *Cell* **150** (5): 909–921. doi:10.1016/j.cell.2012.07.030. PMC 3684241. PMID 22939620.

[30] Hazelbaker DZ, Buratowski S (2012). "Transcription: base J blocks the way". *Curr Biol* **22** (22): R960–2. doi:10.1016/j.cub.2012.10.010. PMC 3648658. PMID 23174300.

[31] Wing R, Drew H, Takano T, Broka C, Tanaka S, Itakura K, Dickerson RE (1980). "Crystal structure analysis of a complete turn of B-DNA". *Nature* **287** (5784): 755–8. Bibcode:1980Natur.287..755W. doi:10.1038/287755a0. PMID 7432492.

[32] Pabo CO, Sauer RT (1984). "Protein-DNA recognition". *Annu Rev Biochem* **53**: 293–321. doi:10.1146/annurev.bi.53.070184.001453. PMID 6236744.

[33] Clausen-Schaumann H, Rief M, Tolksdorf C, Gaub HE (2000). "Mechanical stability of single DNA molecules". *Biophys J* **78** (4): 1997–2007. Bibcode:2000BpJ....78.1997C. doi:10.1016/S0006-3495(00)76747-6. PMC 1300792. PMID 10733978.

[34] Chalikian TV, Völker J, Plum GE, Breslauer KJ (1999). "A more unified picture for the thermodynamics of nucleic acid duplex melting: A characterization by calorimetric and volumetric techniques". *Proc Natl Acad Sci USA* **96** (14): 7853–8. Bibcode:1999PNAS...96.7853C. doi:10.1073/pnas.96.14.7853. PMC 22151. PMID 10393911.

[35] deHaseth PL, Helmann JD (1995). "Open complex formation by Escherichia coli RNA polymerase: the mechanism of polymerase-induced strand separation of double helical DNA". *Mol Microbiol* **16** (5): 817–24. doi:10.1111/j.1365-2958.1995.tb02309.x. PMID 7476180.

[36] Isaksson J, Acharya S, Barman J, Cheruku P, Chattopadhyaya J (2004). "Single-stranded adenine-rich DNA and RNA retain structural characteristics of their respective double-stranded conformations and show directional differences in stacking pattern". *Biochemistry* **43** (51): 15996–6010. doi:10.1021/bi048221v. PMID 15609994.

[37] Designation of the two strands of DNA JCBN/NC-IUB Newsletter 1989. Retrieved 7 May 2008

[38] Hüttenhofer A, Schattner P, Polacek N (2005). "Non-coding RNAs: hope or hype?". *Trends Genet* **21** (5): 289–97. doi:10.1016/j.tig.2005.03.007. PMID 15851066.

[39] Munroe SH (2004). "Diversity of antisense regulation in eukaryotes: multiple mechanisms, emerging patterns". *J Cell Biochem* **93** (4): 664–71. doi:10.1002/jcb.20252. PMID 15389973.

[40] Makalowska I, Lin CF, Makalowski W (2005). "Overlapping genes in vertebrate genomes". *Comput Biol Chem* **29** (1): 1–12. doi:10.1016/j.compbiolchem.2004.12.006. PMID 15680581.

[41] Johnson ZI, Chisholm SW (2004). "Properties of overlapping genes are conserved across microbial genomes". *Genome Res* **14** (11): 2268–72. doi:10.1101/gr.2433104. PMC 525685. PMID 15520290.

[42] Lamb RA, Horvath CM (1991). "Diversity of coding strategies in influenza viruses". *Trends Genet* **7** (8): 261–6. doi:10.1016/0168-9525(91)90326-L. PMID 1771674.

[43] Benham CJ, Mielke SP (2005). "DNA mechanics". *Annu Rev Biomed Eng* **7**: 21–53. doi:10.1146/annurev.bioeng.6.062403.132016. PMID 16004565.

[44] Champoux JJ (2001). "DNA topoisomerases: structure, function, and mechanism". *Annu Rev Biochem* **70**: 369–413. doi:10.1146/annurev.biochem. PMID 11395412.

[45] Wang JC (2002). "Cellular roles of DNA topoisomerases: a molecular perspective". *Nature Reviews Molecular Cell Biology* **3** (6): 430–40. doi:10.1038/nrm831. PMID 12042765.

[46] Basu HS, Feuerstein BG, Zarling DA, Shafer RH, Marton LJ (1988). "Recognition of Z-RNA and Z-DNA determinants by polyamines in solution: experimental and theoretical studies". *J Biomol Struct Dyn* **6** (2): 299–309. doi:10.1080/07391102.1988.10507714. PMID 2482766.

[47] Franklin RE, Gosling RG (6 March 1953). "The Structure of Sodium Thymonucleate Fibres I. The Influence of Water Content" (PDF). *Acta Crystallogr* **6** (8–9): 673–7. doi:10.1107/S0365110X53001939. Archived from the original Check |url= value (help) (PDF) on 2007-06-12.
Franklin RE, Gosling RG (1953). "The structure of sodium thymonucleate fibres. II. The cylindrically symmetrical Patterson function". *Acta Crystallogr* **6** (8–9): 678–85. doi:10.1107/S0365110X53001940.

[48] Franklin RE, Gosling RG (1953). "Molecular Configuration in Sodium Thymonucleate. Franklin R. and Gosling R.G" (PDF). *Nature* **171** (4356): 740–1. Bibcode:1953Natur.171..740F. doi:10.1038/171740a0. PMID 13054694.

[49] Wilkins MH, Stokes AR, Wilson HR (1953). "Molecular Structure of Deoxypentose Nucleic Acids" (PDF). *Nature* **171** (4356): 738–740. Bibcode:1953Natur.171..738W. doi:10.1038/171738a0. PMID 13054693.

[50] Leslie AG, Arnott S, Chandrasekaran R, Ratliff RL (1980). "Polymorphism of DNA double helices". *J. Mol. Biol.* **143** (1): 49–72. doi:10.1016/0022-2836(80)90124-2. PMID 7441761.

[51] Baianu, I.C. (1980). "Structural Order and Partial Disorder in Biological systems". *Bull. Math. Biol.* **42** (4): 137–141. doi:10.1016/s0092-8240(80)80083-8. http://cogprints.org/3822/

[52] Hosemann R., Bagchi R.N., *Direct analysis of diffraction by matter*, North-Holland Publs., Amsterdam – New York, 1962.

[53] Baianu, I.C. (1978). "X-ray scattering by partially disordered membrane systems". *Acta Crystallogr A* **34** (5): 751–753. Bibcode:1978AcCrA..34..751B. doi:10.1107/S0567739478001540.

[54] Wahl MC, Sundaralingam M (1997). "Crystal structures of A-DNA duplexes". *Biopolymers* **44** (1): 45–63. doi:10.1002/(SICI)1097-0282(1997)44:1<45::AID-BIP4>3.0.CO;2-#. PMID 9097733.

[55] Lu XJ, Shakked Z, Olson WK (2000). "A-form conformational motifs in ligand-bound DNA structures". *J. Mol. Biol.* **300** (4): 819–40. doi:10.1006/jmbi.2000.3690. PMID 10891271.

[56] Rothenburg S, Koch-Nolte F, Haag F (2001). "DNA methylation and Z-DNA formation as mediators of quantitative differences in the expression of alleles". *Immunol Rev* **184**: 286–98. doi:10.1034/j.1600-065x.2001.1840125.x. PMID 12086319.

[57] Oh DB, Kim YG, Rich A (2002). "Z-DNA-binding proteins can act as potent effectors of gene expression in vivo". *Proc. Natl. Acad. Sci. U.S.A.* **99** (26): 16666–71. Bibcode:2002PNAS...9916666O. doi:10.1073/pnas.262672699. PMC 139201. PMID 12486233.

[58] Palmer, Jason (2 December 2010). "Arsenic-loving bacteria may help in hunt for alien life". *BBC News.* Retrieved 2 December 2010.

[59] Bortman, Henry (2 December 2010). "Arsenic-Eating Bacteria Opens New Possibilities for Alien Life". *Space.com.* Retrieved 2 December 2010.

[60] Katsnelson, Alla (2 December 2010). "Arsenic-eating microbe may redefine chemistry of life". *Nature News.* doi:10.1038/news.2010.645.

[61] Cressey, Daniel (3 October 2012). "'Arsenic-life' Bacterium Prefers Phosphorus after all". *Nature News.* doi:10.1038/nature.2012.11520.

[62] Greider CW, Blackburn EH (1985). "Identification of a specific telomere terminal transferase activity in Tetrahymena extracts". *Cell* **43** (2 Pt 1): 405–13. doi:10.1016/0092-8674(85)90170-9. PMID 3907856.

[63] Nugent CI, Lundblad V (1998). "The telomerase reverse transcriptase: components and regulation". *Genes Dev* **12** (8): 1073–85. doi:10.1101/gad.12.8.1073. PMID 9553037.

[64] Wright WE, Tesmer VM, Huffman KE, Levene SD, Shay JW (1997). "Normal human chromosomes have long G-rich telomeric overhangs at one end". *Genes Dev* **11** (21): 2801–9. doi:10.1101/gad.11.21.2801. PMC 316649. PMID 9353250.

[65] Created from NDB UD0017

[66] Burge S, Parkinson GN, Hazel P, Todd AK, Neidle S (2006). "Quadruplex DNA: sequence, topology and structure". *Nucleic Acids Res* **34** (19): 5402–15. doi:10.1093/nar/gkl655. PMC 1636468. PMID 17012276.

[67] Parkinson GN, Lee MP, Neidle S (2002). "Crystal structure of parallel quadruplexes from human telomeric DNA". *Nature* **417** (6891): 876–80. Bibcode:2002Natur.417..876P. doi:10.1038/nature755. PMID 12050675.

[68] Griffith JD, Comeau L, Rosenfield S, Stansel RM, Bianchi A, Moss H, de Lange T (1999). "Mammalian telomeres end in a large duplex loop". *Cell* **97** (4): 503–14. doi:10.1016/S0092-8674(00)80760-6. PMID 10338214.

[69] Seeman NC (2005). "DNA enables nanoscale control of the structure of matter". *Q. Rev. Biophys.* **38** (4): 363–71. doi:10.1017/S003358350500$ PMC 3478329. PMID 16515737.

[70] Hu Q, Rosenfeld MG (2012). "Epigenetic regulation of human embryonic stem cells". *Frontiers in Genetics* **3**: 238. doi:10.3389/fgene.2012.002 PMC 3488762. PMID 23133442.

[71] Klose RJ, Bird AP (2006). "Genomic DNA methylation: the mark and its mediators". *Trends Biochem Sci* **31** (2): 89–97. doi:10.1016/j.tibs.2005.12.008. PMID 16403636.

[72] Bird A (2002). "DNA methylation patterns and epigenetic memory". *Genes Dev* **16** (1): 6–21. doi:10.1101/gad.947102. PMID 11782440.

[73] Walsh CP, Xu GL (2006). "Cytosine methylation and DNA repair". *Curr Top Microbiol Immunol.* Current Topics in Microbiology and Immunology **301**: 283–315. doi:10.1007/3-540-31390-7_11. ISBN 3-540-29114-8. PMID 16570853.

[74] Kriaucionis S, Heintz N (2009). "The nuclear DNA base 5-hydroxymethylcytosine is present in Purkinje neurons and the brain". *Science* **324** (5929): 929–30. Bibcode:2009Sci...324..929K. doi:10.1126/science.1169786. PMC 3263819. PMID 19372393.

[75] Ratel D, Ravanat JL, Berger F, Wion D (2006). "N6-methyladenine: the other methylated base of DNA". *BioEssays* **28** (3): 309–15. doi:10.1002/bies.20342. PMC 2754416. PMID 16479578.

[76] Gommers-Ampt JH, Van Leeuwen F, de Beer AL, Vliegenthart JF, Dizdaroglu M, Kowalak JA, Crain PF, Borst P (1993). "beta-D-glucosyl-hydroxymethyluracil: a novel modified base present in the DNA of the parasitic protozoan T. brucei". *Cell* **75** (6): 1129–36. doi:10.1016/0092-8674(93)90322-H. PMID 8261512.

[77] Created from PDB 1JDG

[78] Douki T, Reynaud-Angelin A, Cadet J, Sage E (2003). "Bipyrimidine photoproducts rather than oxidative lesions are the main type of DNA damage involved in the genotoxic effect of solar UVA radiation". *Biochemistry* **42** (30): 9221–6. doi:10.1021/bi034593c. PMID 12885257.

[79] Cadet J, Delatour T, Douki T, Gasparutto D, Pouget JP, Ravanat JL, Sauvaigo S (1999). "Hydroxyl radicals and DNA base damage". *Mutat Res* **424** (1–2): 9–21. doi:10.1016/S0027-5107(99)00004-4. PMID 10064846.

[80] Beckman KB, Ames BN (1997). "Oxidative decay of DNA". *J. Biol. Chem.* **272** (32): 19633–6. doi:10.1074/jbc.272.32.19633. PMID 9289489.

[81] Valerie K, Povirk LF (2003). "Regulation and mechanisms of mammalian double-strand break repair". *Oncogene* **22** (37): 5792–812. doi:10.1038/sj.onc.1206679. PMID 12947387.

[82] Johnson, George (28 December 2010). "Unearthing Prehistoric Tumors, and Debate". *The New York Times*. If we lived long enough, sooner or later we all would get cancer.

[83] Alberts, B, Johnson A, Lewis J; et al. (2002). "The Preventable Causes of Cancer". *Molecular biology of the cell* (4th ed.). New York: Garland Science. ISBN 0-8153-4072-9. A certain irreducible background incidence of cancer is to be expected regardless of circumstances: mutations can never be absolutely avoided, because they are an inescapable consequence of fundamental limitations on the accuracy of DNA replication, as discussed in Chapter 5. If a human could live long enough, it is inevitable that at least one of his or her cells would eventually accumulate a set of mutations sufficient for cancer to develop.

[84] Bernstein H, Payne CM, Bernstein C, Garewal H, Dvorak K (2008). Cancer and aging as consequences of un-repaired DNA damage. In: New Research on DNA Damages (Editors: Honoka Kimura and Aoi Suzuki) Nova Science Publishers, Inc., New York, Chapter 1, pp. 1–47. open access, but read only https://www.novapublishers.com/catalog/product_info.php?products_id=43247 ISBN 978-1604565812

[85] Hoeijmakers JH (October 2009). "DNA damage, aging, and cancer". *N. Engl. J. Med.* **361** (15): 1475–85. doi:10.1056/NEJMra0804615. PMID 19812404.

[86] Freitas AA, de Magalhães JP (2011). "A review and appraisal of the DNA damage theory of ageing". *Mutat. Res.* **728** (1–2): 12–22. doi:10.1016/j.mrrev.2011.05.001. PMID 21600302.

[87] Ferguson LR, Denny WA (1991). "The genetic toxicology of acridines". *Mutat Res* **258** (2): 123–60. doi:10.1016/0165-1110(91)90006-H. PMID 1881402.

[88] Stephens TD, Bunde CJ, Fillmore BJ (2000). "Mechanism of action in thalidomide teratogenesis". *Biochem Pharmacol* **59** (12): 1489–99. doi:10.1016/S0006-2952(99)00388-3. PMID 10799645.

[89] Jeffrey AM (1985). "DNA modification by chemical carcinogens". *Pharmacol Ther* **28** (2): 237–72. doi:10.1016/0163-7258(85)90013-0. PMID 3936066.

[90] Braña MF, Cacho M, Gradillas A, de Pascual-Teresa B, Ramos A (2001). "Intercalators as anticancer drugs". *Curr Pharm Des* **7** (17): 1745–80. doi:10.2174/1381612013397113. PMID 11562309.

[91] Venter JC, Adams MD, Myers EW, Li PW, Mural RJ, Sutton GG; et al. (2001). "The sequence of the human genome". *Science* **291** (5507): 1304–51. Bibcode:2001Sci...291.1304V. doi:10.1126/science.1058040. PMID 11181995.

[92] Thanbichler M, Wang SC, Shapiro L (2005). "The bacterial nucleoid: a highly organized and dynamic structure". *J Cell Biochem* **96** (3): 506–21. doi:10.1002/jcb.20519. PMID 15988757.

[93] Wolfsberg TG, McEntyre J, Schuler GD (2001). "Guide to the draft human genome". *Nature* **409** (6822): 824–6. Bibcode:2001Natur.409..824W. doi:10.1038/35057000. PMID 11236998.

[94] Gregory TR (2005). "The C-value enigma in plants and animals: a review of parallels and an appeal for partnership". *Annals of Botany* **95** (1): 133–46. doi:10.1093/aob/mci009. PMID 15596463.

[95] Birney E, Stamatoyannopoulos JA, Dutta A, Guigó R, Gingeras TR, Margulies EH; et al. (2007). "Identification and analysis of functional elements in 1% of the human genome by the ENCODE pilot project". *Nature* **447** (7146): 799–816. Bibcode:2007Natur.447..799B. doi:10.1038/nature05874. PMC 2212820. PMID 17571346.

[96] Created from PDB 1MSW

[97] Pidoux AL, Allshire RC (2005). "The role of heterochromatin in centromere function". *Philosophical Transactions of the Royal Society B* **360** (1455): 569–79. doi:10.1098/rstb.2004.1611. PMC 1569473. PMID 15905142.

[98] Harrison PM, Hegyi H, Balasubramanian S, Luscombe NM, Bertone P, Echols N, Johnson T, Gerstein M (2002). "Molecular Fossils in the Human Genome: Identification and Analysis of the Pseudogenes in Chromosomes 21 and 22". *Genome Res* **12** (2): 272–80. doi:10.1101/gr.207102. PMC 155275. PMID 11827946.

[99] Harrison PM, Gerstein M (2002). "Studying genomes through the aeons: protein families, pseudogenes and proteome evolution". *J Mol Biol* **318** (5): 1155–74. doi:10.1016/S0022-2836(02)00109-2. PMID 12083509.

[100] Albà M (2001). "Replicative DNA polymerases". *Genome Biol* **2** (1): reviews3002.1–reviews3002.4. doi:10.1186/gb-2001-2-1-reviews3002. PMC 150442. PMID 11178285.

[101] Tani, Katsuji; Nasu, Masao (2010). "Roles of Extracellular DNA in Bacterial Ecosystems". In Kikuchi, Yo; Rykova, Elena Y. *Extracellular Nucleic Acids*. Springer. pp. 25–38. ISBN 978-3-642-12616-1.

[102] Vlassov, V. V.; Laktionov, P. P.; Rykova, E. Y. (2007). "Extracellular nucleic acids". *Bioessays* **29**: 654–667. doi:10.1002/bies.20604.

[103] Finkel, S. E.; Kolter, R. (2001). "DNA as a nutrient: novel role for bacterial competence gene homologs". *J. Bacteriol.* **183**: 6288–6293. doi:10.1128/JB.183.21.6288-6293.2001.

[104] Mulcahy, H.; Charron-Mazenod, L.; Lewenza, S. (2008). "Extracellular DNA chelates cations and induces antibiotic resistance in *Pseudomonas aeruginosa* biofilms". *PLoSPathog* **4**: e1000213. doi:10.1371/journal.ppat.1000213.

[105] Berne, C.; Kysela, D. T.; Brun, Y. V. (2010). "A bacterial extracellular DNA inhibits settling of motile progeny cells within a biofilm". *Mol. Microbiol.* **77**: 815–829. doi:10.1111/j.1365-2958.2010.07267.x.

[106] Whitchurch, C. B.; Tolker-Nielsen, T.; Ragas, P. C.; Mattick, J. S. (2002). "Extracellular DNA required for bacterial biofilm formation" (PDF). *Science* **295**: 1487. doi:10.1126/science.295.5559.1487.

[107] Hu, W.; Li, L.; Sharma, S.; Wang, J.; McHardy, I.; Lux, R.; Yang, Z.; He, X.; Gimzewski, J. K.; Li, Y.; Shi, W. (2012). "DNA Builds and Strengthens the Extracellular Matrix in Myxococcus xanthus Biofilms by Interacting with Exopolysaccharides". *PLoS ONE* **7** (12): e51905. Bibcode:2012PLoSO...751905H. doi:10.1371/journal.pone.0051905.

[108] Sandman K, Pereira SL, Reeve JN (1998). "Diversity of prokaryotic chromosomal proteins and the origin of the nucleosome". *Cell Mol Life Sci* **54** (12): 1350–64. doi:10.1007/s000180050259. PMID 9893710.

[109] Dame RT (2005). "The role of nucleoid-associated proteins in the organization and compaction of bacterial chromatin". *Mol. Microbiol.* **56** (4): 858–70. doi:10.1111/j.1365-2958.2005.04598.x. PMID 15853876.

[110] Luger K, Mäder AW, Richmond RK, Sargent DF, Richmond TJ (1997). "Crystal structure of the nucleosome core particle at 2.8 A resolution". *Nature* **389** (6648): 251–60. Bibcode:1997Natur.389..251L. doi:10.1038/38444. PMID 9305837.

[111] Jenuwein T, Allis CD (2001). "Translating the histone code". *Science* **293** (5532): 1074–80. doi:10.1126/science.1063127. PMID 11498575.

[112] Ito T (2003). "Nucleosome assembly and remodelling". *Curr Top Microbiol Immunol*. Current Topics in Microbiology and Immunology **274**: 1–22. doi:10.1007/978-3-642-55747-7_1. ISBN 978-3-540-44208-0. PMID 12596902.

[113] Thomas JO (2001). "HMG1 and 2: architectural DNA-binding proteins". *Biochem Soc Trans* **29** (Pt 4): 395–401. doi:10.1042/BST0290395. PMID 11497996.

[114] Grosschedl R, Giese K, Pagel J (1994). "HMG domain proteins: architectural elements in the assembly of nucleoprotein structures". *Trends Genet* **10** (3): 94–100. doi:10.1016/0168-9525(94)90232-1. PMID 8178371.

[115] Iftode C, Daniely Y, Borowiec JA (1999). "Replication protein A (RPA): the eukaryotic SSB". *Crit Rev Biochem Mol Biol* **34** (3): 141–80. doi:10.1080/10409239991209255. PMID 10473346.

[116] Created from PDB 1LMB

[117] Myers LC, Kornberg RD (2000). "Mediator of transcriptional regulation". *Annu Rev Biochem* **69**: 729–49. doi:10.1146/annurev.biochem.69.1.729. PMID 10966474.

[118] Spiegelman BM, Heinrich R (2004). "Biological control through regulated transcriptional coactivators". *Cell* **119** (2): 157–67. doi:10.1016/j.cell.2004.09.037. PMID 15479634.

[119] Li Z, Van Calcar S, Qu C, Cavenee WK, Zhang MQ, Ren B (2003). "A global transcriptional regulatory role for c-Myc in Burkitt's lymphoma cells". *Proc Natl Acad Sci USA* **100** (14): 8164–9. Bibcode:2003PNAS..100.8164L. doi:10.1073/pnas.1332764100. PMC 166200. PMID 12808131.

[120] Created from PDB 1RVA

[121] Bickle TA, Krüger DH (1993). "Biology of DNA restriction". *Microbiol Rev* **57** (2): 434–50. PMC 372918. PMID 8336674.

[122] Doherty AJ, Suh SW (2000). "Structural and mechanistic conservation in DNA ligases". *Nucleic Acids Res* **28** (21): 4051–8. doi:10.1093/nar/28.21.4051. PMC 113121. PMID 11058099.

[123] Schoeffler AJ, Berger JM (2005). "Recent advances in understanding structure-function relationships in the type II topoisomerase mechanism". *Biochem Soc Trans* **33** (Pt 6): 1465–70. doi:10.1042/BST20051465. PMID 16246147.

[124] Tuteja N, Tuteja R (2004). "Unraveling DNA helicases. Motif, structure, mechanism and function". *Eur J Biochem* **271** (10): 1849–63. doi:10.1111/j.1432-1033.2004.04094.x. PMID 15128295.

[125] Joyce CM, Steitz TA (1995). "Polymerase structures and function: variations on a theme?". *J Bacteriol* **177** (22): 6321–9. PMC 177480. PMID 7592405.

[126] Hubscher U, Maga G, Spadari S (2002). "Eukaryotic DNA polymerases". *Annu Rev Biochem* **71**: 133–63. doi:10.1146/annurev.biochem.71.090501 PMID 12045093.

[127] Johnson A, O'Donnell M (2005). "Cellular DNA replicases: components and dynamics at the replication fork". *Annu Rev Biochem* **74**: 283–315. doi:10.1146/annurev.biochem.73.011303.073859. PMID 15952889.

[128] Tarrago-Litvak L, Andréola ML, Nevinsky GA, Sarih-Cottin L, Litvak S (1 May 1994). "The reverse transcriptase of HIV-1: from enzymology to therapeutic intervention". *FASEB J* **8** (8): 497–503. PMID 7514143.

[129] Martinez E (2002). "Multi-protein complexes in eukaryotic gene transcription". *Plant Mol Biol* **50** (6): 925–47. doi:10.1023/A:1021258713850. PMID 12516863.

[130] Created from PDB 1M6G

[131] Cremer T, Cremer C (2001). "Chromosome territories, nuclear architecture and gene regulation in mammalian cells". *Nature Reviews Genetics* **2** (4): 292–301. doi:10.1038/35066075. PMID 11283701.

[132] Pál C, Papp B, Lercher MJ (2006). "An integrated view of protein evolution". *Nature Reviews Genetics* **7** (5): 337–48. doi:10.1038/nrg1838. PMID 16619049.

[133] O'Driscoll M, Jeggo PA (2006). "The role of double-strand break repair – insights from human genetics". *Nature Reviews Genetics* **7** (1): 45–54. doi:10.1038/nrg1746. PMID 16369571.

[134] Vispé S, Defais M (1997). "Mammalian Rad51 protein: a RecA homologue with pleiotropic functions". *Biochimie* **79** (9–10): 587–92. doi:10.1016/S0300-9084(97)82007-X. PMID 9466696.

[135] Neale MJ, Keeney S (2006). "Clarifying the mechanics of DNA strand exchange in meiotic recombination". *Nature* **442** (7099): 153–8. Bibcode:2006Natur.442..153N. doi:10.1038/nature04885. PMID 16838012.

[136] Dickman MJ, Ingleston SM, Sedelnikova SE, Rafferty JB, Lloyd RG, Grasby JA, Hornby DP (2002). "The RuvABC resolvasome". *Eur J Biochem* **269** (22): 5492–501. doi:10.1046/j.1432-1033.2002.03250.x. PMID 12423347.

[137] Joyce GF (2002). "The antiquity of RNA-based evolution". *Nature* **418** (6894): 214–21. Bibcode:2002Natur.418..214J. doi:10.1038/418214a. PMID 12110897.

[138] Orgel LE (2004). "Prebiotic chemistry and the origin of the RNA world". *Crit Rev Biochem Mol Biol* **39** (2): 99–123. doi:10.1080/10409230490460765. PMID 15217990.

[139] Davenport RJ (2001). "Ribozymes. Making copies in the RNA world". *Science* **292** (5520): 1278. doi:10.1126/science.292.5520.1278a. PMID 11360970.

[140] Szathmáry E (1992). "What is the optimum size for the genetic alphabet?". *Proc Natl Acad Sci USA* **89** (7): 2614–8. Bibcode:1992PNAS...89.2614S. doi:10.1073/pnas.89.7.2614. PMC 48712. PMID 1372984.

[141] Lindahl T (1993). "Instability and decay of the primary structure of DNA". *Nature* **362** (6422): 709–15. Bibcode:1993Natur.362..709L. doi:10.1038/362709a0. PMID 8469282.

[142] Vreeland RH, Rosenzweig WD, Powers DW (2000). "Isolation of a 250 million-year-old halotolerant bacterium from a primary salt crystal". *Nature* **407** (6806): 897–900. doi:10.1038/35038060. PMID 11057666.

[143] Hebsgaard MB, Phillips MJ, Willerslev E (2005). "Geologically ancient DNA: fact or artefact?". *Trends Microbiol* **13** (5): 212–20. doi:10.1016/j.tim.2005.03.010. PMID 15866038.

[144] Nickle DC, Learn GH, Rain MW, Mullins JI, Mittler JE (2002). "Curiously modern DNA for a "250 million-year-old" bacterium". *J Mol Evol* **54** (1): 134–7. doi:10.1007/s00239-001-0025-x. PMID 11734907.

[145] Callahan MP, Smith KE, Cleaves HJ, Ruzicka J, Stern JC, Glavin DP, House CH, Dworkin JP (August 2011). "Carbonaceous meteorites contain a wide range of extraterrestrial nucleobases". *Proc. Natl. Acad. Sci. U.S.A.* **108** (34): 13995–8. Bibcode:2011PNAS..10813 doi:10.1073/pnas.1106493108. PMC 3161613. PMID 21836052.

[146] Steigerwald, John (8 August 2011). "NASA Researchers: DNA Building Blocks Can Be Made in Space". NASA. Retrieved 10 August 2011.

[147] ScienceDaily Staff (9 August 2011). "DNA Building Blocks Can Be Made in Space, NASA Evidence Suggests". ScienceDaily. Retrieved 9 August 2011.

[148] Marlaire, Ruth (3 March 2015). "NASA Ames Reproduces the Building Blocks of Life in Laboratory". *NASA*. Retrieved 5 March 2015.

[149] Goff SP, Berg P (1976). "Construction of hybrid viruses containing SV40 and lambda phage DNA segments and their propagation in cultured monkey cells". *Cell* **9** (4 PT 2): 695–705. doi:10.1016/0092-8674(76)90133-1. PMID 189942.

[150] Houdebine LM (2007). "Transgenic animal models in biomedical research". *Methods Mol Biol* **360**: 163–202. doi:10.1385/1-59745-165-7:163. ISBN 1-59745-165-7. PMID 17172731.

[151] Daniell H, Dhingra A (2002). "Multigene engineering: dawn of an exciting new era in biotechnology". *Current Opinion in Biotechnology* **13** (2): 136–41. doi:10.1016/S0958-1669(02)00297-5. PMC 3481857. PMID 11950565.

[152] Job D (2002). "Plant biotechnology in agriculture". *Biochimie* **84** (11): 1105–10. doi:10.1016/S0300-9084(02)00013-5. PMID 12595138.

[153] Collins A, Morton NE (1994). "Likelihood ratios for DNA identification". *Proc Natl Acad Sci USA* **91** (13): 6007–11. Bibcode:1994PNAS...91.6007C. doi:10.1073/pnas.91.13.6007. PMC 44126. PMID 8016106.

[154] Weir BS, Triggs CM, Starling L, Stowell LI, Walsh KA, Buckleton J (1997). "Interpreting DNA mixtures". *J Forensic Sci* **42** (2): 213–22. PMID 9068179.

[155] Jeffreys AJ, Wilson V, Thein SL (1985). "Individual-specific 'fingerprints' of human DNA". *Nature* **316** (6023): 76–9. Bibcode:1985Natur.316...76J. doi:10.1038/316076a0. PMID 2989708.

[156] Colin Pitchfork — first murder conviction on DNA evidence also clears the prime suspect Forensic Science Service Accessed 23 December 2006

[157] "DNA Identification in Mass Fatality Incidents". National Institute of Justice. September 2006.

[158] "Paternity Blood Tests That Work Early in a Pregnancy" New York Times June 20, 2012

[159] Breaker, Ronald R.; Joyce, Gerald F. (1994-01-12). "A DNA enzyme that cleaves RNA". *Chemistry & Biology* **1** (4): 223–229. doi:10.1016/1074-5521(94)90014-0. ISSN 1074-5521. PMID 9383394.

[160] Chandra, Madhavaiah; Sachdeva, Amit; Silverman, Scott K. "DNA-catalyzed sequence-specific hydrolysis of DNA". *Nature Chemical Biology* **5** (10): 718–720. doi:10.1038/nchembio.201. PMC 2746877. PMID 19684594.

[161] Carmi, Nir; Shultz, Lisa A.; Breaker, Ronald R. (1996-01-12). "In vitro selection of self-cleaving DNAs". *Chemistry & Biology* **3** (12): 1039–1046. doi:10.1016/S1074-5521(96)90170-2. ISSN 1074-5521. PMID 9000012.

[162] Torabi, Seyed-Fakhreddin; Wu, Peiwen; McGhee, Claire E.; Chen, Lu; Hwang, Kevin; Zheng, Nan; Cheng, Jianjun; Lu, Yi (2015-05-12). "In vitro selection of a sodium-specific DNAzyme and its application in intracellular sensing". *Proceedings of the National Academy of Sciences* **112** (19): 5903–5908. doi:10.1073/pnas.1420361112. ISSN 0027-8424. PMC 4434688. PMID 25918425.

[163] Baldi, Pierre; Brunak, Soren (2001). *Bioinformatics: The Machine Learning Approach.* MIT Press. ISBN 978-0-262-02506-5. OCLC 45951728.

[164] Gusfield, Dan. *Algorithms on Strings, Trees, and Sequences: Computer Science and Computational Biology.* Cambridge University Press, 15 January 1997. ISBN 978-0-521-58519-4.

[165] Sjölander K (2004). "Phylogenomic inference of protein molecular function: advances and challenges". *Bioinformatics* **20** (2): 170–9. doi:10.1093/bioinformatics/bth021. PMID 14734307.

[166] Mount DM (2004). *Bioinformatics: Sequence and Genome Analysis* (2 ed.). Cold Spring Harbor, NY: Cold Spring Harbor Laboratory Press. ISBN 0-87969-712-1. OCLC 55106399.

[167] Rothemund PW (2006). "Folding DNA to create nanoscale shapes and patterns". *Nature* **440** (7082): 297–302. Bibcode:2006Natur.440..297R. doi:10.1038/nature04586. PMID 16541064.

[168] Andersen ES, Dong M, Nielsen MM, Jahn K, Subramani R, Mamdouh W, Golas MM, Sander B, Stark H, Oliveira CL, Pedersen JS, Birkedal V, Besenbacher F, Gothelf KV, Kjems J (2009). "Self-assembly of a nanoscale DNA box with a controllable lid". *Nature* **459** (7243): 73–6. Bibcode:2009Natur.459...73A. doi:10.1038/nature07971. PMID 19424153.

[169] Ishitsuka Y, Ha T (2009). "DNA nanotechnology: a nanomachine goes live". *Nat Nanotechnol* **4** (5): 281–2. Bibcode:2009NatNa...4..281I. doi:10.1038/nnano.2009.101. PMID 19421208.

[170] Aldaye FA, Palmer AL, Sleiman HF (2008). "Assembling materials with DNA as the guide". *Science* **321** (5897): 1795–9. Bibcode:2008Sci...321.1795A. doi:10.1126/science.1154533. PMID 18818351.

[171] Wray GA (2002). "Dating branches on the Tree of Life using DNA". *Genome Biol* **3** (1): reviews0001.1–reviews0001.7. doi:10.1046/j.1525-142X.1999.99010.x. PMC 150454. PMID 11806830.

[172] *Lost Tribes of Israel*, NOVA, PBS airdate: 22 February 2000. Transcript available from PBS.org. Retrieved 4 March 2006.

[173] Kleiman, Yaakov. "The Cohanim/DNA Connection: The fascinating story of how DNA studies confirm an ancient biblical tradition". *aish.com* (13 January 2000). Retrieved 4 March 2006.

[174] Goldman N, Bertone P, Chen S, Dessimoz C, LeProust EM, Sipos B, Birney E (23 January 2013). "Towards practical, high-capacity, low-maintenance information storage in synthesized DNA". *Nature* **494** (7435): 77–80. Bibcode:2013Natur.494...77G. doi:10.1038/nature11875. PMC 3672958. PMID 23354052.

[175] Naik, Gautam (24 January 2013). "Storing Digital Data in DNA". *Wall Street Journal.* Retrieved 24 January 2013.

[176] Miescher, Friedrich (1871) "Ueber die chemische Zusammensetzung der Eiterzellen" (On the chemical composition of pus cells), *Medicinisch-chemische Untersuchungen*, **4** : 441–460. From p. 456: *"Ich habe mich daher später mit meinen Versuchen an die ganzen Kerne gehalten, die Trennung der Körper, die ich einstweilen ohne weiteres Präjudiz als lösliches und unlösliches Nuclein bezeichnen will, einem günstigeren Material überlassend."* (Therefore, in my experiments I subsequently limited myself to the whole nucleus, leaving to a more favorable material the separation of the substances, that for the present, without further prejudice, I will designate as soluble and insoluble nuclear material ("Nuclein").)

[177] Dahm R (2008). "Discovering DNA: Friedrich Miescher and the early years of nucleic acid research". *Hum. Genet.* **122** (6): 565–81. doi:10.1007/s00439-007-0433-0. PMID 17901982.

[178] See:

- Albrect Kossel (1879) "Ueber Nucleïn der Hefe" (On nuclein in yeast) *Zeitschrift für physiologische Chemie*, **3** : 284-291.

- Albrect Kossel (1880) "Ueber Nucleïn der Hefe II" (On nuclein in yeast, Part 2) *Zeitschrift für physiologische Chemie*, **4** : 290-295.

- Albrect Kossel (1881) "Ueber die Verbreitung des Hypoxanthins im Thier- und Pflanzenreich" (On the distribution of hypoxanthins in the animal and plant kingdoms) *Zeitschrift für physiologische Chemie*, **5** : 267-271.

- Albrect Kossel, *Untersuchungen über die Nucleine und ihre Spaltungsprodukte* [Investigations into nuclein and its cleavage products] (Strassburg, Germany: K.J. Trübne, 1881), 19 pages.

- Albrect Kossel (1882) "Ueber Xanthin und Hypoxanthin" (On xanthin and hypoxanthin), *Zeitschrift für physiologische Chemie*, **6** : 422-431.

- Albrect Kossel (1883) "Zur Chemie des Zellkerns" (On the chemistry of the cell nucleus), *Zeitschrift für physiologische Chemie*, **7** : 7-22.

- Albrect Kossel (1886) "Weitere Beiträge zur Chemie des Zellkerns" (Further contributions to the chemistry of the cell nucleus), *Zeitschrift für Physiologische Chemie*, **10** : 248-264. Available on-line at: Max Planck Institute for the History of Science, Berlin, Germany. On p. 264, Kossel remarked presciently: *"Der Erforschung der quantitativen Verhältnisse der vier stickstoffreichen Basen, der Abhängigkeit ihrer Menge von den physiologischen Zuständen der Zelle, verspricht wichtige Aufschlüsse über die elementaren physiologisch-chemischen Vorgänge."* (The study of the quantitative relations of the four nitrogenous bases — [and] of the dependence of their quantity on the physiological states of the cell — promises important insights into the fundamental physiological-chemical processes.)

[179] Jones ME (September 1953). "Albrecht Kossel, A Biographical Sketch". *Yale Journal of Biology and Medicine* (National Center for Biotechnology Information) **26** (1): 80–97. PMC 2599350. PMID 13103145.

[180] Levene P, (1 December 1919). "The structure of yeast nucleic acid". *J Biol Chem* **40** (2): 415–24.

[181] See:

- W. T. Astbury and Florence O. Bell (1938) "Some recent developments in the X-ray study of proteins and related structures," *Cold Spring Harbor Symposia on Quantitative Biology*, **6** : 109-121. Available on-line at: University of Leeds.

- Astbury, W. T., (1947) "X-ray studies of nucleic acids," *Symposia of the Society for Experimental Biology*, **1** : 66-76. Available on-line at: Oregon State University.

[182] Koltsov proposed that a cell's genetic information was encoded in a long chain of amino acids. See:

- Н. К. Кольцов, "Физико-химические основы морфологии" (The physical-chemical basis of morphology) -- speech given at the 3rd All-Union Meeting of Zoologist, Anatomists, and Histologists at Leningrad, U.S.S.R., December 12, 1927.

- Reprinted in: *Успехи экспериментальной биологии* (Advances in Experimental Biology), series B, 7 (1) : ?-? (1928).

- Reprinted in German as: Nikolaj K. Koltzoff (1928) "Physikalisch-chemische Grundlagen der Morphologie" (The physical-chemical basis of morphology), *Biologisches Zentralblatt*, **48** (6) : 345-369.

- In 1934, Koltsov contended that the proteins that contain a cell's genetic information replicate. See: N. K. Koltzoff (October 5, 1934) "The structure of the chromosomes in the salivary glands of Drosophila," *Science*, **80** (2075) : 312-313. From page 313: "I think that the size of the chromosomes in the salivary glands [of Drosophila] is determined through the multiplication of *genonemes*. By this term I designate the axial thread of the chromosome, in which the geneticists locate the linear combination of genes; ... In the normal chromosome there is usually only one genoneme; before cell-division this genoneme has become divided into two strands."

[183] Soyfer VN (2001). "The consequences of political dictatorship for Russian science". *Nature Reviews Genetics* **2** (9): 723–729. doi:10.1038/35088598. PMID 11533721.

[184] Griffith F (January 1928). "The significance of pneumococcal types". *The Journal of Hygiene (London)* **27** (2): 113–59. doi:10.1017/S0022172400031879. PMC 2167760. PMID 20474956.

[185] Lorenz MG, Wackernagel W (1994). "Bacterial gene transfer by natural genetic transformation in the environment". *Microbiol. Rev.* **58** (3): 563–602. PMC 372978. PMID 7968924.

[186] Avery OT, Macleod CM, McCarty M (1944). "Studies on the Chemical Nature of the Substance Inducing Transformation of Pneumococcal Types: Induction of Transformation by a Desoxyribonucleic Acid Fraction Isolated from Pneumococcus Type Iii". *J Exp Med* **79** (2): 137–158. doi:10.1084/jem.79.2.137. PMC 2135445. PMID 19871359.

[187] Hershey AD, Chase M (1952). "Independent Functions of Viral Protein and Nucleic Acid in Growth of Bacteriophage". *J Gen Physiol* **36** (1): 39–56. doi:10.1085/jgp.36.1.39. PMC 2147348. PMID 12981234.

[188] The B-DNA X-ray pattern on the right of this linked image was obtained by Rosalind Franklin and Raymond Gosling in May 1952 at high hydration levels of DNA and it has been labeled as "Photo 51"

[189] Nature Archives Double Helix of DNA: 50 Years

[190] "Original X-ray diffraction image". Osulibrary.oregonstate.edu. Retrieved 6 February 2011.

[191] The Nobel Prize in Physiology or Medicine 1962 Nobelprize .org Accessed 22 December 06

[192] Maddox B (23 January 2003). "The double helix and the 'wronged heroine'" (PDF). *Nature* **421** (6921): 407–408. Bibcode:2003Natur.421..407M. doi:10.1038/nature01399. PMID 12540909.

[193] Crick, F.H.C. On degenerate templates and the adaptor hypothesis (PDF). genome.wellcome.ac.uk (Lecture, 1955). Retrieved 22 December 2006.

[194] Meselson M, Stahl FW (1958). "The replication of DNA in Escherichia coli". *Proc Natl Acad Sci USA* **44** (7): 671–82. Bibcode:1958PNAS...44..671M. doi:10.1073/pnas.44.7.671. PMC 528642. PMID 16590258.

[195] The Nobel Prize in Physiology or Medicine 1968 Nobelprize.org Accessed 22 December 06

8.11 Further reading

- Berry, Andrew; Watson, James. (2003). *DNA: the secret of life.* New York: Alfred A. Knopf. ISBN 0-375-41546-7.

- Calladine, Chris R.; Drew, Horace R.; Luisi, Ben F. and Travers, Andrew A. (2003). *Understanding DNA: the molecule & how it works.* Amsterdam: Elsevier Academic Press. ISBN 0-12-155089-3.

- Dennis, Carina; Julie Clayton (2003). *50 years of DNA.* Basingstoke: Palgrave Macmillan. ISBN 1-4039-1479-6.

- Judson, Horace F. 1979. *The Eighth Day of Creation: Makers of the Revolution in Biology.* Touchstone Books, ISBN 0-671-22540-5. 2nd edition: Cold Spring Harbor Laboratory Press, 1996 paperback: ISBN 0-87969-478-5.

- Olby, Robert C. (1994). *The path to the double helix: the discovery of DNA.* New York: Dover Publications. ISBN 0-486-68117-3., first published in October 1974 by MacMillan, with foreword by Francis Crick;the definitive DNA textbook,revised in 1994 with a 9-page postscript

- Micklas, David. 2003. *DNA Science: A First Course.* Cold Spring Harbor Press: ISBN 978-0-87969-636-8

- Ridley, Matt (2006). *Francis Crick: discoverer of the genetic code.* Ashland, OH: Eminent Lives, Atlas Books. ISBN 0-06-082333-X.

- Olby, Robert C. (2009). *Francis Crick: A Biography.* Plainview, N.Y: Cold Spring Harbor Laboratory Press. ISBN 0-87969-798-9.

- Rosenfeld, Israel. 2010. *DNA: A Graphic Guide to the Molecule that Shook the World.* Columbia University Press: ISBN 978-0-231-14271-7

- Schultz, Mark and Zander Cannon. 2009. *The Stuff of Life: A Graphic Guide to Genetics and DNA.* Hill and Wang: ISBN 0-8090-8947-5

- Stent, Gunther Siegmund; Watson, James. (1980). *The double helix: a personal account of the discovery of the structure of DNA.* New York: Norton. ISBN 0-393-95075-1.

- Watson, James. 2004. *DNA: The Secret of Life.* Random House: ISBN 978-0-09-945184-6

- Wilkins, Maurice (2003). *The third man of the double helix the autobiography of Maurice Wilkins.* Cambridge, Eng: University Press. ISBN 0-19-860665-6.

8.12   External links

- DNA at DMOZ

- DNA binding site prediction on protein

- DNA the Double Helix Game From the official Nobel Prize web site

- DNA under electron microscope

- Dolan DNA Learning Center

- Double Helix: 50 years of DNA, *Nature*

- *Proteopedia DNA*

- *Proteopedia Forms_of_DNA*

- ENCODE threads explorer ENCODE Home page. Nature (journal)

- Double Helix 1953–2003 National Centre for Biotechnology Education

- Genetic Education Modules for Teachers—*DNA from the Beginning* Study Guide

- PDB Molecule of the Month *DNA*

- Rosalind Franklin's contributions to the study of DNA

- U.S. National DNA Day—watch videos and participate in real-time chat with top scientists

- Clue to chemistry of heredity found The New York Times June 1953. First American newspaper coverage of the discovery of the DNA structure

- Olby R (2003). "Quiet debut for the double helix". *Nature* **421** (6921): 402–5. Bibcode:2003Natur.421..402O. doi:10.1038/nature01397. PMID 12540907.

- DNA from the Beginning Another DNA Learning Center site on DNA, genes, and heredity from Mendel to the human genome project.

- The Register of Francis Crick Personal Papers 1938 – 2007 at Mandeville Special Collections Library, University of California, San Diego

- Seven-page, handwritten letter that Crick sent to his 12-year-old son Michael in 1953 describing the structure of DNA. See Crick's medal goes under the hammer, Nature, 5 April 2013.

- 3D map of DNA reveals hidden loops that allow genes to work together (11 December 2014), *Science (Daily News)*

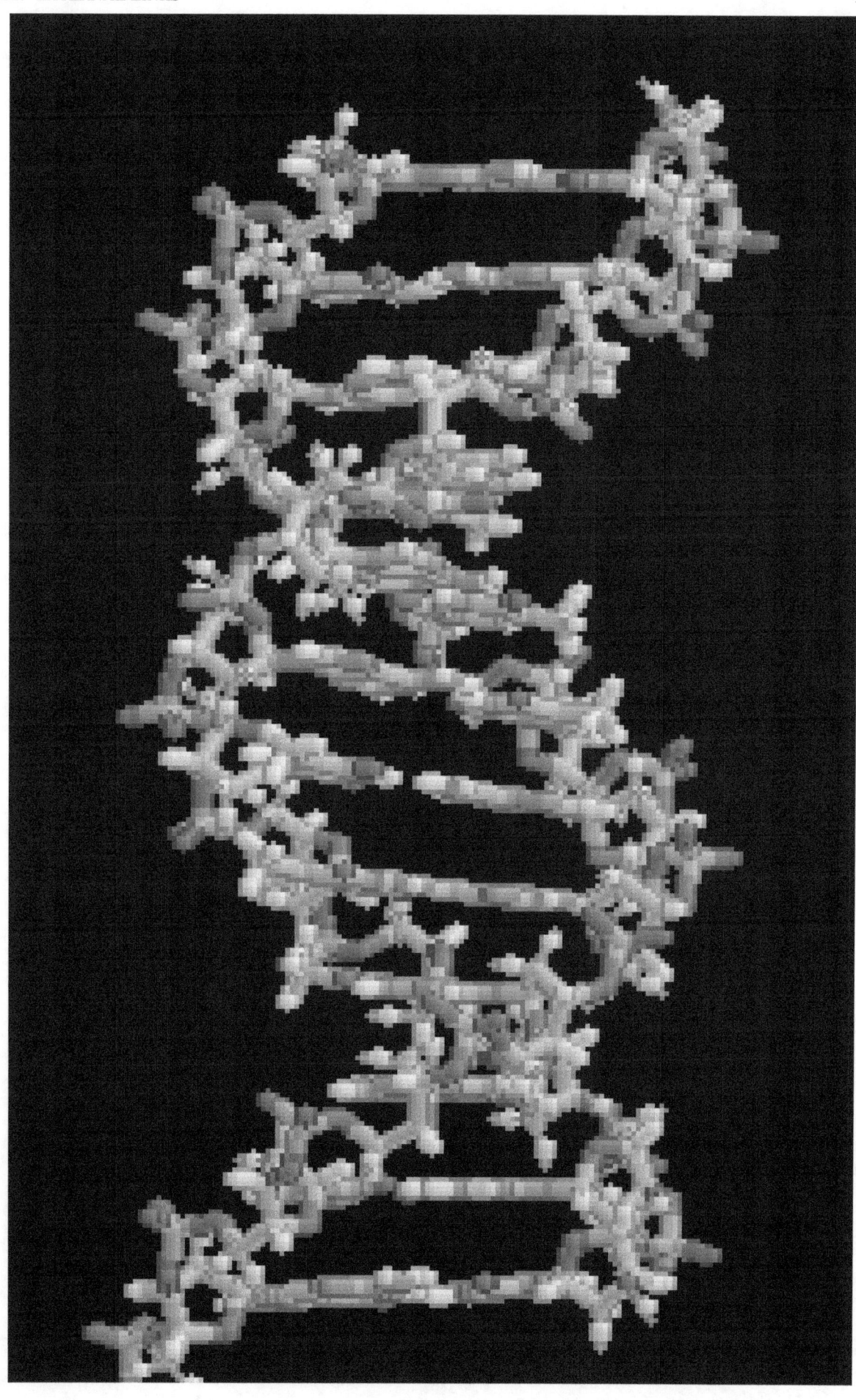

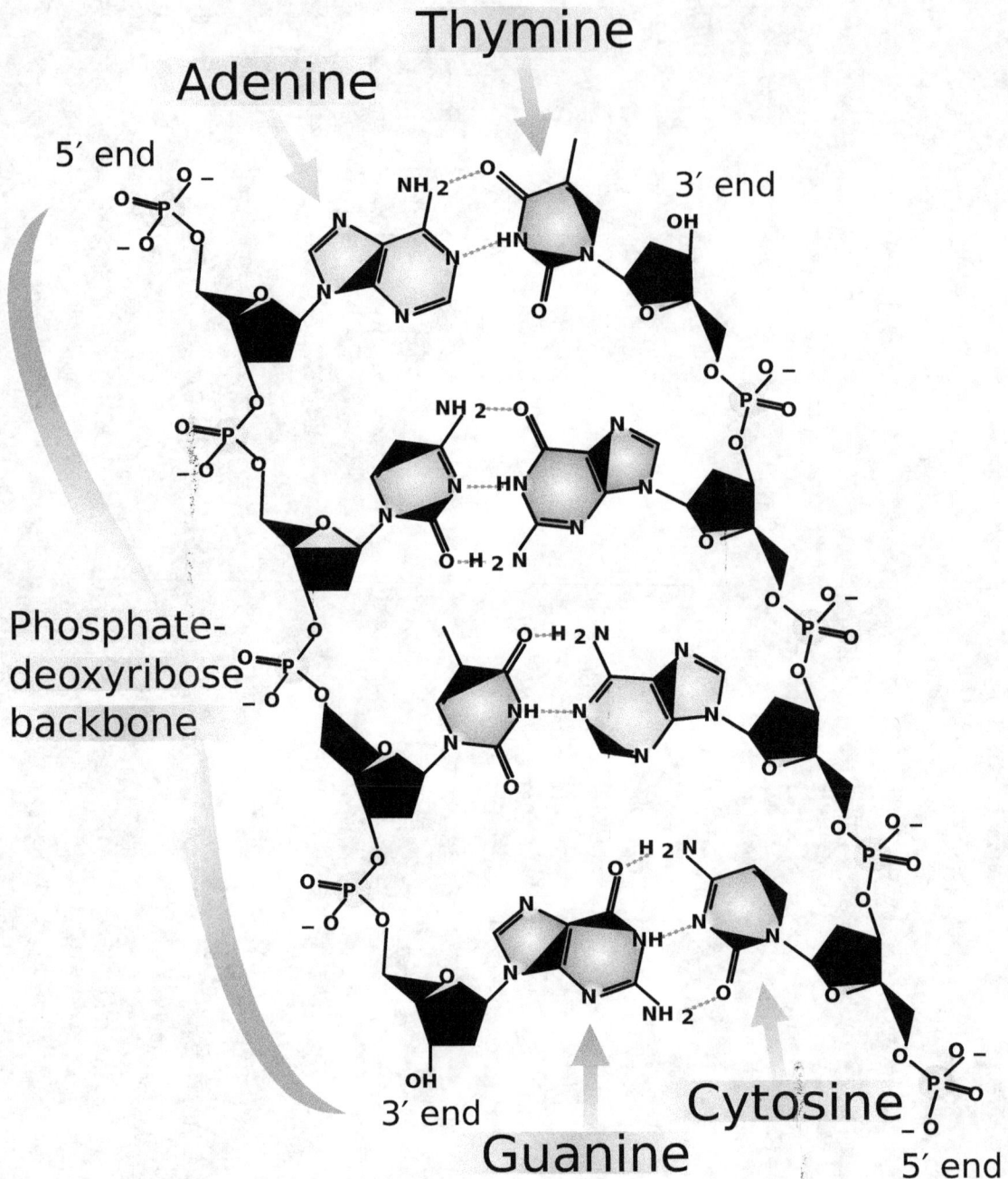

Chemical structure of DNA; hydrogen bonds shown as dotted lines

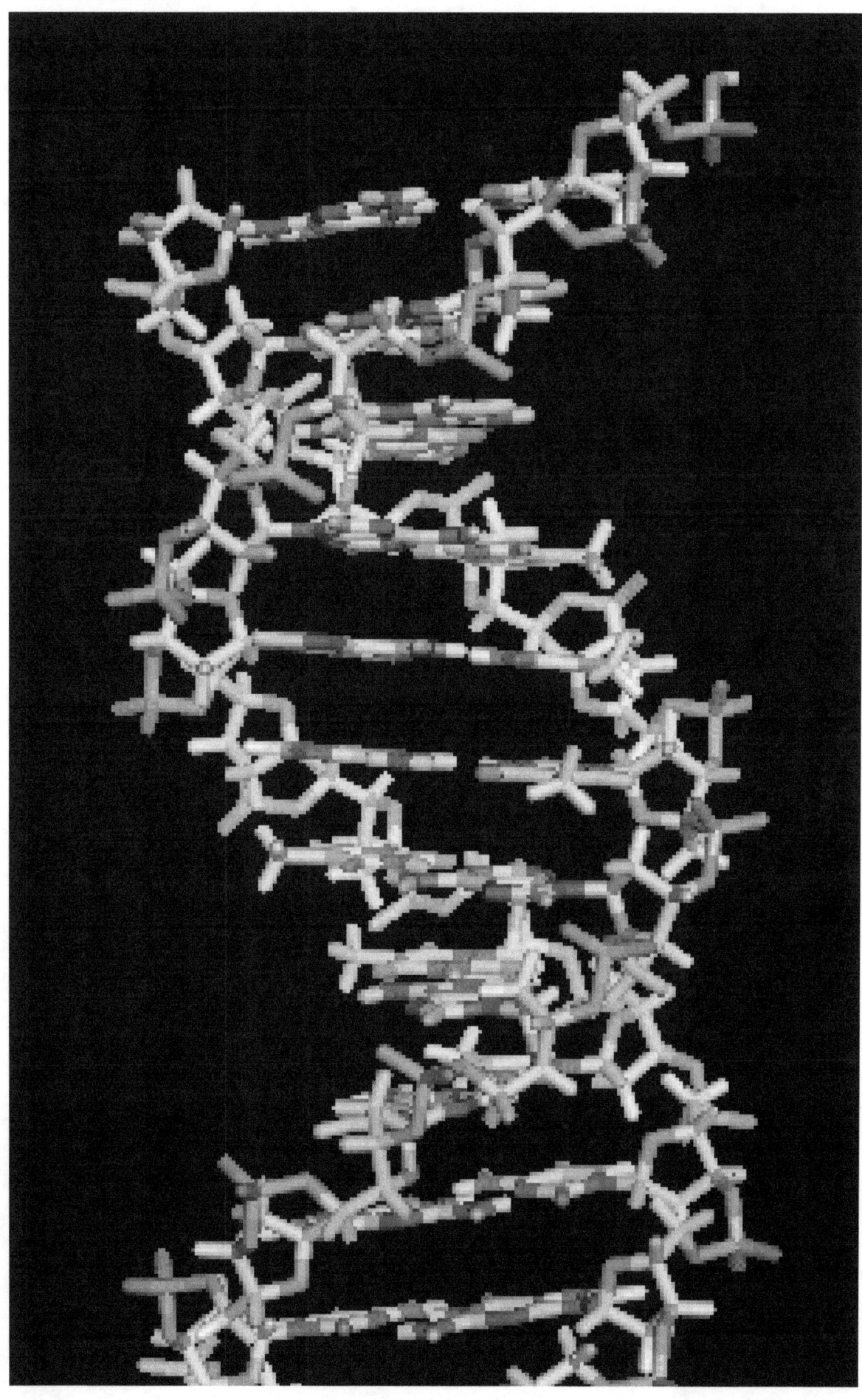

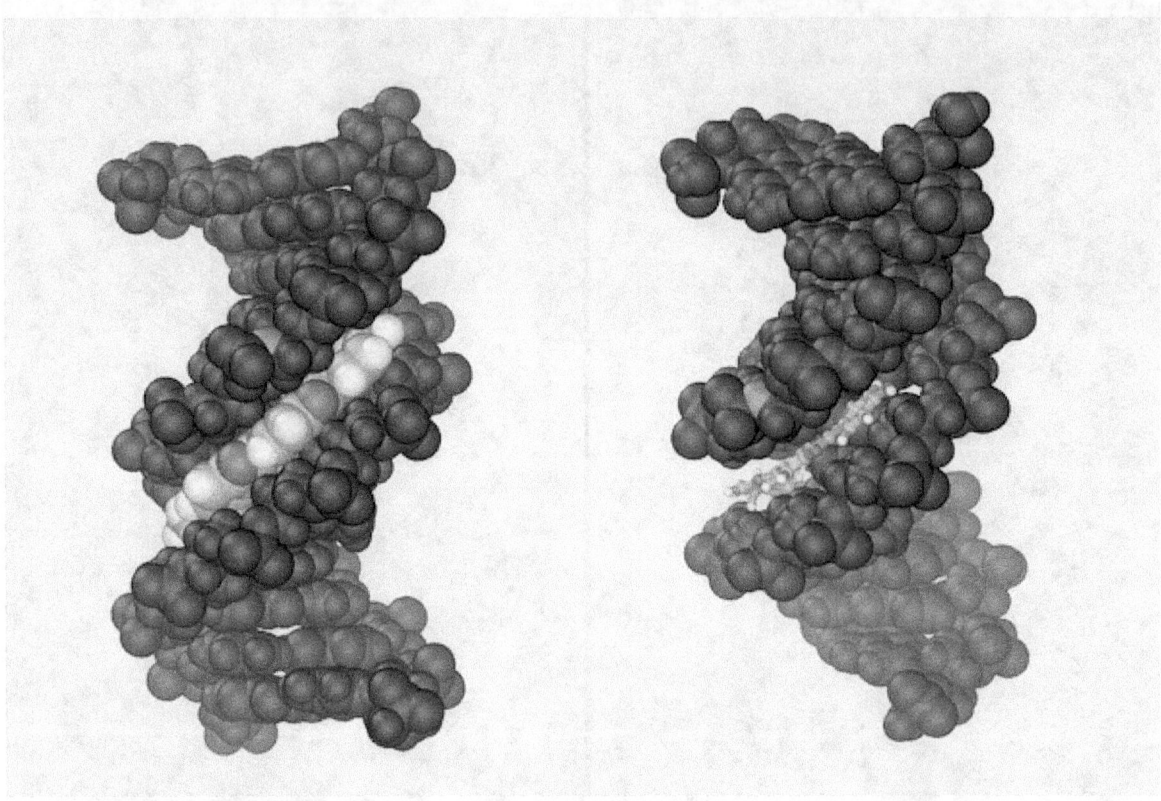

Major and minor grooves of DNA. Minor groove is a binding site for the dye Hoechst 33258.

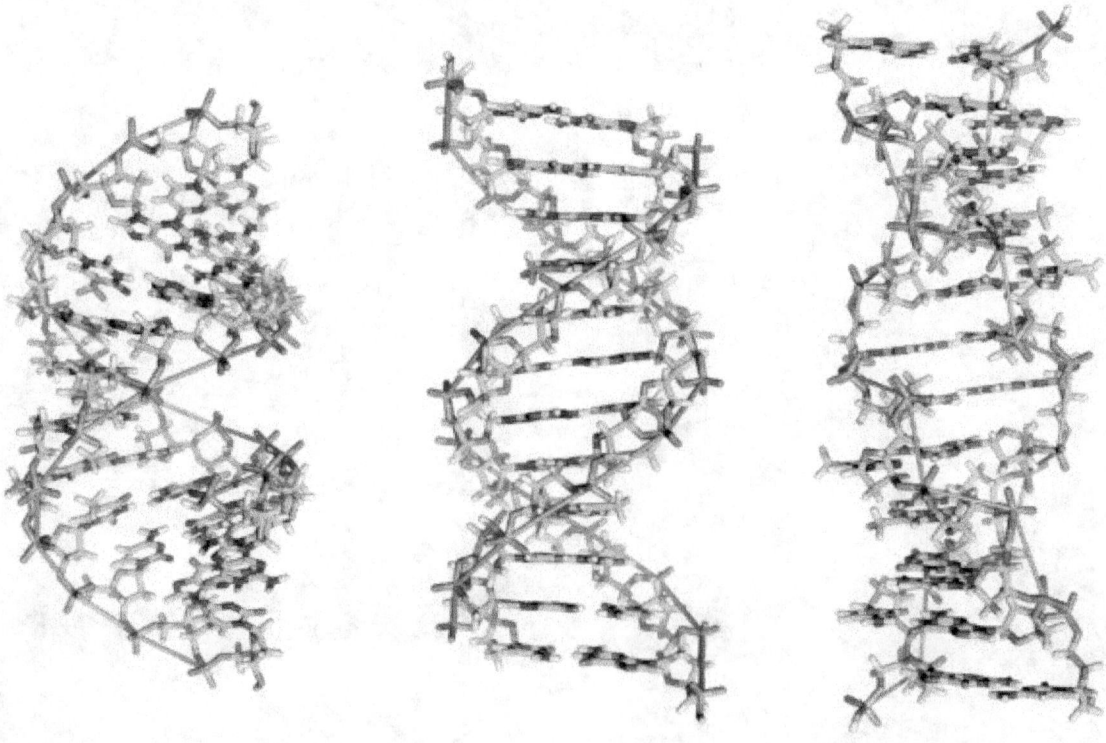

From left to right, the structures of A, B and Z DNA

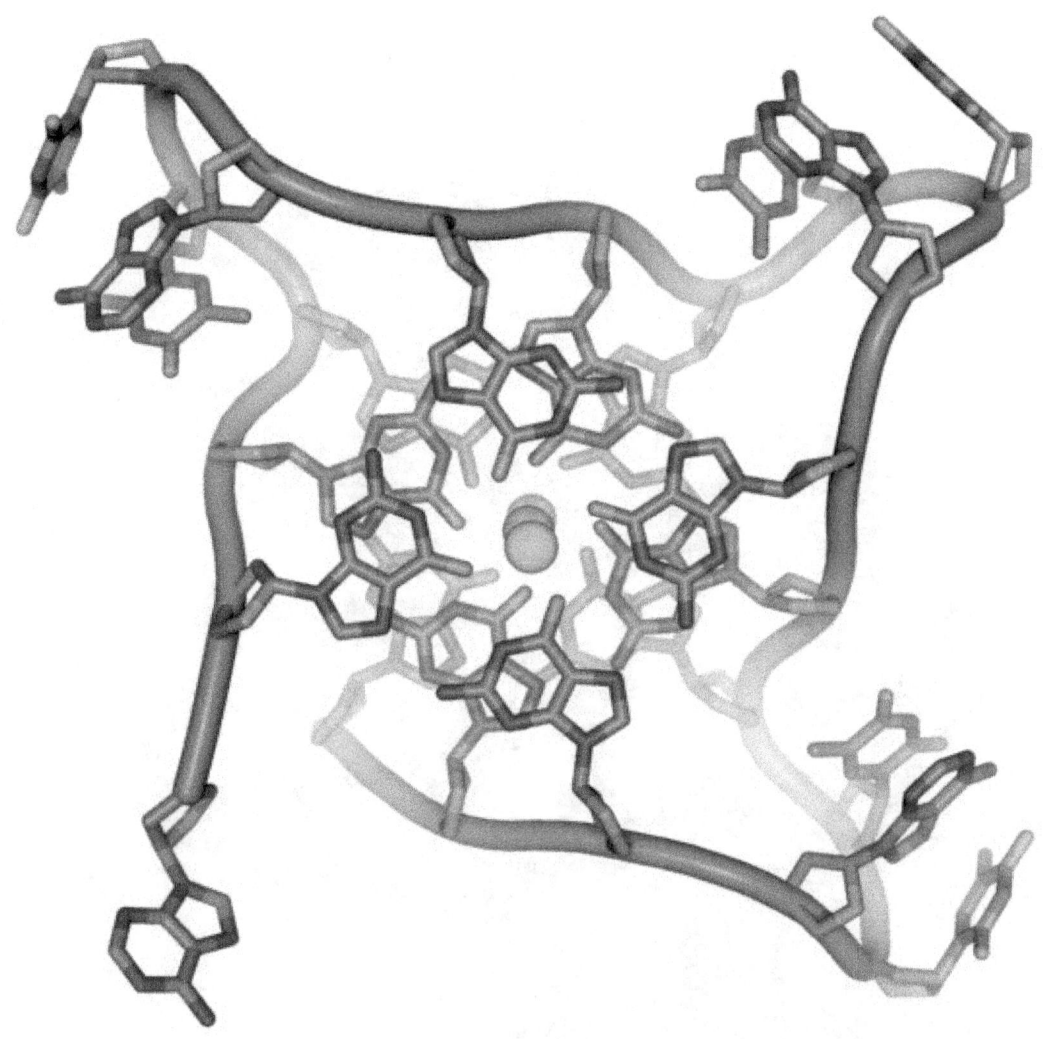

DNA quadruplex formed by telomere repeats. The looped conformation of the DNA backbone is very different from the typical DNA helix.[65]

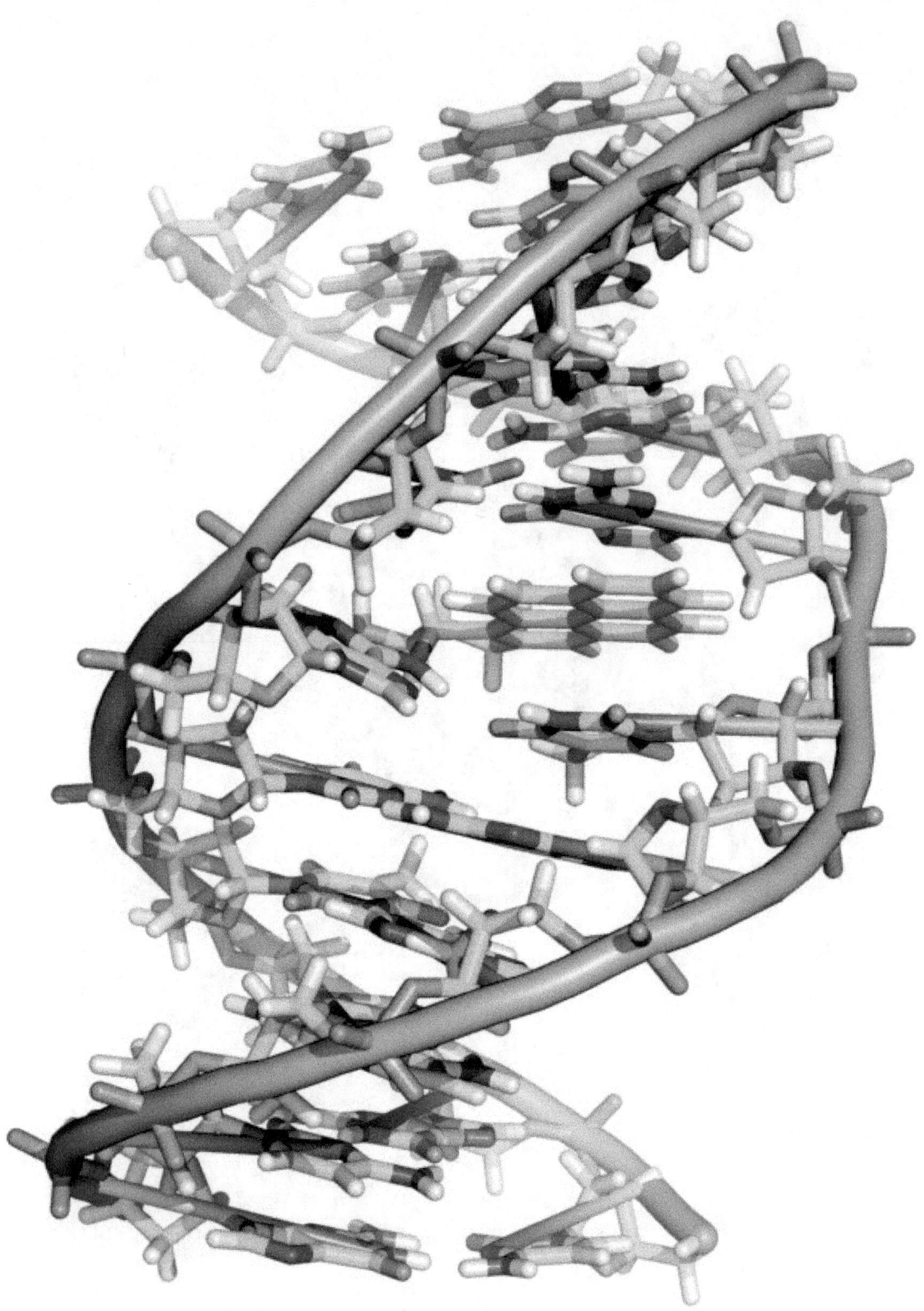

A covalent adduct between a metabolically activated form of benzo[a]pyrene, the major mutagen in tobacco smoke, and DNA[77]

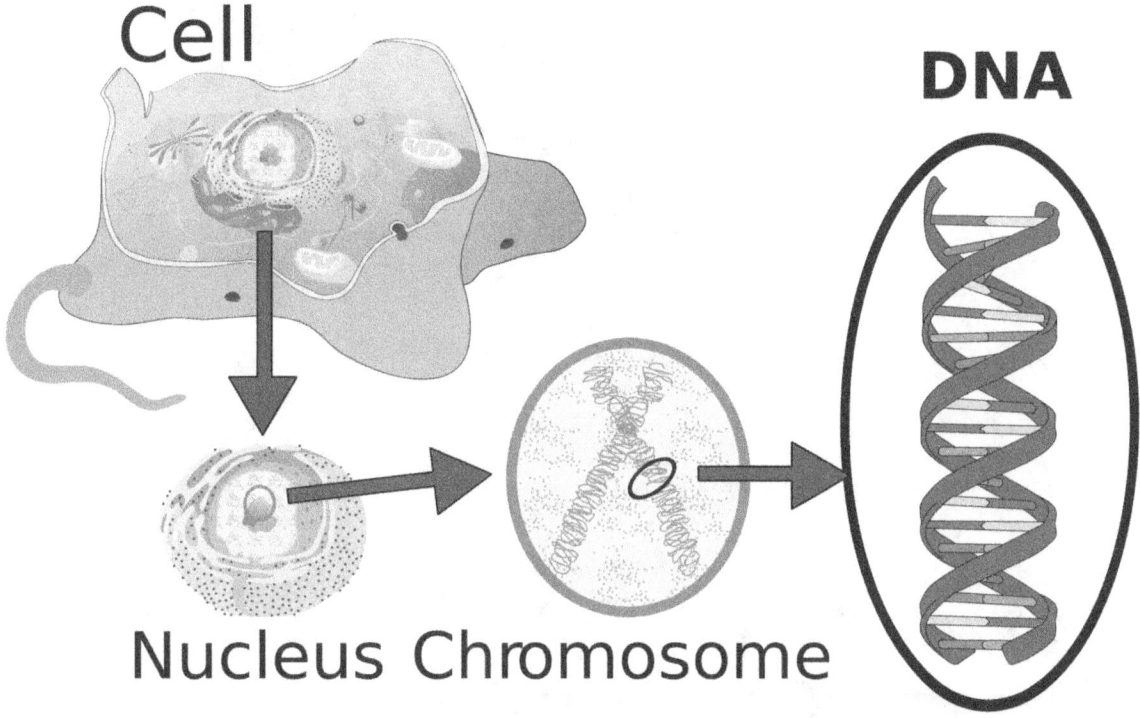

Location of eukaryote nuclear DNA within the chromosomes.

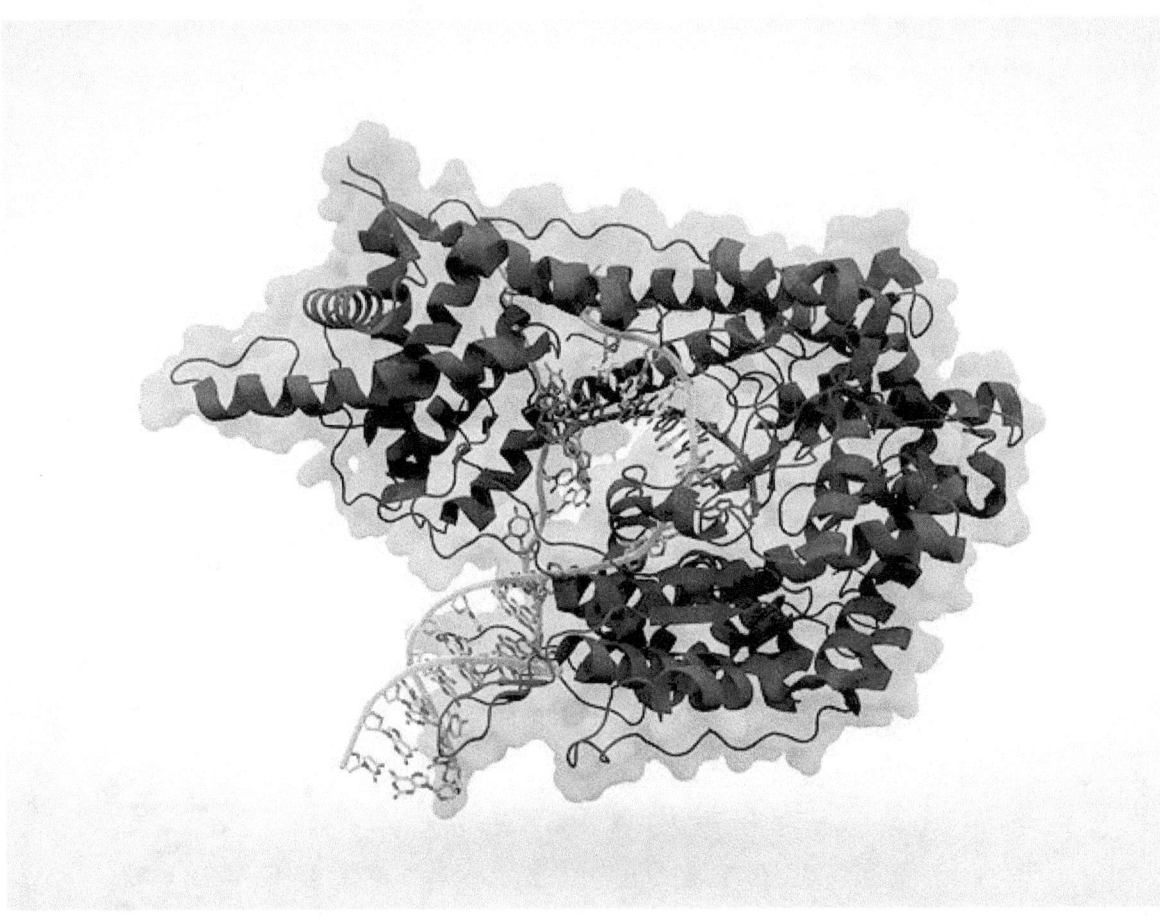

T7 RNA polymerase (blue) producing a mRNA (green) from a DNA template (orange).[96]

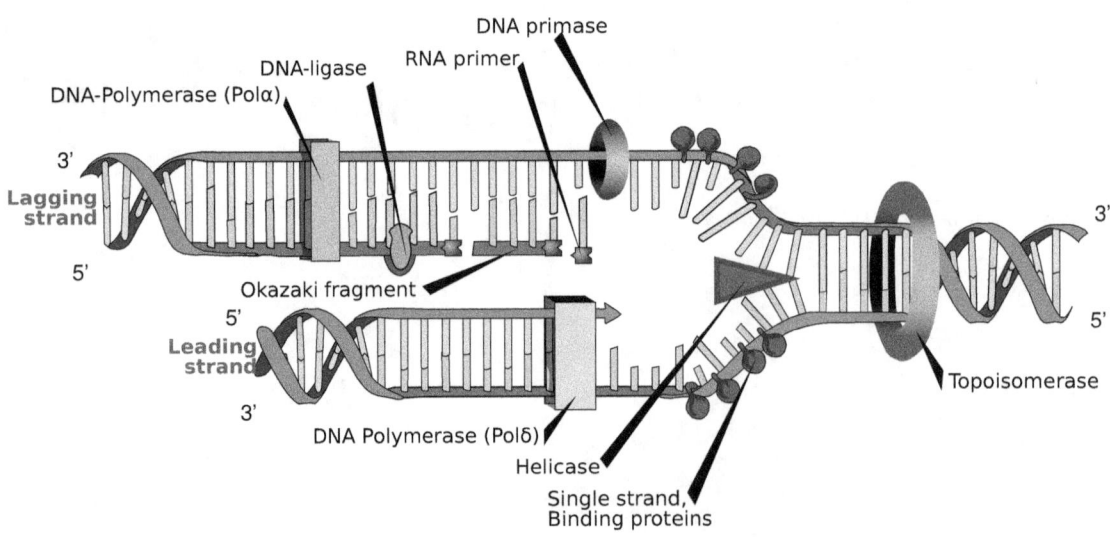

DNA replication. The double helix is unwound by a helicase and topoisomerase. Next, one DNA polymerase produces the leading strand copy. Another DNA polymerase binds to the lagging strand. This enzyme makes discontinuous segments (called Okazaki fragments) before DNA ligase joins them together.

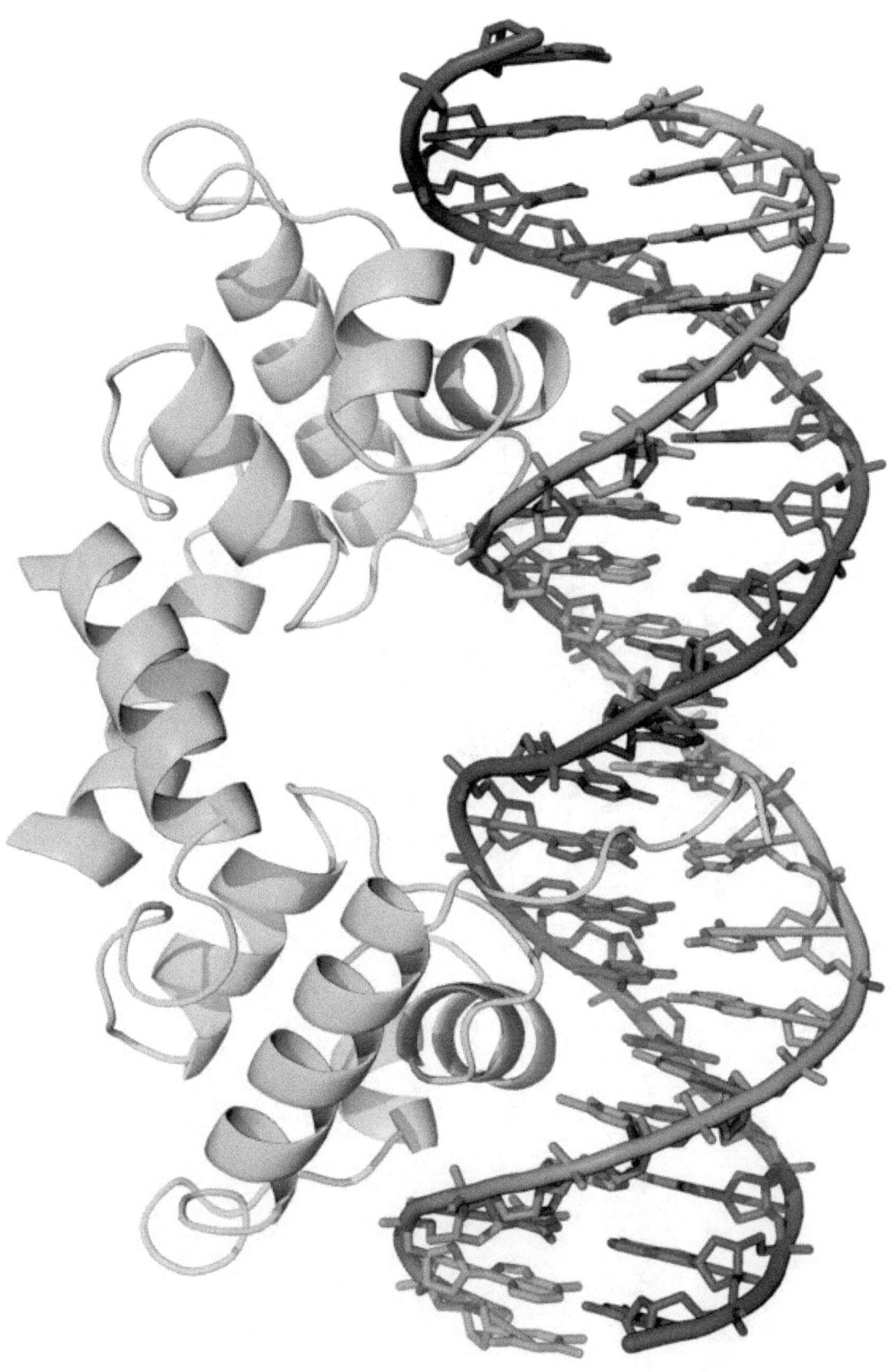

The lambda repressor helix-turn-helix transcription factor bound to its DNA target[116]

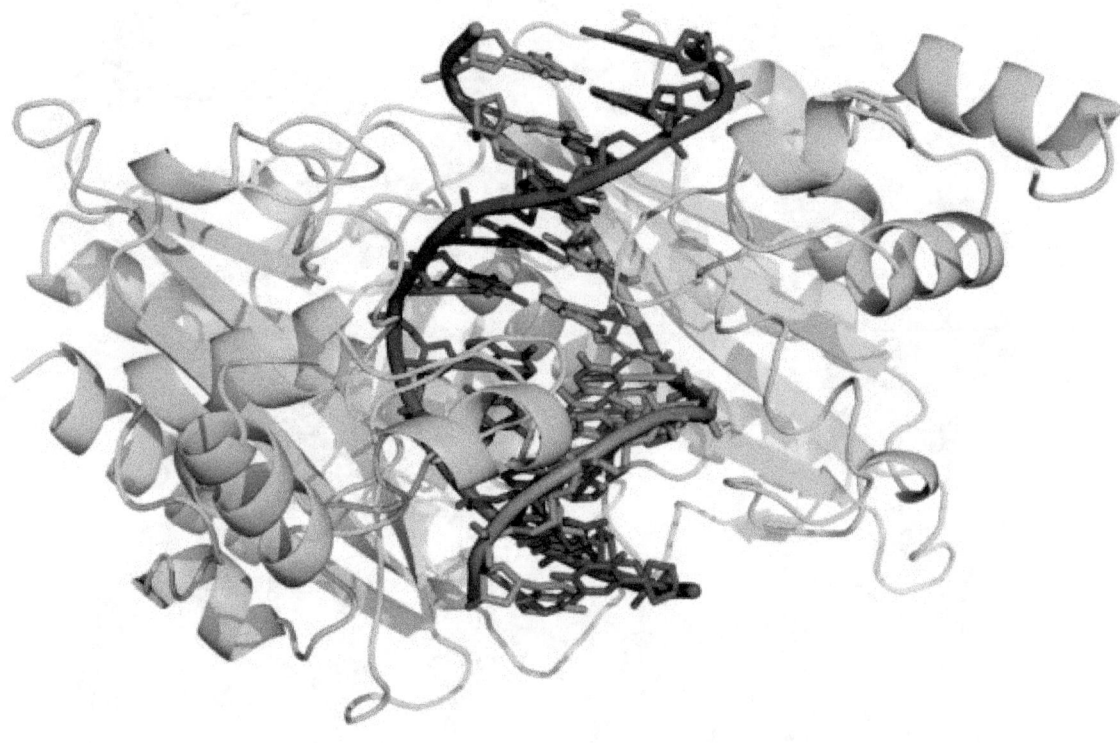

The restriction enzyme EcoRV (green) in a complex with its substrate DNA[120]

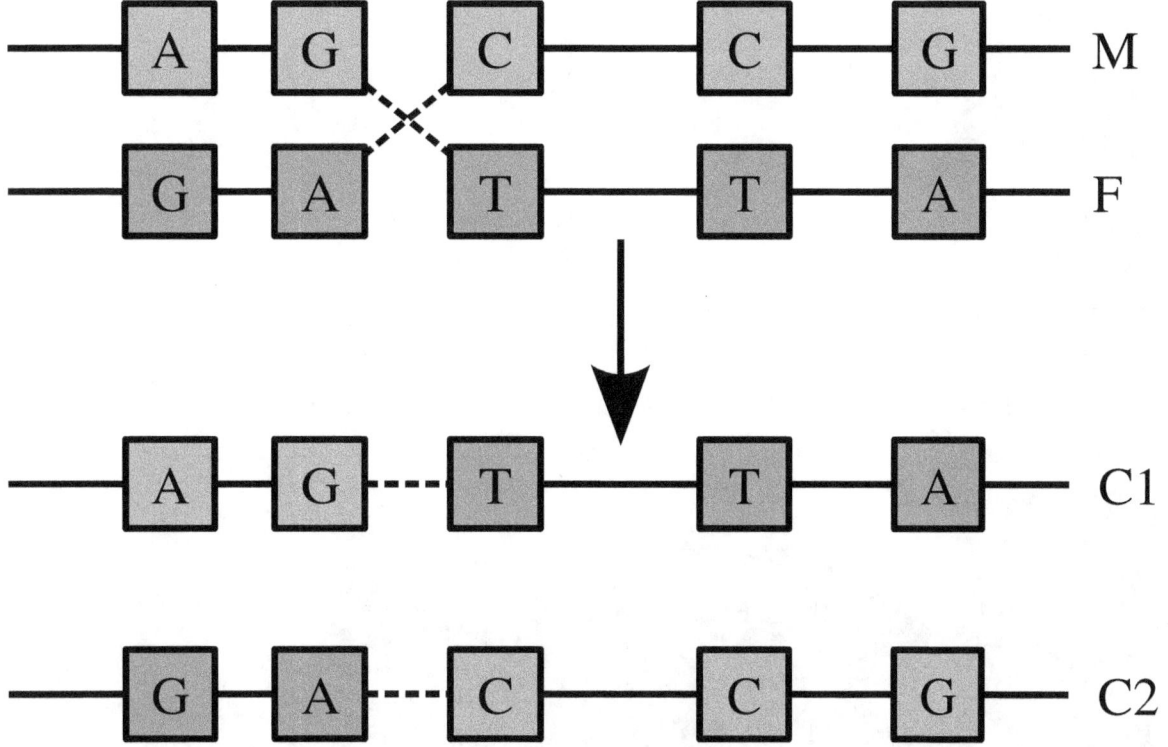

Recombination involves the breakage and rejoining of two chromosomes (M and F) to produce two re-arranged chromosomes (C1 and C2).

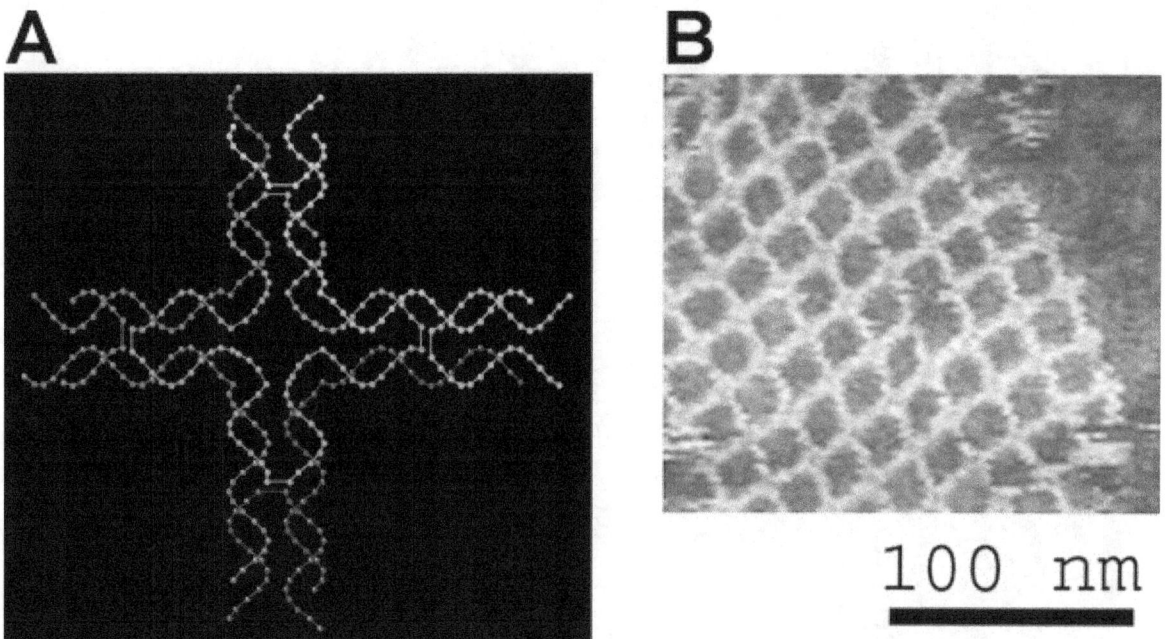

The DNA structure at left (schematic shown) will self-assemble into the structure visualized by atomic force microscopy at right. DNA nanotechnology is the field that seeks to design nanoscale structures using the molecular recognition properties of DNA molecules. Image from Strong, 2004.

James Watson and Francis Crick (right), co-originators of the double-helix model, with Maclyn McCarty (left).

Pencil sketch of the DNA double helix by Francis Crick in 1953

Chapter 9

Mitochondrial DNA

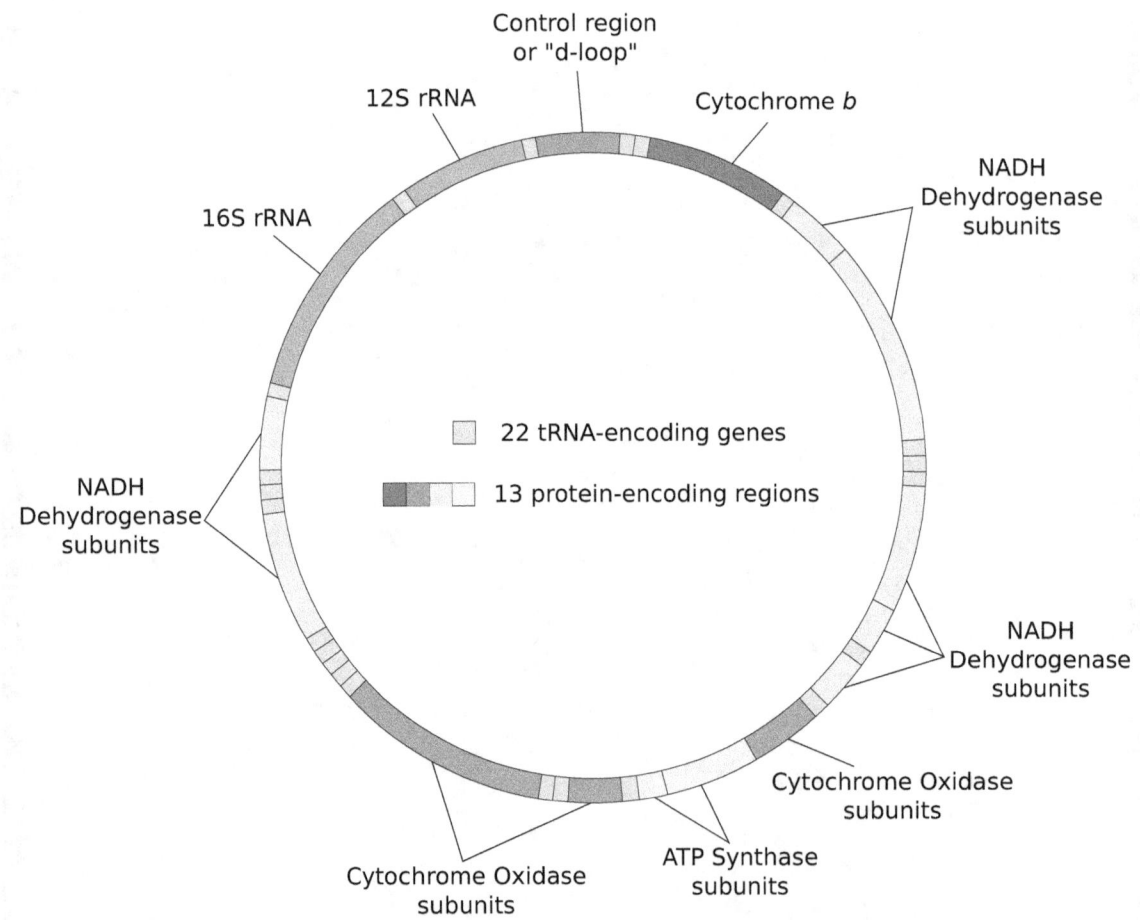

Human mitochondrial DNA.

Mitochondrial DNA (**mtDNA** or **mDNA**)[2] is the DNA located in mitochondria, cellular organelles within eukaryotic cells that convert chemical energy from food into a form that cells can use, adenosine triphosphate (ATP). Mitochondrial DNA is only a small portion of the DNA in a eukaryotic cell; most of the DNA can be found in the cell nucleus and, in plants, in the chloroplast.

In humans, the 16,569 base pairs of mitochondrial DNA encode for only 37 genes. Human mitochondrial DNA was the first significant part of the human genome to be sequenced. In most species, including humans, mtDNA is inherited solely from the mother.[3]

Since mtDNA evolves relatively slowly compared to other genetic markers, it represents a mainstay of phylogenetics and evolutionary biology. It also permits an examination of the relatedness of populations, and so has become important in anthropology and biogeography.

It should not be confused with messenger RNA.

9.1 Origin

Nuclear and mitochondrial DNA are thought to be of separate evolutionary origin, with the mtDNA being derived from the circular genomes of the bacteria that were engulfed by the early ancestors of today's eukaryotic cells. This theory is called the endosymbiotic theory. Each mitochondrion is estimated to contain 2–10 mtDNA copies.[4] In the cells of extant organisms, the vast majority of the proteins present in the mitochondria (numbering approximately 1500 different types in mammals) are coded for by nuclear DNA, but the genes for some of them, if not most, are thought to have originally been of bacterial origin, having since been transferred to the eukaryotic nucleus during evolution.

9.2 Mitochondrial inheritance

In most multicellular organisms, mtDNA is inherited from the mother (maternally inherited). Mechanisms for this include simple dilution (an egg contains on average 200,000 mtDNA molecules, whereas a healthy human sperm was reported to contain on average 5 molecules[5]), degradation of sperm mtDNA in the male genital tract, in the fertilized egg, and, at least in a few organisms, failure of sperm mtDNA to enter the egg. Whatever the mechanism, this single parent (uniparental inheritance) pattern of mtDNA inheritance is found in most animals, most plants and in fungi as well.

9.2.1 Female inheritance

In sexual reproduction, mitochondria are normally inherited exclusively from the mother; the mitochondria in mammalian sperm are usually destroyed by the egg cell after fertilization. Also, most mitochondria are present at the base of the sperm's tail, which is used for propelling the sperm cells; sometimes the tail is lost during fertilization. In 1999 it was reported that paternal sperm mitochondria (containing mtDNA) are marked with ubiquitin to select them for later destruction inside the embryo.[6] Some *in vitro* fertilization techniques, particularly injecting a sperm into an oocyte, may interfere with this.

The fact that mitochondrial DNA is maternally inherited enables genealogical researchers to trace maternal lineage far back in time. (Y-chromosomal DNA, paternally inherited, is used in an analogous way to determine the patrilineal history.) This is accomplished on human mitochondrial DNA by sequencing one or more of the hypervariable control regions (HVR1 or HVR2) of the mitochondrial DNA, as with a genealogical DNA test. HVR1 consists of about 440 base pairs. These 440 base pairs are then compared to the control regions of other individuals (either specific people or subjects in a database) to determine maternal lineage. Most often, the comparison is made to the revised Cambridge Reference Sequence. Vilà *et al.* have published studies tracing the matrilineal descent of domestic dogs to wolves.[7] The concept of the Mitochondrial Eve is based on the same type of analysis, attempting to discover the origin of humanity by tracking the lineage back in time.

mtDNA is highly conserved, and its relatively slow mutation rates (compared to other DNA regions such as microsatellites) make it useful for studying the evolutionary relationships—phylogeny—of organisms. Biologists can determine and then compare mtDNA sequences among different species and use the comparisons to build an evolutionary tree for the species examined. However, due to the slow mutation rates it experiences, it is often hard to distinguish between closely related species to any large degree, so other methods of analysis must be used.

9.2.2   Male inheritance

Main article: Paternal mtDNA transmission

Doubly uniparental inheritance of mtDNA is observed in bivalve mollusks. In those species, females have only one type of mtDNA (F), whereas males have F type mtDNA in their somatic cells, but M type of mtDNA (which can be as much as 30% divergent) in germline cells.[8] Paternally inherited mitochondria have additionally been reported in some insects such as fruit flies,[9] honeybees,[10] and periodical cicadas.[11]

Male mitochondrial inheritance was recently discovered in Plymouth Rock chickens.[12] Evidence supports rare instances of male mitochondrial inheritance in some mammals as well. Specifically, documented occurrences exist for mice,[13][14] where the male-inherited mitochondria were subsequently rejected. It has also been found in sheep,[15] and in cloned cattle.[16] It has been found in a single case in a human male.[17]

Although many of these cases involve cloned embryos or subsequent rejection of the paternal mitochondria, others document *in vivo* inheritance and persistence under lab conditions.

9.2.3   Three-parent inheritance

Main article: Mitochondrial donation

An artificial reproductive process known as Three Parent In Vitro Fertilization (TPIVF) results in offspring containing mtDNA from a donor female, and nuclear DNA from another female and a male. In the process, the nucleus of an egg is inserted into the cytoplasm of an egg from a donor female which has had its nucleus removed, but still contains the donor female's mtDNA. The composite egg is then fertilized with the male's sperm. The procedure is used when a woman with genetically defective mitochondria wishes to procreate and produce offspring with healthy mitochondria.[18]

9.3   Structure

In most multicellular organisms, the mtDNA is organized as a circular, covalently closed, double-stranded DNA. But in many unicellular (e.g. the ciliate *Tetrahymena* or the green alga *Chlamydomonas reinhardtii*) and in rare cases also in multicellular organisms (e.g. in some species of *Cnidaria*) the mtDNA is found as linearly organized DNA. Most of these linear mtDNAs possess telomerase independent telomeres (i.e. the ends of the linear DNA) with different modes of replication, which have made them interesting objects of research, as many of these unicellular organisms with linear mtDNA are known pathogens.[19]

For human mitochondrial DNA (and probably for that of metazoans in general), 100-10,000 separate copies of mtDNA are usually present per cell (egg and sperm cells are exceptions). In mammals, each double-stranded circular mtDNA molecule consists of 15,000-17,000[20] base pairs. The two strands of mtDNA are differentiated by their nucleotide content, with a guanine-rich strand referred to as the heavy strand (or H-strand) and a cytosine-rich strand referred to as the light strand (or L-strand). The heavy strand encodes 28 genes, and the light strand encodes 9 genes for a total of 37 genes. Of the 37 genes, 13 are for proteins (polypeptides), 22 are for transfer RNA (tRNA) and two are for the small and large subunits of ribosomal RNA (rRNA). This pattern is also seen among most metazoans, although in some cases one or more of the 37 genes is absent and the mtDNA size range is greater. Even greater variation in mtDNA gene content and size exists among fungi and plants, although there appears to be a core subset of genes that are present in all eukaryotes (except for the few that have no mitochondria at all). Some plant species have enormous mtDNAs (as many as 2,500,000 base pairs per mtDNA molecule) but, surprisingly, even those huge mtDNAs contain the same number and kinds of genes as related plants with much smaller mtDNAs.[21]

As far as transcription concerns, at least in animals, each strand is transcribed continuously and produces a polycistronic RNA molecule. Mitochondrial genes for ATP8 and ATP6 as well as ND4L and ND4 overlap. Between most (but not all) protein-coding regions, tRNAs are present. During transcription, the tRNAs acquire their characteristic L-shape that gets recognized and cleaved by specific enzymes. Mutations in mitochondrial tRNAs can be responsible for severe diseases like the MELAS and MERRF syndromes.[22]

The genome of the mitochondrion of the cucumber (*Cucumis sativus*) consists of three circular chromosomes (lengths 1556, 84 and 45 kilobases), which are entirely or largely autonomous with regard to their replication.[23]

9.4 Replication

Mitochondrial DNA is replicated by the DNA polymerase gamma complex which is composed of a 140 kDa catalytic DNA polymerase encoded by the *POLG* gene and two 55 kDa accessory subunits encoded by the *POLG2* gene.[24] The replisome machinery is formed by DNA polymerase, TWINKLE and mitochondrial SSB proteins. TWINKLE is a helicase, which unwinds short stretches of dsDNA in the 5′ to 3′ direction.[25]

During embryogenesis, replication of mtDNA is strictly down-regulated from the fertilized oocyte through the preimplantation embryo.[26] At the blastocyst stage, the onset of mtDNA replication is specific to the cells of the trophectoderm.[26] In contrast, the cells of the inner cell mass restrict mtDNA replication until they receive the signals to differentiate to specific cell types.[26]

9.5 Mutations

9.5.1 Susceptibility

The concept that mtDNA is particularly susceptible to reactive oxygen species generated by the respiratory chain due to its proximity remains controversial.[27] mtDNA does not accumulate any more oxidative base damage than nuclear DNA.[28] It has been reported that at least some types of oxidative DNA damage are repaired more efficiently in mitochondria than they are in the nucleus.[29] mtDNA is packaged with proteins which appear to be as protective as proteins of the nuclear chromatin.[30] Moreover, mitochondria evolved a unique mechanism which maintains mtDNA integrity through degradation of excessively damaged genomes followed by replication of intact/repaired mtDNA. This mechanism is not present in the nucleus and is enabled by multiple copies of mtDNA present in mitochondria [31] The outcome of mutation in mtDNA may be an alteration in the coding instructions for some proteins,[32] which may have an effect on organism metabolism and/or fitness.

9.5.2 Genetic illness

Further information: Mitochondrial disease

Mutations of mitochondrial DNA can lead to a number of illnesses including exercise intolerance and Kearns–Sayre syndrome (KSS), which causes a person to lose full function of heart, eye, and muscle movements. Some evidence suggests that they might be major contributors to the aging process and age-associated pathologies.[33]

9.5.3 Use in disease diagnosis

Recently a mutation in mtDNA has been used to help diagnose prostate cancer in patients with negative prostate biopsy.[34][35]

9.5.4 Relationship with aging

Though the idea is controversial, some evidence suggests a link between aging and mitochondrial genome dysfunction.[36] In essence, mutations in mtDNA upset a careful balance of reactive oxygen species (ROS) production and enzymatic ROS scavenging (by enzymes like superoxide dismutase, catalase, glutathione peroxidase and others). However, some mutations that increase ROS production (e.g., by reducing antioxidant defenses)in worms increase, rather than decrease, their longevity.[27] Also, naked mole rats, rodents about the size of mice, live about eight times longer than mice despite having reduced, compared to mice, antioxidant defenses and increased oxidative damage to biomolecules.[37] Once, there was thought to be a positive feedback loop at work (a 'Vicious Cycle'); as mitochondrial DNA accumulates genetic damage caused by free radicals, the mitochondria lose function and leak free radicals into the cytosol. A decrease in mitochondrial function reduces overall metabolic efficiency.[38] However, this concept was conclusively disproved when it was demonstrated that mice, which were genetically altered to accumulate mtDNA mutations at accelerated rate do age prematurely, but their tissues do not produce more ROS as predicted by the 'Vicious Cycle' hypothesis.[39] Supporting a link between longevity and mitochondrial DNA, some studies have found correlations between biochemical properties of the mitochondrial DNA and the longevity of species.[40] Extensive research is being conducted to further investigate this link and methods to combat aging. Presently, gene therapy and nutraceutical supplementation are popular areas of ongoing research.[41][42] Bjelakovic et al. analyzed the results of 78 studies between 1977 and 2012, involving a total of 296,707 participants, and concluded that antioxidant supplements do not reduce all-cause mortality nor extend lifespan, while some of them, such as beta carotene, vitamin E, and higher doses of vitamin A, may actually increase mortality.[43]

9.5.5   Relationship with non-B (non-canonical) DNA structures

Deletion breakpoints frequently occur within or near regions showing non-canonical (non-B) conformations, namely hairpins, cruciforms and cloverleaf-like elements.[44] Moreover, there is data supporting the involvement of helix-distorting intrinsically curved regions and long G-tetrads in eliciting instability events. In addition, higher breakpoint densities were consistently observed within GC-skewed regions and in the close vicinity of the degenerate sequence motif YMMYMNNMMHM.[45]

9.6   Use in identification

For use in human identification, see Human mitochondrial DNA.

Unlike nuclear DNA, which is inherited from both parents and in which genes are rearranged in the process of recombination, there is usually no change in mtDNA from parent to offspring. Although mtDNA also recombines, it does so with copies of itself within the same mitochondrion. Because of this and because the mutation rate of animal mtDNA is higher than that of nuclear DNA,[46] mtDNA is a powerful tool for tracking ancestry through females (matrilineage) and has been used in this role to track the ancestry of many species back hundreds of generations.

The rapid mutation rate (in animals) makes mtDNA useful for assessing genetic relationships of individuals or groups within a species and also for identifying and quantifying the phylogeny (evolutionary relationships; see phylogenetics) among different species. To do this, biologists determine and then compare the mtDNA sequences from different individuals or species. Data from the comparisons is used to construct a network of relationships among the sequences, which provides an estimate of the relationships among the individuals or species from which the mtDNAs were taken. mtDNA can be used to estimate the relationship between both closely related and distantly related species. Due to the high mutation rate of mtDNA in animals, the 3rd positions of the codons change relatively rapidly, and thus provide information about the genetic distances among closely related individuals or species. On the other hand, the substitution rate of mt-proteins is very low, thus amino acid changes accummulate slowly (with corresponding slow changes at 1st and 2nd codon positions) and thus hey provide information about the genetic distances of distantly related species. Statistical models that treat substitution rates among codon positions separately, can thus be used to simultaneously estimate phylogenies that contain both closely and distantly related species[22]

Mitochondrial DNA was admitted into evidence for the first time ever in 1996 during *State of Tennessee v. Paul Ware*.[47]

In the 1998 court case of Commonwealth of Pennsylvania v. Patricia Lynne Rorrer,[48] mitochondrial DNA was admitted into evidence in the State of Pennsylvania for the first time.[49][50] The case was featured in episode 55 of season 5 of the true crime drama series Forensic Files (season 5).

Mitochondrial DNA was first admitted into evidence in California in the successful prosecution of David Westerfield for the 2002 kidnapping and murder of 7-year-old Danielle van Dam in San Diego: it was used for both human and dog identification.[51] This was the first trial in the U.S. to admit canine DNA.[52]

9.7 History

Mitochondrial DNA was discovered in the 1960s by Margit M. K. Nass and Sylvan Nass by electron microscopy as DNase-sensitive threads inside mitochondria,[53] and by Ellen Haslbrunner, Hans Tuppy and Gottfried Schatz by biochemical assays on highly purified mitochondrial fractions.[54]

9.8 Mitochondrial sequence databases

Several specialized databases have been founded to collect mitochondrial genome sequences and other information. Although most of them focus on sequence data, some of them include phylogenetic or functional information.

- **MitoSatPlant**: Mitochondrial microsatellites database of viridiplantae.[55]

- **MitoBreak**: the mitochondrial DNA breakpoints database.[56]

- **MitoFish** and **MitoAnnotator**: a mitochondrial genome database of fish.[57] See also Cawthorn et al.[58]

- **MitoZoa** 2.0: a database for comparative and evolutionary analyses of mitochondrial genomes in Metazoa.[59]

- **InterMitoBase**: an annotated database and analysis platform of protein-protein interactions for human mitochondria.[60]

- **Mitome:** a database for comparative mitochondrial genomics in metazoan animals[61] (no longer available)

- **MitoRes:** a resource of nuclear-encoded mitochondrial genes and their products in metazoa[62] (apparently no longer being updated)

9.9 See also

- Archaeogenetics of the Near East

- Clade

- CoRR hypothesis

- Haplogroup

- Heteroplasmy

- Human mitochondrial DNA haplogroup

- Human mitochondrial genetics

- Mitochondrial disease

- Mitochondrial DNA (journal)

- Mitochondrial Eve

- Mitochondrial rCRS

- Paternal mtDNA transmission

- Single origin theory

- Supercluster (genetic)

- TIM/TOM complex

- Genetic history of Africa

- Genetic history of Europe

- Genetic history of the British Isles

- Genetic history of the Iberian Peninsula

- Genetic history of indigenous peoples of the Americas

- Genetic history of Italy

- Genetic history of North Africa

- Genetics and archaeogenetics of South Asia

9.10 References

[1] Iborra, Francisco J; Kimura, Hiroshi; Cook, Peter R (2004). "The functional organization of mitochondrial genomes in human cells". *BMC Biology* **2**: 9. doi:10.1186/1741-7007-2-9. PMC 425603. PMID 15157274.

[2] Sykes, B (10 September 2003). "Mitochondrial DNA and human history". *The Human Genome*. Wellcome Trust. Retrieved 5 February 2012.

[3] "Mitochondrial DNA: The Eve Gene". *Bradshaw Foundation*. Bradshaw Foundation. Retrieved 5 November 2012.

[4] Wiesner, Rudolf J.; Rüegg, J.Caspar; Morano, Ingo (1992). "Counting target molecules by exponential polymerase chain reaction: Copy number of mitochondrial DNA in rat tissues". *Biochemical and Biophysical Research Communications* **183** (2): 553–9. doi:10.1016/0006-291X(92)90517-O. PMID 1550563.

[5] Gabriel, Maria San; Chan, Sam W.; Alhathal, Naif; Chen, Junjian Z.; Zini, Armand (2012). "Influence of microsurgical varicocelectomy on human sperm mitochondrial DNA copy number: A pilot study". *Journal of Assisted Reproduction and Genetics* **29** (8): 759–64. doi:10.1007/s10815-012-9785-z. PMC 3430774. PMID 22562241.

[6] Schatten, Gerald; Sutovsky, Peter; Moreno, Ricardo D.; Ramalho-Santos, João; Dominko, Tanja; Simerly, Calvin (1999). "Development: Ubiquitin tag for sperm mitochondria". *Nature* **402** (6760): 371–2. Bibcode:1999Natur.402..371S. doi:10.1038/46466. PMID 10586873. Discussed in: Travis, John (2000). "Mom's Eggs Execute Dad's Mitochondria". *Science News* **157**: 5. doi:10.2307/4012086. JSTOR 4012086. Archived from the original on December 19, 2007.

[7] Vila, C.; Savolainen, P; Maldonado, J. E.; Amorim, I. R.; Rice, J. E.; Honeycutt, R. L.; Crandall, K. A.; Lundeberg, J; Wayne, R. K. (1997). "Multiple and Ancient Origins of the Domestic Dog". *Science* **276** (5319): 1687–9. doi:10.1126/science.276.5319.1687. PMID 9180076.

[8] Passamonti, Marco; Ghiselli, Fabrizio (2009). "Doubly Uniparental Inheritance: Two Mitochondrial Genomes, One Precious Model for Organelle DNA Inheritance and Evolution". *DNA and Cell Biology* **28** (2): 79–89. doi:10.1089/dna.2008.0807. PMID 19196051.

[9] Kondo R; Matsuura ET; Chigusa SI (1992). "Further observation of paternal transmission of Drosophila mitochondrial DNA by PCR selective amplification method". *Genet. Res.* **59** (2): 81–4. doi:10.1017/S0016672300030287. PMID 1628820.

[10] Meusel, Michael S.; Moritz, Robin F. A. (1993). "Transfer of paternal mitochondrial DNA during fertilization of honeybee (Apis mellifera L.) eggs". *Current Genetics* **24** (6): 539–43. doi:10.1007/BF00351719. PMID 8299176.

[11] Fontaine, Kathryn M.; Cooley, John R.; Simon, Chris (2007). "Evidence for Paternal Leakage in Hybrid Periodical Cicadas (Hemiptera: Magicicada spp.)". *PLoS ONE* **2** (9): e892. Bibcode:2007PLoSO...2..892F. doi:10.1371/journal.pone.0000892. PMC 1963320. PMID 17849021.

[12] Alexander Michelle; et al. (2015). "Mitogenomic analysis of a 50-generation chicken pedigree reveals a rapid rate of mitochondrial evolution and evidence for paternal mtDNA inheritance". *Biology Letters* **11** (10). doi:10.1098/rsbl.2015.0561.

[13] Gyllensten, Ulf; Wharton, Dan; Josefsson, Agneta; Wilson, Allan C. (1991). "Paternal inheritance of mitochondrial DNA in mice". *Nature* **352** (6332): 255–7. Bibcode:1991Natur.352..255G. doi:10.1038/352255a0. PMID 1857422.

[14] Shitara H; Hayashi JI; Takahama S; Kaneda H; Yonekawa H (1998). "Maternal inheritance of mouse mtDNA in interspecific hybrids: segregation of the leaked paternal mtDNA followed by the prevention of subsequent paternal leakage". *Genetics* **148** (2): 851–7. PMC 1459812. PMID 9504930.

[15] Zhao, X; Li, N; Guo, W; Hu, X; Liu, Z; Gong, G; Wang, A; Feng, J; Wu, C (2004). "Further evidence for paternal inheritance of mitochondrial DNA in the sheep (Ovis aries)". *Heredity* **93** (4): 399–403. doi:10.1038/sj.hdy.6800516. PMID 15266295.

[16] Steinborn, Ralf; Zakhartchenko, Valeri; Jelyazkov, Jivko; Klein, Dieter; Wolf, Eckhard; Müller, Mathias; Brem, Gottfried (1998). "Composition of parental mitochondrial DNA in cloned bovine embryos". *FEBS Letters* **426** (3): 352–6. doi:10.1016/S0014-5793(98)00350-0. PMID 9600265.

[17] Schwartz, Marianne; Vissing, John (2002). "Paternal Inheritance of Mitochondrial DNA". *New England Journal of Medicine* **347** (8): 576–80. doi:10.1056/NEJMoa020350. PMID 12192017.

[18] Frith, Maxine (October 14, 2003). "The Independent".

[19] Nosek, Jozef; Tomáška, L'Ubomír; Fukuhara, Hiroshi; Suyama, Yoshitaka; Kováč, Ladislav (1998). "Linear mitochondrial genomes: 30 years down the line". *Trends in Genetics* **14** (5): 184–8. doi:10.1016/S0168-9525(98)01443-7. PMID 9613202.

[20] Balaresque, Patricia; Bowden, Georgina R.; Adams, Susan M.; Leung, Ho-Yee; King, Turi E.; Rosser, Zoë H.; Goodwin, Jane; Moisan, Jean-Paul; Richard, Christelle; Millward, Ann; Demaine, Andrew G.; Barbujani, Guido; Previderè, Carlo; Wilson, Ian J.; Tyler-Smith, Chris; Jobling, Mark A. (2010). "A Predominantly Neolithic Origin for European Paternal Lineages". *PLoS Biology* **8** (1): e1000285. doi:10.1371/journal.pbio.1000285. PMC 2799514. PMID 20087410.

[21] Ward, Bernard L.; Anderson, Robert S.; Bendich, Arnold J. (1981). "The mitochondrial genome is large and variable in a family of plants (Cucurbitaceae)". *Cell* **25** (3): 793–803. doi:10.1016/0092-8674(81)90187-2. PMID 6269758.

[22] Taylor, Robert W.; Turnbull, Doug M. (29 April 2015). "Mitochondrial DNA Mutations In Human Disease".

[23] Alverson, Andrew J; Rice, Danny W; Dickinson, Stephanie; Barry, Kerrie; Palmer, Jeffrey D (2011). "Origins and Recombination of the Bacterial-Sized Multichromosomal Mitochondrial Genome of Cucumber". *The Plant Cell* **23** (7): 2499–513. doi:10.1105/tpc.111.087189. JSTOR 41433488. PMC 3226218. PMID 21742987.

[24] Yakubovskaya, E.; Chen, Z.; Carrodeguas, J. A.; Kisker, C.; Bogenhagen, D. F. (2005). "Functional Human Mitochondrial DNA Polymerase Forms a Heterotrimer". *Journal of Biological Chemistry* **281** (1): 374–82. doi:10.1074/jbc.M509730200. PMID 16263719.

[25] Jemt, E.; Farge, G.; Backstrom, S.; Holmlund, T.; Gustafsson, C. M.; Falkenberg, M. (2011). "The mitochondrial DNA helicase TWINKLE can assemble on a closed circular template and support initiation of DNA synthesis". *Nucleic Acids Research* **39** (21): 9238–49. doi:10.1093/nar/gkr653. PMC 3241658. PMID 21840902.

[26] St. John, J. C.; Facucho-Oliveira, J.; Jiang, Y.; Kelly, R.; Salah, R. (2010). "Mitochondrial DNA transmission, replication and inheritance: A journey from the gamete through the embryo and into offspring and embryonic stem cells". *Human Reproduction Update* **16** (5): 488–509. doi:10.1093/humupd/dmq002. PMID 20231166.

[27] Alexeyev, Mikhail F. (2009). "Is there more to aging than mitochondrial DNA and reactive oxygen species?". *FEBS Journal* **276** (20): 5768–87. doi:10.1111/j.1742-4658.2009.07269.x. PMC 3097520. PMID 19796285.

[28] Anson, R. M.; Hudson, E; Bohr, V. A. (2000). "Mitochondrial endogenous oxidative damage has been overestimated". *FASEB Journal* **14** (2): 355–60. PMID 10657991.

[29] Thorslund, Tina; Sunesen, Morten; Bohr, Vilhelm A.; Stevnsner, Tinna (2002). "Repair of 8-oxoG is slower in endogenous nuclear genes than in mitochondrial DNA and is without strand bias". *DNA Repair* **1** (4): 261–73. doi:10.1016/S1568-7864(02)00003-4. PMID 12509245.

[30] Guliaeva, N. A.; Kuznetsova, E. A.; Gaziev, A. I. (2006). "Белки, ассоциированные с митохондриальной ДНК, защищают ее от воздействия рентгеновского излучения и перекиси водорода" [Proteins associated with mitochondrial DNA protect it against the action of X-rays and hydrogen peroxide]. *Biofizika* (in Russian) **51** (4): 692–7. PMID 16909848.

[31] Alexeyev, M.; Shokolenko, I.; Wilson, G.; Ledoux, S. (2013). "The Maintenance of Mitochondrial DNA Integrity--Critical Analysis and Update". *Cold Spring Harbor Perspectives in Biology* **5** (5): a012641. doi:10.1101/cshperspect.a012641. PMID 23637283.

[32] Hogan, C. Michael (2010). "Mutation". In Monosson, E.; Cleveland, C. J. *Encyclopedia of Earth*. Washington DC: National Council for Science and the Environment.

[33] Ledoux, Susan P.; Alexeyev, Mikhail F.; Wilson, Glenn L. (2004). "Mitochondrial DNA and aging". *Clinical Science* **107** (4): 355–64. doi:10.1042/CS20040148. PMID 15279618.

[34] Reguly, Brian; Jakupciak, John P.; Parr, Ryan L. (2010). "3.4 kb mitochondrial genome deletion serves as a surrogate predictive biomarker for prostate cancer in histopathologically benign biopsy cores". *Canadian Urological Association Journal* **4** (5): E118–22. PMC 2950771. PMID 20944788.

[35] Robinson, K; Creed, J; Reguly, B; Powell, C; Wittock, R; Klein, D; Maggrah, A; Klotz, L; Parr, R L; Dakubo, G D (2010). "Accurate prediction of repeat prostate biopsy outcomes by a mitochondrial DNA deletion assay". *Prostate Cancer and Prostatic Diseases* **13** (2): 126–31. doi:10.1038/pcan.2009.64. PMID 20084081.

[36] de Grey, Aubrey. *The Mitochondrial Free Radical Theory of Aging* (PDF). ISBN 1-57059-564-X.

[37] Lewis, Kaitlyn N.; Andziak, Blazej; Yang, Ting; Buffenstein, Rochelle (2013). "The Naked Mole-Rat Response to Oxidative Stress: Just Deal with It". *Antioxidants & Redox Signaling* **19** (12): 1388–99. doi:10.1089/ars.2012.4911. PMC 3791056. PMID 23025341.

[38] Shigenaga, M. K.; Hagen, T. M.; Ames, B. N. (1994). "Oxidative damage and mitochondrial decay in aging". *Proceedings of the National Academy of Sciences* **91** (23): 10771–8. Bibcode:1994PNAS...9110771S. doi:10.1073/pnas.91.23.10771. JSTOR 2365473. PMC 45108. PMID 7971961.

[39] Trifunovic, A.; Hansson, A.; Wredenberg, A.; Rovio, A. T.; Dufour, E.; Khvorostov, I.; Spelbrink, J. N.; Wibom, R.; Jacobs, H. T.; Larsson, N.-G. (2005). "Somatic mtDNA mutations cause aging phenotypes without affecting reactive oxygen species production". *Proceedings of the National Academy of Sciences* **102** (50): 17993–8. Bibcode:2005PNAS..10217993T. doi:10.1073/pnas.0508886102. JSTOR 4152716. PMC 1312403. PMID 16332961.

[40] Aledo, Juan Carlos; Li, Yang; De Magalhães, João Pedro; Ruíz-Camacho, Manuel; Pérez-Claros, Juan Antonio (2011). "Mitochondrially encoded methionine is inversely related to longevity in mammals". *Aging Cell* **10** (2): 198–207. doi:10.1111/j.1474-9726.2010.00657.x. PMID 21108730.

[41] Ferrari, Carlos K. B. (2004). "Functional foods, herbs and nutraceuticals: Towards biochemical mechanisms of healthy aging". *Biogerontology* **5** (5): 275–89. doi:10.1007/s10522-004-2566-z. PMID 15547316.

[42] Taylor, Robert W (2005). "Gene therapy for the treatment of mitochondrial DNA disorders". *Expert Opinion on Biological Therapy* **5** (2): 183–94. doi:10.1517/14712598.5.2.183. PMID 15757380.

[43] Bjelakovic, Goran; Nikolova, Dimitrinka; Gluud, Christian (2013). "Antioxidant Supplements to Prevent Mortality". *JAMA* **310** (11): 1178–9. doi:10.1001/jama.2013.277028. PMID 24045742.

[44] Damas, J; Carneiro, J; Goncalves, J; Stewart, JB; Samwels, DC; Amorim, A; Pereira, F (2012). "Mitochondrial DNA deletions are associated with non-B DNA conformations". *Nuclei Acids Res* **40**: 7606–7621. doi:10.1093/nar/gks500.

[45] Oliveira, PH; da Silva, CL; Cabral, JMS (2013). "An Appraisal of Human Mitochondrial DNA Instability: New Insights into the Role of Non-Canonical DNA Structures and Sequence Motifs". *PLoS ONE*. doi:10.1371/journal.pone.0059907.

[46] Brown, W. M.; George, M.; Wilson, A. C. (1979). "Rapid evolution of animal mitochondrial DNA". *Proceedings of the National Academy of Sciences* **76** (4): 1967–71. Bibcode:1979PNAS...76.1967B. doi:10.1073/pnas.76.4.1967. JSTOR 69636. PMC 383514. PMID 109836.

[47] http://www.promega.ca/~{}/media/files/resources/profiles%20in%20dna/103/mitochondrial%20dna%20state%20of%20tennessee%20v%20paul%20ware.pdf[]

[48] Court case name listed in the appeal. Retrieved 17 April 2015.

[49] Defense lawyer. Retrieved 17 April 2015.

[50] Dna Tests Got Rorrer Life In Jail * Those Results, Plus `Mountain' Of Circumstantial Evidence, Convinced The Jury Of Her Guilt. Retrieved 17 April 2015.

[51] "Judge allows DNA in Samantha Runnion case," Associated Press, 18 February 2005. Retrieved 4 April 2007.

[52] "Canine DNA Admitted In California Murder Case," Pit Bulletin Legal News, 5 December 2013. Retrieved 21 January 2014.

[53] Nass, M. M. K.; Nass, S (1963). "INTRAMITOCHONDRIAL FIBERS WITH DNA CHARACTERISTICS: I. Fixation and Electron Staining Reactions". *The Journal of Cell Biology* **19** (3): 593–611. doi:10.1083/jcb.19.3.593. PMC 2106331. PMID 14086138.

[54] Schatz, G.; Haslbrunner, E.; Tuppy, H. (1964). "Deoxyribonucleic acid associated with yeast mitochondria". *Biochemical and Biophysical Research Communications* **15** (2): 127–32. doi:10.1016/0006-291X(64)90311-0.

[55] Kumar, Manjeet; Kapil, Aditi; Shanker, Asheesh (2014). "Mito *Sat Plant*: Mitochondrial microsatellites database of viridiplantae". *Mitochondrion* **19**: 334–7. doi:10.1016/j.mito.2014.02.002. PMID 24561221.

[56] Damas, J.; Carneiro, J.; Amorim, A.; Pereira, F. (2013). "Mito *Break*: The mitochondrial DNA breakpoints database". *Nucleic Acids Research* **42** (Database issue): D1261–8. doi:10.1093/nar/gkt982. PMC 3965124. PMID 24170808.

[57] Iwasaki, W.; Fukunaga, T.; Isagozawa, R.; Yamada, K.; Maeda, Y.; Satoh, T. P.; Sado, T.; Mabuchi, K.; Takeshima, H.; Miya, M.; Nishida, M. (2013). "Mito *Fish* and Mito *Annotator*: A Mitochondrial Genome Database of Fish with an Accurate and Automatic Annotation Pipeline". *Molecular Biology and Evolution* **30** (11): 2531–40. doi:10.1093/molbev/mst141. PMC 3808866. PMID 23955518.

[58] Cawthorn, Donna-Mareè; Steinman, Harris Andrew; Corli Witthuhn, R. (2011). "Establishment of a mitochondrial DNA sequence database for the identification of fish species commercially available in South Africa". *Molecular Ecology Resources* **11** (6): 979–91. doi:10.1111/j.1755-0998.2011.03039.x. PMID 21689383.

[59] d'Onorio De Meo, P.; d'Antonio, M.; Griggio, F.; Lupi, R.; Borsani, M.; Pavesi, G.; Castrignano, T.; Pesole, G.; Gissi, C. (2011). "Mito *Zoa* 2.0: A database resource and search tools for comparative and evolutionary analyses of mitochondrial genomes in Metazoa". *Nucleic Acids Research* **40** (Database issue): D1168–72. doi:10.1093/nar/gkr1144. PMC 3245153. PMID 22123747.

[60] Gu, Zuguang; Li, Jie; Gao, Song; Gong, Ming; Wang, Junling; Xu, Hua; Zhang, Chenyu; Wang, Jin (2011). "Inter *Mito Base*: An annotated database and analysis platform of protein-protein interactions for human mitochondria". *BMC Genomics* **12**: 335. doi:10.1186/1471-2164-12-335. PMC 3142533. PMID 21718467.

[61] Lee, Y. S.; Oh, J.; Kim, Y. U.; Kim, N.; Yang, S.; Hwang, U. W. (2007). "Mitome: Dynamic and interactive database for comparative mitochondrial genomics in metazoan animals". *Nucleic Acids Research* **36** (Database issue): D938–42. doi:10.1093/nar/gkm763. PMC 2238945. PMID 17940090.

[62] Catalano, Domenico; Licciulli, Flavio; Turi, Antonio; Grillo, Giorgio; Saccone, Cecilia; d'Elia, Domenica (2006). "Mito *Res*: A resource of nuclear-encoded mitochondrial genes and their products in Metazoa". *BMC Bioinformatics* **7**: 36. doi:10.1186/1471-2105-7-36. PMC 1395343. PMID 16433928.

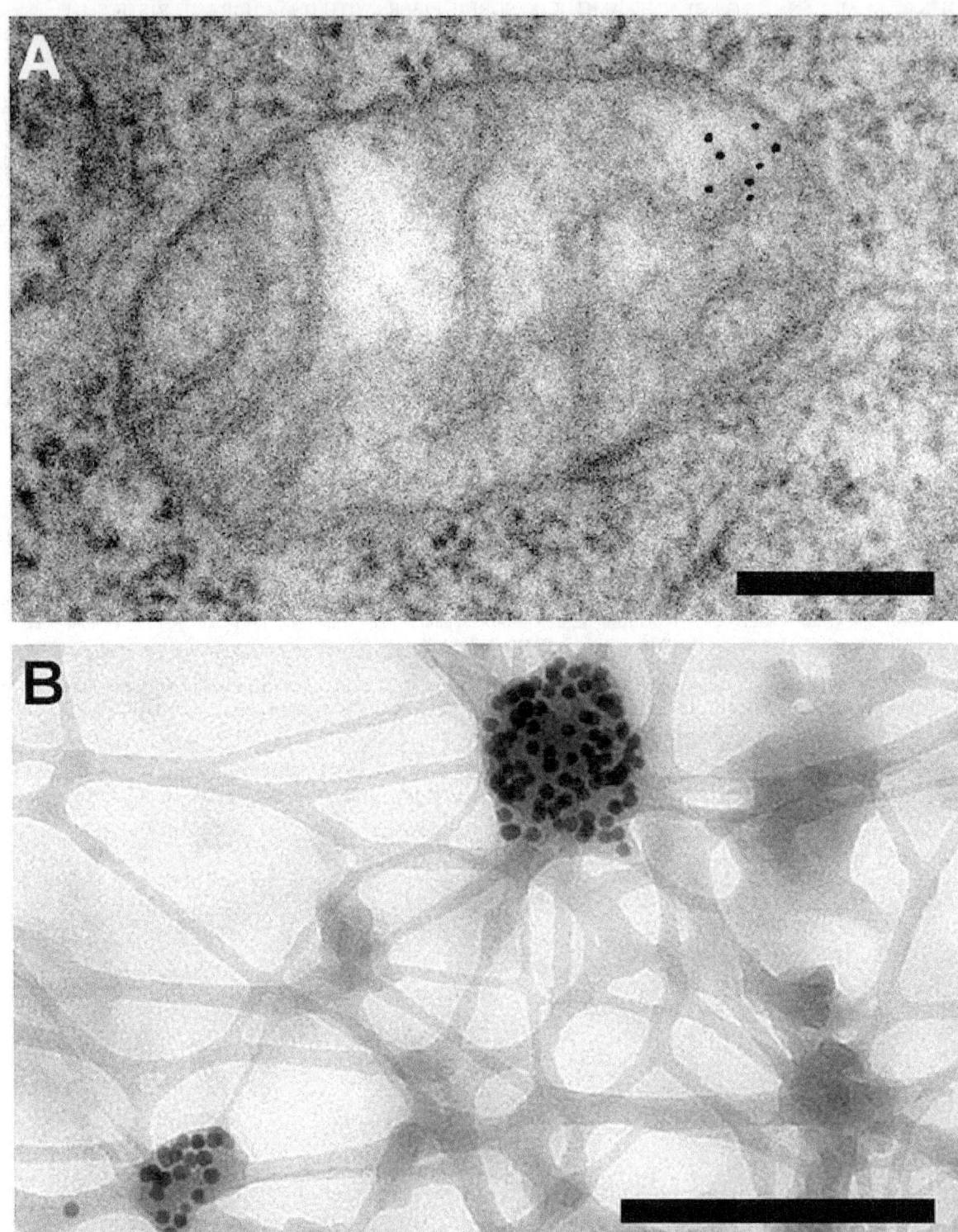

Electron microscopy reveals mitochondrial DNA in discrete foci. Bars: 200 nm. (A) Cytoplasmic section after immunogold labelling with anti-DNA; gold particles marking mtDNA are found near the mitochondrial membrane. (B) Whole mount view of cytoplasm after extraction with CSK buffer and immunogold labelling with anti-DNA; mtDNA (marked by gold particles) resists extraction. From Iborra et al., 2004.[1]

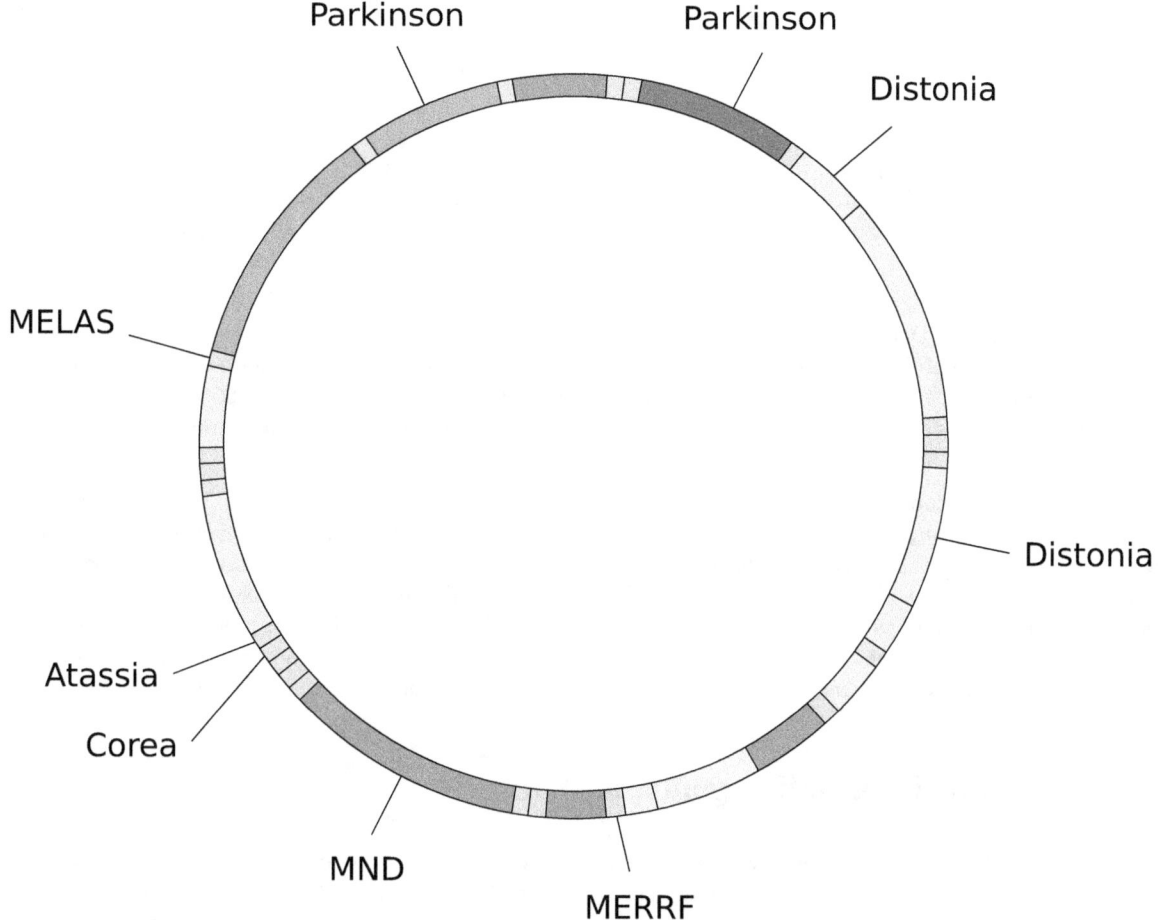

The involvement of mitochondrial DNA in several human diseases.

Chapter 10

Noncoding DNA

In genomics and related disciplines, **noncoding DNA** sequences are components of an organism's DNA that do not encode protein sequences. Some noncoding DNA is transcribed into functional non-coding RNA molecules (e.g. transfer RNA, ribosomal RNA, and regulatory RNAs). Other functions of noncoding DNA include the transcriptional and translational regulation of protein-coding sequences, scaffold attachment regions, origins of DNA replication, centromeres and telomeres.

The amount of noncoding DNA varies greatly among species. For example, over 98% of the human genome is noncoding,[2] while 20% of a typical prokaryote genome is noncoding.[3] When there is much non-coding DNA, a large proportion appears to have no biological function for the organism, as theoretically predicted in the 1960s. Since that time, this non-functional portion has often been referred to as "junk DNA", a term that has elicited strong responses over the years.[4]

The international Encyclopedia of DNA Elements (ENCODE) project uncovered, by direct biochemical approaches, that at least 80% of human genomic DNA has biochemical activity.[5] Though this was not necessarily unexpected due to previous decades of research discovering many functional noncoding regions,[3][6] some scientists criticized the conclusion for conflating biochemical activity with biological function.[7][8][9][10][11] Estimates for the biologically functional fraction of our genome based on comparative genomics range between 8 and 15%.[12][13][14] However, others have argued against relying solely on estimates from comparative genomics due to its limited scope and also because non-coding DNA has been found to be involved in epigenetic activity and making the complexity of species.[6][13][15][16]

10.1   Fraction of noncoding genomic DNA

The amount of total genomic DNA varies widely between organisms, and the proportion of coding and noncoding DNA within these genomes varies greatly as well. More than 98% of the human genome does not encode protein sequences, including most sequences within introns and most intergenic DNA.[2] 20% of a typical prokaryote genome is noncoding.[3]

While overall genome size, and by extension the amount of noncoding DNA, are correlated to organism complexity, there are many exceptions. For example, the genome of the unicellular *Polychaos dubium* (formerly known as *Amoeba dubia*) has been reported to contain more than 200 times the amount of DNA in humans.[17] The pufferfish *Takifugu rubripes* genome is only about one eighth the size of the human genome, yet seems to have a comparable number of genes; approximately 90% of the *Takifugu* genome is noncoding DNA.[2] The extensive variation in nuclear genome size among eukaryotic species is known as the C-value enigma or C-value paradox.[18] Most of the genome size difference appears to lie in the noncoding DNA.

In 2013, a new "record" for the most efficient eukaryotic genome was discovered with *Utricularia gibba*, a bladderwort plant that has only 3% noncoding DNA and 97% of coding DNA. Parts of the noncoding DNA were being deleted by the plant and this suggested that noncoding DNA may not be as critical for plants, even though noncoding DNA is useful for humans.[1] Other studies on plants have discovered crucial functions in portions noncoding DNA that were previously thought to be negligible and have added a new layer to the understanding of gene regulation.[19]

10.2 Types of noncoding DNA sequences

Main article: Conserved non-coding sequence

10.2.1 Noncoding functional RNA

Noncoding RNAs are functional RNA molecules that are not translated into protein. Examples of noncoding RNA include ribosomal RNA, transfer RNA, Piwi-interacting RNA and microRNA.

MicroRNAs are predicted to control the translational activity of approximately 30% of all protein-coding genes in mammals and may be vital components in the progression or treatment of various diseases including cancer, cardiovascular disease, and the immune system response to infection.[20]

10.2.2 *Cis-* and *Trans*-regulatory elements

Cis-regulatory elements are sequences that control the transcription of a nearby gene. Cis-elements may be located in 5' or 3' untranslated regions or within introns. Trans-regulatory elements control the transcription of a distant gene.

Promoters facilitate the transcription of a particular gene and are typically upstream of the coding region. Enhancer sequences may also exert very distant effects on the transcription levels of genes.[21]

10.2.3 Introns

Introns are non-coding sections of a gene, transcribed into the precursor mRNA sequence, but ultimately removed by RNA splicing during the processing to mature messenger RNA. Many introns appear to be mobile genetic elements.[22]

Studies of group I introns from *Tetrahymena* protozoans indicate that some introns appear to be selfish genetic elements, neutral to the host because they remove themselves from flanking exons during RNA processing and do not produce an expression bias between alleles with and without the intron.[22] Some introns appear to have significant biological function, possibly through ribozyme functionality that may regulate tRNA and rRNA activity as well as protein-coding gene expression, evident in hosts that have become dependent on such introns over long periods of time; for example, the *trnL-intron* is found in all green plants and appears to have been vertically inherited for several billions of years, including more than a billion years within chloroplasts and an additional 2–3 billion years prior in the cyanobacterial ancestors of chloroplasts.[22]

10.2.4 Pseudogenes

Pseudogenes are DNA sequences, related to known genes, that have lost their protein-coding ability or are otherwise no longer expressed in the cell. Pseudogenes arise from retrotransposition or genomic duplication of functional genes, and become "genomic fossils" that are nonfunctional due to mutations that prevent the transcription of the gene, such as within the gene promoter region, or fatally alter the translation of the gene, such as premature stop codons or frameshifts.[23] Pseudogenes resulting from the retrotransposition of an RNA intermediate are known as processed pseudogenes; pseudogenes that arise from the genomic remains of duplicated genes or residues of inactivated genes are nonprocessed pseudogenes.[23]

While Dollo's Law suggests that the loss of function in pseudogenes is likely permanent, silenced genes may actually retain function for several million years and can be "reactivated" into protein-coding sequences[24] and a substantial number of pseudogenes are actively transcribed.[23][25] Because pseudogenes are presumed to change without evolutionary constraint, they can serve as a useful model of the type and frequencies of various spontaneous genetic mutations.[26]

10.2.5   Repeat sequences, transposons and viral elements

Transposons and retrotransposons are mobile genetic elements. Retrotransposon repeated sequences, which include long interspersed nuclear elements (LINEs) and short interspersed nuclear elements (SINEs), account for a large proportion of the genomic sequences in many species. Alu sequences, classified as a short interspersed nuclear element, are the most abundant mobile elements in the human genome. Some examples have been found of SINEs exerting transcriptional control of some protein-encoding genes.[27][28][29]

Endogenous retrovirus sequences are the product of reverse transcription of retrovirus genomes into the genomes of germ cells. Mutation within these retro-transcribed sequences can inactivate the viral genome.[30]

Over 8% of the human genome is made up of (mostly decayed) endogenous retrovirus sequences, as part of the over 42% fraction that is recognizably derived of retrotransposons, while another 3% can be identified to be the remains of DNA transposons. Much of the remaining half of the genome that is currently without an explained origin is expected to have found its origin in transposable elements that were active so long ago (> 200 million years) that random mutations have rendered them unrecognizable.[31] Genome size variation in at least two kinds of plants is mostly the result of retrotransposon sequences.[32][33]

10.2.6   Telomeres

Telomeres are regions of repetitive DNA at the end of a chromosome, which provide protection from chromosomal deterioration during DNA replication.

10.3   Junk DNA

The term "junk DNA" became popular in the 1960s.[34][35] According to T. Ryan Gregory, a genomic biologist, the first explicit discussion of the nature of junk DNA was done by David Comings in 1972 and he applied the term to all noncoding DNA.[36] The term was formalized in 1972 by Susumu Ohno,[37] who noted that the mutational load from deleterious mutations placed an upper limit on the number of functional loci that could be expected given a typical mutation rate. Ohno predicted that mammal genomes could not have more than 30,000 loci under selection before the "cost" from the mutational load would cause an inescapable decline in fitness, and eventually extinction. This prediction remains robust, with the human genome containing approximately 20,000 genes. Another source for Ohno's theory was the observation that even closely related species can have widely (orders-of-magnitude) different genome sizes, which had been dubbed the C value paradox in 1971.[8]

Though the fruitfulness of the term "junk DNA" has been questioned on the grounds that it provokes a strong a priori assumption of total non-functionality and though some have recommended using more neutral terminology such as "non-coding DNA" instead;[36] "junk DNA" remains a label for the portions of a genome sequence for which no discernible function has been identified and that through comparative genomics analysis appear under no functional constraint suggesting that the sequence itself has provided no adaptive advantage. Since the late 70s it has become apparent that the majority of non-coding DNA in large genomes finds its origin in the selfish amplification of transposable elements, of which W. Ford Doolittle and Carmen Sapienza in 1980 wrote in the journal *Nature*: "When a given DNA, or class of DNAs, of unproven phenotypic function can be shown to have evolved a strategy (such as transposition) which ensures its genomic survival, then no other explanation for its existence is necessary."[38] The amount of junk DNA can be expected to depend on the rate of amplification of these elements and the rate at which non-functional DNA is lost.[39] In the same issue of *Nature*, Leslie Orgel and Francis Crick wrote that junk DNA has "little specificity and conveys little or no selective advantage to the organism".[40] The term occurs mainly in popular science and in a colloquial way in scientific publications, and it has occasionally been suggested that its connotations may have delayed interest in the biological functions of noncoding DNA.[41]

Several lines of evidence indicate that some "junk DNA" sequences are likely to have unidentified functional activity and that the process of exaptation of fragments of originally selfish or non-functional DNA has been commonplace throughout evolution.[42] In 2012, the ENCODE project, a research program supported by the National Human Genome Research Institute, reported that 76% of the human genome's noncoding DNA sequences were transcribed and that nearly half of the genome was in some way accessible to genetic regulatory proteins such as transcription factors.[4]

However, the suggestion by ENCODE that over 80% of the human genome is biochemically functional has been criticized by other scientists,[7] who argue that neither accessibility of segments of the genome to transcription factors nor their transcription guarantees that those segments have biochemical function and that their transcription is selectively advantageous. Furthermore, the much lower estimates of functionality prior to ENCODE were based on genomic conservation estimates across mammalian lineages.[8][9][10][11]

In response to such views, other scientists argue that the wide spread transcription and splicing that is observed in the human genome directly by biochemical testing is a more accurate indicator of genetic function than genomic conservation because conservation estimates are relative due to incredible variations in genome sizes of even closely related species, it is partially tautological, and these estimates are not based on direct testing for functionality on the genome.[13][15] Conservation estimates may be used to provide clues to identify possible functional elements in the genome, but it does not limit or cap the total amount of functional elements that could possibly exist in the genome since elements that do things at the molecular level can be missed by comparative genomics.[13] Furthermore, much of the apparent junk DNA is involved in epigenetic regulation and appears to be necessary for the development of complex organisms.[6][15][16]

In a 2014 paper, ENCODE researchers tried to address "the question of whether nonconserved but biochemically active regions are truly functional". They noted that in the literature, functional parts of the genome have been identified differently in previous studies depending on the approaches used. There have been three general approaches used to identify functional parts of the human genome: genetic approaches (which rely on changes in phenotype), evolutionary approaches (which rely on conservation) and biochemical approaches (which rely on biochemical testing and was used by ENCODE). All three have limitations: genetic approaches may miss functional elements that do not manifest physically on the organism, evolutionary approaches have difficulties using accurate multispecies sequence alignments since genomes of even closely related species vary considerably, and with biochemical approaches, though having high reproducibility, the biochemical signatures do not always automatically signify a function.[13]

They noted that 70% of the transcription coverage was less than 1 transcript per cell. They noted that this "larger proportion of genome with reproducible but low biochemical signal strength and less evolutionary conservation is challenging to parse between specific functions and biological noise". Furthermore, assay resolution often is much broader than the underlying functional sites so some of the reproducibly "biochemically active but selectively neutral" sequences are unlikely to serve critical functions, especially those with lower-level biochemical signal. To this they added, "However, we also acknowledge substantial limitations in our current detection of constraint, given that some human-specific functions are essential but not conserved and that disease-relevant regions need not be selectively constrained to be functional." On the other hand, they argued that the 12–15% fraction of human DNA under functional constraint, as estimated by a variety of extrapolative evolutionary methods, may still be an underestimate. They concluded that in contrast to evolutionary and genetic evidence, biochemical data offer clues about both the molecular function served by underlying DNA elements and the cell types in which they act. Ultimately genetic, evolutionary, and biochemical approaches can all be used in a complementary way to identify regions that may be functional in human biology and disease.[13]

Some critics have argued that functionality can only be assessed in reference to an appropriate null hypothesis. In this case, the null hypothesis would be that these parts of the genome are non-functional and have properties, be it on the basis of conservation or biochemical activity, that would be expected of such regions based on our general understanding of molecular evolution and biochemistry. According to these critics, until a region in question has been shown to have additional features, beyond what is expected of the null hypothesis, it should provisionally be labelled as non-functional.[43]

10.4 Functions of noncoding DNA

Many noncoding DNA sequences have important biological functions as indicated by comparative genomics studies that report some regions of noncoding DNA that are highly conserved, sometimes on time-scales representing hundreds of millions of years, implying that these noncoding regions are under strong evolutionary pressure and positive selection.[44] For example, in the genomes of humans and mice, which diverged from a common ancestor 65–75 million years ago, protein-coding DNA sequences account for only about 20% of conserved DNA, with the remaining 80% of conserved DNA represented in noncoding regions.[45] Linkage mapping often identifies chromosomal regions associated with a disease with no evidence of functional coding variants of genes within the region, suggesting that disease-causing genetic variants lie in the noncoding DNA.[45] The significance of noncoding DNA mutations in cancer was explored in April 2013.[46]

Noncoding genetic polymorphisms have also been shown to play a role in infectious disease susceptibility, such as hepatitis C.[47] Moreover, noncoding genetic polymorphisms were shown to contribute to susceptibility to Ewing sarcoma - a highly aggressive pediatric bone cancer.[48]

Some specific sequences of noncoding DNA may be features essential to chromosome structure, centromere function and homolog recognition in meiosis.[49]

According to a comparative study of over 300 prokaryotic and over 30 eukaryotic genomes,[50] eukaryotes appear to require a minimum amount of non-coding DNA. This minimum amount can be predicted using a growth model for regulatory genetic networks, implying that it is required for regulatory purposes. In humans the predicted minimum is about 5% of the total genome.

There is evidence that a significant proportion (over 10%) of 32 mammalian genomes may function through the formation of specific RNA secondary structures.[51] The study used comparative genomics to identify compensatory DNA mutations that maintain RNA base-pairings, a distinctive feature of RNA molecules. Over 80% of the genomic regions presenting evolutionary evidence of RNA structure conservation do not present strong DNA sequence conservation.

10.4.1   Protection of the genome

Main article: Mutation

Noncoding DNA separate genes from each other with long gaps, so mutation in one gene or part of a chromosome, for example deletion or insertion, does not have the "frameshift mutation" on the whole chromosome. When genome complexity is relatively high, like in the case of human genome, not only different genes, but also inside one gene there are gaps of introns to protect the entire coding segment to minimise the changes caused by mutation.

It has been suggested that non-coding DNA may serve to decrease the chance of gene disruption during chromosomal crossover.[52]

10.4.2   Genetic switches

Some noncoding DNA sequences are genetic "switches" that regulate when and where genes are expressed.[53]

10.4.3   Regulation of gene expression

Main article: Regulation of gene expression

Some noncoding DNA sequences determine the expression levels of various genes.[54]

10.4.4   Transcription factor sites

Main article: Transcription factor

Some noncoding DNA sequences determine where transcription factors attach.[54] A transcription factor is a protein that binds to specific non-coding DNA sequences, thereby controlling the flow (or transcription) of genetic information from DNA to mRNA. Transcription factors act at very different locations on the genomes of different people.

Operators

Main article: Operator (biology)

An operator is a segment of DNA to which a repressor binds. A repressor is a DNA-binding protein that regulates the expression of one or more genes by binding to the operator and blocking the attachment of RNA polymerase to the promoter, thus preventing transcription of the genes. This blocking of expression is called repression.

Enhancers

Main article: Enhancer (genetics)

An enhancer is a short region of DNA that can be bound with proteins (trans-acting factors), much like a set of transcription factors, to enhance transcription levels of genes in a gene cluster.

Silencers

Main article: Silencer (DNA)

A silencer is a region of DNA that inactivates gene expression when bound by a regulatory protein. It functions in a very similar way as enhancers, only differing in the inactivation of genes.

Promoters

Main article: Promoter (biology)

A promoter is a region of DNA that facilitates transcription of a particular gene. Promoters are typically located near the genes they regulate.

Insulators

Main article: Insulator (genetics)

A genetic insulator is a boundary element that plays two distinct roles in gene expression, either as an enhancer-blocking code, or rarely as a barrier against condensed chromatin. An insulator in a DNA sequence is comparable to a linguistic word divider such as a comma (,) in a sentence, because the insulator indicates where an enhanced or repressed sequence ends.

10.5 Uses of noncoding DNA

10.5.1 Noncoding DNA and evolution

Shared sequences of apparently non-functional DNA are a major line of evidence of common descent.[55]

Pseudogene sequences appear to accumulate mutations more rapidly than coding sequences due to a loss of selective pressure.[26] This allows for the creation of mutant alleles that incorporate new functions that may be favored by natural selection; thus, pseudogenes can serve as raw material for evolution and can be considered "protogenes".[56]

10.5.2 Long range correlations

A statistical distinction between coding and noncoding DNA sequences has been found. It has been observed that nucleotides in non-coding DNA sequences display long range power law correlations while coding sequences do not.[57][58][59]

10.5.3 Forensic anthropology

Police sometimes gather DNA as evidence for purposes of forensic identification. As described in *Maryland v. King*, a 2013 U.S. Supreme Court decision:[60]

> "The current standard for forensic DNA testing relies on an analysis of the chromosomes located within the nucleus of all human cells. "The DNA material in chromosomes is composed of 'coding' and 'noncoding' regions. The coding regions are known as genes and contain the information necessary for a cell to make proteins. . . . Non-protein coding regions . . . are not related directly to making proteins, [and] have been referred to as 'junk' DNA." The adjective "junk" may mislead the lay person, for in fact this is the DNA region used with near certainty to identify a person.

10.6   See also

- Conserved non-coding sequence
- Eukaryotic chromosome fine structure
- Gene-centered view of evolution
- Gene regulatory network
- Intergenic region
- Intragenomic conflict
- Phylogenetic footprinting
- Transcriptome
- Non-coding RNA

10.7   References

[1] "Worlds Record Breaking Plant: Deletes its Noncoding "Junk" DNA". *Design & Trend*. May 12, 2013. Retrieved 2013-06-04.

[2] Elgar G, Vavouri T; Vavouri (July 2008). "Tuning in to the signals: noncoding sequence conservation in vertebrate genomes". *Trends Genet.* **24** (7): 344–52. doi:10.1016/j.tig.2008.04.005. PMID 18514361.

[3] Costa, Fabrico (2012). "7 Non-coding RNAs, Epigenomics, and Complexity in Human Cells". In Morris, Kevin V. *Non-coding RNAs and Epigenetic Regulation of Gene Expression: Drivers of Natural Selection*. Caister Academic Press. ISBN 1904455948.

[4] Pennisi, E. (6 September 2012). "ENCODE Projec t Writes Eulogy for Junk DNA". *Science* **337** (6099): 1159–1161. doi:10.1126/science.337.6099.1159. PMID 22955811.

[5] The ENCODE Project Consortium (2012). "An integrated encyclopedia of DNA elements in the human genome". *Nature* **489** (7414): 57–74. Bibcode:2012Natur.489...57T. doi:10.1038/nature11247. PMC 3439153. PMID 22955616..

[6] Carey, Nessa (2015). *Junk DNA: A Journey Through the Dark Matter of the Genome*. Columbia University Press. ISBN 9780231170840.

[7] Robin McKie (24 February 2013). "Scientists attacked over claim that 'junk DNA' is vital to life". *The Observer.*

[8] Sean Eddy (2012) The C-value paradox, junk DNA, and ENCODE, Curr Biol 22(21):R898–R899.

[9] Doolittle, W. Ford (2013). "Is junk DNA bunk? A critique of ENCODE". *Proc Natl Acad Sci USA* **110** (14): 5294–5300. Bibcode:2013PNAS..110.5294D. doi:10.1073/pnas.1221376110. PMC 3619371. PMID 23479647.

[10] Palazzo, Alexander F.; Gregory, T. Ryan (2014). "The Case for Junk DNA". *PLoS Genetics* **10** (5): e1004351. doi:10.1371/journal.pgen.1004351. ISSN 1553-7404.

[11] Dan Graur, Yichen Zheng, Nicholas Price, Ricardo B. R. Azevedo1, Rebecca A. Zufall and Eran Elhaik (2013). "On the immortality of television sets: "function" in the human genome according to the evolution-free gospel of ENCODE" (PDF). *Genome Biology and Evolution* **5** (3): 578–90. doi:10.1093/gbe/evt028. PMC 3622293. PMID 23431001.

[12] Ponting, CP; Hardison, RC (2011). "What fraction of the human genome is functional?". *Genome Research* **21**: 1769–1776. doi:10.1101/gr.116814.110. PMC 3205562. PMID 21875934.

[13] Kellis, M.; et al. (2014). "Defining functional DNA elements in the human genome". *PNAS* **111** (17): 6131–6138. Bibcode:2014PNAS..111.6131K. doi:10.1073/pnas.1318948111. PMC 4035993. PMID 24753594.

[14] Chris M. Rands, Stephen Meader, Chris P. Ponting and Gerton Lunter (2014). "8.2% of the Human Genome Is Constrained: Variation in Rates of Turnover across Functional Element Classes in the Human Lineage". *PLoS Genet* **10** (7): e1004525. doi:10.1371/journal.pgen.1004525. PMC 4109858. PMID 25057982.

[15] Mattick JS, Dinger ME (2013). "The extent of functionality in the human genome". *The HUGO Journal* **7** (1): 2. doi:10.1186/1877-6566-7-2.

[16] Morris, Kevin, ed. (2012). *Non-Coding RNAs and Epigenetic Regulation of Gene Expression: Drivers of Natural Selection.* Norfolk, UK: Caister Academic Press. ISBN 1904455948.

[17] Gregory TR, Hebert PD; Hebert (April 1999). "The modulation of DNA content: proximate causes and ultimate consequences". *Genome Res.* **9** (4): 317–24. doi:10.1101/gr.9.4.317 (inactive 2015-02-01). PMID 10207154.

[18] Wahls, W.P.; et al. (1990). "Hypervariable minisatellite DNA is a hotspot for homologous recombination in human cells". *Cell* **60** (1): 95–103. doi:10.1016/0092-8674(90)90719-U. PMID 2295091.

[19] Waterhouse, Peter M.; Hellens, Roger P. (25 March 2015). "Plant biology: Coding in non-coding RNAs". *Nature* **520** (7545): 41–42. doi:10.1038/nature14378.

[20] Li M, Marin-Muller C, Bharadwaj U, Chow KH, Yao Q, Chen C; Marin-Muller; Bharadwaj; Chow; Yao; Chen (April 2009). "MicroRNAs: Control and Loss of Control in Human Physiology and Disease". *World J Surg* **33** (4): 667–84. doi:10.1007/s00268-008-9836-x. PMC 2933043. PMID 19030926.

[21] Visel A, Rubin EM, Pennacchio LA (September 2009). "Genomic Views of Distant-Acting Enhancers". *Nature* **461** (7261): 199–205. Bibcode:2009Natur.461..199V. doi:10.1038/nature08451. PMC 2923221. PMID 19741700.

[22] Nielsen H, Johansen SD; Johansen (2009). "Group I introns: Moving in new directions". *RNA Biol* **6** (4): 375–83. doi:10.4161/rna.6.4.9334. PMID 19667762.

[23] Zheng D, Frankish A, Baertsch R; et al. (June 2007). "Pseudogenes in the ENCODE regions: Consensus annotation, analysis of transcription, and evolution". *Genome Res.* **17** (6): 839–51. doi:10.1101/gr.5586307. PMC 1891343. PMID 17568002.

[24] Marshall CR, Raff EC, Raff RA; Raff; Raff (December 1994). "Dollo's law and the death and resurrection of genes". *Proc. Natl. Acad. Sci. U.S.A.* **91** (25): 12283–7. Bibcode:1994PNAS...9112283M. doi:10.1073/pnas.91.25.12283. PMC 45421. PMID 7991619.

[25] Tutar, Y. (2012). "Pseudogenes". *Comp Funct Genomics* **2012**: 424526. doi:10.1155/2012/424526. PMC 3352212. PMID 22611337.

[26] Petrov DA, Hartl DL; Hartl (2000). "Pseudogene evolution and natural selection for a compact genome". *J. Hered.* **91** (3): 221–7. doi:10.1093/jhered/91.3.221. PMID 10833048.

[27] Ponicsan SL, Kugel JF, Goodrich JA; Kugel; Goodrich (February 2010). "Genomic gems: SINE RNAs regulate mRNA production". *Current Opinion in Genetics & Development* **20** (2): 149–55. doi:10.1016/j.gde.2010.01.004. PMC 2859989. PMID 20176473.

[28] Häsler J, Samuelsson T, Strub K; Samuelsson; Strub (July 2007). "Useful 'junk': Alu RNAs in the human transcriptome". *Cell. Mol. Life Sci.* **64** (14): 1793–800. doi:10.1007/s00018-007-7084-0. PMID 17514354.

[29] Walters RD, Kugel JF, Goodrich JA; Kugel; Goodrich (Aug 2009). "InvAluable junk: the cellular impact and function of Alu and B2 RNAs". *IUBMB Life* **61** (8): 831–7. doi:10.1002/iub.227. PMC 4049031. PMID 19621349.

[30] Nelson, PN.; Hooley, P.; Roden, D.; Davari Ejtehadi, H.; Rylance, P.; Warren, P.; Martin, J.; Murray, PG. (Oct 2004). "Human endogenous retroviruses: transposable elements with potential?". *Clin Exp Immunol* **138** (1): 1–9. doi:10.1111/j.1365-2249.2004.02592.x. PMC 1809191. PMID 15373898.

[31] International Human Genome Sequencing Consortium (February 2001). "Initial sequencing and analysis of the human genome". *Nature* **409** (6822): 879–888. Bibcode:2001Natur.409..860L. doi:10.1038/35057062. PMID 11237011.

[32] Piegu, B.; Guyot, R.; Picault, N.; Roulin, A.; Sanyal, A.; Saniyal, A.; Kim, H.; Collura, K.; et al. (Oct 2006). "Doubling genome size without polyploidization: dynamics of retrotransposition-driven genomic expansions in Oryza australiensis, a wild relative of rice". *Genome Res* **16** (10): 1262–9. doi:10.1101/gr.5290206. PMC 1581435. PMID 16963705.

[33] Hawkins, JS.; Kim, H.; Nason, JD.; Wing, RA.; Wendel, JF. (Oct 2006). "Differential lineage-specific amplification of transposable elements is responsible for genome size variation in Gossypium". *Genome Res* **16** (10): 1252–61. doi:10.1101/gr.5282906. PMC 1581434. PMID 16954538.

[34] Ehret CF, De Haller G; De Haller (1963). "Origin, development, and maturation of organelles and organelle systems of the cell surface in Paramecium". *Journal of Ultrastructure Research*. 9 Supplement 1: 1, 3–42. doi:10.1016/S0022-5320(63)80088-X. PMID 14073743.

[35] Dan Graur, The Origin of Junk DNA: A Historical Whodunnit

[36] Gregory, T. Ryan, ed. (2005). *The Evolution of the Genome*. Elsevier. pp. 29–31. ISBN 0123014638. Comings (1972), on the other hand, gave what must be considered the first explicit discussion of the nature of "junk DNA," and was the first to apply the term to all noncoding DNA."; "For this reason, it is unlikely that any one function for noncoding DNA can account for either its sheer mass or its unequal distribution among taxa. However, dismissing it as no more than "junk" in the pejorative sense of "useless" or "wasteful" does little to advance the understanding of genome evolution. For this reason, the far less loaded term "noncoding DNA" is used throughout this chapter and is recommended in preference to "junk DNA" for future treatments of the subject."

[37] Ohno, Susumu (1972). H. H. Smith, ed. *So Much "junk" DNA in Our Genome*. Gordon and Breach, New York. pp. 366–370. Retrieved 2013-05-15.

[38] Doolittle WF, Sapienza C; Sapienza (1980). "Selfish genes, the phenotype paradigm and genome evolution". *Nature* **284** (5757): 601–603. Bibcode:1980Natur.284..601D. doi:10.1038/284601a0. PMID 6245369.

[39] Another source is genome duplication followed by a loss of function due to redundancy.

[40] Orgel LE, Crick FH; Crick (April 1980). "Selfish DNA: the ultimate parasite". *Nature* **284** (5757): 604–7. Bibcode:1980Natur.284..604O. doi:10.1038/284604a0. PMID 7366731.

[41] Khajavinia A, Makalowski W; Makalowski (May 2007). "What is "junk" DNA, and what is it worth?". *Scientific American* **296** (5): 104. doi:10.1038/scientificamerican0307-104. PMID 17503549. The term "junk DNA" repelled mainstream researchers from studying noncoding genetic material for many years

[42] Biémont, Christian; Vieira, C (2006). "Genetics: Junk DNA as an evolutionary force". *Nature* **443** (7111): 521–4. Bibcode:2006Natur.443..52 doi:10.1038/443521a. PMID 17024082.

[43] Palazzo, Alexander F.; Lee, Eliza S. (2015). "Non-coding RNA: what is functional and what is junk?". *Frontiers in Genetics* **6**: 2. doi:10.3389/fgene.2015.00002. ISSN 1664-8021. PMID 25674102.

[44] Ludwig MZ (December 2002). "Functional evolution of noncoding DNA". *Current Opinion in Genetics & Development* **12** (6): 634–9. doi:10.1016/S0959-437X(02)00355-6. PMID 12433575.

[45] Cobb J, Büsst C, Petrou S, Harrap S, Ellis J; Büsst; Petrou; Harrap; Ellis (April 2008). "Searching for functional genetic variants in non-coding DNA". *Clin. Exp. Pharmacol. Physiol.* **35** (4): 372–5. doi:10.1111/j.1440-1681.2008.04880.x. PMID 18307723.

[46] E Khurana; et al. (April 2013). "Integrative annotation of variants from 1092 humans: application to cancer genomics". *Science* **342** (6154): 372–5. doi:10.1126/science.1235587. PMC 3947637. PMID 24092746.

[47] Lu, Yi-Fan; Mauger, David M.; Goldstein, David B.; Urban, Thomas J.; Weeks, Kevin M.; Bradrick, Shelton S. (4 November 2015). "IFNL3 mRNA structure is remodeled by a functional non-coding polymorphism associated with hepatitis C virus clearance". *Scientific Reports* **5**: 16037. doi:10.1038/srep16037.

[48] Grünewald, Thomas G P; Bernard, Virginie; Gilardi-Hebenstreit, Pascale; Raynal, Virginie; Surdez, Didier; Aynaud, Marie-Ming; Mirabeau, Olivier; Cidre-Aranaz, Florencia; Tirode, Franck. "Chimeric EWSR1-FLI1 regulates the Ewing sarcoma susceptibility gene EGR2 via a GGAA microsatellite". *Nature Genetics* **47** (9): 1073–1078. doi:10.1038/ng.3363. PMC 4591073. PMID 26214589.

[49] Subirana JA, Messeguer X; Messeguer (March 2010). "The most frequent short sequences in non-coding DNA". *Nucleic Acids Res.* **38** (4): 1172–81. doi:10.1093/nar/gkp1094. PMC 2831315. PMID 19966278.

[50] S. E. Ahnert; T. M. A. Fink (2008). "How much non-coding DNA do eukaryotes require?" (PDF). *J. Theor. Biol.* **252** (4): 587–592. doi:10.1016/j.jtbi.2008.02.005. PMID 18384817.

[51] Smith MA; et al. (June 2013). "Widespread purifying selection on RNA structure in mammals". *Nucleic Acids Research* **41** (17): 8220–8236. doi:10.1093/nar/gkt596. PMC 3783177. PMID 23847102.

[52] Dileep, V (2009). "The place and function of non-coding DNA in the evolution of variability". *Hypothesis* **7** (1): e7. doi:10.5779/hypothesis.v7i1.146

[53] Carroll, Sean B.; et al. (May 2008). "Regulating Evolution". *Scientific American* **298** (5): 60–67. doi:10.1038/scientificamerican0508-60. PMID 18444326.

[54] Callaway, Ewen (March 2010). "Junk DNA gets credit for making us who we are". *New Scientist.*

[55] "Plagiarized Errors and Molecular Genetics", talkorigins, by Edward E. Max, M.D., Ph.D.

[56] Balakirev ES, Ayala FJ; Ayala (2003). "Pseudogenes: are they "junk" or functional DNA?". *Annu. Rev. Genet.* **37**: 123–51. doi:10.1146/annurev.genet.37.040103.103949. PMID 14616058.

[57] C.-K. Peng, S. V. Buldyrev, A. L. Goldberger, S. Havlin, F. Sciortino, M. Simons, H. E. Stanley; Buldyrev, SV; Goldberger, AL; Havlin, S; Sciortino, F; Simons, M; Stanley, HE (1992). "Long-range correlations in nucleotide sequences". *Nature* **356** (6365): 168–70. Bibcode:1992Natur.356..168P. doi:10.1038/356168a0. PMID 1301010.

[58] W. Li and, K. Kaneko; Kaneko, K (1992). "Long-Range Correlation and Partial 1/f^{alpha} Spectrum in a Non-Coding DNA Sequence" (PDF). *Europhys. Lett* **17** (7): 655–660. Bibcode:1992EL.....17..655L. doi:10.1209/0295-5075/17/7/014.

[59] S. V. Buldyrev, A. L. Goldberger, S. Havlin, R. N. Mantegna, M. Matsa, C.-K. Peng, M. Simons, and H. E. Stanley; Goldberger, A.; Havlin, S.; Mantegna, R.; Matsa, M.; Peng, C.-K.; Simons, M.; Stanley, H. (1995). "Long-range correlations properties of coding and noncoding DNA sequences: GenBank analysis". *Phys. Rev. E* **51** (5): 5084–5091. Bibcode:1995PhRvE..51.5084B. doi:10.1103/PhysRevE.51.5084.

[60] Slip opinion for *Maryland v. King* from the U.S. Supreme Court. Retrieved 2013-06-04.

10.8 Further reading

Bennett, Michael D.; Leitch, Ilia J. (2005). "Genome size evolution in plants". In Gregory, T. Ryan. *The Evolution of the Genome.* San Diego: Elsevier. pp. 89–162. ISBN 978-0-08-047052-8.

Gregory, T.R (2005). "Genome size evolution in animals". In T.R. Gregory (ed.). *The Evolution of the Genome.* San Diego: Elsevier. ISBN 0-12-301463-8.

Shabalina SA, Spiridonov NA; Spiridonov (2004). "The mammalian transcriptome and the function of non-coding DNA sequences". *Genome Biol.* **5** (4): 105. doi:10.1186/gb-2004-5-4-105. PMC 395773. PMID 15059247.

Castillo-Davis CI (October 2005). "The evolution of noncoding DNA: how much junk, how much func?". *Trends Genet.* **21** (10): 533–6. doi:10.1016/j.tig.2005.08.001. PMID 16098630.

10.9   External links

- Plant DNA C-values Database at Royal Botanic Gardens, Kew

- Fungal Genome Size Database at Estonian Institute of Zoology and Botany

- ENCODE: The human encyclopaedia at *Nature* ENCODE

Chapter 11

DNA profiling

For DNA testing for inherited diseases, see Genetic testing.
Not to be confused with DNA barcoding or DNA phenotyping.

DNA profiling (also called **DNA fingerprinting**, **DNA testing**, or **DNA typing**) is a forensic technique used to identify individuals by characteristics of their DNA. A **DNA profile** is a small set of DNA variations that is very likely to be different in all unrelated individuals, thereby being as unique to individuals as are fingerprints (hence the alternate name for the technique). DNA profiling should not be confused with full genome sequencing.[1] First developed and used in 1985,[2] DNA profiling is used in, for example, parentage testing and criminal investigation, to identify a person or to place a person at a crime scene, techniques which are now employed globally in forensic science to facilitate police detective work and help clarify paternity and immigration disputes.[3]

Although 99.9% of human DNA sequences are the same in every person, enough of the DNA is different that it is possible to distinguish one individual from another, unless they are monozygotic ("identical") twins.[4] DNA profiling uses repetitive ("repeat") sequences that are highly variable,[4] called variable number tandem repeats (VNTRs), in particular short tandem repeats (STRs). VNTR loci are very similar between closely related humans, but are so variable that unrelated individuals are extremely unlikely to have the same VNTRs.

The modern process of DNA profiling was developed in 1988.[5][6]

11.1 DNA profiling process

Developed by Professor of Genetics Sir Alec Jeffreys, the process begins with a sample of an individual's DNA (typically called a "reference sample"). The most desirable method of collecting a reference sample is the use of a buccal swab, as this reduces the possibility of contamination. When this is not available (e.g. because a court order is needed but not obtainable) other methods may need to be used to collect a sample of blood, saliva, semen, or other appropriate fluid or tissue from personal items (e.g. a toothbrush, razor) or from stored samples (e.g. banked sperm or biopsy tissue). Samples obtained from blood relatives (related by birth, not marriage) can provide an indication of an individual's profile, as could human remains that had been previously profiled.

A reference sample is then analyzed to create the individual's DNA profile using one of a number of techniques, discussed below. The DNA profile is then compared against another sample to determine whether there is a genetic match.

11.1.1 RFLP analysis

Main article: Restriction fragment length polymorphism

The first methods for finding out genetics used for DNA profiling involved **RFLP analysis**. DNA is collected from cells,

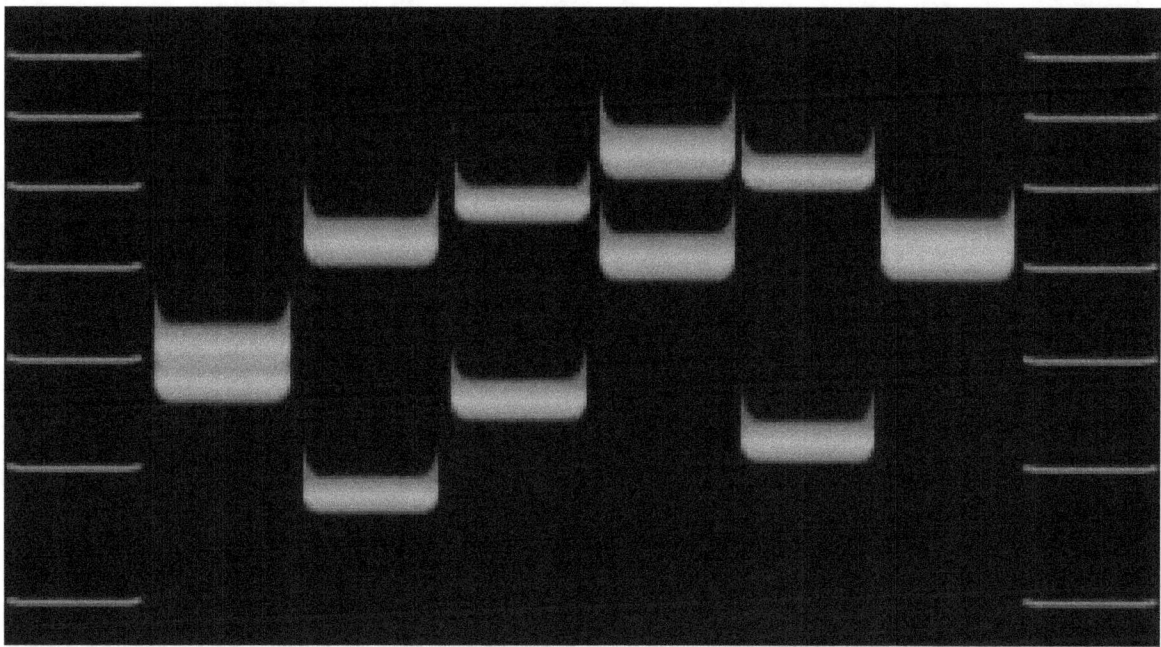

Variations of VNTR allele lengths in 6 individuals.

such as a blood sample, and cut into small pieces using a restriction enzyme (a restriction digest). This generates thousands of DNA fragments of differing sizes as a consequence of variations between DNA sequences of different individuals. The fragments are then separated on the basis of size using gel electrophoresis.

The separated fragments are then transferred to a nitrocellulose or nylon filter; this procedure is called a Southern blot. The DNA fragments within the blot are permanently fixed to the filter, and the DNA strands are denatured. Radiolabeled probe molecules are then added that are complementary to sequences in the genome that contain repeat sequences. These repeat sequences tend to vary in length among different individuals and are called variable number tandem repeat sequences or VNTRs. The probe molecules hybridize to DNA fragments containing the repeat sequences and excess probe molecules are washed away. The blot is then exposed to an X-ray film. Fragments of DNA that have bound to the probe molecules appear as dark bands on the film.

The Southern blot technique is laborious, and requires large amounts of undegraded sample DNA. Also, Karl Brown's original technique looked at many minisatellite loci at the same time, increasing the observed variability, but making it hard to discern individual alleles (and thereby precluding paternity testing). These early techniques have been supplanted by PCR-based assays.

11.1.2   PCR analysis

Main article: polymerase chain reaction

Developed by Kary Mullis in 1983, a process was reported by which specific portions of the sample DNA can be amplified almost indefinitely (Saiki et al. 1985, 1988). This has revolutionized the whole field of DNA study. The process, the polymerase chain reaction (PCR), mimics the biological process of DNA replication, but confines it to specific DNA sequences of interest. With the invention of the PCR technique, DNA profiling took huge strides forward in both discriminating power and the ability to recover information from very small (or degraded) starting samples.

PCR greatly amplifies the amounts of a specific region of DNA. In the PCR process, the DNA sample is denatured into the separate individual polynucleotide strands through heating. Two oligonucleotide DNA primers are used to hybridize

Alec Jeffreys, the pioneer of DNA profiling.

to two corresponding nearby sites on opposite DNA strands in such a fashion that the normal enzymatic extension of the active terminal of each primer (that is, the 3' end) leads toward the other primer. PCR uses replication enzymes that are tolerant of high temperatures, such as the thermostable Taq polymerase. In this fashion, two new copies of the sequence of interest are generated. Repeated denaturation, hybridization, and extension in this fashion produce an exponentially growing number of copies of the DNA of interest. Instruments that perform thermal cycling are now readily available from commercial sources. This process can produce a million-fold or greater amplification of the desired region in 2 hours or less.

Early assays such as the HLA-DQ alpha reverse dot blot strips grew to be very popular due to their ease of use, and the speed with which a result could be obtained. However, they were not as discriminating as RFLP analysis. It was also difficult to determine a DNA profile for mixed samples, such as a vaginal swab from a sexual assault victim.

However, the PCR method was readily adaptable for analyzing VNTR, in particular STR loci. In recent years, research in human DNA quantitation has focused on new "real-time" quantitative PCR (qPCR) techniques. Quantitative PCR methods enable automated, precise, and high-throughput measurements. Interlaboratory studies have demonstrated the importance of human DNA quantitation on achieving reliable interpretation of STR typing and obtaining consistent results across laboratories.

11.1.3   STR analysis

Main article: Short tandem repeats

The system of DNA profiling used today is based on PCR and uses simple sequences[7] or short tandem repeats (STR). This method uses highly polymorphic regions that have short repeated sequences of DNA (the most common is 4 bases repeated, but there are other lengths in use, including 3 and 5 bases). Because unrelated people almost certainly have different numbers of repeat units, STRs can be used to discriminate between unrelated individuals. These STR loci (locations on a chromosome) are targeted with sequence-specific primers and amplified using PCR. The DNA fragments that result are then separated and detected using electrophoresis. There are two common methods of separation and detection, capillary electrophoresis (CE) and gel electrophoresis.

Each STR is polymorphic, but the number of alleles is very small. Typically each STR allele will be shared by around 5 - 20% of individuals. The power of STR analysis comes from looking at multiple STR loci simultaneously. The pattern of alleles can identify an individual quite accurately. Thus STR analysis provides an excellent identification tool. The more STR regions that are tested in an individual the more discriminating the test becomes.

From country to country, different STR-based DNA-profiling systems are in use. In North America, systems that amplify the CODIS 13 core loci are almost universal, whereas in the United Kingdom the SGM+ 11 loci system (which is compatible with The National DNA Database) is in use. Whichever system is used, many of the STR regions used are the same. These DNA-profiling systems are based on multiplex reactions, whereby many STR regions will be tested at the same time.

The true power of STR analysis is in its statistical power of discrimination. Because the 13 loci that are currently used for discrimination in CODIS are independently assorted (having a certain number of repeats at one locus does not change the likelihood of having any number of repeats at any other locus), the product rule for probabilities can be applied. This means that, if someone has the DNA type of ABC, where the three loci were independent, we can say that the probability of having that DNA type is the probability of having type A times the probability of having type B times the probability of having type C. This has resulted in the ability to generate match probabilities of 1 in a quintillion (1×10^{18}) or more. However, DNA database searches showed much more frequent than expected false DNA profile matches.[8] Moreover, since there are about 12 million monozygotic twins on Earth, the theoretical probability is not accurate.

In practice, the risk of contaminated-matching is much greater than matching a distant relative, such as contamination of a sample from nearby objects, or from left-over cells transferred from a prior test. The risk is greater for matching the most common person in the samples: Everything collected from, or in contact with, a victim is a major source of contamination for any other samples brought into a lab. For that reason, multiple control-samples are typically tested in order to ensure that they stayed clean, when prepared during the same period as the actual test samples. Unexpected matches (or variations) in several control-samples indicates a high probability of contamination for the actual test samples. In a relationship test, the full DNA profiles should differ (except for twins), to prove that a person was not actually matched as being related to their own DNA in another sample.

11.1.4 AmpFLP

Main article: Amplified fragment length polymorphism

Another technique, AmpFLP, or amplified fragment length polymorphism was also put into practice during the early 1990s. This technique was also faster than RFLP analysis and used PCR to amplify DNA samples. It relied on variable number tandem repeat (VNTR) polymorphisms to distinguish various alleles, which were separated on a polyacrylamide gel using an allelic ladder (as opposed to a molecular weight ladder). Bands could be visualized by silver staining the gel. One popular locus for fingerprinting was the D1S80 locus. As with all PCR based methods, highly degraded DNA or very small amounts of DNA may cause allelic dropout (causing a mistake in thinking a heterozygote is a homozygote) or other stochastic effects. In addition, because the analysis is done on a gel, very high number repeats may bunch together at the top of the gel, making it difficult to resolve. AmpFLP analysis can be highly automated, and allows for easy creation of phylogenetic trees based on comparing individual samples of DNA. Due to its relatively low cost and ease of set-up and operation, AmpFLP remains popular in lower income countries.

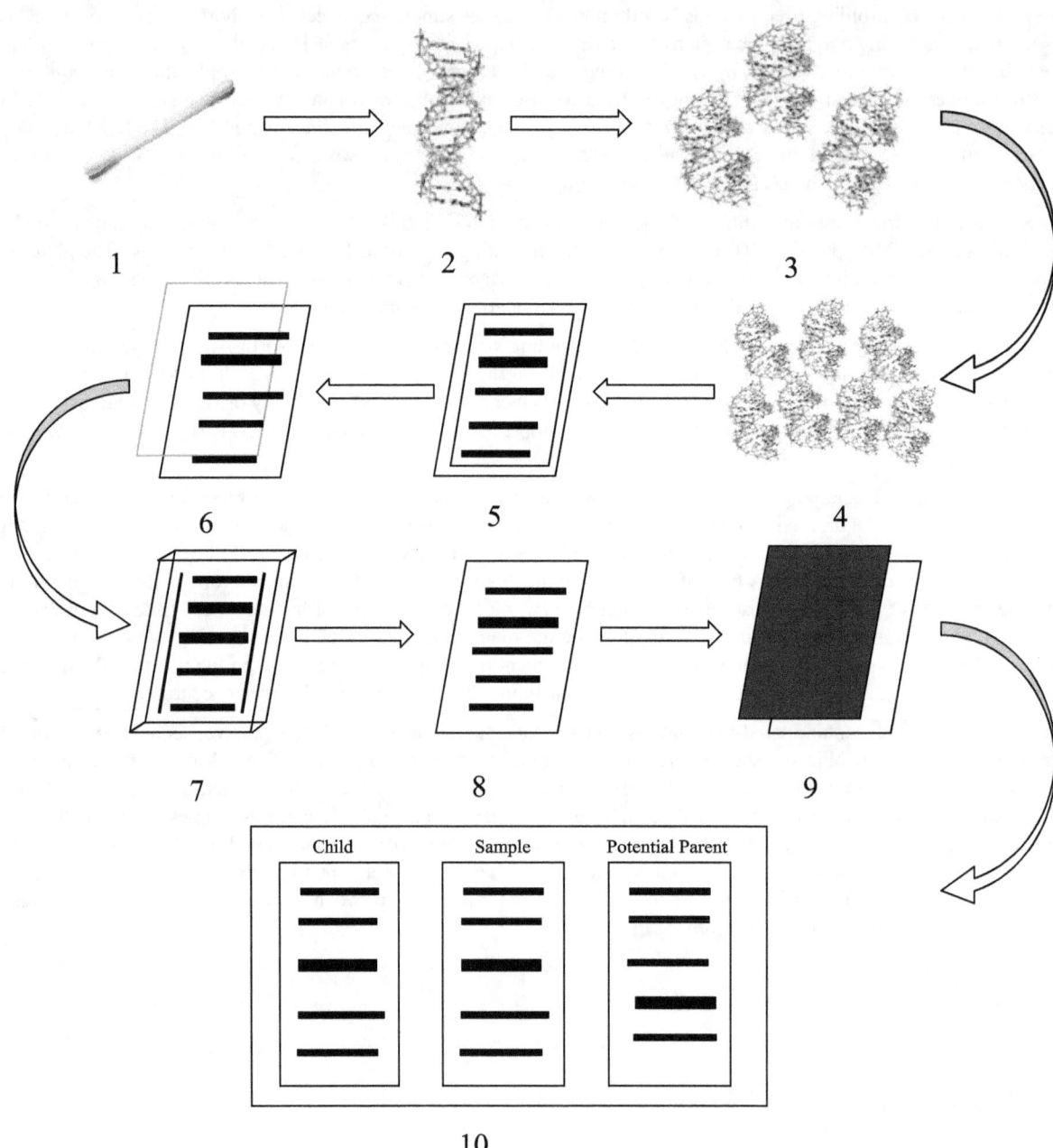

1: A cell sample is taken- usually a cheek swab or blood test
2: DNA is extracted from sample
3: Cleavage of DNA by restriction enzyme- the DNA is broken into small fragments
4: Small fragments are amplified by the Polymerase Chain Reaction- results in many more fragments
5: DNA fragments are separated by electrophoresis
6: The fragments are transferred to an agar plate
7: On the Agar Plate specific DNA fragments are bound to a radioactive DNA probe
8: The Agar Plate is washed free of excess probe
9: An x-ray film is used to detect a radioactive pattern
10: The DNA is compared to other DNA samples

11.1.5 DNA family relationship analysis

Using PCR technology, DNA analysis is widely applied to determine genetic family relationships such as paternity, maternity, siblingship and other kinships.

During conception, the father's sperm cell and the mother's egg cell, each containing half the amount of DNA found in other body cells, meet and fuse to form a fertilized egg, called a zygote. The zygote contains a complete set of DNA molecules, a unique combination of DNA from both parents. This zygote divides and multiplies into an embryo and later, a full human being.

At each stage of development, all the cells forming the body contain the same DNA—half from the father and half from the mother. This fact allows the relationship testing to use all types of all samples including loose cells from the cheeks collected using buccal swabs, blood or other types of samples.

There are predictable inheritance patterns at certain locations (called loci) in the human genome, which have been found to be useful in determining identity and biological relationships. These loci contain specific DNA markers that scientists use to identify individuals. In a routine DNA paternity test, the markers used are Short Tandem Repeats (STRs), short pieces of DNA that occur in highly differential repeat patterns among individuals.

Each person's DNA contains two copies of these markers—one copy inherited from the father and one from the mother. Within a population, the markers at each person's DNA location could differ in length and sometimes sequence, depending on the markers inherited from the parents.

The combination of marker sizes found in each person makes up his/her unique genetic profile. When determining the relationship between two individuals, their genetic profiles are compared to see if they share the same inheritance patterns at a statistically conclusive rate.

For example, the following sample report from this commercial DNA paternity testing laboratory Universal Genetics signifies how relatedness between parents and child is identified on those special markers:

The partial results indicate that the child and the alleged father's DNA match among these five markers. The complete test results show this correlation on 16 markers between the child and the tested man to enable a conclusion to be drawn as to whether or not the man is the biological father.

Each marker is assigned with a Paternity Index (PI), which is a statistical measure of how powerfully a match at a particular marker indicates paternity. The PI of each marker is multiplied with each other to generate the Combined Paternity Index (CPI), which indicates the overall probability of an individual being the biological father of the tested child relative to a randomly selected man from the entire population of the same race. The CPI is then converted into a Probability of Paternity showing the degree of relatedness between the alleged father and child.

The DNA test report in other family relationship tests, such as grandparentage and siblingship tests, is similar to a paternity test report. Instead of the Combined Paternity Index, a different value, such as a Siblingship Index, is reported.

The report shows the genetic profiles of each tested person. If there are markers shared among the tested individuals, the probability of biological relationship is calculated to determine how likely the tested individuals share the same markers due to a blood relationship.

11.1.6 Y-chromosome analysis

Recent innovations have included the creation of primers targeting polymorphic regions on the Y-chromosome (Y-STR), which allows resolution of a mixed DNA sample from a male and female or cases in which a differential extraction is not possible. Y-chromosomes are paternally inherited, so Y-STR analysis can help in the identification of paternally related males. Y-STR analysis was performed in the Sally Hemings controversy to determine if Thomas Jefferson had sired a son with one of his slaves. The analysis of the Y-chromosome yields weaker results than autosomal chromosome analysis. The Y male sex-determining chromosome, as it is inherited only by males from their fathers, is almost identical along the patrilineal line. This leads to a less precise analysis than if autosomal chromosomes were testing, because of the random matching that occurs between pairs of chromosomes as zygotes are being made.[9]

11.1.7   Mitochondrial analysis

Main article: Mitochondrial DNA

For highly degraded samples, it is sometimes impossible to get a complete profile of the 13 CODIS STRs. In these situations, mitochondrial DNA (mtDNA) is sometimes typed due to there being many copies of mtDNA in a cell, while there may only be 1-2 copies of the nuclear DNA. Forensic scientists amplify the HV1 and HV2 regions of the mtDNA, and then sequence each region and compare single-nucleotide differences to a reference. Because mtDNA is maternally inherited, directly linked maternal relatives can be used as match references, such as one's maternal grandmother's daughter's son. In general, a difference of two or more nucleotides is considered to be an exclusion. Heteroplasmy and poly-C differences may throw off straight sequence comparisons, so some expertise on the part of the analyst is required. mtDNA is useful in determining clear identities, such as those of missing people when a maternally linked relative can be found. mtDNA testing was used in determining that Anna Anderson was not the Russian princess she had claimed to be, Anastasia Romanov.

mtDNA can be obtained from such material as hair shafts and old bones/teeth. Control mechanism based on interaction point with data. This can be determined by tooled placement in sample.

11.2   DNA databases

Main article: National DNA database

An early application of a DNA database was the compilation of A Mitochondrial DNA Concordance,[10] prepared by Kevin W. P. Miller and John L. Dawson at the University of Cambridge from 1996 to 1998[11] from data collected as part of Miller's PhD thesis. There are now several DNA databases in existence around the world. Some are private, but most of the largest databases are government controlled. The United States maintains the largest DNA database, with the Combined DNA Index System (CODIS) holding over 5 million records as of 2007.[12] The United Kingdom maintains the National DNA Database (NDNAD), which is of similar size, despite the UK's smaller population. The size of this database, and its rate of growth, is giving concern to civil liberties groups in the UK, where police have wide-ranging powers to take samples and retain them even in the event of acquittal.[13]

The U.S. Patriot Act of the United States provides a means for the U.S. government to get DNA samples from other countries if they are either a division of or a head office of a company operating in the U.S. Under the act; the American offices of the company cannot divulge to their subsidiaries/offices in other countries the reasons that these DNA samples are sought or by whom.

When a match is made from a National DNA Databank to link a crime scene to an offender having provided a DNA Sample to a databank that link is often referred to as a *cold hit*. A cold hit is of value in referring the police agency to a specific suspect but is of less evidential value than a DNA match made from outside the DNA Databank.[14]

FBI agents cannot legally store DNA of a person not convicted of a crime. DNA collected from a suspect not later convicted must be disposed of and not entered into the database. In 1998, a man residing in the UK was arrested on accusation of burglary. His DNA was taken and tested, and he was later released. Nine months later, this man's DNA was accidentally and illegally entered in the DNA database. New DNA is automatically compared to the DNA found at cold cases and, in this case, this man was found to be a match to DNA found at a rape and assault case one year earlier. The government then prosecuted him for these crimes. During the trial the DNA match was requested to be removed from the evidence because it had been illegally entered into the database. The request was carried out.[15]

The DNA collected from victims of rape are often stored for years until matched with the perpetrator's, usually when committing another crime. In 2014, Congress extended a bill that helps states deal with "a backlog" of unexamined evidence.[16]

11.3 Considerations when evaluating DNA evidence

In the early days of the use of genetic fingerprinting as criminal evidence, juries were often swayed by spurious statistical arguments by defense lawyers along these lines: Given a match that had a 1 in 5 million probability of occurring by chance, the lawyer would argue that this meant that in a country of say 60 million people there were 12 people who would also match the profile. This was then translated to a 1 in 12 chance of the suspect's being the guilty one. This argument is not sound unless the suspect was drawn at random from the population of the country. In fact, a jury should consider how likely it is that an individual matching the genetic profile would also have been a suspect in the case for other reasons. Another spurious statistical argument is based on the false assumption that a 1 in 5 million probability of a match automatically translates into a 1 in 5 million probability of innocence and is known as the prosecutor's fallacy.

When using RFLP, the theoretical risk of a coincidental match is 1 in 100 billion (100,000,000,000), although the practical risk is actually 1 in 1000 because monozygotic twins are 0.2% of the human population. Moreover, the rate of laboratory error is almost certainly higher than this, and often actual laboratory procedures do not reflect the theory under which the coincidence probabilities were computed. For example, the coincidence probabilities may be calculated based on the probabilities that markers in two samples have bands in *precisely* the same location, but a laboratory worker may conclude that similar—but not precisely identical—band patterns result from identical genetic samples with some imperfection in the agarose gel. However, in this case, the laboratory worker increases the coincidence risk by expanding the criteria for declaring a match. Recent studies have quoted relatively high error rates, which may be cause for concern.[17] In the early days of genetic fingerprinting, the necessary population data to accurately compute a match probability was sometimes unavailable. Between 1992 and 1996, arbitrary low ceilings were controversially put on match probabilities used in RFLP analysis rather than the higher theoretically computed ones.[18] Today, RFLP has become widely disused due to the advent of more discriminating, sensitive and easier technologies.

Since 1998, the DNA profiling system supported by The National DNA Database in the UK is the SGM+ DNA profiling system that includes 10 STR regions and a sex-indicating test. STRs do not suffer from such subjectivity and provide similar power of discrimination (1 in 10^{13} for unrelated individuals if using a full SGM+ profile). Figures of this magnitude are not considered to be statistically supportable by scientists in the UK; for unrelated individuals with full matching DNA profiles a match probability of 1 in a billion is considered statistically supportable. However, with any DNA technique, the cautious juror should not convict on genetic fingerprint evidence alone if other factors raise doubt. Contamination with other evidence (secondary transfer) is a key source of incorrect DNA profiles and raising doubts as to whether a sample has been adulterated is a favorite defense technique. More rarely, chimerism is one such instance where the lack of a genetic match may unfairly exclude a suspect.

11.3.1 Evidence of genetic relationship

It is also possible to use DNA profiling as evidence of genetic relationship, although such evidence varies in strength from weak to positive. Testing that shows no relationship is absolutely certain.

While almost all individuals have a single and distinct set of genes, ultra-rare individuals, known as "chimeras", have at least two different sets of genes. There have been two cases of DNA profiling that falsely suggested that a mother was unrelated to her children.[19] This happens when two eggs are fertilized at the same time and fuse together to create one individual instead of twins.

11.4 Fake DNA evidence

In one case, a criminal planted fake DNA evidence in his own body: John Schneeberger raped one of his sedated patients in 1992 and left semen on her underwear. Police drew what they believed to be Schneeberger's blood and compared its DNA against the crime scene semen DNA on three occasions, never showing a match. It turned out that he had surgically inserted a Penrose drain into his arm and filled it with foreign blood and anticoagulants.

The functional analysis of genes and their coding sequences (open reading frames [ORFs]) typically requires that each ORF be expressed, the encoded protein purified, antibodies produced, phenotypes examined, intracellular localization determined, and interactions with other proteins sought.[20] In a study conducted by the life science company Nucleix and published in the journal Forensic Science International, scientists found that an In vitro synthesized sample of DNA matching any desired genetic profile can be constructed using standard molecular biology techniques without obtaining any actual tissue from that person. Nucleix claims they can also prove the difference between non-altered DNA and any that was synthesized.[21]

In the case of the Phantom of Heilbronn, police detectives found DNA traces from the same woman on various crime scenes in Austria, Germany, and France—among them murders, burglaries and robberies. Only after the DNA of the "woman" matched the DNA sampled from the burned body of a *male* asylum seeker in France, detectives began to have serious doubts about the DNA evidence. In that case, DNA traces were already present on the cotton swabs used to collect the samples at the crime scene, and the swabs had all been produced at the same factory in Austria. The company's product specification said that the swabs were guaranteed to be sterile, but not DNA-free.

11.5   DNA evidence as evidence in criminal trials

11.5.1   Familial DNA searching

Familial DNA searching (sometimes referred to as "Familial DNA" or "Familial DNA Database Searching") is the practice of creating new investigative leads in cases where DNA evidence found at the scene of a crime (forensic profile) strongly resembles that of an existing DNA profile (offender profile) in a state DNA database but there is not an exact match.[22][23] After all other leads have been exhausted, investigators may use specially developed software to compare the forensic profile to all profiles taken from a state's DNA database to generate a list of those offenders already in the database who are most likely to be a very close relative of the individual whose DNA is in the forensic profile.[24] To eliminate the majority of this list when the forensic DNA is a man's, crime lab technicians conduct Y-STR analysis. Using standard investigative techniques, authorities are then able to build a family tree. The family tree is populated from information gathered from public records and criminal justice records. Investigators rule out family members' involvement in the crime by finding excluding factors such as sex, living out of state or being incarcerated when the crime was committed. They may also use other leads from the case, such as witness or victim statements, to identify a suspect. Once a suspect has been identified, investigators seek to legally obtain a DNA sample from the suspect. This suspect DNA profile is then compared to the sample found at the crime scene to definitively identify the suspect as the source of the crime scene DNA.

Familial DNA database searching was first used in an investigation leading to the conviction of Craig Harman of manslaughter in the United Kingdom on April 19, 2004.[25] Craig Harman was convicted using familial DNA because of the partial matches from Harman's brother. When the police questioned Harman's brother, the police noticed Harman lived very close to the original crime scene. Harman confessed when his DNA isolated from the DNA found on the brick, matched.[26] Currently, familial DNA database searching is not conducted on a national level in the United States. States determine their own policies and decision making processes for how and when to conduct familial searches. The first familial DNA search and subsequent conviction in the United States was conducted in Denver, Colorado, in 2008 using software developed under the leadership of Denver District Attorney Mitch Morrissey and Denver Police Department Crime Lab Director Gregg LaBerge.[27] California was the first state to implement a policy for familial searching under then Attorney General, now Governor, Jerry Brown.[28] In his role as consultant to the Familial Search Working Group of the California Department of Justice, former Alameda County Prosecutor Rock Harmon is widely considered to have been the catalyst in the adoption of familial search technology in California. The technique was used to catch the Los Angeles serial killer known as the "Grim Sleeper" in 2010.[29] It wasn't a witness or informant that tipped off law enforcement to the identity of the "Grim Sleeper" serial killer, who had eluded police for more than two decades, but DNA from the suspect's own son. The suspect's son was arrested and convicted in a felony weapons charge and swabbed for DNA last year. When his DNA was entered into the database of convicted felons, detectives were alerted to a partial match to evidence found at the "Grim Sleeper" crime scenes. David Franklin Jr., also known as the Grim Sleeper, was charged with ten counts of murder and one count of attempted murder.[30] More recently, familial DNA, led to the arrest of 21-year-old Elvis Garcia on charges of sexual assault and false imprisonment of a woman in Santa Cruz in 2008.[31] In March 2011 Virginia Governor Bob McDonnell announced that Virginia would begin using familial DNA searches.[32] Other states are expected to follow.

At a press conference in Virginia on March 7, 2011, regarding the East Coast Rapist, Prince William County prosecutor

Paul Ebert and Fairfax County Police Detective John Kelly said the case would have been solved years ago if Virginia had used familial DNA searching. Aaron Thomas, the suspected East Coast Rapist, was arrested in connection with the rape of 17 women from Virginia to Rhode Island, but familial DNA was not used in the case.[33]

Critics of familial DNA database searches argue that the technique is an invasion of an individual's 4th Amendment rights.[34] Privacy advocates are petitioning for DNA database restrictions, arguing that the only fair way to search for possible DNA matches to relatives of offenders or arrestees would be to have a population-wide DNA database.[15] Some scholars have pointed out that the privacy concerns surrounding familial searching are similar in some respects to other police search techniques,[35] and most have concluded that the practice is constitutional.[36] The Ninth Circuit Court of Appeals in *United States v. Pool* (vacated as moot) suggested that this practice is somewhat analogous to a witness looking at a photograph of one person and stating that it looked like the perpetrator, which leads law enforcement to show the witness photos of similar looking individuals, one of whom is identified as the perpetrator.[37] Regardless of whether familial DNA searching was the method used to identify the suspect, authorities always conduct a normal DNA test to match the suspect's DNA with that of the DNA left at the crime scene.

Critics also claim that racial profiling could occur on account of Familial DNA testing. In the United States, the conviction rates of racial minorities are much higher than that of the overall population. It is unclear whether this is due to discrimination from police officers and the courts, as opposed to a simple higher rate of offence among minorities. Arrest-based databases, which are found in the majority of the United States, lead to an even greater level of racial discrimination. An arrest, as opposed to conviction, relies much more heavily on police discretion.[15]

For instance, investigators with Denver District Attorney's Office successfully identified a suspect in a property theft case using a familial DNA search. In this example, the suspect's blood left at the scene of the crime strongly resembled that of a current Colorado Department of Corrections prisoner.[38] Using publicly available records, the investigators created a family tree. They then eliminated all the family members who were incarcerated at the time of the offense, as well as all of the females (the crime scene DNA profile was that of a male). Investigators obtained a court order to collect the suspect's DNA, but the suspect actually volunteered to come to a police station and give a DNA sample. After providing the sample, the suspect walked free without further interrogation or detainment. Later confronted with an exact match to the forensic profile, the suspect pled guilty to criminal trespass at the first court date and was sentenced to two years probation.

In Italy a familiar DNA search has been done to solve the case of the murder of Yara Gambirasio whose body was found in the bush three months after her disappearance. A DNA trace was found on the underwear of the murdered teenage near and a DNA sample was requested from a person who lived near the municipality of Brembate di Sopra and a common male ancestor was found in the DNA sample of a young man not involved in the murder. After a long investigation the father of the supposed killer was identified in Giuseppe Guerinoni a deceased man but his two sons born from his wife were not related with the DNA samples found on the body of Yara. After 3 and a half years the DNA found on the underwear of the deceased girl was matched with Massimo Giuseppe Bosetti who was arrested and accused of the murder of the 13-year-old girl. Now Bosetti is awaiting in jail his trial.

11.5.2 Partial matches

Partial DNA matches are not searches themselves, but are the result of moderate stringency CODIS searches that produce a potential match that shares at least one allele at every locus.[39] Partial matching does not involve the use of familial search software, such as those used in the UK and United States, or additional Y-STR analysis, and therefore often misses sibling relationships. Partial matching has been used to identify suspects in several cases in the UK and United States,[40] and has also been used as a tool to exonerate the falsely accused. Darryl Hunt was wrongly convicted in connection with the rape and murder of a young woman in 1984 in North Carolina.[41] Hunt was exonerated in 2004 when a DNA database search produced a remarkably close match between a convicted felon and the forensic profile from the case. The partial match led investigators to the felon's brother, Willard E. Brown, who confessed to the crime when confronted by police. A judge then signed an order to dismiss the case against Hunt.

11.5.3   Surreptitious DNA collecting

Police forces may collect DNA samples without the suspects' knowledge, and use it as evidence. Legality of this mode of proceeding has been questioned in Australia.

In the United States, it has been accepted, courts often claiming that there was no expectation of privacy, citing *California v. Greenwood* (1985), in which the Supreme Court held that the Fourth Amendment does not prohibit the warrantless search and seizure of garbage left for collection outside the curtilage of a home. Critics of this practice underline that this analogy ignores that "most people have no idea that they risk surrendering their genetic identity to the police by, for instance, failing to destroy a used coffee cup. Moreover, even if they do realize it, there is no way to avoid abandoning one's DNA in public."[42]

In the UK, the Human Tissue Act 2004 prohibited private individuals from covertly collecting biological samples (hair, fingernails, etc.) for DNA analysis, but excluded medical and criminal investigations from the offence.[43]

The U.S. Supreme Court ruled 5–4 on June 3, 2013, in the case of *Maryland v. King*, that DNA sampling of prisoners arrested for serious crimes is constitutional.[44][45][46]

11.5.4   England and Wales

Evidence from an expert who has compared DNA samples must be accompanied by evidence as to the sources of the samples and the procedures for obtaining the DNA profiles.[47] The judge must ensure that the jury must understand the significance of DNA matches and mismatches in the profiles. The judge must also ensure that the jury does not confuse the 'match probability' (the probability that a person that is chosen at random has a matching DNA profile to the sample from the scene) with the probability that a person with matching DNA committed the crime. In 1996 *R v. Doheny*[48] Phillips LJ gave this example of a summing up, which should be carefully tailored to the particular facts in each case:

> Members of the Jury, if you accept the scientific evidence called by the Crown, this indicates that there are probably only four or five white males in the United Kingdom from whom that semen stain could have come. The Defendant is one of them. If that is the position, the decision you have to reach, on all the evidence, is whether you are sure that it was the Defendant who left that stain or whether it is possible that it was one of that other small group of men who share the same DNA characteristics.

Juries should weigh up conflicting and corroborative evidence, using their own common sense and not by using mathematical formulae, such as Bayes' theorem, so as to avoid "confusion, misunderstanding and misjudgment".[49]

Presentation and evaluation of evidence of partial or incomplete DNA profiles

In *R v Bates*,[50] Moore-Bick LJ said:

> We can see no reason why partial profile DNA evidence should not be admissible provided that the jury are made aware of its inherent limitations and are given a sufficient explanation to enable them to evaluate it. There may be cases where the match probability in relation to all the samples tested is so great that the judge would consider its probative value to be minimal and decide to exclude the evidence in the exercise of his discretion, but this gives rise to no new question of principle and can be left for decision on a case by case basis. However, the fact that there exists in the case of all partial profile evidence the possibility that a "missing" allele might exculpate the accused altogether does not provide sufficient grounds for rejecting such evidence. In many there is a possibility (at least in theory) that evidence that would assist the accused and perhaps even exculpate him altogether exists, but that does not provide grounds for excluding relevant evidence that is available and otherwise admissible, though it does make it important to ensure that the jury are given sufficient information to enable them to evaluate that evidence properly[51]

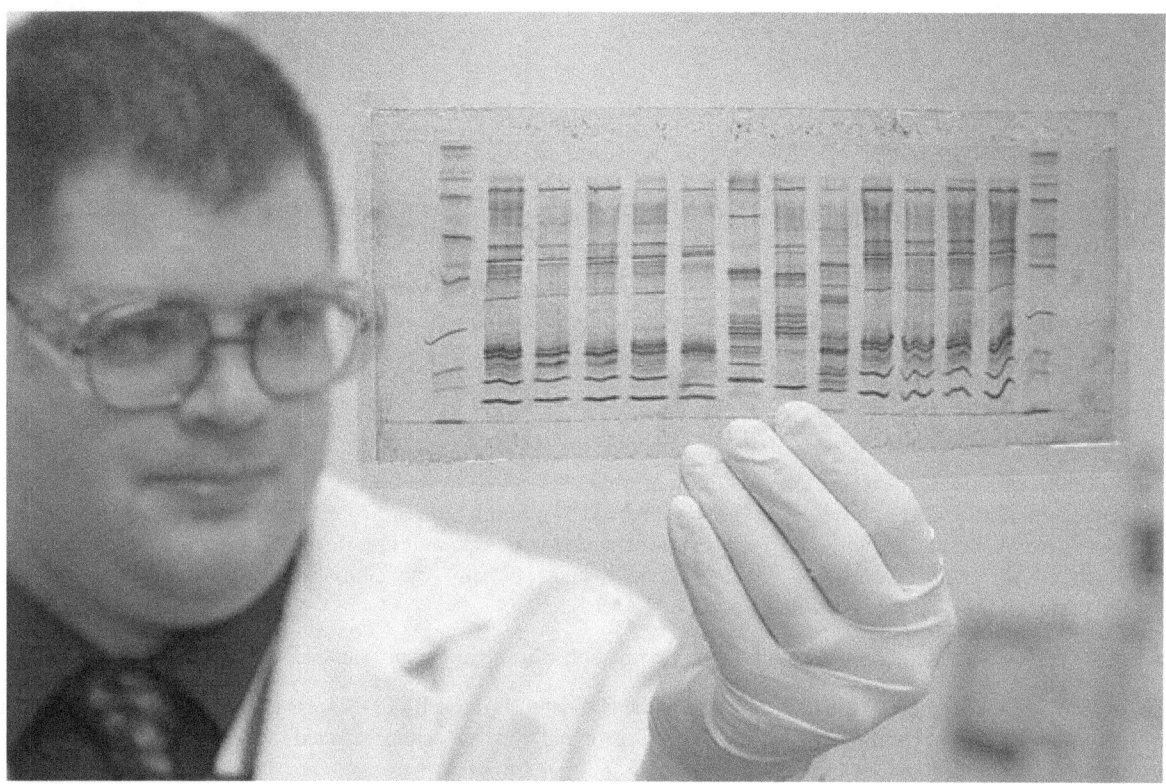

CBP chemist reads a DNA profile to determine the origin of a commodity.

11.5.5 DNA testing in the United States

There are state laws on DNA profiling in all 50 states of the United States.[52] Detailed information on database laws in each state can be found at the National Conference of State Legislatures website.[53]

11.5.6 Development of artificial DNA

In August 2009, scientists in Israel raised serious doubts concerning the use of DNA by law enforcement as the ultimate method of identification. In a paper published in the journal *Forensic Science International: Genetics*, the Israeli researchers demonstrated that it is possible to manufacture DNA in a laboratory, thus falsifying DNA evidence. The scientists fabricated saliva and blood samples, which originally contained DNA from a person other than the supposed donor of the blood and saliva.[54]

The researchers also showed that, using a DNA database, it is possible to take information from a profile and manufacture DNA to match it, and that this can be done without access to any actual DNA from the person whose DNA they are duplicating. The synthetic DNA oligos required for the procedure are common in molecular laboratories.[54]

The New York Times quoted the lead author, Daniel Frumkin, saying, "You can just engineer a crime scene...any biology undergraduate could perform this".[54] Frumkin perfected a test that can differentiate real DNA samples from fake ones. His test detects epigenetic modifications, in particular, DNA methylation. Seventy percent of the DNA in any human genome is methylated, meaning it contains methyl group modifications within a CpG dinucleotide context. Methylation at the promoter region is associated with gene silencing. The synthetic DNA lacks this epigenetic modification, which allows the test to distinguish manufactured DNA from genuine DNA.[54]

It is unknown how many police departments, if any, currently use the test. No police lab has publicly announced that it is using the new test to verify DNA results.[55]

11.6 Cases

- In 1986, Richard Buckland was exonerated, despite having admitted to the rape and murder of a teenager near Leicester, the city where DNA profiling was first developed. This was the first use of DNA fingerprinting in a criminal investigation.[56]

- In 1987, in the same case as Buckland, British baker Colin Pitchfork was the first criminal caught and convicted using DNA fingerprinting.[57]

- In 1987, genetic fingerprinting was used in criminal court for the first time in the trial of a man accused of unlawful intercourse with a mentally handicapped 14-year-old female who gave birth to a baby.[58]

- In 1987, Florida rapist Tommie Lee Andrews was the first person in the United States to be convicted as a result of DNA evidence, for raping a woman during a burglary; he was convicted on November 6, 1987, and sentenced to 22 years in prison.[59][60]

- In 1988, Timothy Wilson Spencer was the first man in Virginia to be sentenced to death through DNA testing, for several rape and murder charges. He was dubbed "The South Side Strangler" because he killed victims on the south side of Richmond, Virginia. He was later charged with rape and first-degree murder and was sentenced to death. He was executed on April 27, 1994. David Vasquez, initially convicted of one of Spencer's crimes, became the first man in America exonerated based on DNA evidence.

- In 1989, Chicago man Gary Dotson was the first person whose conviction was overturned using DNA evidence.

- In 1991, Allan Legere was the first Canadian to be convicted as a result of DNA evidence, for four murders he had committed while an escaped prisoner in 1989. During his trial, his defense argued that the relatively shallow gene pool of the region could lead to false positives.

- In 1992, DNA evidence was used to prove that Nazi doctor Josef Mengele was buried in Brazil under the name Wolfgang Gerhard.

- In 1992, DNA from a palo verde tree was used to convict Mark Alan Bogan of murder. DNA from seed pods of a tree at the crime scene was found to match that of seed pods found in Bogan's truck. This is the first instance of plant DNA admitted in a criminal case.[61][62][63]

- In 1993, Kirk Bloodsworth was the first person to have been convicted of murder and sentenced to death, whose conviction was overturned using DNA evidence.

- The 1993 rape and murder of Mia Zapata, lead singer for the Seattle punk band The Gits was unsolved nine years after the murder. A database search in 2001 failed, but the killer's DNA was collected when he was arrested in Florida for burglary and domestic abuse in 2002.

- The science was made famous in the United States in 1994 when prosecutors heavily relied on DNA evidence allegedly linking O. J. Simpson to a double murder. The case also brought to light the laboratory difficulties and handling procedure mishaps that can cause such evidence to be significantly doubted.

- In 1994, Royal Canadian Mounted Police (RCMP) detectives successfully tested hairs from a cat known as Snowball, and used the test to link a man to the murder of his wife, thus marking for the first time in forensic history the use of non-human animal DNA to identify a criminal (Plant DNA was used in 1992, see above).

- In 1994, the claim that Anna Anderson was Grand Duchess Anastasia Nikolaevna of Russia was tested after her death using samples of her tissue that had been stored at a Charlottesville, Virginia hospital following a medical procedure. The tissue was tested using DNA fingerprinting, and showed that she bore no relation to the Romanovs.[64]

- In 1994, Earl Washington, Jr., of Virginia had his death sentence commuted to life imprisonment a week before his scheduled execution date based on DNA evidence. He received a full pardon in 2000 based on more advanced testing.[65] His case is often cited by opponents of the death penalty.

- In 1995, the British Forensic Science Service carried out its first mass intelligence DNA screening in the investigation of the Naomi Smith murder case.

- In 1998, Richard J. Schmidt was convicted of attempted second-degree murder when it was shown that there was a link between the viral DNA of the human immunodeficiency virus (HIV) he had been accused of injecting in his girlfriend and viral DNA from one of his patients with AIDS. This was the first time viral DNA fingerprinting had been used as evidence in a criminal trial.

- In 1999, Raymond Easton, a disabled man from Swindon, England, was arrested and detained for seven hours in connection with a burglary. He was released due to an inaccurate DNA match. His DNA had been retained on file after an unrelated domestic incident some time previously.[66]

- In 2000 Frank Lee Smith was proved innocent by DNA profiling of the murder of an eight-year-old girl after spending 14 years on death row in Florida, USA. However he had died of cancer just before his innocence was proven.[67] In view of this the Florida state governor ordered that in future any death row inmate claiming innocence should have DNA testing.[65]

- In May 2000 Gordon Graham murdered Paul Gault at his home in Lisburn, Northern Ireland. Graham was convicted of the murder when his DNA was found on a sports bag left in the house as part of an elaborate ploy to suggest the murder occurred after a burglary had gone wrong. Graham was having an affair with the victim's wife at the time of the murder. It was the first time Low Copy Number DNA was used in Northern Ireland.[68]

- In 2001, Wayne Butler was convicted for the murder of Celia Douty. It was the first murder in Australia to be solved using DNA profiling.[69][70]

- In 2002, the body of James Hanratty, hanged in 1962 for the "A6 murder", was exhumed and DNA samples from the body and members of his family were analysed. The results convinced Court of Appeal judges that Hanratty's guilt, which had been strenuously disputed by campaigners, was proved "beyond doubt".[71] Paul Foot and some other campaigners continued to believe in Hanratty's innocence and argued that the DNA evidence could have been contaminated, noting that the small DNA samples from items of clothing, kept in a police laboratory for over 40 years "in conditions that do not satisfy modern evidential standards", had had to be subjected to very new amplification techniques in order to yield any genetic profile.[72] However, no DNA other than Hanratty's was found on the evidence tested, contrary to what would have been expected had the evidence indeed been contaminated.[73]

- In 2002, DNA testing was used to exonerate Douglas Echols, a man who was wrongfully convicted in a 1986 rape case. Echols was the 114th person to be exonerated through post-conviction DNA testing.

- In August 2002, Annalisa Vincenzi was shot dead in Tuscany. Bartender Peter Hamkin, 23, was arrested, in Merseyside, in March 2003 on an extradition warrant heard at Bow Street Magistrates' Court in London to establish whether he should be taken to Italy to face a murder charge. DNA "proved" he shot her, but he was cleared on other evidence.[74]

- In 2003, Welshman Jeffrey Gafoor was convicted of the 1988 murder of Lynette White, when crime scene evidence collected 12 years earlier was re-examined using STR techniques, resulting in a match with his nephew.[75] This may be the first known example of the DNA of an innocent yet related individual being used to identify the actual criminal, via "familial searching".

- In March 2003, Josiah Sutton was released from prison after serving four years of a twelve-year sentence for a sexual assault charge. Questionable DNA samples taken from Sutton were retested in the wake of the Houston Police Department's crime lab scandal of mishandling DNA evidence.

- In June 2003, because of new DNA evidence, Dennis Halstead, John Kogut and John Restivo won a re-trial on their murder conviction, their convictions were struck down and they were released.[76] The three men had already served eighteen years of their thirty-plus-year sentences.

- The trial of Robert Pickton (convicted in December 2003) is notable in that DNA evidence is being used primarily to identify the *victims*, and in many cases to prove their existence.

- In 2004, DNA testing shed new light into the mysterious 1912 disappearance of Bobby Dunbar, a four-year-old boy who vanished during a fishing trip. He was allegedly found alive eight months later in the custody of William Cantwell Walters, but another woman claimed that the boy was her son, Bruce Anderson, whom she had entrusted in Walters' custody. The courts disbelieved her claim and convicted Walters for the kidnapping. The boy was raised and known as Bobby Dunbar throughout the rest of his life. However, DNA tests on Dunbar's son and nephew revealed the two were not related, thus establishing that the boy found in 1912 was not Bobby Dunbar, whose real fate remains unknown.[77]

- In 2005, Gary Leiterman was convicted of the 1969 murder of Jane Mixer, a law student at the University of Michigan, after DNA found on Mixer's pantyhose was matched to Leiterman. DNA in a drop of blood on Mixer's hand was matched to John Ruelas, who was only four years old in 1969 and was never successfully connected to the case in any other way. Leiterman's defense unsuccessfully argued that the unexplained match of the blood spot to Ruelas pointed to cross-contamination and raised doubts about the reliability of the lab's identification of Leiterman.[78][79][80]

- In December 2005, Evan Simmons was proven innocent of a 1981 attack on an Atlanta woman after serving twenty-four years in prison. Mr. Clark is the 164th person in the United States and the fifth in Georgia to be freed using post-conviction DNA testing.

- In March 2009, Sean Hodgson—convicted of 1979 killing of Teresa De Simone, 22, in her car in Southampton—was released after tests proved DNA from the scene was not his. It was later matched to DNA retrieved from the exhumed body of David Lace. Lace had previously confessed to the crime but was not believed by the detectives. He served time in prison for other crimes committed at the same time as the murder and then committed suicide in 1988.[81]

- In November 2008, Anthony Curcio was arrested for masterminding one of the most elaborately planned armored car heists in history. DNA evidence linked Curcio to the crime.[82]

11.7   See also

- DNA barcoding
- DNA database
- National DNA database
- DNA paternity testing
- Capillary electrophoresis (CE)
- Forensic identification
- Full genome sequencing
- Gene mapping
- Genealogical DNA test
- *Harvey v. Horan*
- Identification (biology)
- Kinship analysis
- *Maryland v. King*
- Phantom of Heilbronn
- Project Innocence
- Restriction fragment length polymorphism (RFLP)
- Ribotyping
- Short tandem repeat (STR)
- International Society for Forensic Genetics

11.8 References

[1] *Kijk* magazine, 1 January 2009

[2] "The Guardian Interview With Sir Alec Jeffreys" http://www.theguardian.com/science/2009/may/24/dna-fingerprinting-alec-jeffreys

[3] DNA pioneer's 'eureka' moment BBC. Retrieved 14 October 2011

[4] "Use of DNA in Identification". Accessexcellence.org. Retrieved 2010-04-03.

[5] Tautz, D. (1989). Hypervariability of simple sequences as a general source for polymorphic DNA markers. Nucleic Acids Research, 17, 6463-6471.

[6] Patent Jäckle H & Tautz D (1989) "Process For Analyzing Length Polymorphisms in DNA Regions" europäische Patent Nr. 0 438 512

[7] Tautz, D. (1989). Hypervariability of simple sequences as a general source for polymorphic DNA markers. Nucleic Acids Research, 17, 6463-6471.

[8] Felch, Jason; et al. (July 20, 2008). "FBI resists scrutiny of 'matches'". *Los Angeles Times*. pp. P8.

[9] "STR Analysis"

[10] Miller, Kevin. "Mitochondrial DNA Concordance".

[11] Miller, K.W.P.; Dawson, J.L.; Hagelberg, E. (1996). "A concordance of nucleotide substitutions in the first and second hypervariable segments of the human mtDNA control region". *International Journal of Legal Medicine* (109): 107–113.

[12] "CODIS — National DNA Index System". Fbi.gov. Archived from the original on March 6, 2010. Retrieved 2010-04-03.

[13] "Restrictions on use and destruction of fingerprints and samples". Wikicrimeline.co.uk. 2009-09-01. Retrieved 2010-04-03.

[14] Rose & Goos. *DNA: A Practical Guide*. Toronto: Carswell Publications.

[15] "Double Helix Jeopardy"

[16] "Congress OKs bill to cut rape evidence backlog". Associated Press. Retrieved 18 September 2014.

[17] Nick Paton Walsh False result fear over DNA tests The Observer, Sunday 27 January 2002.

[18] The Evaluation of Forensic DNA Evidence 1996.

[19] "Two Women Don't Match Their Kids' DNA". Abcnews.go.com. 2006-08-15. Retrieved 2010-04-03.

[20] James L. Hartley, Gary F. Temple, and Michael A. Brasch (2000). "DNA Cloning Using In Vitro Site-Specific Recombination". *Cold Spring Harbor Laboratory Press.*

[21] A new test distinguishes between real and fake genetic evidence-Published by MIT Technology Review 2009-08-17, article by Emily Singer; Retrieved 2014-09-03

[22] Diamond, Diane (April 12, 2011). "Searching the Family DNA Tree to Solve Crime". *HuffPost Denver* (Blog). The Huffington Post. Retrieved April 17, 2011.

[23] Bieber Frederick; et al. (2006). "Finding Criminals Through DNA of Their Relatives". *Science* **312** (5778): 1315–16. doi:10.1126/science.1122655.

[24] Staff. "Familial searches allows law enforcement to identify criminals through their family members". *DNA Forensics*. United Kingdom – A Pioneer in Familial Searches. Retrieved December 7, 2015.

[25] Bhattacharya, Shaoni (April 20, 2004). "Killer convicted thanks to relative's DNA". Daily News. *New Scientist*. Retrieved April 17, 2011.

[26] Greely, Henry T.; Riordan, Daniel P.; Garrison, Nanibaa' A.; Mountain, Joanna L. (Summer 2006). "Family Ties: The Use of DNA Offender Databases to Catch Offenders' Kin" (PDF). Symposium. *Journal of Law, Medicine & Ethics* (Wiley for American Society of Law, Medicine & Ethics): 248–62. ISSN 1748-720X.

[27] Pankratz, Howard. "Denver Uses 'Familial DNA Evidence' to Solve Car Break-Ins." The Denver Post accessed April 17, 2011.

[28] Steinhaur, Jennifer. "'Grim Sleeper' Arrest Fans Debate on DNA Use." The New York Times accessed April 17, 2011.

[29] Dolan, Maura. "A New Track in DNA Search." LA Times accessed April 17, 2011.

[30] New DNA Technique Led Police to 'Grim Sleeper' Serial Killer and Will 'Change Policing in America' - ABC News

[31] Dolan, Maura. "Familial DNA Search Used In Grim Sleeper Case Leads to Arrest of Santa Cruz Sex Offender." LA Times accessed April 17, 2011.

[32] Helderman, Rosalind. "McDonnell Approves Familial DNA for VA Crime Fighting." The Washington Post accessed April 17, 2011.

[33] Christoffersen, John and Barakat, Matthew. "Other victims of East Coast Rapist suspect sought." Associated Press. Accessed May 25, 2011.

[34] Murphy, Erin Elizabeth, (2009). "Relative Doubt: Familial Searches of DNA Databases" Michigan Law Review, Vol. 109, 291-348, 2010.

[35] Suter, Sonia. " All in The Family: Privacy and DNA Familial Searching" (2010). Harvard Journal of Law and Technology, Vol. 23,328, 2010.

[36] Kaye, David H., (2013). "The Genealogy Detectives: A Constitutional Analysis of Familial Searching" American Criminal Law Review, Vol. 51, No. 1, 109-163, 2013.

[37] "US v. Pool" Pool 621F .3d 1213.

[38] Pankratz, Howard."Denver Uses 'Familial DNA Evidence' to Solve Car Break-Ins." The Denver Post accessed April 17, 2011.

[39] "Finding Criminals Through DNA Testing of Their Relatives" Technical Bulletin, Chromosomal Laboratories, Inc. accessed April 22, 2011.

[40] "Denver District Attorney DNA Resources" accessed April 20, 2011.

[41] "Darryl Hunt, The Innocence Project".

[42] Amy Harmon, "Lawyers Fight DNA Samples Gained on Sly", *The New York Times*, April 3, 2008.

[43] Human Tissue Act 2004, UK, available in PDF.

[44] "U.S. Supreme Court allows DNA sampling of prisoners". UPI. Retrieved 3 June 2013.

[45] http://www.supremecourt.gov/opinions/12pdf/12-207_d18e.pdf

[46] Samuels, J.E., E.H. Davies, and D.B. Pope. (2013). Collecting DNA at Arrest: Policies, Practices, and Implications, Final Technical Report. Washington, D.C.: Urban Institute, Justice Policy Center.

[47] *R v. Loveridge*, EWCA Crim 734 (2001).

[48] *R v. Doheny* [1996] EWCA Crim 728, [1997] 1 Cr App R 369 (31 July 1996), Court of Appeal

[49] *R v. Adams* [1997] EWCA Crim 2474 (16 October 1997), Court of Appeal

[50] *R v Bates* [2006] EWCA Crim 1395 (7 July 2006), Court of Appeal

[51] "WikiCrimeLine DNA profiling". Wikicrimeline.co.uk. Retrieved 2010-04-03.

[52] "Genelex: The DNA Paternity Testing Site". Healthanddna.com. 1996-01-06. Retrieved 2010-04-03.

[53] Donna Lyons — Posted by Glenda. "State Laws on DNA Data Banks". Ncsl.org. Retrieved 2010-04-03.

[54] Pollack, Andrew (August 18, 2009). "DNA Evidence Can Be Fabricated, Scientists Show". *The New York Times*. Retrieved April 1, 2010.

[55] "Elsevier". Fsigenetics.com. Retrieved 2010-04-03.

[56] "DNA pioneer's 'eureka' moment". *BBC News*. September 9, 2009. Retrieved April 1, 2010.

[57] Joseph Wambaugh, *The Blooding* (New York, New York: A Perigord Press Book, 1989), 369.

[58] Joseph Wambaugh, The Blooding (New York, New York: A Perigord Press Book, 1989), 316.

[59] "Gene Technology — Page 14". Txtwriter.com. 1987-11-06. Retrieved 2010-04-03.

[60] "frontline: the case for innocence: the dna revolution: state and federal dna database laws examined". Pbs.org. Retrieved 2010-04-03.

[61] "Court of Appeals of Arizona: Denial of Bogan's motion to reverse his conviction and sentence" (PDF). Denver DA: www. denverda.org. 2005-04-11. Retrieved 2011-04-21.

[62] "DNA Forensics: Angiosperm Witness for the Prosecution". Human Genome Project. Retrieved 2011-04-21.

[63] "Crime Scene Botanicals". Botanical Society of America. Retrieved 2011-04-21.

[64] Identification of the remains of the Romanov family by DNA analysis by Peter Gill, Central Research and Support Establishment, Forensic Science Service, Aldermaston, Reading, Berkshire, RG7 4PN, UK, Pavel L. Ivanov, Engelhardt Institute of Molecular Biology, Russian Academy of Sciences, 117984, Moscow, Russia, Colin Kimpton, Romelle Piercy, Nicola Benson, Gillian Tully, Ian Evett, Kevin Sullivan, Forensic Science Service, Priory House, Gooch Street North, Birmingham B5 6QQ, UK, Erika Hagelberg, University of Cambridge, Department of Biological Anthropology, Downing Street, Cambridge CB2 3DZ, UK -

[65] Murnaghan, Ian, (28 December 2012) Famous Trials and DNA Testing; Earl Washington Jr. Explore DNA, Retrieved 13 November 2014

[66] Jeffries, Stuart (2006-10-08). "Suspect Nation". London: The Guardian. Retrieved April 1, 2010.

[67] (June 2012) Frank Lee Smith The University of Michigan Law School, National Registry of Exonerations, Retrieved 13 November 2014

[68] Gordon, Stephen (2008-02-17). "Freedom in bag for killer Graham?". Belfasttelegraph.co.uk. Retrieved 2010-06-19.

[69] Dutter, Barbie (2001-06-19). "18 years on, man is jailed for murder of Briton in 'paradise'". London: The Telegraph. Retrieved 2008-06-17.

[70] McCutcheon, Peter (2004-09-08). "DNA evidence may not be infallible: experts". Australian Broadcasting Corporation. Retrieved 2008-06-17.

[71] Joshua Rozenberg,"DNA proves Hanratty guilt 'beyond doubt'", *Daily Telegraph*, London, 11 May 2002.

[72] John Steele, "Hanratty lawyers reject DNA 'guilt'", *Daily Telegraph*, London, 23 June 2001.

[73] "Hanratty: The damning DNA". *BBC News*. 10 May 2002. Retrieved 2011-08-22.

[74] "Mistaken identity claim over murder". *BBC News*. February 15, 2003. Retrieved April 1, 2010.

[75] Satish Sekar. "Lynette White Case: How Forensics Caught the Cellophane Man". Lifeloom.com. Retrieved 2010-04-03.

[76] (18 April 2014) Dennis Halstead The National Registry of Exonerations, University of Michigan Law School, Retrieved 12 January 2015

[77] "DNA clears man of 1914 kidnapping conviction", *USA Today*, (May 5, 2004), by Allen G. Breed, Associated Press.

[78] CBS News story on the Jane Mixer murder case; March 24, 2007.

[79] Another CBS News story on the Mixer case; July 17, 2007.

[80] An advocacy site challenging Leiterman's conviction in the Mixer murder.

[81] Booth, Jenny. "Police name David Lace as true killer of Teresa De Simone". *The Times.*

[82] Doughery, Phil. "D.B. Tuber". History Link.

11.9   Further reading

- Kaye, David H. (2010). *The Double Helix and the Law of Evidence*. Cambridge, Mass.: Harvard University Press. ISBN 9780674035881. OCLC 318876881.

- Koerner, Brendan I. (October 2015). "Family Ties: Your Relatives' DNA Could Turn You Into a Suspect". Argument. *Wired* (paper) (Condé Nast): 35–8. ISSN 1059-1028.

11.10   External links

- McKie, Robin McKie (24 May 2009). "Eureka moment that led to the discovery of DNA fingerprinting". *The Observer* (London).

- Forensic Science, Statistics, and the Law—Blog that tracks scientific and legal developments pertinent to forensic DNA profiling

- Create a DNA Fingerprint—PBS.org

- In silico simulation of Molecular Biology Techniques—A place to learn typing techniques by simulating them

- National DNA Databases in the EU

- The Innocence Record, Winston & Strawn LLP/The Innocence Project

- Making Sense of DNA Backlogs, 2012: Myths vs. Reality United States Department of Justice

- France Tries Mass DNA Test in Hunt for School Rapist

Chapter 12

Repeated sequence (DNA)

Repeated sequences (aka. repetitive elements, or repeats) are patterns of nucleic acids (DNA or RNA) that occur in multiple copies throughout the genome. The functions and descriptions of these sequences are currently being characterized by scientists. Repetitive DNA was first detected because of its rapid reassociation kinetics.

12.1 Types

12.1.1 Main types

There are 3 major categories of **repeated sequence** or **repeats**:

- Terminal repeats

- Tandem repeats: copies which lie adjacent to each other, either directly or inverted

    - Satellite DNA - typically found in centromeres and heterochromatin

    - Minisatellite - repeat units from about 10 to 60 base pairs, found in many places in the genome, including the centromeres

    - Microsatellite - repeat units of less than 10 base pairs; this includes telomeres, which typically have 6 to 8 base pair repeat units

- Interspersed repeats (aka. interspersed nuclear elements)

    - Transposable elements
        - DNA transposons
        - retrotransposons
            - LTR-retrotranposons (HERVs)
            - non LTR-retrotranposons
                - SINEs (**S**hort **I**nterspersed **N**uclear **E**lements)
                - LINEs (**L**ong **I**nterspersed **N**uclear **E**lements)
                - SVAs

In primates, the majority of LINEs are LINE-1 and the majority of SINEs are Alu's. SVAs are hominoid specific.

In prokaryotes, CRISPR are arrays of alternating repeats and spacers.

12.1.2   Other types

Note: The following are covered in detail in "Computing for Comparative Microbial Genomics".[1]

- Direct repeats

  - Global direct repeat
  - Local direct simple repeats
  - Local direct repeats
  - Local direct repeats with spacer

- Inverted repeats

  - Global inverted repeat
  - Local inverted repeat
  - Inverted repeat with spacer
  - Palindromic repeat

- Mirror and everted repeats

12.2   See also

- Eukaryotic chromosome fine structure

- Non-coding DNA

- Intergenic DNA

12.3   References

[1] Ussery, David W.; Wassenaar, Trudy; Borini, Stefano (2008-12-22). "Word Frequencies, Repeats, and Repeat-related Structures in Bacterial Genomes". *Computing for Comparative Microbial Genomics: Bioinformatics for Microbiologists*. Computational Biology **8** (1 ed.). Springer. pp. 133–144. ISBN 978-1-84800-254-8.

12.4   External links

- Function of Repetitive DNA

- DNA Repetitious Region at the US National Library of Medicine Medical Subject Headings (MeSH)

Chapter 13

Nucleic acid sequence

A **nucleic acid sequence** is a succession of letters that indicate the order of nucleotides within a DNA (using GACT) or RNA (GACU) molecule. By convention, sequences are usually presented from the 5' end to the 3' end. For DNA, the sense strand is used. Because nucleic acids are normally linear (unbranched) polymers, specifying the sequence is equivalent to defining the covalent structure of the entire molecule. For this reason, the nucleic acid sequence is also termed the primary structure.

The sequence has capacity to represent information. Biological deoxyribonucleic acid represents the information which directs the functions of a living thing. In that context, the term **genetic sequence** is often used. Sequences can be read from the biological raw material through DNA sequencing methods.

Nucleic acids also have a secondary structure and tertiary structure. Primary structure is sometimes mistakenly referred to as *primary sequence*. Conversely, there is no parallel concept of secondary or tertiary sequence.

13.1   Nucleotides

Main article: Nucleotide

Nucleic acids consist of a chain of linked units called nucleotides. Each nucleotide consists of three subunits: a phosphate group and a sugar (ribose in the case of RNA, deoxyribose in DNA) make up the backbone of the nucleic acid strand, and attached to the sugar is one of a set of nucleobases. The nucleobases are important in base pairing of strands to form higher-level secondary and tertiary structure such as the famed double helix.

The possible letters are *A*, *C*, *G*, and *T*, representing the four nucleotide bases of a DNA strand — adenine, cytosine, guanine, thymine — covalently linked to a phosphodiester backbone. In the typical case, the sequences are printed abutting one another without gaps, as in the sequence AAAGTCTGAC, read left to right in the 5' to 3' direction. With regards to transcription, a sequence is on the coding strand if it has the same order as the transcribed RNA.

One sequence can be complementary to another sequence, meaning that they have the base on each position is the complementary (i.e. A to T, C to G) and in the reverse order. For example, the complementary sequence to TTAC is GTAA. If one strand of the double-stranded DNA is considered the sense strand, then the other strand, considered the antisense strand, will have the complementary sequence to the sense strand.

13.1.1   Notation

Main article: Nucleic acid notation

While A, T, C, and G represent a particular nucleotide at a position, there are also letters that represent ambiguity which are used when more than one kind of nucleotide could occur at that position. The rules of the International Union of Pure and Applied Chemistry (IUPAC) are as follows:[1]

- **A** = adenine

- **C** = cytosine

- **G** = guanine

- **T** = thymine

- **R** = G A (purine)

- **Y** = T C (pyrimidine)

- **K** = G T (keto)

- **M** = A C (amino)

- **S** = G C (strong bonds)

- **W** = A T (weak bonds)

- **B** = G T C (all but A)

- **D** = G A T (all but C)

- **H** = A C T (all but G)

- **V** = G C A (all but T)

- **N** = A G C T (any)

These symbols are also valid for RNA, except with U (uracil) replacing T (thymine).[1]

Apart from adenine (A), cytosine (C), guanine (G), thymine (T) and uracil (U), DNA and RNA also contain bases that have been modified after the nucleic acid chain has been formed. In DNA, the most common modified base is 5-methylcytidine (m5C). In RNA, there are many modified bases, including pseudouridine (Ψ), dihydrouridine (D), inosine (I), ribothymidine (rT) and 7-methylguanosine (m7G).[2][3] Hypoxanthine and xanthine are two of the many bases created through mutagen presence, both of them through deamination (replacement of the amine-group with a carbonyl-group). Hypoxanthine is produced from adenine, xanthine from guanine.[4] Similarly, deamination of cytosine results in uracil.

13.2   Biological significance

Further information: Genetic code and Central dogma of molecular biology

In biological systems, nucleic acids contain information which is used by a living cell to construct specific proteins. The sequence of nucleobases on a nucleic acid strand is translated by cell machinery into a sequence of amino acids making up a protein strand. Each group of three bases, called a codon, corresponds to a single amino acid, and there is a specific genetic code by which each possible combination of three bases corresponds to a specific amino acid.

The central dogma of molecular biology outlines the mechanism by which proteins are constructed using information contained in nucleic acids. DNA is transcribed into mRNA molecules, which travels to the ribosome where the mRNA is used as a template for the construction of the protein strand. Since nucleic acids can bind to molecules with complementary sequences, there is a distinction between "sense" sequences which code for proteins, and the complementary "antisense" sequence which is by itself nonfunctional, but can bind to the sense strand.

13.3 Sequence determination

Main article: DNA sequencing

DNA sequencing is the process of determining the nucleotide sequence of a given DNA fragment. The sequence of the DNA of a living thing encodes the necessary information for that living thing to survive and reproduce. Therefore, determining the sequence is useful in fundamental research into why and how organisms live, as well as in applied subjects. Because of the importance of DNA to living things, knowledge of a DNA sequence may be useful in practically any biological research. For example, in medicine it can be used to identify, diagnose and potentially develop treatments for genetic diseases. Similarly, research into pathogens may lead to treatments for contagious diseases. Biotechnology is a burgeoning discipline, with the potential for many useful products and services.

RNA is not sequenced directly. Instead, it is copied to a DNA by reverse transcriptase, and this DNA is then sequenced.

Current sequencing methods rely on the discriminatory ability of DNA polymerases, and therefore can only distinguish four bases. An inosine (created from adenosine during RNA editing) is read as a G, and 5-methyl-cytosine (created from cytosine by DNA methylation) is read as a C. With current technology, it is difficult to sequence small amounts of DNA, as the signal is too weak to measure. This is overcome by polymerase chain reaction (PCR) amplification.

13.3.1 Digital representation

Once a nucleic acid sequence has been obtained from an organism, it is stored *in silico* in digital format. Digital genetic sequences may be stored in sequence databases, be analyzed (see *Sequence analysis* below), be digitally altered and be used as templates for creating new actual DNA using artificial gene synthesis.

13.4 Sequence analysis

Main article: Sequence analysis

Digital genetic sequences may be analyzed using the tools of bioinformatics to attempt to determine its function.

13.4.1 Genetic testing

Main article: Genetic testing

The DNA in an organism's genome can be analyzed to diagnose vulnerabilities to inherited diseases, and can also be used to determine a child's paternity (genetic father) or a person's ancestry. Normally, every person carries two variations of every gene, one inherited from their mother, the other inherited from their father. The human genome is believed to contain around 20,000 - 25,000 genes. In addition to studying chromosomes to the level of individual genes, genetic testing in a broader sense includes biochemical tests for the possible presence of genetic diseases, or mutant forms of genes associated with increased risk of developing genetic disorders.

Genetic testing identifies changes in chromosomes, genes, or proteins.[5] Usually, testing is used to find changes that are associated with inherited disorders. The results of a genetic test can confirm or rule out a suspected genetic condition or help determine a person's chance of developing or passing on a genetic disorder. Several hundred genetic tests are currently in use, and more are being developed.[6][7]

13.4.2 Sequence alignment

Main article: Sequence alignment

In bioinformatics, a **sequence alignment** is a way of arranging the sequences of DNA, RNA, or protein to identify regions of similarity that may be due to functional, structural, or evolutionary relationships between the sequences.[8] If two sequences in an alignment share a common ancestor, mismatches can be interpreted as point mutations and gaps as insertion or deletion mutations (indels) introduced in one or both lineages in the time since they diverged from one another. In sequence alignments of proteins, the degree of similarity between amino acids occupying a particular position in the sequence can be interpreted as a rough measure of how conserved a particular region or sequence motif is among lineages. The absence of substitutions, or the presence of only very conservative substitutions (that is, the substitution of amino acids whose side chains have similar biochemical properties) in a particular region of the sequence, suggest[9] that this region has structural or functional importance. Although DNA and RNA nucleotide bases are more similar to each other than are amino acids, the conservation of base pairs can indicate a similar functional or structural role.

Computational phylogenetics makes extensive use of sequence alignments in the construction and interpretation of phylogenetic trees, which are used to classify the evolutionary relationships between homologous genes represented in the genomes of divergent species. The degree to which sequences in a query set differ is qualitatively related to the sequences' evolutionary distance from one another. Roughly speaking, high sequence identity suggests that the sequences in question have a comparatively young most recent common ancestor, while low identity suggests that the divergence is more ancient. This approximation, which reflects the "molecular clock" hypothesis that a roughly constant rate of evolutionary change can be used to extrapolate the elapsed time since two genes first diverged (that is, the coalescence time), assumes that the effects of mutation and selection are constant across sequence lineages. Therefore, it does not account for possible difference among organisms or species in the rates of DNA repair or the possible functional conservation of specific regions in a sequence. (In the case of nucleotide sequences, the molecular clock hypothesis in its most basic form also discounts the difference in acceptance rates between silent mutations that do not alter the meaning of a given codon and other mutations that result in a different amino acid being incorporated into the protein.) More statistically accurate methods allow the evolutionary rate on each branch of the phylogenetic tree to vary, thus producing better estimates of coalescence times for genes.

13.4.3   Sequence motifs

Main article: Sequence motif

Frequently the primary structure encodes motifs that are of functional importance. Some examples of sequence motifs are: the C/D[10] and H/ACA boxes[11] of snoRNAs, Sm binding site found in spliceosomal RNAs such as U1, U2, U4, U5, U6, U12 and U3, the Shine-Dalgarno sequence,[12] the Kozak consensus sequence[13] and the RNA polymerase III terminator.[14]

13.4.4   Sequence entropy

Main article: Sequence entropy

In Bioinformatics, a **sequence entropy**, also known as sequence complexity or information profile,[15] is a numerical sequence providing a quantitative measure of the local complexity of a DNA sequence, independently of the direction of processing. The manipulations of the information profiles enable the analysis of the sequences using alignment-free techniques, such as for example in motif and rearrangements detection.[15][16] [17]

13.5   See also

- Single-nucleotide polymorphism (SNP)

- Quaternary numeral system

13.6 References

[1] Nomenclature for Incompletely Specified Bases in Nucleic Acid Sequences, NC-IUB, 1984.

[2] "BIOL2060: Translation". *mun.ca.*

[3] "Research". *uw.edu.pl.*

[4] T Nguyen, D Brunson, C L Crespi, B W Penman, J S Wishnok, and S R Tannenbaum, DNA damage and mutation in human cells exposed to nitric oxide in vitro, Proc Natl Acad Sci U S A. 1992 April 1; 89(7): 3030–3034

[5] "What is genetic testing?". *Genetics Home Reference.* 16 March 2015.

[6] "Genetic Testing". *nih.gov.*

[7] "Definitions of Genetic Testing". *Definitions of Genetic Testing (Jorge Sequeiros and Bárbara Guimarães).* EuroGentest Network of Excellence Project. 2008-09-11. Archived from the original on February 4, 2009. Retrieved 2008-08-10.

[8] Mount DM. (2004). *Bioinformatics: Sequence and Genome Analysis* (2nd ed.). Cold Spring Harbor Laboratory Press: Cold Spring Harbor, NY. ISBN 0-87969-608-7.

[9] Ng, P. C.; Henikoff, S. (2001). "Predicting Deleterious Amino Acid Substitutions". *Genome Research* **11** (5): 863–874. doi:10.1101/gr.176601. PMC 311071. PMID 11337480.

[10] Samarsky, DA; Fournier MJ; Singer RH; Bertrand E (1998). "The snoRNA box C/D motif directs nucleolar targeting and also couples snoRNA synthesis and localization". *EMBO* **17** (13): 3747–3757. doi:10.1093/emboj/17.13.3747. PMC 1170710. PMID 9649444.

[11] Ganot, Philippe; Caizergues-Ferrer, Michèle; Kiss, Tamás (1 April 1997). "The family of box ACA small nucleolar RNAs is defined by an evolutionarily conserved secondary structure and ubiquitous sequence elements essential for RNA accumulation". *Genes & Development* **11** (7): 941–956. doi:10.1101/gad.11.7.941. PMID 9106664.

[12] Shine J, Dalgarno L (1975). "Determinant of cistron specificity in bacterial ribosomes". *Nature* **254** (5495): 34–8. doi:10.1038/254034a0. PMID 803646.

[13] Kozak M (October 1987). "An analysis of 5'-noncoding sequences from 699 vertebrate messenger RNAs". *Nucleic Acids Res.* **15** (20): 8125–8148. doi:10.1093/nar/15.20.8125. PMC 306349. PMID 3313277.

[14] Bogenhagen DF, Brown DD (1981). "Nucleotide sequences in Xenopus 5S DNA required for transcription termination.". *Cell* **24** (1): 261–70. doi:10.1016/0092-8674(81)90522-5. PMID 6263489.

[15] Pinho, A; Garcia, S; Pratas, D; Ferreira, P (Nov 21, 2013). "DNA Sequences at a Glance.". *PLOS ONE* **8** (11): e79922. doi:10.1371/journal.pone.0079922. PMID 24278218.

[16] Pratas, D; Silva, R; Pinho, A; Ferreira, P (May 18, 2015). "An alignment-free method to find and visualise rearrangements between pairs of DNA sequences.". *Scientific Reports (Group Nature)* **5** (10203). doi:10.1038/srep10203. PMID 25984837.

[17] Troyanskaya, O; Arbell, O; Koren, Y; Landau, G; Bolshoy, A (2002). "Sequence complexity profiles of prokaryotic genomic sequences: A fast algorithm for calculating linguistic complexity.". *Bioinformatics* **18** (5). PMID 12050064.

13.7 External links

- A bibliography on features, patterns, correlations in DNA and protein texts

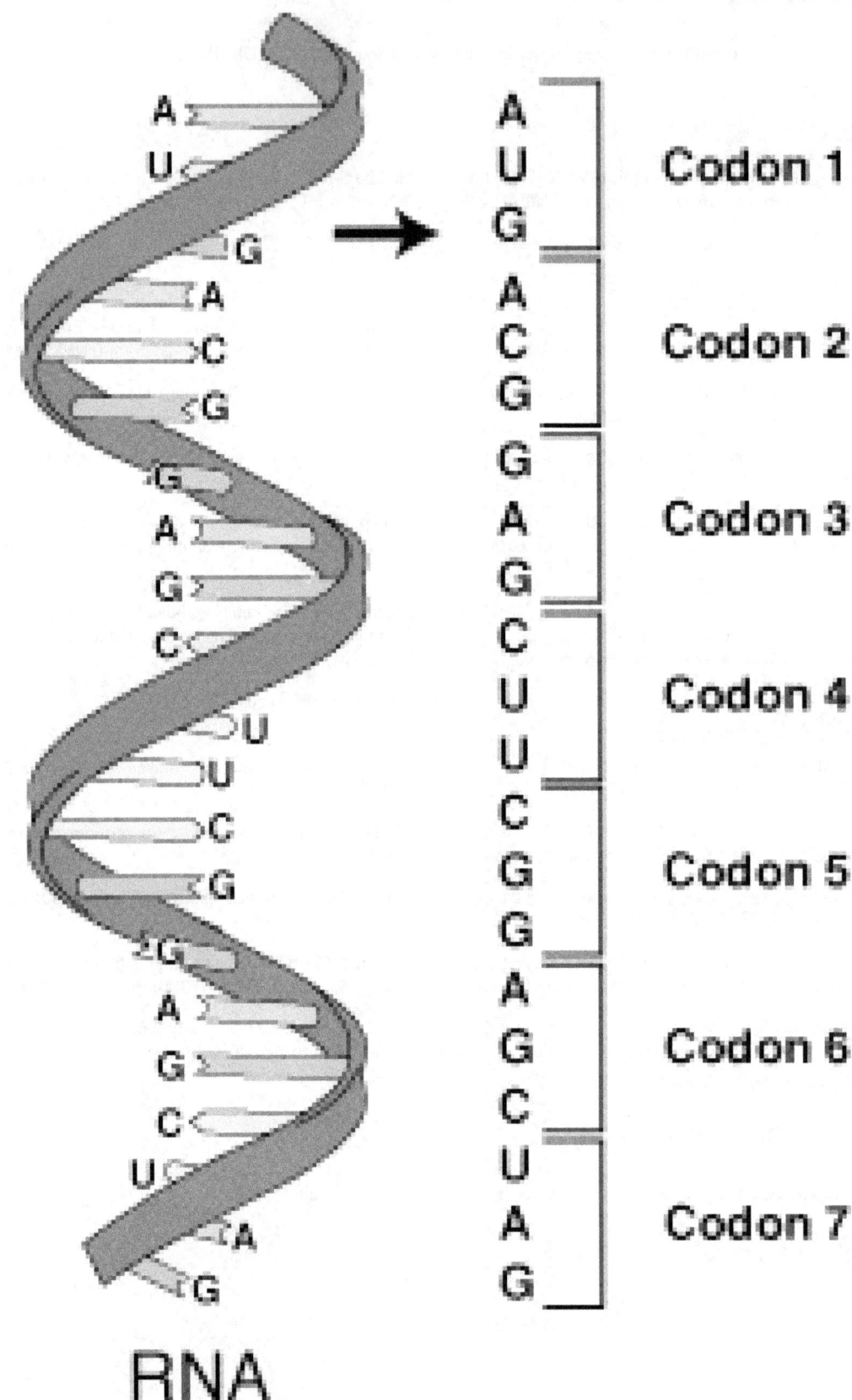

RNA

Ribonucleic acid

Chemical structure of RNA

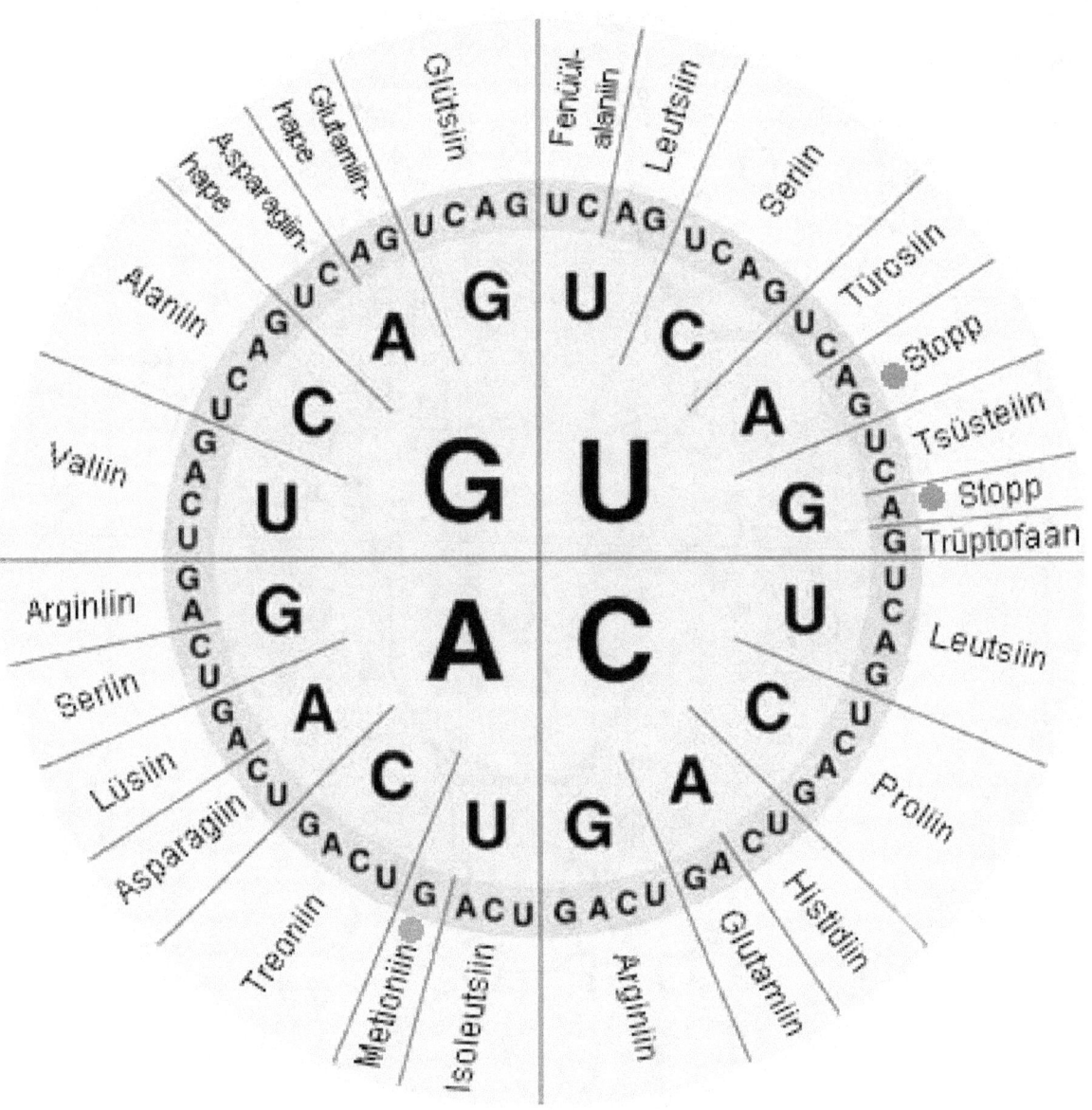

A depiction of the genetic code, by which the information contained in nucleic acids are translated into amino acid sequences in proteins.

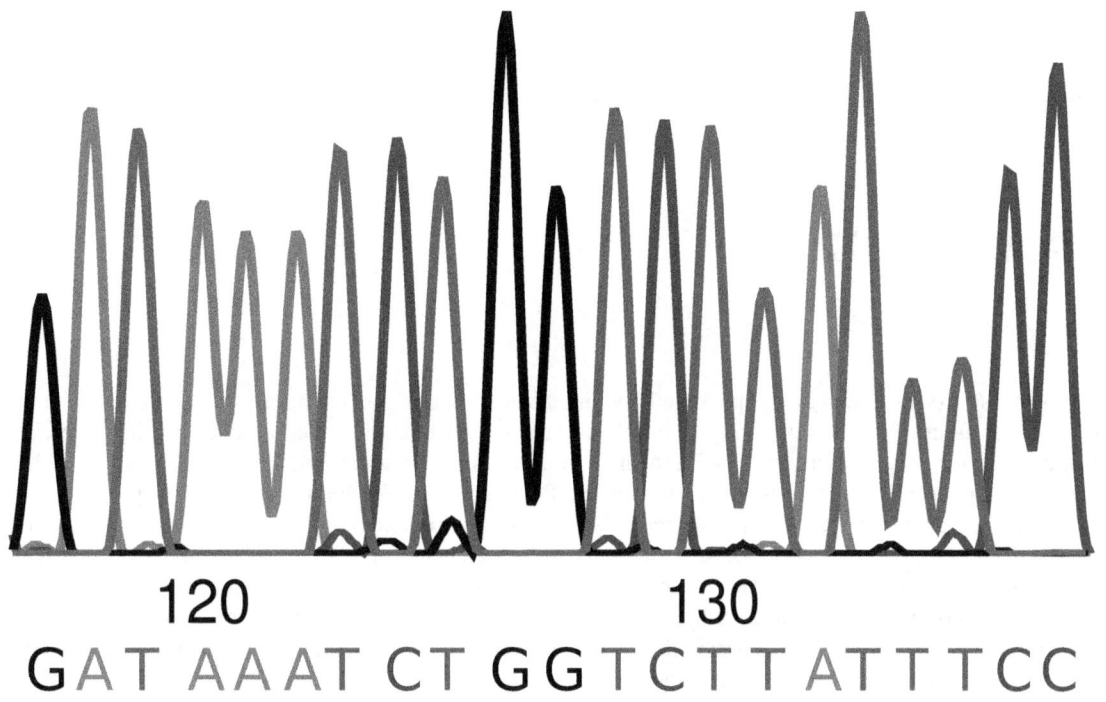

Electropherogram printout from automated sequencer for determining part of a DNA sequence

Genetic sequence in digital format.

Chapter 14

Chromosome

A **chromosome** (*chromo-* + *-some*) is a packaged and organized structure containing most of the DNA of a living organism. It is not usually found on its own, but rather is structured by being wrapped around protein complexes called nucleosomes, which consist of histones. The DNA in chromosomes is also associated with transcription (copying of genetic sequences) factors and several other macromolecules. During most of the duration of the Cell cycle, a chromosome consists of one long double-stranded DNA molecule (with associated proteins). During S phase, the chromosome gets replicated, resulting in an 'X'-shaped structure called a metaphase chromosome. Both the original and the newly copied DNA are now called chromatids. The two "sister" chromatids join together at a protein junction called a centromere. Chromosomes are normally visible under a light microscope only when the cell is undergoing mitosis. Even then, the full chromosome containing both joined sister chromatids becomes visible only during a sequence of mitosis known as metaphase (when chromosomes align together, attached to the mitotic spindle and prepare to divide).[1] This DNA and its associated proteins and macromolecules is collectively known as chromatin, which is further packaged along with its associated molecules into a discrete structure called a nucleosome. Chromatin is present in most cells, with a few exceptions - erythrocytes for example. Occurring only in the nucleus of eukaryotic cells, chromatin composes the vast majority of all DNA, except for a small amount inherited maternally which is found in mitochondria. In prokaryotic cells, chromatin occurs free-floating in cytoplasm, as these cells lack organelles and a defined nucleus. Bacteria also lack histones. The main information-carrying macromolecule is a single piece of coiled double-stranded DNA, containing many genes, regulatory elements and other noncoding DNA.[2] The DNA-bound macromolecules are proteins, which serve to package the DNA and control its functions. Chromosomes vary widely between different organisms. Some species such as certain bacteria also contain plasmids or other extrachromosomal DNA. These are circular structures in the cytoplasm which contain cellular DNA and play a role in horizontal gene transfer.[1]

Compaction of the duplicated chromosomes during cell division (mitosis or meiosis) results either in a four-arm structure (pictured to the right) if the centromere is located in the middle of the chromosome or a two-arm structure if the centromere is located near one of the ends. Chromosomal recombination during meiosis and subsequent sexual reproduction plays a vital role in genetic diversity. If these structures are manipulated incorrectly, through processes known as chromosomal instability and translocation, the cell may undergo mitotic catastrophe and die, or it may unexpectedly evade apoptosis leading to the progression of cancer.

In prokaryotes (see nucleoids) and viruses,[2] the DNA is often densely packed and organized: in the case of archaea, by homologs to eukaryotic histones, and in the case of bacteria, by histone-like proteins. Small circular genomes called plasmids are often found in bacteria and also in mitochondria and chloroplasts, reflecting their bacterial origins.

14.1 History of discovery

The word *chromosome* comes from the Greek χρῶμα (*chroma*, "colour") and σῶμα (*soma*, "body"). Chromatin and chromosomes are both very strongly stained by particular dyes.[3]

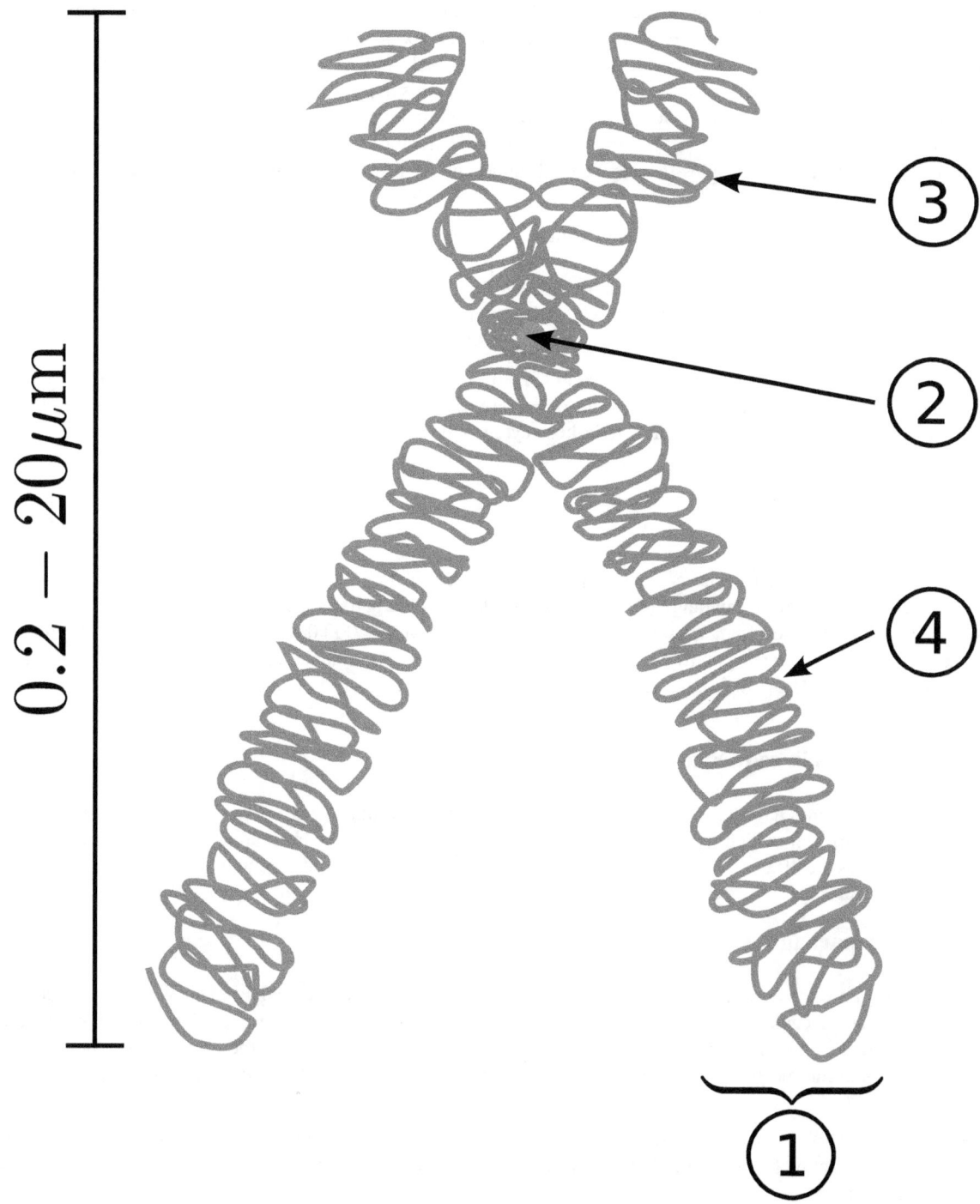

Diagram of a replicated and condensed metaphase eukaryotic chromosome. (1) Chromatid – one of the two identical parts of the chromosome after S phase. (2) Centromere – the point where the two chromatids touch. (3) Short arm. (4) Long arm.

Schleiden,[1] Virchow and Bütschli were among the first scientists who recognized the structures now so familiar to everyone as chromosomes.[4] The term was coined by von Waldeyer-Hartz,[5] referring to the term chromatin, which was introduced by Walther Flemming.

In a series of experiments beginning in the mid-1880s, Theodor Boveri gave the definitive demonstration that chromosomes are the vectors of heredity. His two principles were the *continuity* of chromosomes and the *individuality* of chromosomes. It is the second of these principles that was so original. Wilhelm Roux suggested that each chromosome carries a different genetic load. Boveri was able to test and confirm this hypothesis. Aided by the rediscovery at the start of the 1900s of Gregor Mendel's earlier work, Boveri was able to point out the connection between the rules of inheritance and the behaviour of the chromosomes. Boveri influenced two generations of American cytologists: Edmund Beecher Wilson, Walter Sutton and Theophilus Painter were all influenced by Boveri (Wilson and Painter actually worked with him).

In his famous textbook *The Cell in Development and Heredity*, Wilson linked together the independent work of Boveri and Sutton (both around 1902) by naming the chromosome theory of inheritance the Boveri–Sutton chromosome theory (the names are sometimes reversed).[6] Ernst Mayr remarks that the theory was hotly contested by some famous geneticists: William Bateson, Wilhelm Johannsen, Richard Goldschmidt and T.H. Morgan, all of a rather dogmatic turn of mind. Eventually, complete proof came from chromosome maps in Morgan's own lab.[7]

The number of human chromosomes was published in 1923 by Theophilus Painter. By inspection through the microscope he counted 24 pairs which would mean 48 chromosomes. His error was copied by others and it was not until 1956 that the true number, 46, was determined by Indonesia-born cytogeneticist Joe Hin Tjio.[8]

14.2   Prokaryotes

The prokaryotes – bacteria and archaea – typically have a single circular chromosome, but many variations exist.[9] The chromosomes of most bacteria can range in size from only 130,000 base pairs in the endosymbiotic bacteria *Candidatus Hodgkinia cicadicola*[10] and *Candidatus Tremblaya princeps*,[11] to over 14,000,000 base pairs in the soil-dwelling bacterium *Sorangium cellulosum*.[12] Spirochaetes of the genus *Borrelia* are a notable exception to this arrangement, with bacteria such as *Borrelia burgdorferi*, the cause of Lyme disease, containing a single *linear* chromosome.[13]

14.2.1   Structure in sequences

Prokaryotic chromosomes have less sequence-based structure than eukaryotes. Bacteria typically have a single point (the origin of replication) from which replication starts, whereas some archaea contain multiple replication origins.[14] The genes in prokaryotes are often organized in operons, and do not usually contain introns, unlike eukaryotes.

14.2.2   DNA packaging

Prokaryotes do not possess nuclei. Instead, their DNA is organized into a structure called the nucleoid.[15] The nucleoid is a distinct structure and occupies a defined region of the bacterial cell. This structure is, however, dynamic and is maintained and remodeled by the actions of a range of histone-like proteins, which associate with the bacterial chromosome.[16] In archaea, the DNA in chromosomes is even more organized, with the DNA packaged within structures similar to eukaryotic nucleosomes.[17][18]

Bacterial chromosomes tend to be tethered to the plasma membrane of the bacteria. In molecular biology application, this allows for its isolation from plasmid DNA by centrifugation of lysed bacteria and pelleting of the membranes (and the attached DNA).

Prokaryotic chromosomes and plasmids are, like eukaryotic DNA, generally supercoiled. The DNA must first be released into its relaxed state for access for transcription, regulation, and replication.

14.3   Eukaryotes

See also: Eukaryotic chromosome fine structure

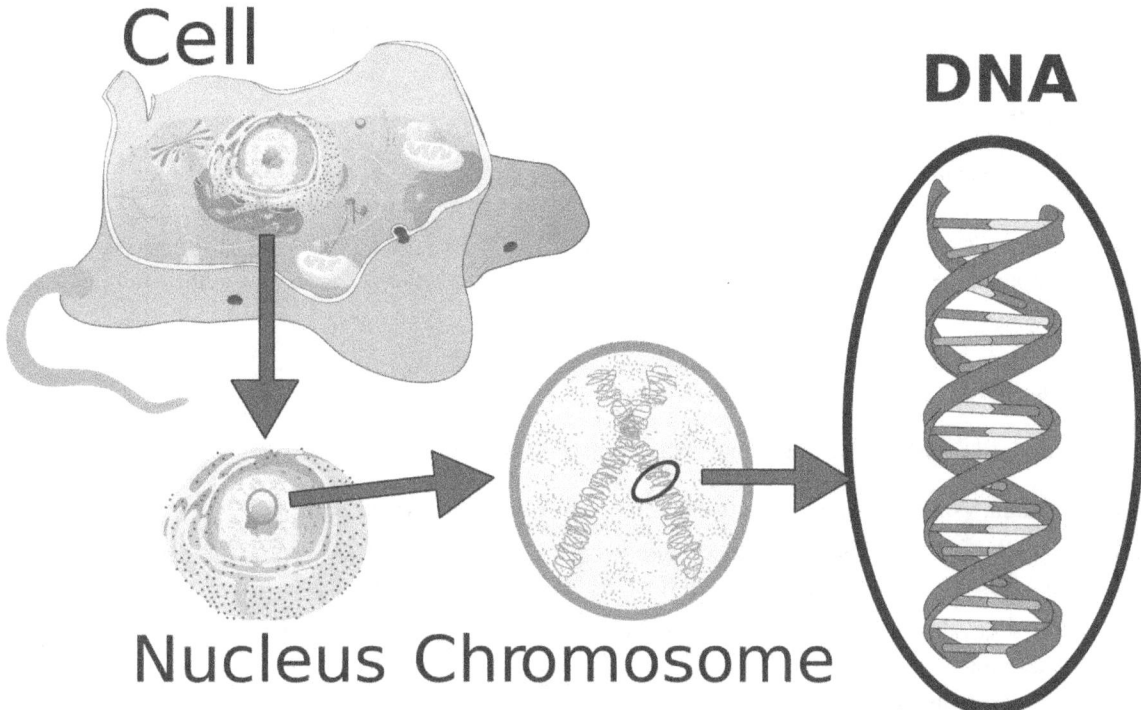

Organization of DNA in a eukaryotic cell.

In eukaryotes, nuclear chromosomes are packaged by proteins into a condensed structure called chromatin. This allows the very long DNA molecules to fit into the cell nucleus. The structure of chromosomes and chromatin varies through the cell cycle. Chromosomes are even more condensed than chromatin and are an essential unit for cellular division. Chromosomes must be replicated, divided, and passed successfully to their daughter cells so as to ensure the genetic diversity and survival of their progeny. Chromosomes may exist as either duplicated or unduplicated. Unduplicated chromosomes are single linear strands, whereas duplicated chromosomes contain two identical copies (called chromatids or sister chromatids) joined by a centromere.

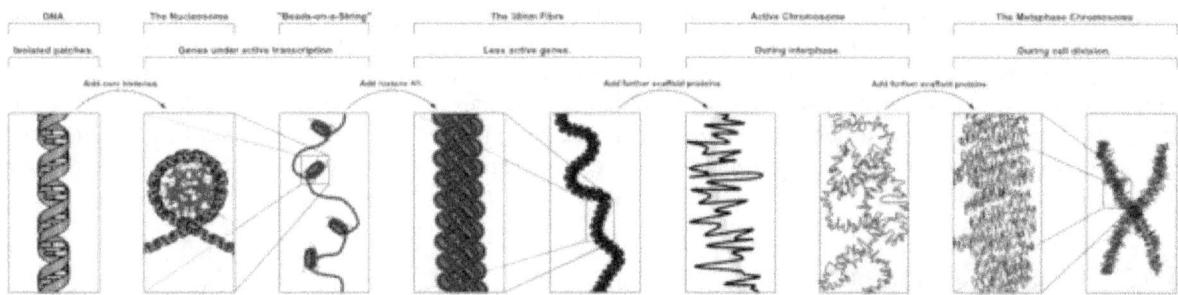

Fig. 2: *The major structures in DNA compaction: DNA, the nucleosome, the 10 nm "beads-on-a-string" fibre, the 30 nm fibre and the metaphase chromosome.*

Eukaryotes (cells with nuclei such as those found in plants, yeast, and animals) possess multiple large linear chromosomes contained in the cell's nucleus. Each chromosome has one centromere, with one or two arms projecting from the centromere, although, under most circumstances, these arms are not visible as such. In addition, most eukaryotes have a small circular mitochondrial genome, and some eukaryotes may have additional small circular or linear cytoplasmic chromosomes.

In the nuclear chromosomes of eukaryotes, the uncondensed DNA exists in a semi-ordered structure, where it is wrapped around histones (structural proteins), forming a composite material called chromatin.

14.3.1   Chromatin

Main article: Chromatin

Chromatin is the complex of DNA and protein found in the eukaryotic nucleus, which packages chromosomes. The structure of chromatin varies significantly between different stages of the cell cycle, according to the requirements of the DNA.

Interphase chromatin

During interphase (the period of the cell cycle where the cell is not dividing), two types of chromatin can be distinguished:

- Euchromatin, which consists of DNA that is active, e.g., being expressed as protein.

- Heterochromatin, which consists of mostly inactive DNA. It seems to serve structural purposes during the chromosomal stages. Heterochromatin can be further distinguished into two types:

  - *Constitutive heterochromatin*, which is never expressed. It is located around the centromere and usually contains repetitive sequences.
  - *Facultative heterochromatin*, which is sometimes expressed.

Metaphase chromatin and division

See also: mitosis and meiosis

In the early stages of mitosis or meiosis (cell division), the chromatin strands become more and more condensed. They cease to function as accessible genetic material (transcription stops) and become a compact transportable form. This compact form makes the individual chromosomes visible, and they form the classic four arm structure, a pair of sister chromatids attached to each other at the centromere. The shorter arms are called *p arms* (from the French *petit*, small) and the longer arms are called *q arms* (*q* follows *p* in the Latin alphabet; q-g "grande"; alternatively it is sometimes said q is short for *queue* meaning tail in French[19]). This is the only natural context in which individual chromosomes are visible with an optical microscope.

During mitosis, microtubules grow from centrosomes located at opposite ends of the cell and also attach to the centromere at specialized structures called kinetochores, one of which is present on each sister chromatid. A special DNA base sequence in the region of the kinetochores provides, along with special proteins, longer-lasting attachment in this region. The microtubules then pull the chromatids apart toward the centrosomes, so that each daughter cell inherits one set of chromatids. Once the cells have divided, the chromatids are uncoiled and DNA can again be transcribed. In spite of their appearance, chromosomes are structurally highly condensed, which enables these giant DNA structures to be contained within a cell nucleus (Fig. 2).

14.3.2   Human chromosomes

Chromosomes in humans can be divided into two types: autosomes and sex chromosomes. Certain genetic traits are linked to a person's sex and are passed on through the sex chromosomes. The autosomes contain the rest of the genetic hereditary information. All act in the same way during cell division. Human cells have 23 pairs of chromosomes (22 pairs of autosomes and one pair of sex chromosomes), giving a total of 46 per cell. In addition to these, human cells have many hundreds of copies of the mitochondrial genome. Sequencing of the human genome has provided a great deal of information about each of the chromosomes. Below is a table compiling statistics for the chromosomes, based on the Sanger Institute's human genome information in the Vertebrate Genome Annotation (VEGA) database.[20] Number of genes is an estimate as it is in part based on gene predictions. Total chromosome length is an estimate as well, based on the estimated size of unsequenced heterochromatin regions.

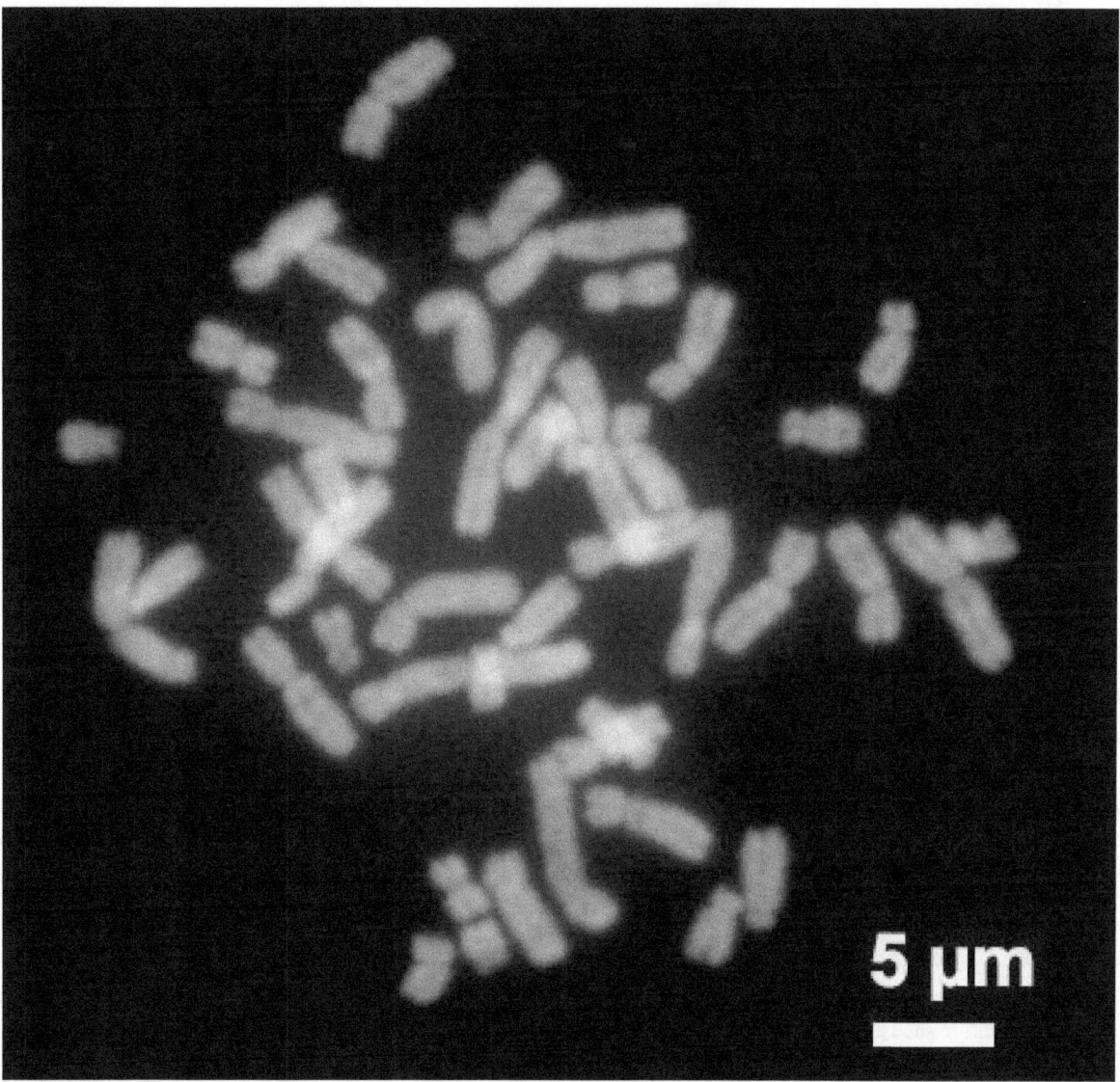

Human chromosomes during metaphase

14.4 Number of chromosomes in various organisms

Main article: List of number of chromosomes of various organisms

14.4.1 In eukaryotes

These tables give the total number of chromosomes (including sex chromosomes) in a cell nucleus. For example, human cells are diploid and have 22 different types of autosome, each present as two copies, and two sex chromosomes. This gives 46 chromosomes in total. Other organisms have more than two copies of their chromosome types, such as bread wheat, which is *hexaploid* and has six copies of seven different chromosome types – 42 chromosomes in total.

Normal members of a particular eukaryotic species all have the same number of nuclear chromosomes (see the table). Other eukaryotic chromosomes, i.e., mitochondrial and plasmid-like small chromosomes, are much more variable in

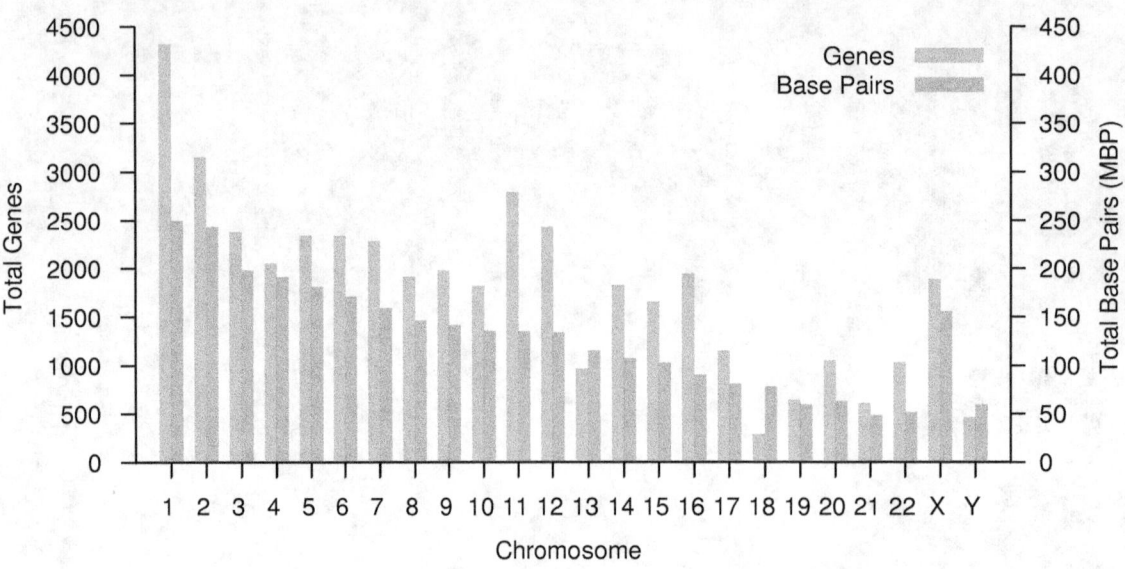

Estimated number of genes and base pairs (in mega base pairs) on each human chromosome

number, and there may be thousands of copies per cell.

Asexually reproducing species have one set of chromosomes, which are the same in all body cells. However, asexual species can be either haploid or diploid.

Sexually reproducing species have somatic cells (body cells), which are diploid [2n] having two sets of chromosomes (23 pairs in humans with one set of 23 chromosomes from each parent), one set from the mother and one from the father. Gametes, reproductive cells, are haploid [n]: They have one set of chromosomes. Gametes are produced by meiosis of a diploid germ line cell. During meiosis, the matching chromosomes of father and mother can exchange small parts of themselves (crossover), and thus create new chromosomes that are not inherited solely from either parent. When a male and a female gamete merge (fertilization), a new diploid organism is formed.

Some animal and plant species are polyploid [Xn]: They have more than two sets of homologous chromosomes. Plants important in agriculture such as tobacco or wheat are often polyploid, compared to their ancestral species. Wheat has a haploid number of seven chromosomes, still seen in some cultivars as well as the wild progenitors. The more-common pasta and bread wheats are polyploid, having 28 (tetraploid) and 42 (hexaploid) chromosomes, compared to the 14 (diploid) chromosomes in the wild wheat.[48]

14.4.2 In prokaryotes

Prokaryote species generally have one copy of each major chromosome, but most cells can easily survive with multiple copies.[49] For example, *Buchnera*, a symbiont of aphids has multiple copies of its chromosome, ranging from 10–400 copies per cell.[50] However, in some large bacteria, such as *Epulopiscium fishelsoni* up to 100,000 copies of the chromosome can be present.[51] Plasmids and plasmid-like small chromosomes are, as in eukaryotes, highly variable in copy number. The number of plasmids in the cell is almost entirely determined by the rate of division of the plasmid – fast division causes high copy number.

14.5 Karyotype

Main article: Karyotype
In general, the **karyotype** is the characteristic chromosome complement of a eukaryote species.[52] The preparation and

study of karyotypes is part of cytogenetics.

Although the replication and transcription of DNA is highly standardized in eukaryotes, *the same cannot be said for their karyotypes*, which are often highly variable. There may be variation between species in chromosome number and in detailed organization. In some cases, there is significant variation within species. Often there is:

1. variation between the two sexes

2. variation between the germ-line and soma (between gametes and the rest of the body)

3. variation between members of a population, due to balanced genetic polymorphism

4. geographical variation between races

5. mosaics or otherwise abnormal individuals.

Also, variation in karyotype may occur during development from the fertilised egg.

The technique of determining the karyotype is usually called *karyotyping*. Cells can be locked part-way through division (in metaphase) in vitro (in a reaction vial) with colchicine. These cells are then stained, photographed, and arranged into a *karyogram*, with the set of chromosomes arranged, autosomes in order of length, and sex chromosomes (here X/Y) at the end: Fig. 3.

Like many sexually reproducing species, humans have special gonosomes (sex chromosomes, in contrast to autosomes). These are XX in females and XY in males.

14.5.1   Historical note

Investigation into the human karyotype took many years to settle the most basic question: *How many chromosomes does a normal diploid human cell contain?* In 1912, Hans von Winiwarter reported 47 chromosomes in spermatogonia and 48 in oogonia, concluding an XX/XO sex determination mechanism.[53] Painter in 1922 was not certain whether the diploid number of man is 46 or 48, at first favouring 46.[54] He revised his opinion later from 46 to 48, and he correctly insisted on humans having an XX/XY system.[55]

New techniques were needed to definitively solve the problem:

1. Using cells in culture

2. Arresting mitosis in metaphase by a solution of colchicine

3. Pretreating cells in a hypotonic solution 0.075 M KCl, which swells them and spreads the chromosomes

4. Squashing the preparation on the slide forcing the chromosomes into a single plane

5. Cutting up a photomicrograph and arranging the result into an indisputable karyogram.

It took until 1954 before the human diploid number was confirmed as 46.[56][57] Considering the techniques of Winiwarter and Painter, their results were quite remarkable.[58] Chimpanzees (the closest living relatives to modern humans) have 48 chromosomes (as well as the other great apes: in humans two chromosomes fused to form chromosome 2).

14.6   Aberrations

Main articles: Chromosome abnormality and aneuploidy

Chromosomal aberrations are disruptions in the normal chromosomal content of a cell and are a major cause of genetic conditions in humans, such as Down syndrome, although most aberrations have little to no effect. Some chromosome abnormalities do not cause disease in carriers, such as translocations, or chromosomal inversions, although they may lead to a higher chance of bearing a child with a chromosome disorder. Abnormal numbers of chromosomes or chromosome sets, called aneuploidy, may be lethal or may give rise to genetic disorders. Genetic counseling is offered for families that may carry a chromosome rearrangement.

The gain or loss of DNA from chromosomes can lead to a variety of genetic disorders. Human examples include:

- Cri du chat, which is caused by the deletion of part of the short arm of chromosome 5. "Cri du chat" means "cry of the cat" in French; the condition was so-named because affected babies make high-pitched cries that sound like those of a cat. Affected individuals have wide-set eyes, a small head and jaw, moderate to severe mental health problems, and are very short.

- Down's syndrome, the most common trisomy, usually caused by an extra copy of chromosome 21 (trisomy 21). Characteristics include decreased muscle tone, stockier build, asymmetrical skull, slanting eyes and mild to moderate developmental disability.[59]

- Edwards syndrome, or trisomy-18, the second most common trisomy. Symptoms include motor retardation, developmental disability and numerous congenital anomalies causing serious health problems. Ninety percent of those affected die in infancy. They have characteristic clenched hands and overlapping fingers.

- Isodicentric 15, also called idic(15), partial tetrasomy 15q, or inverted duplication 15 (inv dup 15).

- Jacobsen syndrome, which is very rare. It is also called the terminal 11q deletion disorder.[60] Those affected have normal intelligence or mild developmental disability, with poor expressive language skills. Most have a bleeding disorder called Paris-Trousseau syndrome.

- Klinefelter syndrome (XXY). Men with Klinefelter syndrome are usually sterile, and tend to be taller and have longer arms and legs than their peers. Boys with the syndrome are often shy and quiet, and have a higher incidence of speech delay and dyslexia. Without testosterone treatment, some may develop gynecomastia during puberty.

- Patau Syndrome, also called D-Syndrome or trisomy-13. Symptoms are somewhat similar to those of trisomy-18, without the characteristic folded hand.

- Small supernumerary marker chromosome. This means there is an extra, abnormal chromosome. Features depend on the origin of the extra genetic material. Cat-eye syndrome and isodicentric chromosome 15 syndrome (or Idic15) are both caused by a supernumerary marker chromosome, as is Pallister-Killian syndrome.

- Triple-X syndrome (XXX). XXX girls tend to be tall and thin and have a higher incidence of dyslexia.

- Turner syndrome (X instead of XX or XY). In Turner syndrome, female sexual characteristics are present but underdeveloped. Females with Turner syndrome often have a short stature, low hairline, abnormal eye features and bone development and a "caved-in" appearance to the chest.

- XYY syndrome. XYY boys are usually taller than their siblings. Like XXY boys and XXX girls, they are more likely to have learning difficulties.

- Wolf-Hirschhorn syndrome, which is caused by partial deletion of the short arm of chromosome 4. It is characterized by growth retardation, delayed motor skills development, "Greek Helmet" facial features, and mild to profound mental health problems.

14.7   Etymology and pronunciation

The word *chromosome* (/ˈkroʊməˌzoʊm/) uses combining forms of *chromo-* and *-some*, yielding "colored body", which describes a chromosome's appearance on microscopy.

14.8 See also

- Genetic deletion

- DNA

- For information about chromosomes in genetic algorithms, see chromosome (genetic algorithm)

- Genetic genealogy

  - Genealogical DNA test

- Lampbrush chromosome

- List of number of chromosomes of various organisms

- Locus (explains gene location nomenclature)

- Maternal influence on sex determination

- Sex-determination system

  - XY sex-determination system
    - X-chromosome
      - X-inactivation
    - Y-chromosome
      - Y-chromosomal Aaron
      - Y-chromosomal Adam

- Polytene chromosome

- Neochromosome

14.9 Notes and references

[1] Schleyden, M.J. (1847). *Microscopical researches into the accordance in the structure and growth of animals and plants.*

[2] Johnson, J; Chiu, W (1 April 2000). "Structures of virus and virus-like particles". *Current Opinion in Structural Biology* **10** (2): 229–235. doi:10.1016/S0959-440X(00)00073-7. PMID 10753814.

[3] Coxx, H. J. (1925). *Biological Stains - A Handbook on the Nature and Uses of the Dyes Employed in the Biological Laboratory.* Commission on Standardization of Biological Stains.

[4] Fokin S.I. (2013). "Otto Bütschli (1848–1920) Where we will genuflect?" (PDF). *Protistology* **8** (1): 22–35.

[5] Waldeyer-Hartz (1888). "Über Karyokinese und ihre Beziehungen zu den Befruchtungsvorgängen". *Archiv für mikroskopische Anatomie und Entwicklungsmechanik* **32**: 27.

[6] Wilson, E.B. (1925). *The Cell in Development and Heredity*, Ed. 3. Macmillan, New York. p. 923.

[7] Mayr, E. (1982). *The growth of biological thought.* Harvard. p. 749.

[8] Matthews, Robert. "The bizarre case of the chromosome that never was" (PDF). Retrieved 13 July 2013.

[9] Thanbichler M; Shapiro L (2006). "Chromosome organization and segregation in bacteria". *J. Struct. Biol.* **156** (2): 292–303. doi:10.1016/j.jsb.2006.05.007. PMID 16860572.

[10] Van Leuven, JT; Meister, RC; Simon, C; McCutcheon, JP (11 September 2014). "Sympatric speciation in a bacterial endosymbiont results in two genomes with the functionality of one.". *Cell* **158** (6): 1270–80. doi:10.1016/j.cell.2014.07.047. PMID 25175626.

[11] McCutcheon, JP; von Dohlen, CD (23 August 2011). "An interdependent metabolic patchwork in the nested symbiosis of mealybugs.". *Current biology : CB* **21** (16): 1366–72. doi:10.1016/j.cub.2011.06.051. PMC 3169327. PMID 21835622.

[12] Han, K; Li, ZF; Peng, R; Zhu, LP; Zhou, T; Wang, LG; Li, SG; Zhang, XB; Hu, W; Wu, ZH; Qin, N; Li, YZ (2013). "Extraordinary expansion of a Sorangium cellulosum genome from an alkaline milieu.". *Scientific reports* **3**: 2101. doi:10.1038/srep02101. PMID 23812535.

[13] Hinnebusch J; Tilly K (1993). "Linear plasmids and chromosomes in bacteria". *Mol Microbiol* **10** (5): 917–22. doi:10.1111/j.1365-2958.1993.tb00963.x. PMID 7934868.

[14] Kelman LM; Kelman Z (2004). "Multiple origins of replication in archaea". *Trends Microbiol.* **12** (9): 399–401. doi:10.1016/j.tim.2004.07.001. PMID 15337158.

[15] Thanbichler M; Wang SC; Shapiro L (2005). "The bacterial nucleoid: a highly organized and dynamic structure". *J. Cell. Biochem.* **96** (3): 506–21. doi:10.1002/jcb.20519. PMID 15988757.

[16] Sandman K; Pereira SL; Reeve JN (1998). "Diversity of prokaryotic chromosomal proteins and the origin of the nucleosome". *Cell. Mol. Life Sci.* **54** (12): 1350–64. doi:10.1007/s000180050259. PMID 9893710.

[17] Sandman K; Reeve JN (2000). "Structure and functional relationships of archaeal and eukaryal histones and nucleosomes". *Arch. Microbiol.* **173** (3): 165–9. doi:10.1007/s002039900122. PMID 10763747.

[18] Pereira SL; Grayling RA; Lurz R; Reeve JN (1997). "Archaeal nucleosomes". *Proc. Natl. Acad. Sci. U.S.A.* **94** (23): 12633–7. Bibcode:1997PNAS...9412633P. doi:10.1073/pnas.94.23.12633. PMC 25063. PMID 9356501.

[19] "Chromosome Mapping: Idiograms" *Nature Education* - August 13, 2013

[20] Vega.sanger.ad.uk, all data in this table was derived from this database, November 11, 2008.

[21] Sequenced percentages are based on fraction of euchromatin portion, as the Human Genome Project goals called for determination of only the euchromatic portion of the genome. Telomeres, centromeres, and other heterochromatic regions have been left undetermined, as have a small number of unclonable gaps. See http://www.ncbi.nlm.nih.gov/genome/seq/ for more information on the Human Genome Project.

[22] Chromosomes - Genetics Home Reference

[23] Armstrong SJ; Jones GH (January 2003). "Meiotic cytology and chromosome behaviour in wild-type Arabidopsis thaliana". *J. Exp. Bot.* **54** (380): 1–10. doi:10.1093/jxb/54.380.1. PMID 12456750.

[24] Gill BS; Kimber G (April 1974). "The Giemsa C-Banded Karyotype of Rye". *Proc. Natl. Acad. Sci. U.S.A.* **71** (4): 1247–9. Bibcode:1974PNAS...71.1247G. doi:10.1073/pnas.71.4.1247. PMC 388202. PMID 4133848.

[25] Kato A; Lamb JC; Birchler JA (September 2004). "Chromosome painting using repetitive DNA sequences as probes for somatic chromosome identification in maize". *Proc. Natl. Acad. Sci. U.S.A.* **101** (37): 13554–9. Bibcode:2004PNAS..10113554K. doi:10.1073/pnas.0403659101. PMC 518793. PMID 15342909.

[26] Dubcovsky J; Luo MC; Zhong GY; et al. (1996). "Genetic Map of Diploid Wheat, Triticum Monococcum L., and Its Comparison with Maps of Hordeum Vulgare L". *Genetics* **143** (2): 983–99. PMC 1207354. PMID 8725244.

[27] Kenton A; Parokonny AS; Gleba YY; Bennett MD (August 1993). "Characterization of the Nicotiana tabacum L. genome by molecular cytogenetics". *Mol. Gen. Genet.* **240** (2): 159–69. doi:10.1007/BF00277053. PMID 8355650.

[28] Leitch IJ; Soltis DE; Soltis PS; Bennett MD (2005). "Evolution of DNA amounts across land plants (embryophyta)". *Ann. Bot.* **95** (1): 207–17. doi:10.1093/aob/mci014. PMID 15596468.

[29] Umeko Semba; Yasuko Umeda; Yoko Shibuya; Hiroaki Okabe; Sumio Tanase & Tetsuro Yamamoto (2004). "Primary structures of guinea pig high- and low-molecular-weight kininogens". *International Immunopharmacology* **4** (10–11): 1391–1400. doi:10.1016/j.intimp.2004.06.003. PMID 15313436.

[30] "The Genetics of the Popular Aquarium Pet - Guppy Fish". Retrieved 2009-12-06.

[31] Vitturi R; Libertini A; Sineo L; et al. (2005). "Cytogenetics of the land snails Cantareus aspersus and C. mazzullii (Mollusca: Gastropoda: Pulmonata)". *Micron* **36** (4): 351–7. doi:10.1016/j.micron.2004.12.010. PMID 15857774.

[32] Vitturi R; Colomba MS; Pirrone AM; Mandrioli M (2002). "rDNA (18S-28S and 5S) colocalization and linkage between ribosomal genes and (TTAGGG)(n) telomeric sequence in the earthworm, Octodrilus complanatus (Annelida: Oligochaeta: Lumbricidae), revealed by single- and double-color FISH". *J. Hered* **93** (4): 279–82. doi:10.1093/jhered/93.4.279. PMID 12407215.

[33] Ambarish, C.N. Sridhar, K.R. (2014). "Cytological and karyological observations of two endemic pill-millipedes Arthrosphaera (Pocock, 1895) (Diplopoda: Sphaerotheriida) of the Western Ghats of India". *Caryologia* **66** (1). doi:10.1080/00087114.

[34] Nie W; Wang J; O'Brien PC; et al. (2002). "The genome phylogeny of domestic cat, red panda and five mustelid species revealed by comparative chromosome painting and G-banding". *Chromosome Res.* **10** (3): 209–22. doi:10.1023/A:1015292005631. PMID 12067210.

[35] Romanenko, Svetlana A.; Perelman, Polina L.; Serdukova, Natalya A.; Trifonov, Vladimir A.; Biltueva, Larisa S.; Wang, Jinhuan; Li, Tangliang; Nie, Wenhui; O'Brien, Patricia C.M.; Volobouev, Vitaly T.; Stanyon, Roscoe; Ferguson-Smith, Malcolm A.; Yang, Fengtang; Graphodatsky, Alexander S. (2006). "Reciprocal chromosome painting between three laboratory rodent species". *Mammalian Genome* **17** (12): 1183–92. doi:10.1007/s00335-006-0081-z. PMID 17143584.

[36] Painter, TS (1928). "A Comparison of the Chromosomes of the Rat and Mouse with Reference to the Question of Chromosome Homology in Mammals". *Genetics* **13** (2): 180–9. PMC 1200977. PMID 17246549.

[37] Hayes, H.; Rogel-Gaillard, C.; Zijlstra, C.; De Haan, N.A.; Urien, C.; Bourgeaux, N.; Bertaud, M.; Bosma, A.A. (2002). "Establishment of an R-banded rabbit karyotype nomenclature by FISH localization of 23 chromosome-specific genes on both G- and R-banded chromosomes". *Cytogenetic and Genome Research* **98** (2–3): 199–205. doi:10.1159/000069807. PMID 12698004.

[38] T.J. Robinson; F. Yang; W.R. Harrison (2002). "Chromosome painting refines the history of genome evolution in hares and rabbits (order Lagomorpha)". *Cytogenic and Genetic Research* **96** (1–4): 223–227. doi:10.1159/000063034. PMID 12438803.

[39] "section 4.W4", *Rabbits, Hares and Pikas. Status Survey and Conservation Action Plan*, pp. 61–94

[40] De Grouchy J (1987). "Chromosome phylogenies of man, great apes, and Old World monkeys". *Genetica* **73** (1–2): 37–52. doi:10.1007/bf00057436. PMID 3333352.

[41] Houck, M.L.; Kumamoto, A.T.; Gallagher, D.S.; Benirschke, K. (2001). "Comparative cytogenetics of the African elephant *(Loxodonta africana)* and Asiatic elephant *(Elephas maximus)*". *Cytogenetic and Genome Research* **93** (3–4): 249–52. doi:10.1159/000056992. PMID 11528120.

[42] Wayne RK; Ostrander EA (1999). "Origin, genetic diversity, and genome structure of the domestic dog". *BioEssays* **21** (3): 247–57. doi:10.1002/(SICI)1521-1878(199903)21:3<247::AID-BIES9>3.0.CO;2-Z. PMID 10333734.

[43] Burt DW (2002). "Origin and evolution of avian microchromosomes". *Cytogenet. Genome Res.* **96** (1–4): 97–112. doi:10.1159/000063018. PMID 12438785.

[44] Ciudad J; Cid E; Velasco A; Lara JM; Aijón J; Orfao A (2002). "Flow cytometry measurement of the DNA contents of G0/G1 diploid cells from three different teleost fish species". *Cytometry* **48** (1): 20–5. doi:10.1002/cyto.10100. PMID 12116377.

[45] Yasukochi Y; Ashakumary LA; Baba K; Yoshido A; Sahara K (2006). "A Second-Generation Integrated Map of the Silkworm Reveals Synteny and Conserved Gene Order Between Lepidopteran Insects". *Genetics* **173** (3): 1319–28. doi:10.1534/genetics.106.055541. PMC 1526672. PMID 16547103.

[46] Itoh, Masahiro; Ikeuchi, Tatsuro; Shimba, Hachiro; Mori, Michiko; Sasaki, Motomichi; Makino, Sajiro (1969). "A Comparative Karyotype Study in Fourteen Species of Birds". *The Japanese journal of genetics* **44** (3): 163–170. doi:10.1266/jjg.44.163.

[47] Smith J; Burt DW (1998). "Parameters of the chicken genome (Gallus gallus)". *Anim. Genet.* **29** (4): 290–4. doi:10.1046/j.1365-2052.1998.00334.x. PMID 9745667.

[48] Sakamura, Tetsu (1918). "Kurze Mitteilung über die Chromosomenzahlen und die Verwandtschaftsverhältnisse der Triticum-Arten". *Shokubutsugaku Zasshi* **32** (379): 150–3. doi:10.15281/jplantres1887.32.379_150.

[49] Charlebois R.L. (ed) 1999. *Organization of the prokaryote genome*. ASM Press, Washington DC.

[50] Komaki K; Ishikawa H (March 2000). "Genomic copy number of intracellular bacterial symbionts of aphids varies in response to developmental stage and morph of their host". *Insect Biochem. Mol. Biol.* **30** (3): 253–8. doi:10.1016/S0965-1748(99)00125-3. PMID 10732993.

[51] Mendell JE; Clements KD; Choat JH; Angert ER (May 2008). "Extreme polyploidy in a large bacterium". *Proc. Natl. Acad. Sci. U.S.A.* **105** (18): 6730–4. Bibcode:2008PNAS..105.6730M. doi:10.1073/pnas.0707522105. PMC 2373351. PMID 18445653.

[52] White, M. J. D. (1973). *The chromosomes* (6th ed.). London: Chapman and Hall, distributed by Halsted Press, New York. p. 28. ISBN 0-412-11930-7.

[53] von Winiwarter H (1912). "Études sur la spermatogenese humaine". *Arch. Biologie* **27** (93): 147–9.

[54] Painter TS (1922). "The spermatogenesis of man". *Anat. Res.* **23**: 129.

[55] Painter TS (1923). "Studies in mammalian spermatogenesis II. The spermatogenesis of man". *J. Exp. Zoology* **37** (3): 291–336. doi:10.1002/jez.1400370303.

[56] Tjio JH; Levan A (1956). "The chromosome number of man". *Hereditas* **42** (1-2): 1–6. doi:10.1111/j.1601-5223.1956.tb03010.x.

[57] Ford C.E; Hamerton J.L (1956). "The Chromosomes of Man". *Nature* **178** (4541): 1020–1023. Bibcode:1956Natur.178.1020F. doi:10.1038/1781020a0. PMID 13378517.

[58] Hsu T.C. *Human and mammalian cytogenetics: a historical perspective.* Springer-Verlag, N.Y. p10: "It's amazing that he [Painter] even came close!"

[59] Miller, Kenneth R. (2000). "Chapter 9-3". *Biology* (5th ed.). Upper Saddle River, New Jersey: Prentice Hall. pp. 194–5. ISBN 0-13-436265-9.

[60] European Chromosome 11 Network

14.10   External links

- An Introduction to DNA and Chromosomes from HOPES: Huntington's Outreach Project for Education at Stanford
- Chromosome Abnormalities at AtlasGeneticsOncology
- On-line exhibition on chromosomes and genome (SIB)
- What Can Our Chromosomes Tell Us?, from the University of Utah's Genetic Science Learning Center
- Try making a karyotype yourself, from the University of Utah's Genetic Science Learning Center
- Kimballs Chromosome pages
- Chromosome News from Genome News Network
- Eurochromnet, European network for Rare Chromosome Disorders on the Internet
- Ensembl.org, Ensembl project, presenting chromosomes, their genes and syntenic loci graphically via the web
- Genographic Project
- Home reference on Chromosomes from the U.S. National Library of Medicine
- Visualisation of human chromosomes and comparison to other species
- Unique - The Rare Chromosome Disorder Support Group Support for people with rare chromosome disorders

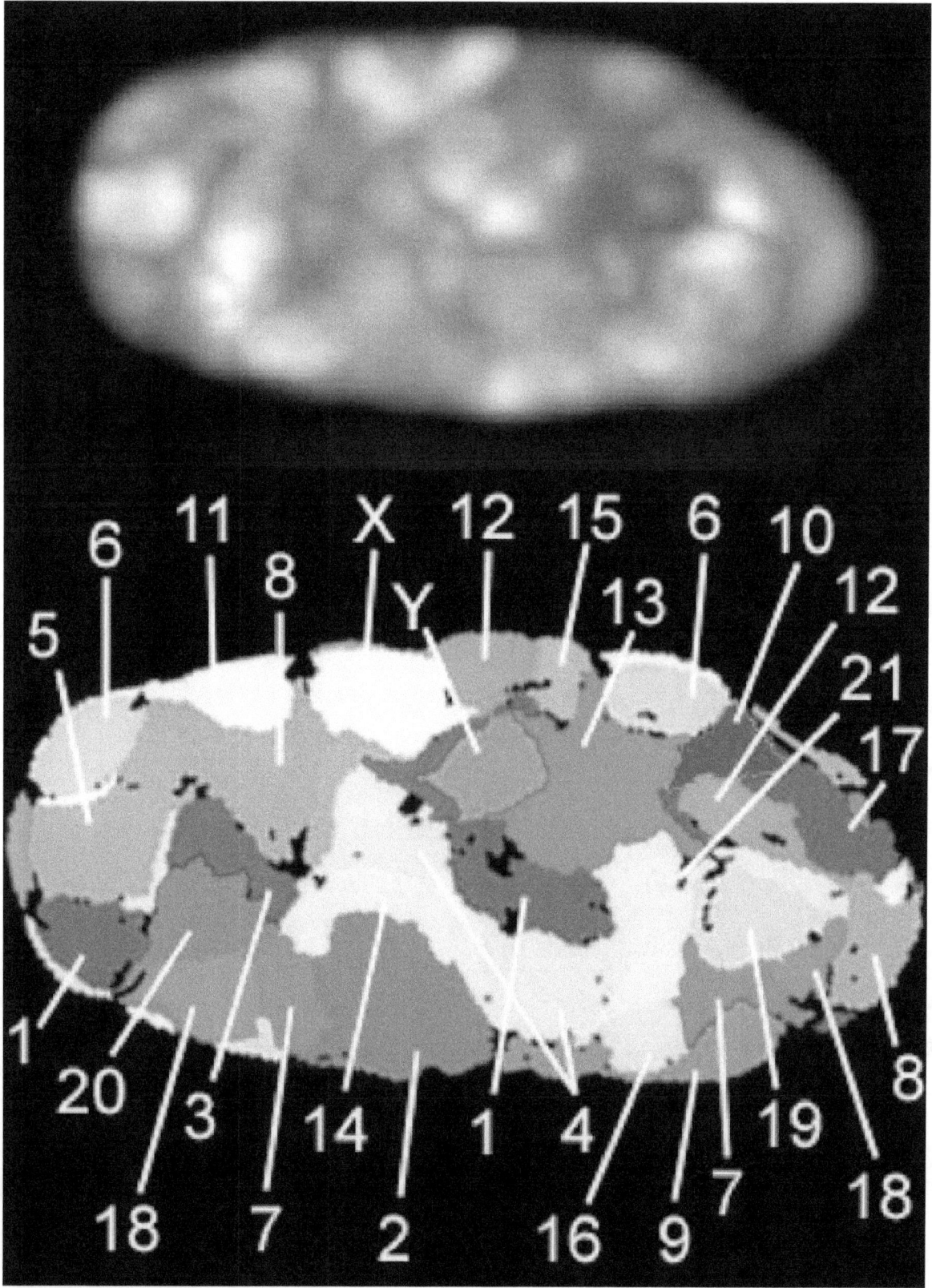

The 23 human chromosome territories during prometaphase in fibroblast cells.

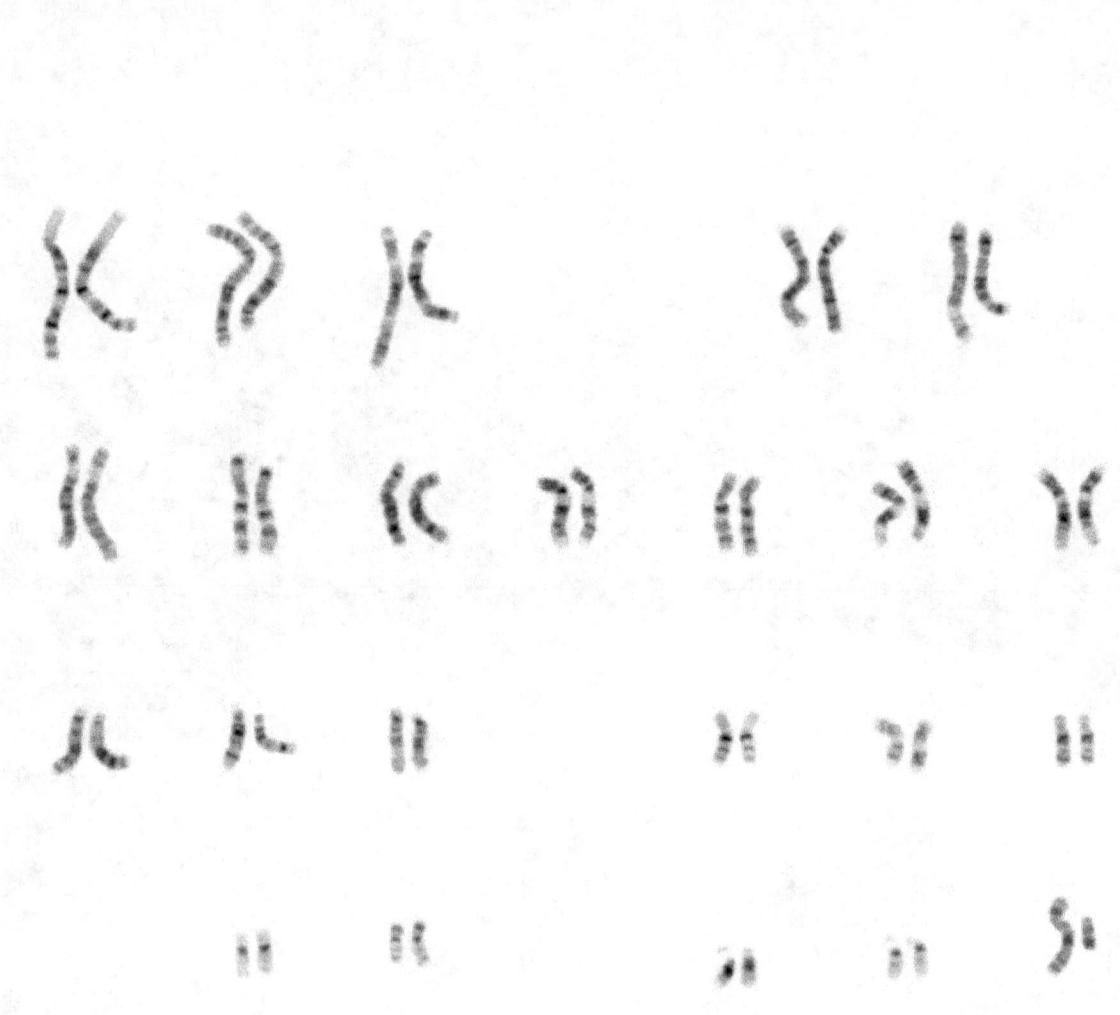

Figure 3: *Karyogram of a human male*

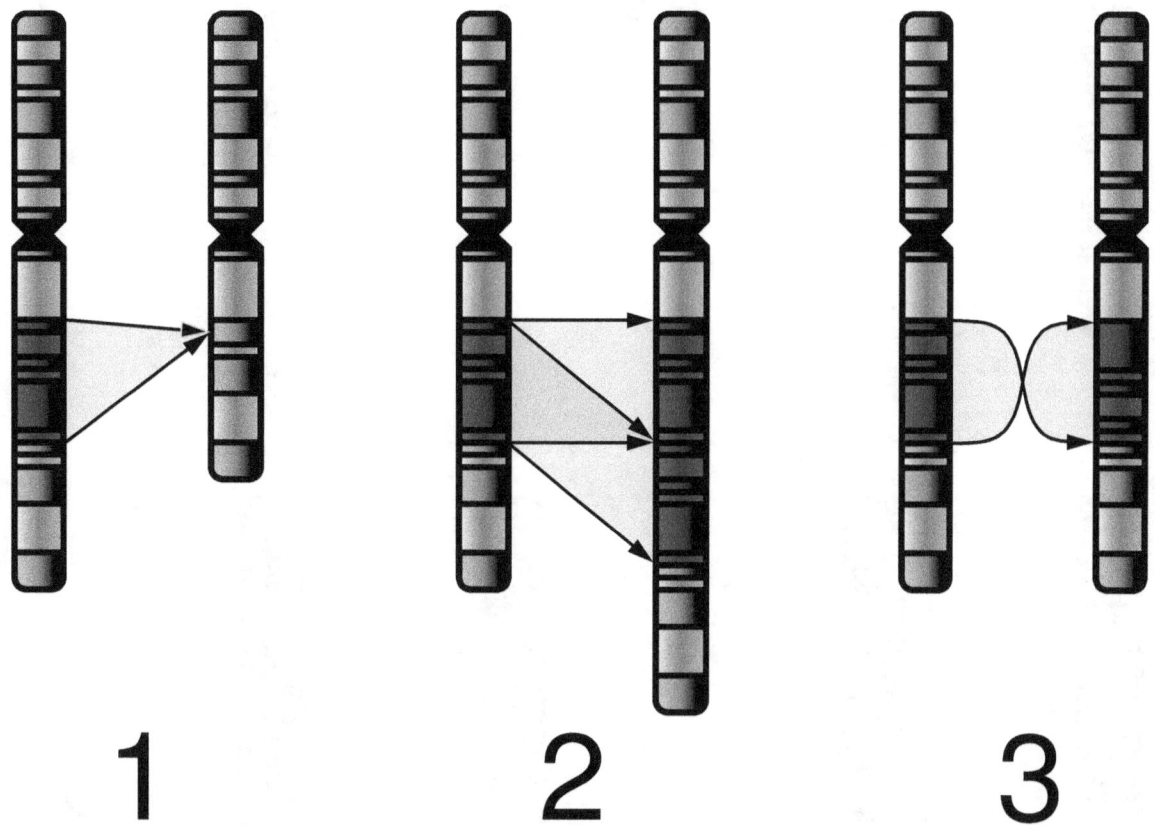

The three major single chromosome mutations; deletion (1), duplication (2) and inversion (3).

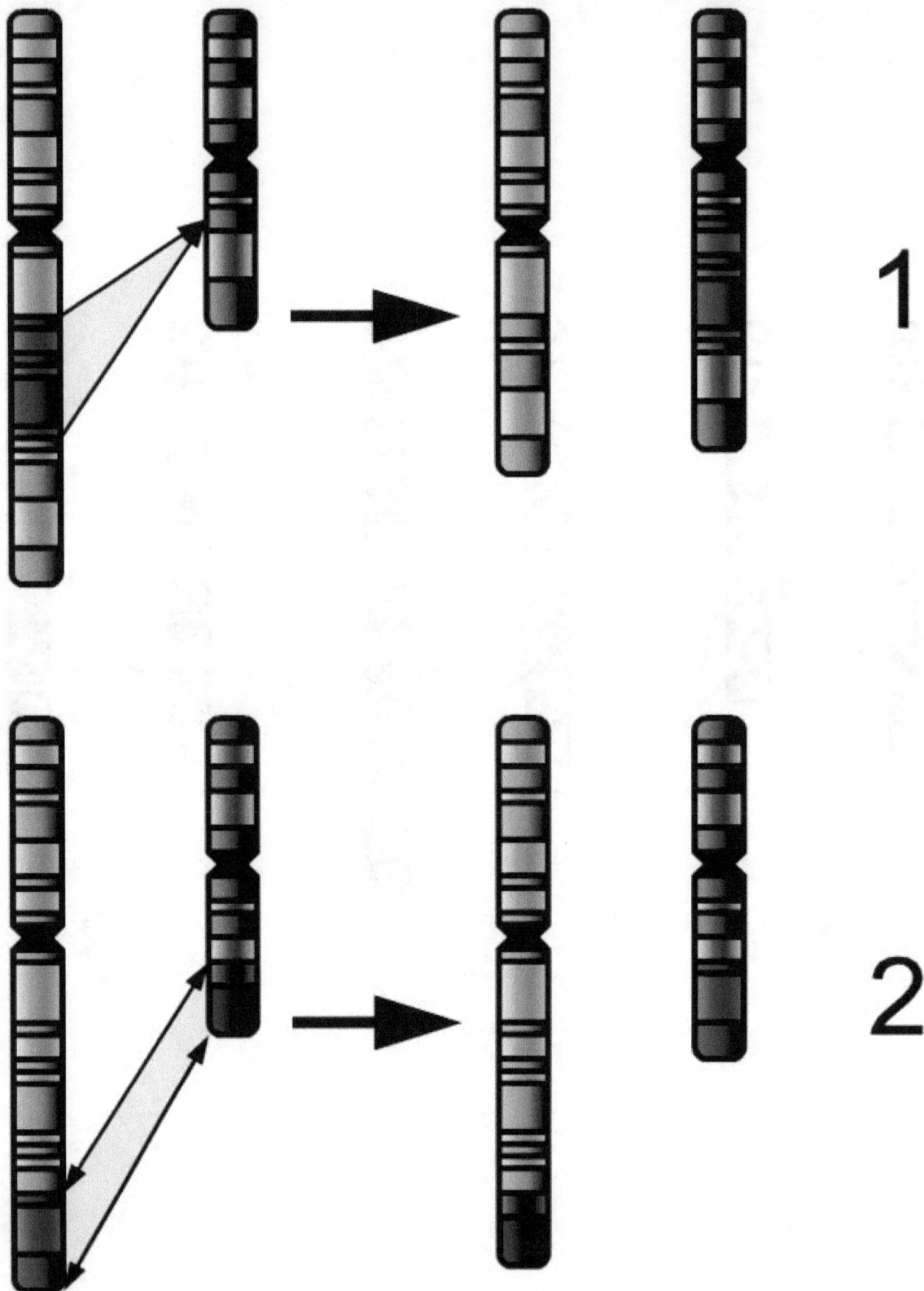

The two major two-chromosome mutations; insertion (1) and translocation (2).

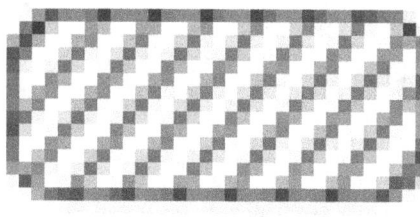

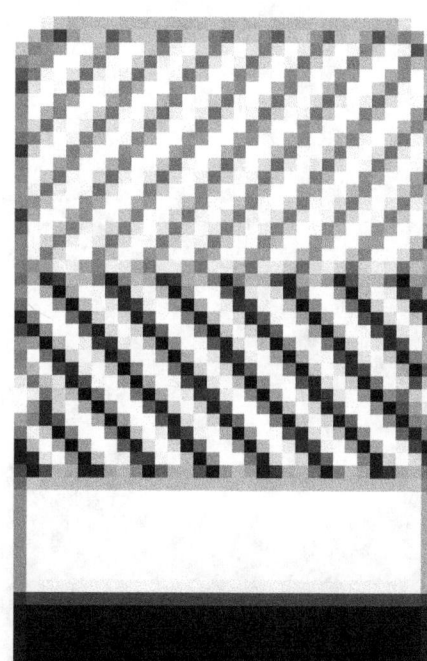

Chapter 15

Cell nucleus

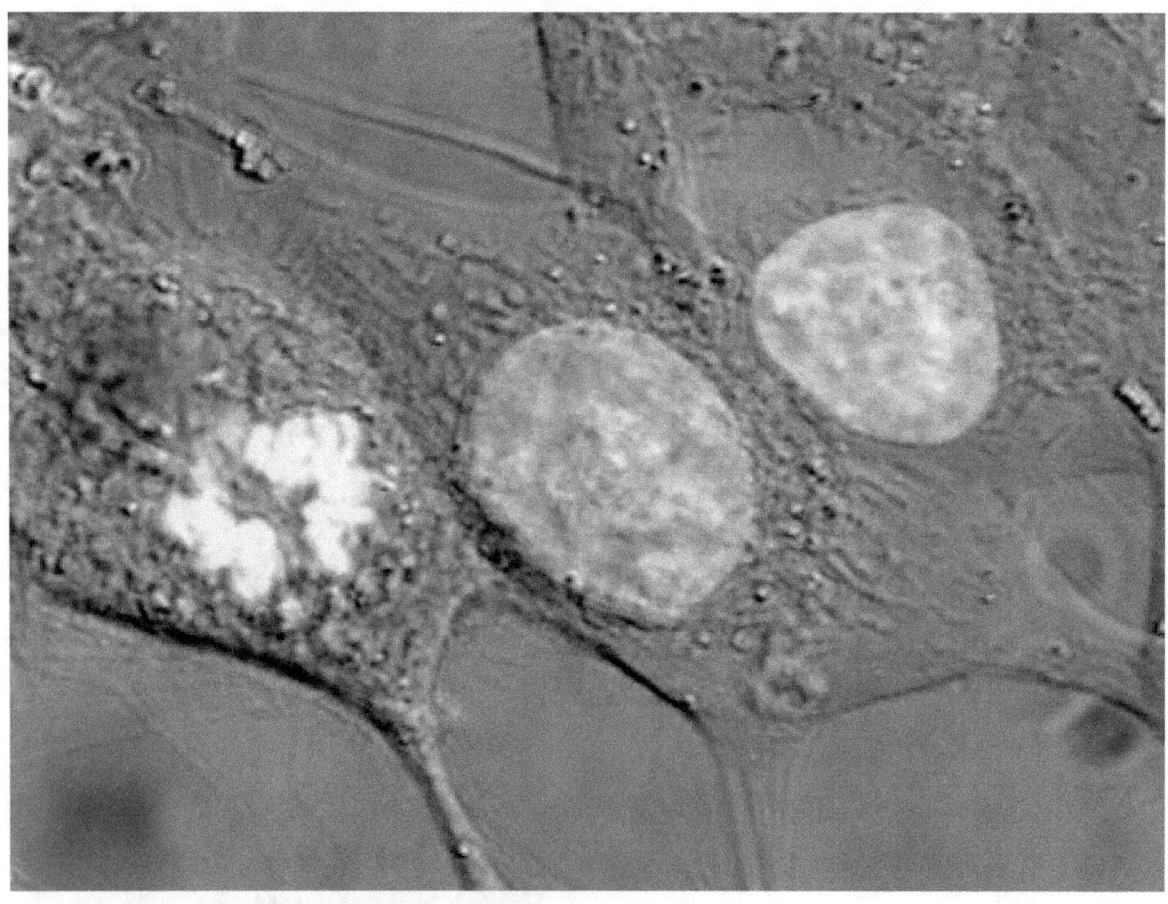

HeLa cells stained for nuclear DNA with the Blue Hoechst dye. The central and rightmost cell are in interphase, thus their entire nuclei are labeled. On the left, a cell is going through mitosis and its DNA has condensed.

In cell biology, the **nucleus** (pl. *nuclei*; from Latin *nucleus* or *nuculeus*, meaning kernel) is a membrane-enclosed organelle found in eukaryotic cells. Eukaryotes usually have a single nucleus, but a few cell types have no nuclei, and a few others have many.

Cell nuclei contain most of the cell's genetic material, organized as multiple long linear DNA molecules in complex with a large variety of proteins, such as histones, to form chromosomes. The genes within these chromosomes are the cell's nuclear genome. The function of the nucleus is to maintain the integrity of these genes and to control the activities of the cell by regulating gene expression—the nucleus is, therefore, the control center of the cell. The main structures making up the nucleus are the nuclear envelope, a double membrane that encloses the entire organelle and isolates its contents from the cellular cytoplasm, and the nucleoskeleton (which includes nuclear lamina), a network within the nucleus that adds mechanical support, much like the cytoskeleton, which supports the cell as a whole.

Because the nuclear membrane is impermeable to large molecules, nuclear pores are required that regulate nuclear transport of molecules across the envelope. The pores cross both nuclear membranes, providing a channel through which larger molecules must be actively transported by carrier proteins while allowing free movement of small molecules and ions. Movement of large molecules such as proteins and RNA through the pores is required for both gene expression and the maintenance of chromosomes. The interior of the nucleus does not contain any membrane-bound sub compartments, its contents are not uniform, and a number of *sub-nuclear bodies* exist, made up of unique proteins, RNA molecules, and particular parts of the chromosomes. The best-known of these is the nucleolus, which is mainly involved in the assembly of ribosomes. After being produced in the nucleolus, ribosomes are exported to the cytoplasm where they translate mRNA.

15.1 History

Oldest known depiction of cells and their nuclei by Antonie van Leeuwenhoek, 1719

The nucleus was the first organelle to be discovered. What is most likely the oldest preserved drawing dates back to the early microscopist Antonie van Leeuwenhoek (1632–1723). He observed a "Lumen", the nucleus, in the red blood cells of salmon.[1] Unlike mammalian red blood cells, those of other vertebrates still contain nuclei.

The nucleus was also described by Franz Bauer in 1804[2] and in more detail in 1831 by Scottish botanist Robert Brown in a talk at the Linnean Society of London. Brown was studying orchids under microscope when he observed an opaque area, which he called the areola or nucleus, in the cells of the flower's outer layer.[3]

He did not suggest a potential function. In 1838, Matthias Schleiden proposed that the nucleus plays a role in generating cells, thus he introduced the name "Cytoblast" (cell builder). He believed that he had observed new cells assembling around "cytoblasts". Franz Meyen was a strong opponent of this view, having already described cells multiplying by division and believing that many cells would have no nuclei. The idea that cells can be generated de novo, by the "cytoblast" or otherwise, contradicted work by Robert Remak (1852) and Rudolf Virchow (1855) who decisively propagated the new paradigm that cells are generated solely by cells ("Omnis cellula e cellula"). The function of the nucleus remained unclear.[4]

Between 1877 and 1878, Oscar Hertwig published several studies on the fertilization of sea urchin eggs, showing that the nucleus of the sperm enters the oocyte and fuses with its nucleus. This was the first time it was suggested that an individual develops from a (single) nucleated cell. This was in contradiction to Ernst Haeckel's theory that the complete phylogeny of a species would be repeated during embryonic development, including generation of the first nucleated cell from a "Monerula", a structureless mass of primordial mucus ("Urschleim"). Therefore, the necessity of the sperm nucleus for

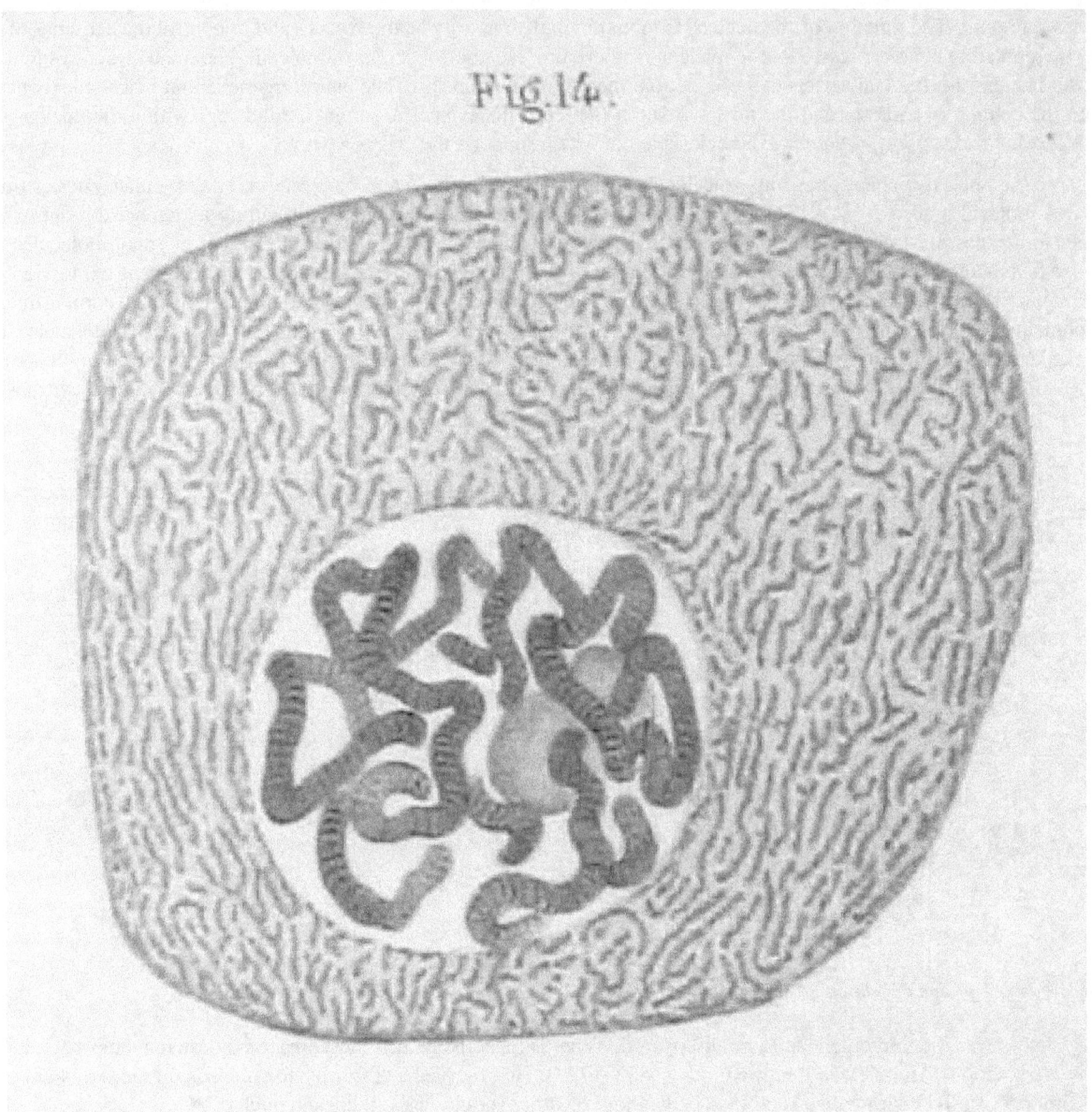

Drawing of a Chironomus *salivary gland cell published by Walther Flemming in 1882. The nucleus contains Polytene chromosomes.*

fertilization was discussed for quite some time. However, Hertwig confirmed his observation in other animal groups, including amphibians and molluscs. Eduard Strasburger produced the same results for plants in 1884. This paved the way to assign the nucleus an important role in heredity. In 1873, August Weismann postulated the equivalence of the maternal and paternal germ *cells* for heredity. The function of the nucleus as carrier of genetic information became clear only later, after mitosis was discovered and the Mendelian rules were rediscovered at the beginning of the 20th century; the chromosome theory of heredity was therefore developed.[4]

15.2 Structures

The nucleus is the largest cellular organelle in animal cells.[5] In mammalian cells, the average diameter of the nucleus is approximately 6 micrometres (μm), which occupies about 10% of the total cell volume.[6] The viscous liquid within it is called nucleoplasm, and is similar in composition to the cytosol found outside the nucleus.[7] It appears as a dense, roughly spherical or irregular organelle.

15.2.1 Nuclear envelope and pores

Main articles: Nuclear envelope and Nuclear pores

The nuclear envelope, otherwise known as nuclear membrane, consists of two cellular membranes, an inner and an outer membrane, arranged parallel to one another and separated by 10 to 50 nanometres (nm). The nuclear envelope completely encloses the nucleus and separates the cell's genetic material from the surrounding cytoplasm, serving as a barrier to prevent macromolecules from diffusing freely between the nucleoplasm and the cytoplasm.[8] The outer nuclear membrane is continuous with the membrane of the rough endoplasmic reticulum (RER), and is similarly studded with ribosomes.[8] The space between the membranes is called the perinuclear space and is continuous with the RER lumen.

Nuclear pores, which provide aqueous channels through the envelope, are composed of multiple proteins, collectively referred to as nucleoporins. The pores are about 125 million daltons in molecular weight and consist of around 50 (in yeast) to several hundred proteins (in vertebrates).[5] The pores are 100 nm in total diameter; however, the gap through which molecules freely diffuse is only about 9 nm wide, due to the presence of regulatory systems within the center of the pore. This size selectively allows the passage of small water-soluble molecules while preventing larger molecules, such as nucleic acids and larger proteins, from inappropriately entering or exiting the nucleus. These large molecules must be actively transported into the nucleus instead. The nucleus of a typical mammalian cell will have about 3000 to 4000 pores throughout its envelope,[9] each of which contains an eightfold-symmetric ring-shaped structure at a position where the inner and outer membranes fuse.[10] Attached to the ring is a structure called the *nuclear basket* that extends into the nucleoplasm, and a series of filamentous extensions that reach into the cytoplasm. Both structures serve to mediate binding to nuclear transport proteins.[5]

Most proteins, ribosomal subunits, and some DNAs are transported through the pore complexes in a process mediated by a family of transport factors known as karyopherins. Those karyopherins that mediate movement into the nucleus are also called importins, whereas those that mediate movement out of the nucleus are called exportins. Most karyopherins interact directly with their cargo, although some use adaptor proteins.[11] Steroid hormones such as cortisol and aldosterone, as well as other small lipid-soluble molecules involved in intercellular signaling, can diffuse through the cell membrane and into the cytoplasm, where they bind nuclear receptor proteins that are trafficked into the nucleus. There they serve as transcription factors when bound to their ligand; in the absence of ligand, many such receptors function as histone deacetylases that repress gene expression.[5]

15.2.2 Nuclear lamina

Main article: Nuclear lamina

In animal cells, two networks of intermediate filaments provide the nucleus with mechanical support: The nuclear lamina forms an organized meshwork on the internal face of the envelope, while less organized support is provided on the cytosolic face of the envelope. Both systems provide structural support for the nuclear envelope and anchoring sites for chromosomes and nuclear pores.[6]

The nuclear lamina is composed mostly of lamin proteins. Like all proteins, lamins are synthesized in the cytoplasm and later transported to the nucleus interior, where they are assembled before being incorporated into the existing network of nuclear lamina.[12][13] Lamins found on the cytosolic face of the membrane, such as emerin and nesprin, bind to the cytoskeleton to provide structural support. Lamins are also found inside the nucleoplasm where they form another regular structure, known as the *nucleoplasmic veil*,[14] that is visible using fluorescence microscopy. The actual function of the veil is not clear, although it is excluded from the nucleolus and is present during interphase.[15] Lamin structures that make up the veil, such as LEM3, bind chromatin and disrupting their structure inhibits transcription of protein-coding genes.[16]

Like the components of other intermediate filaments, the lamin monomer contains an alpha-helical domain used by two monomers to coil around each other, forming a dimer structure called a coiled coil. Two of these dimer structures then join side by side, in an antiparallel arrangement, to form a tetramer called a *protofilament*. Eight of these protofilaments form a lateral arrangement that is twisted to form a ropelike *filament*. These filaments can be assembled or disassembled in a dynamic manner, meaning that changes in the length of the filament depend on the competing rates of filament addition and removal.[6]

Mutations in lamin genes leading to defects in filament assembly cause a group of rare genetic disorders known as *laminopathies*. The most notable laminopathy is the family of diseases known as progeria, which causes the appearance of premature aging in its sufferers. The exact mechanism by which the associated biochemical changes give rise to the aged phenotype is not well understood.[17]

15.2.3 Chromosomes

Main article: Chromosome

The cell nucleus contains the majority of the cell's genetic material in the form of multiple linear DNA molecules organized into structures called chromosomes. Each human cell contains roughly two meters of DNA. During most of the cell cycle these are organized in a DNA-protein complex known as chromatin, and during cell division the chromatin can be seen to form the well-defined chromosomes familiar from a karyotype. A small fraction of the cell's genes are located instead in the mitochondria.

There are two types of chromatin. Euchromatin is the less compact DNA form, and contains genes that are frequently expressed by the cell.[18] The other type, heterochromatin, is the more compact form, and contains DNA that is infrequently transcribed. This structure is further categorized into *facultative* heterochromatin, consisting of genes that are organized as heterochromatin only in certain cell types or at certain stages of development, and *constitutive* heterochromatin that consists of chromosome structural components such as telomeres and centromeres.[19] During interphase the chromatin organizes itself into discrete individual patches,[20] called *chromosome territories*.[21] Active genes, which are generally found in the euchromatic region of the chromosome, tend to be located towards the chromosome's territory boundary.[22]

Antibodies to certain types of chromatin organization, in particular, nucleosomes, have been associated with a number of autoimmune diseases, such as systemic lupus erythematosus.[23] These are known as anti-nuclear antibodies (ANA) and have also been observed in concert with multiple sclerosis as part of general immune system dysfunction.[24] As in the case of progeria, the role played by the antibodies in inducing the symptoms of autoimmune diseases is not obvious.

15.2.4 Nucleolus

Main article: Nucleolus

The nucleolus is a discrete densely stained structure found in the nucleus. It is not surrounded by a membrane, and is sometimes called a *suborganelle*. It forms around tandem repeats of rDNA, DNA coding for ribosomal RNA (rRNA). These regions are called nucleolar organizer regions (NOR). The main roles of the nucleolus are to synthesize rRNA and assemble ribosomes. The structural cohesion of the nucleolus depends on its activity, as ribosomal assembly in the nucleolus results in the transient association of nucleolar components, facilitating further ribosomal assembly, and hence further association. This model is supported by observations that inactivation of rDNA results in intermingling of nucleolar structures.[25]

In the first step of ribosome assembly, a protein called RNA polymerase I transcribes rDNA, which forms a large pre-rRNA precursor. This is cleaved into the subunits 5.8S, 18S, and 28S rRNA.[26] The transcription, post-transcriptional processing, and assembly of rRNA occurs in the nucleolus, aided by small nucleolar RNA (snoRNA) molecules, some of which are derived from spliced introns from messenger RNAs encoding genes related to ribosomal function. The assembled ribosomal subunits are the largest structures passed through the nuclear pores.[5]

When observed under the electron microscope, the nucleolus can be seen to consist of three distinguishable regions: the innermost *fibrillar centers* (FCs), surrounded by the *dense fibrillar component* (DFC), which in turn is bordered by the *granular component* (GC). Transcription of the rDNA occurs either in the FC or at the FC-DFC boundary, and, therefore, when rDNA transcription in the cell is increased, more FCs are detected. Most of the cleavage and modification of rRNAs occurs in the DFC, while the latter steps involving protein assembly onto the ribosomal subunits occur in the GC.[26]

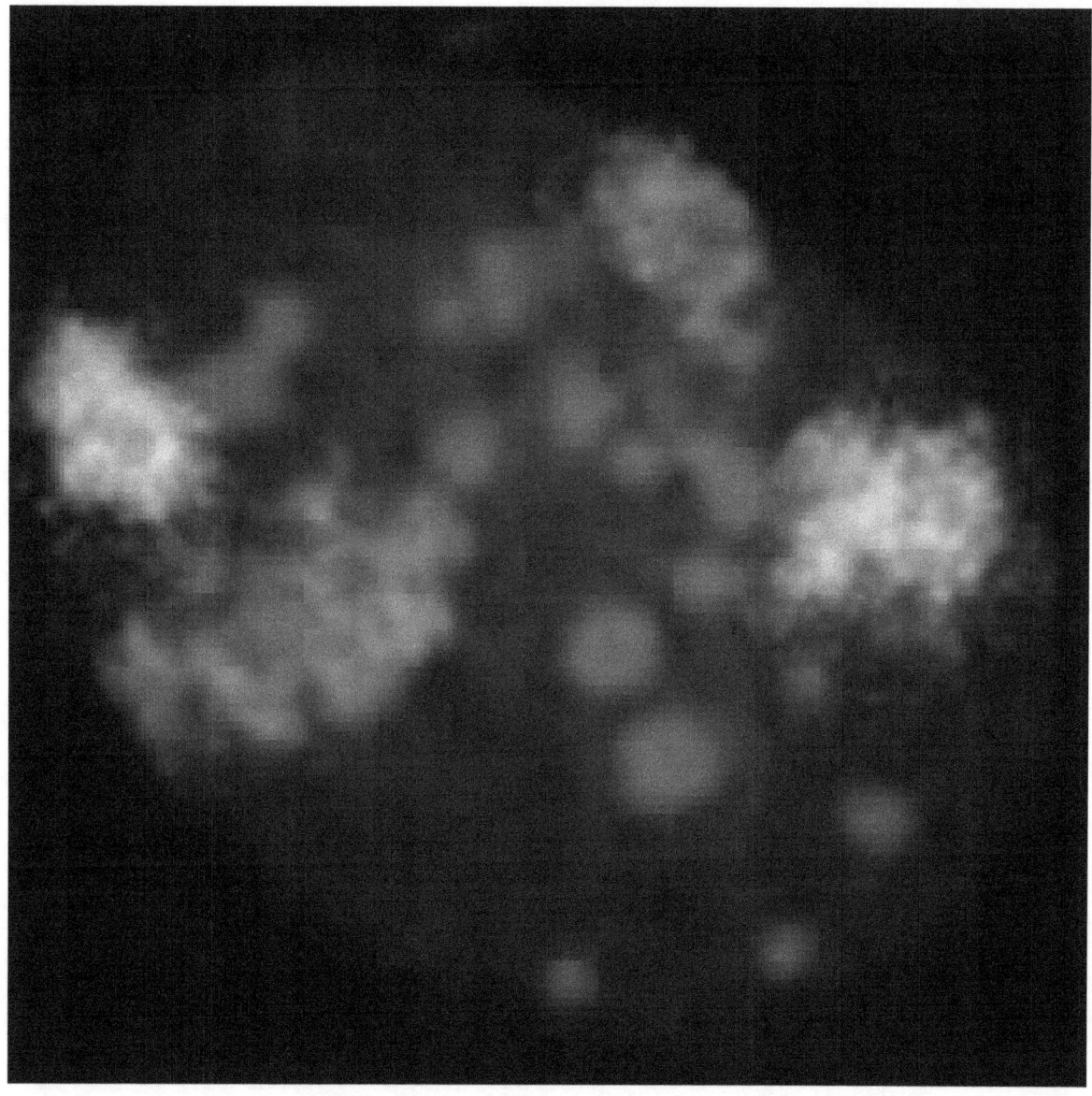

A mouse fibroblast nucleus in which DNA is stained blue. The distinct chromosome territories of chromosome 2 (red) and chromosome 9 (green) are stained with fluorescent in situ hybridization.

15.2.5   Other subnuclear bodies

Besides the nucleolus, the nucleus contains a number of other non-membrane-delineated bodies. These include Cajal bodies, Gemini of coiled bodies, polymorphic interphase karyosomal association (PIKA), promyelocytic leukaemia (PML) bodies, paraspeckles, and splicing speckles. Although little is known about a number of these domains, they are significant in that they show that the nucleoplasm is not a uniform mixture, but rather contains organized functional subdomains.[29]

Other subnuclear structures appear as part of abnormal disease processes. For example, the presence of small intranuclear rods has been reported in some cases of nemaline myopathy. This condition typically results from mutations in actin, and the rods themselves consist of mutant actin as well as other cytoskeletal proteins.[31]

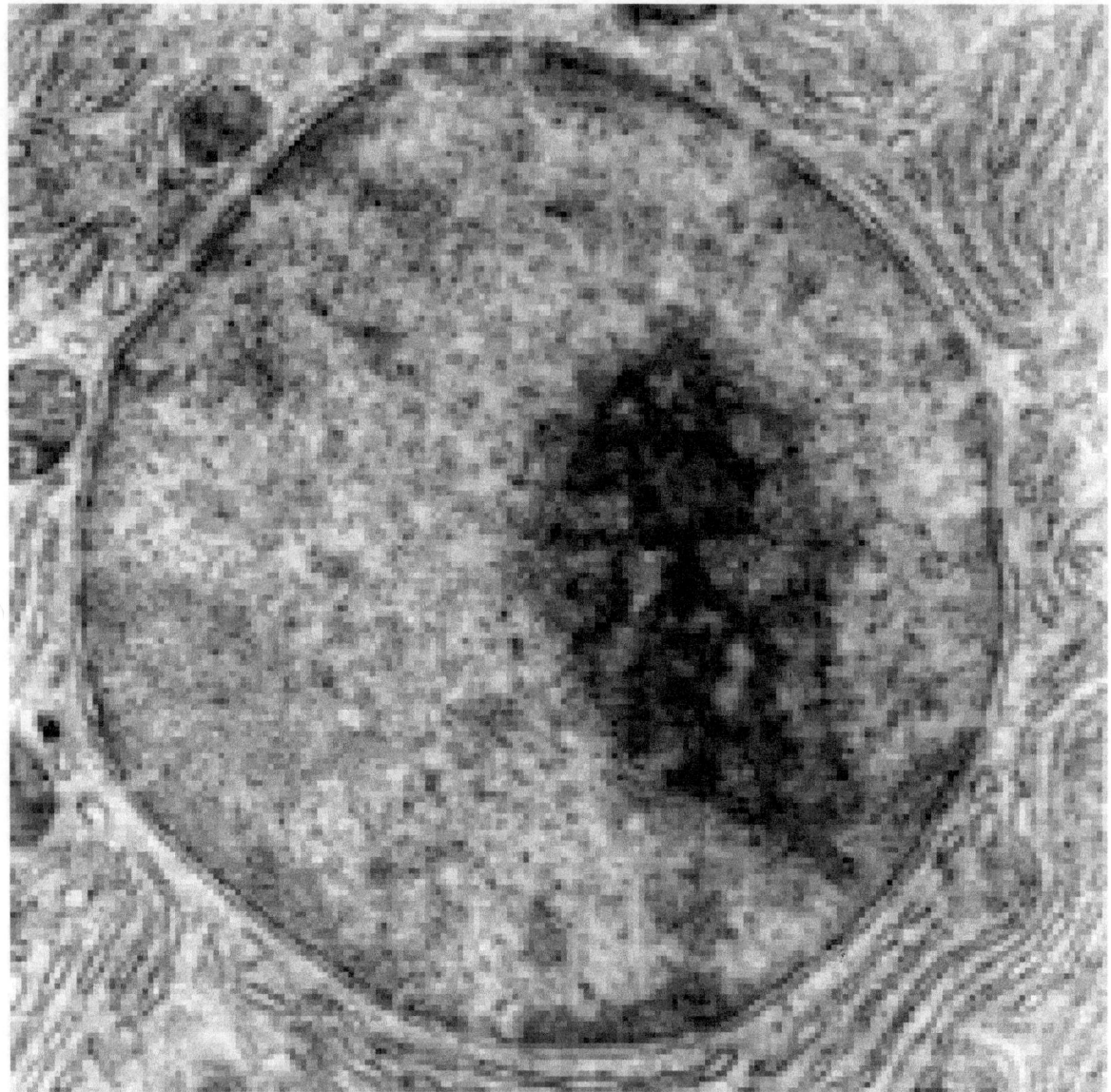

An electron micrograph of a cell nucleus, showing the darkly stained nucleolus

Cajal bodies and gems

A nucleus typically contains between 1 and 10 compact structures called Cajal bodies or coiled bodies (CB), whose diameter measures between 0.2 μm and 2.0 μm depending on the cell type and species.[27] When seen under an electron microscope, they resemble balls of tangled thread[28] and are dense foci of distribution for the protein coilin.[32] CBs are involved in a number of different roles relating to RNA processing, specifically small nucleolar RNA (snoRNA) and small nuclear RNA (snRNA) maturation, and histone mRNA modification.[27]

Similar to Cajal bodies are Gemini of coiled bodies, or gems, whose name is derived from the Gemini constellation in reference to their close "twin" relationship with CBs. Gems are similar in size and shape to CBs, and in fact are virtually indistinguishable under the microscope.[32] Unlike CBs, gems do not contain small nuclear ribonucleoproteins (snRNPs), but do contain a protein called survival of motor neuron (SMN) whose function relates to snRNP biogenesis. Gems are believed to assist CBs in snRNP biogenesis,[33] though it has also been suggested from microscopy evidence that CBs and gems are different manifestations of the same structure.[32]

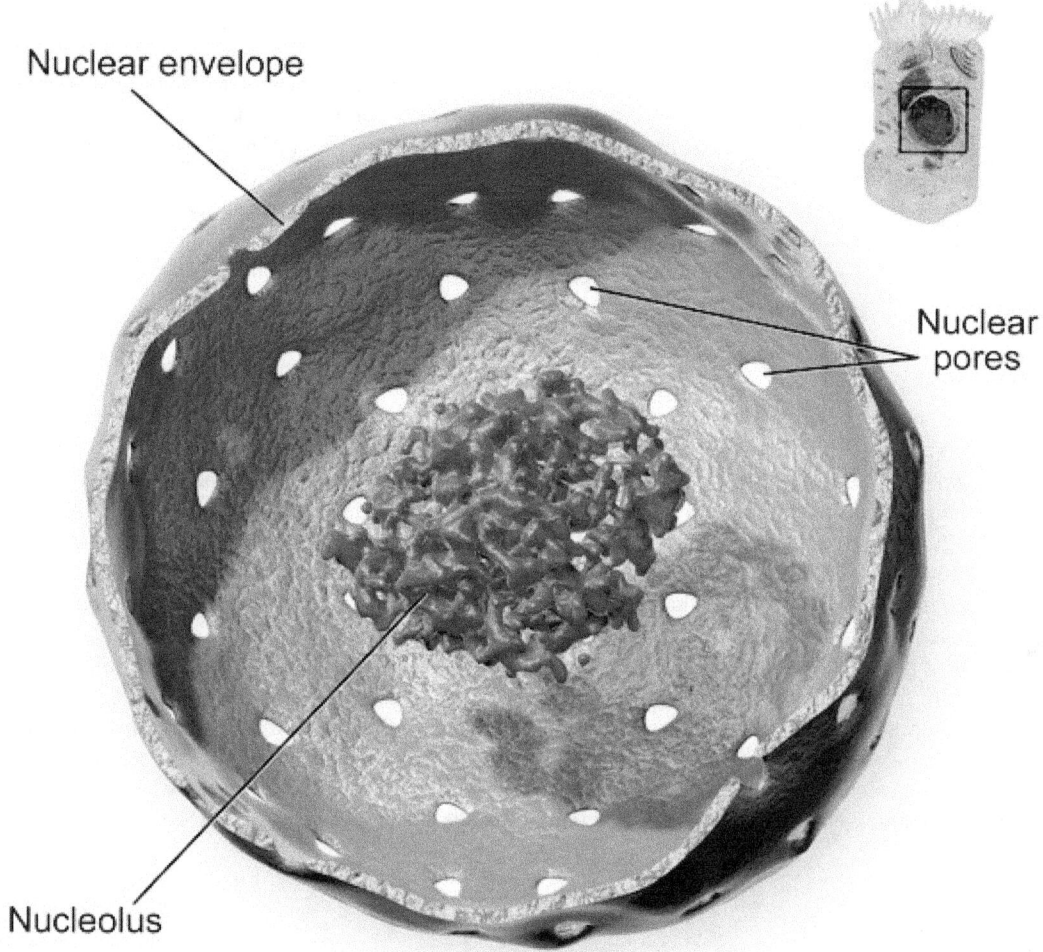

3D rendering of nucleus with location of nucleolus

RAFA and PTF domains

RAFA domains, or polymorphic interphase karyosomal associations, were first described in microscopy studies in 1991. Their function was and remains unclear, though they were not thought to be associated with active DNA replication, transcription, or RNA processing.[34] They have been found to often associate with discrete domains defined by dense localization of the transcription factor PTF, which promotes transcription of small nuclear RNA (snRNA).[35]

PML bodies

Promyelocytic leukaemia bodies (PML bodies) are spherical bodies found scattered throughout the nucleoplasm, measuring around 0.1–1.0 μm. They are known by a number of other names, including nuclear domain 10 (ND10), Kremer bodies, and PML oncogenic domains. PML bodies are named after one of their major components, the promyelocytic leukemia protein (PML). They are often seen in the nucleus in association with Cajal bodies and cleavage bodies.[29] PML bodies belong to the nuclear matrix, an ill-defined super-structure of the nucleus proposed to anchor and regulate many nuclear functions, including DNA replication, transcription, or epigenetic silencing.[36] The PML protein is the key organizer of these domains that recruits an ever-growing number of proteins, whose only common known feature to date is their ability to be SUMOylated. Yet, pml-/- mice (which have their PML gene deleted) cannot assemble nuclear bodies, develop normally and live well, demonstrating that PML bodies are dispensable for most basic biological functions.[36]

Splicing speckles

Speckles are subnuclear structures that are enriched in pre-messenger RNA splicing factors and are located in the interchromatin regions of the nucleoplasm of mammalian cells. At the fluorescence-microscope level they appear as irregular, punctate structures, which vary in size and shape, and when examined by electron microscopy they are seen as clusters of interchromatin granules. Speckles are dynamic structures, and both their protein and RNA-protein components can cycle continuously between speckles and other nuclear locations, including active transcription sites. Studies on the composition, structure and behaviour of speckles have provided a model for understanding the functional compartmentalization of the nucleus and the organization of the gene-expression machinery[37] splicing snRNPs[38][39] and other splicing proteins necessary for pre-mRNA processing.[40] Because of a cell's changing requirements, the composition and location of these bodies changes according to mRNA transcription and regulation via phosphorylation of specific proteins.[41] The splicing speckles are also known as nuclear speckles (nuclear specks), splicing factor compartments (SF compartments), interchromatin granule clusters (IGCs), B snurposomes.[42] B snurposomes are found in the amphibian oocyte nuclei and in *Drosophila melanogaster* embryos. B snurposomes appear alone or attached to the Cajal bodies in the electron micrographs of the amphibian nuclei.[43] IGCs function as storage sites for the splicing factors.[44]

Paraspeckles

Main article: Paraspeckle

Discovered by Fox et al. in 2002, paraspeckles are irregularly shaped compartments in the nucleus' interchromatin space.[45] First documented in HeLa cells, where there are generally 10–30 per nucleus,[46] paraspeckles are now known to also exist in all human primary cells, transformed cell lines, and tissue sections.[47] Their name is derived from their distribution in the nucleus; the "para" is short for parallel and the "speckles" refers to the splicing speckles to which they are always in close proximity.[46]

Paraspeckles are dynamic structures that are altered in response to changes in cellular metabolic activity. They are transcription dependent[45] and in the absence of RNA Pol II transcription, the paraspeckle disappears and all of its associated protein components (PSP1, p54nrb, PSP2, CFI(m)68, and PSF) form a crescent shaped perinucleolar cap in the nucleolus. This phenomenon is demonstrated during the cell cycle. In the cell cycle, paraspeckles are present during interphase and during all of mitosis except for telophase. During telophase, when the two daughter nuclei are formed, there is no RNA Pol II transcription so the protein components instead form a perinucleolar cap.[47]

Perichromatin fibrils

Perichromatin fibrils are visible only under electron microscope. They are located next to the transcriptionally active chromatin and are hypothesized to be the sites of active pre-mRNA processing.[44]

15.3   Function

The nucleus provides a site for genetic transcription that is segregated from the location of translation in the cytoplasm, allowing levels of gene regulation that are not available to prokaryotes. The main function of the cell nucleus is to control gene expression and mediate the replication of DNA during the cell cycle.

15.3.1   Cell compartmentalization

The nuclear envelope allows the nucleus to control its contents, and separate them from the rest of the cytoplasm where necessary. This is important for controlling processes on either side of the nuclear membrane. In most cases where a cytoplasmic process needs to be restricted, a key participant is removed to the nucleus, where it interacts with transcription factors to downregulate the production of certain enzymes in the pathway. This regulatory mechanism occurs in the case of glycolysis, a cellular pathway for breaking down glucose to produce energy. Hexokinase is an enzyme responsible for the first the step of glycolysis, forming glucose-6-phosphate from glucose. At high concentrations of fructose-6-phosphate, a molecule made later from glucose-6-phosphate, a regulator protein removes hexokinase to the nucleus,[48] where it forms a transcriptional repressor complex with nuclear proteins to reduce the expression of genes involved in glycolysis.[49]

In order to control which genes are being transcribed, the cell separates some transcription factor proteins responsible for regulating gene expression from physical access to the DNA until they are activated by other signaling pathways. This prevents even low levels of inappropriate gene expression. For example, in the case of NF-κB-controlled genes, which are involved in most inflammatory responses, transcription is induced in response to a signal pathway such as that initiated by the signaling molecule TNF-α, binds to a cell membrane receptor, resulting in the recruitment of signalling proteins, and eventually activating the transcription factor NF-κB. A nuclear localisation signal on the NF-κB protein allows it to be transported through the nuclear pore and into the nucleus, where it stimulates the transcription of the target genes.[6]

The compartmentalization allows the cell to prevent translation of unspliced mRNA.[50] Eukaryotic mRNA contains introns that must be removed before being translated to produce functional proteins. The splicing is done inside the nucleus before the mRNA can be accessed by ribosomes for translation. Without the nucleus, ribosomes would translate newly transcribed (unprocessed) mRNA, resulting in malformed and nonfunctional proteins.

15.3.2   Gene expression

Main article: Gene expression

Gene expression first involves transcription, in which DNA is used as a template to produce RNA. In the case of genes encoding proteins, that RNA produced from this process is messenger RNA (mRNA), which then needs to be translated by ribosomes to form a protein. As ribosomes are located outside the nucleus, mRNA produced needs to be exported.[51]

Since the nucleus is the site of transcription, it also contains a variety of proteins that either directly mediate transcription or are involved in regulating the process. These proteins include helicases, which unwind the double-stranded DNA molecule to facilitate access to it, RNA polymerases, which synthesize the growing RNA molecule, topoisomerases, which change the amount of supercoiling in DNA, helping it wind and unwind, as well as a large variety of transcription factors that regulate expression.[52]

15.3.3   Processing of pre-mRNA

Main article: Post-transcriptional modification

Newly synthesized mRNA molecules are known as primary transcripts or pre-mRNA. They must undergo post-transcriptional modification in the nucleus before being exported to the cytoplasm; mRNA that appears in the cytoplasm without these modifications is degraded rather than used for protein translation. The three main modifications are 5' capping, 3' polyadenylation, and RNA splicing. While in the nucleus, pre-mRNA is associated with a variety of proteins in complexes known as heterogeneous ribonucleoprotein particles (hnRNPs). Addition of the 5' cap occurs co-transcriptionally and is the first step in post-transcriptional modification. The 3' poly-adenine tail is only added after transcription is complete.

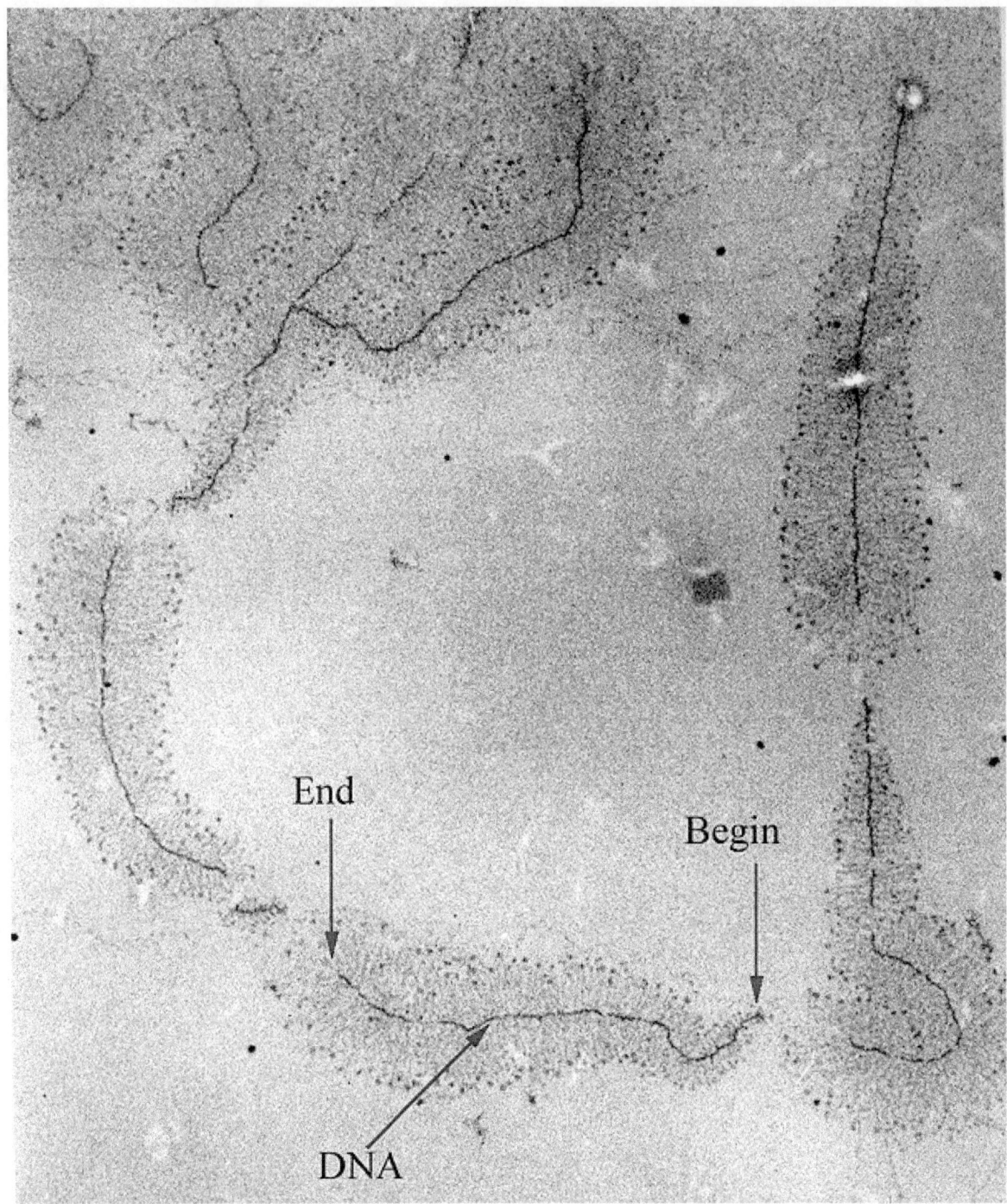

A micrograph of ongoing gene transcription of ribosomal RNA illustrating the growing primary transcripts. "Begin" indicates the 5' end of the DNA, where new RNA synthesis begins; "end" indicates the 3' end, where the primary transcripts are almost complete.

RNA splicing, carried out by a complex called the spliceosome, is the process by which introns, or regions of DNA that do not code for protein, are removed from the pre-mRNA and the remaining exons connected to re-form a single continuous molecule. This process normally occurs after 5' capping and 3' polyadenylation but can begin before synthesis is complete in transcripts with many exons.[5] Many pre-mRNAs, including those encoding antibodies, can be spliced in multiple ways to produce different mature mRNAs that encode different protein sequences. This process is known as alternative splicing, and allows production of a large variety of proteins from a limited amount of DNA.

15.4 Dynamics and regulation

15.4.1 Nuclear transport

Main article: Nuclear transport

The entry and exit of large molecules from the nucleus is tightly controlled by the nuclear pore complexes. Although

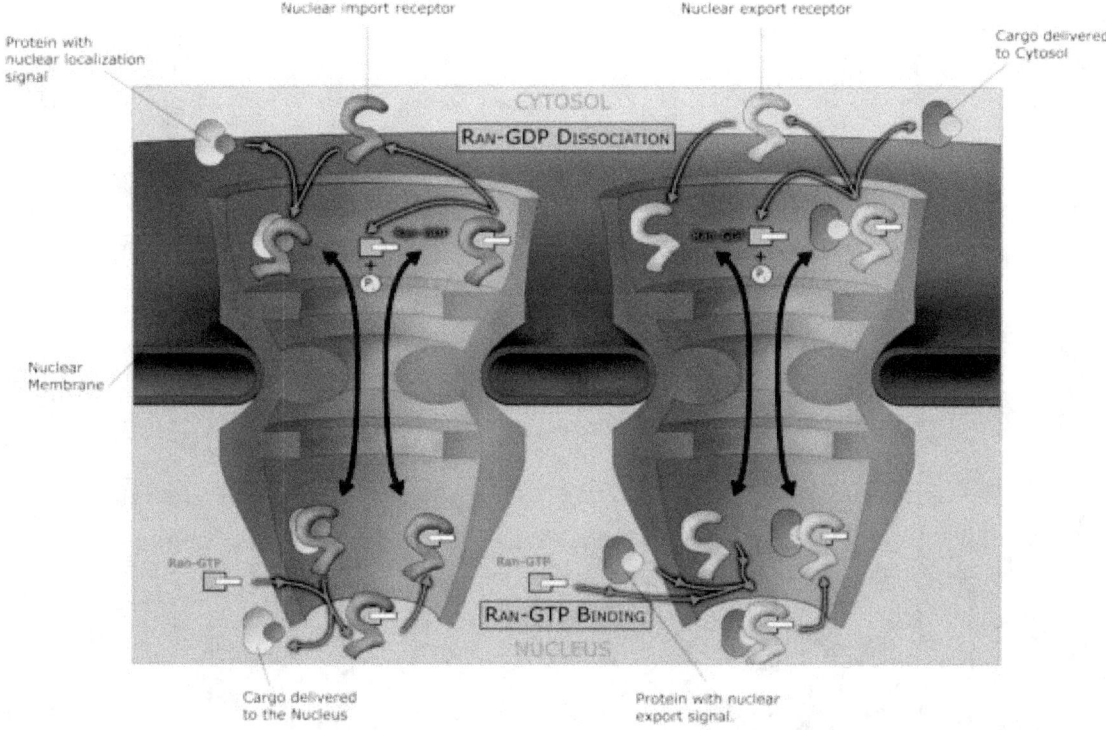

Macromolecules, such as RNA and proteins, are actively transported across the nuclear membrane in a process called the Ran-GTP nuclear transport cycle.

small molecules can enter the nucleus without regulation,[53] macromolecules such as RNA and proteins require association karyopherins called importins to enter the nucleus and exportins to exit. "Cargo" proteins that must be translocated from the cytoplasm to the nucleus contain short amino acid sequences known as nuclear localization signals, which are bound by importins, while those transported from the nucleus to the cytoplasm carry nuclear export signals bound by exportins. The ability of importins and exportins to transport their cargo is regulated by GTPases, enzymes that hydrolyze the molecule guanosine triphosphate to release energy. The key GTPase in nuclear transport is Ran, which can bind either GTP or GDP (guanosine diphosphate), depending on whether it is located in the nucleus or the cytoplasm. Whereas importins depend on RanGTP to dissociate from their cargo, exportins require RanGTP in order to bind to their cargo.[11]

Nuclear import depends on the importin binding its cargo in the cytoplasm and carrying it through the nuclear pore into the nucleus. Inside the nucleus, RanGTP acts to separate the cargo from the importin, allowing the importin to exit the nucleus and be reused. Nuclear export is similar, as the exportin binds the cargo inside the nucleus in a process facilitated by RanGTP, exits through the nuclear pore, and separates from its cargo in the cytoplasm.

Specialized export proteins exist for translocation of mature mRNA and tRNA to the cytoplasm after post-transcriptional modification is complete. This quality-control mechanism is important due to these molecules' central role in protein translation. Mis-expression of a protein due to incomplete excision of exons or mis-incorporation of amino acids could have negative consequences for the cell; thus, incompletely modified RNA that reaches the cytoplasm is degraded rather than used in translation.[5]

15.4.2   Assembly and disassembly

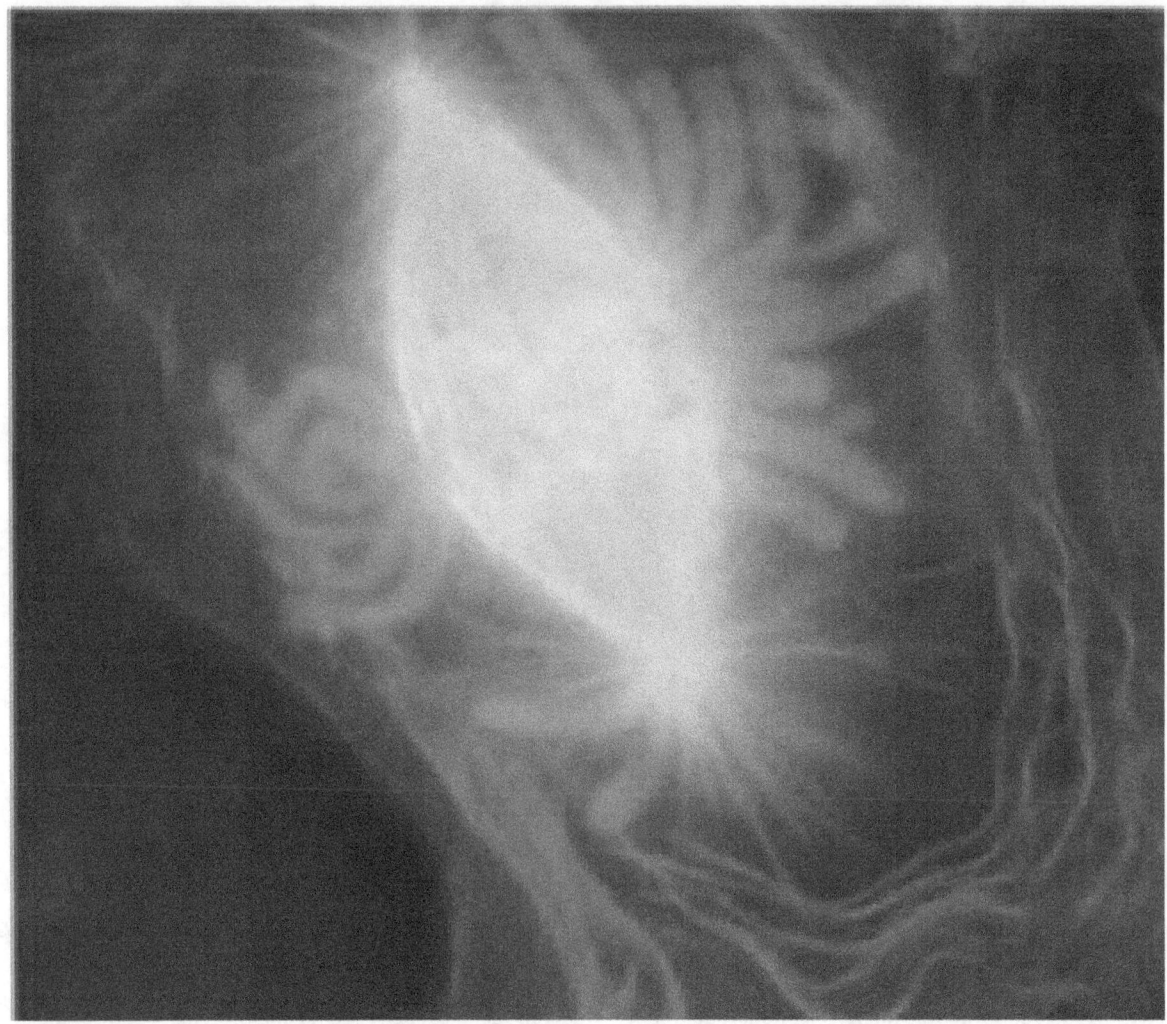

An image of a newt lung cell stained with fluorescent dyes during metaphase. The mitotic spindle can be seen, stained green, attached to the two sets of chromosomes, stained light blue. All chromosomes but one are already at the metaphase plate.

During its lifetime, a nucleus may be broken down or destroyed, either in the process of cell division or as a consequence of apoptosis (the process of programmed cell death). During these events, the structural components of the nucleus — the envelope and lamina — can be systematically degraded. In most cells, the disassembly of the nuclear envelope marks the end of the prophase of mitosis. However, this disassembly of the nucleus is not a universal feature of mitosis and does not occur in all cells. Some unicellular eukaryotes (e.g., yeasts) undergo so-called closed mitosis, in which the nuclear envelope remains intact. In closed mitosis, the daughter chromosomes migrate to opposite poles of the nucleus, which then divides in two. The cells of higher eukaryotes, however, usually undergo open mitosis, which is characterized by breakdown of the nuclear envelope. The daughter chromosomes then migrate to opposite poles of the mitotic spindle, and new nuclei reassemble around them.

At a certain point during the cell cycle in open mitosis, the cell divides to form two cells. In order for this process to be possible, each of the new daughter cells must have a full set of genes, a process requiring replication of the chromosomes as well as segregation of the separate sets. This occurs by the replicated chromosomes, the sister chromatids, attaching to microtubules, which in turn are attached to different centrosomes. The sister chromatids can then be pulled to separate locations in the cell. In many cells, the centrosome is located in the cytoplasm, outside the nucleus; the microtubules would be unable to attach to the chromatids in the presence of the nuclear envelope.[54] Therefore, the early stages in the cell cycle, beginning in prophase and until around prometaphase, the nuclear membrane is dismantled.[14] Likewise, during the same period, the nuclear lamina is also disassembled, a process regulated by phosphorylation of the lamins by protein kinases such as the CDC2 protein kinase.[55] Towards the end of the cell cycle, the nuclear membrane is reformed, and around the same time, the nuclear lamina are reassembled by dephosphorylating the lamins.[55]

However, in dinoflagellates, the nuclear envelope remains intact, the centrosomes are located in the cytoplasm, and the microtubules come in contact with chromosomes, whose centromeric regions are incorporated into the nuclear envelope (the so-called closed mitosis with extranuclear spindle). In many other protists (e.g., ciliates, sporozoans) and fungi, the centrosomes are intranuclear, and their nuclear envelope also does not disassemle during cell division.

Apoptosis is a controlled process in which the cell's structural components are destroyed, resulting in death of the cell. Changes associated with apoptosis directly affect the nucleus and its contents, for example, in the condensation of chromatin and the disintegration of the nuclear envelope and lamina. The destruction of the lamin networks is controlled by specialized apoptotic proteases called caspases, which cleave the lamin proteins and, thus, degrade the nucleus' structural integrity. Lamin cleavage is sometimes used as a laboratory indicator of caspase activity in assays for early apoptotic activity.[14] Cells that express mutant caspase-resistant lamins are deficient in nuclear changes related to apoptosis, suggesting that lamins play a role in initiating the events that lead to apoptotic degradation of the nucleus.[14] Inhibition of lamin assembly itself is an inducer of apoptosis.[56]

The nuclear envelope acts as a barrier that prevents both DNA and RNA viruses from entering the nucleus. Some viruses require access to proteins inside the nucleus in order to replicate and/or assemble. DNA viruses, such as herpesvirus replicate and assemble in the cell nucleus, and exit by budding through the inner nuclear membrane. This process is accompanied by disassembly of the lamina on the nuclear face of the inner membrane.[14]

15.4.3   Disease-related dynamics

Initially, it has been suspected that immunoglobulins in general and autoantibodies in particular do not enter the nucleus. Now there is a body of evidence that under pathological conditions (e.g. lupus erythematosus) IgG can enter the nucleus.[57]

15.5   Nuclei per cell

Most eukaryotic cell types usually have a single nucleus, but some have no nuclei, while others have several. This can result from normal development, as in the maturation of mammalian red blood cells, or from faulty cell division.

15.5.1   Anucleated cells

Anucleated cells contain no nucleus and are, therefore, incapable of dividing to produce daughter cells. The best-known anucleated cell is the mammalian red blood cell, or erythrocyte, which also lacks other organelles such as mitochondria, and serves primarily as a transport vessel to ferry oxygen from the lungs to the body's tissues. Erythrocytes mature through erythropoiesis in the bone marrow, where they lose their nuclei, organelles, and ribosomes. The nucleus is expelled during the process of differentiation from an erythroblast to a reticulocyte, which is the immediate precursor of the mature erythrocyte.[58] The presence of mutagens may induce the release of some immature "micronucleated" erythrocytes into the bloodstream.[59][60] Anucleated cells can also arise from flawed cell division in which one daughter lacks a nucleus and the other has two nuclei.

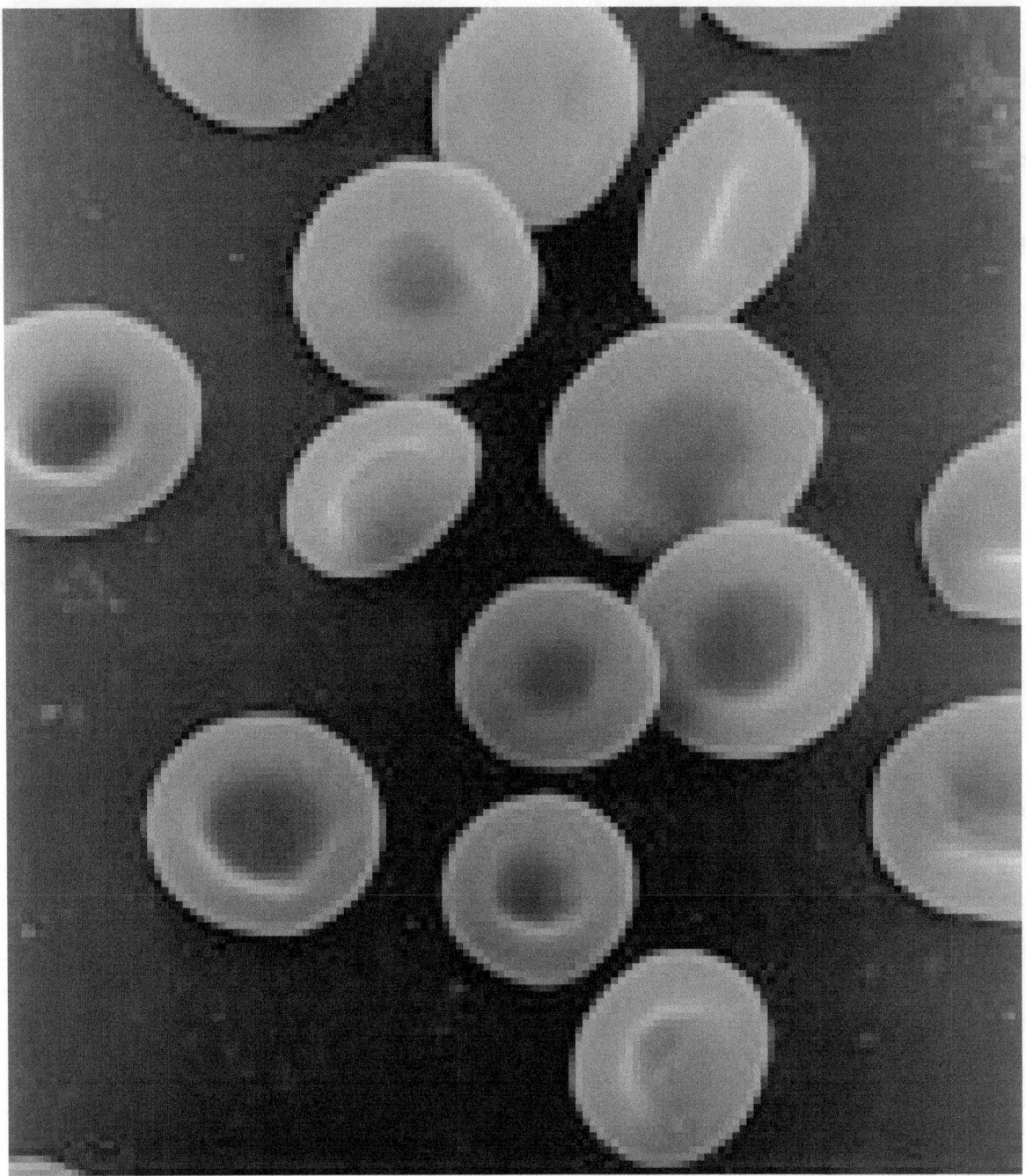

Human red blood cells, like those of other mammals, lack nuclei. This occurs as a normal part of the cells' development.

15.5.2 Multinucleated cells

Multinucleated cells contain multiple nuclei. Most acantharean species of protozoa[61] and some fungi in mycorrhizae[62] have naturally multinucleated cells. Other examples include the intestinal parasites in the genus *Giardia*, which have two nuclei per cell.[63] In humans, skeletal muscle cells, called myocytes and syncytium, become multinucleated during development; the resulting arrangement of nuclei near the periphery of the cells allows maximal intracellular space for myofibrils.[5] Multinucleated and binucleated cells can also be abnormal in humans; for example, cells arising from the fusion of monocytes and macrophages, known as giant multinucleated cells, sometimes accompany inflammation[64] and are also implicated in tumor formation.[65]

A number of dinoflagelates are known to have two nuclei.[66] Unlike other multinucleated cells these nuclei contain two distinct lineages of DNA: one from the dinoflagelate and the other from a symbiotic diatom. Curiously the mitochondrion and the plastid of the diatom remain functional.

15.6 Evolution

As the major defining characteristic of the eukaryotic cell, the nucleus' evolutionary origin has been the subject of much speculation. Four major hypotheses have been proposed to explain the existence of the nucleus, although none have yet earned widespread support.[67]

The first model known as the "syntrophic model" proposes that a symbiotic relationship between the archaea and bacteria created the nucleus-containing eukaryotic cell. (Organisms of the Archaea and Bacteria domain have no cell nucleus.[68]) It is hypothesized that the symbiosis originated when ancient archaea, similar to modern methanogenic archaea, invaded and lived within bacteria similar to modern myxobacteria, eventually forming the early nucleus. This theory is analogous to the accepted theory for the origin of eukaryotic mitochondria and chloroplasts, which are thought to have developed from a similar endosymbiotic relationship between proto-eukaryotes and aerobic bacteria.[69] The archaeal origin of the nucleus is supported by observations that archaea and eukarya have similar genes for certain proteins, including histones. Observations that myxobacteria are motile, can form multicellular complexes, and possess kinases and G proteins similar to eukarya, support a bacterial origin for the eukaryotic cell.[70]

A second model proposes that proto-eukaryotic cells evolved from bacteria without an endosymbiotic stage. This model is based on the existence of modern planctomycetes bacteria that possess a nuclear structure with primitive pores and other compartmentalized membrane structures.[71] A similar proposal states that a eukaryote-like cell, the chronocyte, evolved first and phagocytosed archaea and bacteria to generate the nucleus and the eukaryotic cell.[72]

The most controversial model, known as *viral eukaryogenesis*, posits that the membrane-bound nucleus, along with other eukaryotic features, originated from the infection of a prokaryote by a virus. The suggestion is based on similarities between eukaryotes and viruses such as linear DNA strands, mRNA capping, and tight binding to proteins (analogizing histones to viral envelopes). One version of the proposal suggests that the nucleus evolved in concert with phagocytosis to form an early cellular "predator".[73] Another variant proposes that eukaryotes originated from early archaea infected by poxviruses, on the basis of observed similarity between the DNA polymerases in modern poxviruses and eukaryotes.[74][75] It has been suggested that the unresolved question of the evolution of sex could be related to the viral eukaryogenesis hypothesis.[76]

A more recent proposal, the *exomembrane hypothesis*, suggests that the nucleus instead originated from a single ancestral cell that evolved a second exterior cell membrane; the interior membrane enclosing the original cell then became the nuclear membrane and evolved increasingly elaborate pore structures for passage of internally synthesized cellular components such as ribosomal subunits.[77]

15.7 See also

- Nucleus (neuroanatomy)

15.8 Gallery

- Comparison of human and chimpanzee chromosomes.

- Mouse chromosome territories in different cell types.

- 24 chromosome territories in human cells.

15.9   References

[1] Leeuwenhoek, A. van: Opera Omnia, seu Arcana Naturae ope exactissimorum Microscopiorum detecta, experimentis variis comprobata, Epistolis ad varios illustres viros. J. Arnold et Delphis, A. Beman, Lugdinum Batavorum 1719–1730. Cited after: Dieter Gerlach, Geschichte der Mikroskopie. Verlag Harry Deutsch, Frankfurt am Main, Germany, 2009. ISBN 978-3-8171-1781-9.

[2] Harris, H (1999). *The Birth of the Cell*. New Haven: Yale University Press. ISBN 0-300-07384-4.

[3] Brown, Robert (1866). "On the Organs and Mode of Fecundation of Orchidex and Asclepiadea". *Miscellaneous Botanical Works I*: 511–514.

[4] Cremer, Thomas (1985). *Von der Zellenlehre zur Chromosomentheorie*. Berlin, Heidelberg, New York, Tokyo: Springer Verlag. ISBN 3-540-13987-7. Online Version here

[5] Lodish, H; Berk A; Matsudaira P; Kaiser CA; Krieger M; Scott MP; Zipursky SL; Darnell J. (2004). *Molecular Cell Biology* (5th ed.). New York: WH Freeman. ISBN 0-7167-2672-6.

[6] Bruce Alberts, Alexander Johnson, Julian Lewis, Martin Raff, Keith Roberts, Peter Walter, ed. (2002). *Molecular Biology of the Cell, Chapter 4, pages 191–234* (4th ed.). Garland Science.

[7] Clegg JS (February 1984). "Properties and metabolism of the aqueous cytoplasm and its boundaries". *Am. J. Physiol.* **246** (2 Pt 2): R133–51. PMID 6364846.

[8] Paine P, Moore L, Horowitz S (1975). "Nuclear envelope permeability". *Nature* **254** (5496): 109–114. doi:10.1038/254109a0. PMID 1117994.

[9] Rodney Rhoades, Richard Pflanzer, ed. (1996). "Ch3". *Human Physiology* (3rd ed.). Saunders College Publishing.

[10] Shulga N, Mosammaparast N, Wozniak R, Goldfarb D (2000). "Yeast nucleoporins involved in passive nuclear envelope permeability". *J Cell Biol* **149** (5): 1027–1038. doi:10.1083/jcb.149.5.1027. PMC 2174828. PMID 10831607.

[11] Pemberton L, Paschal B (2005). "Mechanisms of receptor-mediated nuclear import and nuclear export". *Traffic* **6** (3): 187–198. doi:10.1111/j.1600-0854.2005.00270.x. PMID 15702987.

[12] Stuurman N, Heins S, Aebi U (1998). "Nuclear lamins: their structure, assembly, and interactions". *J Struct Biol* **122** (1–2): 42–66. doi:10.1006/jsbi.1998.3987. PMID 9724605.

[13] Goldman A, Moir R, Montag-Lowy M, Stewart M, Goldman R (1992). "Pathway of incorporation of microinjected lamin A into the nuclear envelope". *J Cell Biol* **119** (4): 725–735. doi:10.1083/jcb.119.4.725. PMC 2289687. PMID 1429833.

[14] Goldman R, Gruenbaum Y, Moir R, Shumaker D, Spann T (2002). "Nuclear lamins: building blocks of nuclear architecture". *Genes Dev* **16** (5): 533–547. doi:10.1101/gad.960502. PMID 11877373.

[15] Moir RD, Yoona M, Khuona S, Goldman RD. (2000). "Nuclear Lamins A and B1: Different Pathways of Assembly during Nuclear Envelope Formation in Living Cells". *Journal of Cell Biology* **151** (6): 1155–1168. doi:10.1083/jcb.151.6.1155. PMC 2190592. PMID 11121432.

[16] Spann TP, Goldman AE, Wang C, Huang S, Goldman RD. (2002). "Alteration of nuclear lamin organization inhibits RNA polymerase II–dependent transcription". *Journal of Cell Biology* **156** (4): 603–608. doi:10.1083/jcb.200112047. PMC 2174089. PMID 11854306.

[17] Mounkes LC, Stewart CL (2004). "Aging and nuclear organization: lamins and progeria". *Current Opinion in Cell Biology* **16** (3): 322–327. doi:10.1016/j.ceb.2004.03.009. PMID 15145358.

[18] Ehrenhofer-Murray A (2004). "Chromatin dynamics at DNA replication, transcription and repair". *Eur J Biochem* **271** (12): 2335–2349. doi:10.1111/j.1432-1033.2004.04162.x. PMID 15182349.

[19] Grigoryev S, Bulynko Y, Popova E (2006). "The end adjusts the means: heterochromatin remodelling during terminal cell differentiation". *Chromosome Res* **14** (1): 53–69. doi:10.1007/s10577-005-1021-6. PMID 16506096.

[20] Schardin, Margit; Cremer, T; Hager, HD; Lang, M (December 1985). "Specific staining of human chromosomes in Chinese hamster x man hybrid cell lines demonstrates interphase chromosome territories". *Human Genetics* (Springer Berlin / Heidelberg) **71** (4): 281–287. doi:10.1007/BF00388452. PMID 2416668.

[21] Lamond, Angus I.; William C. Earnshaw (1998-04-24). "Structure and Function in the Nucleus". *Science* **280** (5363): 547–553. doi:10.1126/science.280.5363.547. PMID 9554838.

[22] Kurz, A; Lampel, S; Nickolenko, JE; Bradl, J; Benner, A; Zirbel, RM; Cremer, T; Lichter, P (1996). "Active and inactive genes localize preferentially in the periphery of chromosome territories". *The Journal of Cell Biology* (The Rockefeller University Press) **135** (5): 1195–1205. doi:10.1083/jcb.135.5.1195. PMC 2121085. PMID 8947544.

[23] NF Rothfield, BD Stollar (1967). "The Relation of Immunoglobulin Class, Pattern of Antinuclear Antibody, and Complement-Fixing Antibodies to DNA in Sera from Patients with Systemic Lupus Erythematosus". *J Clin Invest* **46** (11): 1785–1794. doi:10.1172/JCI105669. PMC 292929. PMID 4168731.

[24] S Barned, AD Goodman, DH Mattson (1995). "Frequency of anti-nuclear antibodies in multiple sclerosis". *Neurology* **45** (2): 384–385. doi:10.1212/WNL.45.2.384. PMID 7854544.

[25] Hernandez-Verdun, Daniele (2006). "Nucleolus: from structure to dynamics". *Histochem. Cell. Biol* **125** (1–2): 127–137. doi:10.1007/s00418-005-0046-4. PMID 16328431.

[26] Lamond, Angus I.; Judith E. Sleeman (October 2003). "Nuclear substructure and dynamics". *current biology* **13** (21): R825–828. doi:10.1016/j.cub.2003.10.012. PMID 14588256.

[27] Cioce M, Lamond A (2005). "Cajal bodies: a long history of discovery". *Annu Rev Cell Dev Biol* **21**: 105–131. doi:10.1146/annurev.cellbio.20.01040(PMID 16212489.

[28] Pollard, Thomas D.; William C. Earnshaw (2004). *Cell Biology*. Philadelphia: Saunders. ISBN 0-7216-3360-9.

[29] Dundr, Miroslav; Tom Misteli (2001). "Functional architecture in the cell nucleus". *Biochem. J.* **356** (Pt 2): 297–310. doi:10.1042/0264-6021:3560297. PMC 1221839. PMID 11368755.

[30] Fox, Archa (2007-03-07). *Paraspeckle Size*. Interview with R. Sundby. E-mail Correspondence.

[31] Goebel, H.H.; I Warlow (January 1997). "Nemaline myopathy with intranuclear rods—intranuclear rod myopathy". *Neuromuscular Disorders* **7** (1): 13–19. doi:10.1016/S0960-8966(96)00404-X. PMID 9132135.

[32] Matera AG, Frey MA. (1998). "Coiled Bodies and Gems: Janus or Gemini?". *American Journal of Human Genetics* **63** (2): 317–321. doi:10.1086/301992. PMC 1377332. PMID 9683623.

[33] Matera, A. Gregory (1998). "Of Coiled Bodies, Gems, and Salmon". *Journal of Cellular Biochemistry* **70** (2): 181–192. doi:10.1002/(sici)1097-4644(19980801)70:2<181::aid-jcb4>3.0.co;2-k. PMID 9671224.

[34] Saunders WS, Cooke CA, Earnshaw WC (1991). "Compartmentalization within the nucleus: discovery of a novel subnuclear region.". *Journal of Cellular Biology* **115** (4): 919–931. doi:10.1083/jcb.115.4.919. PMID 1955462

[35] Pombo A, Cuello P, Schul W, Yoon J, Roeder R, Cook P, Murphy S (1998). "Regional and temporal specialization in the nucleus: a transcriptionally active nuclear domain rich in PTF, Oct1 and PIKA antigens associates with specific chromosomes early in the cell cycle". *The EMBO Journal* **17** (6): 1768–1778. doi:10.1093/emboj/17.6.1768. PMC 1170524. PMID 9501098.

[36] Lallemand-Breitenbach, V.; De The, H. (2010). "PML Nuclear Bodies". *Cold Spring Harbor Perspectives in Biology* **2** (5): a000661. doi:10.1101/cshperspect.a000661. PMC 2857171. PMID 20452955.

[37] Lamond AI, Spector DL (August 2003). "Nuclear speckles: a model for nuclear organelles". *Nature Reviews Molecular Cell Biology* **4** (8): 605–12. doi:10.1038/nrm1172. PMID 12923522.

[38] Tripathi K, Parnaik VK (September 2008). "Differential dynamics of splicing factor SC35 during the cell cycle" (PDF). *J. Biosci.* **33** (3): 345–54. doi:10.1007/s12038-008-0054-3. PMID 19005234.

[39] Tripathi, K.; Parnaik, V. K. (2008). "Differential dynamics of splicing factor SC35 during the cell cycle". *Journal of biosciences* **33** (3): 345–354. doi:10.1007/s12038-008-0054-3. PMID 19005234.

[40] Lamond AI, Spector DL (August 2003). "Nuclear speckles: a model for nuclear organelles". *Nature Reviews Molecular Cell Biology* **4** (8): 605–12. doi:10.1038/nrm1172. PMID 12923522.

[41] Handwerger, Korie E.; Joseph G. Gall (January 2006). "Subnuclear organelles: new insights into form and function". *TRENDS in Cell Biology* **16** (1): 19–26. doi:10.1016/j.tcb.2005.11.005. PMID 16325406.

[42] "Cellular component Nucleus speckle". UniProt: UniProtKB. Retrieved 2013-08-30.

[43] Gall, Joseph G.; Bellini, Michel; Wu, Zheng'an; Murphy, Christine (December 1999). "Assembly of the Nuclear Transcription and Processing Machinery: Cajal Bodies (Coiled Bodies) and Transcriptosomes". *Molecular Biology of the Cell* **10** (12): 4385–4402. doi:10.1091/mbc.10.12.4385. ISSN 1059-1524. PMC 25765. PMID 10588665.

[44] Matera, A. Gregory; Rebecca M. Terns; Michael P. Terns (March 2007). "Non-coding RNAs: lessons from the small nuclear and small nucleolar RNAs". *Nature Reviews Molecular Cell Biology* **8** (3): 209–220. doi:10.1038/nrm2124. ISSN 1471-0072. PMID 17318225. Retrieved 2013-08-09.

[45] Fox, Archa; Lam, YW; Leung, AK; Lyon, CE; Andersen, J; Mann, M; Lamond, AI (2002). "Paraspeckles:A Novel Nuclear Domain". *Current Biology* **12** (1): 13–25. doi:10.1016/S0960-9822(01)00632-7. PMID 11790299.

[46] Fox, Archa; Wendy Bickmore (2004). "Nuclear Compartments: Paraspeckles". Nuclear Protein Database. Archived from the original on May 2, 2006. Retrieved 2007-03-06.

[47] Fox, A.; et al. (2005). "P54nrb Forms a Heterodimer with PSP1 That Localizes to Paraspeckles in an RNA-dependent Manner". *Molecular Biology of the Cell* **16** (11): 5304–5315. doi:10.1091/mbc.E05-06-0587. PMC 1266428. PMID 16148043.

[48] Lehninger, Albert L.; Nelson, David L.; Cox, Michael M. (2000). *Lehninger principles of biochemistry* (3rd ed.). New York: Worth Publishers. ISBN 1-57259-931-6.

[49] Moreno F, Ahuatzi D, Riera A, Palomino CA, Herrero P. (2005). "Glucose sensing through the Hxk2-dependent signalling pathway.". *Biochem Soc Trans* **33** (1): 265–268. doi:10.1042/BST0330265. PMID 15667322. PMID 15667322

[50] Görlich, Dirk; Ulrike Kutay (1999). "Transport between the cell nucleus and the cytoplasm". *Ann. Rev. Cell Dev. Biol.* **15** (1): 607–660. doi:10.1146/annurev.cellbio.15.1.607. PMID 10611974.

[51] Nierhaus, Knud H.; Daniel N. Wilson (2004). *Protein Synthesis and Ribosome Structure: Translating the Genome*. Wiley-VCH. ISBN 3-527-30638-2.

[52] Nicolini, Claudio A. (1997). *Genome Structure and Function: From Chromosomes Characterization to Genes Technology*. Springer. ISBN 0-7923-4565-7.

[53] Watson, JD; Baker TA; Bell SP; Gann A; Levine M; Losick R. (2004). "Ch9–10". *Molecular Biology of the Gene* (5th ed.). Peason Benjamin Cummings; CSHL Press. ISBN 0-8053-9603-9.

[54] Lippincott-Schwartz, Jennifer (2002-03-07). "Cell biology: Ripping up the nuclear envelope". *Nature* **416** (6876): 31–32. doi:10.1038/416031a. PMID 11882878.

[55] Boulikas T (1995). "Phosphorylation of transcription factors and control of the cell cycle". *Crit Rev Eukaryot Gene Expr* **5** (1): 1–77. PMID 7549180.

[56] Steen R, Collas P (2001). "Mistargeting of B-type lamins at the end of mitosis: implications on cell survival and regulation of lamins A/C expression". *J Cell Biol* **153** (3): 621–626. doi:10.1083/jcb.153.3.621. PMC 2190567. PMID 11331311.

[57] Böhm I. IgG deposits can be detected in cell nuclei of patients with both lupus erythematosus and malignancy. *Clin Rheumatol* 2007;26(11) 1877-1882

[58] Skutelsky, E.; Danon D. (June 1970). "Comparative study of nuclear expulsion from the late erythroblast and cytokinesis". *J Cell Biol* **60** (60(3)): 625–635. doi:10.1016/0014-4827(70)90536-7. PMID 5422968.

[59] Torous, DK; Dertinger SD; Hall NE; Tometsko CR. (2000). "Enumeration of micronucleated reticulocytes in rat peripheral blood: a flow cytometric study". *Mutat Res* **465** (465(1–2)): 91–99. doi:10.1016/S1383-5718(99)00216-8. PMID 10708974.

[60] Hutter, KJ; Stohr M. (1982). "Rapid detection of mutagen induced micronucleated erythrocytes by flow cytometry". *Histochemistry* **75** (3): 353–362. doi:10.1007/bf00496738. PMID 7141888.

[61] Zettler, LA; Sogin ML; Caron DA (1997). "Phylogenetic relationships between the Acantharea and the Polycystinea: A molecular perspective on Haeckel's Radiolaria". *Proc Natl Acad Sci USA* **94** (21): 11411–11416. doi:10.1073/pnas.94.21.11411. PMC 23483. PMID 9326623.

[62] Horton, TR (2006). "The number of nuclei in basidiospores of 63 species of ectomycorrhizal Homobasidiomycetes". *Mycologia* **98** (2): 233–238. doi:10.3852/mycologia.98.2.233. PMID 16894968.

[63] Adam RD (December 1991). "The biology of Giardia spp". *Microbiol. Rev.* **55** (4): 706–32. PMC 372844. PMID 1779932.

[64] McInnes, A; Rennick DM (1988). "Interleukin 4 induces cultured monocytes/macrophages to form giant multinucleated cells". *J Exp Med* **167** (2): 598–611. doi:10.1084/jem.167.2.598. PMC 2188835. PMID 3258008.

[65] Goldring, SR; Roelke MS; Petrison KK; Bhan AK (1987). "Human giant cell tumors of bone identification and characterization of cell types". *J Clin Invest* **79** (2): 483–491. doi:10.1172/JCI112838. PMC 424109. PMID 3027126.

[66] Imanian, B; Pombert, JF; Dorrell, RG; Burki, F; Keeling, PJ (2012). "Tertiary endosymbiosis in two dinotoms has generated little change in the mitochondrial genomes of their dinoflagellate hosts and diatom endosymbionts". *PLOS ONE* **7** (8): e43763. doi:10.1371/journal.pone.0043763.

[67] Pennisi E. (2004). "Evolutionary biology. The birth of the nucleus". *Science* **305** (5685): 766–768. doi:10.1126/science.305.5685.766. PMID 15297641.

[68] C.Michael Hogan. 2010. *Archaea*. eds. E.Monosson & C.Cleveland, Encyclopedia of Earth. National Council for Science and the Environment, Washington DC.

[69] Margulis, Lynn (1981). *Symbiosis in Cell Evolution*. San Francisco: W. H. Freeman and Company. pp. 206–227. ISBN 0-7167-1256-3.

[70] Lopez-Garcia P, Moreira D. (2006). "Selective forces for the origin of the eukaryotic nucleus". *BioEssays* **28** (5): 525–533. doi:10.1002/bies.20413. PMID 16615090.

[71] Fuerst JA. (2005). "Intracellular compartmentation in planctomycetes". *Annu Rev Microbiol.* **59**: 299–328. doi:10.1146/annurev.micro.59.030804.1 PMID 15910279.

[72] Hartman H, Fedorov A. (2002). "The origin of the eukaryotic cell: a genomic investigation". *Proc Natl Acad Sci U S A.* **99** (3): 1420–1425. doi:10.1073/pnas.032658599. PMC 122206. PMID 11805300.

[73] Bell PJ (September 2001). "Viral eukaryogenesis: was the ancestor of the nucleus a complex DNA virus?". *J. Mol. Evol.* **53** (3): 251–6. doi:10.1007/s002390010215. PMID 11523012.

[74] Takemura M (2001). "Poxviruses and the origin of the eukaryotic nucleus". *J Mol Evol* **52** (5): 419–425. doi:10.1007/s002390010171. PMID 11443345.

[75] Villarreal L, DeFilippis V (2000). "A hypothesis for DNA viruses as the origin of eukaryotic replication proteins". *J Virol* **74** (15): 7079–7084. doi:10.1128/JVI.74.15.7079-7084.2000. PMC 112226. PMID 10888648.

[76] Bell PJ (November 2006). "Sex and the eukaryotic cell cycle is consistent with a viral ancestry for the eukaryotic nucleus". *J. Theor. Biol.* **243** (1): 54–63. doi:10.1016/j.jtbi.2006.05.015. PMID 16846615.

[77] de Roos AD (2006). "The origin of the eukaryotic cell based on conservation of existing interfaces". *Artif Life* **12** (4): 513–523. doi:10.1162/artl.2006.12.4.513. PMID 16953783.

15.10 Further reading

- Goldman, Robert D.; Gruenbaum, Y; Moir, RD; Shumaker, DK; Spann, TP (2002). "Nuclear lamins: building blocks of nuclear architecture". *Genes & Dev.* **16** (5): 533–547. doi:10.1101/gad.960502. PMID 11877373.

  A review article about nuclear lamins, explaining their structure and various roles

- Görlich, Dirk; Kutay, U (1999). "Transport between the cell nucleus and the cytoplasm". *Ann. Rev. Cell Dev. Biol.* **15**: 607–660. doi:10.1146/annurev.cellbio.15.1.607. PMID 10611974.

  A review article about nuclear transport, explains the principles of the mechanism, and the various transport pathways

- Lamond, Angus I.; Earnshaw, WC (1998-04-24). "Structure and Function in the Nucleus". *Science* **280** (5363): 547–553. doi:10.1126/science.280.5363.547. PMID 9554838.

A review article about the nucleus, explaining the structure of chromosomes within the organelle, and describing the nucleolus and other subnuclear bodies

- Pennisi E. (2004). "Evolutionary biology. The birth of the nucleus". *Science* **305** (5685): 766–768. doi:10.1126/science.305.568! PMID 15297641.

A review article about the evolution of the nucleus, explaining a number of different theories

- Pollard, Thomas D.; William C. Earnshaw (2004). *Cell Biology*. Philadelphia: Saunders. ISBN 0-7216-3360-9.

A university level textbook focusing on cell biology. Contains information on nucleus structure and function, including nuclear transport, and subnuclear domains

15.11   External links

- MBInfo - The Nucleus

- cellnucleus.com Website covering structure and function of the nucleus from the Department of Oncology at the University of Alberta.

- http://npd.hgu.mrc.ac.uk/user/?page=compartment The Nuclear Protein Database] Information on nuclear components.

- The Nucleus Collection in the Image & Video Library of The American Society for Cell Biology contains peer-reviewed still images and video clips that illustrate the nucleus.

- Nuclear Envelope and Nuclear Import Section from *Landmark Papers in Cell Biology, Joseph G. Gall, J. Richard McIntosh, eds., contains digitized commentaries and links to seminal research papers on the nucleus. Published online in the Image & Video Library of The American Society for Cell Biology*

- Cytoplasmic patterns generated by human antibodies

Chapter 16

Ploidy

Ploidy is the number of sets of chromosomes in a cell. Usually a gamete (sperm or egg, which fuse into a single cell during the fertilization phase of sexual reproduction) carries a full set of chromosomes that includes a single copy of each chromosome, as aneuploidy generally leads to severe genetic disease in the offspring. The **gametic** or **haploid number** (n) is the number of chromosomes in a gamete. Two gametes form a diploid zygote with twice this number ($2n$, the **zygotic** or **diploid number**) i.e. two copies of autosomal chromosomes. For humans, a diploid species, $n = 23$. A typical human somatic cell contains 46 chromosomes: 2 complete haploid sets, which make up 23 homologous chromosome pairs.

Because chromosome number is generally reduced only by the specialized process of meiosis, the somatic cells of the body inherit and maintain the chromosome number of the zygote. However, in many situations somatic cells double their copy number by means of endoreduplication as an aspect of cellular differentiation. For example, the hearts of two-year-old children contain 85% diploid and 15% tetraploid nuclei, but by 12 years of age the proportions become approximately equal, and adults examined contained 27% diploid, 71% tetraploid and 2% octaploid nuclei.[1]

Cells are described according to the number of sets present (the **ploidy level**): **monoploid** (1 set), **diploid** (2 sets), **triploid** (3 sets), **tetraploid** (4 sets), pentaploid (5 sets), hexaploid (6 sets), heptaploid[2] or septaploid[3] (7 sets), etc. The generic term **polyploid** is frequently used to describe cells with three or more sets of chromosomes (triploid or higher ploidy).

16.1 Etymology

The term *ploidy* is a back-formation from *haploid* and *diploid*. These two terms are from Greek ἁπλόος *haplóos* "single" and διπλόος *diplóos* "double" combined with εἶδος *eîdos* "form" (compare *idol* from Latin *īdōlum*, that from Greek εἴδωλον *eídōlon* derived from εἶδος *eîdos*). Eduard Strasburger, who coined the terms *haploid* and *diploid*, based on Weismann's conception of the id (or germ plasm),[4][5] used diploid to refer to an organism with twice the number of chromosomes of a haploid organism, hence "double" and "single". The two terms were borrowed from German through William Henry Lang's 1908 translation of an 1906 textbook by Strasburger and colleagues.[6]

Technically, ploidy is a description of a nucleus. Though at times authors may report the total ploidy of all nuclei present within the cell membrane of a syncytium,[7] usually the ploidy of the nuclei present will be described. For example, a fungal dikaryon with two haploid nuclei is distinguished from the diploid in which the chromosomes share a nucleus and can be shuffled together.[8] Nonetheless, because in most situations there is only one nucleus, it is commonplace to speak of the ploidy of a cell.

16.2 Case studies

It is also possible on rare occasions for the ploidy to increase in the germline, which can result in polyploid offspring and ultimately polyploid *species*. This is an important evolutionary mechanism in both plants and animals.[9] As a result, it becomes desirable to distinguish between the ploidy of a species or variety as it presently breeds and that of an ancestor. The number of chromosomes in the ancestral (non-homologous) set is called the **monoploid number** (x), and is distinct from the haploid number (n) in the organism as it now reproduces. Both numbers n, and x, apply to every cell of a given organism.

Common wheat is an organism where x and n differ. It has six sets of chromosomes, two sets from each of three different diploid species that are its distant ancestors. The somatic cells are hexaploid, with six sets of chromosomes, $2n = 6x = 42$. The gametes are haploid for their own species, but triploid, with three sets of chromosomes, by comparison to a probable evolutionary ancestor, einkorn wheat. The monoploid number $x = 7$, and the haploid number $n = 21$. Tetraploidy (four sets of chromosomes, $2n = 4x$) is common in plants, and also occurs in amphibians, reptiles, and insects.

Over evolutionary time scales in which chromosomal polymorphisms accumulate, these changes become less apparent by karyotype - for example, humans are generally regarded as diploid, but the 2R hypothesis has confirmed two rounds of whole genome duplication in early vertebrate ancestors.

Ploidy can also differ with life cycle.[10][11] In some insects it differs by caste. In humans, only the gametes are haploid, but in the Australian bulldog ant, *Myrmecia pilosula*, a haplodiploid species, haploid individuals of this species have a single chromosome, and diploid individuals have two chromosomes.[12] In *Entamoeba*, the ploidy level varies from $4n$ to $40n$ in a single population.[13] Alternation of generations occurs in many plants.

Some studies suggest that selection is more likely to favor diploidy in host species and haploidy in parasite species.[14]

16.3 Haploid and monoploid

The nucleus of a eukaryotic cell is **haploid** if it has a single set of chromosomes, each one not being part of a pair. By extension a cell may be called haploid if its nucleus is haploid, and an organism may be called haploid if its body cells (somatic cells) are haploid. The number of chromosomes in a single set is called the **haploid number**, given the symbol n.

Gametes (sperm and ova) are haploid cells. The haploid gametes produced by most organisms combine to form a zygote with n pairs of chromosomes, i.e. $2n$ chromosomes in total. The chromosomes in each pair, one of which comes from the sperm and one from the egg, are said to be **homologous**. Cells and organisms with pairs of homologous chromosomes are called diploid. For example, most animals are diploid and produce haploid gametes. During meiosis, sex cell precursors have their number of chromosomes halved by randomly "choosing" one member of each pair of chromosomes, resulting in haploid gametes. Because homologous chromosomes usually differ genetically, gametes usually differ genetically from one another.

All plants and many fungi and algae switch between a haploid and a diploid state, with one of the stages emphasized over the other. This is called alternation of generations. Most fungi and algae are haploid during the principal stage of their lifecycle, as are plants like mosses. Most animals are diploid, but male bees, wasps, and ants are haploid organisms because they develop from unfertilized, haploid eggs.

In some cases there is evidence that the n chromosomes in a haploid set have resulted from duplications of an originally smaller set of chromosomes. This "base" number – the number of apparently originally unique chromosomes in a haploid set – is called the **monoploid number**,[15] also known as **basic** or **cardinal number**,[16] or **fundamental number**.[17][18] As an example, the chromosomes of common wheat are believed to be derived from three different ancestral species, each of which had 7 chromosomes in its haploid gametes. The monoploid number is thus 7 and the haploid number is $3 \times 7 = 21$. In general n is a multiple of x. The somatic cells in a wheat plant have six sets of 7 chromosomes: three sets from the egg and three sets from the sperm which fused to form the plant, giving a total of 42 chromosomes. As a formula, for wheat $2n = 6x = 42$, so that the haploid number n is 21 and the monoploid number x is 7. The gametes of common wheat are considered to be haploid, since they contain half the genetic information of somatic cells, but they are not monoploid, as they still contain three complete sets of chromosomes ($n = 3x$).[19]

In the case of wheat, the origin of its haploid number of 21 chromosomes from three sets of 7 chromosomes can be demonstrated. In many other organisms, although the number of chromosomes may have originated in this way, this is no longer clear, and the monoploid number is regarded as the same as the haploid number. Thus in humans, $x = n = 23$.

16.4 Diploid

For other uses, see Diploid (crystallography).

Diploid cells have two homologous copies of each chromosome, usually one from the mother and one from the father. Nearly all mammals are diploid organisms (the tetraploid (four sets) plains viscacha rats (*Tympanoctomys barrerae*) and golden vizcacha rat (*Pipanacoctomys aureus*)[20] are the only known exceptions as of 2004[21]), although all individuals have some small fraction of cells that display polyploidy. Human diploid cells have 46 chromosomes and human haploid gametes (egg and sperm) have 23 chromosomes. Retroviruses that contain two copies of their RNA genome in each viral particle are also said to be diploid. Examples include human foamy virus, human T-lymphotropic virus, and HIV.[22]

16.5 Homoploid

"Homoploid" means "at the same ploidy level", i.e. having the same number of homologous chromosomes. For example, homoploid hybridization is hybridization where the offspring have the same ploidy level as the two parental species. This contrasts with a common situation in plants where chromosome doubling accompanies, or happens soon after hybridization. Similarly, homoploid speciation contrasts with polyploid speciation.

16.6 Zygoidy and azygoidy

Zygoidy is the state where the chromosomes are paired and can undergo meiosis. The zygoid state of a species may be diploid or polyploid.[23][24] In the azygoid state the chromosomes are unpaired. It may be the natural state of some asexual species or may occur after meiosis. In diploid organisms the azygoid state is monoploid. (see below for dihaploidy)

16.7 Polyploidy

Main article: Polyploidy

Polyploidy is the state where all cells have multiple sets of chromosomes beyond the basic set, usually 3 or more. Specific terms are **triploid** (3 sets), **tetraploid** (4 sets), pentaploid (5 sets), hexaploid (6 sets), heptaploid[2] or septaploid[3] (7 sets) octoploid (8 sets), nonaploid (9 sets), decaploid (10 sets), undecaploid (11 sets), dodecaploid (12 sets), tridecaploid (13 sets), tetradecaploid (14 sets) etc.[25][26][27][28] Some higher ploidies include hexadecaploid (16 sets), dotriacontaploid (32 sets), and tetrahexacontaploid (64 sets),[29] though Greek terminology may be set aside for readability in cases of higher ploidy (such as "16-ploid").[27] Polytene chromosomes of plants and fruit flies can be 1024-ploid.[30][31] Ploidy of systems such as the salivary gland, elaiosome, endosperm, and trophoblast can exceed this, up to 1048576-ploid in the silk glands of the commercial silkworm *Bombyx mori*.[7]

The chromosome sets may be from the same species or from closely related species. In the latter case, these are known as allopolyploids (or amphidiploids, which are allopolyploids that behave as if they were normal diploids). Allopolyploids are formed from the hybridization of two separate species. In plants, this probably most often occurs from the pairing of meiotically unreduced gametes, and not by diploid–diploid hybridization followed by chromosome doubling.[32] The so-called *Brassica* triangle is an example of allopolyploidy, where three different parent species have hybridized in all possible pair combinations to produce three new species.

Polyploidy occurs commonly in plants, but rarely in animals. Even in diploid organisms, many somatic cells are polyploid due to a process called endoreduplication where duplication of the genome occurs without mitosis (cell division).

The extreme in polyploidy occurs in the fern genus *Ophioglossum*, the adder's-tongues, in which polyploidy results in chromosome counts in the hundreds, or, in at least one case, well over one thousand.

It is also possible for polyploid organisms to revert to lower ploidy by means of haploidisation.

16.8   Variable or indefinite ploidy

Depending on growth conditions, prokaryotes such as bacteria may have a chromosome copy number of 1 to 4, and that number is commonly fractional, counting portions of the chromosome partly replicated at a given time. This is because under exponential growth conditions the cells are able to replicate their DNA faster than they can divide.

In ciliates, the macronucleus is called **ampliploid**, because only part of the genome is amplified.[33]

16.9   Mixoploidy

Mixoploidy refers to the presence of two cell lines, one diploid and one polyploid. Though polyploidy in humans is not viable, mixoploidy has been found in live adults and children.[34] There are two types: diploid-triploid mixoploidy, in which some cells have 46 chromosomes and some have 69, and diploid-tetraploid mixoploidy, in which some cells have 46 and some have 92 chromosomes. It is a major topic of cytology.

16.10   Dihaploidy and polyhaploidy

Not to be confused with haplodiploidy (where diploid and haploid individuals are different sexes).

Dihaploid and polyhaploid cells are formed by haploidisation of polyploids, i.e., by halving the chromosome constitution.

Dihaploids (which are diploid) are important for selective breeding of tetraploid crop plants (notably potatoes), because selection is faster with diploids than with tetraploids. Tetraploids can be reconstituted from the diploids, for example by somatic fusion.

The term "dihaploid" was coined by Bender[35] to combine in one word the number of genome copies (diploid) and their origin (haploid). The term is well established in this original sense,[36][37] but it has also been used for doubled monoploids or doubled haploids, which are homozygous and used for genetic research.[38]

16.11   Euploidy

Euploidy is the state of a cell or organism having the same number of each homologous chromosome, possibly excluding the sex-determining chromosomes. For example, most human cells have 2 of each of the 23 homologous monoploid chromosomes, for a total of 46 chromosomes. A human cell with an abnormal number of 3 of each would also be considered as euploid. **Aneuploidy** is the state of not having euploidy. In humans, examples include having a single extra chromosome (such as Down syndrome), or missing a chromosome (such as Turner syndrome). Aneuploid karyotypes are given names with the suffix -*somy* (rather than -*ploidy*, used for euploid karyotypes), such as trisomy and monosomy.

16.12   Possible adaptive and ecological significance of variation in ploidy

A study comparing the karyotypes of endangered or invasive plants with those of their relatives found that being polyploid as opposed to diploid is associated with a 14% lower risk of being endangered, and a 20% greater chance of being invasive.[39] Polyploidy may be associated with increased vigor and adaptability.[40]

16.13   References

[1] John O. Oberpriller, A Mauro. *The Development and Regenerative Potential of Cardiac Muscle*. Taylor&Francis.

[2] U. R. Murty (1973). "Morphology of pachytene chromosomes and its bearing on the nature of polyploidy in the cytological races of Apluda mutica L.". *Genetica* **44** (2): 234–243. doi:10.1007/bf00119108.

[3] Tuguo Tateoka (May 1975). "A contribution to the taxonomy of the *Agrostis mertensii-flaccida* complex (Poaceae) in Japan". *Journal of Plant Research*. pp. 65–87. doi:10.1007/bf02491243.

[4] Battaglia E (2009). "Caryoneme alternative to chromosome and a new caryological nomenclature" (PDF). *Caryologia* **62**: 1–83.

[5] Haig David (2008). "Homologous versus antithetic alternation of generations and the origin of sporophytes" (PDF). *The Botanical Review* **74** (3): 395–418. doi:10.1007/s12229-008-9012-x.

[6] Strasburger, E.; Noll, F.; Schenck, H.; Karsten, G. 1908. *A Textbook of botany*, 3rd English ed. (1908) , rev. with the 8th German ed. (1906) , translation by W. H. Lang of *Lehrbuch der Botanik für Hochschulen*. Macmillan, London.

[7] Encyclopedia of the Life Sciences (2002) "Polyploidy" Francesco D'Amato and Mauro Durante

[8] James B. Anderson and Linda M Kohn. "Dikaryons, diploids, and evolution" (PDF). University of Toronto.

[9] """Why polyploidy is rarer in animals than in plants': myths and mechanisms" in *Biological relevance of polyploidy: ecology to genomics*" (PDF). *Biological Journal of the Linnean Society* **82**: 453–466. 2004. doi:10.1111/j.1095-8312.2004.00332.x.

[10] Parfrey LW, Lahr DJG, Katz LA. 2008. The dynamic nature of eukaryotic genomes. *Mol Biol Evol*. 25:787–794,

[11] Qiu, Y.-L., A. B. Taylor, H. A. McManus. 2012. Evolution of the life cycle in land plants. *Journal of Systematics and Evolution* 50: 171-194, .

[12] Crosland MW, Crozier RH (1986). "Myrmecia pilosula, an Ant with Only One Pair of Chromosomes". *Science* **231** (4743): 1278. Bibcode:1986Sci...231.1278C. doi:10.1126/science.231.4743.1278. PMID 17839565.

[13] https://bcrc.bio.umass.edu/courses/fall2010/biol/biolh100-03/sites/default/files/vazquez_eukaryotic_diversity_2010.pdf

[14] Nuismer S., Otto S.P. (2004). "Host-parasite interactions and the evolution of ploidy". *Proc. Natl. Acad. Sci. USA* **101** (30): 11036–11039. doi:10.1073/pnas.0403151101.

[15] Langlet, 1927.

[16] Winge, 1917.

[17] Manton, 1932.

[18] Fabbri, F. 1963. Primo supplemento alle tavole cromosomiche delle Pteridophyta di Alberto Chiarugi. *Caryologia* 16: 237–335,
.

[19] http://mcb.berkeley.edu/courses/mcb142/lecture%20topics/Amacher/LECTURE_10_CHROM_F08.pdf

[20] Gallardo MH, González CA, Cebrián I (2006). "Molecular cytogenetics and allotetraploidy in the red vizcacha rat, *Tympanoctomys barrerae* (Rodentia, Octodontidae)]". *Genomics* **88** (2): 214–221. doi:10.1016/j.ygeno.2006.02.010. PMID 16580173.

[21] Gallardo M. H.; et al. (2004). "Whole-genome duplications in South American desert rodents (Octodontidae)". *Biological Journal of the Linnean Society* **82**: 443–451. doi:10.1111/j.1095-8312.2004.00331.x.

[22] http://web.uct.ac.za/depts/mmi/jmoodie/hiv2.html

[23] Books, Elsevier Science & Technology (1950-01-01). *Advances in Genetics*. Academic Press. ISBN 978-0-12-017603-8.

[24] Cosín, Darío J. Díaz, Marta Novo, and Rosa Fernández. "Reproduction of Earthworms: Sexual Selection and Parthenogenesis." In Biology of Earthworms, edited by Ayten Karaca, 24:69-86. Berlin, Heidelberg: Springer Berlin Heidelberg, 2011. http://www.springerlink.com/content/j5j72p2834355w27/.

[25] Dierschke T, Mandáková T, Lysak MA, Mummenhoff K (September 2009). "A bicontinental origin of polyploid Australian/New Zealand *Lepidium* species (Brassicaceae)? Evidence from genomic in situ hybridization". *Annals of Botany* **104** (4): 681–688. doi:10.1093/aob/mcp161. PMC 2729636. PMID 19589857.

[26] Simon Renny-Byfield; et al. (2010). "Flow cytometry and GISH reveal mixed ploidy populations and Spartina nonaploids with genomes of *S. alterniflora* and *S. maritima* origin". *Annals of Botany* **105** (4): 527–533. doi:10.1093/aob/mcq008.

[27] Kim E. Hummer; et al. (March 2009). "Decaploidy in *Fragaria iturupensis* (Rosaceae)". *Am. J. Bot.* pp. 713–716. doi:10.3732/ajb.0800285.

[28] Talyshinskiĭ, G. M. (1990). "Study of the fractional composition of the proteins in the compound fruit of polyploid mulberry". *Shelk* (5): 8–10.

[29] Fujikawa-Yamamoto K (2001). "Temperature dependence in Proliferation of tetraploid Meth-A cells in comparison with the parent diploid cells". *Cell Structure and Function.* pp. 263–269. doi:10.1247/csf.26.263.

[30] Kiichi Fukui, Shigeki Nakayama. *Plant Chromosomes: Laboratory Methods.*

[31] "Genes involved in tissue and organ development: Polytene chromosomes, endoreduplication and puffing". The Interactive Fly.

[32] Ramsey, J.; Schemske, D. W. (2002). "Neopolyploidy in Flowering Plants" (PDF). *Annual Review of Ecology and Systematics* **33**: 589. doi:10.1146/annurev.ecolsys.33.010802.150437.

[33] Schaechter, M. *Eukaryotic microbes.* Amsterdam, Academic Press, 2012, p. 217.

[34] Edwards MJ; et al. "Mixoploidy in humans: two surviving cases of diploid-tetraploid mixoploidy and comparison with diploid-triploid mixoploidy.". Retrieved 17 December 2015.

[35] Bender K (1963). "Über die Erzeugung und Entstehung dihaploider Pflanzen bei *Solanum tuberosum*".". *Zeitschrift für Pflanzenzüchtung* **50**: 141–166.

[36] Nogler, G.A. 1984. Gametophytic apomixis. In *Embryology of angiosperms.* Edited by B.M. Johri. Springer, Berlin, Germany. pp. 475–518.

[37] • Pehu E (1996). "The current status of knowledge on the cellular biology of potato". *Potato Research* **39**: 429–435. doi:10.1007/bf02357948.

[38] • Sprague G.F., Russell W.A., Penny L.H. (1960). "Mutations affecting quantitative traits in the selfed progeny of double monoploid maize stocks". *Genetics* **45** (7): 855–866.

[39] Pandit, M. K.; Pocock, M. J. O.; Kunin, W. E. (2011-03-28). "Ploidy influences rarity and invasiveness in plants". *Journal of Ecology* (Wiley-Blackwell) **99**. doi:10.1111/j.1365-2745.2011.01838.x.

[40] Gilbert, Natasha (2011-04-06). "Ecologists find genomic clues to invasive and endangered plants". *Nature News* (Nature Publishing Group). doi:10.1038/news.2011.213. Retrieved 2011-04-07. External link in |work= (help)

16.14   Bibliography

• Griffiths, A. J. et al. 2000. An introduction to genetic analysis, 7th ed. W. H. Freeman, New York ISBN 0-7167-3520-2

16.15   External links

Some eukaryotic genome-scale or genome size databases and other sources which may contain the ploidy of many organisms:

• Animal genome size database

• Plant genome size database

• Fungal genome size database

• Protist genome-scale database of Ensembl Genomes

• Nuismer S., Otto S.P. (2004). "Host-parasite interactions and the evolution of ploidy". *Proc. Natl. Acad. Sci. USA* **101**: 11036–11039. doi:10.1073/pnas.0403151101. (Supporting Data Set, with information on ploidy level and number of chromosomes of several protists)

• Chromosome number and ploidy mutations YouTube tutorial video

Haploid (N)

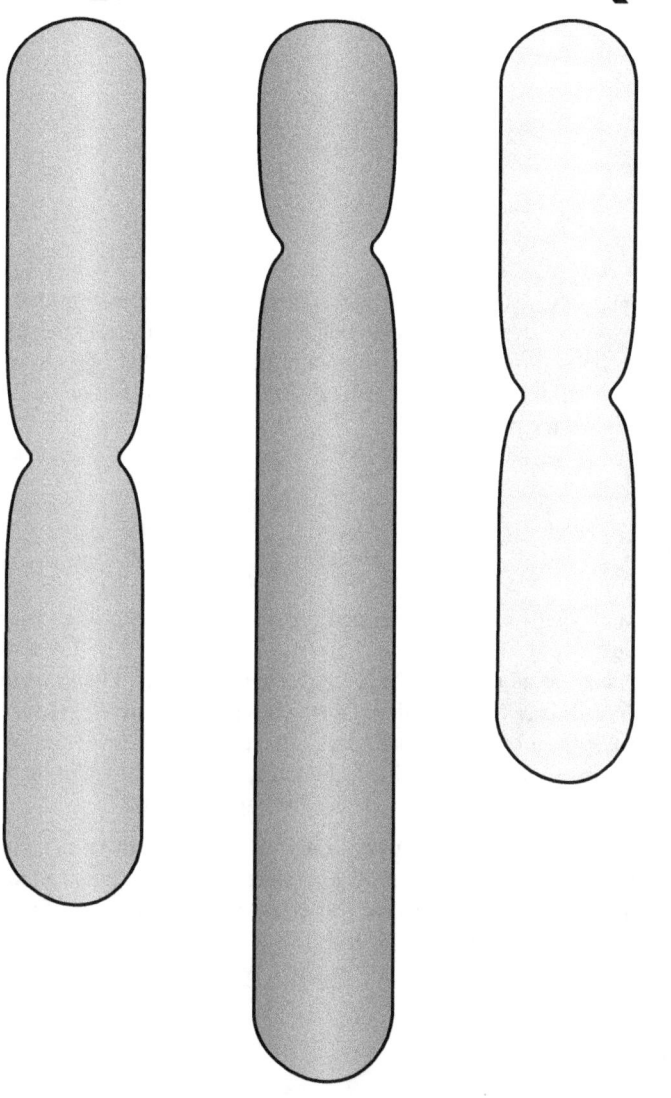

Diploid (2N)

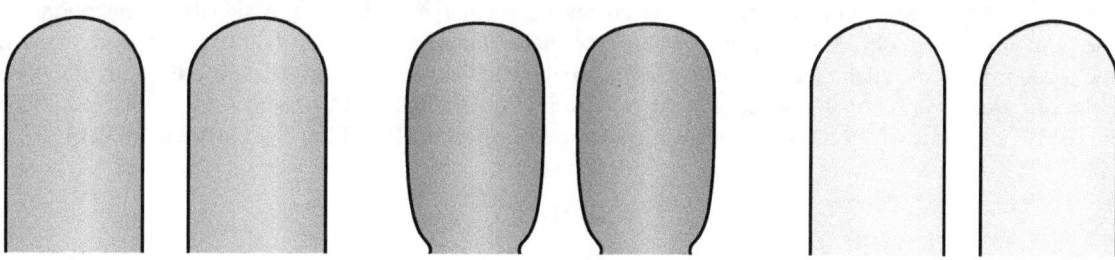

Chapter 17

Germ cell

A **germ cell** is any biological cell that gives rise to the gametes of an organism that reproduces sexually. In many animals, the germ cells originate in the primitive streak and migrate via the gut of an embryo to the developing gonads. There, they undergo meiosis, followed by cellular differentiation into mature gametes, either eggs or sperm. Unlike animals, plants do not have germ cells set aside from in early development. Instead, germ cells can come from somatic cells in the adult (such as the floral meristem of flowering plants).[1][2][3]

17.1   Introduction

Multicellular eukaryotes are made of two fundamental cell types. Germ cells produce gametes and are the only cells that can undergo meiosis as well as mitosis. These cells are sometimes said to be immortal because they are the link between generations. Somatic cells are all the other cells that form the building blocks of the body and they only divide by mitosis. The lineage of germ cells is called germ line. Germ cell specification begins during cleavage in many animals or in the epiblast during gastrulation in birds and mammals. After transport, involving passive movements and active migration, germ cells arrive at the developing gonads. In humans, sexual differentiation starts approximately 6 weeks after conception. The end-products of the germ cell cycle are the egg or sperm.[4]

Under special conditions *in vitro* germ cells can acquire properties similar to those of embryonic stem cells (ES). The underlying mechanism of that change is still unknown. These changed cells are then called embryonic germ cells (EG). Both EG and ES are pluripotent in vitro, but only ES has proven pluripotency in vivo. Recent studies have demonstrated that it is possible to give rise to primordial germ cells from ES.[5]

17.2   Specification

There are two mechanisms to establish the germ cell lineage in the embryo. The first way is called preformistic and involves that the cells destined to become germ cells inherit the specific germ cell determinants present in the germ plasm (specific area of the cytoplasm) of the egg (ovum). The unfertilized egg of most animals is asymmetrical: different regions of the cytoplasm contain different amounts of mRNA and proteins. By this, germ cells obtained by the first divisions of the fertilized egg are characterized by specific molecules of a particular region of the egg cytoplasm.

The second way is found in birds and mammals, where germ cells are not specified by such determinants but by signals controlled by zygotic genes. In mammals, a few cells of the early embryo are induced by signals of neighboring cells to become primordial germ cells. Mammalian eggs are somewhat symmetrical and after the first divisions of the fertilized egg, the produced cells are all totipotent. This means that they can differentiate in any cell type in the body and thus germ cells. Specification of primordial germ cells in the laboratory mouse is initiated by high levels of bone morphogenetic protein (BMP) signaling, which activates expression of the transcription factors Blimp-1/Prdm1 and Prdm14.[6]

17.3 Migration

Primordial germ cells, germ cells that still have to reach the gonads, also known as PGCs, precursor germ cells or gonocytes, divide repeatedly on their migratory route through the gut and into the developing gonads.

17.3.1 Invertebrates

In the model organism *Drosophila*, pole cells passively move from the posterior end of the embryo to the posterior midgut because of the infolding of the blastoderm. Then they actively move through the gut into the mesoderm. Endodermal cells differentiate and together with Wunen proteins they induce the migration through the gut. Wunen proteins are chemorepellents that lead the germ cells away from the endoderm and into the mesoderm. After splitting into two populations, the germ cells continue migrating laterally and in parallel until they reach the gonads. Columbus proteins, chemoattractants, stimulate the migration in the gonadal mesoderm.

17.3.2 Vertebrates

In the *Xenopus* egg, the germ cell determinants are found in the most vegetal blastomeres. These presumptive PGCs are brought to the endoderm of the blastocoel by gastrulation. They are determined as germ cells when gastrulation is completed. Migration from the hindgut along the gut and across the dorsal mesentery then takes place. The germ cells split into two populations and move to the paired gonadal ridges. Migration starts with 3-4 cells that undergo three rounds of cell division so that about 30 PGCs arrive at the gonads. On the migratory path of the PGCs, the orientation of underlying cells and their secreted molecules such as fibronectin play an important role.

Mammals have a migratory path comparable to that in *Xenopus*. Migration begins with 50 gonocytes and about 5,000 PGCs arrive at the gonads. Proliferation occurs also during migration and lasts for 3–4 weeks in humans.

PGCs come from the epiblast and migrate subsequently into the mesoderm, the endoderm and the posterior of the yolk sac. Migration then takes place from the hindgut along the gut and across the dorsal mesentery to reach the gonads (4.5 weeks in human beings). Fibronectin maps here also a polarized network together with other molecules. The somatic cells on the path of germ cells provide them attractive, repulsive, and survival signals. But germ cells also send signals to each other.

In reptiles and birds, germ cells use another path. PGCs come from the epiblast and move to the hypoblast to form the germinal crescent (anterior extraembryonic structure). The gonocytes then squeeze into blood vessels and use the circulatory system for transport. They squeeze out of the vessels when they are at height of the gonadal ridges. Cell adhesion on the endothelium of the blood vessels and molecules such as chemoattractants are probably involved in helping PGCs migrate.

The *Sry* gene of the Y chromosome

The *Sry* gene on the Y chromosome directs male development in mammals by inducing the somatic cells of the gonadal ridge to develop into a testis, rather than an ovary.[7] *Sry* is expressed in a small group of somatic cells of the developing gonad and influence these cells to become Sertoli cells (supporting cells in testis). Sertoli cells are responsible for sexual development along a male pathway in many ways. One of these ways involves stimulation of the arriving primordial cells to differentiate into sperm. In the absence of the *Sry* gene, primordial germ cells differentiate into eggs. Removing genital ridges before they start to develop into testes or ovaries results in the development of a female, independent of the carried sex chromosome.

17.4 Gametogenesis

Gametogenesis, the development of diploid germ cells into either haploid eggs or sperm, (respectively oogenesis and spermatogenesis) is different for each species but the general stages are similar. Oogenesis and spermatogenesis have many features in common, they both involve:

- Meiosis

- Extensive morphological differentiation

- Incapacity of surviving for very long if fertilization does not occur

Despite their homologies they also have major differences:

- Spermatogenesis has equivalent meiotic divisions resulting in four equivalent spermatids while oogenic meiosis is asymmetrical: only one egg is formed together with three polar bodies.

- Different timing of maturation: oogenic meiosis is interrupted at one or more stages (for a long time) while spermatogenic meiosis is rapid and uninterrupted.

17.4.1   Oogenesis

After migration primordial germ cells will become oogonia in the forming gonad (ovary). The oogonia proliferate extensively by mitotic divisions, up to 5-7 million cells in humans. But then many of these oogonia die and about 50,000 remain. These cells differentiate into primary oocytes. In week 11-12 *post coitus* the first meiotic division begins (before birth for most mammals) and remains arrested in prophase I from a few days to many years depending on the species. It is in this period or in some cases at the beginning of sexual maturity that the primary oocytes secrete proteins to form a coat called zona pellucida and they also produce cortical granules containing enzymes and proteins needed for fertilization. Meiosis stands by because of the follicular granulosa cells that send inhibitory signals through gap junctions and the zona pellucida. Sexual maturation is the beginning of periodic ovulation. Ovulation is the regular release of one oocyte from the ovary into the reproductive tract and is preceded by follicular growth. A few follicle cells are stimulated to grow but only one oocyte is ovulated. A primordial follicle consists of an epithelial layer of follicular granulosa cells enclosing an oocyte. The pituitary gland secrete follicle-stimulating hormones (FSHs) that stimulate follicular growth and oocyte maturation. The thecal cells around each follicle secrete estrogen. This hormone stimulates the production of FSH receptors on the follicular granulosa cells and has at the same time a negative feedback on FSH secretion. This results in a competition between the follicles and only the follicle with the most FSH receptors survives and is ovulated. Meiotic division I goes on in the ovulated oocyte stimulated by luteinizing hormones (LHs) produced by the pituitary gland. FSH and LH block the gap junctions between follicle cells and the oocyte therefore inhibiting communication between them. Most follicular granulosa cells stay around the oocyte and so form the cumulus layer. Large non-mammalian oocytes accumulate egg yolk, glycogen, lipids, ribosomes, and the mRNA needed for protein synthesis during early embryonic growth. These intensive RNA biosynthese are mirrored in the structure of the chromosomes, which decondense and form lateral loops giving them a lampbrush appearance (see Lampbrush chromosome). Oocyte maturation is the following phase of oocyte development. It occurs at sexual maturity when hormones stimulate the oocyte to complete meiotic division I. The meiotic division I produces 2 cells differing in size: a small polar body and a large secondary oocyte. The secondary oocyte undergoes meiotic division II and that results in the formation of a second small polar body and a large mature egg, both being haploid cells. The polar bodies degenerate.[8] Oocyte maturation stands by at metaphase II in most vertebrates. During ovulation, the arrested secondary oocyte leaves the ovary and matures rapidly into an egg ready for fertilization. Fertilization will cause the egg to complete meiosis II. In human females there is proliferation of the oogonia in the fetus, meiosis starts then before birth and stands by at meiotic division I up to 50 years, ovulation begins at puberty.

Egg growth

A 10 - 20 μm large somatic cell generally needs 24 hours to double its mass for mitosis. By this way it would take a very long time for that cell to reach the size of a mammalian egg with a diameter of 100 μm (some insects have eggs of about 1,000 μm or greater). Eggs have therefore special mechanisms to grow to their large size. One of these mechanisms is to have extra copies of genes: meiotic division I is paused so that the oocyte grows while it contains two diploid chromosome sets. Some species produce many extra copies of genes, such as amphibians, which may have up to 1 or 2 million copies.

A complementary mechanism is partly dependent on syntheses of other cells. In amphibians, birds, and insects, yolk is made by the liver (or its equivalent) and secreted into the blood. Neighboring accessory cells in the ovary can also provide nutritive help of two types. In some invertebrates some oogonia become nurse cells. These cells are connected by cytoplasmic bridges with oocytes. The nurse cells of insects provide oocytes macromolecules such as proteins and mRNA. Follicular granulosa cells are the second type of accessory cells in the ovary in both invertebrates and vertebrates. They form a layer around the oocyte and nourish them with small molecules, no macromolecules, but eventually their smaller precursor molecules, by gap junctions.

17.4.2 Spermatogenesis

Mammalian spermatogenesis is representative for most animals. In human males, spermatogenesis begins at puberty in seminiferous tubules in the testicles and go on continuously. Spermatogonia are immature germ cells. They proliferate continuously by mitotic divisions around the outer edge of the seminiferous tubules, next to the basal lamina. Some of these cells stop proliferation and differentiate into primary spermatocytes. After they proceed through the first meiotic division, two secondary spermatocytes are produced. The two secondary spermatocytes undergo the second meiotic division to form four haploid spermatids. These spermatids differentiate morphologically into sperm by nuclear condensation, ejection of the cytoplasm and formation of the acrosome and flagellum.

The developing male germ cells do not complete cytokinesis during spermatogenesis. Consequently cytoplasmic bridges assure connection between the clones of differentiating daughter cells to form a syncytium. In this way the haploid cells are supplied with all the products of a complete diploid genome. Sperm that carry a Y chromosome, for example, is supplied with essential molecules that are encoded by genes on the X chromosome.

17.5 Diseases

Germ cell tumor is a rare cancer that can affect people at all ages. 2.4 children out of 1 million suffer the disease, and it counts for 4% of all cancers in children and adolescents younger than 20 years old.

Germ cell tumors are generally located in the gonads but can also appear in the abdomen, pelvis, mediastinum, or brain. Germ cells migrating to the gonads may not reach that intended destination and a tumor can grow wherever they end up, but the exact cause is still unknown. These tumors can be benign or malignant.[9]

17.6 Induced differentiation

Inducing differentiation of certain cells to germ cells has many applications. One implication of induced differentiation is that it may allow for the eradication of male and female factor infertility. Furthermore, it would allow same-sex couples to have biological children if sperm could be produced from female cells or if eggs could be produced from male cells. Efforts to create sperm and eggs from skin and embryonic stem cells were pioneered by Hayashi and Saitou's research group at Kyoto University.[10] These researchers produced primordial germ cell-like cells (PGCs) from embryonic stem cells (ESCs) and skin cells in vitro.

Hayashi and Saitou's group was able to promote the differentiation of embryonic stem cells into PGCs with the use of precise timing and bone morphogenetic protein 4 (Bmp4). The PGCs were then placed in the testes of mice that were initially unable to produce their own sperm cells. As a result the mice were able to produce sperm cells. Upon succeeding with embryonic stem cells, the group was able to successfully promote the differentiation of induced pluripotent stem cells (iPSCs) into PGCs. These primordial germ cell-like cells were then used to create spermatozoa and oocytes.[11]

Efforts for human cells are less advanced due to the fact that the PGCs formed by these experiments are not always viable. In fact Hayashi and Saitou's method is only one third as effective as current in vitro fertilization methods, and the produced PGCs are not always functional. Furthermore, not only are the induced PGCs not as effective as naturally occurring PGCs, but they are also less effective at erasing their epigenetic markers when they differentiate from iPSCs or ESCs to PGCs.

There are also other applications of induced differentiation of germ cells. Another study showed that culture of human embryonic stem cells in mitotically inactivated porcine ovarian fibroblasts (POF) causes differentiation into germ cells, as evidenced by gene expression analysis.[12]

17.7   See also

- Germline

- Germ Line Development

- Germ layer

17.8   References

[1] Alberts B., Johnson A., Lewis J., Raff M., Roberts K., Walter P. (2002). *Molecular biology of the cell*. New York, Garland Science, 1463 p.

[2] Twyman R.M. (2001). *Developmental biology*. Oxford, Bios Scientific Publishers, 451p.

[3] Cinalli R.M., Rangan P., Lehmann R. (2008). "Germ cells are forever". Cell 132:559-562.

[4] Kunwar P.S., Lehmann R. (2003). "Germ-cell attraction". Nature 421:226-227.

[5] Turnpenny L., Spalluto C.M., Perrett R.M., O'Shea M., Hanley K.P., Cameron I.T., Wilson D.I., Hanley N.A. (2006). "Evaluating human embryonic germ cells: concord and conflict as pluripotent stem cells". *Stem Cells* (Stem Cells 24:212-220) **24** (2): 212–20. doi:10.1634/stemcells.2005-0255. PMID 16144875.

[6] Saitou M, Yamaji M (November 1, 2012). "Primordial Germ Cells in Mice". *Cold Spring Harbor Perspectives in Biology* **4** (11): 1–19. doi:10.1101/cshperspect.a008375. PMID 23125014.

[7] Alberts B, Johnson A, Lewis J, et al. (2002). "Primordial Germ Cells and Sex Determination in Mammals". *Molecular Biology of the Cell. 4th edition.* Garland Science.

[8] De Felici M., Scaldaferri M.L., Lobascio M., Iona S., Nazzicone V., Klinger F.G., Farini D. (2004). "Experimental approaches to the study of primordial germ cell lineage and proliferation". *Human Reproduction Update* (Human reproduction update 10:197-206) **10** (3): 197–206. doi:10.1093/humupd/dmh020. PMID 15140867.

[9] Olson T. (2006). "Germ cell tumors". CureSearch.org Medical Editorial Board.

[10] Hayashi K, Ogushi S, Kurimoto K, Shimamoto S, Ohta H, Saitou M (November 2012). "Offspring from Oocytes Derived from in Vitro Primordial Germ Cell–like Cells in Mice". *Science* **338** (6109): 971–975. doi:10.1126/science.1226889. PMID 23042295.

[11] Cyranoski, David (August 2013). "Egg Engineers". *Nature* **500** (7463): 392–94. doi:10.1038/500392a.

[12] Richards M, Fong CY, Bongso A (December 2008). "Comparative evaluation of different in vitro systems that stimulate germ cell differentiation in human embryonic stem cells". *Fertil. Steril.* **93** (3): 986–94. doi:10.1016/j.fertnstert.2008.10.030. PMID 19064262.

17.9   External links

- Germ Cells at the US National Library of Medicine Medical Subject Headings (MeSH)

- Primordial Germ Cell Development

Chapter 18

Egg

This article is about biological eggs. For eggs as food, see Egg (food). For other uses, see Egg (disambiguation).

An **egg** is the organic vessel containing the zygote in which an animal embryo develops until it can survive on its own, at which point the animal hatches. An egg results from fertilization of an ovum. Most arthropods, vertebrates, and mollusks lay eggs, although some do not, such as scorpions and most mammals.

Reptile eggs, bird eggs, and monotreme eggs are laid out of water, and are surrounded by a protective shell, either flexible or inflexible. Eggs laid on land or in nests are usually kept within a favorable temperature range (warm) while the embryo grows. When the embryo is adequately developed it hatches, i.e. breaks out of the egg's shell. Some embryos have a temporary egg tooth with which to crack, pip, or break the eggshell or covering.

The largest recorded egg is from a whaleshark, and was 30 cm × 14 cm × 9 cm (11.8 in × 5.5 in × 3.5 in) in size;[1] whale shark eggs normally hatch within the mother. At 1.5 kg (3.3 lb) and up to 17.8 cm × 14 cm (7.0 in × 5.5 in), the ostrich egg is the largest egg of any living bird,[2] though the extinct elephant bird and some dinosaurs laid larger eggs. The bee hummingbird produces the smallest known bird egg, which weighs half of a gram (around 0.02 oz). The eggs laid by some reptiles and most fish can be even smaller, and those of insects and other invertebrates can be much smaller still.

Reproductive structures similar to the egg in other kingdoms are termed "spores," or in spermatophytes "seeds," or in gametophytes "egg cells".

18.1 Eggs of different animal groups

Further information: egg cell

Several major groups of animals typically have readily distinguishable eggs.

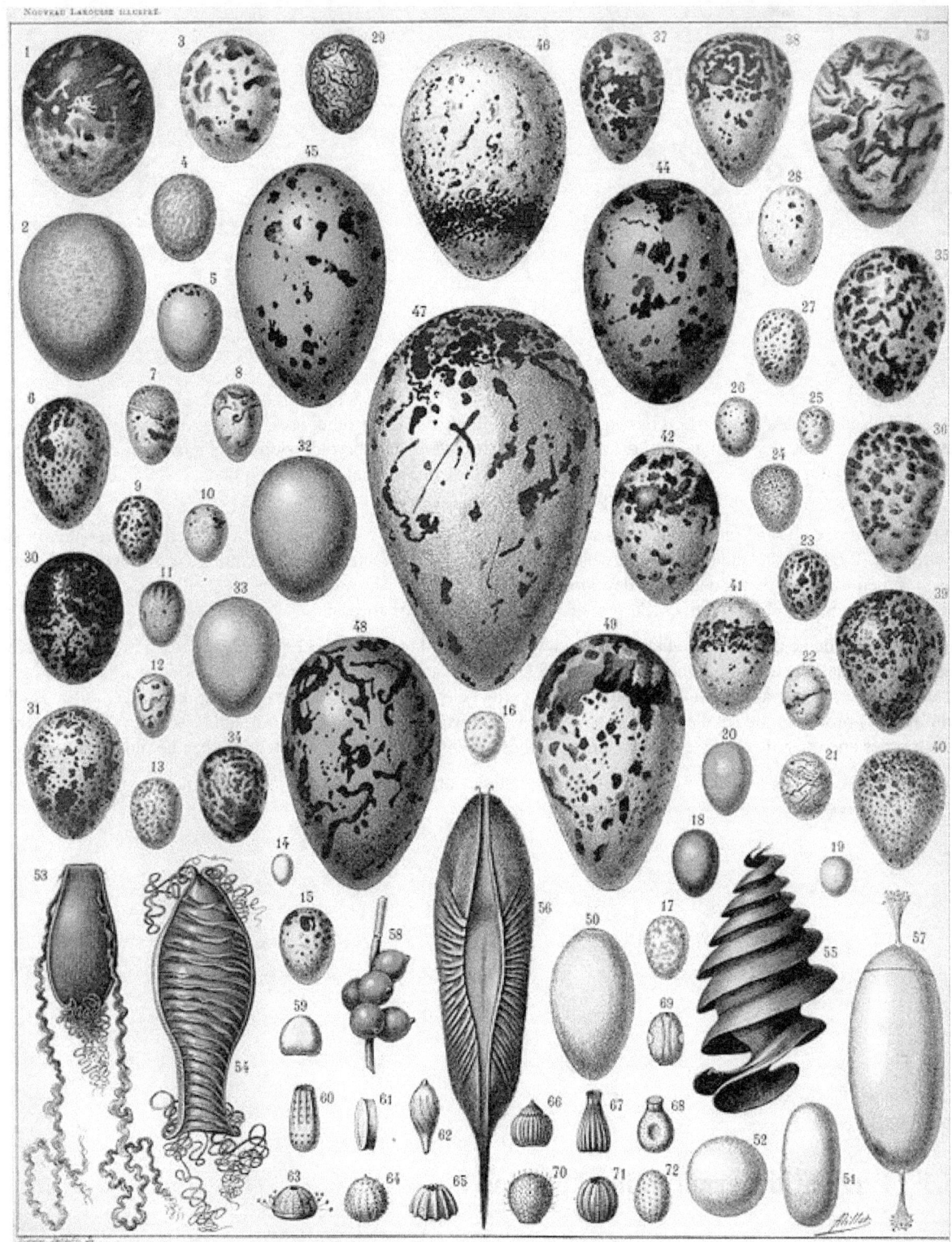

Eggs of various birds, a reptile, various cartilaginous fish, a cuttlefish and various butterflies and moths. (Click on image for key)

18.1.1 Fish and amphibian eggs

See also: Ichthyoplankton and Spawn (biology)

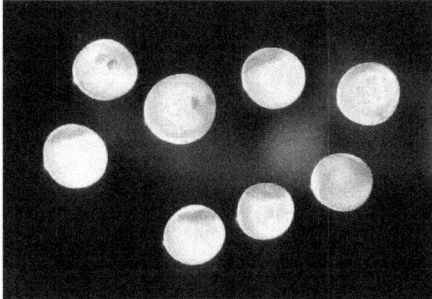

Salmon eggs in different stages of development. In some only a few cells grow on top of the yolk, in the lower right the blood vessels surround the yolk and in the upper left the black eyes are visible.

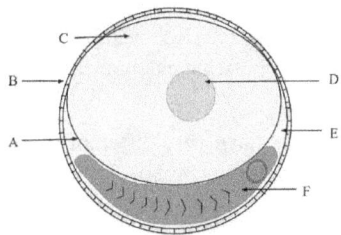

Diagram of a fish egg: A. vitelline membrane B. chorion C. yolk D. oil globule E. perivitelline space F. embryo

The most common reproductive strategy for fish is known as oviparity, in which the female lays undeveloped eggs that are externally fertilized by a male. Typically large numbers of eggs are laid at one time (an adult female cod can produce 4–6 million eggs in one spawning) and the eggs are then left to develop without parental care. When the larvae hatch from the egg, they often carry the remains of the yolk in a yolk sac which continues to nourish the larvae for a few days as they learn how to swim. Once the yolk is consumed, there is a critical point after which they must learn how to hunt and feed or they will die.

A few fish, notably the rays and most sharks use ovoviviparity in which the eggs are fertilized and develop internally. However the larvae still grow inside the egg consuming the egg's yolk and without any direct nourishment from the mother. The mother then gives birth to relatively mature young. In certain instances, the physically most developed offspring will devour its smaller siblings for further nutrition while still within the mother's body. This is known as intrauterine cannibalism.

In certain scenarios, some fish such as the hammerhead shark and reef shark are viviparous, with the egg being fertilized and developed internally, but with the mother also providing direct nourishment.

The eggs of fish and amphibians are jellylike. Cartilagenous fish (sharks, skates, rays, chimaeras) eggs are fertilized internally and exhibit a wide variety of both internal and external embryonic development. Most fish species spawn eggs that are fertilized externally, typically with the male inseminating the eggs after the female lays them. These eggs do not have a shell and would dry out in the air. Even air-breathing amphibians lay their eggs in water, or in protective foam as with the Coast foam-nest treefrog, *Chiromantis xerampelina*.

18.1.2 Bird eggs

Bird eggs are laid by females and incubated for a time that varies according to the species; a single young hatches from each egg. Average clutch sizes range from one (as in condors) to about 17 (the grey partridge). Some birds lay eggs even when not fertilized (e.g. hens); it is not uncommon for pet owners to find their lone bird nesting on a clutch of unfertilized eggs, which are sometimes called wind-eggs.

Colors

The default color of vertebrate eggs is the white of the calcium carbonate from which the shells are made, but some birds, mainly passerines, produce colored eggs. The pigment biliverdin and its zinc chelate give a green or blue ground color, and protoporphyrin produces reds and browns as a ground color or as spotting.

Non-passerines typically have white eggs, except in some ground-nesting groups such as the Charadriiformes, sandgrouse and nightjars, where camouflage is necessary, and some parasitic cuckoos which have to match the passerine host's egg. Most passerines, in contrast, lay colored eggs, even if there is no need of cryptic colors.

However some have suggested that the protoporphyrin markings on passerine eggs actually act to reduce brittleness by acting as a solid state lubricant.[10] If there is insufficient calcium available in the local soil, the egg shell may be thin, especially in a circle around the broad end. Protoporphyrin speckling compensates for this, and increases inversely to the amount of calcium in the soil.[11]

For the same reason, later eggs in a clutch are more spotted than early ones as the female's store of calcium is depleted.

The color of individual eggs is also genetically influenced, and appears to be inherited through the mother only, suggesting that the gene responsible for pigmentation is on the sex determining W chromosome (female birds are WZ, males ZZ).

It used to be thought that color was applied to the shell immediately before laying, but this research shows that coloration is an integral part of the development of the shell, with the same protein responsible for depositing calcium carbonate, or protoporphyrins when there is a lack of that mineral.

In species such as the common guillemot, which nest in large groups,each female's eggs have very different markings, making it easier for females to identify their own eggs on the crowded cliff ledges on which they breed.

Shell

Bird eggshells are diverse. For example:

- cormorant eggs are rough and chalky
- tinamou eggs are shiny
- duck eggs are oily and waterproof
- cassowary eggs are heavily pitted

Tiny pores in bird eggshells allow the embryo to breathe. The domestic hen's egg has around 7500 pores.

Shape

Most bird eggs have an oval shape, with one end rounded (the aerus) and the other more pointed (the taglion). This shape results from the egg being forced through the oviduct. Muscles contract the oviduct behind the egg, pushing it forward. The egg's wall is still shapeable, and the pointy end develops at the back. Cliff-nesting birds often have highly conical eggs. They are less likely to roll off, tending instead to roll around in a tight circle; this trait is likely to have arisen due to evolution via natural selection. In contrast, many hole-nesting birds have nearly spherical eggs.

Predation

Many animals feed on eggs. For example, principal predators of the black oystercatcher's eggs include raccoons, skunks, mink, river and sea otters, gulls, crows and foxes. The stoat (*Mustela erminea*) and long-tailed weasel (*M. frenata*) steal ducks' eggs. Snakes of the genera *Dasypeltis* and *Elachistodon* specialize in eating eggs.

Brood parasitism occurs in birds when one species lays its eggs in the nest of another. In some cases, the host's eggs are removed or eaten by the female, or expelled by her chick. Brood parasites include the cowbirds and many Old World cuckoos.

Various examples

- An average whooping crane egg is 102 mm (4.0 in) long and weighs 208 g (7.3 oz)

- Eurasian oystercatcher eggs camouflaged in the nest

- Egg of a senegal parrot, a bird that nests in tree holes, on a 1 cm (0.39 in) grid

- Eggs of ostrich, emu, kiwi and chicken

- Finch egg next to American dime

- Eggs of duck, goose, guineafowl and chicken

- Eggs of ostrich, cassowary, chicken, flamingo, pigeon and blackbird

- Egg of an emu

18.1.3 Amniote eggs and embryos

Like amphibians, amniotes are air-breathing vertebrates, but they have complex eggs or embryos, including an amniotic membrane. Amniotes include reptiles (including dinosaurs and their descendants, birds) and mammals.

Reptile eggs are often rubbery and are always initially white. They are able to survive in the air. Often the sex of the developing embryo is determined by the temperature of the surroundings, with cooler temperatures favouring males. Not all reptiles lay eggs; some are viviparous ("live birth").

Dinosaurs laid eggs, some of which have been preserved as petrified fossils.

Among mammals, early extinct species laid eggs, as do platypuses and echidnas (spiny anteaters). Platypuses and two genera of echidna are Australian monotremes. Marsupial and placental mammals do not lay eggs, but their unborn young do have the complex tissues that identify amniotes.

18.1.4 Mammalian eggs

The eggs of the egg-laying mammals (the platypus and the echidnas) are macrolecithal eggs very much like those of reptiles. The eggs of marsupials are likewise macrolecithal, but rather small, and develop inside the body of the female, but do not form a placenta. The young are born at a very early stage, and can be classified as a "larva" in the biological sense.[12]

In placental mammals, the egg itself is void of yolk, but develops an umbilical cord from structures that in reptiles would form the yolk sac. Receiving nutrients from the mother, the fetus completes the development while inside the uterus.

18.1.5 Invertebrate eggs

Eggs are common among invertebrates, including insects, spiders, mollusks, and crustaceans.

18.2 Evolution and structure

All sexually reproducing life, including both plants and animals, produces gametes. The male gamete cell, sperm, is usually motile whereas the female gamete cell, the ovum, is generally larger and sessile. The male and female gametes combine to produce the zygote cell. In multicellular organisms the zygote subsequently divides in an organised manner into smaller more specialised cells, so that this new individual develops into an embryo. In most animals the embryo is the sessile initial stage of the individual life cycle, and is followed by the emergence (that is, the hatching) of a motile stage. The zygote or the ovum itself or the sessile organic vessel containing the developing embryo may be called the egg.

A recent proposal suggests that the phylotypic animal body plans originated in cell aggregates before the existence of an egg stage of development. Eggs, in this view, were later evolutionary innovations, selected for their role in ensuring genetic uniformity among the cells of incipient multicellular organisms.[13]

18.3   Scientific classifications

Scientists often classify animal reproduction according to the degree of development that occurs before the new individuals are expelled from the adult body, and by the yolk which the egg provides to nourish the embryo.

18.3.1   Egg size and yolk

Vertebrate eggs can be classified by the relative amount of yolk. Simple eggs with little yolk are called *microlecithal*, medium-sized eggs with some yolk are called *mesolecithal*, and large eggs with a large concentrated yolk are called *macrolecithal*.[7] This classification of eggs is based on the eggs of chordates, though the basic principle extends to the whole animal kingdom.

Microlecithal

Small eggs with little yolk are called microlecithal. The yolk is evenly distributed, so the cleavage of the egg cell cuts through and divides the egg into cells of fairly similar sizes. In sponges and cnidarians the dividing eggs develop directly into a simple larva, rather like a morula with cilia. In cnidarians, this stage is called the planula, and either develops directly into the adult animals or forms new adult individuals through a process of budding.[14]

Microlecithal eggs require minimal yolk mass. Such eggs are found in flatworms, roundworms, annelids, bivalves, echinoderms, the lancelet and in most marine arthropods.[15] In anatomically simple animals, such as cnidarians and flatworms, the fetal development can be quite short, and even microlecithal eggs can undergo direct development. These small eggs can be produced in large numbers. In animals with high egg mortality, microlecithal eggs are the norm, as in bivalves and marine arthropods. However, the latter are more complex anatomically than e.g. flatworms, and the small microlecithal eggs do not allow full development. Instead, the eggs hatch into larvae, which may be markedly different from the adult animal.

In placental mammals, where the egg is nourished from the mother throughout the whole fetal period, the egg is reduced in size to essentially a naked egg cell (zygote).

Mesolecithal

Mesolecithal eggs have comparatively more yolk than the microlecithal eggs. The yolk is concentrated in one part of the egg (the *vegetal pole*), with the cell nucleus and most of the cytoplasm in the other (the *animal pole*). The cell cleavage is uneven, and mainly concentrated in the cytoplasma-rich animal pole.[3]

The larger yolk content of the mesolecithal eggs allows for a longer fetal development. Comparatively anatomically simple animals will be able to go through the full development and leave the egg in a form reminiscent of the adult animal. This is the situation found in hagfish and some snails.[4][15] Animals with smaller size eggs or more advanced anatomy will still have a distinct larval stage, though the larva will be basically similar to the adult animal, as in lampreys, coelacanth and the salamanders.[3]

Macrolecithal

Eggs with a large yolk are called macrolecithal. The eggs are usually few in number, and the embryos have enough food to go through full fetal development in most groups.[7] Macrolecithal eggs are only found in selected representatives of two groups: Cephalopods and vertebrates.[7][16]

Macrolecithal eggs go through a different type of development than other eggs. Due to the large size of the yolk, the cell division can not split up the yolk mass. The fetus instead develops as a plate-like structure on top of the yolk mass, and only envelopes it at a later stage.[7] A portion of the yolk mass is still present as an external or semi-external yolk sac at hatching in many groups. This form of fetal development is common in bony fish, even though their eggs can be quite small. Despite their macrolecithal structure, the small size of the eggs does not allow for direct development, and the eggs hatch to a larval stage ("fry"). In terrestrial animals with macrolecithal eggs, the large volume to surface ratio necessitates structures to aid in transport of oxygen and carbon dioxide, and for storage of waste products so that the embryo does not suffocate or get poisoned from its own waste while inside the egg, see amniote.[9]

In addition to bony fish and cephalopods, macrolecithal eggs are found in cartilaginous fish, reptiles, birds and monotreme mammals.[3] The eggs of the coelacanths can reach a size of 9 cm in diameter, and the young go through full development while in the uterus, living on the copious yolk.[17]

18.3.2 Egg-laying reproduction

Animals are commonly classified by their manner of reproduction, at the most general level distinguishing egg-laying (Latin. *oviparous*) from live-bearing (Latin. *viviparous*).

These classifications are divided into more detail according to the development that occurs before the offspring are expelled from the adult's body. Traditionally:[18]

- **Ovuliparity** means the female spawns unfertilized eggs (ova), which must then be externally fertilised. Ovuliparity is typical of bony fish, anurans, echinoderms, bivalves and cnidarians. Most aquatic organisms are ovuliparous. The term is derived from the diminiutive meaning "little egg".

- **Oviparity** is where fertilisation occurs internally and so the eggs laid by the female are zygotes (or newly developing embryos), often with important outer tissues added (for example, in a chicken egg, no part outside of the yolk originates with the zygote). Oviparity is typical of birds, reptiles, some cartilaginous fish and most arthropods. Terrestrial organisms are typically oviparous, with egg-casings that resist evaporation of moisture.

- **Ovo-viviparity** is where the zygote is retained in the adult's body but there are no *trophic* (feeding) interactions. That is, the embryo still obtains all of its nutrients from inside the egg. Most live-bearing fish, amphibians or reptiles are actually ovoviviparous. Examples include the reptile *Anguis fragilis*, the sea horse (where zygotes are retained in the male's ventral "marsupium"), and the frogs *Rhinoderma darwinii* (where the eggs develop in the vocal sac) and *Rheobatrachus* (where the eggs develop in the stomach).

- **Histotrophic viviparity** means embryos develop in the female's oviducts but obtain nutrients by consuming other ova, zygotes or sibling embryos (oophagy or adelphophagy). This intra-uterine cannibalism occurs in some sharks and in the black salamander *Salamandra atra*. Marsupials excrete an "uterine milk" supplementing the nourishment from the yolk sak.[19]

- **Hemotrophic viviparity** is where nutrients are provided from the female's blood through a designated organ. This most commonly occurs through a placenta, found in most mammals. Similar structures are found in some sharks and in the lizard *Pseudomoia pagenstecheri*.[20][21] In some hylid frogs, the embryo is fed by the mother through specialized gills.[22]

The term hemotropic derives from the Latin for blood-feeding, contrasted with histotrophic for tissue-feeding.[23]

18.4 Human use

18.4.1 Food

Main article: Egg (food)

Eggs laid by many different species, including birds, reptiles, amphibians, and fish, have probably been eaten by mankind for millennia. Popular choices for egg consumption are chicken, duck, roe, and caviar, but by a wide margin the egg most often humanly consumed is the chicken egg, typically unfertilized.

Eggs and Kashrut

See also: Kashrut § Pareve foods and Kosher foods § Eggs

According to the Kashrut, that is the set of Jewish dietary laws, kosher food may be consumed according to *halakha* (Jewish law). Kosher meat and milk (or derivatives) cannot be mixed (Deuteronomy 14:21) or stored together. Eggs are considered *pareve* (neither meat nor dairy) despite being an animal product and can be mixed with either milk or kosher meat. Mayonnaise, for instance, is usually marked "pareve" despite by definition containing egg.[24]

18.4.2   Vaccine manufacture

Many vaccines for infectious diseases are produced in fertile chicken eggs. The basis of this technology was the discovery in 1931 by Alice Miles Woodruff and Ernest William Goodpasture at Vanderbilt University that the rickettsia and viruses that cause a variety of diseases will grow in chicken embryos. This enabled the development of vaccines against influenza, chicken pox, smallpox, yellow fever, typhus, Rocky mountain spotted fever and other diseases.

18.4.3   Culture

A popular Easter tradition in some parts of the world is the decoration of hard-boiled eggs (usually by dyeing, but often by spray-painting). Adults often hide the eggs for children to find, an activity known as an Easter egg hunt. A similar tradition of egg painting exists in areas of the world influenced by the culture of Persia. Before the spring equinox in the Persian New Year tradition (called Norouz), each family member decorates a hard-boiled egg and sets them together in a bowl. The tradition of a dancing egg is held during the feast of Corpus Christi in Barcelona and other Catalan cities since the 16th century. It consists of an emptied egg, positioned over the water jet from a fountain, which starts turning without falling.[25] Although a food item, eggs are sometimes thrown at houses, cars, or people. This act, known commonly as "egging" in the various English-speaking countries, is a minor form of vandalism and, therefore, usually a criminal offense and is capable of damaging property (egg whites can degrade certain types of vehicle paint) as well as causing serious eye injury.[81] On Halloween, for example, trick or treaters have been known to throw eggs (and sometimes flour) at property or people from whom they received nothing. Eggs are also often thrown in protests, as they are inexpensive and nonlethal, yet very messy when broken.

18.5   Gallery

- Insect eggs, in this case those of the Emperor Gum Moth, are often laid on the underside of leaves.

- Fish eggs, such as these herring eggs are often transparent and fertilized after laying.

- Skates and some sharks have a uniquely shaped egg case called a mermaid's purse.

- A *Testudo hermanni* emerging fully developed from a reptilian egg.

- A *Schistosoma mekongi* egg.

- Eggs of *Huffmanela hamo*, a Nematode parasite in a fish

- Eggs of various parasites (mainly nematodes) from wild primates

18.6 See also

- List of egg topics
- Animal shell
- Bird egg
- Butterfly eggs
- Egg (food)
- Egg yolk and Egg white
- Eggshell
- Fossil egg
- Haugh unit
- Oology
- Ovary
- Ovulation
- Oviparous
- Trophic egg

18.7 References

[1] "Whale Shark – Cartilaginous Fish". SeaWorld Parks & Entertainment. Retrieved 2014-06-26.

[2] D.R. Khanna (1 January 2005). *Biology of Birds.* Discovery Publishing House. p. 130. ISBN 978-81-7141-933-3.

[3] Hildebrand, M. & Gonslow, G. (2001): Analysis of Vertebrate Structure. 5th edition. *John Wiley & Sons, Inc.* New York

[4] Gorbman, A. (June 1997). "Hagfish development". *Zoological journal* **14** (3): 375–390. doi:10.2108/zsj.14.375.

[5] Hardisty, M. W., and Potter, I. C. (1971). The Biology of Lampreys 1st ed. (Academic Press Inc.).

[6] Leonard J. V. Compagno (1984). Sharks of the World: An annotated and illustrated catalogue of shark species known to date. Food and Agriculture Organization of the United Nations. ISBN 92-5-104543-7. OCLC 156157504.

[7] Romer, A. S. & Parsons, T. S. (1985): *The Vertebrate Body.* (6th ed.) Saunders, Philadelphia.

[8] Peter Scott: Livebearing Fishes, p. 13. Tetra Press 1997. ISBN 1-56465-193-2

[9] Stewart J. R. (1997): *Morphology and evolution of the egg of oviparous amniotes.* In: S. Sumida and K. Martin (ed.) Amniote Origins-Completing the Transition to Land (1): 291–326. London: Academic Press.

[10] Solomon, S.E. (1987). Egg shell pigmentation. In Egg Quality : Current Problems and Recent Advances (eds R.G. Wells & C.G. Belyarin). Butterworths, London, pp. 147–157.

[11] Gosler, Andrew G.; James P. Higham; S. James Reynolds (2005). "Why are birds' eggs speckled?". *Ecology Letters* **8**: 1105–1113. doi:10.1111/j.1461-0248.2005.00816.x.

[12] Colbert, H.E & Morales, M. (1991): Evolution of the Vertebrates – A History of Backboned Animals Through Time. 4. utgave. John Wiely & sons inc, New York. 470 pages ISBN 0-471-85074-8

[13] Newman, S.A. (2011). "Animal egg as evolutionary innovation: a solution to the 'embryonic hourglass' puzzle". *Journal of Experimental Zoology (Molecular and Developmental Evolution)* **316**: 467–483. doi:10.1002/jez.b.21417.

[14] Reitzel, A.M.; Sullivan, J.C; Finnery, J.R (2006). "Qualitative shift to indirect development in the parasitic sea anemone *Edwardsiella lineata*". *Integrative and Comparative Biology* **46** (6): 827–837. doi:10.1093/icb/icl032.

[15] Barns, R.D. (1968): Invertebrate Zoology. W. B. Saunders Company, Philadelphia. 743 pages

[16] Nixon, M. & Messenger, J.B (eds) (1977): The Biology of Cephalopods. Symposium of the Zoological Society of London, pp 38–615

[17] Fricke, H.W. & Frahm, J. (1992): Evidence for lecithotrophic viviparity in the living coelacanth. *Naturwissenschaften* no 79: pp 476–479

[18] Thierry Lodé 2001. Les stratégies de reproduction des animaux (reproduction strategies in animal kingdom). Eds Dunod Sciences, Paris

[19] USA, David O. Norris, Ph.D., Professor Emeritus, Department of Integrative Physiology, University of Colorado at Boulder, Colorado, USA, James A. Carr, Ph.D., faculty director, Joint Admission Medical Program, Department of Biological Sciences, Texas Tech University, Lubbock, Texas, (2013). *Vertebrate endocrinology.* (Fifth ed.). p. 349. ISBN 0123948150. Retrieved 25 November 2014.

[20] Hamlett, William C. (1989). "Evolution and morphogenesis of the placenta in sharks". *Journal of Experimental Zoology* **252** (S2): 35–52. doi:10.1002/jez.1402520406. Retrieved 25 November 2014.

[21] Jerez, Adriana; Ramírez-Pinilla, Martha Patricia (November 2003). "Morphogenesis of extraembryonic membranes and placentation inMabuya mabouya (Squamata, Scincidae)". *Journal of Morphology* **258** (2): 158–178. doi:10.1002/jmor.10138. Retrieved 25 November 2014.

[22] Gorbman, edited by Peter K.T. Pang, Martin P. Schreibman ; consulting editor, Aubrey (1986). *Vertebrate endocrinology : fundamentals and biomedical implications.* Orlando: Academic Press. p. 237. ISBN 0125449011. Retrieved 25 November 2014.

[23] "Online Etymology Dictionary". Etymonline.com. Retrieved 2013-07-27.

[24] Jewish Virtual Library Kashrut: Jewish Dietary Laws

[25] L'ou com balla, Barcelona Cathedral.

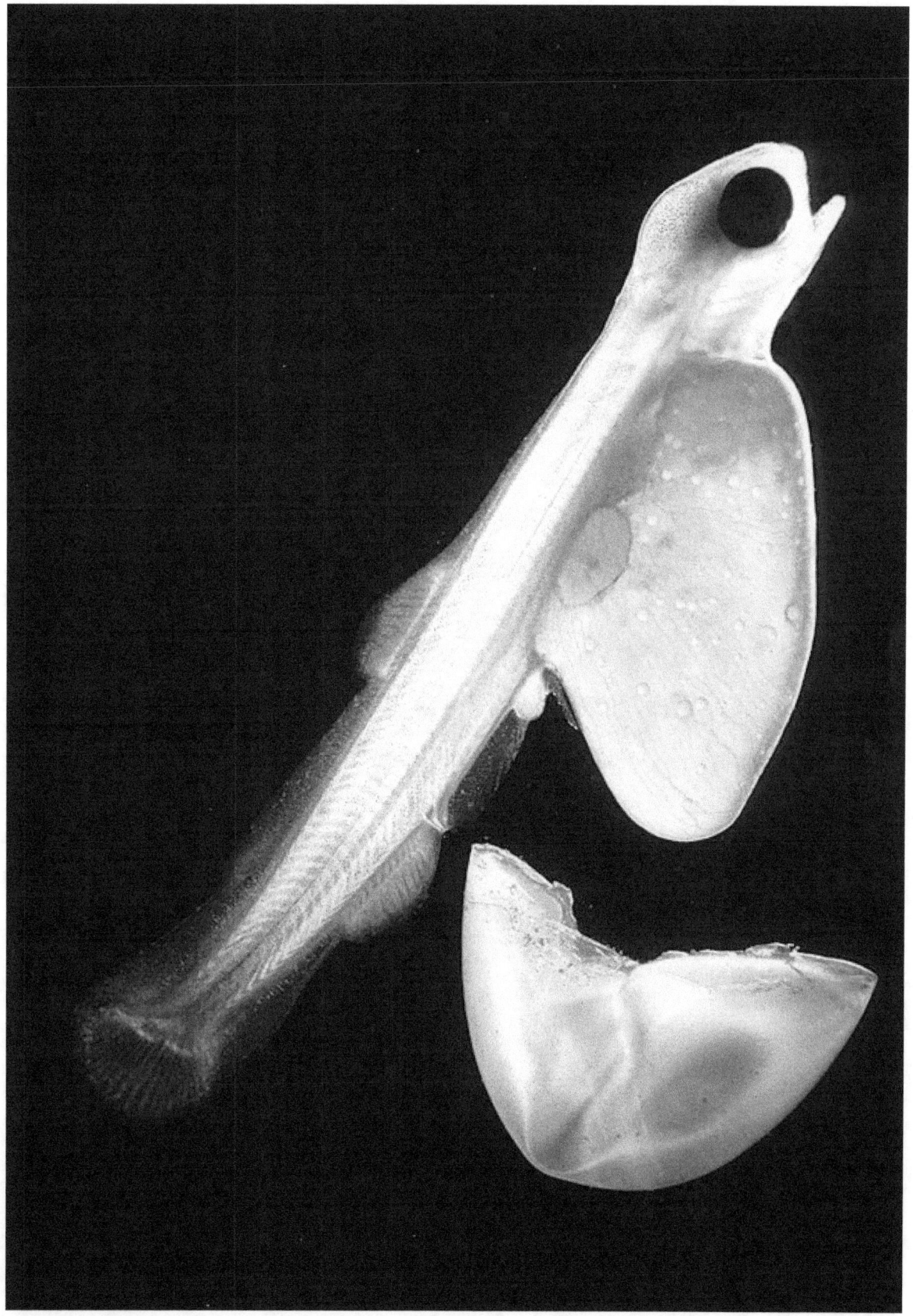

Salmon fry hatching. The larva has grown around the remains of the yolk and the remains of the soft, transparent egg are discarded.

Guillemot eggs

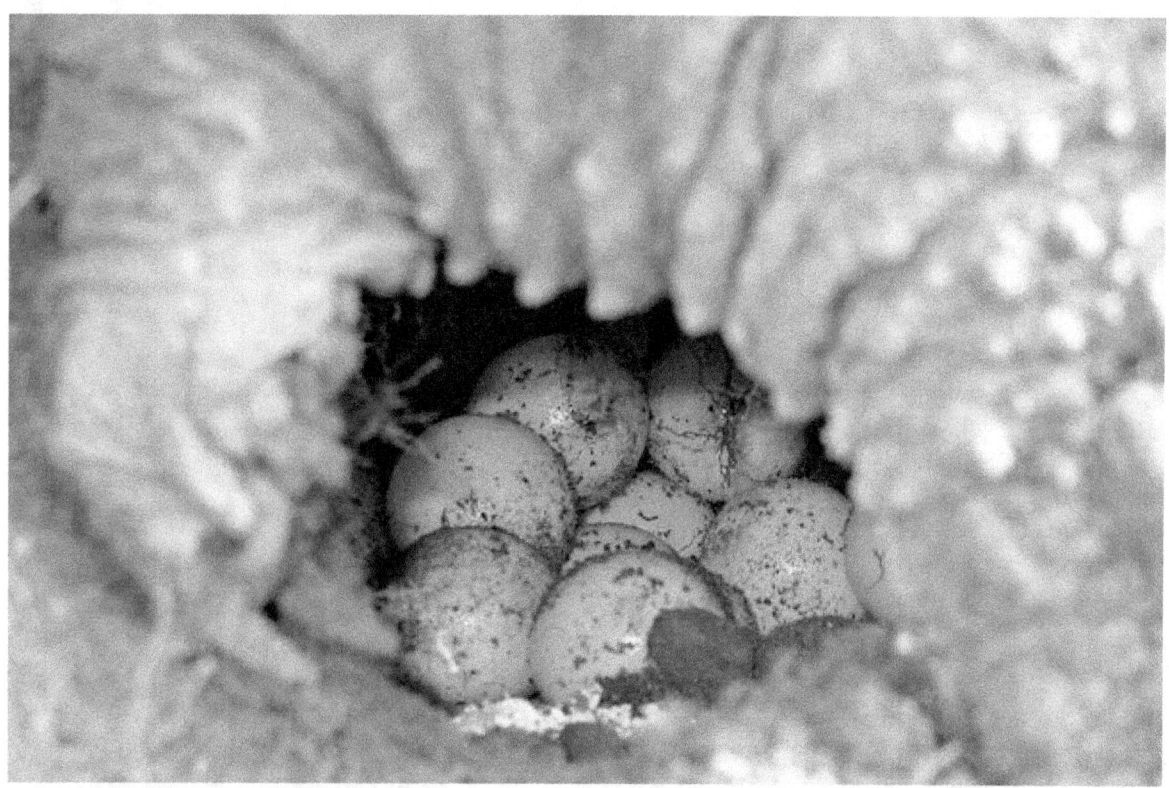

Turtle eggs in a nest dug by a female common snapping turtle (Chelydra serpentina)

Nudibranch Orange-peel doris Acanthodoris lutea *in tide pool laying eggs*

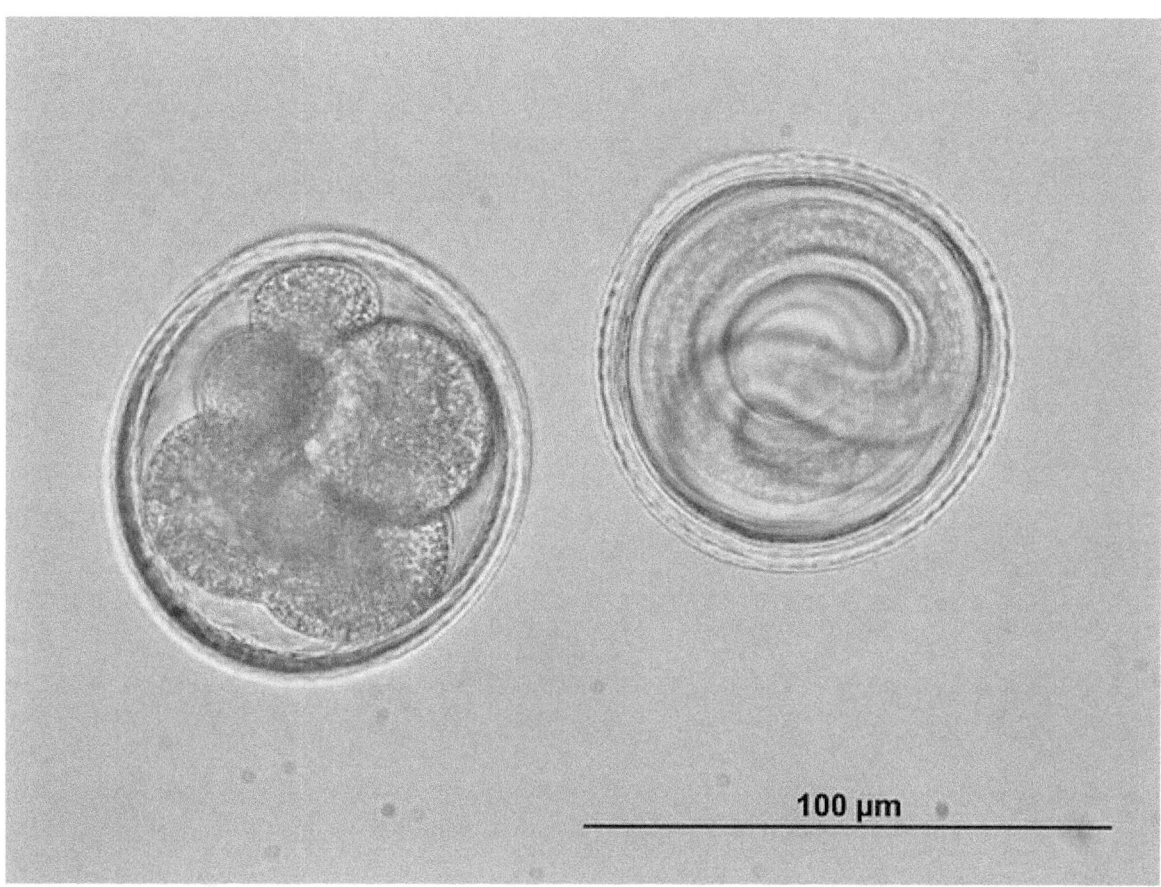

Microlecithal eggs from the roundworm Toxocara

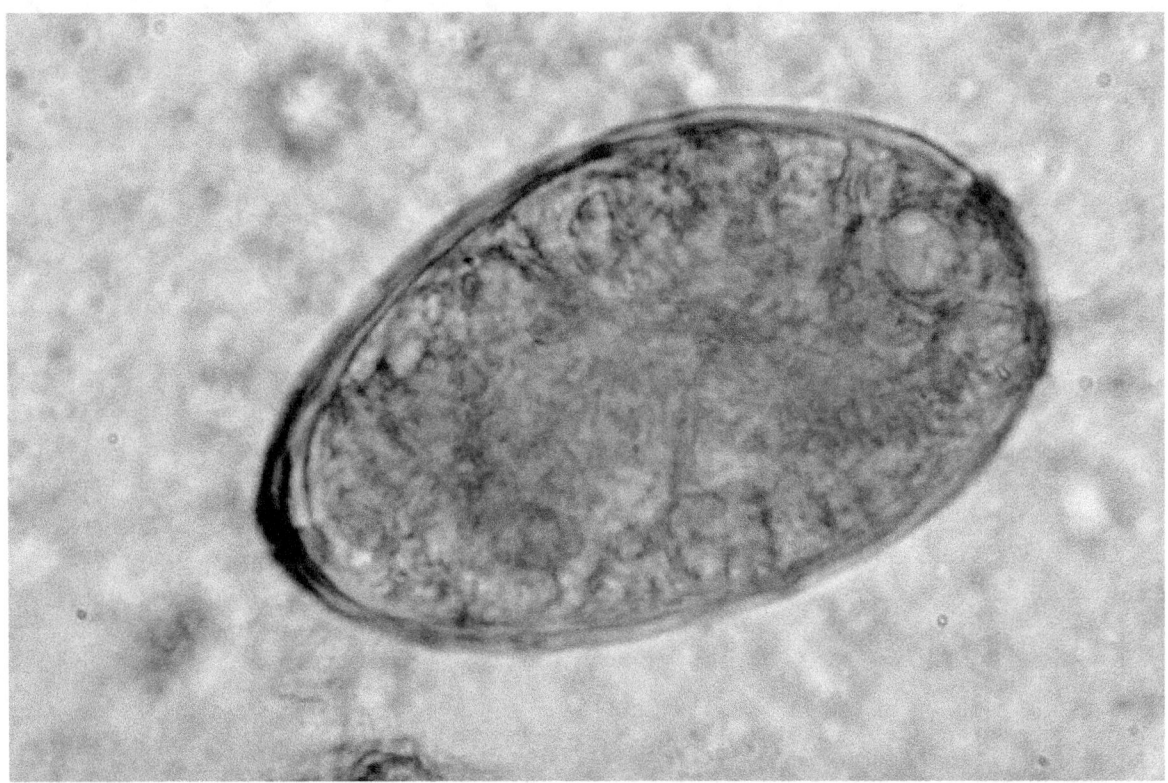

Microlecithal eggs from the flatworm Paragonimus westermani

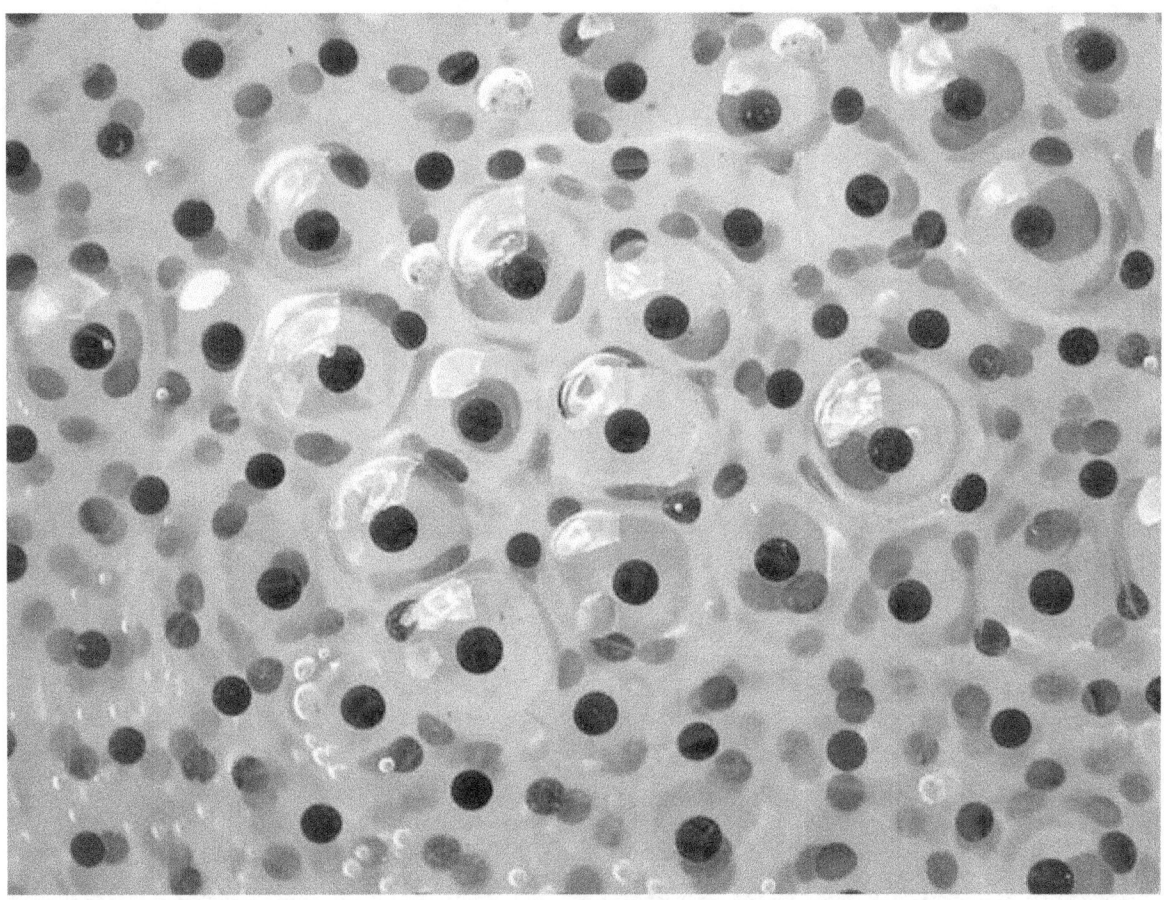

Frogspawn is mesolecithal.

A baby tortoise begins to emerge "fully developed" from its macrolecithal egg.

Chapter 19

Sperm

For other uses, see Sperm (disambiguation).

Sperm is the male reproductive cell and is derived from the Greek word (σπέρμα) *sperma* (meaning "seed"). In the types of sexual reproduction known as anisogamy and its subtype oogamy, there is a marked difference in the size of the gametes with the smaller one being termed the "male" or sperm cell. A uniflagellar sperm cell that is motile is referred to as a **spermatozoon**, whereas a non-motile sperm cell is referred to as a **spermatium**. Sperm cells cannot divide and have a limited life span, but after fusion with egg cells during fertilization, a new organism begins developing, starting as a totipotent zygote. The human sperm cell is haploid, so that its 23 chromosomes can join the 23 chromosomes of the female egg to form a diploid cell. In mammals, sperm develops in the testicles and is released from the penis. It is also possible to extract sperm through TESE. Some sperm banks hold up to 170 litres (37 imp gal; 45 US gal) of sperm.[1]

19.1 Sperm in animals

Further information: Spermatozoon

19.1.1 Anatomy

The mammalian sperm cell consists of a head, a midpiece and a tail. The head contains the nucleus with densely coiled chromatin fibres, surrounded anteriorly by an acrosome, which contains enzymes used for penetrating the female egg. The midpiece has a central filamentous core with many mitochondria spiralled around it, used for ATP production for the journey through the female cervix, uterus and uterine tubes. The tail or "flagellum" executes the lashing movements that propel the spermatocyte.

During fertilization, the sperm provides three essential parts to the oocyte: (1) a signalling or activating factor, which causes the metabolically dormant oocyte to activate; (2) the haploid paternal genome; (3) the centrosome, which is responsible for maintaining the microtubule system.[2]

Although semen contains millions of sperm, the egg will admit only one. The other ones will soon die and be absorbed.

19.1.2 Origin

The spermatozoa of animals are produced through spermatogenesis inside the male gonads (testicles) via meiotic division. The initial spermatozoon process takes around 70 days to complete. The spermatid stage is where the sperm develops the familiar tail. The next stage where it becomes fully mature takes around 60 days when it is called a spermatozoan.[3] Sperm cells are carried out of the male body in a fluid known as semen. Human sperm cells can survive within the

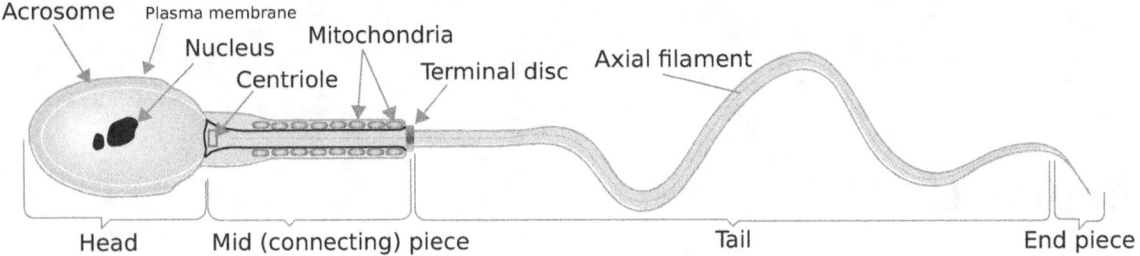

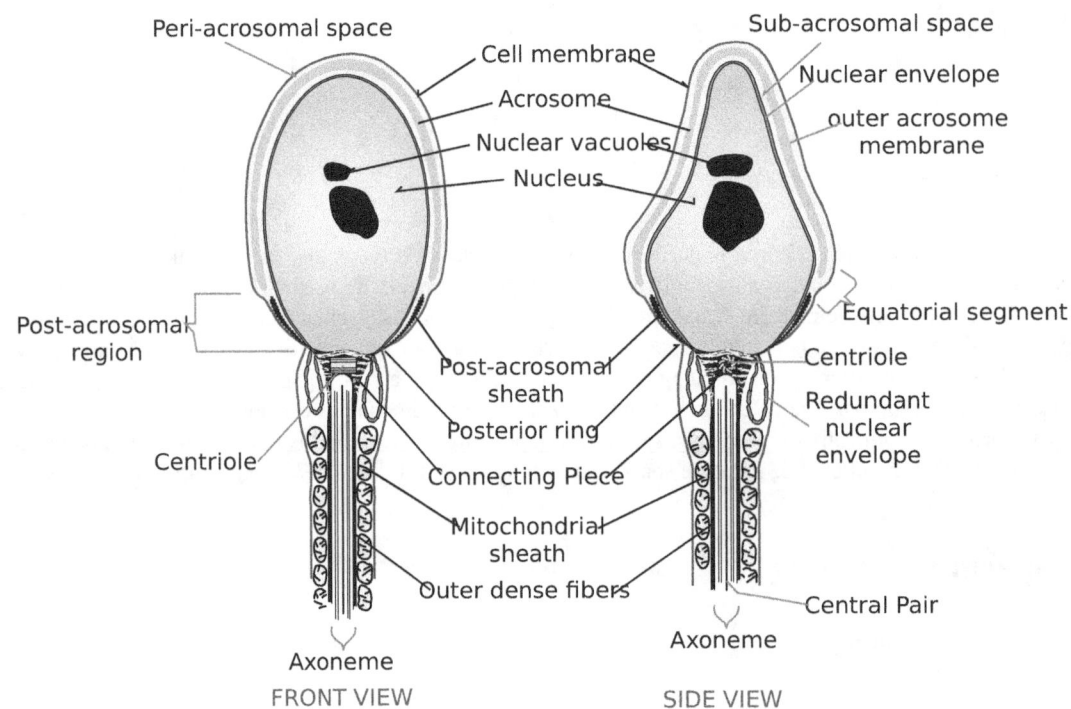

Diagram of a human sperm cell

female reproductive tract for more than 5 days post coitus.[4] Semen is produced in the seminal vesicles, prostate gland and urethral glands.

19.1.3   Sperm quality

Main article: Semen quality

Sperm quantity and quality are the main parameters in semen quality, which is a measure of the ability of semen to accomplish fertilization. Thus, in humans, it is a measure of fertility in a man. The genetic quality of sperm, as well as its volume and motility, all typically decrease with age.[5] (See paternal age effect.)

DNA damages present in sperm cells in the period after meiosis but before fertilization may be repaired in the fertilized egg, but if not repaired, can have serious deleterious effects on fertility and the developing embryo. Human sperm cells are particularly vulnerable to free radical attack and the generation of oxidative DNA damage.[6] (see e.g. 8-Oxo-2'-deoxyguanosine)

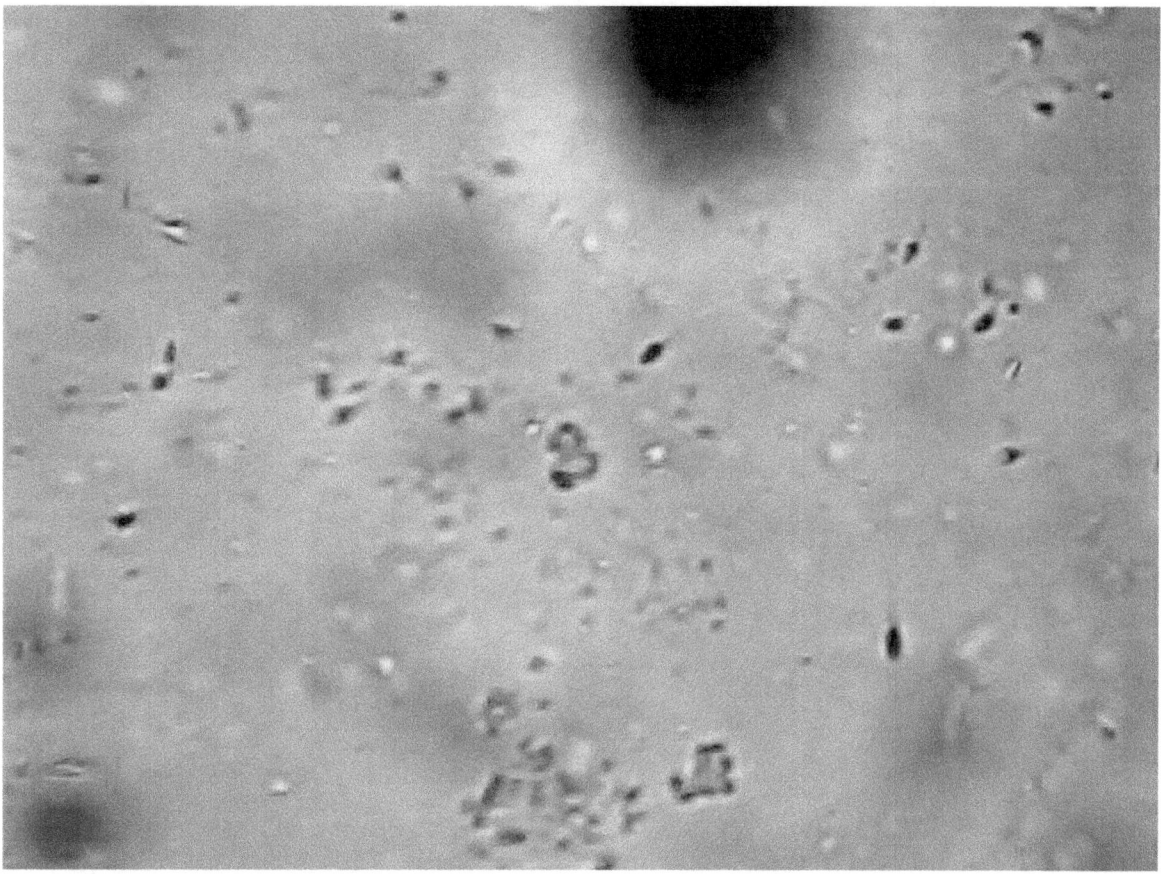

Video of human sperm cells recorded by an affordable home microscope.

The postmeiotic phase of mouse spermatogenesis is very sensitive to environmental genotoxic agents, because as male germ cells form mature sperm they progressively lose the ability to repair DNA damage.[7] Irradiation of male mice during late spermatogenesis can induce damage that persists for at least 7 days in the fertilizing sperm cells, and disruption of maternal DNA double-strand break repair pathways increases sperm cell-derived chromosomal aberrations.[8] Treatment of male mice with melphalan, a bifunctional alkylating agent frequently employed in chemotherapy, induces DNA lesions during meiosis that may persist in an unrepaired state as germ cells progress though DNA repair-competent phases of spermatogenic development.[9] Such unrepaired DNA damages in sperm cells, after fertilization, can lead to offspring with various abnormalities.

19.1.4   Market for human sperm

Further information: Sperm donation

On the global market, Denmark has a well-developed system of human sperm export. This success mainly comes from the reputation of Danish sperm donors for being of high quality[10] and, in contrast with the law in the other Nordic countries, gives donors the choice of being either anonymous or non-anonymous to the receiving couple.[10] Furthermore, Nordic sperm donors tend to be tall and highly educated[11] and have altruistic motives for their donations,[11] partly due to the relatively low monetary compensation in Nordic countries. More than 50 countries worldwide are importers of Danish sperm, including Paraguay, Canada, Kenya, and Hong Kong.[10] However, the Food and Drug Administration (FDA) of the US has banned import of any sperm, motivated by a risk of transmission of Creutzfeldt-Jakob disease, although such a risk is insignificant, since artificial insemination is very different from the route of transmission of Creutzfeldt-Jakob

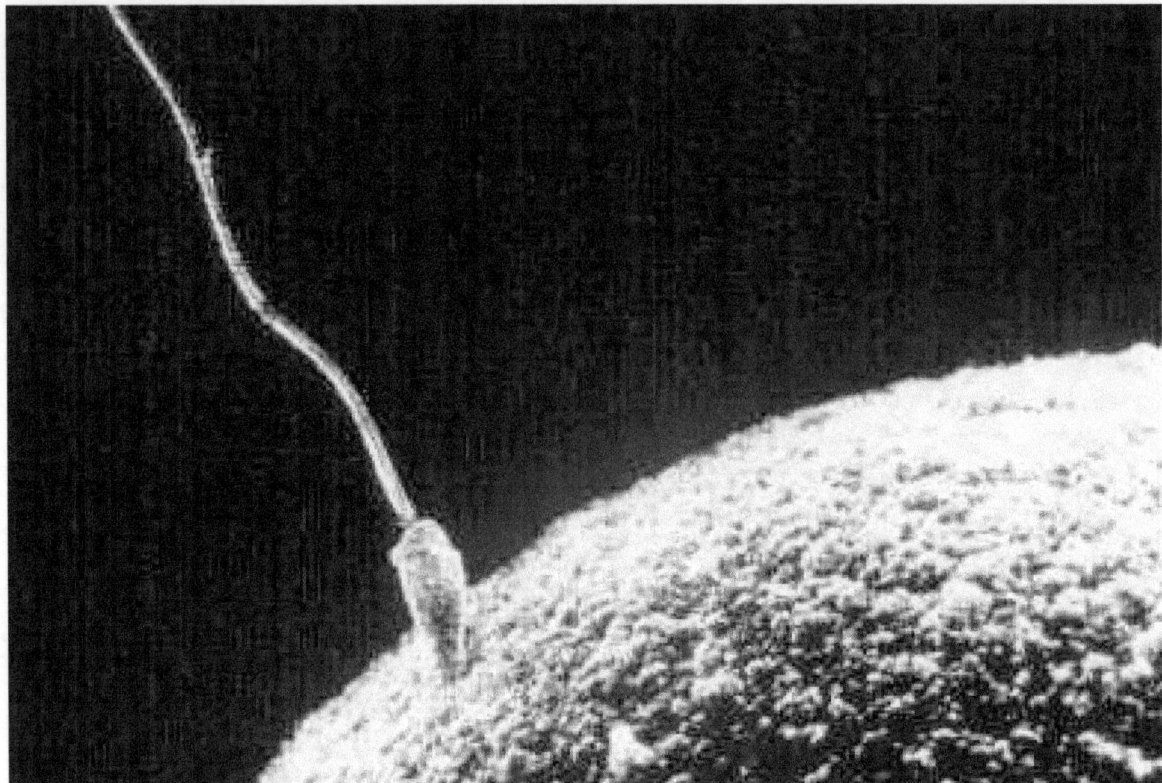

Sperm and egg fusing

disease.[12] The prevalence of Creutzfeldt-Jakob disease for donors is at most one in a million, and if the donor was a carrier, the infectious proteins would still have to cross the blood-testis barrier to make transmission possible.[12]

19.1.5   History

See also: Homunculus § Homunculus of spermists

Sperm were first observed in 1677 by Antonie van Leeuwenhoek[13] using a microscope, he described them as being animalcules (little animals), probably due to his belief in preformationism, which thought that each sperm contained a fully formed but small human.

19.1.6   Forensic analysis

Ejaculated fluids are detected by ultraviolet light, irrespective of the structure or colour of the surface.[14] Sperm heads, e.g. from vaginal swabs, are still detected by microscopy using the "Christmas Tree Stain" method, i.e., Kernechtrot-Picroindigocarmine (KPIC) staining.[15][16]

19.2   Sperm in plants

Sperm cells in algal and many plant gametophytes are produced in male gametangia (antheridia) via mitotic division. In flowering plants, sperm nuclei are produced inside pollen.

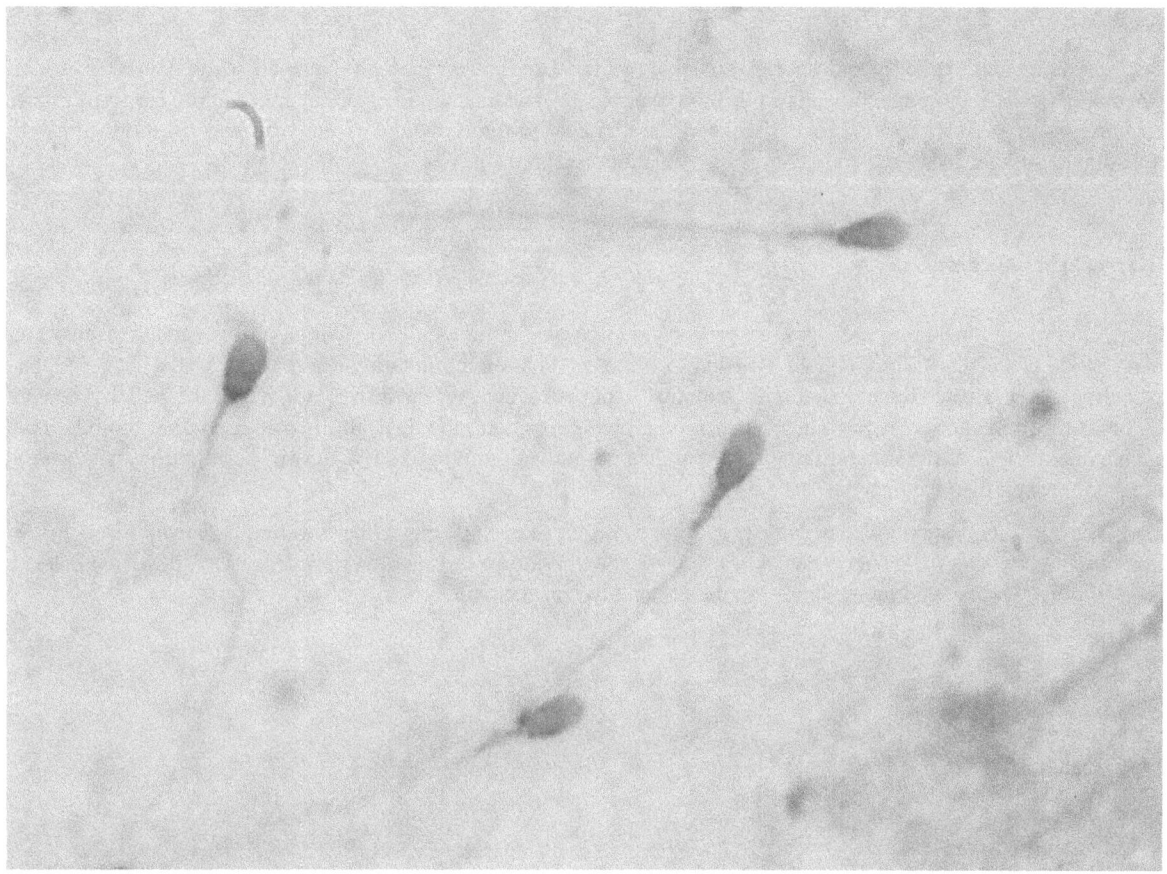

Human sperm stained for semen quality testing.

19.3 Motile sperm cells

Motile sperm cells typically move via flagella and require a water medium in order to swim toward the egg for fertilization. In animals most of the energy for sperm motility is derived from the metabolism of fructose carried in the seminal fluid. This takes place in the mitochondria located in the sperm's midpiece (at the base of the sperm head). These cells cannot swim backwards due to the nature of their propulsion. The uniflagellated sperm cells (with one flagellum) of animals are referred to as **spermatozoa**, and are known to vary in size.

Motile sperm are also produced by many protists and the gametophytes of bryophytes, ferns and some gymnosperms such as cycads and ginkgo. The sperm cells are the only flagellated cells in the life cycle of these plants. In many ferns and lycophytes, they are multi-flagellated (carrying more than one flagellum).[17]

In nematodes, the sperm cells are amoeboid and crawl, rather than swim, towards the egg cell.[18]

19.4 Non-motile sperm cells

Non-motile sperm cells called **spermatia** lack flagella and therefore cannot swim. Spermatia are produced in a spermatangium.[17]

Because spermatia cannot swim, they depend on their environment to carry them to the egg cell. Some red algae, such as *Polysiphonia*, produce non-motile spermatia that are spread by water currents after their release.[17] The spermatia of rust fungi are covered with a sticky substance. They are produced in flask-shaped structures containing nectar, which attract flies that transfer the spermatia to nearby hyphae for fertilization in a mechanism similar to insect pollination in flowering plants.[19]

Fungal spermatia (also called pycniospores, especially in the Uredinales) may be confused with conidia. Conidia are spores that germinate independently of fertilization, whereas spermatia are gametes that are required for fertilization. In some fungi, such as *Neurospora crassa*, spermatia are identical to microconidia as they can perform both functions of fertilization as well as giving rise to new organisms without fertilization.[20]

19.5   Sperm nuclei

In many land plants, including most gymnosperms and all angiosperms, the male gametophytes (pollen grains) are the primary mode of dispersal, for example via wind or insect pollination, eliminating the need for water to bridge the gap between male and female. Each pollen grain contains a spermatogenous (generative) cell. Once the pollen lands on the stigma of a receptive flower, it germinates and starts growing a pollen tube through the carpel. Before the tube reaches the ovule, the nucleus of the generative cell in the pollen grain divides and gives rise to two sperm nuclei, which are then discharged through the tube into the ovule for fertilization.[17]

In some protists, fertilization also involves sperm nuclei, rather than cells, migrating toward the egg cell through a fertilization tube. Oomycetes form sperm nuclei in a syncytical antheridium surrounding the egg cells. The sperm nuclei reach the eggs through fertilization tubes, similar to the pollen tube mechanism in plants.[17]

19.6   See also

- Ejaculation
- Female sperm
- Female sperm storage
- Polyspermy
- Semen
- Sperm competition
- Sperm donation
- Sperm granuloma
- Spermatogenesis
- Spermatozoon

19.7   References

[1] Sarfraz Manzoor (2 November 2012). "Come inside: the world's biggest sperm bank". *The Guardian*. Retrieved 4 August 2013.

[2] Hewitson, Laura & Schatten, Gerald P. (2003). "The biology of fertilization in humans". In Patrizio, Pasquale et al. *A color atlas for human assisted reproduction: laboratory and clinical insights*. Lippincott Williams & Wilkins. p. 3. ISBN 978-0-7817-3769-2. Retrieved 2013-11-09.

[3] Semen and sperm quality

[4] Gould JE, Overstreet JW and Hanson FW (1984) "Assessment of human sperm function after recovery from the female reproductive tract". *Biology of Reproduction* 31, 888–894.

[5] Gurevich, Rachel (2008-06-10). "Does Age Affect Male Fertility?". About.com. Retrieved 14 February 2010.

[6] Gavriliouk D, Aitken RJ (2015). "Damage to Sperm DNA Mediated by Reactive Oxygen Species: Its Impact on Human Reproduction and the Health Trajectory of Offspring". *Advances in Experimental Medicine and Biology* **868**: 23–47. doi:10.1007/978-3-319-18881-2_2. PMID 26178844.

[7] Marchetti F, Wyrobek AJ (2008). "DNA repair decline during mouse spermiogenesis results in the accumulation of heritable DNA damage". *DNA Repair* **7** (4): 572–81. doi:10.1016/j.dnarep.2007.12.011. PMID 18282746.

[8] Marchetti F, Essers J, Kanaar R, Wyrobek AJ (2007). "Disruption of maternal DNA repair increases sperm-derived chromosomal aberrations". *Proceedings of the National Academy of Sciences of the United States of America* **104** (45): 17725–9. doi:10.1073/pnas.0705257104. PMC 2077046. PMID 17978187.

[9] Marchetti F, Bishop J, Gingerich J, Wyrobek AJ (2015). "Meiotic interstrand DNA damage escapes paternal repair and causes chromosomal aberrations in the zygote by maternal misrepair". *Scientific Reports* **5**: 7689. doi:10.1038/srep07689. PMC 4286742. PMID 25567288.

[10] Assisted Reproduction in the Nordic Countries ncbio.org

[11] FDA Rules Block Import of Prized Danish Sperm Posted Aug 13, 08 7:37 AM CDT in World, Science & Health

[12] Steven Kotler (26 September 2007). "The God of Sperm".

[13] "Timeline: Assisted reproduction and birth control". *CBC News*. Retrieved 2006-04-06.

[14] Anja Fiedler, Mark Benecke; et al. "Detection of Semen (Human and Boar) and Saliva on Fabrics by a Very High Powered UV-/VIS-Light Source". Retrieved 2015-10-22.

[15] Allery, J. P.; Telmon, N.; Mieusset, R.; Blanc, A.; Rougé, D. (2001). "Cytological detection of spermatozoa: Comparison of three staining methods". *Journal of forensic sciences* **46** (2): 349–351. PMID 11305439.

[16] Illinois State Police/President's DNA Initiative. "The Presidents's DNA Initiative: Semen Stain Identification: Kernechtrot" (PDF). Retrieved 2009-12-10.

[17] Raven, Peter H.; Ray F. Evert; Susan E. Eichhorn (2005). *Biology of Plants, 7th Edition*. New York: W.H. Freeman and Company Publishers. ISBN 0-7167-1007-2.

[18] Bottino D, Mogilner A, Roberts T, Stewart M, Oster G (2002). "How nematode sperm crawl". *Journal of Cell Science* **115** (Pt 2): 367–84. PMID 11839788.

[19] Sumbali, Geeta (2005). *The Fungi*. Alpha Science Int'l Ltd. ISBN 1-84265-153-6.

[20] Maheshwari R (1999). "Microconidia of Neurospora crassa". *Fungal Genetics and Biology* **26** (1): 1–18. doi:10.1006/fgbi.1998.1103. PMID 10072316.

19.8 External links

- The Great Sperm Race pdf

- Human Sperm Under a Microscope

Plant & algae sperm cells

Lycopodium

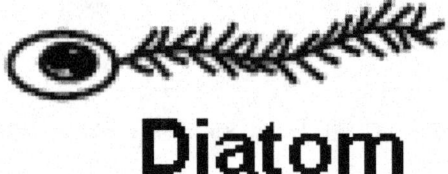

Diatom

Selaginella

Laminaria

Fucus

Psilotum

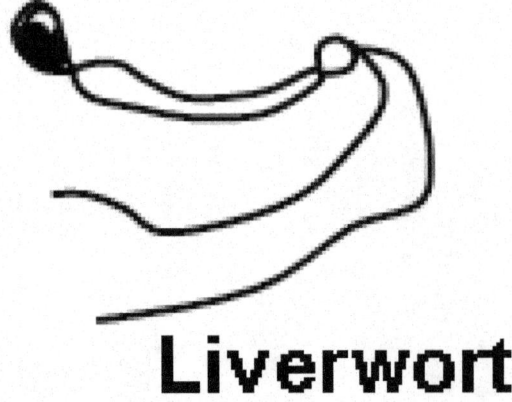

Liverwort

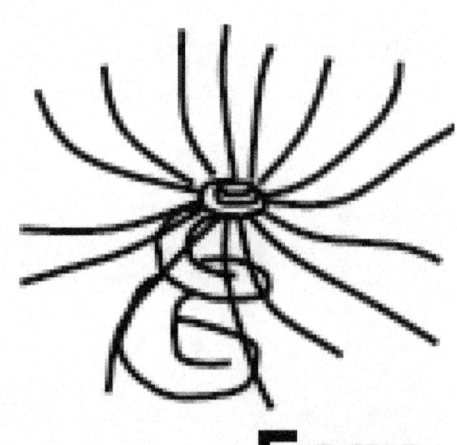

Fern

Chapter 20

Meiosis

Not to be confused with Mitosis, Miosis, Myositis. Note that *mitosis* is the process by which most cells divide to produce two copies of themselves.

For the figure of speech, see Meiosis (figure of speech).

Meiosis 🔊i /maɪˈoʊsɪs/ is a specialized type of cell division that reduces the chromosome number by half. This process

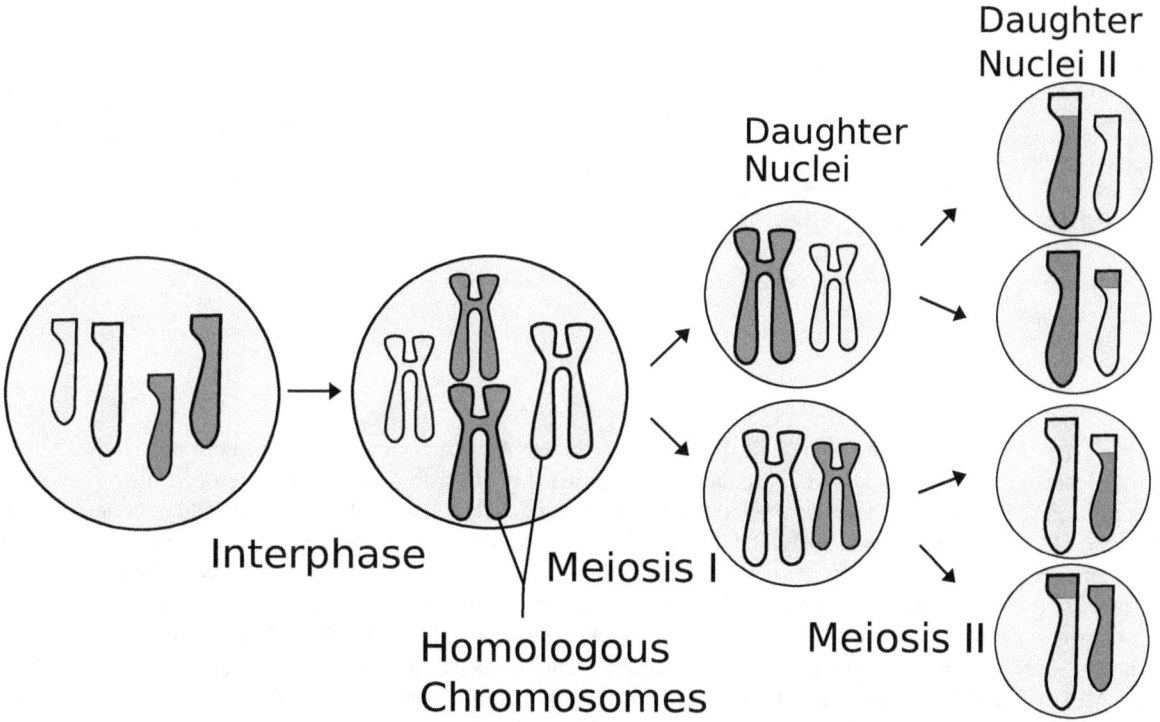

In meiosis, the chromosomes duplicate (during interphase) and homologous chromosomes exchange genetic information (chromosomal crossover) before a first division, called meiosis I. The daughter cells divide again in meiosis II, splitting up sister chromatids to form haploid gametes. Male and female gametes fuse during fertilization, creating a diploid cell with a complete set of paired chromosomes.

occurs in all sexually reproducing single-celled and multicellular eukaryotes, including animals, plants, and fungi.[1][2][3][4] Errors in meiosis resulting in aneuploidy are the leading known cause of miscarriage and the most frequent genetic cause of developmental disabilities. [5]

In meiosis, DNA replication is followed by two rounds of cell division to produce four potential daughter cells, each with half the number of chromosomes as the original parent cell. The two meiotic divisions are known as *Meiosis I* and *Meiosis*

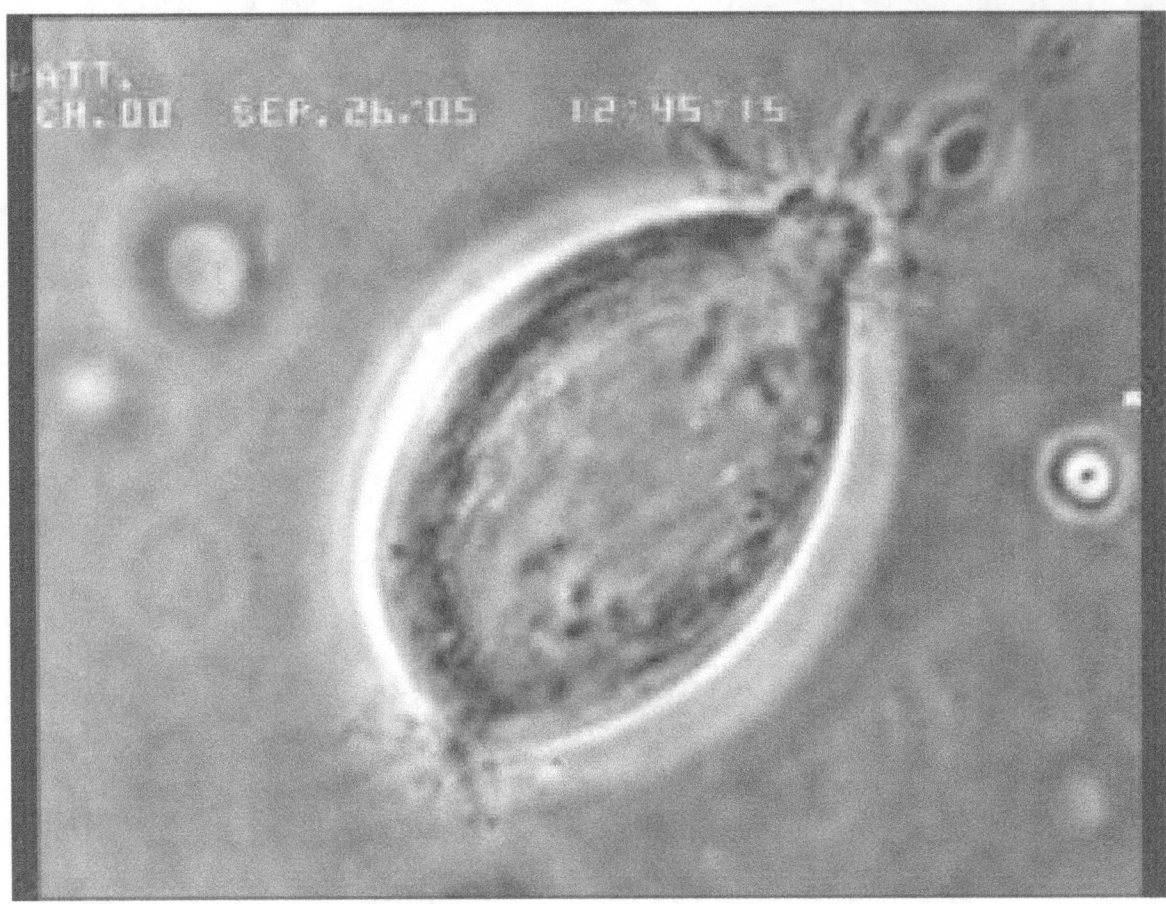

A video of meiosis I in a crane-fly spermatocyte, played back at 120× the recorded speed.

II. Before meiosis begins, during S phase of the cell cycle, the DNA of each chromosome is replicated so that it consists of two identical sister chromatids, which remain held together through sister chromatid cohesion. This S-phase can be referred to as "premeiotic S-phase" or "meiotic S-phase." Immediately following DNA replication, meiotic cells enter a prolonged G2-like stage known as meiotic prophase. During this time, homologous chromosomes pair with each other and undergo genetic recombination, a programmed process in which DNA is cut and then repaired, which allows them to exchange some of their genetic information. A subset of recombination events results in crossovers, which create physical links known as chiasmata (singular:chiasma, for the Greek letter Chi) between the homologous chromosomes. In most organisms, these links are essential to direct each pair of homologous chromosomes to segregate away from each other during Meiosis I, resulting in two haploid cells that half the number of chromosomes as the parent cell. During Meiosis II, the cohesion between sister chromatids is released and they segregate from one another, as during mitosis. In some cases all four of the meiotic products form gametes such as sperm, spores, or pollen. In female animals, three of the four meiotic products are typically eliminated by extrusion into polar bodies, and only one cell develops to produce an ovum.

Because the number of chromosomes is halved during meiosis, gametes can fuse (i.e. fertilization) to form a diploid zygote that contains two copies of each chromosome, one from each parent. Thus, alternating cycles of meiosis and fertilization enable sexual reproduction, with successive generations maintaining the same number of chromosomes. For example, diploid human cells contain 23 pairs of chromosomes (46 total), half of maternal origin and half of paternal origin. Meiosis produces haploid gametes (ova or sperm) that contain one set of 23 chromosomes. When two gametes (an egg and a sperm) fuse, the resulting zygote is once again diploid, with the mother and father each contributing 23 chromosomes. This same pattern, but not the same number of chromosomes, occurs in all organisms that utilize meiosis.

20.1   Overview

While the process of meiosis is related to the more general cell division process of mitosis, it differs in two important respects:

Meiosis begins with a diploid cell, which contains two copies of each chromosome, termed homologs. First, the cell undergoes DNA replication, so each homolog now consists of two identical sister chromatids. Then each set of homologs pair with each other and exchange DNA by homologous recombination leading to physical connections (crossovers) between the homologs. In the first meiotic division, the homologs are segregated to separate daughter cells by the spindle apparatus. The cells then proceed to a second division without an intervening round of DNA replication. The sister chromatids are segregated to separate daughter cells to produce a total of four haploid cells. Female animals employ a slight variation on this pattern and produce one large ovum and two small polar bodies. Because of recombination, an individual chromatid can consist of a new combination of maternal and paternal DNA, resulting in offspring that are genetically distinct from either parent. Furthermore, an individual gamete can include an assortment of maternal, paternal, and recombinant chromatids. This genetic diversity resulting from sexual reproduction contributes to the variation in traits upon which natural selection can act.

Meiosis uses many of the same mechanisms as mitosis, the type of cell division used by eukaryotes to divide one cell into two identical daughter cells. In some plants, fungi, and protists meiosis results in the formation of spores: haploid cells that can divide vegetatively without undergoing fertilization. Some eukaryotes, like bdelloid rotifers, do not have the ability to carry out meiosis and have acquired the ability to reproduce by parthenogenesis.

Meiosis does not occur in archaea or bacteria, which generally reproduce via asexual processes such as binary fission. However, a "sexual" process known as horizontal gene transfer involves the transfer of DNA from one bacterium or archaeon to another and recombination of these DNA molecules of different parental origin.

20.2   History

Meiosis was discovered and described for the first time in sea urchin eggs in 1876 by the German biologist Oscar Hertwig. It was described again in 1883, at the level of chromosomes, by the Belgian zoologist Edouard Van Beneden, in *Ascaris* roundworm eggs. The significance of meiosis for reproduction and inheritance, however, was described only in 1890 by German biologist August Weismann, who noted that two cell divisions were necessary to transform one diploid cell into four haploid cells if the number of chromosomes had to be maintained. In 1911 the American geneticist Thomas Hunt Morgan detected crossovers in meiosis in the fruit fly *Drosophila melanogaster*, which helped to establish that genetic traits are transmitted on chromosomes.

The term meiosis (originally spelled "maiosis") was introduced to biology by J.B. Farmer and J.E.S. Moore in 1905:

> We propose to apply the terms Maiosis or Maiotic phase to cover the whole series of nuclear changes included in the two divisions that were designated as Heterotype and Homotype by Flemming.[6]

It is derived from the Greek word μείωσις, meaning 'lessening'.

20.3   Occurrence in eukaryotic life cycles

Main article: Biological life cycle

Meiosis occurs in eukaryotic life cycles involving sexual reproduction, consisting of the constant cyclical process of meiosis and fertilization. This takes place alongside normal mitotic cell division. In multicellular organisms, there is an intermediary step between the diploid and haploid transition where the organism grows. At certain stages of the life cycle, germ cells produce gametes. Somatic cells make up the body of the organism and are not involved in gamete production.

Cycling meiosis and fertilization events produces a series of transitions back and forth between alternating haploid and diploid states. The organism phase of the life cycle can occur either during the diploid state (*gametic* or *diploid* life cycle),

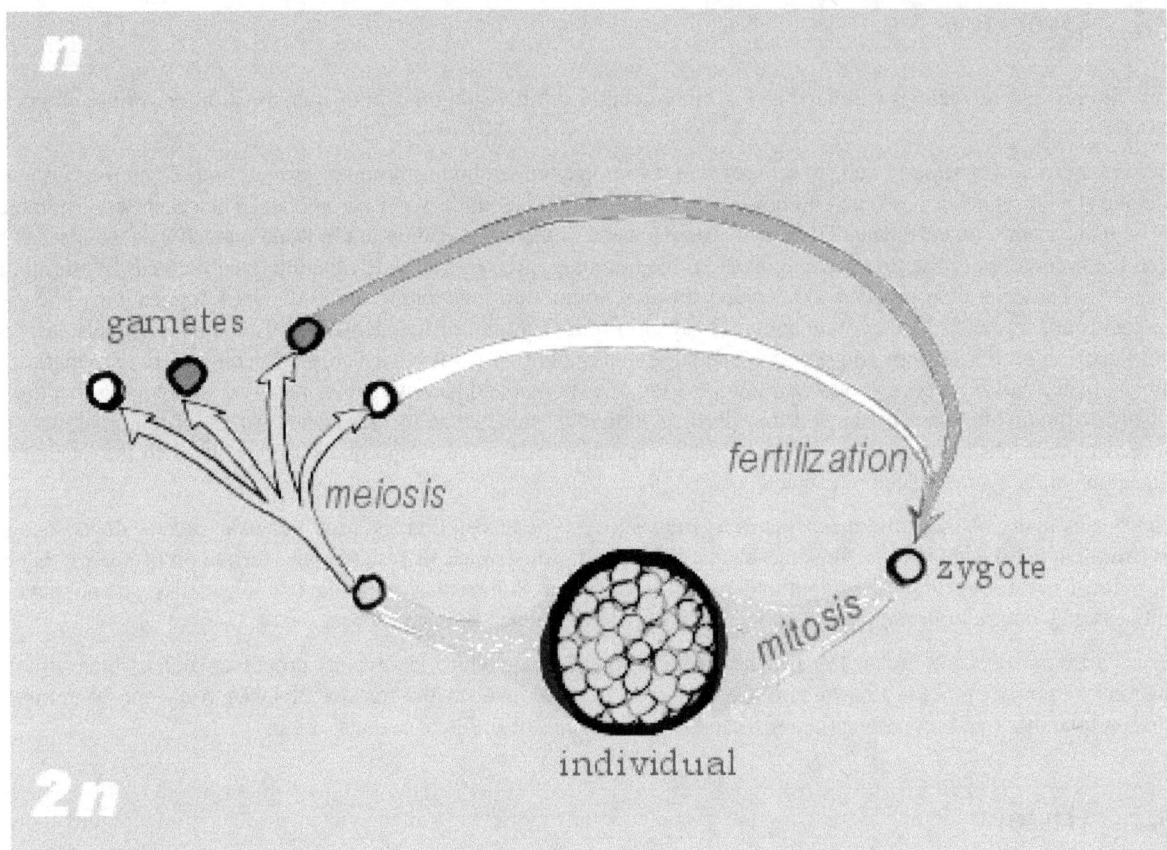

Gametic life cycle.

during the haploid state (*zygotic* or *haploid* life cycle), or both (*sporic* or *haplodiploid* life cycle, in which there are two distinct organism phases, one during the haploid state and the other during the diploid state). In this sense there are three types of life cycles that utilize sexual reproduction, differentiated by the location of the organism phase(s).

In the *gametic life cycle* or " diplontic life cycle", of which humans are a part, the organism is diploid, grown from a diploid cell called the zygote. The organism's diploid germ-line stem cells undergo meiosis to create haploid gametes (the spermatozoa for males and ova for females), which fertilize to form the zygote. The diploid zygote undergoes repeated cellular division by mitosis to grow into the organism.

In the *zygotic life cycle* the organism is haploid instead, spawned by the proliferation and differentiation of a single haploid cell called the gamete. Two organisms of opposing gender contribute their haploid gametes to form a diploid zygote. The zygote undergoes meiosis immediately, creating four haploid cells. These cells undergo mitosis to create the organism. Many fungi and many protozoa utilize the zygotic life cycle.

Finally, in the *sporic life cycle*, the living organism alternates between haploid and diploid states. Consequently, this cycle is also known as the alternation of generations. The diploid organism's germ-line cells undergo meiosis to produce spores. The spores proliferate by mitosis, growing into a haploid organism. The haploid organism's gamete then combines with another haploid organism's gamete, creating the zygote. The zygote undergoes repeated mitosis and differentiation to become a diploid organism again. The sporic life cycle can be considered a fusion of the gametic and zygotic life cycles.

20.4   Process

The preparatory steps that lead up to meiosis are identical in pattern and name to interphase of the mitotic cell cycle.

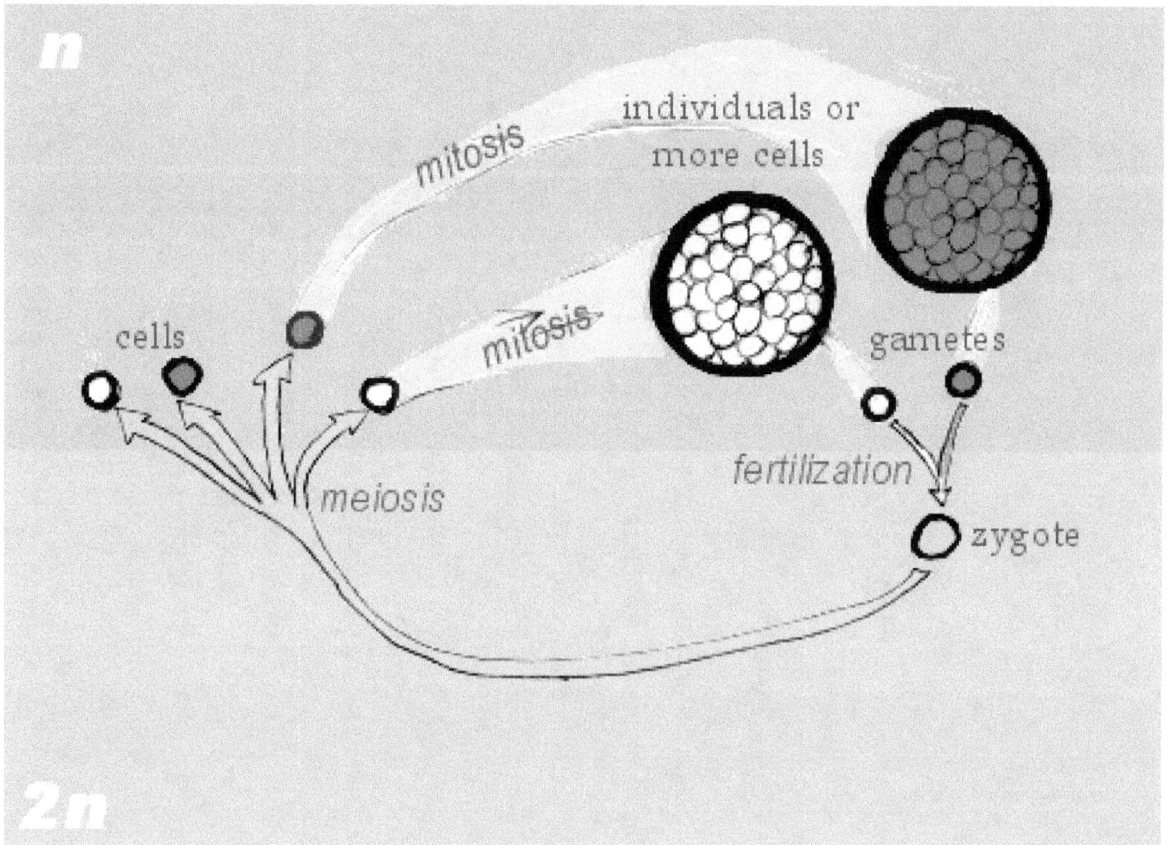

Zygotic life cycle.

Interphase is divided into three phases:

- Growth 1 (G_1) phase: In this very active phase, the cell synthesizes its vast array of proteins, including the enzymes and structural proteins it will need for growth. In G_1, each of the chromosomes consists of a single linear molecule of DNA.

- Synthesis (S) phase: The genetic material is replicated; each of the cell's chromosomes duplicates to become two identical sister chromatids attached at a centromere. This replication does not change the ploidy of the cell since the centromere number remains the same. The identical sister chromatids have not yet condensed into the densely packaged chromosomes visible with the light microscope. This will take place during prophase I in meiosis.

- Growth 2 (G_2) phase: G_2 phase as seen before mitosis is not present in meiosis. Meiotic prophase corresponds most closely to the G_2 phase of mitotic cell cycle.

Interphase is followed by meiosis I and then meiosis II. Meiosis I separates homologous chromosome, each still made up of two sister chromatids, into two daughter cells, thus reducing the chromosome number by half. During meiosis II, sister chromatids decouple and the resultant daughter chromosomes are segregated into four daughter cells. For diploid organisms, the daughter cells resulting from meiosis are haploid and contain only one copy of each chromosome. In some species, cells enter a resting phase known as interkinesis between meiosis I and meiosis II.

Meiosis I and II are each divided into prophase, metaphase, anaphase, and telophase stages, similar in purpose to their analogous subphases in the mitotic cell cycle. Therefore, meiosis includes the stages of meiosis I (prophase I, metaphase I, anaphase I, telophase I) and meiosis II (prophase II, metaphase II, anaphase II, telophase II).

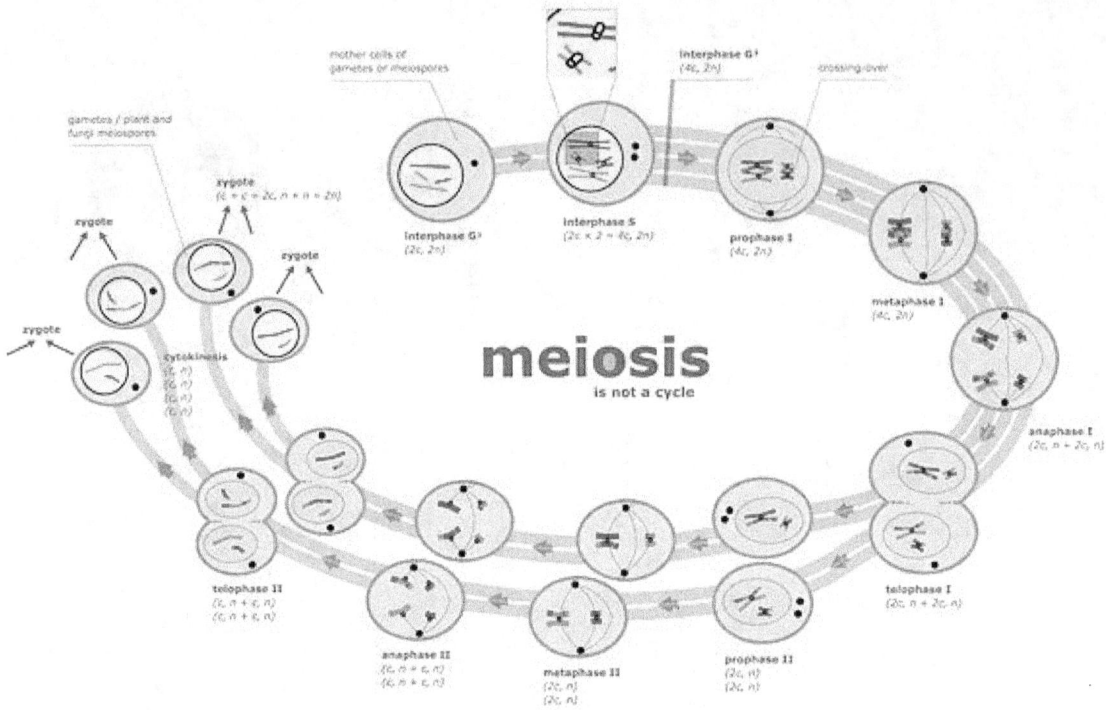

A diagram of the meiotic phases

Meiosis generates gamete genetic diversity in two ways: (1) the independent orientation of homologous chromosome pairs along the metaphase plate during metaphase I and the subsequent separation of homologs during anaphase I allows a random and independent distribution of chromosomes to each daughter cell (and ultimately to gametes); and (2) physical exchange of homologous chromosomal regions by homologous recombination during prophase I results in new combinations of DNA within chromosomes.

During meiosis, specific genes are more highly transcribed.[7] In addition to strong meiotic stage-specific expression of mRNA, there are also pervasive translational controls (e.g. selective usage of preformed mRNA), regulating the ultimate meiotic stage-specific protein expression of genes during meiosis.[8] Thus, both transcriptional and translational controls determine the broad restructuring of meiotic cells needed to carry out meiosis.

20.5   Phases

Meiosis is divided into meiosis I and meiosis II which are further divided into Karyokinesis I and Cytokinesis I & Karyokinesis II and Cytokinesis II respectively.

20.5.1   Meiosis I

Meiosis I segregates homologous chromosomes, producing two haploid cells (n chromosomes, 23 in humans). Because the ploidy is reduced from diploid to haploid, meiosis I is referred to as a *reductional division*. Meiosis II is an *equational division* analogous to mitosis, in which the sister chromatids are segregated, creating four haploid daughter cells .[9]

Prophase I

Prophase I is typically the longest phase of meiosis. During prophase I, homologous chromosomes pair and exchange DNA in a process called homologous recombination. This often results in chromosomal crossover. This process is critical for pairing between homologous chromosomes and hence for accurate segregation of the chromosomes at the first meiosis division. The new combinations of DNA created during crossover are a significant source of genetic variation, and result in new combinations of alleles, which may be beneficial. The paired and replicated chromosomes are called bivalents or tetrads, which have two chromosomes and four chromatids, with one chromosome coming from each parent. The process of pairing the homologous chromosomes is called synapsis. At this stage, non-sister chromatids may cross-over at points called chiasmata (plural; singular chiasma).[10] Prophase I has historically been divided into a series of substages which are named according to the appearance of chromosomes.

Leptotene   The first stage of prophase I is the *leptotene* stage, also known as *leptonema*, from Greek words meaning "thin threads".[11]:27 In this stage of prophase I, individual chromosomes—each consisting of two sister chromatids—become "individualized" to form visible strands within the nucleus.[11]:27[12]:353 The two sister chromatids closely associated and are visually indistinguishable from one another. During leptotene, lateral elements of the synaptonemal complex assemble. Leptotene is of very short duration and progressive condensation and coiling of chromosome fibers takes place.

Zygotene   The *zygotene* stage, also known as *zygonema*, from Greek words meaning "paired threads",[11]:27 occurs as the chromosomes approximately line up with each other into homologous chromosome pairs. In some organisms, this is called the bouquet stage because of the way the telomeres cluster at one end of the nucleus. At this stage, the synapsis (pairing/coming together) of homologous chromosomes takes place, facilitated by assembly of central element of the synaptonemal complex. Pairing is brought about in a zipper-like fashion and may start at the centromere (procentric), at the chromosome ends (proterminal), or at any other portion (intermediate). Individuals of a pair are equal in length and in position of the centromere. Thus pairing is highly specific and exact. The paired chromosomes are called bivalent or tetrad chromosomes.

Pachytene   The *pachytene* (pronounced /ˈpækɪtiːn/ *PAK-ə-teen*) stage, also known as *pachynema*, from Greek words meaning "thick threads",.[11]:27 At this point a tetrad of the chromosomes has formed known as a bivalent. This is the stage when chromosomal crossover (crossing over) occurs. Nonsister chromatids of homologous chromosomes may exchange segments over regions of homology. Sex chromosomes, however, are not wholly identical, and only exchange information over a small region of homology. At the sites where exchange happens, chiasmata form. The exchange of information between the non-sister chromatids results in a recombination of information; each chromosome has the complete set of information it had before, and there are no gaps formed as a result of the process. Because the chromosomes cannot be distinguished in the synaptonemal complex, the actual act of crossing over is not perceivable through the microscope, and chiasmata are not visible until the next stage.

Diplotene   During the *diplotene* stage, also known as *diplonema*, from Greek words meaning "two threads",[11]:30 the synaptonemal complex degrades and homologous chromosomes separate from one another a little. The chromosomes themselves uncoil a bit, allowing some transcription of DNA. However, the homologous chromosomes of each bivalent remain tightly bound at chiasmata, the regions where crossing-over occurred. The chiasmata remain on the chromosomes until they are severed at the transition to anaphase I.

In mammalian and human fetal oogenesis all developing oocytes develop to this stage and are arrested before birth. This suspended state is referred to as the *dictyotene stage* or dictyate. It lasts until meiosis is resumed to prepare the oocyte for ovulation, which happens at puberty or even later.

Diakinesis   Chromosomes condense further during the *diakinesis* stage, from Greek words meaning "moving through".[11]:30 This is the first point in meiosis where the four parts of the tetrads are actually visible. Sites of crossing over entangle together, effectively overlapping, making chiasmata clearly visible. Other than this observation, the rest of the stage closely resembles prometaphase of mitosis; the nucleoli disappear, the nuclear membrane disintegrates into vesicles, and the meiotic spindle begins to form.

Synchronous processes   During these stages, two centrosomes, containing a pair of centrioles in animal cells, migrate to the two poles of the cell. These centrosomes, which were duplicated during S-phase, function as microtubule organizing centers nucleating microtubules, which are essentially cellular ropes and poles. The microtubules invade the nuclear region after the nuclear envelope disintegrates, attaching to the chromosomes at the kinetochore. The kinetochore functions as a motor, pulling the chromosome along the attached microtubule toward the originating centrosome, like a train on a track. There are four kinetochores on each tetrad, but the pair of kinetochores on each sister chromatid fuses and functions as a unit during meiosis I.[13][14]

Microtubules that attach to the kinetochores are known as *kinetochore microtubules*. Other microtubules will interact with microtubules from the opposite centrosome: these are called *nonkinetochore microtubules* or *polar microtubules*. A third type of microtubules, the aster microtubules, radiates from the centrosome into the cytoplasm or contacts components of the membrane skeleton.

Metaphase I

Homologous pairs move together along the metaphase plate: As *kinetochore microtubules* from both centrosomes attach to their respective kinetochores, the paired homologous chromosomes align along an equatorial plane that bisects the spindle, due to continuous counterbalancing forces exerted on the bivalents by the microtubules emanating from the two kinetochores of homologous chromosomes. The physical basis of the independent assortment of chromosomes is the random orientation of each bivalent along the metaphase plate, with respect to the orientation of the other bivalents along the same equatorial line.[10] The protein complex cohesin holds sister chromatids together from the time of their replication until anaphase. In mitosis, the force of kinetochore microtubules pulling in opposite directions creates tension. The cell senses this tension and does not progress with anaphase until all the chromosomes are properly bi-oriented. In meiosis, establishing tension requires at least one crossover per chromosome pair in addition to cohesin between sister chromatids.

Anaphase I

Kinetochore microtubules shorten, pulling homologous chromosomes (which consist of a pair of sister chromatids) to opposite poles. Nonkinetochore microtubules lengthen, pushing the centrosomes farther apart. The cell elongates in preparation for division down the center.[10] Unlike in mitosis, only the cohesin from the chromosome arms is degraded while the cohesin surrounding the centromere remains protected. This allows the sister chromatids to remain together while homologs are segregated.

Telophase I

The first meiotic division effectively ends when the chromosomes arrive at the poles. Each daughter cell now has half the number of chromosomes but each chromosome consists of a pair of chromatids. The microtubules that make up the spindle network disappear, and a new nuclear membrane surrounds each haploid set. The chromosomes uncoil back into chromatin. Cytokinesis, the pinching of the cell membrane in animal cells or the formation of the cell wall in plant cells, occurs, completing the creation of two daughter cells. Sister chromatids remain attached during telophase I.

Cells may enter a period of rest known as interkinesis or interphase II. No DNA replication occurs during this stage.

20.5.2   Meiosis II

Meiosis II is the second meiotic division, and usually involves equational segregation, or separation of sister chromatids. Mechanically, the process is similar to mitosis, though its genetic results are fundamentally different. The end result is production of four haploid cells (n chromosomes, 23 in humans) from the two haploid cells (with n chromosomes, each consisting of two sister chromatids) produced in meiosis I. The four main steps of Meiosis II are: Prophase II, Metaphase II, Anaphase II, and Telophase II.

In **prophase II** we see the disappearance of the nucleoli and the nuclear envelope again as well as the shortening and thickening of the chromatids. Centrosomes move to the polar regions and arrange spindle fibers for the second meiotic division.

In **metaphase II**, the centromeres contain two kinetochores that attach to spindle fibers from the centrosomes at opposite poles. The new equatorial metaphase plate is rotated by 90 degrees when compared to meiosis I, perpendicular to the previous plate.

This is followed by **anaphase II**, in which the remaining centromeric cohesin is cleaved allowing the sister chromatids to segregate. The sister chromatids by convention are now called sister chromosomes as they move toward opposing poles.

The process ends with **telophase II**, which is similar to telophase I, and is marked by decondensation and lengthening of the chromosomes and the disassembly of the spindle. Nuclear envelopes reform and cleavage or cell wall formation eventually produces a total of four daughter cells, each with a haploid set of chromosomes.

Meiosis is now complete and ends up with four new daughter cells.

20.6 Origin and function

Main article: Origin and function of meiosis

The **origin and function of meiosis** are fundamental to understanding the evolution of sexual reproduction in Eukaryotes. There is no current consensus among biologists on the questions of how sex in Eukaryotes arose in evolution, what basic function sexual reproduction serves, and why it is maintained, given the basic two-fold cost of sex. It is clear that it evolved over 1.2 billion years ago, and that almost all species which are descendents of the original sexually reproducing species are still sexual reproducers, including plants, fungi, and animals.

Meiosis is ubiquitous among eukaryotes. It occurs in single-celled organisms such as yeast, as well as in multicellular organisms, such as humans. Eukaryotes arose from prokaryotes more than 1.5 billion years ago,[15] and the earliest eukaryotes were likely single-celled organisms. To understand sex in eukaryotes, it is necessary to understand (1) how meiosis arose in single celled eukaryotes, and (2) the function of meiosis.

20.7 Nondisjunction

Main article: Nondisjunction

The normal separation of chromosomes in meiosis I or sister chromatids in meiosis II is termed *disjunction*. When the segregation is not normal, it is called *nondisjunction*. This results in the production of gametes which have either too many or too few of a particular chromosome, and is a common mechanism for trisomy or monosomy. Nondisjunction can occur in the meiosis I or meiosis II, phases of cellular reproduction, or during mitosis.

Most monosomic and trisomic human embryos are not viable, but some aneuploidies can be tolerated, such as trisomy for the smallest chromosome, chromosome 21. Phenotypes of these aneuploidies range from severe developmental disorders to asymptomatic. Medical conditions include but are not limited to:

- Down Syndrome - trisomy of chromosome 21

- Patau Syndrome - trisomy of chromosome 13

- Edward Syndrome - trisomy of chromosome 18

- Klinefelter Syndrome - extra X chromosomes in males - i.e. XXY, XXXY, XXXXY, etc.

- Turner Syndrome - lacking of one X chromosome in females - i.e. X0

- Triple X syndrome - an extra X chromosome in females

- XYY Syndrome - an extra Y chromosome in males.

The probability of nondisjunction in human oocytes increases with increasing maternal age,[16] presumably due to loss of cohesin over time.[17]

20.8   Meiosis in plants and animals

Meiosis occurs in all animals and plants. The end result, the production of gametes with half the number of chromosomes as the parent cell, is the same, but the detailed process is different. In animals, meiosis produces gametes directly. In land plants and some algae, there is an alternation of generations such that meiosis in the diploid sporophyte generation produces haploid spores. These spores multiply by mitosis, developing into the haploid gametophyte generation, which then gives rise to gametes directly (i.e. without further meiosis). In both animals and plants, the final stage is for the gametes to fuse, restoring the original number of chromosomes.[18]

20.9   Meiosis in mammals

In females, meiosis occurs in cells known as oocytes (singular: oocyte). Each oocyte that initiates meiosis divides twice, unequally in each case. The first division produces a daughter cell that will undergo a second division, and a much smaller "polar body" that is extruded from the surface of the cell and does not divide further. Following Meiosis II, a "second polar body" is extruded, and the single remaining haploid cell enlarges to become an ovum. Since the first polar body normally disintegrates rather than dividing again, meiosis in female mammals results in three products, the oocyte and two polar bodies. However, before these divisions occur, these cells stop at the diplotene stage of meiosis I and lie dormant within a protective shell of somatic cells called the follicle. Follicles begin growth at a steady pace in a process known as folliculogenesis, and a small number enter the menstrual cycle. Menstruated oocytes continue meiosis I and arrest at meiosis II until fertilization. The process of meiosis in females occurs during oogenesis, and differs from the typical meiosis in that it features a long period of meiotic arrest known as the dictyate stage and lacks the assistance of centrosomes.[19][20]

In males, meiosis occurs during spermatogenesis in the seminiferous tubules of the testicles. Meiosis during spermatogenesis is specific to a type of cell called spermatocytes that will later mature to become spermatozoa.

In female mammals, meiosis begins immediately after primordial germ cells migrate to the ovary in the embryo, but in the males, meiosis begins later, at the time of puberty. It is retinoic acid, derived from the primitive kidney (mesonephros) that stimulates meiosis in ovarian oogonia. Tissues of the male testis suppress meiosis by degrading retinoic acid, a stimulator of meiosis. This is overcome at puberty when cells within seminiferous tubules called Sertoli cells start making their own retinoic acid. Sensitivity to retinoic acid is also adjusted by proteins called nanos and DAZL.[21][22]

20.10   Meiosis vs. mitosis

In order to understand meiosis, a comparison to mitosis is helpful. The table below shows the differences between meiosis and mitosis.[23]

20.11   See also

- Coefficient of coincidence

- DNA repair

- Evolution of sexual reproduction

- Fertilization

- Genetic recombination

- Mitosis

- Oxidative stress

- Synizesis (biology)

20.12 References

[1] name="Letunic I and Bork P" Letunic, I; Bork, P (2006). "Interactive Tree of Life". Retrieved 23 July 2011.

[2] Bernstein H, Bernstein C, Michod RE (2011). "Meiosis as an evolutionary adaptation for DNA repair." In "DNA Repair", Intech Publ (Inna Kruman, editor), Chapter 19: 357-382 DOI: 10.5772/1751 ISBN 978-953-307-697-3 Available online from: http://www.intechopen.com/books/dna-repair/meiosis-as-an-evolutionary-adaptation-for-dna-repair

[3] Bernstein H, Bernstein C (2010). "Evolutionary origin of recombination during meiosis". *BioScience* **60** (7): 498–505. doi:10.1525/bio.2010.60.7.5.

[4] LODÉ T (2011). "Sex is not a solution for reproduction: the libertine bubble theory". *BioEssays* **33** (6): 419–422. doi:10.1002/bies.201000125. PMID 21472739.

[5] http://www.nature.com/nrg/journal/v2/n4/abs/nrg0401_280a.html

[6] J.B. Farmer and J.E.S. Moore, *Quarterly Journal of Microscopic Science* **48**:489 (1905) as quoted in the Oxford English Dictionary, Third Edition, June 2001, *s.v.*

[7] Zhou, A.; Pawlowski, W.P. (August 2014). "Regulation of meiotic gene expression in plants". *Frontiers in Plant Science* **5**: Article 413. doi:10.3389/fpls.2014.00413. PMID 25202317.

[8] Brar GA, Yassour M, Friedman N, Regev A, Ingolia NT, Weissman JS (February 2012). "High-resolution view of the yeast meiotic program revealed by ribosome profiling". *Science* **335** (6068): 552–7. doi:10.1126/science.1215110.

[9] Freeman 2005, pp. 244–45

[10] Freeman 2005, pp. 249–250

[11] Snustad, DP; Simmons, MJ (December 2008). *Principles of Genetics* (5th ed.). Wiley. ISBN 978-0-470-38825-9.

[12] Krebs, JE; Goldstein, ES; Kilpatrick, ST (November 2009). *Lewin's Genes X* (10th ed.). Jones & Barlett Learning. ISBN 978-0-7637-6632-0.

[13] Raven, Peter H.; Johnson, George B.; Mason, Kenneth A.; Losos, Jonathan & Singer, Susan. Biology, 8th ed. McGraw-Hill 2007.

[14] Petronczki M, Siomos MF, Nasmyth K (February 2003). "Un ménage à quatre: the molecular biology of chromosome segregation in meiosis". *Cell* **112** (4): 423–40. doi:10.1016/S0092-8674(03)00083-7. PMID 12600308.

[15] Javaux EJ, Knoll AH, Walter MR (July 2001). "Morphological and ecological complexity in early eukaryotic ecosystems". *Nature* **412** (6842): 66–9. doi:10.1038/35083562. PMID 11452306.

[16] Hassold, T.; Jacobs, P.; Kline, J.; Stein, Z.; Warburton, D. (July 1980). "Effect of maternal age on autosomal trisomies". *Annals of Human Genetics* **44** (1): 29–36. doi:10.1111/j.1469-1809.1980.tb00943.x. PMID 7198887.

[17] Tsutsumi, M.; Fujiwara, R.; Nishizawa, H.; Ito, M.; Kogo, H.; Inagaki, H.; Ohye, T.; Kato, T.; Fujii, T.; Kurahashi, H. (May 2014). "Age-related decrease of meiotic cohesins in human oocytes". *PLoS One* **9** (5): Article e96710. doi:10.1371/journal.pone.0096710. PMID 24806359.

[18] Bidlack, James E. (2011). *Introductory Plant Biology*. New York, NY: McGraw HIll. pp. 214–29.

[19] Brunet, S.; Verlhac, M. H. (2010). "Positioning to get out of meiosis: The asymmetry of division". *Human Reproduction Update* **17** (1): 68–75. doi:10.1093/humupd/dmq044. PMID 20833637.

[20] Rosenbusch B (November 2006). "The contradictory information on the distribution of non-disjunction and pre-division in female gametes". *Hum. Reprod.* **21** (11): 2739–42. doi:10.1093/humrep/del122. PMID 16982661.

[21] Lin Y, Gill ME, Koubova J, Page DC (December 2008). "Germ cell-intrinsic and -extrinsic factors govern meiotic initiation in mouse embryos". *Science* **322** (5908): 1685–7. doi:10.1126/science.1166340. PMID 19074348.

[22] Suzuki A, Saga Y (February 2008). "Nanos2 suppresses meiosis and promotes male germ cell differentiation". *Genes Dev.* **22** (4): 430–5. doi:10.1101/gad.1612708. PMC 2238665. PMID 18281459.

[23] "How Cells Divide". *PBS*. Public Broadcasting Service. Retrieved 6 December 2012.

[24] Heywood, P.; Magee, P.T. (1976). "Meiosis in protists. Some structural and physiological aspects of meiosis in algae, fungi, and protozoa". *Bacteriological Reviews* **40** (1): 190–240.

[25] Raikov, I. B. (1995). "Meiosis in protists: recent advances and persisting problems". *Europ J Protistol* **31**: 1–7. doi:10.1016/s0932-4739(11)80349-4.

20.12.1 Cited texts

- Freeman, Scott (2005). *Biological Science* (3rd ed.). Upper Saddle River, NJ: Pearson Prentice Hall.

20.13 External links

- Meiosis Flash Animation

- Animations from the U. of Arizona Biology Dept.

- Meiosis at Kimball's Biology Pages

- Khan Academy, video lecture

- CCO The Cell-Cycle Ontology

- Stages of Meiosis animation

- *"Abby Denburg seminar: Chromosome Dynamics During Meiosis"

Chapter 21

Zygote

For other uses, see Zygote (disambiguation).
"Fertilized egg" redirects here. For the food product, see Balut (egg).

A **zygote** (from Greek ζυγωτός *zygōtos* "joined" or "yoked", from ζυγοῦν *zygoun* "to join" or "to yoke"),[1] is a eukaryotic cell formed by a fertilization event between two gametes. The zygote's genome is a combination of the DNA in each gamete, and contains all of the genetic information necessary to form a new individual. In multicellular organisms, the zygote is the earliest developmental stage. In single-celled organisms, the zygote can divide asexually by mitosis to produce identical offspring.

Oscar Hertwig and Richard Hertwig made some of the first discoveries on animal zygote formation.

21.1 Fungi

In fungi, the sexual fusion of haploid cells is called karyogamy. The result of karyogamy is a diploid cell called a zygote or zygospore. This cell may then enter meiosis or mitosis depending on the life cycle of the species.

21.2 Plants

In plants, the zygote may be polyploid if fertilization occurs between meiotically unreduced gametes.

In land plants, the zygote is formed within a chamber called the archegonium. In seedless plants, the archegonium is usually flask-shaped, with a long hollow neck through which the sperm cell enters. As the zygote divides and grows, it does so inside the archegonium.

21.3 Humans

In human fertilization, two 1n haploid cells—an ovum (female gamete) and a sperm cell (male gamete)—combine to form a single 2n diploid cell called the zygote. DNA is then replicated in the two separate pronuclei derived from the sperm and ovum, making the zygote's chromosome number temporarily 4n diploid. After approximately 30 hours, fusion of the pronuclei and subsequent mitotic division produce two 2n diploid daughter cells called blastomeres.[2]

Between the stages of fertilization and implantation, the developing human is called the *preimplantation conceptus* or the proembryo. It is not correct to call the conceptus an *embryo*, because it will later differentiate into both intraembryonic and extraembryonic tissues,[3] and can even split to produce multiple embryos (identical twins).

After fertilization, the conceptus travels down the oviduct towards the uterus while continuing to divide[4] mitotically without actually increasing in size, in a process called cleavage.[5] After four divisions, the conceptus consists of 16 blastomeres, and it is known as the morula.[6] Through the processes of compaction, cell division, and blastulation, the conceptus takes the form of the blastocyst by the fifth day of development, just as it approaches the site of implantation.[7] When the blastocyst hatches from the zona pellucida, it can implant in the endometrial lining of the uterus and begin the embryonic stage of development.

The human zygote has been genetically edited in experiments designed to cure inherited diseases.[8]

21.4   In other species

A Chlamydomonas zygote that contains chloroplast DNA (cpDNA) from both parents, such cells generally are rare since normally cpDNA is inherited uniparental from the mt+ mating type parent.These rare biparental zygotes allowed mapping of chloroplast genes by recombination.

21.5   In Protozoa

In the Amoeba, reproduction occurs by cell division of the parent cell: first the nucleus of the parent divides into two and then the cell membrane also cleaves, becoming two "daughter" Amoebae.

21.6   See also

- Proembryo

21.7   References

[1] "English etymology of zygote". *myetymology.com*.

[2] Blastomere Encyclopædia Britannica. Encyclopædia Britannica Online. Encyclopædia Britannica Inc., 2012. Web. 06 Feb. 2012.

[3] Larsen's Human Embryology. 4th Ed. Page 4.

[4] O'Reilly, Deirdre. "Fetal development". *MedlinePlus Medical Encyclopedia* (2007-10-19). Retrieved 2009-02-15.

[5] Klossner, N. Jayne and Hatfield, Nancy. *Introductory Maternity & Pediatric Nursing,* p. 107 (Lippincott Williams & Wilkins, 2006).

[6] Neas, John F. "Human Development". *Embryology Atlas*

[7] Blackburn, Susan. *Maternal, Fetal, & Neonatal Physiology,* p. 80 (Elsevier Health Sciences 2007).

[8] Human zygote edited genetically

Chapter 22

Bioinformatics

For the journal, see Bioinformatics (journal).

Bioinformatics ◀◢ⁱ/ˌbaɪ.oʊˌɪnfərˈmætɪks/ is an interdisciplinary field that develops methods and software tools for understanding biological data. As an interdisciplinary field of science, bioinformatics combines computer science, statistics, mathematics, and engineering to analyze and interpret biological data.

Bioinformatics is both an umbrella term for the body of biological studies that use computer programming as part of their methodology, as well as a reference to specific analysis "pipelines" that are repeatedly used, particularly in the field of genomics. Common uses of bioinformatics include the identification of candidate genes and nucleotides (SNPs). Often, such identification is made with the aim of better understanding the genetic basis of disease, unique adaptations, desirable properties (esp. in agricultural species), or differences between populations. In a less formal way, bioinformatics also tries to understand the organisational principles within nucleic acid and protein sequences.

22.1 Introduction

Bioinformatics has become an important part of many areas of biology. In experimental molecular biology, bioinformatics techniques such as image and signal processing allow extraction of useful results from large amounts of raw data. In the field of genetics and genomics, it aids in sequencing and annotating genomes and their observed mutations. It plays a role in the text mining of biological literature and the development of biological and gene ontologies to organize and query biological data. It also plays a role in the analysis of gene and protein expression and regulation. Bioinformatics tools aid in the comparison of genetic and genomic data and more generally in the understanding of evolutionary aspects of molecular biology. At a more integrative level, it helps analyze and catalogue the biological pathways and networks that are an important part of systems biology. In structural biology, it aids in the simulation and modeling of DNA, RNA, and protein structures as well as molecular interactions.

22.1.1 History

Historically, the term *bioinformatics* did not mean what it means today. Paulien Hogeweg and Ben Hesper coined it in 1970 to refer to the study of information processes in biotic systems.[1][2][3] This definition placed bioinformatics as a field parallel to biophysics (the study of physical processes in biological systems) or biochemistry (the study of chemical processes in biological systems).[1]

Sequences

Computers became essential in molecular biology when protein sequences became available after Frederick Sanger determined the sequence of insulin in the early 1950s. Comparing multiple sequences manually turned out to be impractical. A pioneer in the field was Margaret Oakley Dayhoff, who has been hailed by David Lipman, director of the National

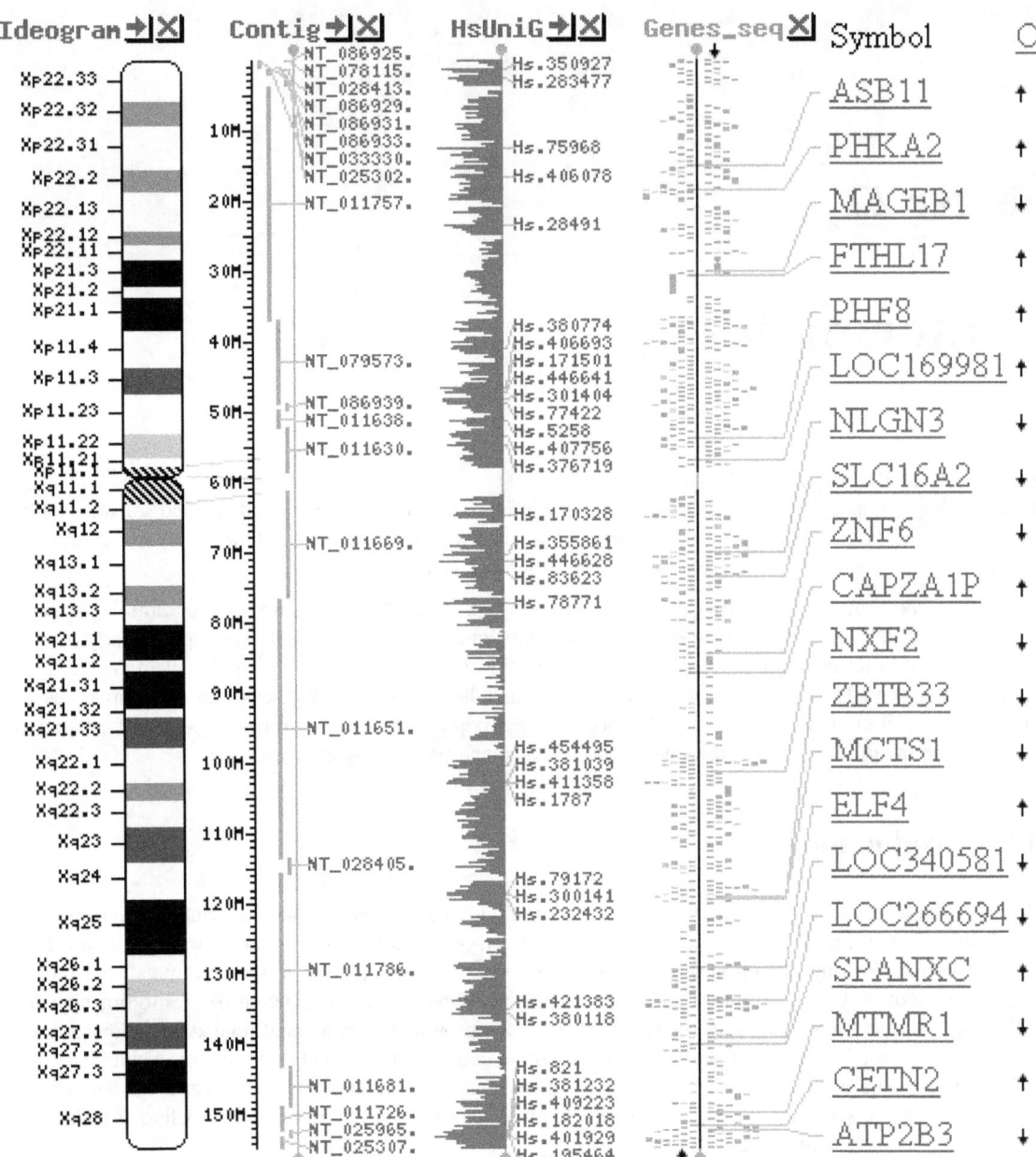

Map of the human X chromosome *(from the National Center for Biotechnology Information website).*

Center for Biotechnology Information, as the "mother and father of bioinformatics."[4] Dayhoff compiled one of the first protein sequence databases, initially published as books[5] and pioneered methods of sequence alignment and molecular evolution.[6] Another early contributor to bioinformatics was Elvin A. Kabat, who pioneered biological sequence analysis in 1970 with his comprehensive volumes of antibody sequences released with Tai Te Wu between 1980 and 1991.[7]

22.1.2   Goals

To study how normal cellular activities are altered in different disease states, the biological data must be combined to form a comprehensive picture of these activities. Therefore, the field of bioinformatics has evolved such that the most

5' A T G A C G T G G G G A 3'

3' T A C T G C A C C C C T 5'

Sequences of genetic material are frequently used in bioinformatics and are easier to manage using computers than manually.

pressing task now involves the analysis and interpretation of various types of data. This includes nucleotide and amino acid sequences, protein domains, and protein structures.[8] The actual process of analyzing and interpreting data is referred to as computational biology. Important sub-disciplines within bioinformatics and computational biology include:

- Development and implementation of computer programs that enable efficient access to, use and management of, various types of information

- Development of new algorithms (mathematical formulas) and statistical measures that assess relationships among members of large data sets. For example, there are methods to locate a gene within a sequence, to predict protein structure and/or function, and to cluster protein sequences into families of related sequences.

The primary goal of bioinformatics is to increase the understanding of biological processes. What sets it apart from other approaches, however, is its focus on developing and applying computationally intensive techniques to achieve this goal. Examples include: pattern recognition, data mining, machine learning algorithms, and visualization. Major research efforts in the field include sequence alignment, gene finding, genome assembly, drug design, drug discovery, protein structure alignment, protein structure prediction, prediction of gene expression and protein–protein interactions, genome-wide association studies, the modeling of evolution and cell division/mitosis.

Bioinformatics now entails the creation and advancement of databases, algorithms, computational and statistical techniques, and theory to solve formal and practical problems arising from the management and analysis of biological data.

Over the past few decades, rapid developments in genomic and other molecular research technologies and developments in information technologies have combined to produce a tremendous amount of information related to molecular biology. Bioinformatics is the name given to these mathematical and computing approaches used to glean understanding of biological processes.

Common activities in bioinformatics include mapping and analyzing DNA and protein sequences, aligning DNA and protein sequences to compare them, and creating and viewing 3-D models of protein structures.

22.1.3 Relation to other fields

Bioinformatics is a science field that is similar to but distinct from biological computation and computational biology. Biological computation uses bioengineering and biology to build biological computers, whereas bioinformatics uses computation to better understand biology. Bioinformatics and computational biology have similar aims and approaches, but they differ in scale: bioinformatics organizes and analyzes basic biological data, whereas computational biology builds theoretical models of biological systems, just as mathematical biology does with mathematical models.

Analyzing biological data to produce meaningful information involves writing and running software programs that use algorithms from graph theory, artificial intelligence, soft computing, data mining, image processing, and computer simulation. The algorithms in turn depend on theoretical foundations such as discrete mathematics, control theory, system theory, information theory, and statistics.

22.2   Sequence analysis

Main articles: Sequence alignment and Sequence database

Since the Phage Φ-X174 was sequenced in 1977,[9] the DNA sequences of thousands of organisms have been decoded

```
A5ASC3.1  14 SIKLWPPSQTTRLLLVERMANNLST..PSIFTRK..YGSLSKEEARENAKQIEEVACSTANQ.....HYEKEPDGDGGSAVQLYAKECSKLILEVLK 101
B4F917.1  13 SIKLWPPSESTRIMLVDRMTNNLST..ESIFSRK..YRLLGQEAHENAKTIEELCFALADE.....HFREEPDGDGSSAVQLYAKETSKRMALEVLK 100
A9S1V2.1  23 VFKLWPPSQGTREAVRQKMALKLSS..ACFESQS..FARIELADAQEHARAIEEVAFGAAQE.....ADSGGDKTGSAVWVYAKHASKLMLETLR 109
B9GSN7.1  13 SVKLWPPGQSTRLMLVERMTKNFIT..PSFISRK..YGLLSKEEAEEDAKKIEEVAFAAANQ.....HYEKQPDGDGGSAVQIYAKESSRLMLEVLK 100
Q8H056.1  30 SFSIWPPTQRTRDAVVRRLVQTLGG..DTILCKR..YGAVPAADAEPAARGIEAEAFDAAAA..SGEAAATASVEEGIKALQLYSKEVSRRLLDFVK 120
QOD4Z3.2  44 SLSIWPPSQRTRDAVVRRLVQTLVA..PSILSQR..YGAVPEAEAGRAARAAVEAEAYAAVTES,SSAAAAPASVEDGIEVLQRYSKEVSRRLLELAK 135
B9MVW8.1  56 SFSIWPPTQRTRDAIISRLIETLST..TSVLSKR..YGTIPKEEASEASRRIEEEAFSGAST.......VASSEKDGLEVLQLYSKEISKRMLETVK 141
QOIYC5.1  29 SFAVWPPTRRTRDAVVRRLVAVLSGDTTTALPKRYRYGAVPAADAERAARAVEAQAFDAASA....SSSSSSSVEDGIETLQLYSREVSNRLLAFVR 121
A9NWJ46.1  13 SIKLWPPSESTRLMLVERMTDNLSS..VSFFSRK..YGLLSKEEAAENAKRIEETAFLAAND.....HEAKEPNLDDSSVVQFYAREASKLMLEALK 100
Q9C500.1  57 SLRIWPPTQKTRDAVLNRLIETLST..ESILSKR..YGTLKSDDATTVAKLIEEEAYGVASN.......AVSSDDDGIKILELYSKEISKRMLESVK 142
Q2HRI7.1  25 NYSIWPPKQRTRDAVXNRLIETLST..PSVLTKR..YGTMSADEASAAAIQIEDEAFSVANA.......SSSTSNDNVTILEVYSKEISKRMIETVK 110
Q9M7N3.1  28 SFKIWPPTQRTREAVVRRLVETLTS..QSVLSKR..YGVIPEEDATSAARIIEEEAFSVASV.ASAASTGGRPEDEWIEVLHIYSQEIXQRVVESAK 119
Q9M7N6.1  25 SFSIWPPTQRTRDAVINRLIESLST..PSILSKR..YGTLPQDEASETARLIEEEAFAAAGS.......TASDADDGIEILQVYSKEISKRMIDTVK 110
Q9LE82.1  14 SVKMWPPSKSTRLMLVERMTKNITT..PSIFSRK..YGLLSVEEAEQDAKRIEDLAFATANK.....HFQHEPDGDGTSAVHVYAKESSKLMLDVIK 101
Q9M651.2  13 SIKLWPPSLPTRKALIEPITNNFSS..KTIFTEK..YGSLTKDQATENAKRIEDIAFSTANQ.....QFEREPDGDGGSAVQLYAKECSKLILEVLK 100
B9R748.1  48 SLSIWPPTQRTRDAVITRLIETLSS..PSVLSKR..YGTISHDEAESAARRIEDEAFGVANT.......ATSAEDDGLEILQLYSKEISRRMLDTVK 133
```

The sequences of different genes or proteins may be aligned side-by-side to measure their similarity. This alignment compares protein sequences containing WPP domains.

and stored in databases. This sequence information is analyzed to determine genes that encode proteins, RNA genes, regulatory sequences, structural motifs, and repetitive sequences. A comparison of genes within a species or between different species can show similarities between protein functions, or relations between species (the use of molecular systematics to construct phylogenetic trees). With the growing amount of data, it long ago became impractical to analyze DNA sequences manually. Today, computer programs such as BLAST are used daily to search sequences from more than 260 000 organisms, containing over 190 billion nucleotides.[10] These programs can compensate for mutations (exchanged, deleted or inserted bases) in the DNA sequence, to identify sequences that are related, but not identical. A variant of this sequence alignment is used in the sequencing process itself. The so-called shotgun sequencing technique (which was used, for example, by The Institute for Genomic Research to sequence the first bacterial genome, *Haemophilus influenzae*)[11] does not produce entire chromosomes. Instead it generates the sequences of many thousands of small DNA fragments (ranging from 35 to 900 nucleotides long, depending on the sequencing technology). The ends of these fragments overlap and, when aligned properly by a genome assembly program, can be used to reconstruct the complete genome. Shotgun sequencing yields sequence data quickly, but the task of assembling the fragments can be quite complicated for larger genomes. For a genome as large as the human genome, it may take many days of CPU time on large-memory, multiprocessor computers to assemble the fragments, and the resulting assembly usually contains numerous gaps that must be filled in later. Shotgun sequencing is the method of choice for virtually all genomes sequenced today, and genome assembly algorithms are a critical area of bioinformatics research.

Following the goals that the Human Genome Project left to achieve after its closure in 2003, a new project developed by the National Human Genome Research Institute in the U.S appeared. The so-called ENCODE project is a collaborative data collection of the functional elements of the human genome that uses next-generation DNA-sequencing technologies and genomic tiling arrays, technologies able to generate automatically large amounts of data with lower research costs but with the same quality and viability.

Another aspect of bioinformatics in sequence analysis is annotation. This involves computational gene finding to search for protein-coding genes, RNA genes, and other functional sequences within a genome. Not all of the nucleotides within a genome are part of genes. Within the genomes of higher organisms, large parts of the DNA do not serve any obvious purpose.

See also: sequence analysis, sequence mining, sequence profiling tool and sequence motif

22.2.1   Genome annotation

Main article: Gene prediction

In the context of genomics, annotation is the process of marking the genes and other biological features in a DNA sequence.

This process needs to be automated because most genomes are too large to annotate by hand, not to mention the desire to annotate as many genomes as possible, as the rate of sequencing has ceased to pose a bottleneck. Annotation is made possible by the fact that genes have recognisable start and stop regions, although the exact sequence found in these regions can vary between genes.

The first genome annotation software system was designed in 1995 by Owen White, who was part of the team at The Institute for Genomic Research that sequenced and analyzed the first genome of a free-living organism to be decoded, the bacterium *Haemophilus influenzae*.[11] White built a software system to find the genes (fragments of genomic sequence that encode proteins), the transfer RNAs, and to make initial assignments of function to those genes. Most current genome annotation systems work similarly, but the programs available for analysis of genomic DNA, such as the GeneMark program trained and used to find protein-coding genes in *Haemophilus influenzae*, are constantly changing and improving.

22.2.2 Computational evolutionary biology

Evolutionary biology is the study of the origin and descent of species, as well as their change over time. Informatics has assisted evolutionary biologists by enabling researchers to:

- trace the evolution of a large number of organisms by measuring changes in their DNA, rather than through physical taxonomy or physiological observations alone,

- more recently, compare entire genomes, which permits the study of more complex evolutionary events, such as gene duplication, horizontal gene transfer, and the prediction of factors important in bacterial speciation,

- build complex computational models of populations to predict the outcome of the system over time[12]

- track and share information on an increasingly large number of species and organisms

Future work endeavours to reconstruct the now more complex tree of life.

The area of research within computer science that uses genetic algorithms is sometimes confused with computational evolutionary biology, but the two areas are not necessarily related.

22.2.3 Comparative genomics

Main article: Comparative genomics

The core of comparative genome analysis is the establishment of the correspondence between genes (orthology analysis) or other genomic features in different organisms. It is these intergenomic maps that make it possible to trace the evolutionary processes responsible for the divergence of two genomes. A multitude of evolutionary events acting at various organizational levels shape genome evolution. At the lowest level, point mutations affect individual nucleotides. At a higher level, large chromosomal segments undergo duplication, lateral transfer, inversion, transposition, deletion and insertion.[13] Ultimately, whole genomes are involved in processes of hybridization, polyploidization and endosymbiosis, often leading to rapid speciation. The complexity of genome evolution poses many exciting challenges to developers of mathematical models and algorithms, who have recourse to a spectra of algorithmic, statistical and mathematical techniques, ranging from exact, heuristics, fixed parameter and approximation algorithms for problems based on parsimony models to Markov Chain Monte Carlo algorithms for Bayesian analysis of problems based on probabilistic models.

Many of these studies are based on the homology detection and protein families computation.[14]

22.2.4 Pan genomics

Main article: Pan-genome

Pan genomics is a concept introduced in 2005 by Tettelin and Medini which eventually took root in bioinformatics. Pan genome is the complete gene repertoire of a particular taxonomic group: although initially applied to closely related strains of a species, it can be applied to a larger context like genus, phylum etc. It is divided in two parts- The Core genome: Set of genes common to all the genomes under study (These are often housekeeping genes vital for survival) and The Dispensable/Flexible Genome: Set of genes not present in all but one or some genomes under study.

22.2.5   Genetics of disease

Main article: Genome-wide association studies

With the advent of next-generation sequencing we are obtaining enough sequence data to map the genes of complex diseases such as diabetes,[15] infertility,[16] breast cancer[17] or Alzheimer's Disease.[18] Genome-wide association studies are a useful approach to pinpoint the mutations responsible for such complex diseases.[19] Through these studies, thousands of DNA variants have been identified that are associated with similar diseases and traits.[20] Furthermore, the possibility for genes to be used at prognosis, diagnosis or treatment is one of the most essential applications. Many studies are discussing both the promising ways to choose the genes to be used and the problems and pitfalls of using genes to predict disease presence or prognosis.[21]

22.2.6   Analysis of mutations in cancer

Main article: Oncogenomics

In cancer, the genomes of affected cells are rearranged in complex or even unpredictable ways. Massive sequencing efforts are used to identify previously unknown point mutations in a variety of genes in cancer. Bioinformaticians continue to produce specialized automated systems to manage the sheer volume of sequence data produced, and they create new algorithms and software to compare the sequencing results to the growing collection of human genome sequences and germline polymorphisms. New physical detection technologies are employed, such as oligonucleotide microarrays to identify chromosomal gains and losses (called comparative genomic hybridization), and single-nucleotide polymorphism arrays to detect known *point mutations*. These detection methods simultaneously measure several hundred thousand sites throughout the genome, and when used in high-throughput to measure thousands of samples, generate terabytes of data per experiment. Again the massive amounts and new types of data generate new opportunities for bioinformaticians. The data is often found to contain considerable variability, or noise, and thus Hidden Markov model and change-point analysis methods are being developed to infer real copy number changes.

However, with the breakthroughs that the next-generation sequencing technology is providing to the field of Bioinformatics, cancer genomics may be drastically change. This new methods and software allow bioinformaticians to sequence in a rapid and affordable way many cancer genomes. This could mean a more flexible process to classify types of cancer by analysis of cancer driven mutations in the genome. Furthermore, individual tracking of patients during the progression of the disease may be possible in the future with the sequence of cancer samples.[22]

Another type of data that requires novel informatics development is the analysis of lesions found to be recurrent among many tumors.

22.3   Gene and protein expression

22.3.1   Analysis of gene expression

The expression of many genes can be determined by measuring mRNA levels with multiple techniques including microarrays, expressed cDNA sequence tag (EST) sequencing, serial analysis of gene expression (SAGE) tag sequencing, massively parallel signature sequencing (MPSS), RNA-Seq, also known as "Whole Transcriptome Shotgun Sequencing" (WTSS), or various applications of multiplexed in-situ hybridization. All of these techniques are extremely noise-prone and/or subject to bias in the biological measurement, and a major research area in computational biology involves developing statistical tools to separate signal from noise in high-throughput gene expression studies. Such studies are often used to determine the genes implicated in a disorder: one might compare microarray data from cancerous epithelial cells to data from non-cancerous cells to determine the transcripts that are up-regulated and down-regulated in a particular population of cancer cells.

22.3.2 Analysis of protein expression

Protein microarrays and high throughput (HT) mass spectrometry (MS) can provide a snapshot of the proteins present in a biological sample. Bioinformatics is very much involved in making sense of protein microarray and HT MS data; the former approach faces similar problems as with microarrays targeted at mRNA, the latter involves the problem of matching large amounts of mass data against predicted masses from protein sequence databases, and the complicated statistical analysis of samples where multiple, but incomplete peptides from each protein are detected.

22.3.3 Analysis of regulation

Regulation is the complex orchestration of events starting with an extracellular signal such as a hormone and leading to an increase or decrease in the activity of one or more proteins. Bioinformatics techniques have been applied to explore various steps in this process. For example, promoter analysis involves the identification and study of sequence motifs in the DNA surrounding the coding region of a gene. These motifs influence the extent to which that region is transcribed into mRNA. Expression data can be used to infer gene regulation: one might compare microarray data from a wide variety of states of an organism to form hypotheses about the genes involved in each state. In a single-cell organism, one might compare stages of the cell cycle, along with various stress conditions (heat shock, starvation, etc.). One can then apply clustering algorithms to that expression data to determine which genes are co-expressed. For example, the upstream regions (promoters) of co-expressed genes can be searched for over-represented regulatory elements. Examples of clustering algorithms applied in gene clustering are k-means clustering, self-organizing maps (SOMs), hierarchical clustering, and consensus clustering methods such as the Bi-CoPaM. The later, namely Bi-CoPaM, has been actually proposed to address various issues specific to gene discovery problems such as consistent co-expression of genes over multiple microarray datasets.[23][24]

22.4 Structural bioinformatics

Main articles: Structural bioinformatics and Protein structure prediction
See also: Structural motif and Structural domain
Protein structure prediction is another important application of bioinformatics. The amino acid sequence of a protein, the so-called primary structure, can be easily determined from the sequence on the gene that codes for it. In the vast majority of cases, this primary structure uniquely determines a structure in its native environment. (Of course, there are exceptions, such as the bovine spongiform encephalopathy – a.k.a. Mad Cow Disease – prion.) Knowledge of this structure is vital in understanding the function of the protein. Structural information is usually classified as one of *secondary*, *tertiary* and *quaternary* structure. A viable general solution to such predictions remains an open problem. Most efforts have so far been directed towards heuristics that work most of the time.

One of the key ideas in bioinformatics is the notion of homology. In the genomic branch of bioinformatics, homology is used to predict the function of a gene: if the sequence of gene *A*, whose function is known, is homologous to the sequence of gene *B*, whose function is unknown, one could infer that B may share A's function. In the structural branch of bioinformatics, homology is used to determine which parts of a protein are important in structure formation and interaction with other proteins. In a technique called homology modeling, this information is used to predict the structure of a protein once the structure of a homologous protein is known. This currently remains the only way to predict protein structures reliably.

One example of this is the similar protein homology between hemoglobin in humans and the hemoglobin in legumes (leghemoglobin). Both serve the same purpose of transporting oxygen in the organism. Though both of these proteins

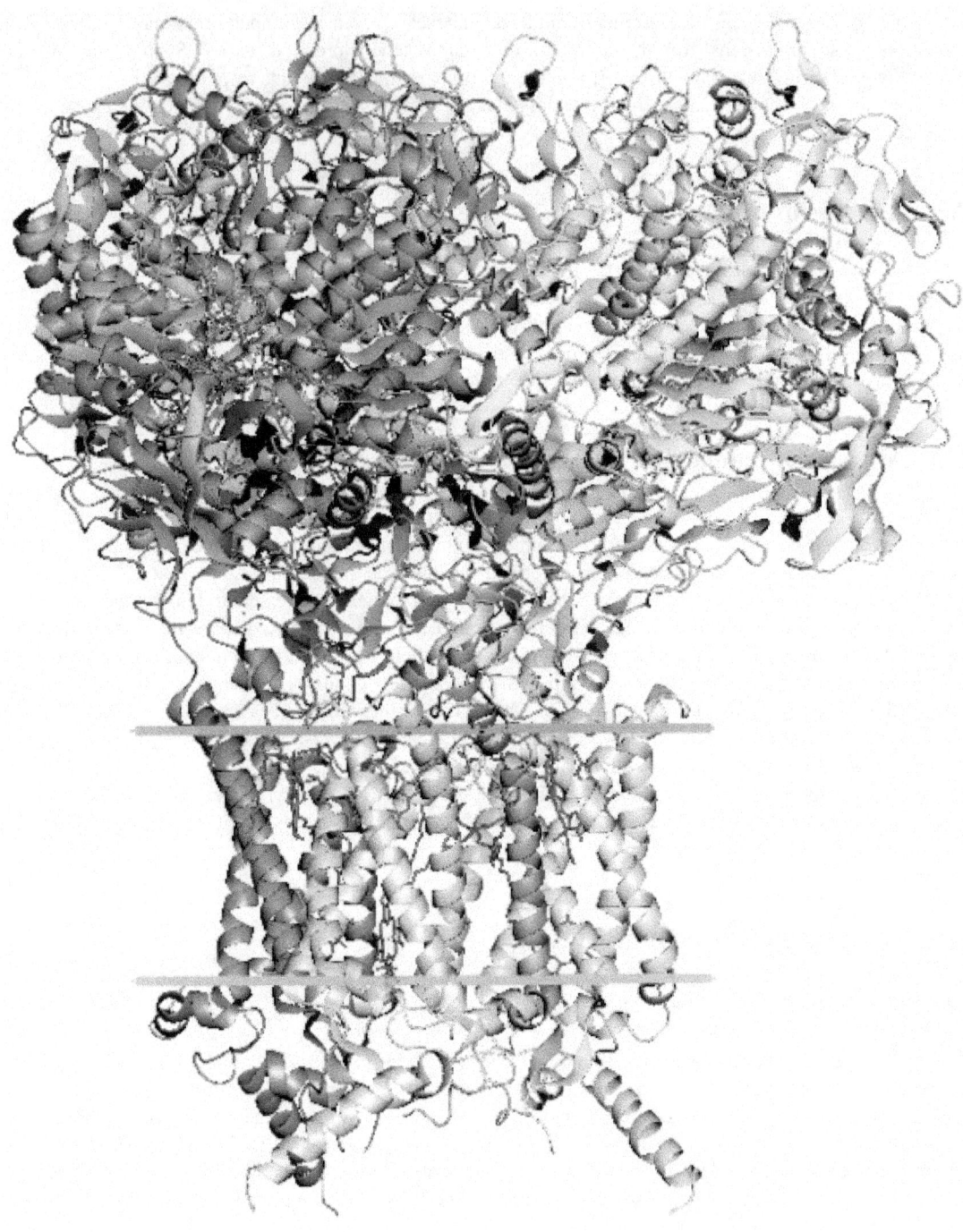

3-dimensional protein structures such as this one are common subjects in bioinformatic analyses.

have completely different amino acid sequences, their protein structures are virtually identical, which reflects their near identical purposes.[25]

Other techniques for predicting protein structure include protein threading and *de novo* (from scratch) physics-based modeling.

22.5 Network and systems biology

Main articles: Computational systems biology, Biological network and Interactome

Network analysis seeks to understand the relationships within biological networks such as metabolic or protein-protein interaction networks. Although biological networks can be constructed from a single type of molecule or entity (such as genes), network biology often attempts to integrate many different data types, such as proteins, small molecules, gene expression data, and others, which are all connected physically, functionally, or both.

Systems biology involves the use of computer simulations of cellular subsystems (such as the networks of metabolites and enzymes that comprise metabolism, signal transduction pathways and gene regulatory networks) to both analyze and visualize the complex connections of these cellular processes. Artificial life or virtual evolution attempts to understand evolutionary processes via the computer simulation of simple (artificial) life forms.

22.5.1 Molecular interaction networks

Main articles: Protein–protein interaction prediction and interactome

Tens of thousands of three-dimensional protein structures have been determined by X-ray crystallography and protein nuclear magnetic resonance spectroscopy (protein NMR) and a central question in structural bioinformatics is whether it is practical to predict possible protein–protein interactions only based on these 3D shapes, without performing protein–protein interaction experiments. A variety of methods have been developed to tackle the protein–protein docking problem, though it seems that there is still much work to be done in this field.

Other interactions encountered in the field include Protein–ligand (including drug) and protein–peptide. Molecular dynamic simulation of movement of atoms about rotatable bonds is the fundamental principle behind computational algorithms, termed docking algorithms, for studying molecular interactions.

22.6 Others

22.6.1 Literature analysis

Main articles: Text mining and Biomedical text mining

The growth in the number of published literature makes it virtually impossible to read every paper, resulting in disjointed sub-fields of research. Literature analysis aims to employ computational and statistical linguistics to mine this growing library of text resources. For example:

- Abbreviation recognition – identify the long-form and abbreviation of biological terms
- Named entity recognition – recognizing biological terms such as gene names
- Protein-protein interaction – identify which proteins interact with which proteins from text

The area of research draws from statistics and computational linguistics.

22.6.2 High-throughput image analysis

Computational technologies are used to accelerate or fully automate the processing, quantification and analysis of large amounts of high-information-content biomedical imagery. Modern image analysis systems augment an observer's ability to make measurements from a large or complex set of images, by improving accuracy, objectivity, or speed. A fully

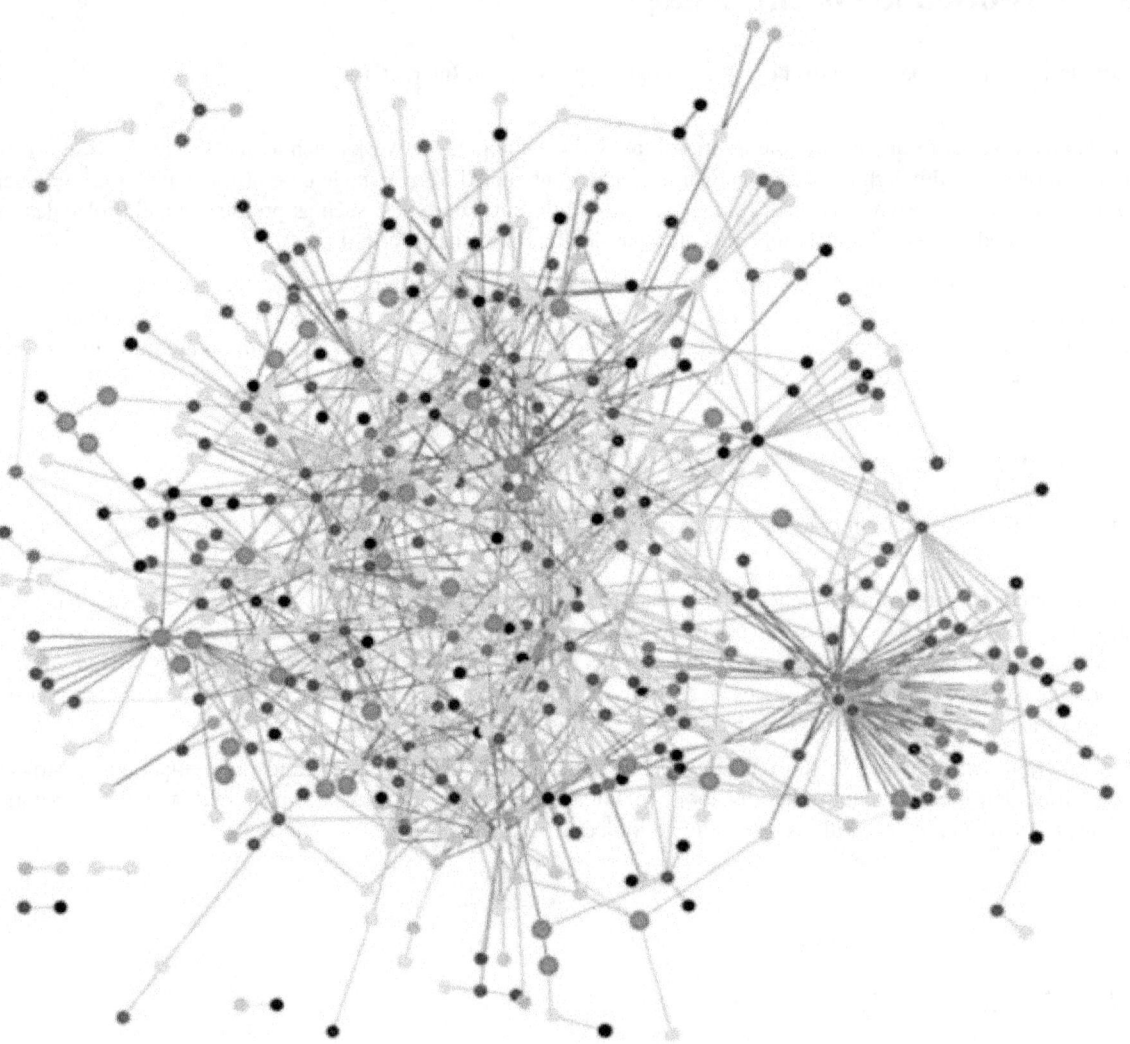

Interactions between proteins are frequently visualized and analyzed using networks. This network is made up of protein-protein interactions from Treponema pallidum, *the causative agent of syphilis and other diseases.*

developed analysis system may completely replace the observer. Although these systems are not unique to biomedical imagery, biomedical imaging is becoming more important for both diagnostics and research. Some examples are:

- high-throughput and high-fidelity quantification and sub-cellular localization (high-content screening, cytohistopathology, Bioimage informatics)

- morphometrics

- clinical image analysis and visualization

- determining the real-time air-flow patterns in breathing lungs of living animals

- quantifying occlusion size in real-time imagery from the development of and recovery during arterial injury

- making behavioral observations from extended video recordings of laboratory animals

- infrared measurements for metabolic activity determination

- inferring clone overlaps in DNA mapping, e.g. the Sulston score

22.6.3   High-throughput single cell data analysis

Main article: Flow cytometry bioinformatics

Computational techniques are used to analyse high-throughput, low-measurement single cell data, such as that obtained from flow cytometry. These methods typically involve finding populations of cells that are relevant to a particular disease state or experimental condition.

22.6.4   Biodiversity informatics

Main article: Biodiversity informatics

Biodiversity informatics deals with the collection and analysis of biodiversity data, such as taxonomic databases, or microbiome data. Examples of such analyses include phylogenetics, niche modelling, species richness mapping, or species identification tools.

22.7   Databases

Main articles: List of biological databases and Biological database

Databases are essential for bioinformatics research and applications. There is a huge number of available databases covering almost everything from DNA and protein sequences, molecular structures, to phenotypes and biodiversity. Databases generally fall into one of three types. Some contain data resulting directly from empirical methods such as gene knockouts. Others consist of predicted data, and most contain data from both sources. There are meta-databases that incorporate data compiled from multiple other databases. Some others are specialized, such as those specific to an organism. These databases vary in their format, way of accession and whether they are public or not. Some of the most commonly used databases are listed below. For a more comprehensive list, please check the link at the beginning of the subsection.

- Used in Motif Finding: GenomeNet MOTIF Search

- Used in Gene Ontology: ToppGene FuncAssociate, Enrichr, GATHER

- Used in Gene Finding: Hidden Markov Model

- Used in finding Protein Structures/Family: PFAM

- Used for Next Generation Sequencing: (Not database but data format), FASTQ Format

- Used in Gene Expression Analysis: GEO, ArrayExpress

- Used in Network Analysis: Interaction Analysis Databases(BioGRID, MINT, HPRD, Curated Human Signaling Network), Functional Networks (STRING, KEGG)

- Used in design of synthetic genetic circuits: GenoCAD

Please keep in mind that this is a quick sampling and generally most computation data is supported by wet lab data as well.

22.8   Software and tools

Software tools for bioinformatics range from simple command-line tools, to more complex graphical programs and standalone web-services available from various bioinformatics companies or public institutions.

22.8.1   Open-source bioinformatics software

Many free and open-source software tools have existed and continued to grow since the 1980s.[26] The combination of a continued need for new algorithms for the analysis of emerging types of biological readouts, the potential for innovative *in silico* experiments, and freely available open code bases have helped to create opportunities for all research groups to contribute to both bioinformatics and the range of open-source software available, regardless of their funding arrangements. The open source tools often act as incubators of ideas, or community-supported plug-ins in commercial applications. They may also provide *de facto* standards and shared object models for assisting with the challenge of bioinformation integration.

The range of open-source software packages includes titles such as Bioconductor, BioPerl, Biopython, BioJava, BioJS, BioRuby, Bioclipse, EMBOSS, .NET Bio, Apache Taverna, UGENE and GenoCAD. To maintain this tradition and create further opportunities, the non-profit Open Bioinformatics Foundation[26] have supported the annual Bioinformatics Open Source Conference (BOSC) since 2000.[27]

An alternative method to build public bioinformatics databases is to use the MediaWiki engine with the *WikiOpener* extension. This system allows the database to be accessed and updated by all experts in the field.[28]

22.8.2   Web services in bioinformatics

SOAP- and REST-based interfaces have been developed for a wide variety of bioinformatics applications allowing an application running on one computer in one part of the world to use algorithms, data and computing resources on servers in other parts of the world. The main advantages derive from the fact that end users do not have to deal with software and database maintenance overheads.

Basic bioinformatics services are classified by the EBI into three categories: SSS (Sequence Search Services), MSA (Multiple Sequence Alignment), and BSA (Biological Sequence Analysis).[29] The availability of these service-oriented bioinformatics resources demonstrate the applicability of web-based bioinformatics solutions, and range from a collection of standalone tools with a common data format under a single, standalone or web-based interface, to integrative, distributed and extensible bioinformatics workflow management systems.

22.8.3   Bioinformatics workflow management systems

Main article: Bioinformatics workflow management systems

A Bioinformatics workflow management system is a specialized form of a workflow management system designed specifically to compose and execute a series of computational or data manipulation steps, or a workflow, in a Bioinformatics application. Such systems are designed to

- provide an easy-to-use environment for individual application scientists themselves to create their own workflows

- provide interactive tools for the scientists enabling them to execute their workflows and view their results in real-time

- simplify the process of sharing and reusing workflows between the scientists.

- enable scientists to track the provenance of the workflow execution results and the workflow creation steps.

Some of the platforms giving this service: Galaxy, Kepler, Taverna, UGENE, Anduril.

22.9 Education platforms

Software platforms designed to teach bioinformatics concepts and methods include Rosalind and online courses offered through the Swiss Institute of Bioinformatics Training Portal. The Canadian Bioinformatics Workshops provides videos and slides from training workshops on their website under a Creative Commons license.

22.10 Conferences

There are several large conferences that are concerned with bioinformatics. Some of the most notable examples are Intelligent Systems for Molecular Biology (ISMB), European Conference on Computational Biology (ECCB), and Research in Computational Molecular Biology (RECOMB).

22.11 See also

- Biodiversity informatics

- Bioinformatics companies

- Computational biology

- Computational biomodeling

- Computational genomics

- Functional genomics

- Health informatics

- International Society for Computational Biology

- Jumping library

- List of open-source bioinformatics software

- List of scientific journals in bioinformatics

- Margaret Oakley Dayhoff

- Metabolomics

- Nucleic acid sequence

- Phylogenetics

- Proteomics

- Structural bioinformatics

- Gene Disease Database

22.12 References

[1] Hogeweg P (2011). Searls, David B., ed. "The Roots of Bioinformatics in Theoretical Biology". *PLoS Computational Biology* **7** (3): e1002021. Bibcode:2011PLSCB...7E0020H. doi:10.1371/journal.pcbi.1002021. PMC 3068925. PMID 21483479.

[2] Hesper B, Hogeweg P (1970). "Bioinformatica: een werkconcept" **1** (6). Kameleon: 28–29.

[3] Hogeweg P (1978). "Simulating the growth of cellular forms". *Simulation* **31** (3): 90–96. doi:10.1177/003754977803100305.

[4] Moody, Glyn (2004). *Digital Code of Life: How Bioinformatics is Revolutionizing Science, Medicine, and Business*. ISBN 978-0-471-32788-2.

[5] Dayhoff, M.O. (1966) Atlas of protein sequence and structure. National Biomedical Research Foundation, 215 pp.

[6] Eck RV, Dayhoff MO (1966). "Evolution of the structure of ferredoxin based on living relics of primitive amino Acid sequences". *Science* **152** (3720): 363–6. Bibcode:1966Sci...152..363E. doi:10.1126/science.152.3720.363. PMID 17775169.

[7] Johnson G, Wu TT (January 2000). "Kabat Database and its applications: 30 years after the first variability plot". *Nucleic Acids Res* **28** (1): 214–218. doi:10.1093/nar/28.1.214. PMC 102431. PMID 10592229.

[8] Attwood TK, Gisel A, Eriksson N-E, Bongcam-Rudloff E (2011). "Concepts, Historical Milestones and the Central Place of Bioinformatics in Modern Biology: A European Perspective". *Bioinformatics – Trends and Methodologies*. InTech. Retrieved 8 Jan 2012.

[9] Sanger F, Air GM, Barrell BG, Brown NL, Coulson AR, Fiddes CA, Hutchison CA, Slocombe PM, Smith M (February 1977). "Nucleotide sequence of bacteriophage phi X174 DNA". *Nature* **265** (5596): 687–95. Bibcode:1977Natur.265..687S. doi:10.1038/265687a0. PMID 870828.

[10] Benson DA, Karsch-Mizrachi I, Lipman DJ, Ostell J, Wheeler DL (January 2008). "GenBank". *Nucleic Acids Res.* **36** (Database issue): D25–30. doi:10.1093/nar/gkm929. PMC 2238942. PMID 18073190.

[11] Fleischmann RD, Adams MD, White O, Clayton RA, Kirkness EF, Kerlavage AR, Bult CJ, Tomb JF, Dougherty BA, Merrick JM (July 1995). "Whole-genome random sequencing and assembly of Haemophilus influenzae Rd". *Science* **269** (5223): 496–512. Bibcode:1995Sci...269..496F. doi:10.1126/science.7542800. PMID 7542800.

[12] Carvajal-Rodríguez A (2012). "Simulation of Genes and Genomes Forward in Time". *Current Genomics* (Bentham Science Publishers Ltd.) **11** (1): 58–61. doi:10.2174/138920210790218007. PMC 2851118. PMID 20808525.

[13] Brown, TA (2002). "Mutation, Repair and Recombination". *Genomes* (2nd ed.). Manchester (UK): Oxford.

[14] Carter, N. P.; Fiegler, H.; Piper, J. (2002). "Comparative analysis of comparative genomic hybridization microarray technologies: Report of a workshop sponsored by the Wellcome trust". *Wiley Subscription Services, Inc* **49** (2): 43–8. doi:10.1002/cyto.10153.

[15] Ionescu-Tîrgovişte, Constantin; Gagniuc, Paul Aurelian; Guja, Cristian. "Structural Properties of Gene Promoters Highlight More than Two Phenotypes of Diabetes". *PLOS ONE* **10** (9): e0137950. doi:10.1371/journal.pone.0137950. PMC 4574929. PMID 26379145.

[16] Aston KI (2014). "Genetic susceptibility to male infertility: News from genome-wide association studies". *Andrology* **2** (3): 315–21. doi:10.1111/j.2047-2927.2014.00188.x. PMID 24574159.

[17] Véron A, Blein S, Cox DG (2014). "Genome-wide association studies and the clinic: A focus on breast cancer". *Biomarkers in Medicine* **8** (2): 287–96. doi:10.2217/bmm.13.121. PMID 24521025.

[18] Tosto G, Reitz C (2013). "Genome-wide association studies in Alzheimer's disease: A review". *Current Neurology and Neuroscience Reports* **13** (10): 381. doi:10.1007/s11910-013-0381-0. PMC 3809844. PMID 23954969.

[19] Londin E, Yadav P, Surrey S, Kricka LJ, Fortina P (2013). "Use of Linkage Analysis, Genome-Wide Association Studies, and Next-Generation Sequencing in the Identification of Disease-Causing Mutations". *Pharmacogenomics*. Methods in Molecular Biology **1015**: 127–46. doi:10.1007/978-1-62703-435-7_8. ISBN 978-1-62703-434-0. PMID 23824853.

[20] Hindorff, L.A.,; et al. (2009). "Potential etiologic and functional implications of genome-wide association loci for human diseases and traits.". *Proc. Natl Acad. Sci. USA* **106**: 9362–9367. doi:10.1073/pnas.0903103106. PMC 2687147. PMID 19474294.

[21] Hall, L.O. (2010). "Finding the right genes for disease and prognosis prediction.". *System Science and Engineering (ICSSE),2010 International Conference*: 1–2. doi:10.1109/ICSSE.2010.5551766.

[22] Hye-Jung, E.C.; Jaswinder, K.; Martin, K.; Samuel, A.A; Marco, A.M (2014). ""Second-Generation Sequencing for Cancer Genome Analysis"". In Dellaire, Graham; Berman, Jason N.; Arceci, Robert J. *Cancer Genomics*. Boston (US): Academic Press. pp. 13–30. doi:10.1016/B978-0-12-396967-5.00002-5. ISBN 9780123969675.

[23] Abu-Jamous B, Fa R, Roberts DJ, Nandi AK (11 February 2013). "Paradigm of Tunable Clustering Using Binarization of Consensus Partition Matrices (Bi-CoPaM) for Gene Discovery". *PLoS ONE* **8** (2): e56432. Bibcode:2013PLoSO...856432A. doi:10.1371/journal.pone.0056432. PMC 3569426. PMID 23409186.

[24] Abu-Jamous B, Fa R, Roberts DJ, Nandi AK (2013). "Yeast gene CMR1/YDL156W is consistently co-expressed with genes participating in DNA-metabolic processes in a variety of stringent clustering experiments". *J R Soc Interface* **10** (81): 20120990. doi:10.1098/rsif.2012.0990. PMC 3627109. PMID 23349438.

[25] Hoy, JA; Robinson, H; Trent JT, 3rd; Kakar, S; Smagghe, BJ; Hargrove, MS (3 August 2007). "Plant hemoglobins: a molecular fossil record for the evolution of oxygen transport.". *Journal of Molecular Biology* **371** (1): 168–79. doi:10.1016/j.jmb.2007.05.029. PMID 17560601.

[26] "Open Bioinformatics Foundation: About us". *Official website*. Open Bioinformatics Foundation. Retrieved 10 May 2011.

[27] "Open Bioinformatics Foundation: BOSC". *Official website*. Open Bioinformatics Foundation. Retrieved 10 May 2011.

[28] Brohée, Sylvain; Barriot, Roland; Moreau, Yves. "Biological knowledge bases using Wikis: combining the flexibility of Wikis with the structure of databases". *Bioinformatics*. Oxford Journals. Retrieved 5 May 2015.

[29] Nisbet, Robert (14 May 2009). "BIOINFORMATICS". *Handbook of Statistical Analysis and Data Mining Applications*. John Elder IV, Gary Miner. Academic Press. p. 328. Retrieved 9 May 2014.

22.13 Further reading

- Raul Isea The Present-Day Meaning Of The Word Bioinformatics, Global Journal of Advanced Research, 2015

- Ilzins, O., Isea, R. and Hoebeke, J. Can Bioinformatics Be Considered as an Experimental Biological Science 2015

- Achuthsankar S Nair Computational Biology & Bioinformatics – A gentle Overview, Communications of Computer Society of India, January 2007

- Aluru, Srinivas, ed. *Handbook of Computational Molecular Biology*. Chapman & Hall/Crc, 2006. ISBN 1-58488-406-1 (Chapman & Hall/Crc Computer and Information Science Series)

- Baldi, P and Brunak, S, *Bioinformatics: The Machine Learning Approach*, 2nd edition. MIT Press, 2001. ISBN 0-262-02506-X

- Barnes, M.R. and Gray, I.C., eds., *Bioinformatics for Geneticists*, first edition. Wiley, 2003. ISBN 0-470-84394-2

- Baxevanis, A.D. and Ouellette, B.F.F., eds., *Bioinformatics: A Practical Guide to the Analysis of Genes and Proteins*, third edition. Wiley, 2005. ISBN 0-471-47878-4

- Baxevanis, A.D., Petsko, G.A., Stein, L.D., and Stormo, G.D., eds., *Current Protocols in Bioinformatics*. Wiley, 2007. ISBN 0-471-25093-7

- Cristianini, N. and Hahn, M. *Introduction to Computational Genomics*, Cambridge University Press, 2006. (ISBN 9780521671910 | ISBN 0-521-67191-4)

- Durbin, R., S. Eddy, A. Krogh and G. Mitchison, *Biological sequence analysis*. Cambridge University Press, 1998. ISBN 0-521-62971-3

- Gilbert D (2004). "Bioinformatics software resources". *Briefings in Bioinformatics* **5** (3): 300–304. doi:10.1093/bib/5.3.300. PMID 15383216.

- Keedwell, E., *Intelligent Bioinformatics: The Application of Artificial Intelligence Techniques to Bioinformatics Problems*. Wiley, 2005. ISBN 0-470-02175-6

- Kohane, et al. *Microarrays for an Integrative Genomics.* The MIT Press, 2002. ISBN 0-262-11271-X

- Lund, O. et al. *Immunological Bioinformatics.* The MIT Press, 2005. ISBN 0-262-12280-4

- Pachter, Lior and Sturmfels, Bernd. "Algebraic Statistics for Computational Biology" Cambridge University Press, 2005. ISBN 0-521-85700-7

- Pevzner, Pavel A. *Computational Molecular Biology: An Algorithmic Approach* The MIT Press, 2000. ISBN 0-262-16197-4

- Soinov, L. Bioinformatics and Pattern Recognition Come Together Journal of Pattern Recognition Research (JPRR), Vol 1 (1) 2006 p. 37–41

- Stevens, Hallam, *Life Out of Sequence: A Data-Driven History of Bioinformatics*, Chicago: The University of Chicago Press, 2013, ISBN 9780226080208

- Tisdall, James. "Beginning Perl for Bioinformatics" O'Reilly, 2001. ISBN 0-596-00080-4

- Dedicated issue of *Philosophical Transactions B* on Bioinformatics freely available

- Catalyzing Inquiry at the Interface of Computing and Biology (2005) CSTB report

- Calculating the Secrets of Life: Contributions of the Mathematical Sciences and computing to Molecular Biology (1995)

- Foundations of Computational and Systems Biology MIT Course

- Computational Biology: Genomes, Networks, Evolution Free MIT Course

22.14   External links

- Bioinformatics Resource Portal (SIB)

Chapter 23

Regulation of gene expression

"Gene modulation" redirects here. For information on therapeutic regulation of gene expression, see therapeutic gene modulation.

For vocabulary, see Glossary of gene expression terms.

Regulation of gene expression includes a wide range of mechanisms that are used by cells to increase or decrease

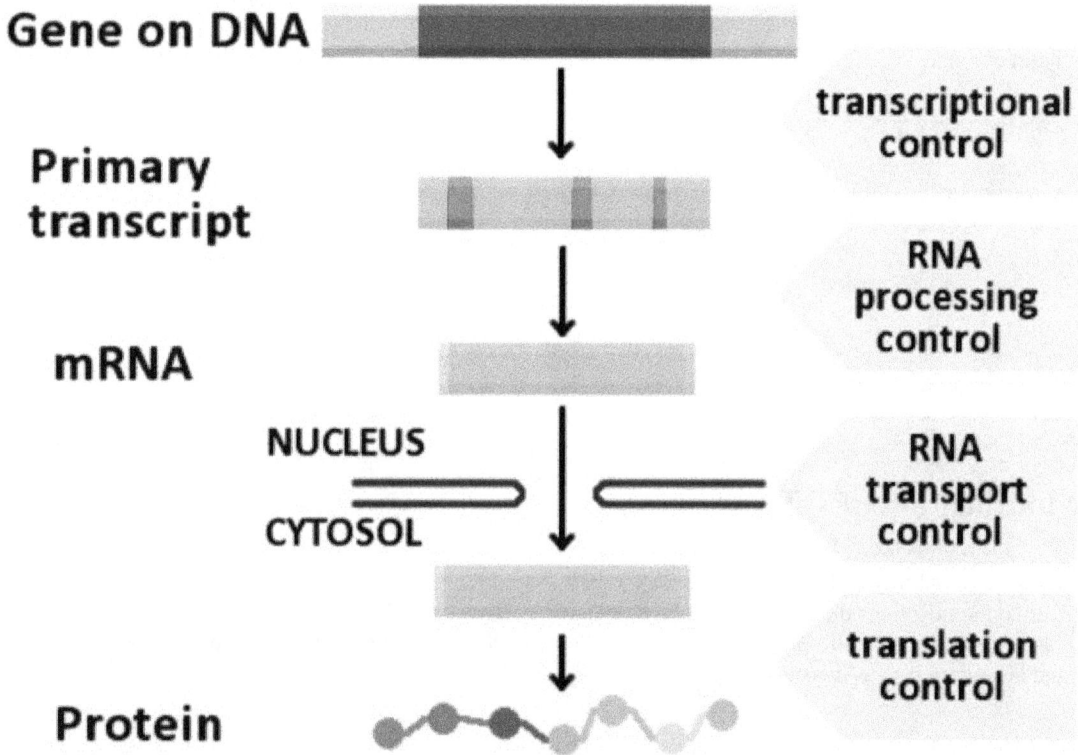

Diagram showing at which stages in the DNA-mRNA-protein pathway expression can be controlled

the production of specific gene products (protein or RNA), and is informally termed *gene regulation*. Sophisticated programs of gene expression are widely observed in biology, for example to trigger developmental pathways, respond to environmental stimuli, or adapt to new food sources. Virtually any step of gene expression can be modulated, from transcriptional initiation, to RNA processing, and to the post-translational modification of a protein.

Gene regulation is essential for viruses, prokaryotes and eukaryotes as it increases the versatility and adaptability of an organism by allowing the cell to express protein when needed. Although as early as 1951, Barbara McClintock showed interaction between two genetic loci, Activator (*Ac*) and Dissociator (*Ds*), in the color formation of maize seeds, the first discovery of a gene regulation system is widely considered to be the identification in 1961 of the *lac* operon, discovered by Jacques Monod, in which some enzymes involved in lactose metabolism are expressed by *E. coli* only in the presence of lactose and absence of glucose.

Furthermore, in multicellular organisms, gene regulation drives the processes of cellular differentiation and morphogenesis, leading to the creation of different cell types that possess different gene expression profiles, and hence produce different proteins/have different ultrastructures that suit them to their functions (though they all possess the genotype, which follows the same genome sequence).

The initiating event leading to a change in gene expression include activation or deactivation of receptors. Also, there is evidence that changes in a cell's choice of catabolism leads to altered gene expressions.[1]

23.1   Regulated stages of gene expression

Any step of gene expression may be modulated, from the DNA-RNA transcription step to post-translational modification of a protein. The following is a list of stages where gene expression is regulated, the most extensively utilised point is Transcription Initiation:

- Chromatin domains

- Transcription

- Post-transcriptional modification

- RNA transport

- Translation

- mRNA degradation

23.2   Modification of DNA

In eukaryotes, the accessibility of large regions of DNA can depend on its chromatin structure, which can be altered as a result of histone modifications directed by DNA methylation, ncRNA, or DNA-binding protein. Hence these modifications may up or down regulate the expression of a gene. Some of these modifications that regulate gene expression are inheritable and are referred to as epigenetic regulation.

23.2.1   Structural

Transcription of DNA is dictated by its structure. In general, the density of its packing is indicative of the frequency of transcription. Octameric protein complexes called nucleosomes are responsible for the amount of supercoiling of DNA, and these complexes can be temporarily modified by processes such as phosphorylation or more permanently modified by processes such as methylation. Such modifications are considered to be responsible for more or less permanent changes in gene expression levels.[2]

23.2.2 Chemical

Methylation of DNA is a common method of gene silencing. DNA is typically methylated by methyltransferase enzymes on cytosine nucleotides in a CpG dinucleotide sequence (also called "CpG islands" when densely clustered). Analysis of the pattern of methylation in a given region of DNA (which can be a promoter) can be achieved through a method called bisulfite mapping. Methylated cytosine residues are unchanged by the treatment, whereas unmethylated ones are changed to uracil. The differences are analyzed by DNA sequencing or by methods developed to quantify SNPs, such as Pyrosequencing (Biotage) or MassArray (Sequenom), measuring the relative amounts of C/T at the CG dinucleotide. Abnormal methylation patterns are thought to be involved in oncogenesis.[3]

Histone acetylation is also an important process in transcription. Histone acetyltransferase enzymes (HATs) such as CREB-binding protein also dissociate the DNA from the histone complex, allowing transcription to proceed. Often, DNA methylation and histone deacetylation work together in gene silencing. The combination of the two seems to be a signal for DNA to be packed more densely, lowering gene expression.

23.3 Regulation of transcription

Main article: Transcriptional regulation
Regulation of transcription thus controls when transcription occurs and how much RNA is created. Transcription of a

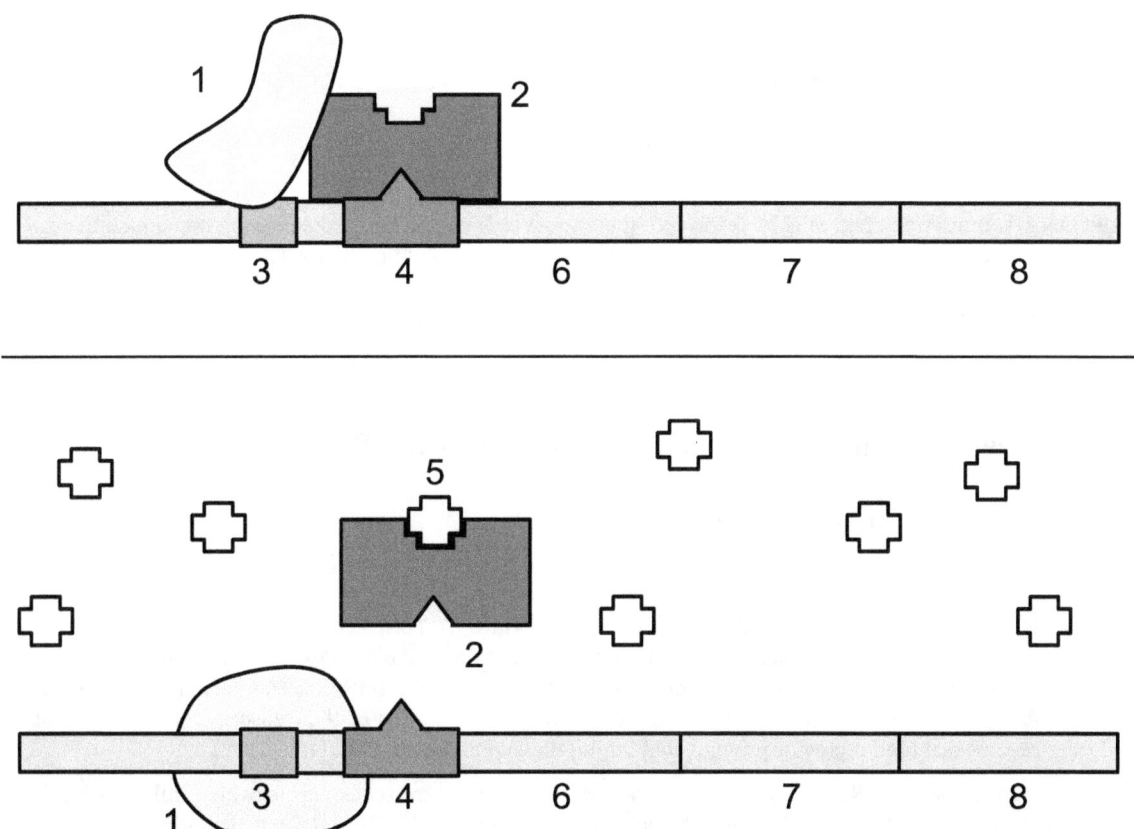

1: *RNA Polymerase,* **2**: *Repressor,* **3**: *Promoter,* **4**: *Operator,* **5**: *Lactose,* **6**: *lacZ,* **7**: *lacY,* **8**: *lacA.* **Top**: *The gene is essentially turned off. There is no lactose to inhibit the repressor, so the repressor binds to the operator, which obstructs the RNA polymerase from binding to the promoter and making lactase.* **Bottom**: *The gene is turned on. Lactose is inhibiting the repressor, allowing the RNA polymerase to bind with the promoter, and express the genes, which synthesize lactase. Eventually, the lactase will digest all of the lactose, until there is none to bind to the repressor. The repressor will then bind to the operator, stopping the manufacture of lactase.*

gene by RNA polymerase can be regulated by at least five mechanisms:

- **Specificity factors** alter the specificity of RNA polymerase for a given promoter or set of promoters, making it more or less likely to bind to them (i.e., sigma factors used in prokaryotic transcription).

- **Repressors** bind to the **Operator**, coding sequences on the DNA strand that are close to or overlapping the promoter region, impeding RNA polymerase's progress along the strand, thus impeding the expression of the gene. The image to the right demonstrates regulation by a repressor in the lac operon.

- **General transcription factors** position RNA polymerase at the start of a protein-coding sequence and then release the polymerase to transcribe the mRNA.

- **Activators** enhance the interaction between RNA polymerase and a particular promoter, encouraging the expression of the gene. Activators do this by increasing the attraction of RNA polymerase for the promoter, through interactions with subunits of the RNA polymerase or indirectly by changing the structure of the DNA.

- **Enhancers** are sites on the DNA helix that are bound by activators in order to loop the DNA bringing a specific promoter to the initiation complex. Enhancers are much more common in eukaryotes than prokaryotes, where only a few examples exist (to date).[4]

- **Silencers** are regions of DNA sequences that, when bound by particular transcription factors, can silence expression of the gene.

23.4   Post-transcriptional regulation

Main article: Post-transcriptional regulation

After the DNA is transcribed and mRNA is formed, there must be some sort of regulation on how much the mRNA is translated into proteins. Cells do this by modulating the capping, splicing, addition of a Poly(A) Tail, the sequence-specific nuclear export rates, and, in several contexts, sequestration of the RNA transcript. These processes occur in eukaryotes but not in prokaryotes. This modulation is a result of a protein or transcript that, in turn, is regulated and may have an affinity for certain sequences.

23.5   Three prime untranslated regions and microRNAs

Main article: Three prime untranslated region
Main article: MicroRNA

Three prime untranslated regions (3'-UTRs) of messenger RNAs (mRNAs) often contain regulatory sequences that post-transcriptionally influence gene expression. Such 3'-UTRs often contain both binding sites for microRNAs (miRNAs) as well as for regulatory proteins. By binding to specific sites within the 3'-UTR, miRNAs can decrease gene expression of various mRNAs by either inhibiting translation or directly causing degradation of the transcript. The 3'-UTR also may have silencer regions that bind repressor proteins that inhibit the expression of a mRNA.

The 3'-UTR often contains miRNA response elements (MREs). MREs are sequences to which miRNAs bind. These are prevalent motifs within 3'-UTRs. Among all regulatory motifs within the 3'-UTRs (e.g. including silencer regions), MREs make up about half of the motifs.

As of 2014, the miRBase web site,[5] an archive of miRNA sequences and annotations, listed 28,645 entries in 233 biologic species. Of these, 1,881 miRNAs were in annotated human miRNA loci. miRNAs were predicted to have an average of about four hundred target mRNAs (affecting expression of several hundred genes).[6] Freidman et al.[6] estimate that >45,000 miRNA target sites within human mRNA 3'-UTRs are conserved above background levels, and >60% of human protein-coding genes have been under selective pressure to maintain pairing to miRNAs.

Direct experiments show that a single miRNA can reduce the stability of hundreds of unique mRNAs.[7] Other experiments show that a single miRNA may repress the production of hundreds of proteins, but that this repression often is relatively mild (less than 2-fold).[8][9]

The effects of miRNA dysregulation of gene expression seem to be important in cancer.[10] For instance, in gastrointestinal cancers, a 2015 paper identified nine miRNAs as epigenetically altered and effective in down-regulating DNA repair enzymes.[11]

The effects of miRNA dysregulation of gene expression also seem to be important in neuropsychiatric disorders, such as schizophrenia, bipolar disorder, major depressive disorder, Parkinson's disease, Alzheimer's disease and autism spectrum disorders.[12][13][14]

23.6 Regulation of translation

Main article: Translational regulation

The translation of mRNA can also be controlled by a number of mechanisms, mostly at the level of initiation. Recruitment of the small ribosomal subunit can indeed be modulated by mRNA secondary structure, antisense RNA binding, or protein binding. In both prokaryotes and eukaryotes, a large number of RNA binding proteins exist, which often are directed to their target sequence by the secondary structure of the transcript, which may change depending on certain conditions, such as temperature or presence of a ligand (aptamer). Some transcripts act as ribozymes and self-regulate their expression.

23.7 Examples of gene regulation

- Enzyme induction is a process in which a molecule (e.g., a drug) induces (i.e., initiates or enhances) the expression of an enzyme.

- The induction of heat shock proteins in the fruit fly *Drosophila melanogaster*.

- The Lac operon is an interesting example of how gene expression can be regulated.

- Viruses, despite having only a few genes, possess mechanisms to regulate their gene expression, typically into an early and late phase, using collinear systems regulated by anti-terminators (lambda phage) or splicing modulators (HIV).

- GAL4 is a transcriptional activator that controls the expression of GAL1, GAL7, and GAL10 (all of which code for the metabolic of galactose in yeast). The GAL4/UAS system has been used in a variety of organisms across various phyla to study gene expression.[15]

23.7.1 Developmental biology

Main article: morphogen

A large number of studied regulatory systems come from developmental biology. Examples include:

- The colinearity of the Hox gene cluster with their nested antero-posterior patterning

- It has been speculated that pattern generation of the hand (digits - interdigits) The gradient of Sonic hedgehog (secreted inducing factor) from the zone of polarizing activity in the limb, which creates a gradient of active Gli3, which activates Gremlin, which inhibits BMPs also secreted in the limb, resulting in the formation of an alternating pattern of activity as a result of this reaction-diffusion system.

- Somitogenesis is the creation of segments (somites) from a uniform tissue (Pre-somitic Mesoderm, PSM). They are formed sequentially from anterior to posterior. This is achieved in amniotes possibly by means of two opposing gradients, Retinoic acid in the anterior (wavefront) and Wnt and Fgf in the posterior, coupled to an oscillating pattern (segmentation clock) composed of FGF + Notch and Wnt in antiphase.[16]

- Sex determination in the soma of a Drosophila requires the sensing of the ratio of autosomal genes to sex chromosome-encoded genes, which results in the production of sexless splicing factor in females, resulting in the female isoform of doublesex.[17]

23.8   Circuitry

Main article: Gene regulatory network

23.8.1   Up-regulation and down-regulation

Up-regulation is a process that occurs within a cell triggered by a signal (originating internal or external to the cell), which results in increased expression of one or more genes and as a result the protein(s) encoded by those genes. On the converse, **down-regulation** is a process resulting in decreased gene and corresponding protein expression.

- Up-regulation occurs, for example, when a cell is deficient in some kind of receptor. In this case, more receptor protein is synthesized and transported to the membrane of the cell and, thus, the sensitivity of the cell is brought back to normal, reestablishing homeostasis.

- Down-regulation occurs, for example, when a cell is overstimulated by a neurotransmitter, hormone, or drug for a prolonged period of time, and the expression of the receptor protein is decreased in order to protect the cell (see also tachyphylaxis).

23.8.2   Inducible vs. repressible systems

Gene Regulation can be summarized by the response of the respective system:

- Inducible systems - An inducible system is off unless there is the presence of some molecule (called an inducer) that allows for gene expression. The molecule is said to "induce expression". The manner by which this happens is dependent on the control mechanisms as well as differences between prokaryotic and eukaryotic cells.

- Repressible systems - A repressible system is on except in the presence of some molecule (called a corepressor) that suppresses gene expression. The molecule is said to "repress expression". The manner by which this happens is dependent on the control mechanisms as well as differences between prokaryotic and eukaryotic cells.

The GAL4/UAS system is an example of both an inducible and repressible system. GAL4 binds an upstream activation sequence (UAS) to activate the transcription of the GAL1/GAL7/GAL10 cassette. On the other hand, a MIG1 response to the presence of glucose can inhibit GAL4 and therefore stop the expression of the GAL1/GAL7/GAL10 cassette.[18]

23.8.3   Theoretical circuits

- Repressor/Inducer: an activation of a sensor results in the change of expression of a gene

- negative feedback: the gene product downregulates its own production directly or indirectly, which can result in

  - keeping transcript levels constant/proportional to a factor

- inhibition of run-away reactions when coupled with a positive feedback loop

    - creating an oscillator by taking advantage in the time delay of transcription and translation, given that the mRNA and protein half-life is shorter

- positive feedback: the gene product upregulates its own production directly or indirectly, which can result in

    - signal amplification

    - bistable switches when two genes inhibit each other and both have positive feedback

    - pattern generation

23.9   Study methods

For DNA and RNA methods, see nucleic acid methods.
For protein methods, see protein methods.

In general, most experiments investigating differential expression used whole cell extracts of RNA, called steady-state levels, to determine which genes changed and by how much they did. These are, however, not informative of where the regulation has occurred and may actually mask conflicting regulatory processes (*see post-transcriptional regulation*), but it is still the most commonly analysed (quantitative PCR and DNA microarray).

When studying gene expression, there are several methods to look at the various stages. In eukaryotes these include:

- The local chromatin environment of the region can be determined by ChIP-chip analysis by pulling down RNA Polymerase II, Histone 3 modifications, Trithorax-group protein, Polycomb-group protein, or any other DNA-binding element to which a good antibody is available.

- Epistatic interactions can be investigated by synthetic genetic array analysis

- Due to post-transcriptional regulation, transcription rates and total RNA levels differ significantly. To measure the transcription rates nuclear run-on assays can be done and newer high-throughput methods are being developed, using thiol labelling instead of radioactivity.[19]

- Only 5% of the RNA polymerised in the nucleus actually exits,[20] and not only introns, abortive products, and non-sense transcripts are degradated. Therefore, the differences in nuclear and cytoplasmic levels can be see by separating the two fractions by gentle lysis.[21]

- Alternative splicing can be analysed with a splicing array or with a tiling array (*see DNA microarray*).

- All in vivo RNA is complexed as RNPs. The quantity of transcripts bound to specific protein can be also analysed by RIP-Chip. For example, DCP2 will give an indication of sequestered protein; ribosome-bound gives and indication of transcripts active in transcription (although it should be noted that a more dated method, called polysome fractionation, is still popular in some labs)

- Protein levels can be analysed by Mass spectrometry, which can be compared only to quantitative PCR data, as microarray data is relative and not absolute.

- RNA and protein degradation rates are measured by means of transcription inhibitors (actinomycin D or α-amanitin) or translation inhibitors (Cycloheximide), respectively.

23.10   See also

- Enhancer (genetics)

- Artificial transcription factors (small molecules that mimic transcription factor protein)

- Cellular model

- Conserved non-coding DNA sequence

- Spatiotemporal gene expression

23.11   Notes and references

[1] Pereira SL, Rodrigues AS, Sousa MI, Correia M, Perestrelo T, Ramalho-Santos J (2014). "From gametogenesis and stem cells to cancer: common metabolic themes". *Human Reproduction Update* **20** (6): 924–43. doi:10.1093/humupd/dmu034. PMID 25013216.

[2] Bell JT, Pai AA, Pickrell JK, Gaffney DJ, Pique-Regi R, Degner JF, Gilad Y, Pritchard JK (2011). "DNA methylation patterns associate with genetic and gene expression variation in HapMap cell lines". *Genome Biology* **12** (1): R10. doi:10.1186/gb-2011-12-1-r10. PMC 3091299. PMID 21251332.

[3] Vertino PM, Spillare EA, Harris CC, Baylin SB (Apr 1993). "Altered chromosomal methylation patterns accompany oncogene-induced transformation of human bronchial epithelial cells" (PDF). *Cancer Research* **53** (7): 1684–9. PMID 8453642.

[4] Austin S, Dixon R (Jun 1992). "The prokaryotic enhancer binding protein NTRC has an ATPase activity which is phosphorylation and DNA dependent". *The EMBO Journal* **11** (6): 2219–28. PMC 556689. PMID 1534752.

[5] miRBase.org

[6] Friedman RC, Farh KK, Burge CB, Bartel DP (Jan 2009). "Most mammalian mRNAs are conserved targets of microRNAs". *Genome Research* **19** (1): 92–105. doi:10.1101/gr.082701.108. PMC 2612969. PMID 18955434.

[7] Lim LP, Lau NC, Garrett-Engele P, Grimson A, Schelter JM, Castle J, Bartel DP, Linsley PS, Johnson JM (Feb 2005). "Microarray analysis shows that some microRNAs downregulate large numbers of target mRNAs". *Nature* **433** (7027): 769–73. Bibcode:2005Natur.433..769L. doi:10.1038/nature03315. PMID 15685193.

[8] Selbach M, Schwanhäusser B, Thierfelder N, Fang Z, Khanin R, Rajewsky N (Sep 2008). "Widespread changes in protein synthesis induced by microRNAs". *Nature* **455** (7209): 58–63. doi:10.1038/nature07228. PMID 18668040.

[9] Baek D, Villén J, Shin C, Camargo FD, Gygi SP, Bartel DP (Sep 2008). "The impact of microRNAs on protein output". *Nature* **455** (7209): 64–71. doi:10.1038/nature07242. PMC 2745094. PMID 18668037.

[10] Palmero EI, de Campos SG, Campos M, de Souza NC, Guerreiro ID, Carvalho AL, Marques MM (Jul 2011). "Mechanisms and role of microRNA deregulation in cancer onset and progression". *Genetics and Molecular Biology* **34** (3): 363–70. doi:10.1590/S1415-47572011000300001. PMC 3168173. PMID 21931505.

[11] Bernstein C, Bernstein H (May 2015). "Epigenetic reduction of DNA repair in progression to gastrointestinal cancer". *World Journal of Gastrointestinal Oncology* **7** (5): 30–46. doi:10.4251/wjgo.v7.i5.30. PMC 4434036. PMID 25987950.

[12] Maffioletti E, Tardito D, Gennarelli M, Bocchio-Chiavetto L (2014). "Micro spies from the brain to the periphery: new clues from studies on microRNAs in neuropsychiatric disorders". *Frontiers in Cellular Neuroscience* **8**: 75. doi:10.3389/fncel.2014.00075. PMC 3949217. PMID 24653674.

[13] Mellios N, Sur M (2012). "The Emerging Role of microRNAs in Schizophrenia and Autism Spectrum Disorders". *Frontiers in Psychiatry* **3**: 39. doi:10.3389/fpsyt.2012.00039. PMC 3336189. PMID 22539927.

[14] Geaghan M, Cairns MJ (Aug 2015). "MicroRNA and Posttranscriptional Dysregulation in Psychiatry". *Biological Psychiatry* **78** (4): 231–9. doi:10.1016/j.biopsych.2014.12.009. PMID 25636176.

[15] Barnett, J. A. (2004), A history of research on yeasts 7: enzymic adaptation and regulation. Yeast, 21: 703–746. doi: 10.1002/yea.1113

[16] Dequéant ML, Pourquié O. Segmental patterning of the vertebrate embryonic axis. Nat Rev Genet. 2008 May;9(5):370-82. PMID 18414404

[17] Gilbert SF (2003). Developmental biology, 7th ed., Sunderland, Mass: Sinauer Associates, 65–6. ISBN 0-87893-258-5.

[18] Nehlin JO, Carlberg M, Ronne H (1991). "Control of yeast GAL genes by MIG1 repressor: a transcriptional cascade in the glucose response". *EMBO J.* **10** (11): 3373–7. PMC 453065. PMID 1915298.

[19] Cheadle C, Fan J, Cho-Chung YS, Werner T, Ray J, Do L, Gorospe M, Becker KG (2005). "Control of gene expression during T cell activation: alternate regulation of mRNA transcription and mRNA stability". *BMC Genomics* **6**: 75. doi:10.1186/1471-2164-6-75. PMC 1156890. PMID 15907206.

[20] Jackson DA, Pombo A, Iborra F (Feb 2000). "The balance sheet for transcription: an analysis of nuclear RNA metabolism in mammalian cells". *FASEB Journal* **14** (2): 242–54. PMID 10657981.

[21] Schwanekamp JA, Sartor MA, Karyala S, Halbleib D, Medvedovic M, Tomlinson CR (2006). "Genome-wide analyses show that nuclear and cytoplasmic RNA levels are differentially affected by dioxin". *Biochimica Et Biophysica Acta* **1759** (8-9): 388–402. doi:10.1016/j.bbaexp.2006.07.005. PMID 16962184.

23.12 Bibliography

• Latchman, David S. (2005). *Gene regulation: a eukaryotic perspective*. Psychology Press. ISBN 978-0-415-36510-9.

23.13 External links

Regulation of Gene Expression at the US National Library of Medicine Medical Subject Headings (MeSH)

• Cellular Darwinism

Chapter 24

Chromatin

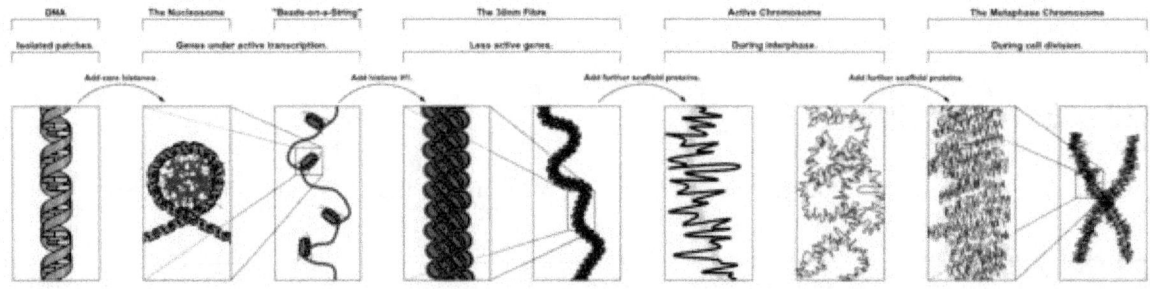

The major structures in DNA compaction: DNA, the nucleosome, the 10 nm "beads-on-a-string" fibre, the 30 nm chromatin fibre and the metaphase chromosome.

Chromatin is a complex of macromolecules found in cells, consisting of DNA, protein, and RNA. The primary functions of chromatin are 1) to package DNA into a smaller volume to fit in the cell, 2) to reinforce the DNA macromolecule to allow mitosis, 3) to prevent DNA damage, and 4) to control gene expression and DNA replication. The primary protein components of chromatin are histones that compact the DNA. Chromatin is only found in eukaryotic cells (cells with defined nuclei). Prokaryotic cells have a different organization of their DNA (the prokaryotic chromosome equivalent is called genophore and is localized within the nucleoid region).

The structure of chromatin depends on several factors. The overall structure depends on the stage of the cell cycle. During interphase, the chromatin is structurally loose to allow access to RNA and DNA polymerases that transcribe and replicate the DNA. The local structure of chromatin during interphase depends on the genes present on the DNA: DNA coding genes that are actively transcribed ("turned on") are more loosely packaged and are found associated with RNA polymerases (referred to as euchromatin) while DNA coding inactive genes ("turned off") are found associated with structural proteins and are more tightly packaged (heterochromatin).[1][2] Epigenetic chemical modification of the structural proteins in chromatin also alters the local chromatin structure, in particular chemical modifications of histone proteins by methylation and acetylation. As the cell prepares to divide, i.e. enters mitosis or meiosis, the chromatin packages more tightly to facilitate segregation of the chromosomes during anaphase. During this stage of the cell cycle this makes the individual chromosomes in many cells visible by optical microscope.

In general terms, there are three levels of chromatin organization:

1. DNA wraps around histone proteins forming nucleosomes; the "beads on a string" structure (euchromatin).

2. Multiple histones wrap into a 30 nm fibre consisting of nucleosome arrays in their most compact form (heterochromatin). (Definitively established to exist in vitro, the 30-nanometer fibre was not seen in recent X-ray studies of human mitotic chromosomes.[3])

3. Higher-level DNA packaging of the 30 nm fibre into the metaphase chromosome (during mitosis and meiosis).

There are, however, many cells that do not follow this organisation. For example, spermatozoa and avian red blood cells have more tightly packed chromatin than most eukaryotic cells, and trypanosomatid protozoa do not condense their chromatin into visible chromosomes for mitosis.

24.1 During interphase

The structure of chromatin during interphase of mitosis is optimized to allow simple access of transcription and DNA repair factors to the DNA while compacting the DNA into the nucleus. The structure varies depending on the access required to the DNA. Genes that require regular access by RNA polymerase require the looser structure provided by euchromatin.

24.2 Dynamic chromatin structure and hierarchy

Chromatin undergoes various structural changes during a cell cycle. Histone proteins are the basic packer and arranger of chromatin and can be modified by various post-translational modifications to alter chromatin packing (Histone modification). Most of the modifications occur on the histone tail. The consequences in terms of chromatin accessibility and compaction depend both on the amino-acid that is modified and the type of modification. For example, Histone acetylation results in loosening and increased accessibility of chromatin for replication and transcription. Lysine tri-methylation can either be correlated with transcriptional activity (tri-methylation of histone H3 Lysine 4) or transcriptional repression and chromatin compaction (tri-methylation of histone H3 Lysine 9 or 27). Several studies suggested that different modifications could occur simultaneously. For example, it was proposed that a bivalent structure (with tri-methylation of both Lysine 4 and 27 on histone H3) was involved in mammalian early development.[4]

Polycomb-group proteins play a role in regulating genes through modulation of chromatin structure.[5]

For additional information, see Histone modifications in chromatin regulation and RNA polymerase control by chromatin structure.

24.2.1 DNA structure

Main articles: Mechanical properties of DNA and Z-DNA

In nature, DNA can form three structures, A-, B-, and Z-DNA. A- and B-DNA are very similar, forming right-handed helices, whereas Z-DNA is a left-handed helix with a zig-zag phosphate backbone. Z-DNA is thought to play a specific role in chromatin structure and transcription because of the properties of the junction between B- and Z-DNA.

At the junction of B- and Z-DNA, one pair of bases is flipped out from normal bonding. These play a dual role of a site of recognition by many proteins and as a sink for torsional stress from RNA polymerase or nucleosome binding.

24.2.2 Nucleosomes and beads-on-a-string

Main articles: Nucleosome, Chromatosome and Histone

The basic repeat element of chromatin is the nucleosome, interconnected by sections of linker DNA, a far shorter arrangement than pure DNA in solution.

In addition to the core histones, there is the linker histone, H1, which contacts the exit/entry of the DNA strand on the nucleosome. The nucleosome core particle, together with histone H1, is known as a chromatosome. Nucleosomes,

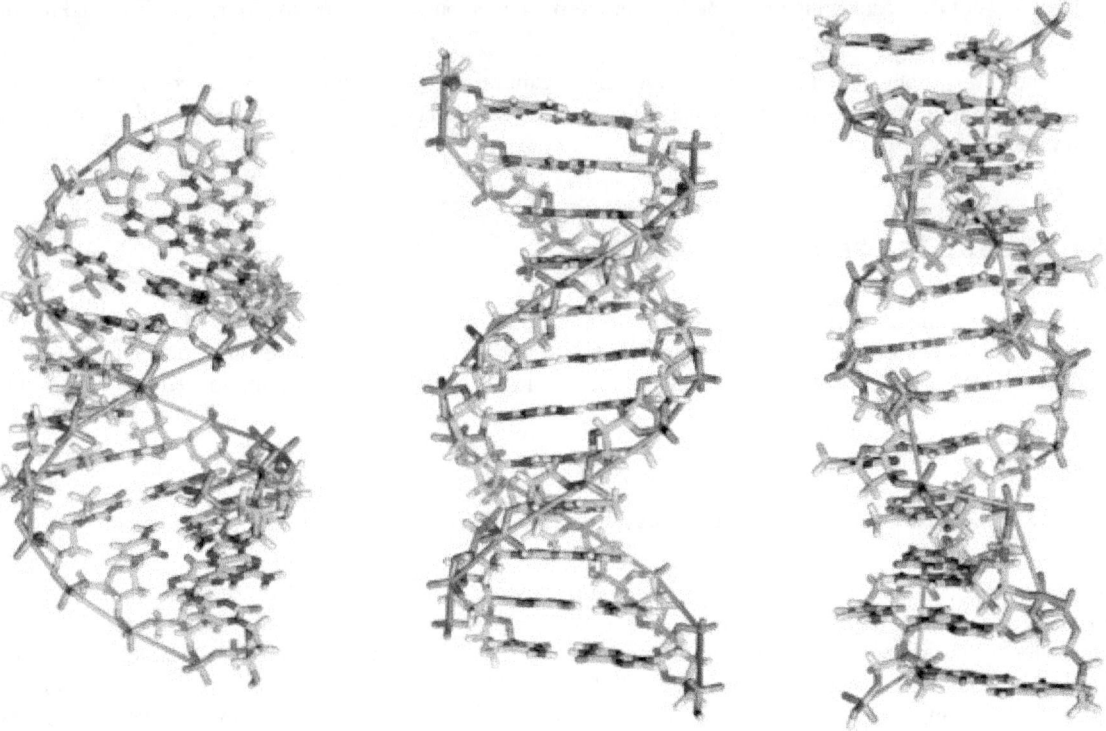

The structures of A-, B-, and Z-DNA.

with about 20 to 60 base pairs of linker DNA, can form, under non-physiological conditions, an approximately 10 nm "beads-on-a-string" fibre. (Fig. 1-2). .

The nucleosomes bind DNA non-specifically, as required by their function in general DNA packaging. There are, however, large DNA sequence preferences that govern nucleosome positioning. This is due primarily to the varying physical properties of different DNA sequences: For instance, adenine and thymine are more favorably compressed into the inner minor grooves. This means nucleosomes can bind preferentially at one position approximately every 10 base pairs (the helical repeat of DNA)- where the DNA is rotated to maximise the number of A and T bases that will lie in the inner minor groove. (See mechanical properties of DNA.)

24.2.3   30 nanometer chromatin fibre

With addition of H1, the beads-on-a-string structure in turn coils into a 30 nm diameter helical structure known as the 30 nm fibre or filament. The precise structure of the chromatin fibre in the cell is not known in detail, and there is still some debate over this.[6]

This level of chromatin structure is thought to be the form of euchromatin, which contains actively transcribed genes. EM studies have demonstrated that the 30 nm fibre is highly dynamic such that it unfolds into a 10 nm fiber ("beads-on-a-string") structure when transversed by an RNA polymerase engaged in transcription.

The existing models commonly accept that the nucleosomes lie perpendicular to the axis of the fibre, with linker histones arranged internally. A stable 30 nm fibre relies on the regular positioning of nucleosomes along DNA. Linker DNA is relatively resistant to bending and rotation. This makes the length of linker DNA critical to the stability of the fibre, requiring nucleosomes to be separated by lengths that permit rotation and folding into the required orientation without excessive stress to the DNA. In this view, different lengths of the linker DNA should produce different folding topologies of the chromatin fiber. Recent theoretical work, based on electron-microscopy images[7] of reconstituted fibers supports this view.[8]

A cartoon representation of the nucleosome structure. From PDB: 1KX5.

24.2.4   Spatial organization of chromatin in the cell nucleus

The spatial arrangement of the chromatin within the nucleus is not random - specific regions of the chromatin can be found in certain territories. Territories are, for example, the lamina-associated domains (LADs), and the topological association domains (TADs), which are bound together by protein complexes.[9] Currently, polymer models such as the Strings & Binders Switch (SBS) model[10] and the Dynamic Loop (DL) model[11] are used to describe the folding of chromatin within the nucleus.

24.3   Chromatin and bursts of transcription

Chromatin and its interaction with enzymes has been researched, and a conclusion being made is that it is relevant and an important factor in gene expression. Vincent G. Allfrey, a professor at Rockefeller University, stated that RNA synthesis

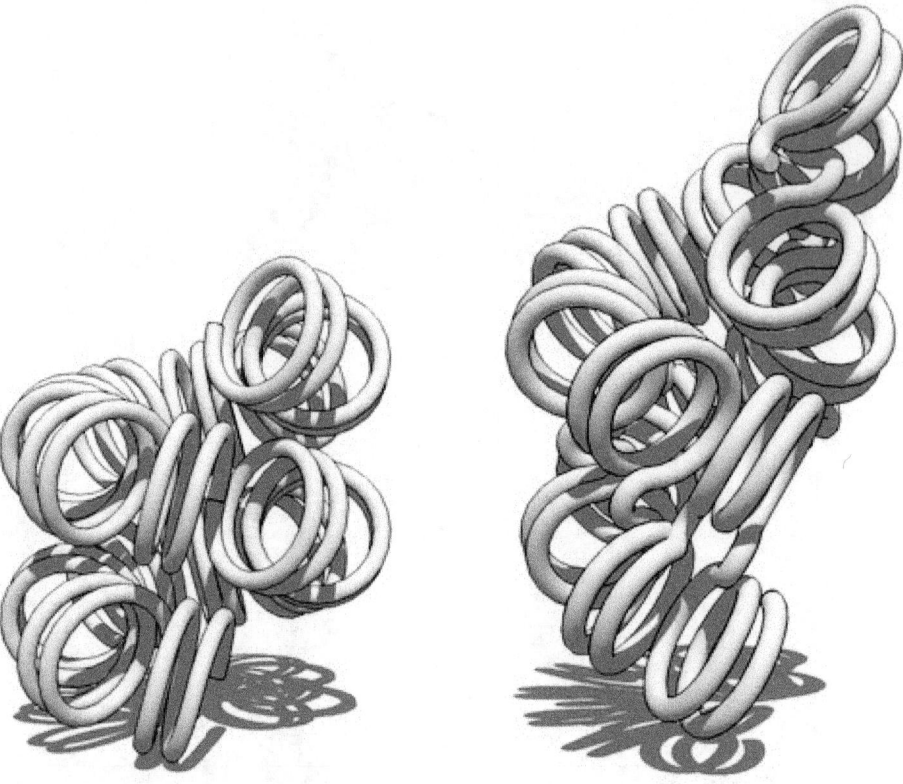

Two proposed structures of the 30nm chromatin filament.
Left: 1 start helix "solenoid" structure.
Right: 2 start loose helix structure.
Note: the histones are omitted in this diagram - only the DNA is shown.

is related to histone acetylation.[12] The lysine amino acid attached to the end of the histones is positively charged. The acetylation of these tails would make the chromatin ends neutral, allowing for DNA access.

When the chromatin decondenses, the DNA is open to entry of molecular machinery. Fluctuations between open and closed chromatin may contribute to the discontinuity of transcription, or transcriptional bursting. Other factors are probably involved, such as the association and dissociation of transcription factor complexes with chromatin. The phenomenon, as opposed to simple probabilistic models of transcription, can account for the high variability in gene expression occurring between cells in isogenic populations[13]

24.4   Metaphase chromatin (chromosomes)

The metaphase structure of chromatin differs vastly to that of interphase. It is optimised for physical strength and manageability, forming the classic chromosome structure seen in karyotypes. The structure of the condensed chromatin is thought to be loops of 30 nm fibre to a central scaffold of proteins. It is, however, not well-characterised.

The physical strength of chromatin is vital for this stage of division to prevent shear damage to the DNA as the daughter chromosomes are separated. To maximise strength the composition of the chromatin changes as it approaches the centromere, primarily through alternative histone H1 anologues.

It should also be noted that, during mitosis, while most of the chromatin is tightly compacted, there are small regions that

Four proposed structures of the 30 nm chromatin filament for DNA repeat length per nucleosomes ranging from 177 to 207 bp. Linker DNA in yellow and nucleosomal DNA in pink.

are not as tightly compacted. These regions often correspond to promoter regions of genes that were active in that cell type prior to entry into chromitosis. The lack of compaction of these regions is called bookmarking, which is an epigenetic mechanism believed to be important for transmitting to daughter cells the "memory" of which genes were active prior to entry into mitosis.[14] This bookmarking mechanism is needed to help transmit this memory because transcription ceases during mitosis.

24.5 Chromatin: alternative definitions

1. **Simple and concise definition:** Chromatin is a macromolecular complex of a DNA macromolecule and protein macromolecules (and RNA). The proteins package and arrange the DNA and control its functions within the cell nucleus.

2. **A biochemists' operational definition:** Chromatin is the DNA/protein/RNA complex extracted from eukaryotic lysed interphase nuclei. Just which of the multitudinous substances present in a nucleus will constitute a part of the extracted material partly depends on the technique each researcher uses. Furthermore, the composition and properties of chromatin vary from one cell type to the another, during development of a specific cell type, and at different stages in the cell cycle.

3. **The *DNA + histone = chromatin* definition:** The DNA double helix in the cell nucleus is packaged by special proteins termed histones. The formed protein/DNA complex is called chromatin. The basic structural unit of chromatin is the nucleosome.

Karyogram of human male using Giemsa staining, showing the classic metaphase chromatin structure.

24.6   Alternative chromatin organizations

During metazoan spermiogenesis, the spermatid's chromatin is remodelled into a more spaced-packaged, widened, almost crystal-like structure. This process is associated with the cessation of transcription and involves nuclear protein exchange. The histones are mostly displaced, and replaced by protamines (small, arginine-rich proteins).[15]

24.7   Nobel Prizes

The following scientists were recognized for their contributions to chromatin research with Nobel Prizes:

24.8   See also

- Chromatid

- Epigenetics

- Histone-Modifying Enzymes

- Position-effect variegation

- Salt-and-pepper chromatin

- Transcriptional bursting

24.9 References

[1] "Chromatin Network Home Page.". Retrieved 2008-11-18.

[2] Dame, R.T. (May 2005). "The role of nucleoid-associated proteins in the organization and compaction of bacterial chromatin". *Molecular Microbiology* **56** (4): 858–870. doi:10.1111/j.1365-2958.2005.04598.x. PMID 15853876.

[3] Hansen, Jeffrey (March 2012). "Human mitotic chromosome structure: what happened to the 30-nm fibre?". *The EMBO Journal* **31** (7): 1621–1623. doi:10.1038/emboj.2012.66. PMC 3321215. PMID 22415369.

[4] Bernstein, B.E., T.S. Mikkelsen, X. Xie, M. Kamal, D.J. Huebert, J. Cuff, B. Fry, A. Meissner, M. Wernig, K. Plath, R. Jaenisch, A. Wagschal, R. Feil, S.L. Schreiber & E.S. Lander (April 2006). "A bivalent chromatin structure marks key developmental genes in embryonic stem cells". *Cell* **125** (2): 315–26. doi:10.1016/j.cell.2006.02.041. ISSN 0092-8674. PMID 16630819.

[5] Portoso M and Cavalli G (2008). "The Role of RNAi and Noncoding RNAs in Polycomb Mediated Control of Gene Expression and Genomic Programming". *RNA and the Regulation of Gene Expression: A Hidden Layer of Complexity*. Caister Academic Press. isbn=978-1-904455-25-7.

[6] Annunziato, Anthony T. "DNA Packaging: Nucleosomes and Chromatin". *Scitable*. Nature Education. Retrieved 2015-10-29.

[7] Robinson DJ, Fairall L, Huynh VA, Rhodes D. (April 2006). "EM measurements define the dimensions of the "30-nm" chromatin fiber: Evidence for a compact, interdigitated structure". *PNAS* **103** (17): 6506–11. doi:10.1073/pnas.0601212103. PMC 1436021. PMID 16617109.

[8] Wong H, Victor JM, Mozziconacci J. (September 2007). Chen, Pu, ed. "An All-Atom Model of the Chromatin Fiber Containing Linker Histones Reveals a Versatile Structure Tuned by the Nucleosomal Repeat Length". *PLoS ONE* **2** (9): e877. doi:10.1371/journal.pone.0000877. PMC 1963316. PMID 17849006.

[9] Nicodemi M, Pombo A (June 2014). "Models of chromosome structure". *Curr. Opin. Cell Biol.* **28**: 90–5. doi:10.1016/j.ceb.2014.04.004. PMID 24804566.

[10] Nicodemi M, Panning B, Prisco A (May 2008). "A thermodynamic switch for chromosome colocalization". *Genetics* **179** (1): 717–21. doi:10.1534/genetics.107.083154. PMC 2390650. PMID 18493085.

[11] Bohn M, Heermann DW (2010). "Diffusion-driven looping provides a consistent framework for chromatin organization". *PLoS ONE* **5** (8): e12218. doi:10.1371/journal.pone.0012218. PMC 2928267. PMID 20811620.

[12] ALLFREY VG, FAULKNER R, MIRSKY AE (May 1964). "ACETYLATION AND METHYLATION OF HISTONES AND THEIR POSSIBLE ROLE IN THE REGULATION OF RNA SYNTHESIS". *Proc. Natl. Acad. Sci. U.S.A.* **51** (5): 786–94. doi:10.1073/pnas.51.5.786. PMC 300163. PMID 14172992.

[13] Kaochar S, Tu BP (November 2012). "Gatekeepers of chromatin: Small metabolites elicit big changes in gene expression". *Trends Biochem. Sci.* **37** (11): 477–83. doi:10.1016/j.tibs.2012.07.008. PMC 3482309. PMID 22944281.

[14] Xing H, Vanderford NL, Sarge KD (November 2008). "The TBP-PP2A mitotic complex bookmarks genes by preventing condensin action". *Nat. Cell Biol.* **10** (11): 1318–23. doi:10.1038/ncb1790. PMC 2577711. PMID 18931662.

[15] De Vries M, Ramos L, Housein Z, De Boer P (May 2012). "Chromatin remodelling initiation during human spermiogenesis". *Biol Open* **1** (5): 446–57. doi:10.1242/bio.2012844. PMC 3507207. PMID 23213436.

[16] "Thomas Hunt Morgan and His Legacy". Nobelprize.org. 7 Sep 2012

24.10   Other references

- Cooper, Geoffrey M. 2000. The Cell, 2nd edition, A Molecular Approach. Chapter 4.2 Chromosomes and Chromatin.

- Corces, V. G. (1995). "Chromatin insulators. Keeping enhancers under control". *Nature* **376** (6540): 462–463. doi:10.1038/376462a0.

- Cremer, T. 1985. Von der Zellenlehre zur Chromosomentheorie: Naturwissenschaftliche Erkenntnis und Theorienwechsel in der frühen Zell- und Vererbungsforschung, Veröffentlichungen aus der Forschungsstelle für Theoretische Pathologie der Heidelberger Akademie der Wissenschaften. Springer-Vlg., Berlin, Heidelberg.

- Elgin, S. C. R. (ed.). 1995. Chromatin Structure and Gene Expression, vol. 9. IRL Press, Oxford, New York, Tokyo.

- Gerasimova, T. I.; Corces, V. G. (1996). "Boundary and insulator elements in chromosomes". *Current Op. Genet. and Dev.* **6**: 185–192. doi:10.1016/s0959-437x(96)80049-9.

- Gerasimova, T. I.; Corces, V. G. (1998). "Polycomb and Trithorax group proteins mediate the function of a chromatin insulator". *Cell* **92**: 511–521. doi:10.1016/s0092-8674(00)80944-7.

- Gerasimova, T. I.; Corces, V. G. (2001). "CHROMATIN INSULATORS AND BOUNDARIES: Effects on Transcription and Nuclear Organization". *Annu Rev Genet* **35**: 193–208.

- Gerasimova, T. I.; Byrd, K.; Corces, V. G. (2000). "A chromatin insulator determines the nuclear localization of DNA [In Process Citation]". *Mol Cell* **6**: 1025–35. doi:10.1016/s1097-2765(00)00101-5.

- Ha, S. C.; Lowenhaupt, K.; Rich, A.; Kim, Y. G.; Kim, K. K. (2005). "Crystal structure of a junction between B-DNA and Z-DNA reveals two extruded bases". *Nature* **437**: 1183–6. doi:10.1038/nature04088. PMID 16237447.

- Pollard, T., and W. Earnshaw. 2002. Cell Biology. Saunders.

- Saumweber, H. 1987. Arrangement of Chromosomes in Interphase Cell Nuclei, p. 223-234. In W. Hennig (ed.), Structure and Function of Eucaryotic Chromosomes, vol. 14. Springer-Verlag, Berlin, Heidelberg.

- Sinden, R. R. (2005). "Molecular biology: DNA twists and flips". *Nature* **437**: 1097–8. doi:10.1038/4371097a.

- Van Holde KE. 1989. Chromatin. New York: Springer-Verlag. ISBN 0-387-96694-3.

- Van Holde, K., J. Zlatanova, G. Arents, and E. Moudrianakis. 1995. Elements of chromatin structure: histones, nucleosomes, and fibres, p. 1-26. In S. C. R. Elgin (ed.), Chromatin structure and gene expression. IRL Press at Oxford University Press, Oxford.

24.11   External links

- Chromatin, Histones & Cathepsin; PMAP The Proteolysis Map-animation

- Recent chromatin publications and news

- Protocol for *in vitro* Chromatin Assembly

- ENCODE threads Explorer Chromatin patterns at transcription factor binding sites. Nature (journal)

Chapter 25

Epigenetics

For the unfolding of an organism or the hypothesis that plants and animals (including humans) develop in this way, see epigenesis (biology). For epigenetics in robotics, see developmental robotics. For the scientific journal, see Epigenetics (journal). For earth science concepts labelled *epigenetic*, see epigenetic (earth sciences).

Epigenetics is the study, in the field of genetics, of cellular and physiological phenotypic trait variations that are caused

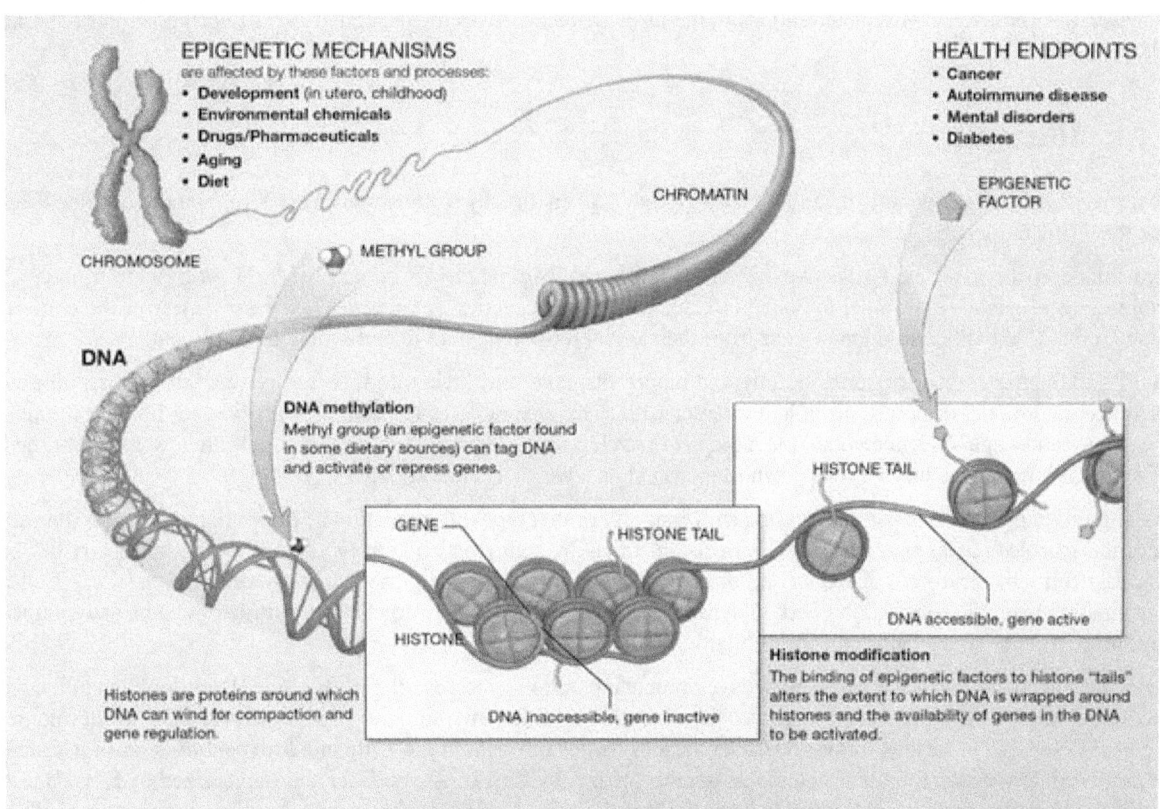

Epigenetic mechanisms

by external or environmental factors that switch genes on and off and affect how cells *read* genes instead of being caused by changes in the DNA sequence.[1][2] Hence, epigenetic research seeks to describe dynamic alterations in the transcriptional potential of a cell. These alterations may or may not be heritable, although the use of the term "epigenetic" to describe processes that are not heritable is controversial.[3] Unlike genetics based on changes to the DNA sequence (the genotype), the changes in gene expression or cellular phenotype of epigenetics have other causes, thus use of the prefix *epi*- (Greek:

ἐπί- over, outside of, around).[4][5]

The term also refers to the changes themselves: functionally relevant changes to the genome that do not involve a change in the nucleotide sequence. Examples of mechanisms that produce such changes are DNA methylation and histone modification, each of which alters how genes are expressed without altering the underlying DNA sequence. Gene expression can be controlled through the action of repressor proteins that attach to silencer regions of the DNA. These epigenetic changes may last through cell divisions for the duration of the cell's life, and may also last for multiple generations even though they do not involve changes in the underlying DNA sequence of the organism;[6] instead, non-genetic factors cause the organism's genes to behave (or "express themselves") differently.[7]

One example of an epigenetic change in eukaryotic biology is the process of cellular differentiation. During morphogenesis, totipotent stem cells become the various pluripotent cell lines of the embryo, which in turn become fully differentiated cells. In other words, as a single fertilized egg cell – the zygote – continues to divide, the resulting daughter cells change into all the different cell types in an organism, including neurons, muscle cells, epithelium, endothelium of blood vessels, etc., by activating some genes while inhibiting the expression of others.[8]

25.1   Definitions

The term *epigenetics* in its contemporary usage emerged in the 1990s, but for some years has been used in somewhat variable meanings. A consensus definition of the concept of *epigenetic trait* as "stably heritable phenotype resulting from changes in a chromosome without alterations in the DNA sequence" was formulated at a Cold Spring Harbor meeting in 2008.

25.1.1   Historical

The term *epigenesis* has a generic meaning "extra growth", taken directly from Koine Greek ἐπιγέννησις, used in English since the 17th century.[9]

From this, and the associated adjective *epigenetic*, the term *epigenetics* was coined by C. H. Waddington in 1942 as pertaining to *epigenesis* in parallel to Valentin Haecker's 'phenogenetics' (*Pänogenetik*).[10] *Epigenesis* in the context of biology refers to the differentiation of cells from their initial totipotent state in embryonic development.[11]

When Waddington coined the term the physical nature of genes and their role in heredity was not known; he used it as a conceptual model of how genes might interact with their surroundings to produce a phenotype; he used the phrase "epigenetic landscape" as a metaphor for biological development. Waddington held that cell fates were established in development much like a marble rolls down to the point of lowest local elevation.[12]

Waddington suggested visualising increasing irreversibility of cell type differentiation as ridges rising between the valleys where the marbles (cells) are travelling.[13] In recent times Waddington's notion of the epigenetic landscape has been rigorously formalized in the context of the systems dynamics state approach to the study of cell-fate.[14][15] Cell-fate determination is predicted to exhibit certain dynamics, such as attractor-convergence (the attractor can be an equilibrium point, limit cycle or strange attractor) or oscillatory.[15]

The term "epigenetic" has also been used in developmental psychology to describe psychological development as the result of an ongoing, bi-directional interchange between heredity and the environment.[16] Interactivist ideas of development have been discussed in various forms and under various names throughout the 19th and 20th centuries. An early version was proposed, among the founding statements in embryology, by Karl Ernst von Baer and popularized by Ernst Haeckel. A radical epigenetic view (physiological epigenesis) was developed by Paul Wintrebert. Another variation, probabilistic epigenesis, was presented by Gilbert Gottlieb in 2003.[17] This view encompasses all of the possible developing factors on an organism and how they not only influence the organism and each other, but how the organism also influences its own development.

The developmental psychologist Erik Erikson used the term *epigenetic principle* in his book *Identity: Youth and Crisis* (1968), and used it to encompass the notion that we develop through an unfolding of our personality in predetermined stages, and that our environment and surrounding culture influence how we progress through these stages. This biological unfolding in relation to our socio-cultural settings is done in stages of psychosocial development, where "progress through each stage is in part determined by our success, or lack of success, in all the previous stages."[18][19][20]

25.1.2 Contemporary

Robin Holliday defined epigenetics as "the study of the mechanisms of temporal and spatial control of gene activity during the development of complex organisms."[21] Thus *epigenetic* can be used to describe anything other than DNA sequence that influences the development of an organism.

The more recent usage of the word in science has a stricter definition. It is, as defined by Arthur Riggs and colleagues, "the study of mitotically and/or meiotically heritable changes in gene function that cannot be explained by changes in DNA sequence."[22] The Greek prefix *epi-* in *epigenetics* implies features that are "on top of" or "in addition to" genetics; thus *epigenetic* traits exist on top of or in addition to the traditional molecular basis for inheritance.[23]

The term "epigenetics", however, has been used to describe processes which have not been demonstrated to be heritable such as histone modification; there are therefore attempts to redefine it in broader terms that would avoid the constraints of requiring heritability. For example, Sir Adrian Bird defined epigenetics as *"the structural adaptation of chromosomal regions so as to register, signal or perpetuate altered activity states."*[6] This definition would be inclusive of transient modifications associated with DNA repair or cell-cycle phases as well as stable changes maintained across multiple cell generations, but exclude others such as templating of membrane architecture and prions unless they impinge on chromosome function. Such redefinitions however are not universally accepted and are still subject to dispute.[3] The NIH "Roadmap Epigenomics Project," ongoing as of 2013, uses the following definition: *"...For purposes of this program, epigenetics refers to both heritable changes in gene activity and expression (in the progeny of cells or of individuals) and also stable, long-term alterations in the transcriptional potential of a cell that are not necessarily heritable."*[24]

In 2008, a consensus definition of the epigenetic trait, "stably heritable phenotype resulting from changes in a chromosome without alterations in the DNA sequence", was made at a Cold Spring Harbor meeting.[25]

The similarity of the word to "genetics" has generated many parallel usages. The "epigenome" is a parallel to the word "genome", referring to the overall epigenetic state of a cell, and epigenomics refers to more global analyses of epigenetic changes across the entire genome.[24] The phrase "genetic code" has also been adapted—the "epigenetic code" has been used to describe the set of epigenetic features that create different phenotypes in different cells. Taken to its extreme, the "epigenetic code" could represent the total state of the cell, with the position of each molecule accounted for in an *epigenomic map*, a diagrammatic representation of the gene expression, DNA methylation and histone modification status of a particular genomic region. More typically, the term is used in reference to systematic efforts to measure specific, relevant forms of epigenetic information such as the histone code or DNA methylation patterns.

25.2 Molecular basis

Epigenetic changes modify the activation of certain genes, but not the genetic code sequence of DNA. The microstructure (not code) of DNA itself or the associated chromatin proteins may be modified, causing activation or silencing. This mechanism enables differentiated cells in a multicellular organism to express only the genes that are necessary for their own activity. Epigenetic changes are preserved when cells divide. Most epigenetic changes only occur within the course of one individual organism's lifetime; however, if gene inactivation occurs in a sperm or egg cell that results in fertilization, then some epigenetic changes can be transferred to the next generation.[26] This raises the question of whether or not epigenetic changes in an organism can alter the basic structure of its DNA (see Evolution, below), a form of Lamarckism.

Specific epigenetic processes include paramutation, bookmarking, imprinting, gene silencing, X chromosome inactivation, position effect, reprogramming, transvection, maternal effects, the progress of carcinogenesis, many effects of teratogens, regulation of histone modifications and heterochromatin, and technical limitations affecting parthenogenesis and cloning.

DNA damage can also cause epigenetic changes.[27][28][29] DNA damages are very frequent, occurring on average about 60,000 times a day per cell of the human body (see DNA damage (naturally occurring)). These damages are largely repaired, but at the site of a DNA repair, epigenetic changes can remain.[30] In particular, a double strand break in DNA can initiate unprogrammed epigenetic gene silencing both by causing DNA methylation as well as by promoting silencing types of histone modifications (chromatin remodeling) (see next section).[31] In addition, the enzyme Parp1 (poly(ADP)-

ribose polymerase) and its product poly(ADP)-ribose (PAR) accumulate at sites of DNA damage as part of a repair process.[32] This accumulation, in turn, directs recruitment and activation of the chromatin remodeling protein ALC1 that can cause nucleosome remodeling.[33] Nucleosome remodeling has been found to cause, for instance, epigenetic silencing of DNA repair gene MLH1.[22][34] DNA damaging chemicals, such as benzene, hydroquinone, styrene, carbon tetrachloride and trichloroethylene, cause considerable hypomethylation of DNA, some through the activation of oxidative stress pathways.[35]

Foods are known to alter the epigenetics of rats on different diets.[36] Some food components epigenetically increase the levels of DNA repair enzymes such as MGMT and MLH1[37] and p53.[38][39] Other food components can reduce DNA damage, such as soy isoflavones[40][41] and bilberry anthocyanins.[42]

Epigenetic research uses a wide range of molecular biologic techniques to further our understanding of epigenetic phenomena, including chromatin immunoprecipitation (together with its large-scale variants ChIP-on-chip and ChIP-Seq), fluorescent in situ hybridization, methylation-sensitive restriction enzymes, DNA adenine methyltransferase identification (DamID) and bisulfite sequencing. Furthermore, the use of bioinformatic methods is playing an increasing role (computational epigenetics).

Computer simulations and molecular dynamics approaches revealed the atomistic motions associated with the molecular recognition of the histone tail through an allosteric mechanism.[43]

25.3 Mechanisms

Several types of epigenetic inheritance systems may play a role in what has become known as cell memory,[44] note however that not all of these are universally accepted to be examples of epigenetics.

25.3.1 Covalent modifications

Covalent modifications of either DNA (e.g. cytosine methylation and hydroxymethylation) or of histone proteins (e.g. lysine acetylation, lysine and arginine methylation, serine and threonine phosphorylation, and lysine ubiquitination and sumoylation) play central roles in many types of epigenetic inheritance. Therefore, the word "epigenetics" is sometimes used as a synonym for these processes. However, this can be misleading. Chromatin remodeling is not always inherited, and not all epigenetic inheritance involves chromatin remodeling.[45]

Because the phenotype of a cell or individual is affected by which of its genes are transcribed, heritable transcription states can give rise to epigenetic effects. There are several layers of regulation of gene expression. One way that genes are regulated is through the remodeling of chromatin. Chromatin is the complex of DNA and the histone proteins with which it associates. If the way that DNA is wrapped around the histones changes, gene expression can change as well. Chromatin remodeling is accomplished through two main mechanisms:

1. The first way is post translational modification of the amino acids that make up histone proteins. Histone proteins are made up of long chains of amino acids. If the amino acids that are in the chain are changed, the shape of the histone might be modified. DNA is not completely unwound during replication. It is possible, then, that the modified histones may be carried into each new copy of the DNA. Once there, these histones may act as templates, initiating the surrounding new histones to be shaped in the new manner. By altering the shape of the histones around them, these modified histones would ensure that a lineage-specific transcription program is maintained after cell division.

2. The second way is the addition of methyl groups to the DNA, mostly at CpG sites, to convert cytosine to 5-methylcytosine. 5-Methylcytosine performs much like a regular cytosine, pairing with a guanine in double-stranded DNA. However, some areas of the genome are methylated more heavily than others, and highly methylated areas tend to be less transcriptionally active, through a mechanism not fully understood. Methylation of cytosines can also persist from the germ line of one of the parents into the zygote, marking the chromosome as being inherited from one parent or the other (genetic imprinting).

DNA associates with histone proteins to form chromatin.

Mechanisms of heritability of histone state are not well understood; however, much is known about the mechanism of heritability of DNA methylation state during cell division and differentiation. Heritability of methylation state depends on certain enzymes (such as DNMT1) that have a higher affinity for 5-methylcytosine than for cytosine. If this enzyme reaches a "hemimethylated" portion of DNA (where 5-methylcytosine is in only one of the two DNA strands) the enzyme will methylate the other half.

Although histone modifications occur throughout the entire sequence, the unstructured N-termini of histones (called histone tails) are particularly highly modified. These modifications include acetylation, methylation, ubiquitylation, phosphorylation, sumoylation, ribosylation and citrullination. Acetylation is the most highly studied of these modifications. For example, acetylation of the K14 and K9 lysines of the tail of histone H3 by histone acetyltransferase enzymes (HATs) is generally related to transcriptional competence.

One mode of thinking is that this tendency of acetylation to be associated with "active" transcription is biophysical in nature. Because it normally has a positively charged nitrogen at its end, lysine can bind the negatively charged phosphates of the DNA backbone. The acetylation event converts the positively charged amine group on the side chain into a neutral amide linkage. This removes the positive charge, thus loosening the DNA from the histone. When this occurs, complexes like SWI/SNF and other transcriptional factors can bind to the DNA and allow transcription to occur. This is the "cis" model of epigenetic function. In other words, changes to the histone tails have a direct effect on the DNA itself.

Another model of epigenetic function is the "trans" model. In this model, changes to the histone tails act indirectly on the DNA. For example, lysine acetylation may create a binding site for chromatin-modifying enzymes (or transcription machinery as well). This chromatin remodeler can then cause changes to the state of the chromatin. Indeed, a bromodomain — a protein domain that specifically binds acetyl-lysine — is found in many enzymes that help activate transcription, including the SWI/SNF complex. It may be that acetylation acts in this and the previous way to aid in transcriptional activation.

The idea that modifications act as docking modules for related factors is borne out by histone methylation as well. Methylation of lysine 9 of histone H3 has long been associated with constitutively transcriptionally silent chromatin (constitutive heterochromatin). It has been determined that a chromodomain (a domain that specifically binds methyl-lysine) in the transcriptionally repressive protein HP1 recruits HP1 to K9 methylated regions. One example that seems to refute this biophysical model for methylation is that tri-methylation of histone H3 at lysine 4 is strongly associated with (and required for full) transcriptional activation. Tri-methylation in this case would introduce a fixed positive charge on the tail.

It has been shown that the histone lysine methyltransferase (KMT) is responsible for this methylation activity in the pattern of histones H3 & H4. This enzyme utilizes a catalytically active site called the SET domain (Suppressor of variegation, Enhancer of zeste, Trithorax). The SET domain is a 130-amino acid sequence involved in modulating gene activities. This domain has been demonstrated to bind to the histone tail and causes the methylation of the histone.[46]

Differing histone modifications are likely to function in differing ways; acetylation at one position is likely to function differently from acetylation at another position. Also, multiple modifications may occur at the same time, and these modifications may work together to change the behavior of the nucleosome. The idea that multiple dynamic modifications regulate gene transcription in a systematic and reproducible way is called the histone code, although the idea that histone state can be read linearly as a digital information carrier has been largely debunked. One of the best-understood systems that orchestrates chromatin-based silencing is the SIR protein based silencing of the yeast hidden mating type loci HML and HMR.

DNA methylation frequently occurs in repeated sequences, and helps to suppress the expression and mobility of 'transposable elements':[47] Because 5-methylcytosine can be spontaneously deaminated (replacing nitrogen by oxygen) to thymidine, CpG sites are frequently mutated and become rare in the genome, except at CpG islands where they remain unmethylated. Epigenetic changes of this type thus have the potential to direct increased frequencies of permanent genetic mutation. DNA methylation patterns are known to be established and modified in response to environmental factors by a complex interplay of at least three independent DNA methyltransferases, DNMT1, DNMT3A, and DNMT3B, the loss of any of which is lethal in mice.[48] DNMT1 is the most abundant methyltransferase in somatic cells,[49] localizes to replication foci,[50] has a 10–40-fold preference for hemimethylated DNA and interacts with the proliferating cell nuclear antigen (PCNA).[51]

By preferentially modifying hemimethylated DNA, DNMT1 transfers patterns of methylation to a newly synthesized strand after DNA replication, and therefore is often referred to as the 'maintenance' methyltransferase.[52] DNMT1 is essential for proper embryonic development, imprinting and X-inactivation.[48][53] To emphasize the difference of this molecular mechanism of inheritance from the canonical Watson-Crick base-pairing mechanism of transmission of genetic information, the term 'Epigenetic templating' was introduced.[54] Furthermore, in addition to the maintenance and transmission of methylated DNA states, the same principle could work in the maintenance and transmission of histone modifications and even cytoplasmic (structural) heritable states.[55]

Histones H3 and H4 can also be manipulated through demethylation using histone lysine demethylase (KDM). This recently identified enzyme has a catalytically active site called the Jumonji domain (JmjC). The demethylation occurs when JmjC utilizes multiple cofactors to hydroxylate the methyl group, thereby removing it. JmjC is capable of demethylating mono-, di-, and tri-methylated substrates.[56]

Chromosomal regions can adopt stable and heritable alternative states resulting in bistable gene expression without changes to the DNA sequence. Epigenetic control is often associated with alternative covalent modifications of histones.[57] The stability and heritability of states of larger chromosomal regions are suggested to involve positive feedback where modified nucleosomes recruit enzymes that similarly modify nearby nucleosomes.[58] A simplified stochastic model for this type of epigenetics is found here.[59][60]

It has been suggested that chromatin-based transcriptional regulation could be mediated by the effect of small RNAs. Small interfering RNAs can modulate transcriptional gene expression via epigenetic modulation of targeted promoters.[61]

25.3.2 RNA transcripts

Sometimes a gene, after being turned on, transcribes a product that (directly or indirectly) maintains the activity of that gene. For example, Hnf4 and MyoD enhance the transcription of many liver- and muscle-specific genes, respectively, including their own, through the transcription factor activity of the proteins they encode. RNA signalling includes differential recruitment of a hierarchy of generic chromatin modifying complexes and DNA methyltransferases to specific loci by RNAs during differentiation and development.[62] Other epigenetic changes are mediated by the production of different splice forms of RNA, or by formation of double-stranded RNA (RNAi). Descendants of the cell in which the gene was turned on will inherit this activity, even if the original stimulus for gene-activation is no longer present. These genes are often turned on or off by signal transduction, although in some systems where syncytia or gap junctions are important, RNA may spread directly to other cells or nuclei by diffusion. A large amount of RNA and protein is contributed to the zygote by the mother during oogenesis or via nurse cells, resulting in maternal effect phenotypes. A smaller quantity of sperm RNA is transmitted from the father, but there is recent evidence that this epigenetic information can lead to visible changes in several generations of offspring.[63]

25.3.3 MicroRNAs

MicroRNAs (miRNAs) are members of non-coding RNAs that range in size from 17 to 25 nucleotides. miRNAs regulate a large variety of biological functions in plants and animals.[64] So far, in 2013, about 2000 miRNAs have been discovered in humans and these can be found online in an miRNA database.[65] Each miRNA expressed in a cell may target about 100 to 200 messenger RNAs that it downregulates.[66] Most of the downregulation of mRNAs occurs by causing the decay of the targeted mRNA, while some downregulation occurs at the level of translation into protein.[67]

It appears that about 60% of human protein coding genes are regulated by miRNAs.[68] Many miRNAs are epigenetically regulated. About 50% of miRNA genes are associated with CpG islands,[64] that may be repressed by epigenetic methylation. Transcription from methylated CpG islands is strongly and heritably repressed.[69] Other miRNAs are epigenetically regulated by either histone modifications or by combined DNA methylation and histone modification.[64]

25.3.4 mRNA

In 2011, it was demonstrated that the methylation of mRNA plays a critical role in human energy homeostasis. The obesity-associated FTO gene is shown to be able to demethylate N6-methyladenosine in RNA.[70][71]

25.3.5 sRNAs

sRNAs are small (50–250 nucleotides), highly structured, non-coding RNA fragments found in bacteria. They control gene expression including virulence genes in pathogens and are viewed as new targets in the fight against drug-resistant bacteria.[72] They play an important role in many biological processes, binding to mRNA and protein targets in prokaryotes. Their phylogenetic analyses, for example through sRNA–mRNA target interactions or protein binding properties, are used to build comprehensive databases.[73] sRNA-gene maps based on their targets in microbial genomes are also constructed.[74]

25.3.6 Prions

For more details on this topic, see Fungal prions.

Prions are infectious forms of proteins. In general, proteins fold into discrete units that perform distinct cellular functions, but some proteins are also capable of forming an infectious conformational state known as a prion. Although often viewed in the context of infectious disease, prions are more loosely defined by their ability to catalytically convert other native state versions of the same protein to an infectious conformational state. It is in this latter sense that they can be viewed as epigenetic agents capable of inducing a phenotypic change without a modification of the genome.[75]

Fungal prions are considered by some to be epigenetic because the infectious phenotype caused by the prion can be inherited without modification of the genome. PSI+ and URE3, discovered in yeast in 1965 and 1971, are the two best studied of this type of prion.[76][77] Prions can have a phenotypic effect through the sequestration of protein in aggregates, thereby reducing that protein's activity. In PSI+ cells, the loss of the Sup35 protein (which is involved in termination of translation) causes ribosomes to have a higher rate of read-through of stop codons, an effect that results in suppression of nonsense mutations in other genes.[78] The ability of Sup35 to form prions may be a conserved trait. It could confer an adaptive advantage by giving cells the ability to switch into a PSI+ state and express dormant genetic features normally terminated by stop codon mutations.[79][80][81][82]

25.3.7   Structural inheritance

For more details on this topic, see Structural inheritance.

In ciliates such as *Tetrahymena* and *Paramecium*, genetically identical cells show heritable differences in the patterns of ciliary rows on their cell surface. Experimentally altered patterns can be transmitted to daughter cells. It seems existing structures act as templates for new structures. The mechanisms of such inheritance are unclear, but reasons exist to assume that multicellular organisms also use existing cell structures to assemble new ones.[83][84][85]

25.3.8   Nucleosome positioning

Eukaryotic genomes have numerous nucleosomes. Nucleosome position is not random, and determine the accessibility of DNA to regulatory proteins. This determines differences in gene expression and cell differentiation. It has been shown that at least some nucleosomes are retained in sperm cells (where most but not all histones are replaced by protamines). Thus nucleosome positioning is to some degree inheritable. Recent studies have uncovered connections between nucleosome positioning and other epigenetic factors, such as DNA methylation and hydroxymethylation [86]

25.4   Functions and consequences

25.4.1   Development

Developmental epigenetics can be divided into predetermined and probabilistic epigenesis. Predetermined epigenesis is a unidirectional movement from structural development in DNA to the functional maturation of the protein. "Predetermined" here means that development is scripted and predictable. Probabilistic epigenesis on the other hand is a bidirectional structure-function development with experiences and external molding development.[87]

Somatic epigenetic inheritance, particularly through DNA and histone covalent modifications and nucleosome repositioning, is very important in the development of multicellular eukaryotic organisms.[86] The genome sequence is static (with some notable exceptions), but cells differentiate into many different types, which perform different functions, and respond differently to the environment and intercellular signalling. Thus, as individuals develop, morphogens activate or silence genes in an epigenetically heritable fashion, giving cells a memory. In mammals, most cells terminally differentiate, with only stem cells retaining the ability to differentiate into several cell types ("totipotency" and "multipotency"). In mammals, some stem cells continue producing new differentiated cells throughout life, such as in neurogenesis, but mammals are not able to respond to loss of some tissues, for example, the inability to regenerate limbs, which some other animals are capable of. Epigenetic modifications regulate the transition from neural neural stem cells to glial progenitor cells (for example, differentiation into oligodendrocytes is regulated by the deacetylation and methylation of histones.[88]

Unlike animals, plant cells do not terminally differentiate, remaining totipotent with the ability to give rise to a new individual plant. While plants do utilise many of the same epigenetic mechanisms as animals, such as chromatin remodeling, it has been hypothesised that some kinds of plant cells do not use or require "cellular memories", resetting their gene expression patterns using positional information from the environment and surrounding cells to determine their fate.[89]

Epigenetic changes can occur in response to environmental exposure—for example, mice given some dietary supplements have epigenetic changes affecting expression of the agouti gene, which affects their fur color, weight, and propensity to develop cancer.[90][91]

Controversial results from one study suggested that traumatic experiences might produce an epigenetic signal that is capable of being passed to future generations. Mice were trained, using foot shocks, to fear a cherry blossom odor. The investigators reported that the mouse offspring had an increased aversion to this specific odor.[92][93] They suggested epigenetic changes that increase gene expression, rather than in DNA itself, in a gene, M71, that governs the functioning of an odor receptor in the nose that responds specifically to this cherry blossom smell. There were physical changes that correlated with olfactory (smell) function in the brains of the trained mice and their descendants. Several criticisms were reported, including the study's low statistical power as evidence of some irregularity such as bias in reporting results.[94] Due to limits of sample size, there is a probability that an effect will not be demonstrated to within statistical significance even if it exists. The criticism suggested that the probability that all the experiments reported would show positive results if an identical protocol was followed, assuming the claimed effects exist, is merely 0.4%. The authors also did not indicate which mice were siblings, and treated all of the mice as statistically independent.[95] The original researchers pointed out negative results in the paper's appendix that the criticism omitted in its calculations, and undertook to track which mice were siblings in the future.[96]

25.4.2 Transgenerational

Main article: Transgenerational epigenetics

Epigenetics can impact evolution when epigenetic changes are heritable.[1] A sequestered germ line or Weismann barrier is specific to animals, and epigenetic inheritance is more common in plants and microbes. Eva Jablonka, Marion J. Lamb and Étienne Danchin have argued that these effects may require enhancements to the standard conceptual framework of the modern synthesis and have called for an extended evolutionary synthesis.[97][98][99] Other evolutionary biologists have incorporated epigenetic inheritance into population genetics models and are openly skeptical, stating that epigenetic mechanisms such as DNA methylation and histone modification are genetically inherited under the control of natural selection.[100][101][102]

Two important ways in which epigenetic inheritance can be different from traditional genetic inheritance, with important consequences for evolution, are that rates of epimutation can be much faster than rates of mutation[103] and the epimutations are more easily reversible.[104] In plants heritable DNA methylation mutations are 100.000 times more likely to occur compared to DNA mutations.[105] An epigenetically inherited element such as the PSI+ system can act as a "stop-gap", good enough for short-term adaptation that allows the lineage to survive for long enough for mutation and/or recombination to genetically assimilate the adaptive phenotypic change.[106] The existence of this possibility increases the evolvability of a species.

More than 100 cases of transgenerational epigenetic inheritance phenomena have been reported in a wide range of organisms, including prokaryotes, plants, and animals.[107] For instance, Mourning Cloak butterflies will change color through hormone changes in response to experimentation of varying temperatures.[108] Bacteria make widespread use of postreplicative DNA methylation for the epigenetic control of DNA-protein interactions. Bacteria make use of DNA adenine methylation (rather than DNA cytosine methylation) as an epigenetic signal. DNA adenine methylation is important in bacteria virulence in organisms such as *Escherichia coli*, *Salmonella*, *Vibrio*, *Yersinia*, *Haemophilus*, and *Brucella*. In *Alphaproteobacteria*, methylation of adenine regulates the cell cycle and couples gene transcription to DNA replication. In *Gammaproteobacteria*, adenine methylation provides signals for DNA replication, chromosome segregation, mismatch repair, packaging of bacteriophage, transposase activity and regulation of gene expression.[109][110]

The filamentous fungus *Neurospora crassa* is a prominent model system for understanding the control and function of cytosine methylation. In this organisms, DNA methylation is associated with relics of a genome defense system called RIP (repeat-induced point mutation) and silences gene expression by inhibiting transcription elongation.[111]

The yeast prion PSI is generated by a conformational change of a translation termination factor, which is then inherited by daughter cells. This can provide a survival advantage under adverse conditions. This is an example of epigenetic regulation enabling unicellular organisms to respond rapidly to environmental stress. Prions can be viewed as epigenetic agents capable of inducing a phenotypic change without modification of the genome.[110]

Direct detection of epigenetic marks in microorganisms is possible with single molecule real time sequencing, in which polymerase sensitivity allows for measuring methylation and other modifications as a DNA molecule is being sequenced.[112] Several projects have demonstrated the ability to collect genome-wide epigenetic data in bacteria.[113][114][115][116]

25.5 Medicine

Epigenetics has many and varied potential medical applications.[117] In 2008, the National Institutes of Health announced that $190 million had been earmarked for epigenetics research over the next five years. In announcing the funding, government officials noted that epigenetics has the potential to explain mechanisms of aging, human development, and the origins of cancer, heart disease, mental illness, as well as several other conditions. Some investigators, like Randy Jirtle, PhD, of Duke University Medical Center, think epigenetics may ultimately turn out to have a greater role in disease than genetics.[118]

25.5.1 Twins

Direct comparisons of identical twins constitute an optimal model for interrogating environmental epigenetics. In the case of humans with different environmental exposures, monozygotic (identical) twins were epigenetically indistinguishable during their early years, while older twins had remarkable differences in the overall content and genomic distribution of 5-methylcytosine DNA and histone acetylation.[1] The twin pairs who had spent less of their lifetime together and/or had greater differences in their medical histories were those who showed the largest differences in their levels of 5-methylcytosine DNA and acetylation of histones H3 and H4.[119]

Dizygotic (fraternal) and monozygotic (identical) twins show evidence of epigenetic influence in humans.[119][120][121] DNA sequence differences that would be abundant in a singleton-based study do not interfere with the analysis. Environmental differences can produce long-term epigenetic effects, and different developmental monozygotic twin subtypes may be different with respect to their susceptibility to be discordant from an epigenetic point of view.[122]

A high-throughput study, which denotes technology that looks at extensive genetic markers, focused on epigenetic differences between monozygotic twins to compare global and locus-specific changes in DNA methylation and histone modifications in a sample of 40 monozygotic twin pairs.[119] In this case, only healthy twin pairs were studied, but a wide range of ages was represented, between 3 and 74 years. One of the major conclusions from this study was that there is an age-dependent accumulation of epigenetic differences between the two siblings of twin pairs. This accumulation suggests the existence of epigenetic "drift".

A more recent study, where 114 monozygotic twins and 80 dizygotic twins were analyzed for the DNA methylation status of around 6000 unique genomic regions, concluded that epigenetic similarity at the time of blastocyst splitting may also contribute to phenotypic similarities in monozygotic co-twins. This supports the notion that microenvironment at early stages of embryonic development can be quite important for the establishment of epigenetic marks.[123] Congenital genetic disease is well understood and it is clear that epigenetics can play a role, for example, in the case of Angelman syndrome and Prader-Willi syndrome. These are normal genetic diseases caused by gene deletions or inactivation of the genes, but are unusually common because individuals are essentially hemizygous because of genomic imprinting, and therefore a single gene knock out is sufficient to cause the disease, where most cases would require both copies to be knocked out.[124]

25.5.2 Genomic imprinting

Some human disorders are associated with genomic imprinting, a phenomenon in mammals where the father and mother contribute different epigenetic patterns for specific genomic loci in their germ cells.[125] The best-known case of imprinting in human disorders is that of Angelman syndrome and Prader-Willi syndrome—both can be produced by the same genetic mutation, chromosome 15q partial deletion, and the particular syndrome that will develop depends on whether the mutation is inherited from the child's mother or from their father.[126] This is due to the presence of genomic imprinting in the region. Beckwith-Wiedemann syndrome is also associated with genomic imprinting, often caused by abnormalities in maternal genomic imprinting of a region on chromosome 11.

Rett syndrome is underlied by mutations in the MECP2 gene despite no large-scale changes in expression of MeCP2 being found in microarray analyses. BDNF is downregulated in the MECP2 mutant resulting in Rett syndrome.

In the Överkalix study, paternal (but not maternal) grandsons[127] of Swedish men who were exposed during preadolescence to famine in the 19th century were less likely to die of cardiovascular disease. If food was plentiful, then diabetes mortality in the grandchildren increased, suggesting that this was a transgenerational epigenetic inheritance.[128] The opposite effect was observed for females—the paternal (but not maternal) granddaughters of women who experienced famine while in the womb (and therefore while their eggs were being formed) lived shorter lives on average.[129]

25.5.3 Cancer

For more details on this topic, see Cancer epigenetics.

A variety of epigenetic mechanisms can be perturbed in different types of cancer. Epigenetic alterations of DNA repair genes or cell cycle control genes are very frequent in sporadic (non-germ line) cancers, being significantly more common than germ line (familial) mutations in these sporadic cancers.[130][131] Epigenetic alterations are important in cellular transformation to cancer, and their manipulation holds great promise for cancer prevention, detection, and therapy.[132][133] Several medications which have epigenetic impact are used in several of these diseases. These aspects of epigenetics are addressed in cancer epigenetics.

25.6 Research

The two forms of heritable information, namely genetic and epigenetic, are collectively denoted as dual inheritance. Members of the APOBEC/AID family of cytosine deaminases may concurrently influence genetic and epigenetic inheritance using similar molecular mechanisms, and may be a point of crosstalk between these conceptually compartmentalized processes.[134]

Fluoroquinolone antibiotics induce epigenetic changes in mammalian cells through iron chelation. This leads to epigenetic effects through inhibition of α-ketoglutarate-dependent dioxygenases that require iron as a co-factor.[135]

Various pharmacological agents are applied for the production of induced pluripotent stem cells (iPSC) or maintain the embryonic stem cell (ESC) phenotypic via epigenetic approach. Adult stem cells like bone marrow stem cells have also shown a potential to differentiate into cardiac competent cells when treated with G9a histone methyltransferase inhibitor BIX01294.[136][137]

25.7 Caution

Due to the early stages of epigenetics as a science and to the sensationalism surrounding it, surgical oncologist David Gorski and geneticist Adam Rutherford caution against the drawing and proliferation of false and pseudoscientific conclusions from new age authors such as Deepak Chopra and Bruce Lipton.[138][139] Lysenkoism also refutes these theories.

25.8 In popular culture

In Neal Stephenson's 2015 novel Seveneves, survivors of a worldwide holocaust are tasked with seeding new life on a dormant Earth. Rather than create specific breeds of animals to be hunters, scavengers, or prey, species like "canids" are developed with mutable epigenetic traits, with the intention that the animals would quickly transform into the necessary roles that would be required for an ecosystem to quickly evolve. Additionally, a race of humans, "Moirans," are created to survive in space, with the hope that this subspecies of human would be able to adapt to unforeseeable dangers and circumstances, via an epigenetic process called "going epi".

25.9   See also

- B chromosome

- Baldwin effect

- Behavioral epigenetics

- Computational epigenetics

- Emergenesis

- Epigenetic therapy

- Evolutionary capacitance

- Extranuclear inheritance

- Hologenome theory of evolution

- Human genome

- Lamarckism

- Lysenkoism

- Molecular biology

- Molecular pathology

- Nutriepigenomics

- Position-effect variegation

- Preformationism

- Somatic epitype

- Synthetic genetic array

- Weismann barrier

25.10   References

[1] Moore, David S. (2015). *The Developing Genome: An Introduction to Behavioral Epigenetics* (1st ed.). Oxford University Press. ISBN 978-0199922345.

[2] "Epigenetics". Icahn School of Medicine at Mount Sinai. Retrieved 26 May 2015.

[3] Ledford H (2008). "Disputed definitions". *Nature* **455** (7216): 1023–8. doi:10.1038/4551023a. PMID 18948925.

[4] Spector, Tim (2012). *Identically Different: Why You Can Change Your Genes*. London: Weidenfeld & Nicolson. p. 8. Just over ten years ago researchers found that the diets of pregnant mothers could alter the behaviour of genes in their children and that these changes could last a lifetime and then be passed on in turn to their children. The genes were literally being switched on or off by a new mechanism we call epigenetics – meaning in Greek 'around the gene'. Contrary to traditional genetic dogma, these changes could be transferred to the next generation. In this case the mothers just happened to be rats, but recent similar findings in humans have created a revolution in our thinking.

[5] Carey N. (2011): Epigenetics revolution: How modern biology is rewriting our understanding of genetics, disease and inheritance. Icon Books, London, ISBN 978-1-84831-315-6; ISBN 978-1-84831-316-3.

[6] Bird A (May 2007). "Perceptions of epigenetics". *Nature* **447** (7143): 396–8. Bibcode:2007Natur.447..396B. doi:10.1038/nature05913. PMID 17522671.

[7] "Special report: 'What genes remember' by Philip Hunter | Prospect Magazine May 2008 issue 146". Web.archive.org. 1 May 2008. Retrieved 26 July 2012.

[8] Reik W (May 2007). "Stability and flexibility of epigenetic gene regulation in mammalian development". *Nature* **447** (7143): 425–32. Bibcode:2007Natur.447..425R. doi:10.1038/nature05918. PMID 17522676.

[9] Oxford English Dictionary: "The word is used by W. Harvey, *Exercitationes* 1651, p. 148, and in the *English Anatomical Exercitations* 1653, p. 272. It is explained to mean 'partium super-exorientium additamentum', 'the additament of parts budding one out of another'."

[10] Waddington CH (1942). "The epigenotype". *Endeavour* **1**: 18–20. "For the purpose of a study of inheritance, the relation between phenotypes and genotypes [...] is, from a wider biological point of view, of crucial importance, since it is the kernel of the whole problem of development. Many geneticists have recognized this and attempted to discover the processes involved in the mechanism by which the genes of the genotype bring about phenotypic effects. The first step in such an enterprise is – or rather should be, since it is often omitted by those with an undue respect for the powers of reason – to describe what can be seen of the developmental processes. For enquiries of this kind, the word 'phenogenetics' was coined by Haecker [1918, *Phänogenetik*]. The second and more important part of the task is to discover the causal mechanisms at work, and to relate them as far as possible to what experimental embryology has already revealed of the mechanics of development. We might use the name 'epigenetics' for such studies, thus emphasizing their relation to the concepts, so strongly favourable to the classical theory of epigenesis, which have been reached by the experimental embryologists. We certainly need to remember that between genotype and phenotype, and connecting them to each other, there lies a whole complex of developmental processes. It is convenient to have a name for this complex: 'epigenotype' seems suitable."

[11] See *preformationism* for historical background. Oxford English Dictionary: "the theory that the germ is brought into existence (by successive accretions), and not merely developed, in the process of reproduction. [...] The opposite theory was formerly known as the 'theory of evolution'; to avoid the ambiguity of this name, it is now spoken of chiefly as the 'theory of preformation', sometimes as that of 'encasement' or 'emboîtement'."

[12] C. H. Waddington (1953). *The Epigenetics of Birds*. Cambridge University Press. pp. 1–. ISBN 978-1-107-44047-0. (2014 edition)

[13] Hall BK (15 January 2004). "In search of evolutionary developmental mechanisms: the 30-year gap between 1944 and 1974". *302* **302** (1): 5–18. doi:10.1002/jez.b.20002. PMID 14760651.

[14] Alvarez-Buylla ER, Chaos A, Aldana M, Benítez M, Cortes-Poza Y, Espinosa-Soto C, Hartasánchez DA, Lotto RB, Malkin D, Escalera Santos GJ, Padilla-Longoria P (November 3, 2008). "Floral Morphogenesis: Stochastic Explorations of a Gene Network Epigenetic Landscape.". *PLoS ONE* **3**: e3626. Bibcode:2008PLoSO...3.3626A. doi:10.1371/journal.pone.0003626. PMID 18978941.

[15] Rabajante JF, Babierra AL (January 30, 2015). "Branching and oscillations in the epigenetic landscape of cell-fate determination". *Progress in Biophysics and Molecular Biology*. doi:10.1016/j.pbiomolbio.2015.01.006. PMID 25641423.

[16] Gottlieb G (1991). "Epigenetic systems view of human development". *Developmental Psychology* **27** (1): 33–34. doi:10.1037/0012-1649.27.1.33.

[17] Gilbert Gottlieb. Probabilistic epigenesis, Developmental Science 10:1 (2007), 1–11

[18] Boeree, C. George, (1997/2006), *Personality Theories, Erik Erikson*

[19] Erikson, Erik (1968). *Identity: Youth and Crisis*. Chapter 3: W.W. Norton and Company, Inc. p. 92.

[20] "Epigenetics". Bio-Medicine.org. Retrieved 21 May 2011.

[21] Holliday R (Jan 30, 1990). "DNA Methylation and Epigenetic Inheritance". *Philosophical Transactions of the Royal Society of London. Series B, Biological Sciences* **326** (1235): 329–338. Bibcode:1990RSPTB.326..329H. doi:10.1098/rstb.1990.0015.

[22] Riggs AD, Russo VEA, Martienssen RA (1996). *Epigenetic mechanisms of gene regulation*. Plainview, N.Y.: Cold Spring Harbor Laboratory Press. ISBN 0-87969-490-4.

[23] Beware the pseudo gene genies Beware the pseudo gene genies The Guardian

[24] "Overview". *NIH Roadmap Epigenomics Project.*

[25] Berger SL, Kouzarides T, Shiekhattar R, Shilatifard A (2009). "An operational definition of epigenetics". *Genes Dev.* **23** (7): 781–3. doi:10.1101/gad.1787609. PMC 3959995. PMID 19339683.

[26] Chandler VL (February 2007). "Paramutation: from maize to mice". *Cell* **128** (4): 641–5. doi:10.1016/j.cell.2007.02.007. PMID 17320501.

[27] Kovalchuk O, Baulch JE (January 2008). "Epigenetic changes and nontargeted radiation effects—is there a link?". *Environ. Mol. Mutagen.* **49** (1): 16–25. doi:10.1002/em.20361. PMID 18172877.

[28] Ilnytskyy Y, Kovalchuk O (September 2011). "Non-targeted radiation effects-an epigenetic connection". *Mutat. Res.* **714** (1–2): 113–25. doi:10.1016/j.mrfmmm.2011.06.014. PMID 21784089.

[29] Friedl AA, Mazurek B, Seiler DM (2012). "Radiation-induced alterations in histone modification patterns and their potential impact on short-term radiation effects". *Front Oncol* **2**: 117. doi:10.3389/fonc.2012.00117. PMC 3445916. PMID 23050241.

[30] Cuozzo C, Porcellini A, Angrisano T, Morano A, Lee B, Di Pardo A, Messina S, Iuliano R, Fusco A, Santillo MR, Muller MT, Chiariotti L, Gottesman ME, Avvedimento EV (July 2007). "DNA damage, homology-directed repair, and DNA methylation". *PLoS Genet.* **3** (7): e110. doi:10.1371/journal.pgen.0030110. PMC 1913100. PMID 17616978.

[31] O'Hagan HM, Mohammad HP, Baylin SB (2008). Lee JT, ed. "Double strand breaks can initiate gene silencing and SIRT1-dependent onset of DNA methylation in an exogenous promoter CpG island". *PLoS Genet.* **4** (8): e1000155. doi:10.1371/journal.pgen.1000155 PMC 2491723. PMID 18704159.

[32] Malanga M, Althaus FR (2005). "The role of poly(ADP-ribose) in the DNA damage signaling network". *Biochem Cell Biol* **83** (3): 354–364. doi:10.1139/o05-038. PMID 15959561.

[33] Gottschalk AJ, Timinszky G, Kong SE, Jin J, Cai Y, Swanson SK, Washburn MP, Florens L, Ladurner AG, Conaway JW, Conaway RC (August 2009). "Poly(ADP-ribosyl)ation directs recruitment and activation of an ATP-dependent chromatin remodeler". *Proc. Natl. Acad. Sci. U.S.A.* **106** (33): 13770–4. Bibcode:2009PNAS..10613770G. doi:10.1073/pnas.0906920106. PMC 2722505. PMID 19666485.

[34] Lin JC, Jeong S, Liang G, Takai D, Fatemi M, Tsai YC, Egger G, Gal-Yam EN, Jones PA (November 2007). "Role of nucleosomal occupancy in the epigenetic silencing of the MLH1 CpG island". *Cancer Cell* **12** (5): 432–44. doi:10.1016/j.ccr.2007.10.014. PMID 17996647.

[35] Tabish AM, Poels K, Hoet P, Godderis L (2012). Chiariotti L, ed. "Epigenetic factors in cancer risk: effect of chemical carcinogens on global DNA methylation pattern in human TK6 cells". *PLoS ONE* **7** (4): e34674. Bibcode:2012PLoSO...734674T. doi:10.1371/journal.pone.0034674. PMC 3324488. PMID 22509344.

[36] Burdge GC, Hoile SP, Uller T, Thomas NA, Gluckman PD, Hanson MA, Lillycrop KA (2011). Imhof A, ed. "Progressive, transgenerational changes in offspring phenotype and epigenotype following nutritional transition". *PLoS ONE* **6** (11): e28282. Bibcode:2011PLoSO...628282B. doi:10.1371/journal.pone.0028282. PMC 3227644. PMID 22140567.

[37] Fang M, Chen D, Yang CS (January 2007). "Dietary polyphenols may affect DNA methylation". *J. Nutr.* **137** (1 Suppl): 223S–228S. PMID 17182830.

[38] Olaharski AJ, Rine J, Marshall BL, Babiarz J, Zhang L, Verdin E, Smith MT (December 2005). "The flavoring agent dihydrocoumarin reverses epigenetic silencing and inhibits sirtuin deacetylases". *PLoS Genet.* **1** (6): e77. doi:10.1371/journal.pgen.0010077. PMC 1315280. PMID 16362078.

[39] Kikuno N, Shiina H, Urakami S, Kawamoto K, Hirata H, Tanaka Y, Majid S, Igawa M, Dahiya R (August 2008). "Genistein mediated histone acetylation and demethylation activates tumor suppressor genes in prostate cancer cells". *Int. J. Cancer* **123** (3): 552–60. doi:10.1002/ijc.23590. PMID 18431742.

[40] Davis JN, Kucuk O, Djuric Z, Sarkar FH (June 2001). "Soy isoflavone supplementation in healthy men prevents NF-kappa B activation by TNF-alpha in blood lymphocytes". *Free Radic. Biol. Med.* **30** (11): 1293–302. doi:10.1016/S0891-5849(01)00535-4. PMID 11368927.

[41] Djuric Z, Chen G, Doerge DR, Heilbrun LK, Kucuk O (October 2001). "Effect of soy isoflavone supplementation on markers of oxidative stress in men and women". *Cancer Lett.* **172** (1): 1–6. doi:10.1016/S0304-3835(01)00627-9. PMID 11595123.

[42] Kropat C, Mueller D, Boettler U, Zimmermann K, Heiss EH, Dirsch VM, Rogoll D, Melcher R, Richling E, Marko D (March 2013). "Modulation of Nrf2-dependent gene transcription by bilberry anthocyanins in vivo". *Mol Nutr Food Res* **57** (3): 545–50. doi:10.1002/mnfr.201200504. PMID 23349102.

[43] Baron R, Vellore NA (2012). "LSD1/CoREST is an allosteric nanoscale clamp regulated by H3-histone-tail molecular recognition". *Proc Natl Acad Sci U S A.* **109** (31): 12509–14. Bibcode:2012PNAS..10912509B. doi:10.1073/pnas.1207892109. PMC 3411975. PMID 22802671.

[44] Jablonka E, Lamb MJ, Lachmann M (September 1992). "Evidence, mechanisms and models for the inheritance of acquired characteristics". *J. Theor. Biol.* **158** (2): 245–268. doi:10.1016/S0022-5193(05)80722-2.

[45] Ptashne M (April 2007). "On the use of the word 'epigenetic'". *Curr. Biol.* **17** (7): R233–6. doi:10.1016/j.cub.2007.02.030. PMID 17407749.

[46] Jenuwein T, Laible G, Dorn R, Reuter G (January 1998). "SET domain proteins modulate chromatin domains in eu- and heterochromatin". *Cell. Mol. Life Sci.* **54** (1): 80–93. doi:10.1007/s000180050127. PMID 9487389.

[47] Slotkin RK, Martienssen R (April 2007). "Transposable elements and the epigenetic regulation of the genome". *Nature Reviews Genetics* **8** (4): 272–85. doi:10.1038/nrg2072. PMID 17363976.

[48] Li E, Bestor TH, Jaenisch R (June 1992). "Targeted mutation of the DNA methyltransferase gene results in embryonic lethality". *Cell* **69** (6): 915–26. doi:10.1016/0092-8674(92)90611-F. PMID 1606615.

[49] Robertson KD, Uzvolgyi E, Liang G, Talmadge C, Sumegi J, Gonzales FA, Jones PA (June 1999). "The human DNA methyltransferases (DNMTs) 1, 3a and 3b: coordinate mRNA expression in normal tissues and overexpression in tumors". *Nucleic Acids Res.* **27** (11): 2291–8. doi:10.1093/nar/27.11.2291. PMC 148793. PMID 10325416.

[50] Leonhardt H, Page AW, Weier HU, Bestor TH (November 1992). "A targeting sequence directs DNA methyltransferase to sites of DNA replication in mammalian nuclei". *Cell* **71** (5): 865–73. doi:10.1016/0092-8674(92)90561-P. PMID 1423634.

[51] Chuang LS, Ian HI, Koh TW, Ng HH, Xu G, Li BF (September 1997). "Human DNA-(cytosine-5) methyltransferase-PCNA complex as a target for p21WAF1". *Science* **277** (5334): 1996–2000. doi:10.1126/science.277.5334.1996. PMID 9302295.

[52] Robertson KD, Wolffe AP (October 2000). "DNA methylation in health and disease". *Nature Reviews Genetics* **1** (1): 11–9. doi:10.1038/35049533. PMID 11262868.

[53] Li E, Beard C, Jaenisch R (November 1993). "Role for DNA methylation in genomic imprinting". *Nature* **366** (6453): 362–5. Bibcode:1993Natur.366..362L. doi:10.1038/366362a0. PMID 8247133.

[54] Viens A, Mechold U, Brouillard F, Gilbert C, Leclerc P, Ogryzko V (July 2006). "Analysis of human histone H2AZ deposition in vivo argues against its direct role in epigenetic templating mechanisms". *Mol. Cell. Biol.* **26** (14): 5325–35. doi:10.1128/MCB.00584-06. PMC 1592707. PMID 16809769.

[55] Ogryzko VV (2008). "Erwin Schroedinger, Francis Crick and epigenetic stability". *Biol. Direct* **3**: 15. doi:10.1186/1745-6150-3-15. PMC 2413215. PMID 18419815.

[56] Nottke A, Colaiácovo MP, Shi Y (March 2009). "Developmental roles of the histone lysine demethylases". *Development* **136** (6): 879–89. doi:10.1242/dev.020966. PMC 2692332. PMID 19234061.

[57] Rosenfeld JA, Wang Z, Schones DE, Zhao K, DeSalle R, Zhang MQ (2009). "Determination of enriched histone modifications in non-genic portions of the human genome". *BMC Genomics* **10**: 143. doi:10.1186/1471-2164-10-143. PMC 2667539. PMID 19335899.

[58] Sneppen K, Micheelsen MA, Dodd IB (April 15, 2008). "Ultrasensitive gene regulation by positive feedback loops in nucleosome modification". *Molecular systems biology* **4** (1): 182. doi:10.1038/msb.2008.21. PMC 2387233. PMID 18414483. Retrieved 5 May 2014.

[59] "Epigenetic cell memory". Cmol.nbi.dk. Retrieved 26 July 2012.

[60] Dodd IB, Micheelsen MA, Sneppen K, Thon G (May 2007). "Theoretical analysis of epigenetic cell memory by nucleosome modification". *Cell* **129** (4): 813–22. doi:10.1016/j.cell.2007.02.053. PMID 17512413.

[61] Morris KL (2008). "Epigenetic Regulation of Gene Expression". *RNA and the Regulation of Gene Expression: A Hidden Layer of Complexity*. Norfolk, England: Caister Academic Press. ISBN 1-904455-25-5.

[62] Mattick JS, Amaral PP, Dinger ME, Mercer TR, Mehler MF (January 2009). "RNA regulation of epigenetic processes". *BioEssays* **31** (1): 51–9. doi:10.1002/bies.080099. PMID 19154003.

[63] Choi CQ (25 May 2006). "The Scientist: RNA can be hereditary molecule". The Scientist. Retrieved 2006.

[64] Bernal JE, Duran C, Papiha SS (2012). "Transcriptional and epigenetic regulation of human microRNAs". *Cancer Lett* **331** (1): 1–10. doi:10.1016/j.canlet.2012.12.006. PMID 3246373.

[65] Browse miRBase by species

[66] Lim LP, Lau NC, Garrett-Engele P, Grimson A, Schelter JM, Castle J, Bartel DP, Linsley PS, Johnson JM (2005). "Microarray analysis shows that some microRNAs downregulate large numbers of target mRNAs". *Nature* **433** (7027): 769–773. Bibcode:2005Natur.433..769L. doi:10.1038/nature03315. PMID 15685193.

[67] Lee D, Shin C (2012). MicroRNA-target interactions: new insights from genome-wide approaches" *Ann N Y Acad Sci* 1271:118-28. doi: 10.1111/j.1749-6632.2012.06745.x. Review. PMID 23050973

[68] Friedman RC, Farh KK, Burge CB, Bartel DP (2009). "Most mammalian mRNAs are conserved targets of microRNAs". *Genome Res* **19** (1): 92–105. doi:10.1101/gr.082701.108. PMC 2612969. PMID 18955434.

[69] Goll MG, Bestor TH (2005). "Eukaryotic cytosine methyltransferases". *Annu Rev Biochem* **74**: 481–514. doi:10.1146/annurev.biochem.74.010 PMID 15952895.

[70] Guifang Jia; Ye Fu; Xu Zhao; Qing Dai; Guanqun Zheng; Ying Yang; Chengqi Yi; Lindahl, Tomas; Tao Pan; Yun-Gui Yang; Chuan He (16 October 2011). "N6-Methyladenosine in nuclear RNA is a major substrate of the obesity-associated FTO". *Nature Chemical Biology* **7** (12): 885–887. doi:10.1038/nchembio.687. PMC 3218240. PMID 22002720.

[71] "New research links common RNA modification to obesity". Physorg.com. Retrieved 26 July 2012.

[72] Howden BP, Beaume M, Harrison PF, Hernandez D, Schrenzel J, Seemann T, Francois P, Stinear TP (August 2013). "Analysis of the Small RNA Transcriptional Response in Multidrug-Resistant Staphylococcus aureus after Antimicrobial Exposure". *Antimicrob. Agents Chemother.* **57** (8): 3864–74. doi:10.1128/AAC.00263-13. PMC 3719707. PMID 23733475.

[73] sRNATarBase 2.0 A comprehensive database of bacterial SRNA targets verified by experiments

[74] Genomics maps for small non-coding RNA's and their targets in microbial genomes

[75] Yool A, Edmunds WJ (1998). "Epigenetic inheritance and prions". *Journal of Evolutionary Biology* **11** (2): 241–242. doi:10.1007/s000360050

[76] Cox BS (1965). "[PSI], a cytoplasmic suppressor of super-suppression in yeast". *Heredity* **20** (4): 505–521. doi:10.1038/hdy.1965.65.

[77] Lacroute F (May 1971). "Non-Mendelian mutation allowing ureidosuccinic acid uptake in yeast". *J. Bacteriol.* **106** (2): 519–22. PMC 285125. PMID 5573734.

[78] Liebman SW, Sherman F (September 1979). "Extrachromosomal psi+ determinant suppresses nonsense mutations in yeast". *J. Bacteriol.* **139** (3): 1068–71. PMC 218059. PMID 225301.

[79] True HL, Lindquist SL (September 2000). "A yeast prion provides a mechanism for genetic variation and phenotypic diversity". *Nature* **407** (6803): 477–83. doi:10.1038/35035005. PMID 11028992.

[80] Shorter J, Lindquist S (June 2005). "Prions as adaptive conduits of memory and inheritance". *Nature Reviews Genetics* **6** (6): 435–50. doi:10.1038/nrg1616. PMID 15931169.

[81] Giacomelli MG, Hancock AS, Masel J (2007). "The conversion of 3′ UTRs into coding regions". *Molecular Biology & Evolution* **24** (2): 457–464. doi:10.1093/molbev/msl172. PMC 1808353. PMID 17099057.

[82] Lancaster AK, Bardill JP, True HL, Masel J (2010). "The Spontaneous Appearance Rate of the Yeast Prion PSI+ and Its Implications for the Evolution of the Evolvability Properties of the PSI+ System". *Genetics* **184** (2): 393–400. doi:10.1534/genetics.109.110213. PMC 2828720. PMID 19917766.

[83] Sapp J (1991). "Concepts of organization. The leverage of ciliate protozoa". *Dev. Biol. (NY)* **7**: 229–58. doi:10.1007/978-1-4615-6823-0_11. PMID 1804215.

[84] Sapp J (2003). *Genesis: the evolution of biology*. Oxford [Oxfordshire]: Oxford University Press. ISBN 0-19-515619-6.

[85] Gray RD, Oyama S, Griffiths PE (2003). *Cycles of Contingency: Developmental Systems and Evolution (Life and Mind: Philosophical Issues in Biology and Psychology)*. Cambridge, Mass: The MIT Press. ISBN 0-262-65063-0.

[86] Teif VB, Beshnova DA, Vainshtein Y, Marth C, Mallm JP, Höfer T, Rippe K (8 May 2014). "Nucleosome repositioning links DNA (de)methylation and differential CTCF binding during stem cell development". *Genome Research* **24**: 1285–1295. doi:10.1101/gr.164418.113. PMID 24812327.

[87] Griesemer J, Haber MH, Yamashita G, Gannett L (March 2005). "Critical Notice: Cycles of Contingency – Developmental Systems and Evolution". *Biology & Philosophy* **20** (2–3): 517–544. doi:10.1007/s10539-004-0836-4.

[88] Chapter: "Nervous System Development" in "Epigenetics," by Benedikt Hallgrimsson and Brian Hall

[89] Costa S, Shaw P (March 2007). "'Open minded' cells: how cells can change fate" (PDF). *Trends Cell Biol.* **17** (3): 101–6. doi:10.1016/j.tcb.2006.12.005. PMID 17194589. This might suggest that plant cells do not use or require a cellular memory mechanism and just respond to positional information. However, it has been shown that plants do use cellular memory mechanisms mediated by PcG proteins in several processes, ... (p.104)

[90] Cooney CA, Dave AA, Wolff GL (August 2002). "Maternal methyl supplements in mice affect epigenetic variation and DNA methylation of offspring". *J. Nutr.* **132** (8 Suppl): 2393S–2400S. PMID 12163699.

[91] Waterland RA, Jirtle RL (August 2003). "Transposable elements: targets for early nutritional effects on epigenetic gene regulation". *Mol. Cell. Biol.* **23** (15): 5293–300. doi:10.1128/MCB.23.15.5293-5300.2003. PMC 165709. PMID 12861015.

[92] Fearful Memories Passed Down to Mouse Descendants: Genetic imprint from traumatic experiences carries through at least two generations, By Ewen Callaway and Nature magazine | Sunday, 1 December 2013.

[93] Mice can 'warn' sons, grandsons of dangers via sperm, by Mariette Le Roux, 12/1/13.

[94] G. Francis, "Too Much Success for Recent Groundbreaking Epigenetic Experiments" http://www.genetics.org/content/198/2/449.abstract

[95] http://www.ncbi.nlm.nih.gov/pubmed/24292232 (see comment by Gonzalo Otazu)

[96] http://www.the-scientist.com/?articles.view/articleNo/41239/title/Epigenetics-Paper-Raises-Questions/

[97] Lamb MJ, Jablonka E (2005). *Evolution in four dimensions: genetic, epigenetic, behavioral, and symbolic variation in the history of life*. Cambridge, Mass: MIT Press. ISBN 0-262-10107-6.

[98] See also Denis Noble *The Music of Life* see esp pp. 93–8 and p. 48 where he cites Jablonka & Lamb and Massimo Pigliucci's review of Jablonka and Lamb in Nature **435**, 565–566 (2 June 2005)

[99] Danchin É, Charmantier A, Champagne FA, Mesoudi A, Pujol B, Blanchet S (2011). "Beyond DNA: integrating inclusive inheritance into an extended theory of evolution". *Nature Reviews Genetics* **12**: 475–486. doi:10.1038/nrg3028. PMID 21681209.

[100] Maynard Smith J (1990). "Models of a Dual Inheritance System". *Journal of Theoretical Biology* **143** (1): 41–53. doi:10.1016/S0022-5193(05)80287-5. PMID 2359317.

[101] Lynch M (2007). "The frailty of adaptive hypotheses for the origins of organismal complexity". *PNAS* **104** (suppl. 1): 8597–8604. Bibcode:2007PNAS..104.8597L. doi:10.1073/pnas.0702207104. PMC 1876435. PMID 17494740.

[102] Thomas Dickens, Qazi Rahman. (2012). *The extended evolutionary synthesis and the role of soft inheritance in evolution.* Proceedings of the Royal Society: B biological sciences, 279 (1740). pp. 2913-2921.

[103] Rando OJ, Verstrepen KJ (February 2007). "Timescales of genetic and epigenetic inheritance". *Cell* **128** (4): 655–68. doi:10.1016/j.cell.2007.01.023. PMID 17320504.

[104] Lancaster AK, Masel J (1 September 2009). "The evolution of reversible switches in the presence of irreversible mimics". *Evolution* **63** (9): 2350–2362. doi:10.1111/j.1558-5646.2009.00729.x. PMC 2770902. PMID 19486147.

[105] Graaf, Adriaan van der; Wardenaar, René; Neumann, Drexel A.; Taudt, Aaron; Shaw, Ruth G.; Jansen, Ritsert C.; Schmitz, Robert J.; Colomé-Tatché, Maria; Johannes, Frank (2015-05-11). "Rate, spectrum, and evolutionary dynamics of spontaneous epimutations". *Proceedings of the National Academy of Sciences* **112**: 201424254. doi:10.1073/pnas.1424254112. ISSN 0027-8424. Retrieved 2015-05-12.

[106] Griswold CK, Masel J (2009). Úbeda F, ed. "Complex Adaptations Can Drive the Evolution of the Capacitor PSI+, Even with Realistic Rates of Yeast Sex". *PLoS Genetics* **5** (6): e1000517. doi:10.1371/journal.pgen.1000517. PMC 2686163. PMID 19521499.

[107] Jablonka E, Raz G (June 2009). "Transgenerational epigenetic inheritance: prevalence, mechanisms, and implications for the study of heredity and evolution". *Q Rev Biol* **84** (2): 131–76. doi:10.1086/598822. PMID 19606595.

[108] Davies, Hazel (2008). Do Butterflies Bite?: Fascinating Answers to Questions about Butterflies and Moths (Animals Q&A). Rutgers University Press.

[109] Casadesús J, Low D (September 2006). "Epigenetic gene regulation in the bacterial world". *Microbiol. Mol. Biol. Rev.* **70** (3): 830–56. doi:10.1128/MMBR.00016-06. PMC 1594586. PMID 16959970.

[110] Jorg Tost (2008). *Epigenetics*. Norfolk, England: Caister Academic Press. ISBN 1-904455-23-9.

[111] Lewis ZA, Honda S, Khlafallah TK, Jeffress JK, Freitag M, Mohn F, Schübeler D, Selker EU (March 2009). "Relics of repeat-induced point mutation direct heterochromatin formation in Neurospora crassa". *Genome Res.* **19** (3): 427–37. doi:10.1101/gr.086231.108. PMC 2661801. PMID 19092133.

[112] Schadt EE, Banerjee O, Fang G, Feng Z, Wong WH, Zhang X, Kislyuk A, Clark TA, Luong K, Keren-Paz A, Chess A, Kumar V, Chen-Plotkin A, Sondheimer N, Korlach J, Kasarskis A (2012). "Modeling kinetic rate variation in third generation DNA sequencing data to detect putative modifications to DNA bases". *Genome Research* **23** (1): 129–41. doi:10.1101/gr.136739.111. PMC 3530673. PMID 23093720.

[113] Davis BM, Chao MC, Waldor MK (2013). "Entering the era of bacterial epigenomics with single molecule real time DNA sequencing". *Current Opinion in Microbiology* **16** (2): 192–8. doi:10.1016/j.mib.2013.01.011. PMC 3646917. PMID 23434113.

[114] Lluch-Senar M, Luong K, Lloréns-Rico V, Delgado J, Fang G, Spittle K, Clark TA, Schadt E, Turner SW, Korlach J, Serrano L (2013). Richardson PM, ed. "Comprehensive Methylome Characterization of Mycoplasma genitalium and Mycoplasma pneumoniae at Single-Base Resolution". *PLoS Genetics* **9** (1): e1003191. doi:10.1371/journal.pgen.1003191. PMC 3536716. PMID 23300489.

[115] Murray IA, Clark TA, Morgan RD, Boitano M, Anton BP, Luong K, Fomenkov A, Turner SW, Korlach J, Roberts RJ (2012). "The methylomes of six bacteria". *Nucleic Acids Research* **40** (22): 11450–62. doi:10.1093/nar/gks891. PMC 3526280. PMID 23034806.

[116] Fang G, Munera D, Friedman DI, Mandlik A, Chao MC, Banerjee O, Feng Z, Losic B, Mahajan MC, Jabado OJ, Deikus G, Clark TA, Luong K, Murray IA, Davis BM, Keren-Paz A, Chess A, Roberts RJ, Korlach J, Turner SW, Kumar V, Waldor MK, Schadt EE (2012). "Genome-wide mapping of methylated adenine residues in pathogenic Escherichia coli using single-molecule real-time sequencing". *Nature Biotechnology* **30** (12): 1232–9. doi:10.1038/nbt.2432. PMID 23138224.

[117] Chahwan R, Wontakal SN, Roa S (March 2011). "The multidimensional nature of epigenetic information and its role in disease". *Discov Med* **11** (58): 233–43. PMID 21447282.

[118] Beil, Laura (Winter 2008). "Medicine's New Epicenter? Epigenetics: New field of epigenetics may hold the secret to flipping cancer's "off" switch.". CURE (Cancer Updates, Research and Education).

[119] Fraga MF, Ballestar E, Paz MF, Ropero S, Setien F, Ballestar ML, Heine-Suñer D, Cigudosa JC, Urioste M, Benitez J, Boix-Chornet M, Sanchez-Aguilera A, Ling C, Carlsson E, Poulsen P, Vaag A, Stephan Z, Spector TD, Wu YZ, Plass C, Esteller M (July 2005). "Epigenetic differences arise during the lifetime of monozygotic twins". *Proc. Natl. Acad. Sci. U.S.A.* **102** (30): 10604–9. Bibcode:2005PNAS..10210604F. doi:10.1073/pnas.0500398102. PMC 1174919. PMID 16009939.

[120] Kaminsky ZA, Tang T, Wang SC, Ptak C, Oh GH, Wong AH, Feldcamp LA, Virtanen C, Halfvarson J, Tysk C, McRae AF, Visscher PM, Montgomery GW, Gottesman II, Martin NG, Petronis A (February 2009). "DNA methylation profiles in monozygotic and dizygotic twins". *Nat. Genet.* **41** (2): 240–5. doi:10.1038/ng.286. PMID 19151718.

[121] O'Connor, Anahad (11 March 2008). "The Claim: Identical Twins Have Identical DNA". New York Times. Retrieved 2 May 2010.

[122] Ballestar E (2010). "Epigenetics lessons from twins: prospects for autoimmune disease". *Clin Rev Allergy Immunol* **39** (1): 30–41. doi:10.1007/s12016-009-8168-4. PMID 19653134.

[123] Kaminsky ZA, Tang T, Wang SC, Ptak C, Oh GH, Wong AH, Feldcamp LA, Virtanen C, Halfvarson J, Tysk C, McRae AF, Visscher PM, Montgomery GW, Gottesman II, Martin NG, Petronis A (2009). "DNA methylation profiles in monozygotic and dizygotic twins". *Nat Genet* **41** (2): 240–245. doi:10.1038/ng.286. PMID 19151718.

[124] Online 'Mendelian Inheritance in Man' (OMIM) 105830

[125] Wood AJ, Oakey RJ (November 2006). "Genomic imprinting in mammals: emerging themes and established theories". *PLoS Genet.* **2** (11): e147. doi:10.1371/journal.pgen.0020147. PMC 1657038. PMID 17121465.

[126] Knoll JH, Nicholls RD, Magenis RE, Graham JM, Lalande M, Latt SA (February 1989). "Angelman and Prader-Willi syndromes share a common chromosome 15 deletion but differ in parental origin of the deletion". *Am. J. Med. Genet.* **32** (2): 285–90. doi:10.1002/ajmg.1320320235. PMID 2564739.

[127] A person's paternal grandson is the son of a son of that person; a maternal grandson is the son of a daughter.

[128] Pembrey ME, Bygren LO, Kaati G, Edvinsson S, Northstone K, Sjöström M, Golding J (February 2006). "Sex-specific, male-line transgenerational responses in humans". *Eur. J. Hum. Genet.* **14** (2): 159–66. doi:10.1038/sj.ejhg.5201538. PMID 16391557. Robert Winston refers to this study in a lecture; see also discussion at Leeds University, here

[129] "NOVA | Transcripts | Ghost in Your Genes". PBS. 16 October 2007. Retrieved 26 July 2012.

[130] Wood LD, Parsons DW, Jones S, Lin J, Sjöblom T, Leary RJ, Shen D, Boca SM, Barber T, Ptak J, Silliman N, Szabo S, Dezso Z, Ustyanksky V, Nikolskaya T, Nikolsky Y, Karchin R, Wilson PA, Kaminker JS, Zhang Z, Croshaw R, Willis J, Dawson D, Shipitsin M, Willson JK, Sukumar S, Polyak K, Park BH, Pethiyagoda CL, Pant PV, Ballinger DG, Sparks AB, Hartigan J, Smith DR, Suh E, Papadopoulos N, Buckhaults P, Markowitz SD, Parmigiani G, Kinzler KW, Velculescu VE, Vogelstein B (2007). "The genomic landscapes of human breast and colorectal cancers". *Science* **318** (5853): 1108–1113. Bibcode:2007Sci...318.1108W. doi:10.1126/science.1145720. PMID 17932254.

[131] Jasperson KW, Tuohy TM, Neklason DW, Burt RW (2010). "Hereditary and familial colon cancer". *Gastroenterology* **138** (6): 2044–2058. doi:10.1053/j.gastro.2010.01.054. PMID 20420945.

[132] Novak, Kris (20 December 2004). "Epigenetics Changes in Cancer Cells". *Medscape General Medicine* **6** (4): 17. PMC 1480584. PMID 15775844. Retrieved 31 May 2012.

[133] Banno K, Kisu I, Yanokura M, Tsuji K, Masuda K, Ueki A, Kobayashi Y, Yamagami W, Nomura H, Tominaga E, Susumu N, Aoki D (2012). "Epimutation and cancer: a new carcinogenic mechanism of Lynch syndrome (Review)". *Int. J. Oncol.* **41** (3): 793–7. doi:10.3892/ijo.2012.1528. PMC 3582986. PMID 22735547.

[134] Chahwan R, Wontakal SN, Roa S (October 2010). "Crosstalk between genetic and epigenetic information through cytosine deamination". *Trends Genet.* **26** (10): 443–8. doi:10.1016/j.tig.2010.07.005. PMID 20800313.

[135] "Nonantibiotic Effects of Fluoroquinolones in Mammalian Cells.". *J Biol Chem* **290**: 22287–97. Sep 2015. doi:10.1074/jbc.M115.671222. PMID 26205818.

[136] mezentseva, nadejda; yang, jinpu; kaur, keerat; eisenberg, carol; eisenberg, leonard (2012). "The histone methyltransferase inhibitor BIX01294 enhances the cardiac potential of bone marrow cells.". *Stem Cells Dev* **22**: 654–67. doi:10.1089/scd.2012.0181. PMID 22994322.

[137] yang, jinpu; kaur, keerat; ong, lilin; eisenberg, carol; eisenberg, leonard (2015). "Inhibition of G9a Histone Methyltransferase Converts Bone Marrow Mesenchymal Stem Cells to Cardiac Competent Progenitors.". *stem cell international* **2015**: 270428. doi:10.1155/2015/270428. PMID 26089912.

[138] "Beware the pseudo gene genies". The Guardian.

[139] "Epigenetics: It doesn't mean what quacks think it means". Science-Based Medicine.

25.11 External links

- Haque FN, Gottesman II, Wong AH (May 2009). "Not really identical: epigenetic differences in monozygotic twins and implications for twin studies in psychiatry". *Am J Med Genet C Semin Med Genet* **151C** (2): 136–41. doi:10.1002/ajmg.c.30206. PMID 19378334.

- The Human Epigenome Project (HEP)

- The Epigenome Network of Excellence (NoE)

- Canadian Epigenetics, Environment and Health Research Consortium (CEEHRC)

- The Epigenome Network of Excellence (NoE)- public international site

- DNA Is Not Destiny – Discover Magazine cover story

- BBC – Horizon – 2005 – The Ghost In Your Genes

- Epigenetics article at Hopkins Medicine

- Towards a global map of epigenetic variation

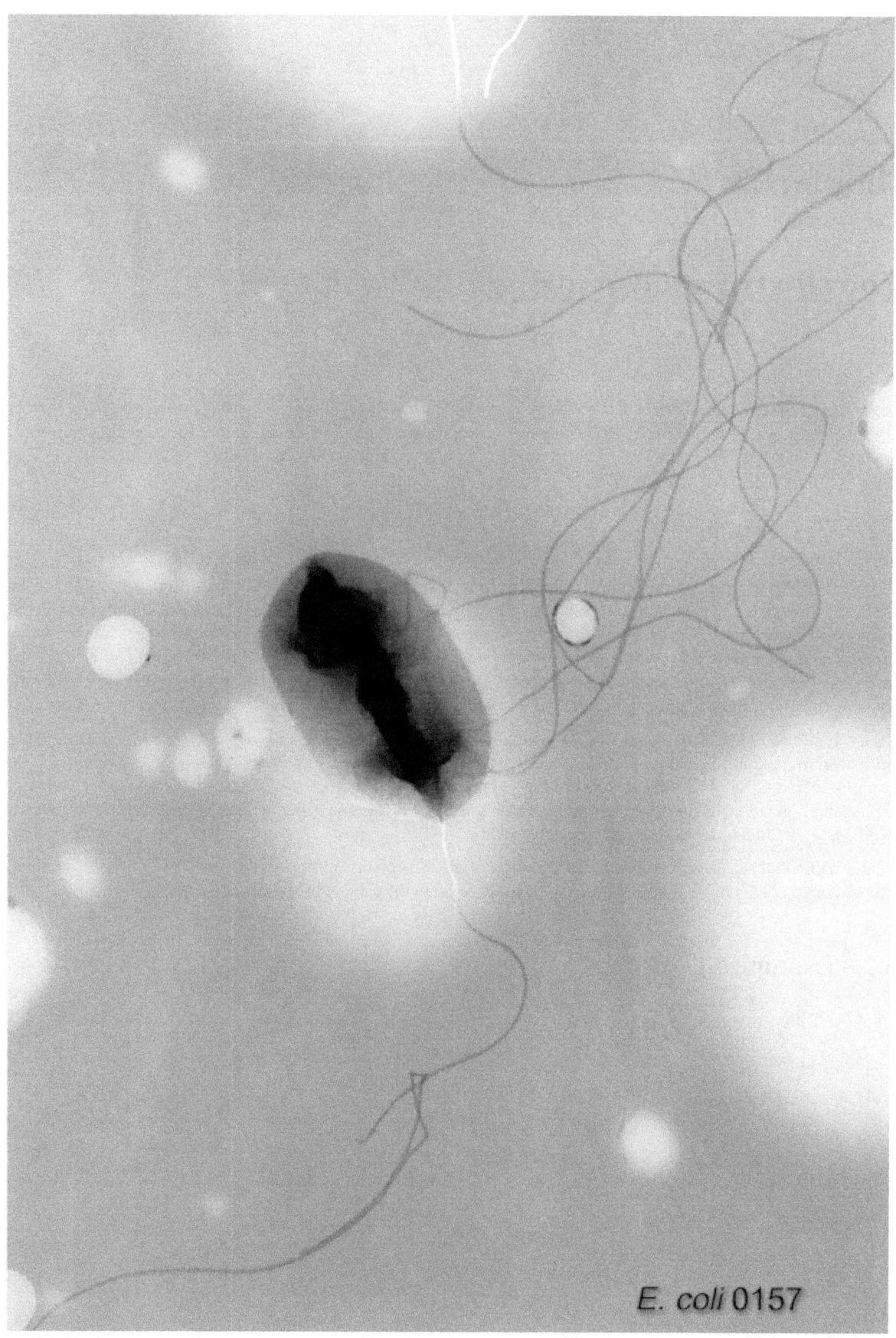

E. coli 0157

Escherichia coli bacteria

Chapter 26

Regulatory sequence

A **regulatory sequence** is a segment of a nucleic acid molecule which is capable of increasing or decreasing the expression of specific genes within an organism. Regulation of gene expression is an essential feature of all living organisms and viruses.

26.1 Description

In DNA, regulation of gene expression normally happens at the level of RNA biosynthesis (transcription), and is accomplished through the sequence-specific binding of proteins (transcription factors) that activate or inhibit transcription. Transcription factors may act as activators, repressors, or both. Repressors often act by preventing RNA polymerase from forming a productive complex with the transcriptional initiation region (promoter), while activators facilitate formation of a productive complex. Furthermore, DNA motifs have been shown to be predictive of epigenomic modifications, suggesting that transcription factors play a role in regulating the epigenome.[1]

In RNA, regulation may occur at the level of protein biosynthesis (translation), RNA cleavage, RNA splicing, or transcriptional termination. Regulatory sequences are frequently associated with messenger RNA (mRNA) molecules, where they are used to control mRNA biogenesis or translation. A variety of biological molecules may bind to the RNA to accomplish this regulation, including proteins (e.g. translational repressors and splicing factors), other RNA molecules (e.g. miRNA) and small molecules, in the case of riboswitches.

Research to find all regulatory regions in the genomes of all sorts of organisms is under way.[2] Conserved non-coding sequences often contain regulatory regions, and so they are often the subject of these analyses.

26.2 Examples

- CAAT box
- CCAAT box
- Operator (biology)
- Pribnow box
- TATA box
- SECIS element, mRNA
- Polyadenylation signals, mRNA
- A-box

- Z-box

- C-box

- E-box

- G-box

26.3 Insulin gene

Regulatory sequences for the insulin gene are:[3]

- A5

- Z

- negative regulatory element (NRE)[4]

- C2

- E2

- A3

- cAMP response element

- A2

- CAAT enhancer binding (CEB)

- C1

- E1

- G1

26.4 See also

- Regulator gene

- Regulation of gene expression

- Cis-acting element

- Gene regulatory network

- Operon

- DNA binding site

- Promoter

- Trans-acting factor

- ORegAnno

26.5   References

[1] Whitaker JW, Zhao Chen, Wei Wang. (2014) Predicting the Human Epigenome from DNA Motifs. Nature Methods. doi: 10.1038/nmeth.3065

[2] Stepanova et al., Bioinformatics, 21(9): 1789-96, year 2005. A comparative analysis of relative occurrence of transcription factor binding sites in vertebrate genomes and gene promoter areas

[3] Melloul et al., Diabetologica, 45, 309-326, year 2002. Regulation of insulin gene transcription

[4] Biochemical and Biophysical Research Communications ...

26.6   External links

- ORegAnno - Open Regulatory Annotation Database

Chapter 27

Retrotransposon

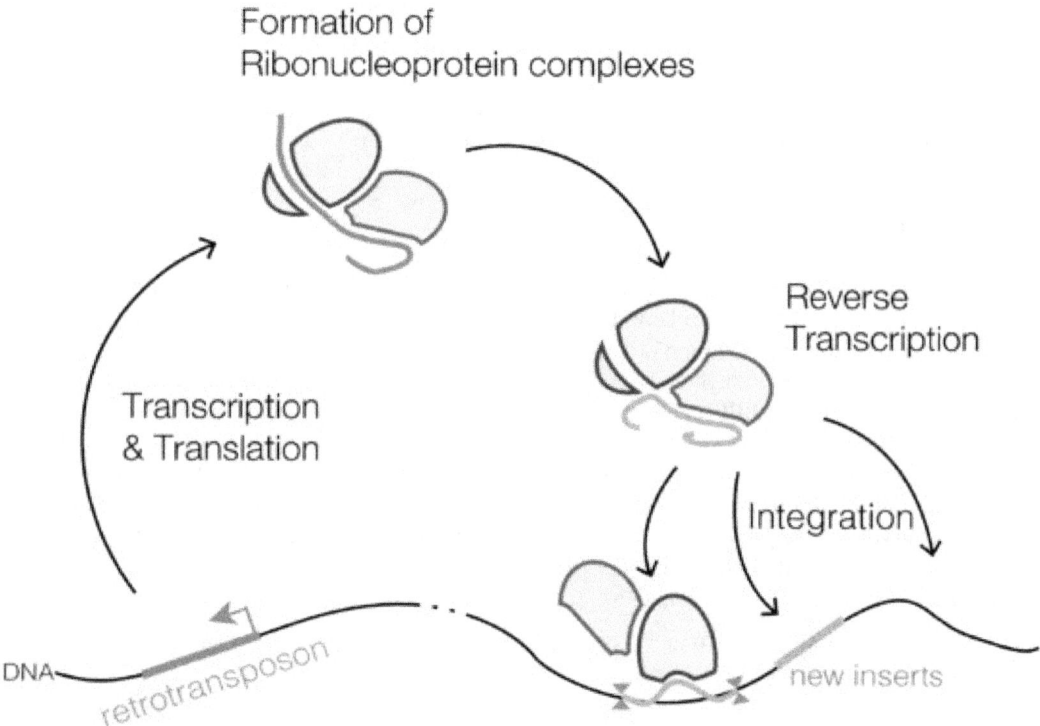

Formation of
Ribonucleoprotein complexes

Reverse
Transcription

Integration

Transcription
& Translation

DNA

retrotransposon

new inserts

Simplified representation of the life cycle of a retrotransposon

Retrotransposons (also called transposons via RNA intermediates) are genetic elements that can amplify themselves in a genome and are ubiquitous components of the DNA of many eukaryotic organisms. They are one of the two subclasses of transposon, where the other is DNA transposon, which does not involve an RNA intermediate. They are particularly abundant in plants, where they are often a principal component of nuclear DNA. In maize, 49–78% of the genome is made up of retrotransposons.[1] In wheat, about 90% of the genome consists of repeated sequences and 68% of transposable elements.[2] In mammals, almost half the genome (45% to 48%) is transposons or remnants of transposons. Around 42% of the human genome is made up of retrotransposons, while DNA transposons account for about 2–3%.[3]

27.1   Biological activity

The retrotransposons' replicative mode of transposition by means of an RNA intermediate rapidly increases the copy numbers of elements and thereby can increase genome size. Like DNA transposable elements (class II transposons), retrotransposons can induce mutations by inserting near or within genes. Furthermore, retrotransposon-induced mutations are relatively stable, because the sequence at the insertion site is retained as they transpose via the replication mechanism.

Retrotransposons copy themselves to RNA and then back to DNA that may integrate back to the genome. The second step of forming DNA may be carried out by a reverse transcriptase, which the retrotransposon encodes.[4] Transposition and survival of retrotransposons within the host genome are possibly regulated both by retrotransposon- and host-encoded factors, to avoid deleterious effects on host and retrotransposon as well, in a relationship that has existed for many millions of years between retrotransposons and their hosts. The understanding of how retrotransposons and their hosts' genomes have co-evolved mechanisms to regulate transposition, insertion specificities, and mutational outcomes in order to optimize each other's survival is still in its infancy.

Because of accumulated mutations, most retrotransposons are no longer able to retrotranspose.

27.2   Types of retrotransposons

Retrotransposons, also known as class I transposable elements, consist of two subclasses, the long terminal repeat (LTR) and the non-LTR retrotransposons. Classification into these subclasses is based on the phylogeny of the reverse transcriptase,[5] which goes in line with structural differences, such as presence/absence of long terminal repeats as well as number and types of open reading frames, encoding domains and target site duplication lengths.

27.2.1   LTR retrotransposons

LTR retrotransposons have direct LTRs that range from ~100 bp to over 5 kb in size. LTR retrotransposons are further sub-classified into the Ty1-*copia*-like (Pseudoviridae), Ty3-*gypsy*-like (Metaviridae), and BEL-Pao-like groups based on both their degree of sequence similarity and the order of encoded gene products. Ty1-*copia* and Ty3-*gypsy* groups of retrotransposons are commonly found in high copy number (up to a few million copies per haploid nucleus) in animals, fungi, protista, and plants genomes. BEL-Pao like elements have so far only been found in animals.[6][7] Although Retroviruses are often classified separately, they share many features with LTR retrotransposons. A major difference with Ty1-*copia* and Ty3-*gypsy* retrotransposons is that retroviruses have an Envelope protein (ENV). A retrovirus can be transformed into an LTR retrotransposon through inactivation or deletion of the domains that enable extracellular mobility. If such a retrovirus infects and subsequently inserts itself in the genome in germ line cells, it may become transmitted vertically and become an Endogenous Retrovirus (ERV).[7] Endogenous retroviruses make up about 8% of the human genome and approximately 10% of the mouse genome.[8]

In plant genomes, LTR retrotransposons are the major repetitive sequence class, e.g. able to constitute more than 75% of the maize genome.[9]

Ty1-*copia* retrotransposons

Ty1-*copia* retrotransposons are abundant in species ranging from single-cell algae to bryophytes, gymnosperms, and angiosperms. They encode four protein domains in the following order: protease, integrase, reverse transcriptase, and ribonuclease H.

At least two classification systems exist for the subdivision of Ty1-*copia* retrotransposons into five lineages:[10][11] *Sireviruses*/Maximus, Oryco/Ivana, Retrofit/Ale, TORK (subdivided in Angela/Sto, TAR/Fourf, GMR/Tork), and Bianca.

Sireviruses/Maximus retrotransposons contain an additional putative envelope gene. This lineage is named for the founder element SIRE1 in the *Glycine max* genome,[12] and was later described in many species such as *Zea mays*,[13] *Arabidopsis thaliana*,[14] *Beta vulgaris*,[15] and *Pinus pinaster*.[16] Plant *Sireviruses* of many sequenced plant genomes are summarized at the MASIVEdb *Sirevirus* database.[17]

Ty3-*gypsy* retrotransposons

Ty3-*gypsy* retrotransposons (*Metaviridae*) are widely distributed in the plant kingdom, including both gymnosperms and angiosperms. They encode at least four protein domains in the order: protease, reverse transcriptase, ribonuclease H, and integrase. Based on structure, presence/absence of specific protein domains, and conserved protein sequence motifs, they can be subdivided into several lineages:

Errantiviruses contain an additional defective envelope ORF with similarities to the retroviral envelope gene. First described as Athila-elements in *Arabidopsis thaliana*,[18][19] they have been later identified in many species, such as *Glycine max* [20] and *Beta vulgaris*.[21]

Chromoviruses contain an additional chromodomain (chromatin organization modifier domain) at the C-terminus of their integrase protein.[22][23] They are widespread in plants and fungi, probably retaining protein domains during evolution of these two kingdoms.[24] It is thought that the chromodomain directs retrotransposon integration to specific target sites.[25] According to sequence and structure of the chromodomain, chromoviruses are subdivided into the four clades CRM, Tekay, Reina and Galadriel. Chromoviruses from each clade show distinctive integration patterns, e.g. into centromeres or into the rRNA genes.[26][27]

Ogre-elements are gigantic Ty3-*gypsy* retrotransposons reaching lengths up to 25 kb.[28] Ogre elements have been first described in *Pisum sativum*.[29]

Metaviruses describe conventional Ty3-*gypsy* retrotransposons that do not contain additional domains or ORFs.

Endogenous retroviruses (ERV)

Main article: Endogenous retrovirus

Endogenous retroviruses are the most important LTR retrotransposons in mammals, including human where the Human ERVs make up 8% of the genome.

27.2.2 Non-LTR retrotransposons

Non-LTR retrotransposons consist of two sub-types, long interspersed elements (LINEs) and short interspersed elements (SINEs). They can also be found in high copy numbers, as shown in the plant species.[30] Non-long terminal repeat (LTR) retroposons are widespread in eukaryotic genomes. LINEs possess two ORFs, which encode all the functions needed for retrotransposition. These functions include reverse transcriptase and endonuclease activities, in addition to a nucleic acid-binding property needed to form a ribonucleoprotein particle.[31] SINEs, on the other hand, co-opt the LINE machinery and function as nonautonomous retroelements.

LINEs

Main article: Long interspersed nuclear element

Long Interspersed Nuclear Elements[32] (**LINE**) are a group of genetic elements that are found in large numbers in eukaryotic genomes, comprising 17% of the human genome (99.9% of which is no longer capable of mobilization).[33] Among the LINE, there are several subgroups, such as L1, L2 and L3. Human coding L1 begin with an untranslated region (UTR) that includes an RNA polymerase II promoter, two non-overlapping open reading frames (ORF1 and ORF2), and ends with another UTR.[33] Recently, a new open reading frame in the 5' end of the LINE elements has been identified in the reverse strand. It is shown to be transcribed and endogenous proteins are observed. The name ORF0 is coined due to its position with respect to ORF1 and ORF2.[34] ORF1 encodes an RNA binding protein and ORF2 encodes a protein having an endonuclease (e.g. RNase H) as well as a reverse transcriptase. The reverse transcriptase has a higher specificity for the LINE RNA than other RNA, and makes a DNA copy of the RNA that can be integrated into the genome at a new site.[35] The endonuclease encoded by non-LTR retroposons may be AP (Apurinic/Pyrimidinic) type or REL (Restriction Endonuclease Like) type. Elements in the R2 group have REL type endonuclease, which shows site specificity in insertion.[36]

The 5' UTR contains the promoter sequence, while the 3' UTR contains a polyadenylation signal (AATAAA) and a poly-A tail.[37] Because LINEs (and other class I transposons, e.g. LTR retrotransposons and SINEs) move by copying themselves (instead of moving by a cut and paste like mechanism, as class II transposons do), they enlarge the genome. The human genome, for example, contains about 500,000 LINEs, which is roughly 17% of the genome.[38] Of these, approximately 7,000 are full-length, a small subset of which are capable of retrotransposition.[39][40]

Interestingly, it was recently found that specific LINE-1 retroposons in the human genome are actively transcribed and the associated LINE-1 RNAs are tightly bound to nucleosomes and essential in the establishment of local chromatin environment.[41]

SINEs

Short Interspersed Nuclear Elements[32] are short DNA sequences (<500 bases[42]) that represent reverse-transcribed RNA molecules originally transcribed by RNA polymerase III into tRNA, 5S ribosomal RNA, and other small nuclear RNAs. The mechanism of retrotransposition of these elements is more complicated than LINEs, and less dependent solely on the actual elements that they encode. SINEs do not encode a functional reverse transcriptase protein and rely on other mobile elements for transposition. In some cases they may have their own endonuclease that will allow them to cleave their way into the genome, but the majority of SINEs integrate at chromosomal breaks by using random DNA breaks to prime reverse transcriptase.[32]

The most common SINEs in primates are called Alu sequences. Alu elements are approximately 350 base pairs long, do not contain any coding sequences, and can be recognized by the restriction enzyme AluI (hence the name). With about 1,500,000 copies, SINEs make up about 11% of the human genome.[38] While historically viewed as "junk DNA", recent research suggests that, in some rare cases, both LINEs and SINEs were incorporated into novel genes so as to evolve new functionality.[43][44] The distribution of these elements has been implicated in some genetic diseases and cancers. Although sequence analysis of human Alu subfamilies shows the existence of mosaic (recombinant) elements, experimental evidence is lacking. In the primitive eukaryote *Entamoeba histolytica*, the frequent exchange of sequence during retrotransposition has been reported; this results in a mosaic pattern in its SINE sequences.[45]

27.3 See also

- Transposon

- Endogenous retrovirus

- Paleogenetics

- Paleovirology

- Insertion sequences

- Copy-number variation

- Genomic organization

- Interspersed repeat

- Retrotransposon markers, a powerful method of reconstructing phylogenies.

- RetrOryza

27.4 References

[1] SanMiguel P, Bennetzen JL (1998). "Evidence that a recent increase in maize genome size was caused by the massive amplification of intergene retrotranposons" (PDF). *Annals of Botany* **82** (Suppl A): 37–44. doi:10.1006/anbo.1998.0746.

[2] Li W, Zhang P, Fellers JP, Friebe B, Gill BS (November 2004). "Sequence composition, organization, and evolution of the core Triticeae genome". *Plant J.* **40** (4): 500–11. doi:10.1111/j.1365-313X.2004.02228.x. PMID 15500466.

[3] Lander ES, Linton LM, Birren B, et al. (February 2001). "Initial sequencing and analysis of the human genome". *Nature* **409** (6822): 860–921. doi:10.1038/35057062. PMID 11237011.

[4] Dombroski BA, Feng Q, Mathias SL, et al. (July 1994). "An in vivo assay for the reverse transcriptase of human retrotransposon L1 in Saccharomyces cerevisiae". *Mol. Cell. Biol.* **14** (7): 4485–92. doi:10.1128/mcb.14.7.4485. PMC 358820. PMID 7516468.

[5] Xiong, Y; Eickbush, TH (October 1990). "Origin and evolution of retroelements based upon their reverse transcriptase sequences.". *The EMBO Journal* **9** (10): 3353–62. PMID 1698615.

[6] Copeland CS, Mann VH, Morales ME, Kalinna BH, Brindley PJ (2005). "The Sinbad retrotransposon from the genome of the human blood fluke, Schistosoma mansoni, and the distribution of related Pao-like elements". *BMC Evol. Biol.* **5** (1): 20. doi:10.1186/1471-2148-5-20. PMC 554778. PMID 15725362.

[7] Wicker T, Sabot F, Hua-Van A, et al. (December 2007). "A unified classification system for eukaryotic transposable elements". *Nat. Rev. Genet.* **8** (12): 973–82. doi:10.1038/nrg2165. PMID 17984973.

[8] McCarthy EM, McDonald JF (2004). "Long terminal repeat retrotransposons of Mus musculus". *Genome Biol.* **5** (3): R14. doi:10.1186/gb-2004-5-3-r14. PMC 395764. PMID 15003117.

[9] Baucom, RS; Estill, JC; Chaparro, C; Upshaw, N; Jogi, A; Deragon, JM; Westerman, RP; Sanmiguel, PJ; Bennetzen, JL (November 2009). "Exceptional diversity, non-random distribution, and rapid evolution of retroelements in the B73 maize genome.". *PLoS Genetics* **5** (11): e1000732. doi:10.1371/journal.pgen.1000732. PMID 19936065.

[10] Wicker, T; Keller, B (July 2007). "Genome-wide comparative analysis of copia retrotransposons in Triticeae, rice, and Arabidopsis reveals conserved ancient evolutionary lineages and distinct dynamics of individual copia families.". *Genome Research* **17** (7): 1072–81. doi:10.1101/gr.6214107. PMID 17556529.

[11] Llorens, C; Muñoz-Pomer, A; Bernad, L; Botella, H; Moya, A (2 November 2009). "Network dynamics of eukaryotic LTR retroelements beyond phylogenetic trees.". *Biology Direct* **4**: 41. doi:10.1186/1745-6150-4-41. PMID 19883502.

[12] Laten, HM; Majumdar, A; Gaucher, EA (9 June 1998). "SIRE-1, a copia/Ty1-like retroelement from soybean, encodes a retroviral envelope-like protein.". *Proceedings of the National Academy of Sciences of the United States of America* **95** (12): 6897–902. doi:10.1073/pnas.95.12.6897. PMID 9618510.

[13] Bousios, A; Kourmpetis, YA; Pavlidis, P; Minga, E; Tsaftaris, A; Darzentas, N (February 2012). "The turbulent life of Sirevirus retrotransposons and the evolution of the maize genome: more than ten thousand elements tell the story.". *The Plant journal* **69** (3): 475–88. doi:10.1111/j.1365-313x.2011.04806.x. PMID 21967390.

[14] Kapitonov, VV; Jurka, J (1999). "Molecular paleontology of transposable elements from Arabidopsis thaliana.". *Genetica* **107** (1-3): 27–37. PMID 10952195.

[15] Weber, B; Wenke, T; Frömmel, U; Schmidt, T; Heitkam, T (February 2010). "The Ty1-copia families SALIRE and Cotzilla populating the Beta vulgaris genome show remarkable differences in abundance, chromosomal distribution, and age.". *Chromosome Research* **18** (2): 247–63. doi:10.1007/s10577-009-9104-4. PMID 20039119.

[16] Miguel, C; Simões, M; Oliveira, MM; Rocheta, M (November 2008). "Envelope-like retrotransposons in the plant kingdom: evidence of their presence in gymnosperms (Pinus pinaster).". *Journal of Molecular Evolution* **67** (5): 517–25. doi:10.1007/s00239-008-9168-3. PMID 18925379.

[17] Bousios, A; Minga, E; Kalitsou, N; Pantermali, M; Tsaballa, A; Darzentas, N (30 April 2012). "MASiVEdb: the Sirevirus Plant Retrotransposon Database.". *BMC Genomics* **13**: 158. doi:10.1186/1471-2164-13-158. PMID 22545773.

[18] Pélissier, T; Tutois, S; Deragon, JM; Tourmente, S; Genestier, S; Picard, G (November 1995). "Athila, a new retroelement from Arabidopsis thaliana.". *Plant Molecular Biology* **29** (3): 441–52. PMID 8534844.

[19] Wright, DA; Voytas, DF (June 1998). "Potential retroviruses in plants: Tat1 is related to a group of Arabidopsis thaliana Ty3/gypsy retrotransposons that encode envelope-like proteins.". *Genetics* **149** (2): 703–15. PMID 9611185.

[20] Wright, DA; Voytas, DF (January 2002). "Athila4 of Arabidopsis and Calypso of soybean define a lineage of endogenous plant retroviruses.". *Genome Research* **12** (1): 122–31. doi:10.1101/gr.196001. PMID 11779837.

[21] Wollrab, C; Heitkam, T; Holtgräwe, D; Weisshaar, B; Minoche, AE; Dohm, JC; Himmelbauer, H; Schmidt, T (November 2012). "Evolutionary reshuffling in the Errantivirus lineage Elbe within the Beta vulgaris genome.". *The Plant Journal* **72** (4): 636–51. doi:10.1111/j.1365-313x.2012.05107.x. PMID 22804913.

[22] Marín, I; Lloréns, C (July 2000). "Ty3/Gypsy retrotransposons: description of new Arabidopsis thaliana elements and evolutionary perspectives derived from comparative genomic data.". *Molecular Biology and Evolution* **17** (7): 1040–9. doi:10.1093/oxfordjournals.molbev PMID 10889217.

[23] Gorinsek, B; Gubensek, F; Kordis, D (May 2004). "Evolutionary genomics of chromoviruses in eukaryotes.". *Molecular Biology and Evolution* **21** (5): 781–98. doi:10.1093/molbev/msh057. PMID 14739248.

[24] Novikova, O; Smyshlyaev, G; Blinov, A (8 April 2010). "Evolutionary genomics revealed interkingdom distribution of Tcn1-like chromodomain-containing Gypsy LTR retrotransposons among fungi and plants.". *BMC Genomics* **11**: 231. doi:10.1186/1471-2164-11-231. PMID 20377908.

[25] Gao, X; Hou, Y; Ebina, H; Levin, HL; Voytas, DF (March 2008). "Chromodomains direct integration of retrotransposons to heterochromatin.". *Genome Research* **18** (3): 359–69. doi:10.1101/gr.7146408. PMID 18256242.

[26] Neumann, P; Navrátilová, A; Koblížková, A; Kejnovský, E; Hřibová, E; Hobza, R; Widmer, A; Doležel, J; Macas, J (3 March 2011). "Plant centromeric retrotransposons: a structural and cytogenetic perspective.". *Mobile DNA* **2** (1): 4. doi:10.1186/1759-8753-2-4. PMID 21371312.

[27] Weber, B; Heitkam, T; Holtgräwe, D; Weisshaar, B; Minoche, AE; Dohm, JC; Himmelbauer, H; Schmidt, T (1 March 2013). "Highly diverse chromoviruses of Beta vulgaris are classified by chromodomains and chromosomal integration.". *Mobile DNA* **4** (1): 8. doi:10.1186/1759-8753-4-8. PMID 23448600.

[28] Macas, J; Neumann, P (1 April 2007). "Ogre elements--a distinct group of plant Ty3/gypsy-like retrotransposons.". *Gene* **390** (1-2): 108–16. doi:10.1016/j.gene.2006.08.007. PMID 17052864.

[29] Neumann, P; Pozárková, D; Macas, J (October 2003). "Highly abundant pea LTR retrotransposon Ogre is constitutively transcribed and partially spliced.". *Plant Molecular Biology* **53** (3): 399–410. doi:10.1023/b:plan.0000006945.77043.ce. PMID 14750527.

[30] Schmidt, Thomas (1999-08-01). "LINEs, SINEs and repetitive DNA: non-LTR retrotransposons in plant genomes". *Plant Molecular Biology* **40** (6): 903–910. doi:10.1023/A:1006212929794. ISSN 0167-4412.

[31] Yadav, VP; Mandal, PK; Rao, DN; Bhattacharya, S (December 2009). "Characterization of the restriction enzyme-like endonuclease encoded by the Entamoeba histolytica non-long terminal repeat retroposon EhLINE1". *The FEBS journal* **276** (23): 7070–82. doi:10.1111/j.1742-4658.2009.07419.x. PMID 19878305.

[32] Singer MF (March 1982). "SINEs and LINEs: highly repeated short and long interspersed sequences in mammalian genomes". *Cell* **28** (3): 433–4. doi:10.1016/0092-8674(82)90194-5. PMID 6280868.

[33] Doucet AJ, Hulme AE, Sahinovic E, Kulpa DA, Moldovan JB, Kopera HC, Athanikar JN, Hasnaoui M, Bucheton A, Moran JV, Gilbert N (October 7, 2010). "Characterization of LINE-1 ribonucleoprotein particles". *PLOS Genetics* **6** (10): e1001150. doi:10.1371/journal.pgen.1001150. PMC 2951350. PMID 20949108.

[34] Denli, AM; Narvaiza, I; Kerman, BE; Pena, M; Benner, C; Marchetto, MC; Diedrich, JK; Aslanian, A; Ma, J; Moresco, JJ; Moore, L; Hunter, T; Saghatelian, A; Gage, FH (22 October 2015). "Primate-Specific ORF0 Contributes to Retrotransposon-Mediated Diversity.". *Cell* **163** (3): 583–93. doi:10.1016/j.cell.2015.09.025. PMID 26496605.

[35] Ohshima K, Okada N (2005). "SINEs and LINEs: symbionts of eukaryotic genomes with a common tail". *Cytogenet. Genome Res.* **110** (1–4): 475–90. doi:10.1159/000084981. PMID 16093701.

[36] Yadav, VP; Mandal, PK; Rao, DN; Bhattacharya, S (December 2009). "Characterization of the restriction enzyme-like endonuclease encoded by the Entamoeba histolytica non-long terminal repeat retrotransposon EhLINE1". *The FEBS journal* **276** (23): 7070–82. doi:10.1111/j.1742-4658.2009.07419.x. PMID 19878305.

[37] Deininger PL, Batzer MA (October 2002). "Mammalian retroelements". *Genome Res.* **12** (10): 1455–65. doi:10.1101/gr.282402. PMID 12368238.

[38] Richard Cordaux and Mark Batzer (October 2009). "The impact of retrotransposons on human genome evolution". *Nature Reviews Genetics* **10** (10): 691–703. doi:10.1038/nrg2640. PMC 2884099. PMID 19763152.

[39] Griffiths, Anthony J. (2008). *Introduction to genetic analysis* (9th ed.). New York: W.H. Freeman. p. 505. ISBN 0-7167-6887-9.

[40] Rangwala S, Kazazian HH (2009). "Many LINE1 elements contribute to the transcriptome of human somatic cells". *Genome Biology* **10** (9): R100. doi:10.1186/gb-2009-10-9-r100. PMC 2768975. PMID 19772661.

[41] Chueh, A.C.; Northrop, Emma L.; Brettingham-Moore, Kate H.; Choo, K. H. Andy; Wong, Lee H. (Jan 2009). Bickmore, Wendy A., ed. "LINE Retrotransposon RNA Is an Essential Structural and Functional Epigenetic Component of a Core Neocentromeric Chromatin". *PLoS Genetics* **5** (1): e1000354. doi:10.1371/journal.pgen.1000354. PMC 2625447. PMID 19180186.

[42] Stansfield, William D.; King, Robert C. (1997). *A dictionary of genetics* (5th ed.). Oxford [Oxfordshire]: Oxford University Press. ISBN 0-19-509441-7.

[43] Santangelo, Andrea; de Souza, Flavio; Franchini, Lucia; Bumaschny, Viviana; Low, Malcolm; Rubinstein,Marcelo (October 2007). "Ancient Exaptation of a CORE-SINE Retroposon into a Highly Conserved Mammalian Neuronal Enhancer of the Proopiomelanocortin Gene". *PLoS Genetics* (Public Library of Science) **3** (10): 1813–26. doi:10.1371/journal.pgen.0030166. PMC 2000970. PMID 17922573. Retrieved 2007-12-31. Cite uses deprecated parameter |coauthors= (help)

[44] Liang, Kung-Hao; Yeh, Chau-Ting (2013). "A gene expression restriction network mediated by sense and antisense Alu sequences located on protein-coding messenger RNAs.". *BMC Genomics* **14**: 325. doi:10.1186/1471-2164-14-325. PMC 3655826. PMID 23663499. Retrieved 2013-05-11.

[45] Yadav, Vijay Pal; Mandal, Prabhat Kumar; Bhattacharya, Alok; Bhattacharya, Sudha (21 May 2012). "Recombinant SINEs are formed at high frequency during induced retrotransposition in vivo". *Nature Communications* **3**: 854. doi:10.1038/ncomms1855.

Chapter 28

Intron

For the interferon-based drug used in viral and cancer treatments, see Intron A. For the album by LCD Soundsystem, see Introns (album).

An **intron** is any nucleotide sequence within a gene that is removed by RNA splicing during maturation of the final

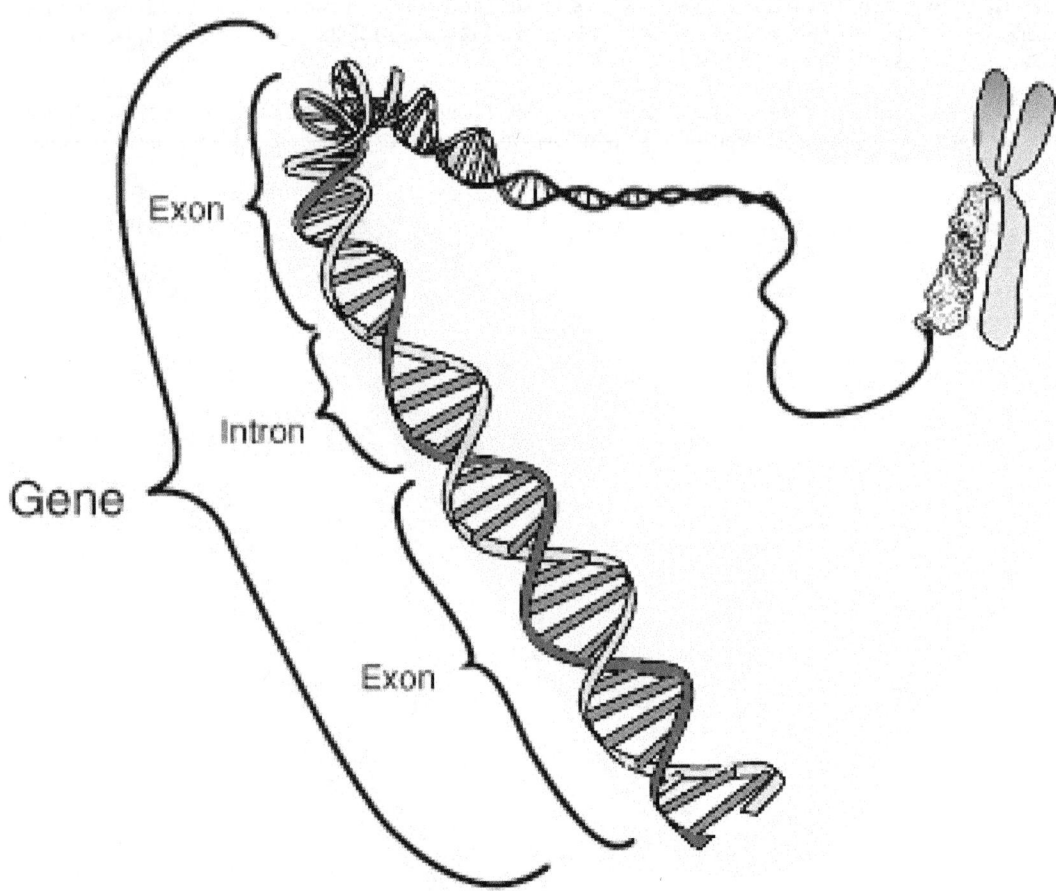

Representation of intron and exons within a simple gene containing a single intron.

RNA product.[1][2] The term *intron* refers to both the DNA sequence within a gene and the corresponding sequence in

RNA transcripts.[3] Sequences that are joined together in the final mature RNA after RNA splicing are exons. Introns are found in the genes of most organisms and many viruses, and can be located in a wide range of genes, including those that generate proteins, ribosomal RNA (rRNA), and transfer RNA (tRNA). When proteins are generated from intron-containing genes, RNA splicing takes place as part of the RNA processing pathway that follows transcription and precedes translation.

The word *intron* is derived from the term *intragenic region*, i.e. a region inside a gene. Although introns are sometimes called *intervening sequences*, the term "intervening sequence" can refer to any of several families of internal nucleic acid sequences that are not present in the final gene product, including inteins, untranslated sequences (UTR), and nucleotides removed by RNA editing, in addition to introns.

28.1 Introduction

Introns were first discovered in protein-coding genes of adenovirus,[4][5] and were subsequently identified in genes encoding transfer RNA and ribosomal RNA genes. Introns are now known to occur within a wide variety of genes throughout organisms and viruses within all of the biological kingdoms.

The fact that genes were split or interrupted by introns was discovered independently in 1977 by Phillip Allen Sharp and Richard J. Roberts, for which they shared the Nobel Prize in Physiology or Medicine in 1993.[6] The term *intron* was introduced by American biochemist Walter Gilbert:[7]

> "The notion of the cistron [...] must be replaced by that of a transcription unit containing regions which will be lost from the mature messenger - which I suggest we call introns (for intragenic regions) - alternating with regions which will be expressed - exons." (Gilbert 1978)

The frequency of introns within different genomes is observed to vary widely across the spectrum of biological organisms. For example, introns are extremely common within the nuclear genome of higher vertebrates (e.g. humans and mice), where protein-coding genes almost always contain multiple introns, while introns are rare within the nuclear genes of some eukaryotic microorganisms,[8] for example baker's/brewer's yeast (*Saccharomyces cerevisiae*). In contrast, the mitochondrial genomes of vertebrates are entirely devoid of introns, while those of eukaryotic microorganisms may contain many introns. Introns are well known in bacterial and archaeal genes, but occur more rarely than in most eukaryotic genomes.

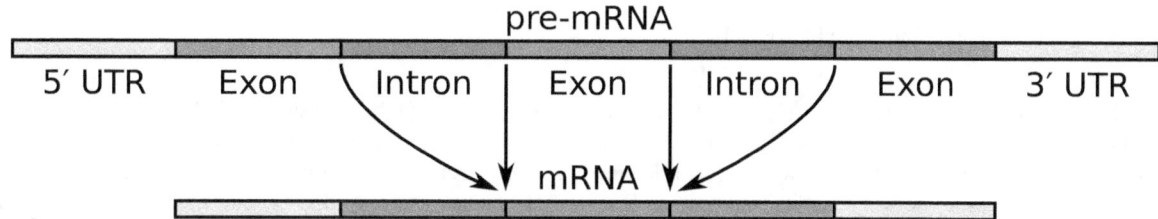

Simple illustration of an unspliced mRNA precursor, with two introns and three exons (top). After the introns have been removed via splicing, the mature mRNA sequence is ready for translation (bottom).

A particularly extreme case is the *Drosophila dhc7* gene containing a ≥3.6 Mb intron, which takes roughly three days to transcribe.[9][10]

28.2 Classification

Splicing of all intron-containing RNA molecules is superficially similar, as described above. However, different types of introns were identified through the examination of intron structure by DNA sequence analysis, together with genetic and biochemical analysis of RNA is splicing reactions.

At least four distinct classes of introns have been identified.[1]

- Introns in nuclear protein-coding genes that are removed by spliceosomes (spliceosomal introns)

- Introns in nuclear and archaeal transfer RNA genes that are removed by proteins (tRNA introns)

- Self-splicing group I introns that are removed by RNA catalysis.

- Self-splicing group II introns that are removed by RNA catalysis

Group III introns are proposed to be a fifth family, but little is known about the biochemical apparatus that mediates their splicing. They appear to be related to group II introns, and possibly to spliceosomal introns.[11]

28.2.1   Spliceosomal introns

See also: RNA splicing § Spliceosomal

Nuclear pre-mRNA introns (spliceosomal introns) are characterized by specific intron sequences located at the boundaries between introns and exons.[12] These sequences are recognized by spliceosomal RNA molecules when the splicing reactions are initiated.[13] In addition, they contain a branch point, a particular nucleotide sequence near the 3' end of the intron that becomes covalently linked to the 5' end of the intron during the splicing process, generating a branched (*lariat*) intron. Apart from these three short conserved elements, nuclear pre-mRNA intron sequences are highly variable. Nuclear pre-mRNA introns are often much longer than their surrounding exons.

28.2.2   tRNA introns

Transfer RNA introns that depend upon proteins for removal occur at a specific location within the anticodon loop of unspliced tRNA precursors, and are removed by a tRNA splicing endonuclease. The exons are then linked together by a second protein, the tRNA splicing ligase.[14] Note that self-splicing introns are also sometimes found within tRNA genes.[15]

28.2.3   Group I and group II introns

See also: Group I catalytic intron and Group II intron

Group I and group II introns are found in genes encoding proteins (messenger RNA), transfer RNA and ribosomal RNA in a very wide range of living organisms.,[16][17] Following transcription into RNA, group I and group II introns also make extensive internal interactions that allow them to fold into a specific, complex three-dimensional architecture. These complex architectures allow some group I and group II introns to be *self-splicing*, that is, the intron-containing RNA molecule can rearrange its own covalent structure so as to precisely remove the intron and link the exons together in the correct order. In some cases, particular intron-binding proteins are involved in splicing, acting in such a way that they assist the intron in folding into the three-dimensional structure that is necessary for self-splicing activity. Group I and group II introns are distinguished by different sets of internal conserved sequences and folded structures, and by the fact that splicing of RNA molecules containing group II introns generates branched introns (like those of spliceosomal RNAs), while group I introns use a non-encoded guanosine nucleotide (typically GTP) to initiate splicing, adding it on to the 5'-end of the excised intron.

28.3   Biological functions and evolution

While introns do not encode protein products, they are integral to gene expression regulation. Some introns themselves encode functional RNAs through further processing after splicing to generate noncoding RNA molecules.[18] Alternative splicing is widely used to generate multiple proteins from a single gene. Furthermore, some introns play essential roles in a wide range of gene expression regulatory functions such as non-sense mediated decay[19] and mRNA export.[20]

The biological origins of introns are obscure. After the initial discovery of introns in protein-coding genes of the eukaryotic nucleus, there was significant debate as to whether introns in modern-day organisms were inherited from a common ancient ancestor (termed the introns-early hypothesis), or whether they appeared in genes rather recently in the evolutionary process (termed the introns-late hypothesis). Another theory is that the spliceosome and the intron-exon structure of genes is a relic of the RNA world (the introns-first hypothesis).[21] There is still considerable debate about the extent to which of these hypotheses is most correct. The popular consensus at the moment is that introns arose within the eukaryote lineage as selfish elements .

Early studies of genomic DNA sequences from a wide range of organisms show that the intron-exon structure of homologous genes in different organisms can vary widely.[22] More recent studies of entire eukaryotic genomes have now shown that the lengths and density (introns/gene) of introns varies considerably between related species. For example, while the human genome contains an average of 8.4 introns/gene (139,418 in the genome), the unicellular fungus *Encephalitozoon cuniculi* contains only 0.0075 introns/gene (15 introns in the genome).[23] Since eukaryotes arose from a common ancestor (Common descent), there must have been extensive gain or loss of introns during evolutionary time.[24][25] This process is thought to be subject to selection, with a tendency towards intron gain in larger species due to their smaller population sizes, and the converse in smaller (particularly unicellular) species.[26] Biological factors also influence which genes in a genome lose or accumulate introns.[27][28][29]

Alternative splicing of introns within a gene acts to introduce greater variability of protein sequences translated from a single gene, allowing multiple related proteins to be generated from a single gene and a single precursor mRNA transcript. The control of alternative RNA splicing is performed by a complex network of signaling molecules that respond to a wide range of intracellular and extracellular signals.

Introns contain several short sequences that are important for efficient splicing, such as acceptor and donor sites at either end of the intron as well as a branch point site, which are required for proper splicing by the spliceosome. Some introns are known to enhance the expression of the gene that they are contained in by a process known as intron-mediated enhancement (IME).

28.4 Introns as mobile genetic elements

Introns may be lost or gained over evolutionary time, as shown by many comparative studies of orthologous genes. Subsequent analyses have identified thousands of examples of intron loss and gain events, and it has been proposed that the emergence of eukaryotes, or the initial stages of eukaryotic evolution, involved an intron invasion.[30] Two definitive mechanisms of intron loss, Reverse Transcriptase-Mediated Intron Loss (RTMIL) and genomic deletions, have been identified, and are known to occur.[31] The definitive mechanisms of intron gain, however, remain elusive and controversial. At least seven mechanisms of intron gain have been reported thus far: Intron Transposition, Transposon Insertion, Tandem Genomic Duplication, Intron Transfer, Intron Gain during Double-Strand Break Repair (DSBR), Insertion of a Group II Intron, and Intronization. In theory it should be easiest to deduce the origin of recently gained introns due to the lack of host-induced mutations, yet even introns gained recently did not arise from any of the aforementioned mechanisms. These findings thus raise the question of whether or not the proposed mechanisms of intron gain fail to describe the mechanistic origin of many novel introns because they are not accurate mechanisms of intron gain, or if there are other, yet to be discovered, processes generating novel introns.[32]

In intron transposition, the most commonly purported intron gain mechanism, a spliced intron is thought to reverse splice into either its own mRNA or another mRNA at a previously intron-less position. This intron-containing mRNA is then reverse transcribed and the resulting intron-containing cDNA may then cause intron gain via complete or partial recombination with its original genomic locus. Transposon insertions can also result in intron creation. Such an insertion could intronize the transposon without disrupting the coding sequence when a transposon inserts into the sequence AGGT, resulting in the duplication of this sequence on each side of the transposon. It is not yet understood why these elements are spliced, whether by chance, or by some preferential action by the transposon. In tandem genomic duplication, due to the similarity between consensus donor and acceptor splice sites, which both closely resemble AGGT, the tandem genomic duplication of an exonic segment harboring an AGGT sequence generates two potential splice sites. When recognized by the spliceosome, the sequence between the original and duplicated AGGT will be spliced, resulting in the creation of an intron without alteration of the coding sequence of the gene. Double-stranded break repair via non-homologous end joining was recently identified as a source of intron gain when researchers identified short direct repeats flanking 43% of gained introns in Daphnia.[32] These numbers must be compared to the number of conserved introns flanked by repeats in other organisms, though, for statistical relevance. For group II intron insertion, the retrohoming of a group II intron into a nuclear gene was proposed to cause recent spliceosomal intron gain.

Intron transfer has been hypothesized to result in intron gain when a paralog or pseudogene gains an intron and then transfers this intron via recombination to an intron-absent location in its sister paralog. Intronization is the process by which mutations create novel introns from formerly exonic sequence. Thus, unlike other proposed mechanisms of intron gain, this mechanism does not require the insertion or generation of DNA to create a novel intron.[32]

The only hypothesized mechanism of recent intron gain lacking any direct evidence is that of group II intron insertion, which when demonstrated in vivo, abolishes gene expression.[33] Group II introns are therefore likely the presumed ancestors of spliceosomal introns, acting as site-specific retroelements, and are no longer responsible for intron gain.[34][35] Tandem genomic duplication is the only proposed mechanism with supporting in vivo experimental evidence: a short intragenic tandem duplication can insert a novel intron into a protein-coding gene, leaving the corresponding peptide sequence unchanged.[36] This mechanism also has extensive indirect evidence lending support to the idea that tandem genomic duplication is a prevalent mechanism for intron gain. The testing of other proposed mechanisms in vivo, particularly intron gain during DSBR, intron transfer, and intronization, is possible, although these mechanisms must be demonstrated in vivo to solidify them as actual mechanisms of intron gain. Further genomic analyses, especially when executed at the population level, may then quantify the relative contribution of each mechanism, possibly identifying species-specific biases that may shed light on varied rates of intron gain amongst different species.[32]

28.5   See also

Structure:

- Exon

- mRNA

- Eukaryotic chromosome fine structure

- Small t intron

Splicing:

- Alternative splicing

- Exitron

- Minor spliceosome

Function

- MicroRNA

Others:

- Intein

- Interrupted gene

- Noncoding DNA

- Noncoding RNA

- Selfish DNA

- Twintron

28.6 References

[1] Alberts, Bruce (2008). *Molecular biology of the cell*. New York: Garland Science. ISBN 0-8153-4105-9.

[2] Stryer, Lubert; Berg, Jeremy Mark; Tymoczko, John L. (2007). *Biochemistry*. San Francisco: W.H. Freeman. ISBN 0-7167-6766-X.

[3] Kinniburgh, Alan; mertz, j; Ross, J. (July 1978). "The precursor of mouse β-globin messenger RNA contains two intervening RNA sequences". *Cell* **14** (3): 681–693. doi:10.1016/0092-8674(78)90251-9. PMID 688388.

[4] Chow LT, Gelinas RE, Broker TR, Roberts RJ (September 1977). "An amazing sequence arrangement at the 5' ends of adenovirus 2 messenger RNA". *Cell* **12** (1): 1–8. doi:10.1016/0092-8674(77)90180-5. PMID 902310.

[5] Berget SM, Moore C, Sharp PA (August 1977). "Spliced segments at the 5' terminus of adenovirus 2 late mRNA". *Proc. Natl. Acad. Sci. U.S.A.* **74** (8): 3171–5. doi:10.1073/pnas.74.8.3171. PMC 431482. PMID 269380.

[6] http://www.nobelprize.org/nobel_prizes/medicine/laureates/1993/press.html

[7] Gilbert, Walter (1978). "Why genes in pieces". *Nature* **271** (5645): 501–501. doi:10.1038/271501a0. PMID 622185.

[8] Stajich JE, Dietrich FS, Roy SW (2007). "Comparative genomic analysis of fungal genomes reveals intron-rich ancestors". *Genome Biol.* **8** (10): R223. doi:10.1186/gb-2007-8-10-r223. PMC 2246297. PMID 17949488.

[9] Tollervey, David; Caceres, Javier F (November 2000). "RNA Processing Marches on". *Cell* **103** (5): 703–709. doi:10.1016/S0092-8674(00)00174-4. Retrieved 12 December 2014.

[10] Reugels, AM; Kurek, R; Lammermann, U; Bünemann, H (February 2000). "Mega-introns in the dynein gene DhDhc7(Y) on the heterochromatic Y chromosome give rise to the giant threads loops in primary spermatocytes of Drosophila hydei.". *Genetics* **154** (2): 759–69. PMID 10655227. Retrieved 12 December 2014.

[11] Copertino DW, Hallick RB (December 1993). "Group II and group III introns of twintrons: potential relationships with nuclear pre-mRNA introns". *Trends Biochem. Sci.* **18** (12): 467–71. doi:10.1016/0968-0004(93)90008-b. PMID 8108859.

[12] Padgett RA, Grabowski PJ, Konarska MM, Seiler S, Sharp PA (1986). "Splicing of messenger RNA precursors". *Annu. Rev. Biochem.* **55**: 1119–50. doi:10.1146/annurev.bi.55.070186.005351. PMID 2943217.

[13] Guthrie C, Patterson B (1988). "Spliceosomal snRNAs". *Annu. Rev. Genet.* **22**: 387–419. doi:10.1146/annurev.ge.22.120188.002131. PMID 2977088.

[14] Greer CL, Peebles CL, Gegenheimer P, Abelson J (February 1983). "Mechanism of action of a yeast RNA ligase in tRNA splicing". *Cell* **32** (2): 537–46. doi:10.1016/0092-8674(83)90473-7. PMID 6297798.

[15] Reinhold-Hurek B, Shub DA (May 1992). "Self-splicing introns in tRNA genes of widely divergent bacteria". *Nature* **357** (6374): 173–6. doi:10.1038/357173a0. PMID 1579169.

[16] Cech TR (1990). "Self-splicing of group I introns". *Annu. Rev. Biochem.* **59**: 543–68. doi:10.1146/annurev.bi.59.070190.002551. PMID 2197983.

[17] Michel F, Ferat JL (1995). "Structure and activities of group II introns". *Annu. Rev. Biochem.* **64**: 435–61. doi:10.1146/annurev.bi.64.070195.0022: PMID 7574489.

[18] Rearick D, Prakash A, McSweeny A, Shepard SS, Fedorova L, Fedorov A (March 2011). "Critical association of ncRNA with introns". *Nucleic Acids Res.* **39** (6): 2357–66. doi:10.1093/nar/gkq1080. PMC 3064772. PMID 21071396.

[19] Bicknell AA, Cenik C, Chua HN, Roth FP, Moore MJ (Dec 2012). "Introns in UTRs: why we should stop ignoring them.". *Bioessays* **34** (12): 1025–34. doi:10.1002/bies.201200073. PMID 23108796.

[20] Cenik, Can; Chua, Hon Nian; Zhang, Hui; Tarnawsky, Stefan P.; Akef, Abdalla; Derti, Adnan; Tasan, Murat; Moore, Melissa J.; Palazzo, Alexander F.; Roth, Frederick P. (2011). Snyder, Michael, ed. "Genome Analysis Reveals Interplay between 5′UTR Introns and Nuclear mRNA Export for Secretory and Mitochondrial Genes". *PLoS Genetics* **7** (4): e1001366. doi:10.1371/journal.pgen.100136(ISSN 1553-7404. PMC 3077370. PMID 21533221.

[21] Penny D, Hoeppner MP, Poole AM, Jeffares DC (November 2009). "An overview of the introns-first theory". *Journal of Molecular Evolution* **69** (5): 527–40. doi:10.1007/s00239-009-9279-5. PMID 19777149.

[22] Rodríguez-Trelles F, Tarrío R, Ayala FJ (2006). "Origins and evolution of spliceosomal introns". *Annu. Rev. Genet* **40**: 47–76. doi:10.1146/annurev.genet.40.110405.090625. PMID 17094737.

[23] Mourier T, Jeffares DC (May 2003). "Eukaryotic intron loss". *Science* **300** (5624): 1393–1393. doi:10.1126/science.1080559. PMID 12775832.

[24] Roy SW, Gilbert W (March 2006). "The evolution of spliceosomal introns: patterns, puzzles and progress". *Nature Reviews Genetics* **7** (3): 211–21. doi:10.1038/nrg1807. PMID 16485020.

[25] de Souza SJ (July 2003). "The emergence of a synthetic theory of intron evolution". *Genetica* **118** (2–3): 117–21. doi:10.1023/A:102419332339 PMID 12868602.

[26] Lynch M (April 2002). "Intron evolution as a population-genetic process". *Proceedings of the National Academy of Sciences* **99** (9): 6118–23. doi:10.1073/pnas.092595699. PMC 122912. PMID 11983904.

[27] Jeffares DC, Mourier T, Penny D (January 2006). "The biology of intron gain and loss". *Trends in Genetics* **22** (1): 16–22. doi:10.1016/j.tig.2005.10.006. PMID 16290250.

[28] Jeffares DC, Penkett CJ, Bähler J (August 2008). "Rapidly regulated genes are intron poor". *Trends in Genetics* **24** (8): 375–8. doi:10.1016/j.tig.2008.05.006. PMID 18586348.

[29] Castillo-Davis CI, Mekhedov SL, Hartl DL, Koonin EV, Kondrashov FA (August 2002). "Selection for short introns in highly expressed genes". *Nature Genetics* **31** (4): 415–8. doi:10.1038/ng940. PMID 12134150.

[30] Rogozin, I. B.; Carmel, L.; Csuros, M.; Koonin, E. V. (2012). "Origin and evolution of spliceosomal introns". *Biology Direct* **7**: 11. doi:10.1186/1745-6150-7-11. PMC 3488318. PMID 22507701.

[31] Derr, L. K.; Strathern, J. N. (1993). "A role for reverse transcripts in gene conversion". *Nature* **361** (6408): 170–173. doi:10.1038/361170a0. PMID 8380627.

[32] Yenerall, P.; Zhou, L. (2012). "Identifying the mechanisms of intron gain: Progress and trends". *Biology Direct* **7**: 29. doi:10.1186/1745-6150-7-29. PMC 3443670. PMID 22963364.

[33] Chalamcharla, V. R.; Curcio, M. J.; Belfort, M. (2010). "Nuclear expression of a group II intron is consistent with spliceosomal intron ancestry". *Genes & Development* **24** (8): 827–836. doi:10.1101/gad.1905010. PMC 2854396. PMID 20351053.

[34] Cech, T. R. (1986). "The generality of self-splicing RNA: Relationship to nuclear mRNA splicing". *Cell* **44** (2): 207–210. doi:10.1016/0092-8674(86)90751-8. PMID 2417724.

[35] Dickson, L.; Huang, H. -R.; Liu, L.; Matsuura, M.; Lambowitz, A. M.; Perlman, P. S. (2001). "Retrotransposition of a yeast group II intron occurs by reverse splicing directly into ectopic DNA sites". *Proceedings of the National Academy of Sciences* **98** (23): 13207. doi:10.1073/pnas.231494498.

[36] Hellsten, U.; Aspden, J. L.; Rio, D. C.; Rokhsar, D. S. (2011). "A segmental genomic duplication generates a functional intron". *Nature Communications* **2**: 454–. doi:10.1038/ncomms1461. PMC 3265369. PMID 21878908.

28.7 External links

- A search engine for exon/intron sequences defined by NCBI

- Bruce Alberts, Alexander Johnson, Julian Lewis, Martin Raff, Keith Roberts, and Peter Walter *Molecular Biology of the Cell*, 2007, ISBN 978-0-8153-4105-5. Fourth edition is available online through the NCBI Bookshelf: link

- Jeremy M Berg, John L Tymoczko, and Lubert Stryer, *Biochemistry* 5th edition, 2002, W H Freeman. Available online through the NCBI Bookshelf: link

- Intron finding tool for plant genomic sequences

- Exon-intron graphic maker

Chapter 29

Non-coding RNA

A **non-coding RNA** (**ncRNA**) is an RNA molecule that is not translated into a protein. Less-frequently used synonyms are **non-protein-coding RNA** (npcRNA), **non-messenger RNA** (nmRNA) and **functional RNA** (fRNA). The DNA sequence from which a functional non-coding RNA is transcribed is often called an RNA gene.

Non-coding RNA genes include highly abundant and functionally important RNAs such as transfer RNAs (tRNAs) and ribosomal RNAs (rRNAs), as well as RNAs such as snoRNAs, microRNAs, siRNAs, snRNAs, exRNAs, piRNAs and scaRNAs and the long ncRNAs that include examples such as Xist and HOTAIR (see here for a more complete list of ncRNAs). The number of ncRNAs encoded within the human genome is unknown; however, recent transcriptomic and bioinformatic studies suggest the existence of thousands of ncRNAs.[1][2][3][4], but see [5] Since many of the newly identified ncRNAs have not been validated for their function, it is possible that many are non-functional.[6] It is also likely that many ncRNAs are non functional (sometimes referred to as **Junk RNA**), and are the product of spurious transcription.[7]

29.1   History and discovery

Further information: History of molecular biology

Nucleic acids were first discovered in 1868 by Friedrich Miescher[8] and by 1939 RNA had been implicated in protein synthesis.[9] Two decades later, Francis Crick predicted a functional RNA component which mediated translation; he reasoned that RNA is better suited to base-pair with an mRNA transcript than a pure polypeptide.[10]

The first non-coding RNA to be characterised was an alanine tRNA found in baker's yeast, its structure was published in 1965.[11] To produce a purified alanine tRNA sample, Robert W. Holley *et al.* used 140kg of commercial baker's yeast to give just 1g of purified tRNA^{Ala} for analysis.[12] The 80 nucleotide tRNA was sequenced by first being digested with Pancreatic ribonuclease (producing fragments ending in Cytosine or Uridine) and then with takadiastase ribonuclease Tl (producing fragments which finished with Guanosine). Chromatography and identification of the 5' and 3' ends then helped arrange the fragments to establish the RNA sequence.[12] Of the three structures originally proposed for this tRNA,[11] the 'cloverleaf' structure was independently proposed in several following publications.[13][14][15][16] The cloverleaf secondary structure was finalised following X-ray crystallography analysis performed by two independent research groups in 1974.[17][18]

Ribosomal RNA was next to be discovered, followed by URNA in the early 1980s. Since then, the discovery of new non-coding RNAs has continued with snoRNAs, Xist, CRISPR and many more.[19] Recent notable additions include riboswitches and miRNA; the discovery of the RNAi mechanism associated with the latter earned Craig C. Mello and Andrew Fire the 2006 Nobel Prize in Physiology or Medicine.[20]

The cloverleaf structure of Yeast tRNA^{Phe} (inset) and the 3D structure determined by X-ray analysis.

29.2 Biological roles of ncRNA

Noncoding RNAs belong to several groups and are involved in many cellular processes. These range from ncRNAs of central importance that are conserved across all or most cellular life through to more transient ncRNAs specific to one or a few closely related species. The more conserved ncRNAs are thought to be molecular fossils or relics from LUCA and the RNA world and their current roles remain mostly in regulation of information flow from DNA to protein.[21][22][23]

29.2.1 ncRNAs in translation

Many of the conserved, essential and abundant ncRNAs are involved in translation. Ribonucleoprotein (RNP) particles called ribosomes are the 'factories' where translation takes place in the cell. The ribosome consists of more than 60% ribosomal RNA; these are made up of 3 ncRNAs in prokaryotes and 4 ncRNAs in eukaryotes. Ribosomal RNAs catalyse the translation of nucleotide sequences to protein. Another set of ncRNAs, Transfer RNAs, form an 'adaptor molecule'

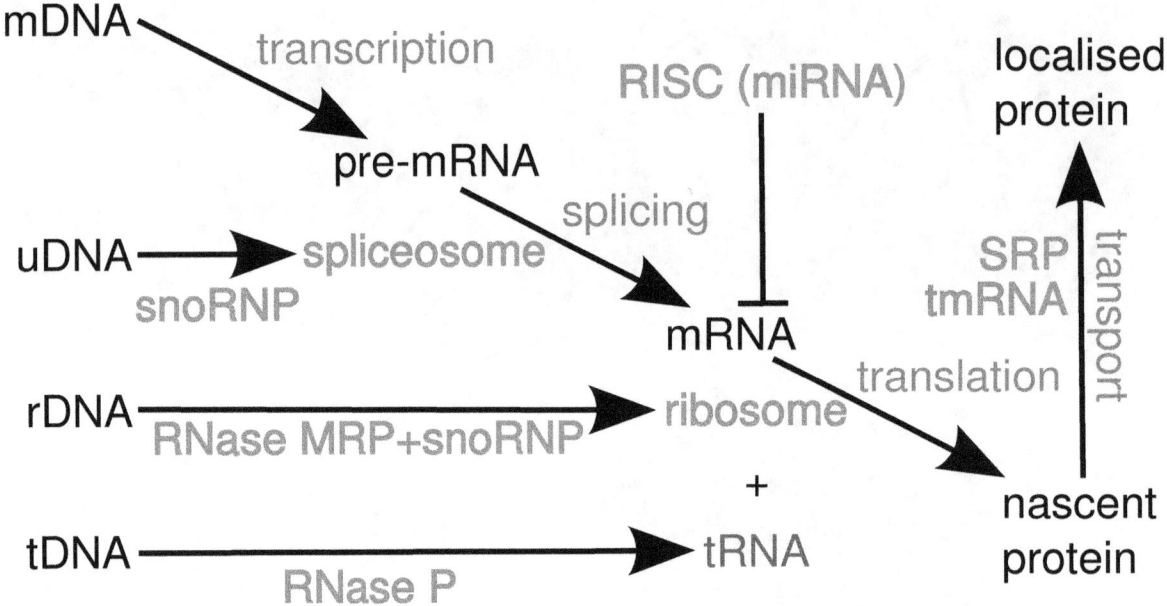

An illustration of the central dogma of molecular biology annotated with the processes ncRNAs are involved in. RNPs are shown in red, ncRNAs are shown in blue.

between mRNA and protein. The H/ACA box and C/D box snoRNAs are ncRNAs found in archaea and eukaryotes. RNase MRP is restricted to eukaryotes. Both groups of ncRNA are involved in the maturation of rRNA. The snoRNAs guide covalent modifications of rRNA, tRNA and snRNAs; RNase MRP cleaves the internal transcribed spacer 1 between 18S and 5.8S rRNAs.[25] The ubiquitous ncRNA, RNase P, is an evolutionary relative of RNase MRP.[25] RNase P matures tRNA sequences by generating mature 5'-ends of tRNAs through cleaving the 5'-leader elements of precursor-tRNAs. Another ubiquitous RNP called SRP recognizes and transports specific nascent proteins to the endoplasmic reticulum in eukaryotes and the plasma membrane in prokaryotes. In bacteria Transfer-messenger RNA (tmRNA) is an RNP involved in rescuing stalled ribosomes, tagging incomplete polypeptides and promoting the degradation of aberrant mRNA.

29.2.2   ncRNAs in RNA splicing

In eukaryotes the spliceosome performs the splicing reactions essential for removing intron sequences, this process is required for the formation of mature mRNA. The spliceosome is another RNP often also known as the snRNP or tri-snRNP. There are two different forms of the spliceosome, the major and minor forms. The ncRNA components of the major spliceosome are U1, U2, U4, U5, and U6. The ncRNA components of the minor spliceosome are U11, U12, U5, U4atac and U6atac.

Another group of introns can catalyse their own removal from host transcripts; these are called self-splicing RNAs. There are two main groups of self-splicing RNAs: group I catalytic intron and group II catalytic intron. These ncRNAs catalyze their own excision from mRNA, tRNA and rRNA precursors in a wide range of organisms.

In mammals it has been found that snoRNAs can also regulate the alternative splicing of mRNA, for example snoRNA HBII-52 regulates the splicing of serotonin receptor 2C.[26]

In nematodes, the SmY ncRNA appears to be involved in mRNA trans-splicing.

29.2.3   ncRNAs in DNA replication

Y RNAs are stem loops, necessary for DNA replication through interactions with chromatin and initiation proteins (including the origin recognition complex).[28][29] They are also components of the Ro60 ribonucleoprotein particle[30] which

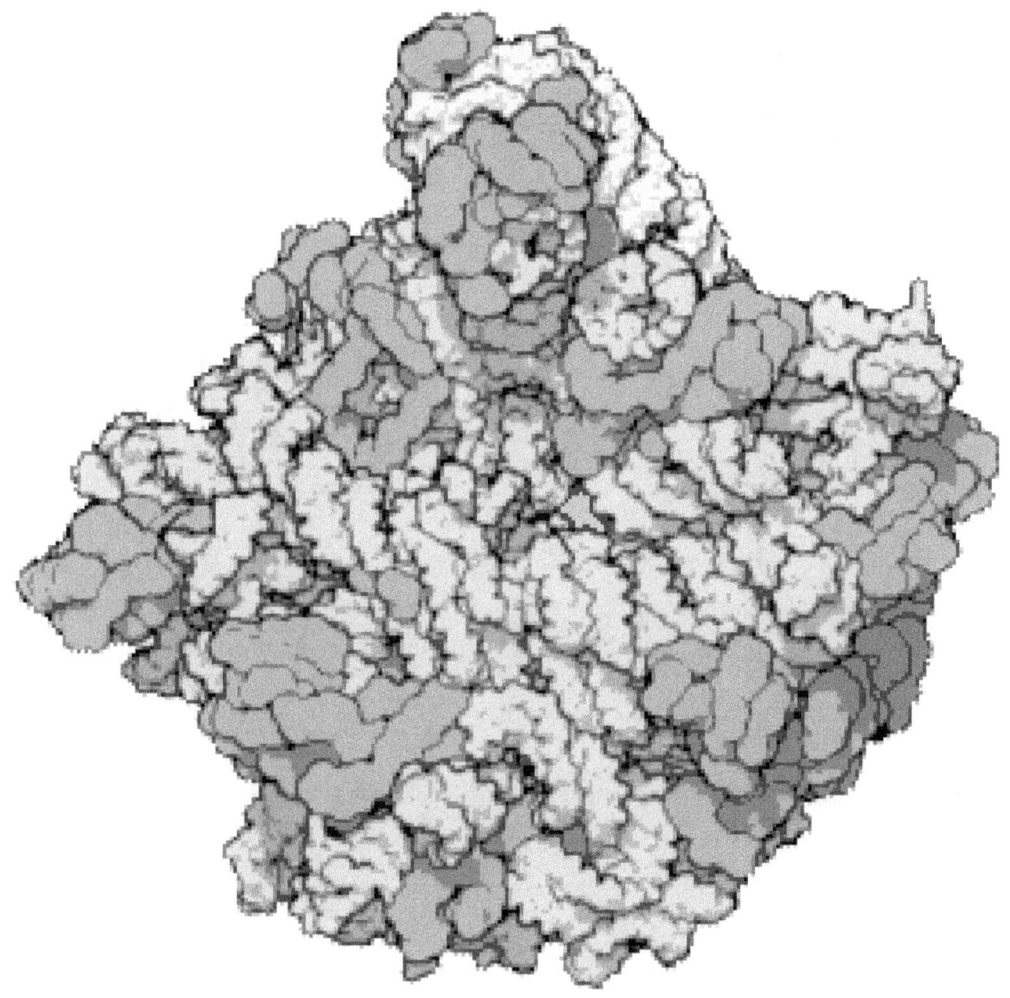

Atomic structure of the 50S Subunit from Haloarcula marismortui. *Proteins are shown in blue and the two RNA strands in orange and yellow.*[24] *The small patch of green in the center of the subunit is the active site.*

is a target of autoimmune antibodies in patients with systemic lupus erythematosus.[31]

29.2.4 ncRNAs in gene regulation

The expression of many thousands of genes are regulated by ncRNAs. This regulation can occur in trans or in cis.

Trans-acting ncRNAs

In higher eukaryotes microRNAs regulate gene expression. A single miRNA can reduce the expression levels of hundreds of genes. The mechanism by which mature miRNA molecules act is through partial complementary to one or more messenger RNA (mRNA) molecules, generally in 3' UTRs. The main function of miRNAs is to down-regulate gene

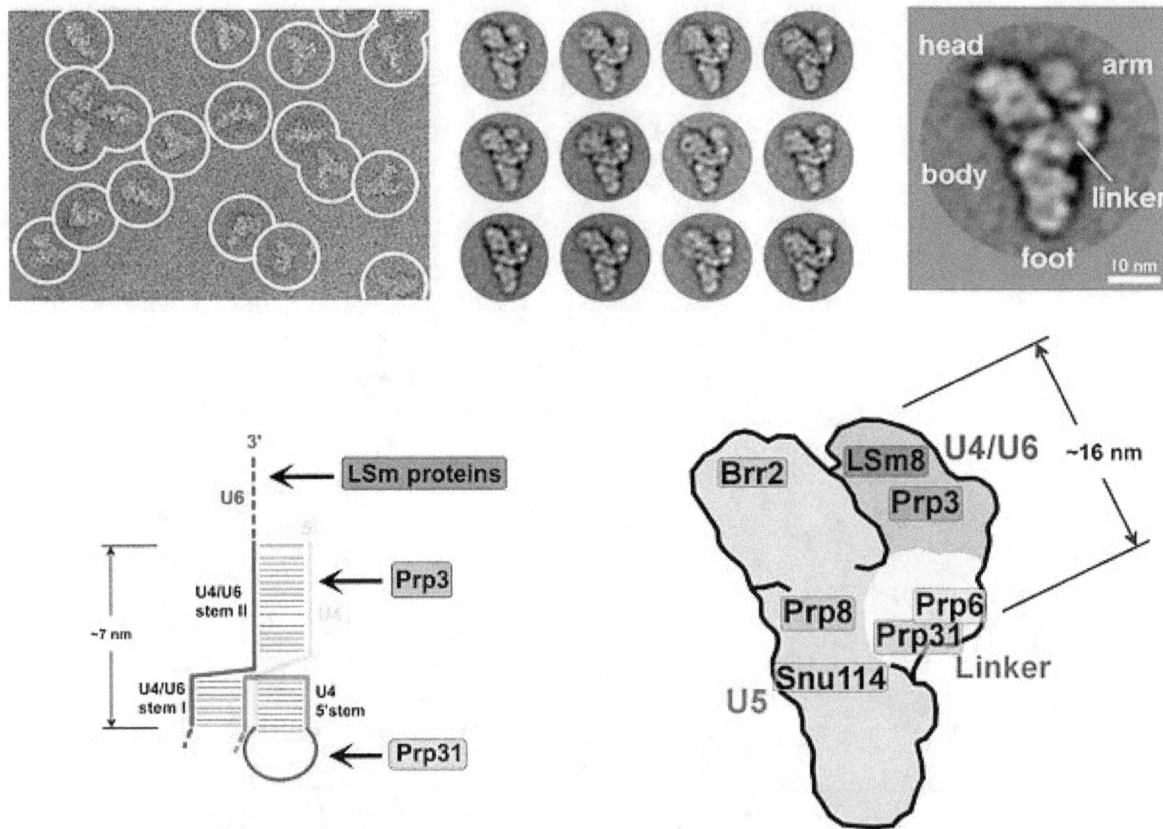

Electron microscopy images of the yeast spliceosome. Note the bulk of the complex is in fact ncRNA.

expression.

The ncRNA RNase P has also been shown to influence gene expression. In the human nucleus RNase P is required for the normal and efficient transcription of various ncRNAs transcribed by RNA polymerase III. These include tRNA, 5S rRNA, SRP RNA, and U6 snRNA genes. RNase P exerts its role in transcription through association with Pol III and chromatin of active tRNA and 5S rRNA genes.[32]

It has been shown that 7SK RNA, a metazoan ncRNA, acts as a negative regulator of the RNA polymerase II elongation factor P-TEFb, and that this activity is influenced by stress response pathways.

The bacterial ncRNA, 6S RNA, specifically associates with RNA polymerase holoenzyme containing the sigma70 specificity factor. This interaction represses expression from a sigma70-dependent promoter during stationary phase.

Another bacterial ncRNA, OxyS RNA represses translation by binding to Shine-Dalgarno sequences thereby occluding ribosome binding. OxyS RNA is induced in response to oxidative stress in Escherichia coli.

The B2 RNA is a small noncoding RNA polymerase III transcript that represses mRNA transcription in response to heat shock in mouse cells. B2 RNA inhibits transcription by binding to core Pol II. Through this interaction, B2 RNA assembles into preinitiation complexes at the promoter and blocks RNA synthesis.[33]

A recent study has shown that just the act of transcription of ncRNA sequence can have an influence on gene expression. RNA polymerase II transcription of ncRNAs is required for chromatin remodelling in the Schizosaccharomyces pombe. Chromatin is progressively converted to an open configuration, as several species of ncRNAs are transcribed.[34]

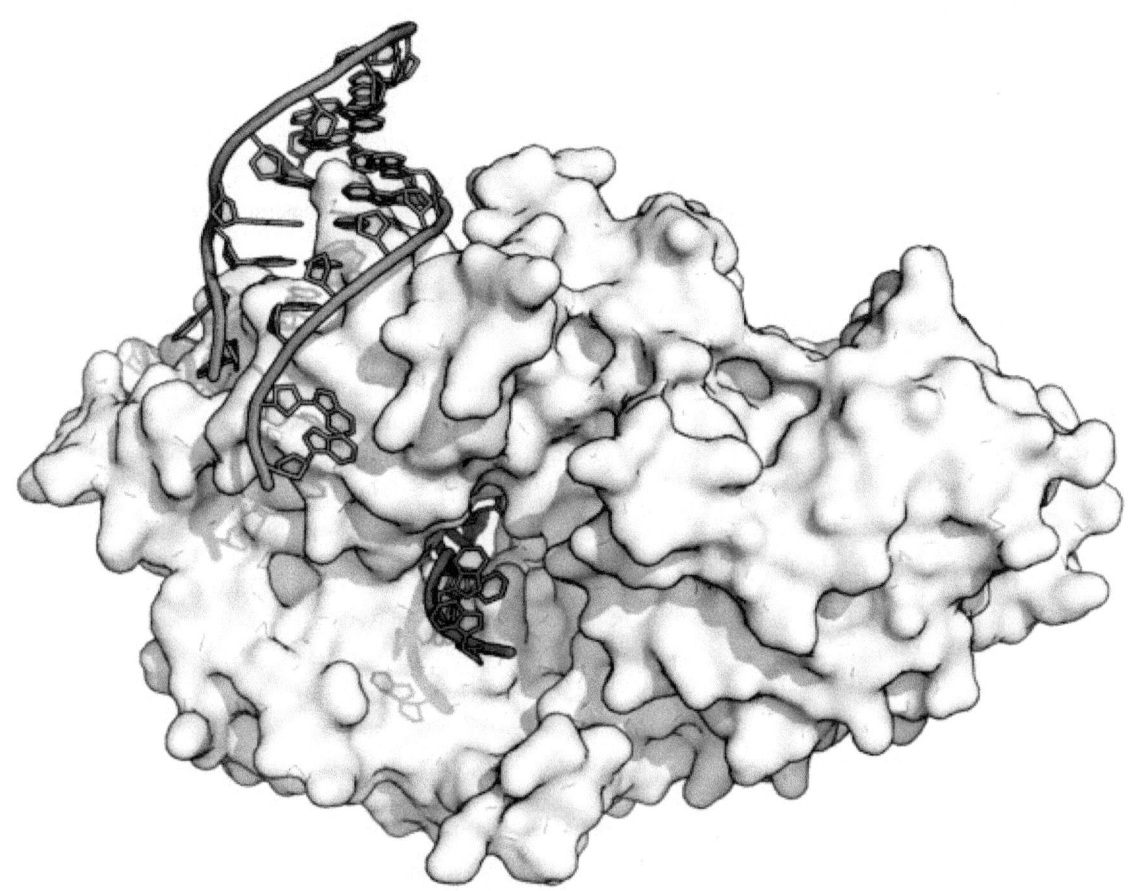

The Ro autoantigen protein (white) binds the end of a double-stranded Y RNA (red) and a single stranded RNA (blue). (PDB: 1YVP).[27]

Cis-acting ncRNAs

Main articles: Five prime untranslated region and Three prime untranslated region

A number of ncRNAs are embedded in the 5' UTRs (Untranslated Regions) of protein coding genes and influence their expression in various ways. For example, a riboswitch can directly bind a small target molecule; the binding of the target affects the gene's activity.

RNA leader sequences are found upstream of the first gene of amino acid biosynthetic operons. These RNA elements form one of two possible structures in regions encoding very short peptide sequences that are rich in the end product amino acid of the operon. A terminator structure forms when there is an excess of the regulatory amino acid and ribosome movement over the leader transcript is not impeded. When there is a deficiency of the charged tRNA of the regulatory amino acid the ribosome translating the leader peptide stalls and the antiterminator structure forms. This allows RNA polymerase to transcribe the operon. Known RNA leaders are Histidine operon leader, Leucine operon leader, Threonine operon leader and the Tryptophan operon leader.

Iron response elements (IRE) are bound by iron response proteins (IRP). The IRE is found in UTRs of various mRNAs whose products are involved in iron metabolism. When iron concentration is low, IRPs bind the ferritin mRNA IRE leading to translation repression.

Internal ribosome entry sites (IRES) are RNA structures that allow for translation initiation in the middle of a mRNA sequence as part of the process of protein synthesis.

29.2.5   ncRNAs and genome defense

Piwi-interacting RNAs (piRNAs) expressed in mammalian testes and somatic cells form RNA-protein complexes with Piwi proteins. These piRNA complexes (piRCs) have been linked to transcriptional gene silencing of retrotransposons and other genetic elements in germ line cells, particularly those in spermatogenesis.

Clustered Regularly Interspaced Short Palindromic Repeats (CRISPR) are repeats found in the DNA of many bacteria and archaea. The repeats are separated by spacers of similar length. It has been demonstrated that these spacers can be derived from phage and subsequently help protect the cell from infection.

29.2.6   ncRNAs and chromosome structure

Telomerase is an RNP enzyme that adds specific DNA sequence repeats ("TTAGGG" in vertebrates) to telomeric regions, which are found at the ends of eukaryotic chromosomes. The telomeres contain condensed DNA material, giving stability to the chromosomes. The enzyme is a reverse transcriptase that carries Telomerase RNA, which is used as a template when it elongates telomeres, which are shortened after each replication cycle.

Xist (X-inactive-specific transcript) is a long ncRNA gene on the X chromosome of the placental mammals that acts as major effector of the X chromosome inactivation process forming Barr bodies. An antisense RNA, Tsix, is a negative regulator of Xist. X chromosomes lacking Tsix expression (and thus having high levels of Xist transcription) are inactivated more frequently than normal chromosomes. In drosophilids, which also use an XY sex-determination system, the roX (RNA on the X) RNAs are involved in dosage compensation.[35] Both Xist and roX operate by epigenetic regulation of transcription through the recruitment of histone-modifying enzymes.

29.2.7   Bifunctional RNA

Bifunctional RNAs, or *dual-function RNAs*, are RNAs that have two distinct functions.[36][37] The majority of the known bifunctional RNAs are mRNAs that encode both a protein and ncRNAs. However, a growing number of ncRNAs fall into two different ncRNA categories; e.g., H/ACA box snoRNA and miRNA.[38][39]

Two well known examples of bifunctional RNAs are SgrS RNA and RNAIII. However, a handful of other bifunctional RNAs are known to exist (e.g., steroid receptor activator/SRA,[40] VegT RNA,[41][42] Oskar RNA,[43] ENOD40,[44] p53 RNA[45] and SR1 RNA.[46] Bifunctional RNAs have recently been the subject of a special issue of Biochimie.[47]

29.3   ncRNAs and disease

See also: Long noncoding RNA § Long non-coding RNAs in disease

As with proteins, mutations or imbalances in the ncRNA repertoire within the body can cause a variety of diseases.

29.3.1   Cancer

Many ncRNAs show abnormal expression patterns in cancerous tissues. These include miRNAs,[48] long mRNA-like ncRNAs,[49][50] GAS5,[51] SNORD50,[52] telomerase RNA and Y RNAs.[53] The miRNAs are involved in the large scale regulation of many protein coding genes,[54][55] the Y RNAs are important for the initiation of DNA replication,[28] telomerase RNA that serves as a primer for telomerase, an RNP that extends telomeric regions at chromosome ends (see telomeres and disease for more information). The direct function of the long mRNA-like ncRNAs is less clear.

Germ-line mutations in miR-16-1 and miR-15 primary precursors have been shown to be much more frequent in patients with chronic lymphocytic leukemia compared to control populations.[56][57]

It has been suggested that a rare SNP (rs11614913) that overlaps has-mir-196a2 has been found to be associated with non-small cell lung carcinoma.[58] Likewise, a screen of 17 miRNAs that have been predicted to regulate a number of breast cancer associated genes found variations in the microRNAs miR-17 and miR-30c-1of patients; these patients were noncarriers of BRCA1 or BRCA2 mutations, lending the possibility that familial breast cancer may be caused by variation in these miRNAs.[59]

The p53 tumor suppressor is arguably the most important player in preventing tumor formation and progression. The p53 protein functions as a transcription factor with a crucial role in orchestrating the cellular stress response. In addition to its crucial role in cancer, p53 has been implicated in other diseases including diabetes, cell death after ischemia, and various neurodegenerative diseases such as Huntington, Parkinson, and Alzheimer. Studies have suggested that p53 expression is subject to regulation by non-coding RNA.[4]

29.3.2 Prader–Willi syndrome

The deletion of the 48 copies of the C/D box snoRNA SNORD116 has been shown to be the primary cause of Prader–Willi syndrome.[60][61][62][63] Prader–Willi is a developmental disorder associated with over-eating and learning difficulties. SNORD116 has potential target sites within a number of protein-coding genes, and could have a role in regulating alternative splicing.[64]

29.3.3 Autism

The chromosomal locus containing the small nucleolar RNA SNORD115 gene cluster has been duplicated in approximately 5% of individuals with autistic traits.[65][66] A mouse model engineered to have a duplication of the SNORD115 cluster displays autistic-like behaviour.[67] A recent small study of post-mortem brain tissue demonstrated altered expression of long non-coding RNAs in the prefrontal cortex and cerebellum of autistic brains as compared to controls.[68]

29.3.4 Cartilage-hair hypoplasia

Mutations within RNase MRP have been shown to cause cartilage-hair hypoplasia, a disease associated with an array of symptoms such as short stature, sparse hair, skeletal abnormalities and a suppressed immune system that is frequent among Amish and Finnish.[69][70][71] The best characterised variant is an A-to-G transition at nucleotide 70 that is in a loop region two bases 5' of a conserved pseudoknot. However, many other mutations within RNase MRP also cause CHH.

29.3.5 Alzheimer's disease

The antisense RNA, BACE1-AS is transcribed from the opposite strand to BACE1 and is upregulated in patients with Alzheimer's disease.[72] BACE1-AS regulates the expression of BACE1 by increasing BACE1 mRNA stability and generating additional BACE1 through a post-transcriptional feed-forward mechanism. By the same mechanism it also raises concentrations of beta amyloid, the main constituent of senile plaques. BACE1-AS concentrations are elevated in subjects with Alzheimer's disease and in amyloid precursor protein transgenic mice.

29.3.6 miR-96 and hearing loss

Variation within the seed region of mature miR-96 has been associated with autosomal dominant, progressive hearing loss in humans and mice. The homozygous mutant mice were profoundly deaf, showing no cochlear responses. Heterozygous mice and humans progressively lose the ability to hear.[73][74][75]

29.4   Distinction between functional RNA (fRNA) and ncRNA

Several publications[76][77][78] have started using the term *functional RNA* (*fRNA*), as opposed to ncRNA, to describe regions functional at the RNA level that may or may not be stand-alone RNA transcripts. According to this terminology, there exist fRNA (such as riboswitches, SECIS elements, and other cis-regulatory regions) that are not ncRNA. Yet the term fRNA could also include mRNA as this is RNA coding for protein and hence is functional. Additionally artificially evolved RNAs also fall under the fRNA umbrella term. Some publications[19] state that the terms *ncRNA* and *fRNA* are nearly synonymous, however others have pointed out that a large proportion of annotated ncRNAs likely have no function.[7]

29.5   See also

- Extracellular RNA

- List of RNAs

- Nucleic acid structure

- Rfam

- Riboswitch

- Ribozyme

- RNAs present in environmental samples

29.6   References

[1] Cheng J, Kapranov P, Drenkow J, Dike S, Brubaker S, Patel S, Long J, Stern D, Tammana H, Helt G, Sementchenko V, Piccolboni A, Bekiranov S, Bailey DK, Ganesh M, Ghosh S, Bell I, Gerhard DS, Gingeras TR; Kapranov; Drenkow; Dike; Brubaker; Patel; Long; Stern; Tammana; Helt; Sementchenko; Piccolboni; Bekiranov; Bailey; Ganesh; Ghosh; Bell; Gerhard; Gingeras (2005). "Transcriptional maps of 10 human chromosomes at 5-nucleotide resolution". *Science* **308** (5725): 1149–54. Bibcode:2005Sci...308.1149C. doi:10.1126/science.1108625. PMID 15790807.

[2] ENCODE Project Consortium; Birney, E; Stamatoyannopoulos, JA; Dutta, A; Guigó, R; Gingeras, TR; Margulies, EH; Weng, Z; Snyder, M; Dermitzakis, John A.; Thurman, Robert E.; Kuehn, Michael S.; Taylor, Christopher M.; Neph, Shane; Koch, Christoph M.; Asthana, Saurabh; Malhotra, Ankit; Adzhubei, Ivan; Greenbaum, Jason A.; Andrews, Robert M.; Flicek, Paul; Boyle, Patrick J.; Cao, Hua; Carter, Nigel P.; Clelland, Gayle K.; Davis, Sean; Day, Nathan; Dhami, Pawandeep; Dillon, Shane C.; et al. (2007). "Identification and analysis of functional elements in 1% of the human genome by the ENCODE pilot project". *Nature* **447** (7146): 799–816. Bibcode:2007Natur.447..799B. doi:10.1038/nature05874. PMC 2212820. PMID 17571346.

[3] Washietl S, Pedersen JS, Korbel JO, Stocsits C, Gruber AR, Hackermüller J, Hertel J, Lindemeyer M, Reiche K, Tanzer A, Ucla C, Wyss C, Antonarakis SE, Denoeud F, Lagarde J, Drenkow J, Kapranov P, Gingeras TR, Guigó R, Snyder M, Gerstein MB, Reymond A, Hofacker IL, Stadler PF; Pedersen; Korbel; Stocsits; Gruber; Hackermüller; Hertel; Lindemeyer; Reiche; Tanzer; Ucla; Wyss; Antonarakis; Denoeud; Lagarde; Drenkow; Kapranov; Gingeras; Guigó; Snyder; Gerstein; Reymond; Hofacker; Stadler (2007). "Structured RNAs in the ENCODE selected regions of the human genome". *Genome Res* **17** (6): 852–64. doi:10.1101/gr.5650707. PMC 1891344. PMID 17568003.

[4] Morris, KV (editor) (2012). *Non-coding RNAs and Epigenetic Regulation of Gene Expression: Drivers of Natural Selection*. Caister Academic Press. ISBN 978-1-904455-94-3.

[5] van Bakel H, Nislow C, Blencowe BJ, Hughes TR; Nislow; Blencowe; Hughes (2010). Eddy, Sean R., ed. "Most "dark matter" transcripts are associated with known genes". *PLoS Biol* **8** (5): e1000371. doi:10.1371/journal.pbio.1000371. PMC 2872640. PMID 20502517.

[6] Hüttenhofer A, Schattner P, Polacek N; Schattner; Polacek (2005). "Non-coding RNAs: hope or hype?". *Trends Genet* **21** (5): 289–97. doi:10.1016/j.tig.2005.03.007. PMID 15851066.

[7] Palazzo, Alexander F.; Lee, Eliza S. (2015). "Non-coding RNA: what is functional and what is junk?". *Frontiers in Genetics* **6**: 2. doi:10.3389/fgene.2015.00002. ISSN 1664-8021. PMID 25674102.

[8] Dahm R (February 2005). "Friedrich Miescher and the discovery of DNA". *Dev. Biol.* **278** (2): 274–88. doi:10.1016/j.ydbio.2004.11.028. PMID 15680349.

[9] Caspersson T, Schultz J; Schultz (1939). "Pentose nucleotides in the cytoplasm of growing tissues". *Nature* **143** (3623): 602–3. Bibcode:1939Natur.143..602C. doi:10.1038/143602c0.

[10] CRICK FH (1958). "On protein synthesis". *Symp. Soc. Exp. Biol.* **12**: 138–63. PMID 13580867.

[11] HOLLEY RW, APGAR J, EVERETT GA; et al. (March 1965). "STRUCTURE OF A RIBONUCLEIC ACID". *Science* **147** (3664): 1462–5. Bibcode:1965Sci...147.1462H. doi:10.1126/science.147.3664.1462. PMID 14263761.

[12] "The Nobel Prize in Physiology or Medicine 1968". Nobel Foundation. Retrieved 2007-07-28.

[13] Madison JT, Everett GA, Kung H; Everett; Kung (1966). "Nucleotide sequence of a yeast tyrosine transfer RNA". *Science* **153** (3735): 531–4. Bibcode:1966Sci...153..531M. doi:10.1126/science.153.3735.531. PMID 5938777.

[14] Zachau HG, Dütting D, Feldmann H, Melchers F, Karau W; Dütting; Feldmann; Melchers; Karau (1966). "Serine specific transfer ribonucleic acids. XIV. Comparison of nucleotide sequences and secondary structure models". *Cold Spring Harb. Symp. Quant. Biol.* **31**: 417–24. doi:10.1101/SQB.1966.031.01.054. PMID 5237198.

[15] Dudock BS, Katz G, Taylor EK, Holley RW; Katz; Taylor; Holley (March 1969). "Primary structure of wheat germ phenylalanine transfer RNA". *Proc. Natl. Acad. Sci. U.S.A.* **62** (3): 941–5. Bibcode:1969PNAS...62..941D. doi:10.1073/pnas.62.3.941. PMC 223689. PMID 5257014.

[16] Cramer F, Doepner H, Haar F VD, Schlimme E, Seidel H; Doepner; Haar f; Schlimme; Seidel (December 1968). "On the conformation of transfer RNA". *Proc. Natl. Acad. Sci. U.S.A.* **61** (4): 1384–91. Bibcode:1968PNAS...61.1384C. doi:10.1073/pnas.61.4.1384. PMC 225267. PMID 4884685.

[17] Ladner JE, Jack A, Robertus JD; et al. (November 1975). "Structure of yeast phenylalanine transfer RNA at 2.5 A resolution". *Proc. Natl. Acad. Sci. U.S.A.* **72** (11): 4414–8. Bibcode:1975PNAS...72.4414L. doi:10.1073/pnas.72.11.4414. PMC 388732. PMID 1105583.

[18] Kim SH, Quigley GJ, Suddath FL; et al. (1973). "Three-dimensional structure of yeast phenylalanine transfer RNA: folding of the polynucleotide chain". *Science* **179** (4070): 285–8. Bibcode:1973Sci...179..285K. doi:10.1126/science.179.4070.285. PMID 4566654.

[19] Eddy SR (December 2001). "Non-coding RNA genes and the modern RNA world". *Nat. Rev. Genet.* **2** (12): 919–29. doi:10.1038/35103511. PMID 11733745.

[20] Daneholt, Bertil. "Advanced Information: RNA interference". *The Nobel Prize in Physiology or Medicine 2006*. Archived from the original on 2007-01-20. Retrieved 2007-01-25.

[21] Jeffares DC, Poole AM, Penny D; Poole; Penny (1998). "Relics from the RNA world". *J Mol Evol* **46** (1): 18–36. doi:10.1007/PL00006280. PMID 9419222.

[22] Poole AM, Jeffares DC, Penny D; Jeffares; Penny (1998). "The path from the RNA world". *J Mol Evol* **46** (1): 1–17. doi:10.1007/PL00006275. PMID 9419221.

[23] Poole A, Jeffares D, Penny D; Jeffares; Penny (1999). "Early evolution: prokaryotes, the new kids on the block". *BioEssays* **21** (10): 880–9. doi:10.1002/(SICI)1521-1878(199910)21:10<880::AID-BIES11>3.0.CO;2-P. PMID 10497339.

[24] Ban N, Nissen P, Hansen J, Moore P, Steitz T; Nissen; Hansen; Moore; Steitz (2000). "The complete atomic structure of the large ribosomal subunit at 2.4 ångström resolution". *Science* **289** (5481): 905–20. Bibcode:2000Sci...289..905B. doi:10.1126/science.289.5481.905. PMID 10937989.

[25] Zhu Y, Stribinskis V, Ramos KS, Li Y; Stribinskis; Ramos; Li (2006). "Sequence analysis of RNase MRP RNA reveals its origination from eukaryotic RNase P RNA". *RNA* **12** (5): 699–706. doi:10.1261/rna.2284906. PMC 1440897. PMID 16540690.

[26] Kishore S, Stamm S; Stamm (2006). "The snoRNA HBII-52 regulates alternative splicing of the serotonin receptor 2C". *Science* **311** (5758): 230–231. Bibcode:2006Sci...311..230K. doi:10.1126/science.1118265. PMID 16357227.

[27] Stein, AJ; Fuchs G; Fu C; Wolin SL; Reinisch KM (2005). "Structural insights into RNA quality control: The Ro autoantigen binds misfolded RNAs via its central cavity". *Cell* **121** (4): 529–537. doi:10.1016/j.cell.2005.03.009. PMC 1769319. PMID 15907467.

[28] Christov CP, Gardiner TJ, Szüts D, Krude T; Gardiner; Szüts; Krude (2006). "Functional Requirement of Noncoding Y RNAs for Human Chromosomal DNA Replication". *Mol. Cell. Biol.* **26** (18): 6993–7004. doi:10.1128/MCB.01060-06. PMC 1592862. PMID 16943439.

[29] Zhang, AT; Langley, AR; Christov, CP; Kheir, E; Shafee, T; Gardiner, TJ; Krude, T (Jun 15, 2011). "Dynamic interaction of Y RNAs with chromatin and initiation proteins during human DNA replication.". *Journal of Cell Science* **124** (Pt 12): 2058–69. doi:10.1242/jcs.086561. PMC 3104036. PMID 21610089.

[30] Hall, Adam E.; Turnbull, Carly; Dalmay, Tamas (2013). "Y RNAs: recent developments". *Biomolecular Concepts* **4** (2): 103–110. doi:10.1515/bmc-2012-0050.

[31] Lerner, MR; Boyle JA; Hardin JA; Steitz JA (1981). "Two novel classes of small ribonucleoproteins detected by antibodies associated with lupus erythematosus". *Science* **211** (4480): 400–402. Bibcode:1981Sci...211..400L. doi:10.1126/science.6164096. PMID 6164096.

[32] Reiner R, Ben-Asouli Y, Krilovetzky I, Jarrous N; Ben-Asouli; Krilovetzky; Jarrous (2006). "A role for the catalytic ribonucleoprotein RNase P in RNA polymerase III transcription". *Genes Dev* **20** (12): 1621–35. doi:10.1101/gad.386706. PMC 1482482. PMID 16778078.

[33] Espinoza CA, Allen TA, Hieb AR, Kugel JF, Goodrich JA; Allen; Hieb; Kugel; Goodrich (2004). "B2 RNA binds directly to RNA polymerase II to repress transcript synthesis". *Nat Struct Mol Biol* **11** (9): 822–9. doi:10.1038/nsmb812. PMID 15300239.

[34] Hirota K, Miyoshi T, Kugou K, Hoffman CS, Shibata T, Ohta K; Miyoshi; Kugou; Hoffman; Shibata; Ohta (2008). "Stepwise chromatin remodelling by a cascade of transcription initiation of non-coding RNAs". *Nature* **456** (7218): 130–4. Bibcode:2008Natur.456..130 doi:10.1038/nature07348. PMID 18820678.

[35] Park Y, Kelley RL, Oh H, Kuroda MI, Meller VH; Kelley; Oh; Kuroda; Meller (2002). "Extent of chromatin spreading determined by roX RNA recruitment of MSL proteins". *Science* **298** (5598): 1620–3. Bibcode:2002Sci...298.1620P. doi:10.1126/science.1076686 PMID 12446910.

[36] Wadler CS, Vanderpool CK; Vanderpool (2007). "A dual function for a bacterial small RNA: SgrS performs base pairing-dependent regulation and encodes a functional polypeptide". *Proc Natl Acad Sci USA* **104** (51): 20454–9. Bibcode:2007PNAS..10420454W. doi:10.1073/pnas.0708102104. PMC 2154452. PMID 18042713.

[37] Dinger ME, Pang KC, Mercer TR, Mattick JS; Pang; Mercer; Mattick (2008). McEntyre, Johanna, ed. "Differentiating protein-coding and noncoding RNA: challenges and ambiguities". *PLoS Comput Biol* **4** (11): e1000176. Bibcode:2008PLSCB...4E0176D. doi:10.1371/journal.pcbi.1000176. PMC 2518207. PMID 19043537.

[38] Saraiya AA, Wang CC; Wang (2008). Goldberg, Daniel Eliot, ed. "snoRNA, a novel precursor of microRNA in Giardia lamblia". *PLoS Pathog* **4** (11): e1000224. doi:10.1371/journal.ppat.1000224. PMC 2583053. PMID 19043559.

[39] Ender C, Krek A, Friedländer MR, Beitzinger M, Weinmann L, Chen W, Pfeffer S, Rajewsky N, Meister G; Krek; Friedländer; Beitzinger; Weinmann; Chen; Pfeffer; Rajewsky; Meister (2008). "A human snoRNA with microRNA-like functions". *Mol Cell* **32** (4): 519–28. doi:10.1016/j.molcel.2008.10.017. PMID 19026782.

[40] Leygue E (2007). "Steroid receptor RNA activator (SRA1): unusual bifaceted gene products with suspected relevance to breast cancer". *Nucl Recept Signal* **5**: e006. doi:10.1621/nrs.05006. PMC 1948073. PMID 17710122.

[41] Zhang J, King ML; King (1996). "Xenopus VegT RNA is localized to the vegetal cortex during oogenesis and encodes a novel T-box transcription factor involved in mesodermal patterning". *Development* **122** (12): 4119–29. PMID 9012531.

[42] Kloc M, Wilk K, Vargas D, Shirato Y, Bilinski S, Etkin LD; Wilk; Vargas; Shirato; Bilinski; Etkin (2005). "Potential structural role of non-coding and coding RNAs in the organization of the cytoskeleton at the vegetal cortex of Xenopus oocytes". *Development* **132** (15): 3445–57. doi:10.1242/dev.01919. PMID 16000384.

[43] Jenny A, Hachet O, Závorszky P, Cyrklaff A, Weston MD, Johnston DS, Erdélyi M, Ephrussi A; Hachet; Závorszky; Cyrklaff; Weston; Johnston; Erdélyi; Ephrussi (2006). "A translation-independent role of oskar RNA in early Drosophila oogenesis". *Development* **133** (15): 2827–33. doi:10.1242/dev.02456. PMID 16835436.

[44] Gultyaev AP, Roussis A; Roussis (2007). "Identification of conserved secondary structures and expansion segments in enod40 RNAs reveals new enod40 homologues in plants". *Nucleic Acids Res* **35** (9): 3144–52. doi:10.1093/nar/gkm173. PMC 1888808. PMID 17452360.

[45] Candeias MM, Malbert-Colas L, Powell DJ, Daskalogianni C, Maslon MM, Naski N, Bourougaa K, Calvo F, Fahraeus R; Malbert-Colas; Powell; Daskalogianni; Maslon; Naski; Bourougaa; Calvo; Fåhraeus (2008). "p53 mRNA controls p53 activity by managing Mdm2 functions". *Nature Cell Biology* **10** (9): 1098–1105. doi:10.1038/ncb1770. PMID 19160491.

[46] Gimpel, M; Preis, H; Barth, E; Gramzow, L; Brantl, S (Oct 13, 2012). "SR1--a small RNA with two remarkably conserved functions". *Nucleic Acids Research* **40** (22): 11659–72. doi:10.1093/nar/gks895. PMC 3526287. PMID 23034808.

[47] Francastel, Claire; Hubé, Florent (2011). "Coding or non-coding: Need they be exclusive?". *Biochimie* **93** (11): vi–vii. doi:10.1016/S0300-9084(11)00322-1. PMID 21963143.

[48] Mraz, M.; Pospisilova, S. (2012). "MicroRNAs in chronic lymphocytic leukemia: From causality to associations and back". *Expert Review of Hematology* **5** (6): 579–581. doi:10.1586/ehm.12.54. PMID 23216588.

[49] Pibouin L, Villaudy J, Ferbus D, Muleris M, Prospéri MT, Remvikos Y, Goubin G; Villaudy; Ferbus; Muleris; Prospéri; Remvikos; Goubin (2002). "Cloning of the mRNA of overexpression in colon carcinoma-1: a sequence overexpressed in a subset of colon carcinomas". *Cancer Genet Cytogenet* **133** (1): 55–60. doi:10.1016/S0165-4608(01)00634-3. PMID 11890990.

[50] Fu X, Ravindranath L, Tran N, Petrovics G, Srivastava S; Ravindranath; Tran; Petrovics; Srivastava (2006). "Regulation of apoptosis by a prostate-specific and prostate cancer-associated noncoding gene, PCGEM1". *DNA Cell Biol* **25** (3): 135–41. doi:10.1089/dna.2006.25.135. PMID 16569192.

[51] Mourtada-Maarabouni M, Pickard MR, Hedge VL, Farzaneh F, Williams GT; Pickard; Hedge; Farzaneh; Williams (2009). "GAS5, a non-protein-coding RNA, controls apoptosis and is downregulated in breast cancer". *Oncogene* **28** (2): 195–208. doi:10.1038/onc.2008.373. PMID 18836484.

[52] Dong XY, Guo P, Boyd J, Sun X, Li Q, Zhou W, Dong JT; Guo; Boyd; Sun; Li; Zhou; Dong (2009). "Implication of snoRNA U50 in human breast cancer". *J Genet Genomics* **36** (8): 447–54. doi:10.1016/S1673-8527(08)60134-4. PMC 2854654. PMID 19683667.

[53] Christov CP, Trivier E, Krude T; Trivier; Krude (2008). "Noncoding human Y RNAs are overexpressed in tumours and required for cell proliferation". *Br J Cancer* **98** (5): 981–8. doi:10.1038/sj.bjc.6604254. PMC 2266855. PMID 18283318.

[54] Farh KK, Grimson A, Jan C, Lewis BP, Johnston WK, Lim LP, Burge CB, Bartel DP; Grimson; Jan; Lewis; Johnston; Lim; Burge; Bartel (2005). "The widespread impact of mammalian MicroRNAs on mRNA repression and evolution". *Science* **310** (5755): 1817–21. Bibcode:2005Sci...310.1817F. doi:10.1126/science.1121158. PMID 16308420.

[55] Lim LP, Lau NC, Garrett-Engele P, Grimson A, Schelter JM, Castle J, Bartel DP, Linsley PS, Johnson JM; Lau; Garrett-Engele; Grimson; Schelter; Castle; Bartel; Linsley; Johnson (2005). "Microarray analysis shows that some microRNAs downregulate large numbers of target mRNAs". *Nature* **433** (7027): 769–73. Bibcode:2005Natur.433..769L. doi:10.1038/nature03315. PMID 15685193.

[56] Calin GA, Ferracin M, Cimmino A; et al. (October 2005). "A MicroRNA signature associated with prognosis and progression in chronic lymphocytic leukemia". *N. Engl. J. Med.* **353** (17): 1793–801. doi:10.1056/NEJMoa050995. PMID 16251535.

[57] Calin, GA; Dumitru CD, Shimizu M, Bichi R, Zupo S, Noch E, Aldler H, Rattan S, Keating M, Rai K, Rassenti L, Kipps T, Negrini M, Bullrich F, Croce CM; Shimizu, Masayoshi; Bichi, Roberta; Zupo, Simona; Noch, Evan; Aldler, Hansjuerg; Rattan, Sashi; Keating, Michael; Rai, Kanti; Rassenti, Laura; Kipps, Thomas; Negrini, Massimo; Bullrich, Florencia; Croce, Carlo M. (2002). "Frequent deletions and down-regulation of micro-RNA genes miR15 and miR16 at 13q14 in chronic lymphocytic leukemia". *Proc Natl Acad Sci USA* **99** (24): 15524–15529. Bibcode:2002PNAS...9915524C. doi:10.1073/pnas.242606799. PMC 137750. PMID 12434020.

[58] Hu Z, Chen J, Tian T, Zhou X, Gu H, Xu L, Zeng Y, Miao R, Jin G, Ma H, Chen Y, Shen H; Chen; Tian; Zhou; Gu; Xu; Zeng; Miao; Jin; Ma; Chen; Shen (2008). "Genetic variants of miRNA sequences and non-small cell lung cancer survival". *J Clin Invest* **118** (7): 2600–8. doi:10.1172/JCI34934. PMC 2402113. PMID 18521189.

[59] Shen J, Ambrosone CB, Zhao H; Ambrosone; Zhao (2009). "Novel genetic variants in microRNA genes and familial breast cancer". *Int J Cancer* **124** (5): 1178–82. doi:10.1002/ijc.24008. PMID 19048628.

[60] Sahoo T, del Gaudio D, German JR, Shinawi M, Peters SU, Person RE, Garnica A, Cheung SW, Beaudet AL; Del Gaudio; German; Shinawi; Peters; Person; Garnica; Cheung; Beaudet (2008). "Prader–Willi phenotype caused by paternal deficiency for the HBII-85 C/D box small nucleolar RNA cluster". *Nat Genet* **40** (6): 719–21. doi:10.1038/ng.158. PMC 2705197. PMID 18500341.

[61] Skryabin BV, Gubar LV, Seeger B, Pfeiffer J, Handel S, Robeck T, Karpova E, Rozhdestvensky TS, Brosius J; Gubar; Seeger; Pfeiffer; Handel; Robeck; Karpova; Rozhdestvensky; Brosius (2007). "Deletion of the MBII-85 snoRNA gene cluster in mice results in postnatal growth retardation". *PLoS Genet* **3** (12): e235. doi:10.1371/journal.pgen.0030235. PMC 2323313. PMID 18166085.

[62] Ding F, Li HH, Zhang S, Solomon NM, Camper SA, Cohen P, Francke U; Li; Zhang; Solomon; Camper; Cohen; Francke (2008). Akbarian, Schahram, ed. "SnoRNA Snord116 (Pwcr1/MBII-85) deletion causes growth deficiency and hyperphagia in mice". *PLoS ONE* **3** (3): e1709. Bibcode:2008PLoSO...3.1709D. doi:10.1371/journal.pone.0001709. PMC 2248623. PMID 18320030.

[63] Ding F, Prints Y, Dhar MS, Johnson DK, Garnacho-Montero C, Nicholls RD, Francke U; Prints; Dhar; Johnson; Garnacho-Montero; Nicholls; Francke (2005). "Lack of Pwcr1/MBII-85 snoRNA is critical for neonatal lethality in Prader–Willi syndrome mouse models". *Mamm Genome* **16** (6): 424–31. doi:10.1007/s00335-005-2460-2. PMID 16075369.

[64] Bazeley PS, Shepelev V, Talebizadeh Z, Butler MG, Fedorova L, Filatov V, Fedorov A; Shepelev; Talebizadeh; Butler; Fedorova; Filatov; Fedorov (2008). "snoTARGET shows that human orphan snoRNA targets locate close to alternative splice junctions". *Gene* **408** (1–2): 172–9. doi:10.1016/j.gene.2007.10.037. PMID 18160232.

[65] Bolton PF, Veltman MW, Weisblatt E; et al. (September 2004). "Chromosome 15q11-13 abnormalities and other medical conditions in individuals with autism spectrum disorders". *Psychiatr. Genet.* **14** (3): 131–7. doi:10.1097/00041444-200409000-00002. PMID 15318025.

[66] Cook EH, Scherer SW; Scherer (October 2008). "Copy-number variations associated with neuropsychiatric conditions". *Nature* **455** (7215): 919–23. Bibcode:2008Natur.455..919C. doi:10.1038/nature07458. PMID 18923514.

[67] Nakatani J, Tamada K, Hatanaka F; et al. (June 2009). "Abnormal behavior in a chromosome-engineered mouse model for human 15q11-13 duplication seen in autism". *Cell* **137** (7): 1235–46. doi:10.1016/j.cell.2009.04.024. PMC 3710970. PMID 19563756.

[68] Ziats, Mark N.; Rennert, Owen M. (2012). "Aberrant Expression of Long Noncoding RNAs in Autistic Brain". *Journal of Molecular Neuroscience* **49** (3): 589–93. doi:10.1007/s12031-012-9880-8. PMC 3566384. PMID 22949041.

[69] Ridanpää M, van Eenennaam H, Pelin K, Chadwick R, Johnson C, Yuan B, vanVenrooij W, Pruijn G, Salmela R, Rockas S, Mäkitie O, Kaitila I, de la Chapelle A; Van Eenennaam; Pelin; Chadwick; Johnson; Yuan; Vanvenrooij; Pruijn; Salmela; Rockas; Mäkitie; Kaitila; de la Chapelle (2001). "Mutations in the RNA component of RNase MRP cause a pleiotropic human disease, cartilage-hair hypoplasia". *Cell* **104** (2): 195–203. doi:10.1016/S0092-8674(01)00205-7. PMID 11207361.

[70] Martin AN, Li Y; Li (2007). "RNase MRP RNA and human genetic diseases". *Cell Res* **17** (3): 219–26. doi:10.1038/sj.cr.7310120. PMID 17189938.

[71] Kavadas FD, Giliani S, Gu Y, Mazzolari E, Bates A, Pegoiani E, Roifman CM, Notarangelo LD; Giliani; Gu; Mazzolari; Bates; Pegoiani; Roifman; Notarangelo (2008). "Variability of clinical and laboratory features among patients with ribonuclease mitochondrial RNA processing endoribonuclease gene mutations". *J Allergy Clin Immunol* **122** (6): 1178–84. doi:10.1016/j.jaci.2008.07.036. PMID 18804272.

[72] Faghihi MA, Modarresi F, Khalil AM, Wood DE, Sahagan BG, Morgan TE, Finch CE, St Laurent G, Kenny PJ, Wahlestedt C; Modarresi; Khalil; Wood; Sahagan; Morgan; Finch; St Laurent g; Kenny; Wahlestedt (2008). "Expression of a noncoding RNA is elevated in Alzheimer's disease and drives rapid feed-forward regulation of beta-secretase". *Nat Med* **14** (7): 723–30. doi:10.1038/nm1784. PMC 2826895. PMID 18587408.

[73] Mencía A, Modamio-Høybjør S, Redshaw N, Morín M, Mayo-Merino F, Olavarrieta L, Aguirre LA, del Castillo I, Steel KP, Dalmay T, Moreno F, Moreno-Pelayo MA; Modamio-Høybjør; Redshaw; Morín; Mayo-Merino; Olavarrieta; Aguirre; Del Castillo; Steel; Dalmay; Moreno; Moreno-Pelayo (2009). "Mutations in the seed region of human miR-96 are responsible for nonsyndromic progressive hearing loss". *Nat Genet* **41** (5): 609–13. doi:10.1038/ng.355. PMID 19363479.

[74] Lewis MA, Quint E, Glazier AM, Fuchs H, De Angelis MH, Langford C, van Dongen S, Abreu-Goodger C, Piipari M, Redshaw N, Dalmay T, Moreno-Pelayo MA, Enright AJ, Steel KP; Quint; Glazier; Fuchs; De Angelis; Langford; Van Dongen; Abreu-Goodger; Piipari; Redshaw; Dalmay; Moreno-Pelayo; Enright; Steel (2009). "An ENU-induced mutation of miR-96 associated with progressive hearing loss in mice". *Nat Genet* **41** (5): 614–8. doi:10.1038/ng.369. PMC 2705913. PMID 19363478.

[75] Soukup GA (2009). "Little but loud: Small RNAs have a resounding affect on ear development". *Brain Res* **1277**: 104–14. doi:10.1016/j.brainres.2009.02.027. PMC 2700218. PMID 19245798.

[76] Richard J. Carter, Inna Dubchak, Stephen R. Holbrook; Dubchak; Holbrook (2001). "A computational approach to identify genes for functional RNAs in genomic sequences". *Nucleic Acids Research* **29** (19): 3928–3938. doi:10.1093/nar/29.19.3928 (inactive 2015-02-01). PMC 60242. PMID 11574674.

[77] Jakob Skou Pedersen, Gill Bejerano, Adam Siepel, Kate Rosenbloom, Kerstin Lindblad-Toh, Eric S. Lander, Jim Kent, Webb Miller, David Haussler; Bejerano; Siepel; Rosenbloom; Lindblad-Toh; Lander; Kent; Miller; Haussler (2006). "Identification and Classification of Conserved RNA Secondary Structures in the Human Genome". *PLOS Computational Biology* **2** (4): e33. Bibcode:2006PLSCB...2...33P. doi:10.1371/journal.pcbi.0020033. PMC 1440920. PMID 16628248.

[78] Tomas Babak, Benjamin J Blencowe, Timothy R Hughes; Horspool; Brown; Tcherepanov; Upton (2007). "Considerations in the identification of functional RNA structural elements in genomic alignments". *BMC Bioinformatics* **8**: 33. doi:10.1186/1471-2105-8-21. PMC 1783863. PMID 17244370.

29.7   External links

- Comprehensive database of mammalian ncRNAs

- The Rfam Database — a curated list of hundreds of families of related ncRNAs.

- NONCODE.org — a free database of all kinds of noncoding RNAs (except tRNAs and rRNAs).

- Joint ncRNA Database — over 30,000 individual sequences from 99 species of bacteria, archaea and eukaryota

- RNAcon Prediction and classification of ncRNA BMC Genomics 2014, 15:127

- ENCODE threads explorer Non-coding RNA characterization. Nature (journal)

- The Non-coding RNA Databases Resource (NRDR) — a curated source of data related to over non-coding RNA databases available over the internet.

Chapter 30

Precursor mRNA

See main article Primary transcript for more details

Precursor mRNA (**pre-mRNA**) is an immature single strand of messenger ribonucleic acid (mRNA). Pre-mRNA is synthesized from a DNA template in the cell nucleus by transcription. Pre-mRNA comprises the bulk of **heterogeneous nuclear RNA** (**hnRNA**). The term hnRNA is often used as a synonym for pre-mRNA, although, in the strict sense, hnRNA may include nuclear RNA transcripts that do not end up as cytoplasmic mRNA.

Once pre-mRNA has been completely processed, it is termed "mature messenger RNA", "mature mRNA", or simply "mRNA".

30.1 Processing

Eukaryotic pre-mRNA exists only briefly before it is fully processed into mRNA. Pre-mRNAs include two different types of segments, exons and introns. Exons are segments that are retained in the final mRNA, whereas introns are removed in a process called splicing, which is performed by the spliceosome (except for self-splicing introns).

Additional processing steps attach modifications to the 5' and 3' ends of Eukaryotic pre-mRNA. These include a 5' cap of 7-methylguanosine and a poly-A tail. In addition, eukaryotic pre-mRNAs have their introns spliced out by spliceosomes made up of small nuclear ribonucleoproteins.[1][2]

When a pre-mRNA strand has been properly processed to an mRNA sequence, it is exported out of the nucleus and eventually translated into a protein – a process accomplished in conjunction with ribosomes.

30.2 References

[1] Weaver, Robert F. (2005). *Molecular Biology*, p.432-448. McGraw-Hill, New York, NY. ISBN 0-07-284611-9.

[2] Wahl, M. C.; Will, C. L.; Lührmann, R. (2009). "The Spliceosome: Design Principles of a Dynamic RNP Machine". *Cell* **136** (4): 701–718. doi:10.1016/j.cell.2009.02.009. PMID 19239890.

30.3 External links

- Scienceden.com RNA Article

Chapter 31

Ribosomal RNA

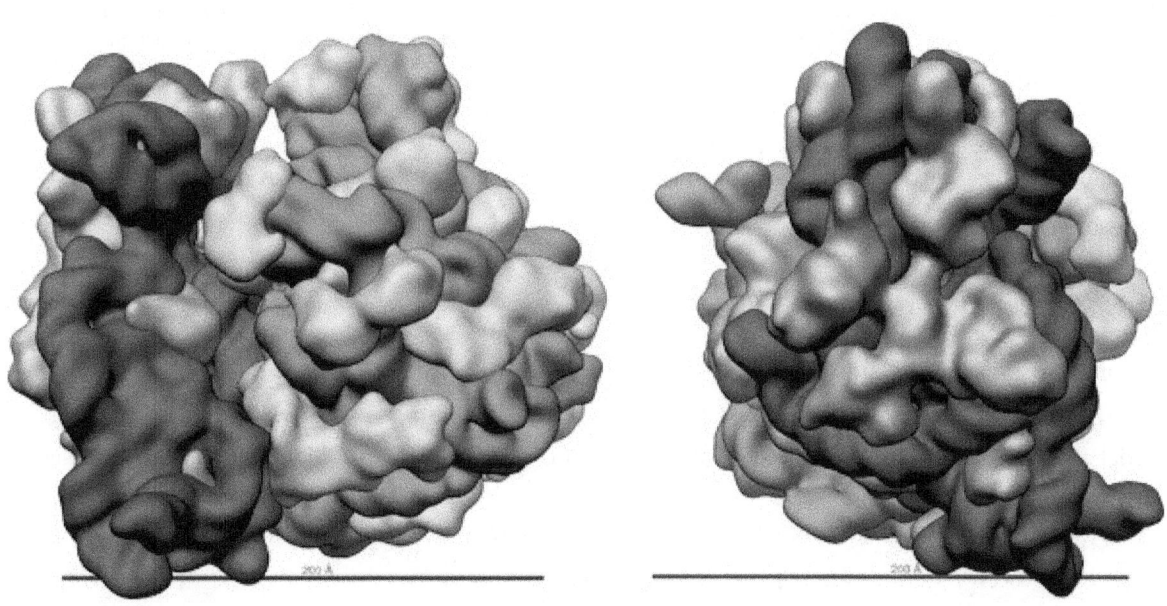

Three-dimensional views of the ribosome, showing rRNA in dark blue (small subunit) and dark red (large subunit). Lighter colors represent ribosomal proteins.

In molecular biology, **ribosomal ribonucleic acid** (**rRNA**) is the RNA component of the ribosome, and is essential for protein synthesis in all living organisms. It constitutes the predominant material within the ribosome, which is approximately 60% rRNA and 40% protein by weight. Ribosomes contain two major rRNAs and 50 or more proteins. The ribosomal RNAs form two subunits, the large subunit (LSU) and small subunit (SSU). The LSU rRNA acts as a ribozyme, catalyzing peptide bond formation. rRNA sequences are widely used for working out evolutionary relationships among organisms, since they are of ancient origin and are found in all known forms of life.

31.1   Structure

The ribosomal RNAs complex with proteins to form two subunits, the large subunit (LSU) and small subunit (SSU). mRNA is sandwiched between the small and large subunits, and the ribosome catalyzes the formation of a peptide bond between the two amino acids that are contained in the rRNA.

A ribosome also has three binding sites called A, P, and E.

- The A site in the ribosome binds to an aminoacyl-tRNA (a tRNA bound to an amino acid).

- The amino (NH$_2$) group of the aminoacyl-tRNA, which contains the new amino acid, attacks the ester linkage of peptidyl-tRNA (contained within the P site), which contains the last amino acid of the growing chain, forming a new peptide bond. This reaction is catalyzed by peptidyl transferase.

- The tRNA that was holding onto the last amino acid is moved to the E site, and what used to be the aminoacyl-tRNA is the peptidyl-tRNA.

A single mRNA can be translated simultaneously by multiple ribosomes..

31.2   Prokaryotes vs. eukaryotes

Both prokaryotic and eukaryotic ribosomes can be broken down into two subunits (the S in 16S represents Svedberg units), nt= length in nucleotides of the respective rRNAs, for exemplary species *Escherichia coli* (prokaryote) and human (eukaryote):

Note that the S units of the subunits (or the rRNAs) cannot simply be added because they represent measures of sedimentation rate rather than of mass. The sedimentation rate of each subunit is affected by its shape, as well as by its mass. The nt units can be added as these represent the integer number of units in the linear rRNA polymers (for example, the total length of the human rRNA = 7216 nt).

31.2.1   Prokaryotes

In prokaryotes a small 30S ribosomal subunit contains the 16S ribosomal RNA.

The large 50S ribosomal subunit contains two rRNA species (the 5S and 23S ribosomal RNAs).

Bacterial 16S ribosomal RNA, 23S ribosomal RNA, and 5S rRNA genes are typically organized as a co-transcribed operon.

There may be one or more copies of the operon dispersed in the genome (for example, *Escherichia coli* has seven).

Archaea contains either a single rDNA operon or multiple copies of the operon.

The 3' end of the 16S ribosomal RNA (in a ribosome) binds to a sequence on the 5' end of mRNA called the Shine-Dalgarno sequence.

31.2.2   Eukaryotes

In contrast, eukaryotes generally have many copies of the rRNA genes organized in tandem repeats; in humans approximately 300–400 repeats are present in five clusters (on chromosomes 13, 14, 15, 21 and 22). Because of their special structure and transcription behaviour, rRNA gene clusters are commonly called "ribosomal DNA" (note that the term seems to imply that ribosomes contain DNA, which is not the case).

The 18S rRNA in most eukaryotes is in the small ribosomal subunit, and the large subunit contains three rRNA species (the 5S, 5.8S and 28S in mammals, 25S in plants, rRNAs).

Mammalian cells have 2 mitochondrial (12S and 16S) rRNA molecules and 4 types of cytoplasmic rRNA (the 28S, 5.8S, 18S, and 5S subunits). The 28S, 5.8S, and 18S rRNAs are encoded by a single transcription unit (45S) separated by 2 internally transcribed spacers. The 45S rDNA is organized into 5 clusters (each has 30-40 repeats) on chromosomes 13, 14, 15, 21, and 22. These are transcribed by RNA polymerase I. 5S occurs in tandem arrays (~200-300 true 5S genes and many dispersed pseudogenes), the largest one on the chromosome 1q41-42. 5S rRNA is transcribed by RNA polymerase III.

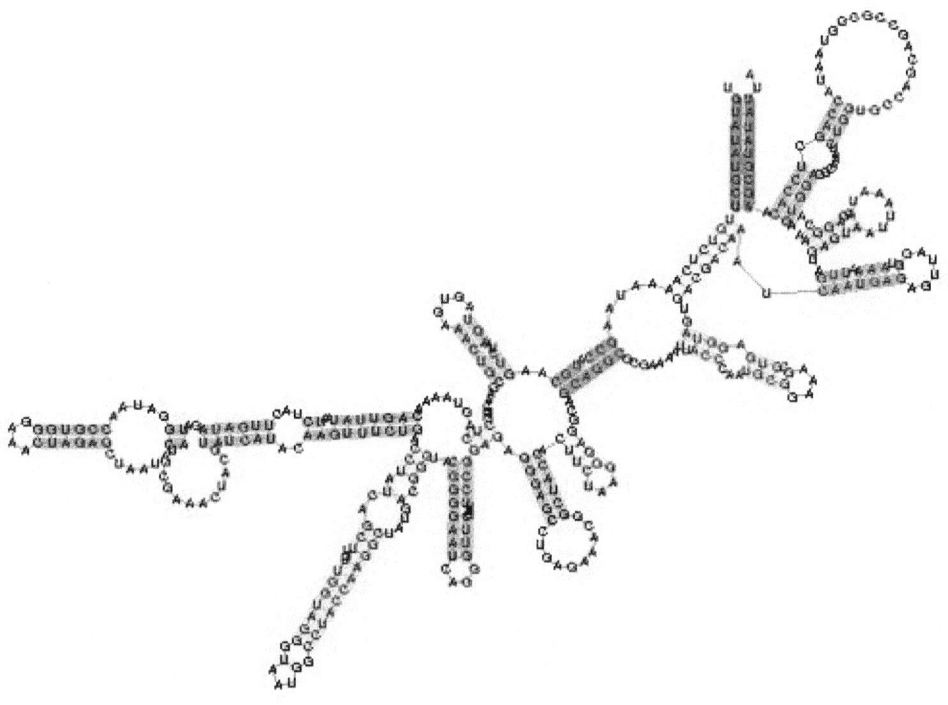

Small subunit ribosomal RNA, 5' domain taken from the Rfam database. This example is RF00177

The tertiary structure of the small subunit ribosomal RNA (SSU rRNA) has been resolved by X-ray crystallography.[5] The secondary structure of SSU rRNA contains 4 distinct domains — the 5', central, 3' major and 3' minor domains. A model of the secondary structure for the 5' domain (500-800 nucleotides) is shown.

31.3 Translation

Translation is the net effect of proteins being synthesized by ribosomes, from a copy (mRNA) of the DNA template in the nucleus. One of the components of the ribosome (16S rRNA) base pairs complementary to a Shine–Dalgarno sequence upstream of the start codon in mRNA.

31.4   Importance of rRNA

Ribosomal RNA characteristics are important in evolution, thus taxonomy, and medicine.

- rRNA is one of only a few gene products present in all cells.[6] For this reason, genes that encode the rRNA (rDNA) are sequenced to identify an organism's taxonomic group, calculate related groups, and estimate rates of species divergence. As a result, many thousands of rRNA sequences are known and stored in specialized databases such as RDP-II[7] and SILVA.[8]

- rRNA is the target of numerous clinically relevant antibiotics: chloramphenicol, erythromycin, kasugamycin, micrococcin, paromomycin, ricin, sarcin, spectinomycin, streptomycin, and thiostrepton.

- rRNA have been shown to be the origin of species-specific microRNAs, like miR-663 in humans and miR-712 in mouse. These miRNAs originate from the Internal Transcribed Spacers of the rRNA.[9]

31.5   Genes

- RPL1, RPL2, RPL3, RPL4, RPL5, RPL6, RPL7, RPL8, RPL9, RPL10, RPL11, RPL12, RPL13, RPL14, RPL15, RPL16, RPL17, RPL18, RPL19, RPL20, RPL21, RPL22, RPL23, RPL24, RPL25, RPL26, RPL27, RPL28, RPL29, RPL30, RPL31, RPL32, RPL33, RPL34, RPL35, RPL36, RPL37, RPL38, RPL39, RPL40, RPL41

- MRPL1, MRPL2, MRPL3, MRPL4, MRPL5, MRPL6, MRPL7, MRPL8, MRPL9, MRPL10, MRPL11, MRPL12, MRPL13, MRPL14, MRPL15, MRPL16, MRPL17, MRPL18, MRPL19, MRPL20, MRPL21, MRPL22, MRPL23, MRPL24, MRPL25, MRPL26, MRPL27, MRPL28, MRPL29, MRPL30, MRPL31, MRPL32, MRPL33, MRPL34, MRPL35, MRPL36, MRPL37, MRPL38, MRPL39, MRPL40, MRPL41, MRPL42

- RPS1, RPS2, RPS3, RPS4, RPS5, RPS6, RPS7, RPS8, RPS9, RPS10, RPS11, RPS12, RPS13, RPS14, RPS15, RPS16, RPS17, RPS18, RPS19, RPS20, RPS21, RPS22, RPS23, RPS24, RPS25, RPS26, RPS27, RPS28, RPS29

- MRPS1, MRPS2, MRPS3, MRPS4, MRPS5, MRPS6, MRPS7, MRPS8, MRPS9, MRPS10, MRPS11, MRPS12, MRPS13, MRPS14, MRPS15, MRPS16, MRPS17, MRPS18, MRPS19, MRPS20, MRPS21, MRPS22, MRPS23, MRPS24, MRPS25, MRPS26, MRPS27, MRPS28, MRPS29, MRPS30, MRPS31, MRPS32, MRPS33, MRPS34, MRPS35

These denote genes encoding for the proteins of the ribosome and are transcribed as mRNA, not rRNA.

31.6   See also

- Ribotyping

31.7   References

[1] "*Homo sapiens* 5S ribosomal RNA".

[2] "*Homo sapiens* 5.8S ribosomal RNA".

[3] "*Homo sapiens* 28S ribosomal RNA".

[4] "*Homo sapiens* 18S ribosomal RNA".

[5] Yusupov MM, Yusupova GZ, Baucom A, et al. (2001). "Crystal structure of the ribosome at 5.5 A resolution". *Science* **292** (5518): 883–96. doi:10.1126/science.1060089. PMID 11283358.

[6] Smit S, Widmann J, Knight R (2007). "Evolutionary rates vary among rRNA structural elements". *Nucleic Acids Res* **35** (10): 3339–54. doi:10.1093/nar/gkm101. PMC 1904297. PMID 17468501.

[7] Cole, JR; Chai B; Marsh TL; Farris RJ; Wang Q; Kulam SA; Chandra S; McGarrell DM; Schmidt TM; Garrity GM; Tiedje JM (2003). "The Ribosomal Database Project (RDP-II): previewing a new autoaligner that allows regular updates and the new prokaryotic taxonomy". *Nucleic Acids Res* **31** (1): 442–3. doi:10.1093/nar/gkg039. PMC 165486. PMID 12520046.

[8] Pruesse, E; Quast C; Knittel K; Fuchs BM; Ludwig W; Peplies J; Gloeckner FO (2007). "SILVA: a comprehensive online resource for quality checked and aligned ribosomal RNA sequence data compatible with ARB". *Nucleic Acids Res* **35** (1): 7188–7196. doi:10.1093/nar/gkm864. PMC 2175337. PMID 17947321.

[9] The atypical mechanosensitive microRNA-712 derived from pre-ribosomal RNA induces endothelial inflammation and atherosclerosis Nature Communications, 2013 doi:10.1038/ncomms4000

31.8 External links

- SILVA rRNA Database Project (also includes Eukaryotes (18S) and LSU (23S/28S))

- European database on small subunit ribosomal RNA

- Ribosomal Database Project II

- 16S rRNA, BioMineWiki

-

- Page for Small subunit ribosomal RNA, 5' domain at Rfam

- Ribosomal RNA at the US National Library of Medicine Medical Subject Headings (MeSH)

Chapter 32

microRNA

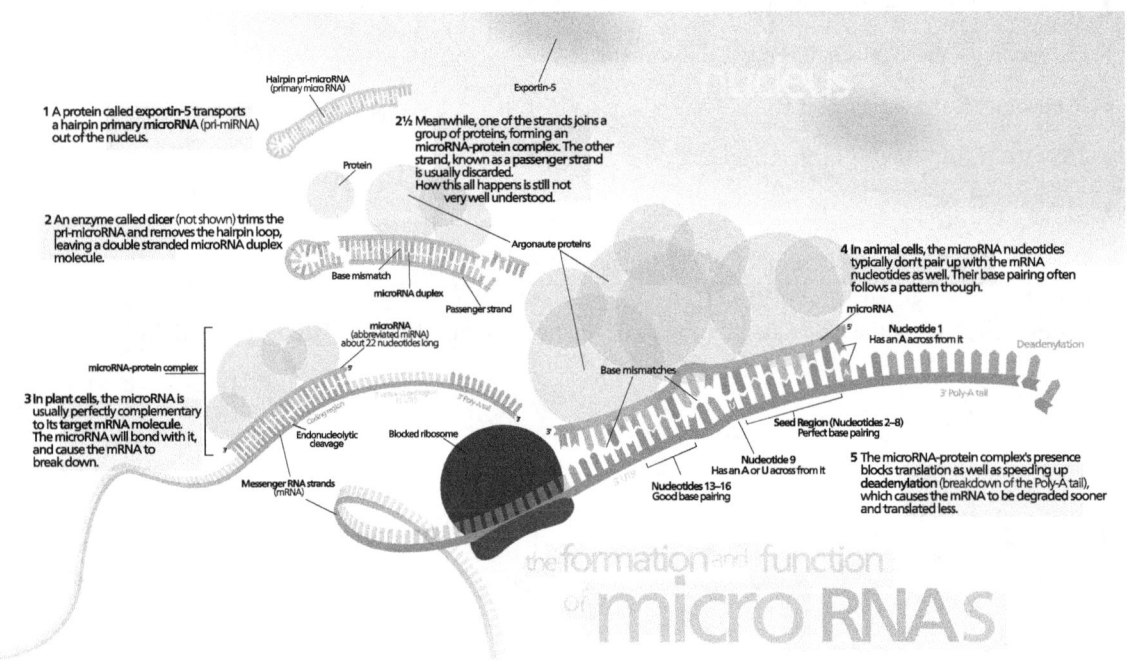

Diagram of micro RNA (miRNA) action with mRNA

A **micro RNA** (abbreviated **miRNA**) is a small non-coding RNA molecule (containing about 22 nucleotides) found in plants, animals, and some viruses, which functions in RNA silencing and post-transcriptional regulation of gene expression.[1][2]

Encoded by eukaryotic nuclear DNA in plants and animals and by viral DNA in certain viruses whose genome is based on DNA, miRNAs function via base-pairing with complementary sequences within mRNA molecules.[3] As a result, these mRNA molecules are silenced by one or more of the following processes: 1) cleavage of the mRNA strand into two pieces, 2) destabilization of the mRNA through shortening of its poly(A) tail, and 3) less efficient translation of the mRNA into proteins by ribosomes.[3][4] miRNAs resemble the small interfering RNAs (siRNAs) of the RNA interference (RNAi) pathway, except miRNAs derive from regions of RNA transcripts that fold back on themselves to form short hairpins, whereas siRNAs derive from longer regions of double-stranded RNA.[2] The human genome may encode over 1000 miRNAs,[5][6] which are abundant in many mammalian cell types[7][8] and appear to target about 60% of the genes of humans and other mammals.[9][10]

miRNAs are well conserved in both plants and animals, and are thought to be a vital and evolutionarily ancient com-

Examples of miRNA stem-loops, with the mature miRNAs shown in red

ponent of genetic regulation.[11][12][13][14][15] While core components of the microRNA pathway are conserved between plants and animals, miRNA repertoires in the two kingdoms appear to have emerged independently with different primary modes of action.[16][17] Plant miRNAs usually have near-perfect pairing with their mRNA targets, which induces gene repression through cleavage of the target transcripts.[18] In contrast, animal miRNAs are able to recognize their target mRNAs by using as little as 6–8 nucleotides (the seed region) at the 5' end of the miRNA,[9][19][20] which is not enough pairing to induce cleavage of the target mRNAs.[3] Combinatorial regulation is a feature of miRNA regulation in animals.[3][21] A given miRNA may have hundreds of different mRNA targets, and a given target might be regulated by multiple miRNAs.[10][22]

The first miRNA was discovered in the early 1990s.[23][24] However, miRNAs were not recognized as a distinct class of biological regulators until the early 2000s.[25][26][27][28][29] Since then, miRNA research has revealed different sets of miRNAs expressed in different cell types and tissues[8][30] and has revealed multiple roles for miRNAs in plant and animal development and in many other biological processes.[18][31][32][33][34][35][36] Aberrant expression of miRNAs has been implicated in numerous disease states, and miRNA-based therapies are under investigation.[37][38][39][40]

Estimates of the average number of unique messenger RNAs that are targets for repression by a typical microRNA vary, depending on the method used to make the estimate,[41] but several approaches show that mammalian miRNAs can have many unique targets. For example, an analysis of the miRNAs highly conserved in vertebrate animals shows that each of these miRNAs has, on average, roughly 400 conserved targets.[10] Likewise, experiments show that a single miRNA can reduce the stability of hundreds of unique messenger RNAs,[42] and other experiments show that a single miRNA may repress the production of hundreds of proteins, but that this repression often is relatively mild (less than 2-fold).[43][44] The first human disease associated with deregulation of miRNAs was chronic lymhocytic leukemia, and other B cell malignancies followed. ".[45]

32.1 History

The first miRNA was discovered in 1993 by a group led by Victor Ambros and including Rosalind Lee and Rhonda Feinbaum; but full understanding of its mode of action was dependent on simultaneously published work by Gary Ruvkun's team, including Bruce Wightman and Ilho Ha.[23][24] These groups published back-to-back papers on the *lin-4* gene, which was known to control the timing of *C. elegans* larval development by repressing the *lin-14* gene. When Lee et al. isolated the *lin-4* gene, they found that instead of producing an mRNA encoding a protein, it produced short noncoding RNAs, one of which was a ~22-nucleotide RNA that contained sequences partially complementary to multiple sequences in the 3' UTR of the *lin-14* mRNA.[23] This complementarity was proposed to inhibit the translation of the *lin-14* mRNA into the LIN-14 protein. At the time, the *lin-4* small RNA was thought to be a nematode idiosyncrasy. Only in 2000 was a second small RNA characterized: *let-7* RNA, which represses *lin-41* to promote a later developmental transition in *C. elegans*.[25]

The *let-7* RNA was soon found to be conserved in many species, leading to the suggestion that *let-7* RNA and additional "small temporal RNAs" might regulate the timing of development in diverse animals, including humans.[26] A year later, the *lin-4* and *let-7* RNAs were found to be part of a very large class of small RNAs present in *C. elegans*, *Drosophila* and human cells.[27][28][29] The many newly discovered RNAs of this class resembled the *lin-4* and *let-7* RNAs, except their expression patterns were usually inconsistent with a role in regulating the timing of development, which suggested that most might function in other types of regulatory pathways. At this point, researchers started using the term "microRNA" to refer to this class of small regulatory RNAs.[27][28][29] The first human diasease associated with deregulaton of miRNAs was chronic lymphocytic leukemia ".[45]

32.2   Nomenclature

Under a standard nomenclature system, names are assigned to experimentally confirmed miRNAs before publication of their discovery.[46][47] The prefix "miR" is followed by a dash and a number, the latter often indicating order of naming. For example, miR-124 was named and likely discovered prior to miR-456. A capitalized "miR-" refers to the mature form of the miRNA, while the uncapitalized "mir-" refers to the pre-miRNA and the pri-miRNA, and "MIR" refers to the gene that encodes them.[48] miRNAs with nearly identical sequences except for one or two nucleotides are annotated with an additional lower case letter. For example, miR-124a is closely related to miR-124b. Pre-miRNAs, pri-miRNAs and genes that lead to 100% identical mature miRNAs but that are located at different places in the genome are indicated with an additional dash-number suffix. For example, the pre-miRNAs hsa-mir-194-1 and hsa-mir-194-2 lead to an identical mature miRNA (hsa-miR-194) but are from genes located in different regions of the genome. Species of origin is designated with a three-letter prefix, e.g., hsa-miR-124 is a human (*Homo sapiens*) miRNA and oar-miR-124 is a sheep (*Ovis aries*) miRNA. Other common prefixes include 'v' for viral (miRNA encoded by a viral genome) and 'd' for *Drosophila* miRNA (a fruit fly commonly studied in genetic research). When two mature microRNAs originate from opposite arms of the same pre-miRNA and are found in roughly similar amounts, they are denoted with a −3p or −5p suffix. (In the past, this distinction was also made with 's' (sense) and 'as' (antisense)). However, the mature microRNA found from one arm of the hairpin is usually much more abundant than that found from the other arm,[2] in which case, an asterisk following the name indicates the mature species found at low levels from the opposite arm of a hairpin. For example, miR-124 and miR-124* share a pre-miRNA hairpin, but much more miR-124 is found in the cell.

32.3   Biogenesis

MicroRNAs are produced from either their own genes or from introns. A video of this process can be found here.

The majority of the characterized miRNA genes are intergenic or oriented antisense to neighboring genes and are therefore suspected to be transcribed as independent units.[27][28][28][29][49] However, in some cases a microRNA gene is transcribed together with its host gene; this provides a means for coupled regulation of miRNA and protein-coding gene.[50][51] As much as 40% of miRNA genes may lie in the introns of protein and non-protein coding genes or even in exons of long nonprotein-coding transcripts.[52] These are usually, though not exclusively, found in a sense orientation,[53][54] and thus usually are regulated together with their host genes.[52][55][56] Other miRNA genes showing a common promoter include the 42-48% of all miRNAs originating from polycistronic units containing multiple discrete loops from which mature miRNAs are processed,[49][57] although this does not necessarily mean the mature miRNAs of a family will be homologous in structure and function. The promoters mentioned have been shown to have some similarities in their motifs to promoters of other genes transcribed by RNA polymerase II such as protein coding genes.[49][58] The DNA template is not the final word on mature miRNA production: 6% of human miRNAs show RNA editing (IsomiRs), the site-specific modification of RNA sequences to yield products different from those encoded by their DNA. This increases the diversity and scope of miRNA action beyond that implicated from the genome alone.

32.3.1   Transcription

miRNA genes are usually transcribed by RNA polymerase II (Pol II).[49][58] The polymerase often binds to a promoter found near the DNA sequence encoding what will become the hairpin loop of the pre-miRNA. The resulting transcript is

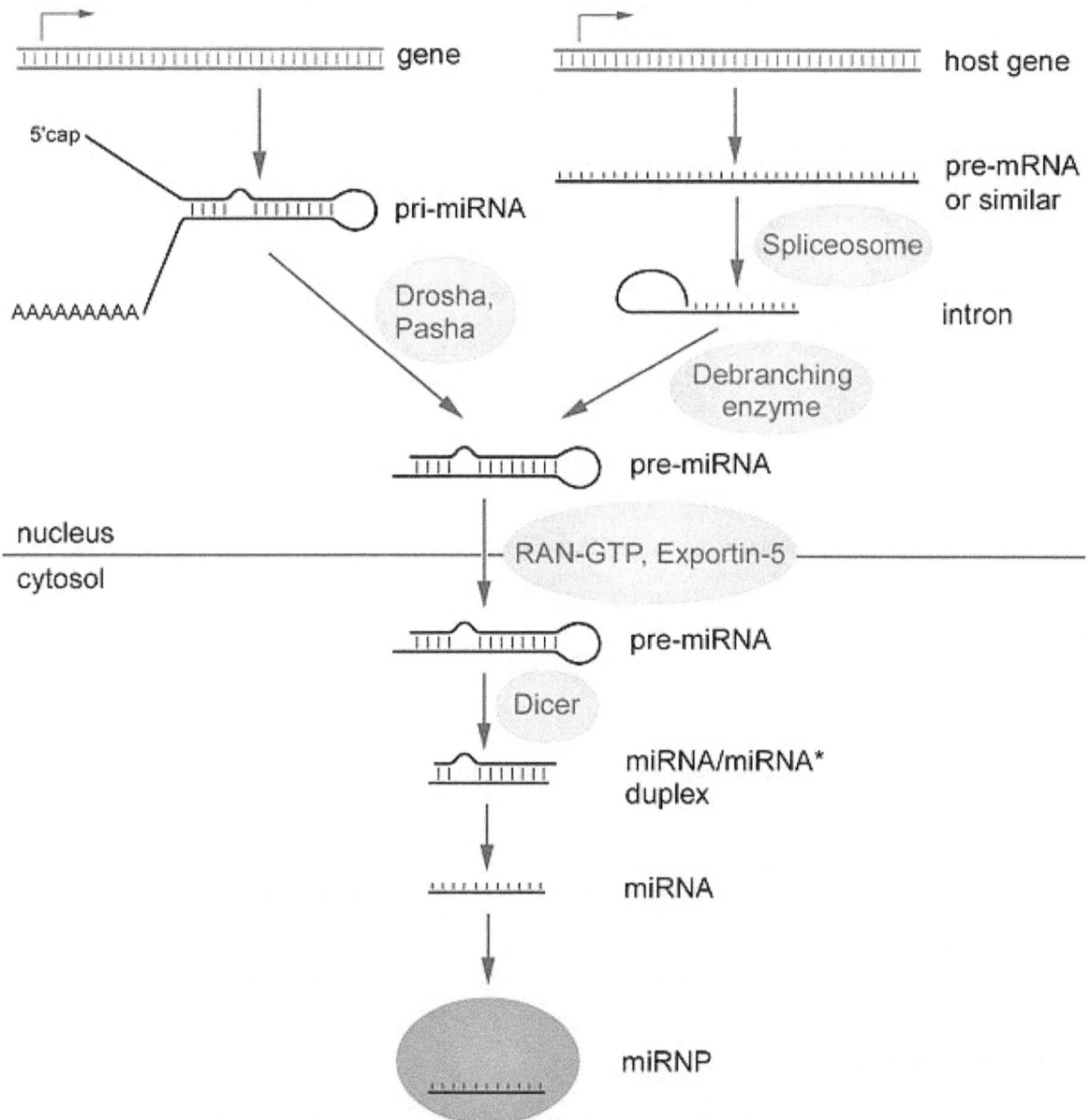

capped with a specially modified nucleotide at the 5' end, polyadenylated with multiple adenosines (a poly(A) tail),[49][53] and spliced. Animal miRNAs are initially transcribed as part of one arm of an ~80 nucleotide RNA stem-loop that in turn forms part of a several hundred nucleotides long miRNA precursor termed a primary miRNA (pri-miRNA)s.[49][53] When a stem-loop precursor is found in the 3' UTR, a transcript may serve as a pri-miRNA and a mRNA.[53] RNA polymerase III (Pol III) transcribes some miRNAs, especially those with upstream Alu sequences, transfer RNAs (tRNAs), and mammalian wide interspersed repeat (MWIR) promoter units.[59]

32.3.2 Nuclear processing

A single pri-miRNA may contain from one to six miRNA precursors. These hairpin loop structures are composed of about 70 nucleotides each. Each hairpin is flanked by sequences necessary for efficient processing. The double-stranded

RNA structure of the hairpins in a pri-miRNA is recognized by a nuclear protein known as DiGeorge Syndrome Critical Region 8 (DGCR8 or "Pasha" in invertebrates), named for its association with DiGeorge Syndrome. DGCR8 associates with the enzyme Drosha, a protein that cuts RNA, to form the "Microprocessor" complex.[60][61] In this complex, DGCR8 orients the catalytic RNase III domain of Drosha to liberate hairpins from pri-miRNAs by cleaving RNA about eleven nucleotides from the hairpin base (one helical dsRNA turn into the stem).[62][63] The product resulting has a two-nucleotide overhang at its 3' end; it has 3' hydroxyl and 5' phosphate groups. It is often termed as a pre-miRNA (precursor-miRNA). Sequence motifs downstream of the pre-miRNA that are important for efficient processing have been identified.[64][65][66]

Pre-miRNAs that are spliced directly out of introns, bypassing the Microprocessor complex, are known as "Mirtrons." Originally thought to exist only in *Drosophila* and *C. elegans*, mirtrons have now been found in mammals.[67]

Perhaps as many as 16% of pre-miRNAs may be altered through nuclear RNA editing.[68][69][70] Most commonly, enzymes known as adenosine deaminases acting on RNA (ADARs) catalyze adenosine to inosine (A to I) transitions. RNA editing can halt nuclear processing (for example, of pri-miR-142, leading to degradation by the ribonuclease Tudor-SN) and alter downstream processes including cytoplasmic miRNA processing and target specificity (e.g., by changing the seed region of miR-376 in the central nervous system).[68]

32.3.3 Nuclear export

Pre-miRNA hairpins are exported out of the nucleus in a process involving the nucleocytoplasmic shuttler Exportin-5. This protein, a member of the *karyopherin* family, recognizes a two-nucleotide overhang left by the RNase III enzyme Drosha at the 3' end of the pre-miRNA hairpin. Exportin-5-mediated transport to the cytoplasm is energy-dependent, using GTP bound to the Ran protein.[71]

32.3.4 Cytoplasmic processing

In the cytoplasm, the pre-miRNA hairpin is cleaved by the RNase III enzyme Dicer.[72] This endoribonuclease interacts with 5' and 3' ends of the hairpin[73] and cuts away the loop joining the 3' and 5' arms, yielding an imperfect miRNA: miRNA* duplex about 22 nucleotides in length.[72] Overall hairpin length and loop size influence the efficiency of Dicer processing, and the imperfect nature of the miRNA:miRNA* pairing also affects cleavage.[72][74] Some of the G-rich pre-miRNAs can potentially adopt the G-quadruplex structure as an alternative to the canonical stem-loop structure. For example, human pre-miRNA 92b adopts a G-quadruplex structure which is resistant to the Dicer mediated cleavage in the cytoplasm.[75] Although either strand of the duplex may potentially act as a functional miRNA, only one strand is usually incorporated into the RNA-induced silencing complex (RISC) where the miRNA and its mRNA target interact.

32.3.5 Biogenesis in plants

miRNA biogenesis in plants differs from animal biogenesis mainly in the steps of nuclear processing and export. Instead of being cleaved by two different enzymes, once inside and once outside the nucleus, both cleavages of the plant miRNA are performed by a Dicer homolog, called Dicer-like1 (DL1). DL1 is only expressed in the nucleus of plant cells, which indicates that both reactions take place inside the nucleus. Before plant miRNA:miRNA* duplexes are transported out of the nucleus, its 3' overhangs are methylated by a RNA methyltransferaseprotein called Hua-Enhancer1 (HEN1). The duplex is then transported out of the nucleus to the cytoplasm by a protein called Hasty (HST), an Exportin 5 homolog, where they disassemble and the mature miRNA is incorporated into the RISC.[76]

32.4 The RNA-induced silencing complex

Main article: RNA-induced silencing complex

The mature miRNA is part of an active RNA-induced silencing complex (RISC) containing Dicer and many associated proteins.[77] RISC is also known as a microRNA ribonucleoprotein complex (miRNP);[78] RISC with incorporated miRNA is sometimes referred to as "miRISC."

Dicer processing of the pre-miRNA is thought to be coupled with unwinding of the duplex. Generally, only one strand is incorporated into the miRISC, selected on the basis of its thermodynamic instability and weaker base-pairing relative to the other strand.[79][80][81] The position of the stem-loop may also influence strand choice.[82] The other strand, called the passenger strand due to its lower levels in the steady state, is denoted with an asterisk (*) and is normally degraded. In some cases, both strands of the duplex are viable and become functional miRNA that target different mRNA populations.[83]

Members of the Argonaute (Ago) protein family are central to RISC function. Argonautes are needed for miRNA-induced silencing and contain two conserved RNA binding domains: a PAZ domain that can bind the single stranded 3' end of the mature miRNA and a PIWI domain that structurally resembles ribonuclease-H and functions to interact with the 5' end of the guide strand. They bind the mature miRNA and orient it for interaction with a target mRNA. Some argonautes, for example human Ago2, cleave target transcripts directly; argonautes may also recruit additional proteins to achieve translational repression.[84] The human genome encodes eight argonaute proteins divided by sequence similarities into two families: AGO (with four members present in all mammalian cells and called E1F2C/hAgo in humans), and PIWI (found in the germ line and hematopoietic stem cells).[78][84]

Additional RISC components include TRBP [human immunodeficiency virus (HIV) transactivating response RNA (TAR) binding protein],[85] PACT (protein activator of the interferon induced protein kinase), the SMN complex, fragile X mental retardation protein (FMRP), Tudor staphylococcal nuclease-domain-containing protein (Tudor-SN), the putative DNA helicase MOV10, and the RNA recognition motif containing protein TNRC6B.[71][86][87]

32.4.1 Mode of silencing and regulatory loops

Gene silencing may occur either via mRNA degradation or preventing mRNA from being translated. For example, miR16 contains a sequence complementary to the AU-rich element found in the 3'UTR of many unstable mRNAs, such as TNF alpha or GM-CSF.[88] It has been demonstrated that if there is complete complementation between the miRNA and target mRNA sequence, Ago2 can cleave the mRNA and lead to direct mRNA degradation. Yet, if there isn't complete complementation the silencing is achieved by preventing translation.[42] The relation of miRNA and its target mRNA(s) can be based on the simple negative regulation of a target mRNA, but it seems that a common scenario is the use of a "coherent feed-forward loop" (Fig. 1C), "mutual negative feedback loop" (also termed double negative loop) and "positive feedback/feed-forward loop".[45] It was also demonstrated that some miRNAs work as buffers of random gene expression changes arising due to stochastic events in transcription, translation and protein stability.[45] Such regulation is typically achieved by the virtue of negative feedback loops or incoherent feed-forward loop uncoupling protein output from mRNA transcription.[45]

32.5 miRNA turnover

Turnover of mature miRNA is needed for rapid changes in miRNA expression profiles. During miRNA maturation in the cytoplasm, uptake by the Argonaute protein is thought to stabilize the guide strand, while the opposite (* or "passenger") strand is preferentially destroyed. In what has been called a "Use it or lose it" strategy, Argonaute may preferentially retain miRNAs with many targets over miRNAs with few or no targets, leading to degradation of the non-targeting molecules.[89]

Decay of mature miRNAs in *Caenorhabditis elegans* is mediated by the 5′-to-3′ exoribonuclease XRN2, also known as Rat1p.[90] In plants, SDN (small RNA degrading nuclease) family members degrade miRNAs in the opposite (3'-to-5') direction. Similar enzymes are encoded in animal genomes, but their roles have not yet been described.[89]

Several miRNA modifications affect miRNA stability. As indicated by work in the model organism *Arabidopsis thaliana* (thale cress), mature plant miRNAs appear to be stabilized by the addition of methyl moieties at the 3' end. The 2'-O-conjugated methyl groups block the addition of uracil (U) residues by uridyltransferase enzymes, a modification that may be associated with miRNA degradation. However, uridylation may also protect some miRNAs; the consequences of this modification are incompletely understood. Uridylation of some animal miRNAs has also been reported. Both plant and animal miRNAs may be altered by addition of adenine (A) residues to the 3' end of the miRNA. An extra A added to the end of mammalian miR-122, a liver-enriched miRNA important in Hepatitis C, stabilizes the molecule, and plant miRNAs ending with an adenine residue have slower decay rates.[89]

32.6 Cellular functions

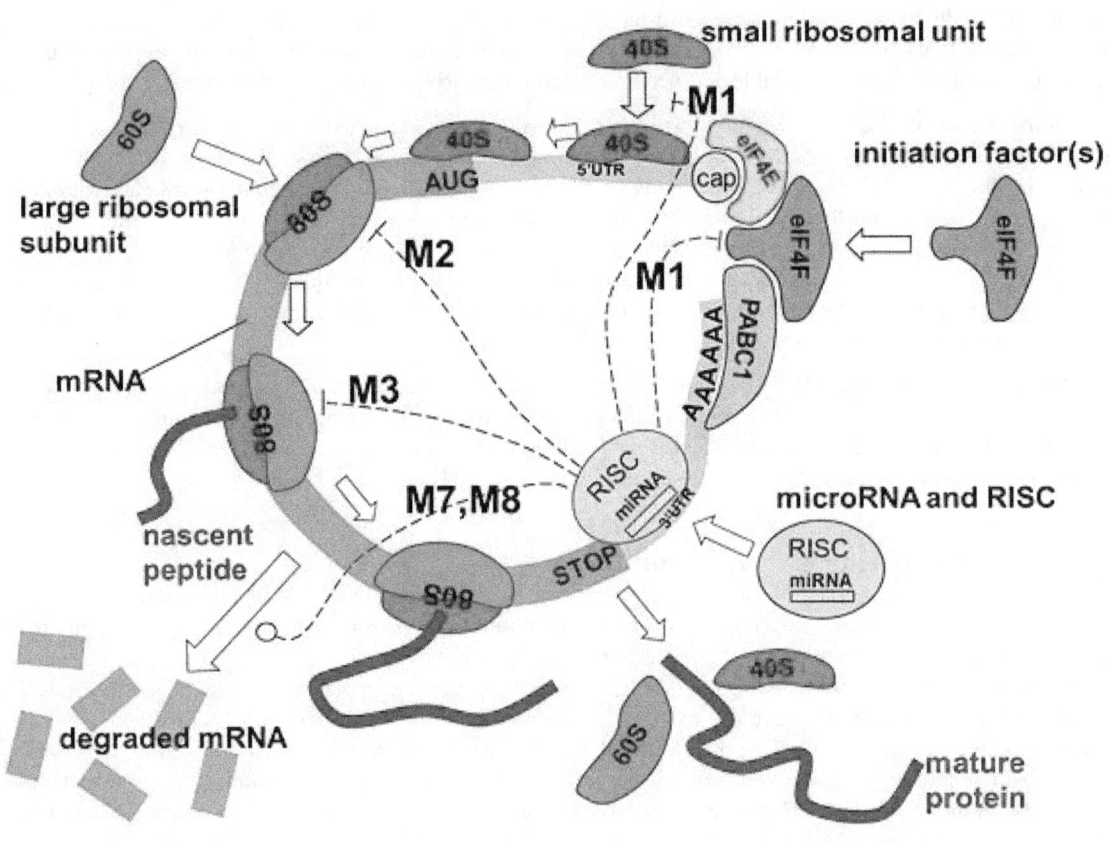

Interaction of microRNA with protein translation process. Several (from nine documented) mechanisms of translation repression are shown: M1) on the initiation process, preventing assembling of the initiation complex or recruiting the 40S ribosomal subunit; M2) on the ribosome assembly; M3) on the translation process; M7, M8) on the degradation of mRNA. There exist other mechanisms of microRNA action on protein translation (transcriptional, transport to P-bodies, ribosome drop-off, co-translational protein degradation and others) that are not visualized here.[91] Here, 40S and 60S are light and heavy components of the ribosome, 80S is the assembled ribosome bound to mRNA, eIF4F is an translation initiation factor, PABC1 is the Poly-A binding protein, and "cap" is the mRNA cap structure needed for mRNA circularization (which can be the normal m7G-cap or artificial modified A-cap). The initiation of mRNA can proceed in a cap-independent manner, through recruiting 40S to IRES (Internal Ribosome Entry Site) located in 5'UTR region. The actual work of RNA silencing is performed by RISC (RNA-induced silencing complex) in which the main catalytic subunit is one of the Argonaute proteins (AGO), and miRNA serves as a template for recognizing specific mRNA sequences.

The function of miRNAs appears to be in gene regulation. For that purpose, a miRNA is complementary to a part of one or more messenger RNAs (mRNAs). Animal miRNAs are usually complementary to a site in the 3' UTR whereas plant miRNAs are usually complementary to coding regions of mRNAs.[92] Perfect or near perfect base pairing with the target RNA promotes cleavage of the RNA.[93] This is the primary mode of plant miRNAs.[94] In animals miRNAs more often have only partly the right sequence of nucleotides to bond with the target mRNA. The match-ups are imperfect. For partially complementary microRNAs to recognise their targets, nucleotides 2–7 of the miRNA (its 'seed region'[9][19]) still have to be perfectly complementary.[95] Animal miRNAs inhibit protein translation of the target mRNA[96] (this exists in plants as well but is less common).[94] MicroRNAs that are partially complementary to a target can also speed up deadenylation, causing mRNAs to be degraded sooner.[97] While degradation of miRNA-targeted mRNA is well documented, whether or not translational repression is accomplished through mRNA degradation, translational inhibition, or a combination of the two is hotly debated. Recent work on miR-430 in zebrafish, as well as on bantam-miRNA and miR-9 in *Drosophila* cultured cells, shows that translational repression is caused by the disruption of translation initiation, independent of mRNA deadenylation.[98][99]

miRNAs occasionally also cause histone modification and DNA methylation of promoter sites, which affects the expression of target genes.[100][101]

Nine mechanisms of miRNA action are described and assembled in a unified mathematical model:[91]

1. Cap-40S initiation inhibition;

2. 60S Ribosomal unit joining inhibition;

3. Elongation inhibition;

4. Ribosome drop-off (premature termination);

5. Co-translational nascent protein degradation;

6. Sequestration in P-bodies;

7. mRNA Decay (destabilisation);

8. mRNA Cleavage;

9. Transcriptional inhibition through microRNA-mediated chromatin reorganization followed by gene silencing.

It is often impossible to discern these mechanisms using the experimental data about stationary reaction rates. Nevertheless, they are differentiated in dynamics and have different *kinetic signatures*.[91]

Unlike plant microRNAs, the animal microRNAs target a diverse set of genes.[19] However, genes involved in functions common to all cells, such as gene expression, have relatively fewer microRNA target sites and seem to be under selection to avoid targeting by microRNAs.[102]

dsRNA can also activate gene expression, a mechanism that has been termed "small RNA-induced gene activation" or RNAa. dsRNAs targeting gene promoters can induce potent transcriptional activation of associated genes. This was demonstrated in human cells using synthetic dsRNAs termed small activating RNAs (saRNAs),[103] but has also been demonstrated for endogenous microRNA.[104]

Interactions between microRNAs and complementary sequences on genes and even pseudogenes that share sequence homology are thought to be a back channel of communication regulating expression levels between paralogous genes. Given the name "competing endogenous RNAs" (ceRNAs), these microRNAs bind to "microRNA response elements" on genes and pseudogenes and may provide another explanation for the persistence of non-coding DNA.[105]

32.7 Evolution

MicroRNAs are useful phylogenetic markers because of their apparently low rate of evolution.[106] MicroRNAs' origin as a regulatory mechanism developed from previous RNAi machinery which was initially used as a defense against exogenous genetic material such as viruses.[107] Their origin may have permitted the development of morphological innovation, and by making gene expression more specific and 'fine-tunable', permitted the genesis of complex organs[108] and perhaps, ultimately, complex life.[109] Indeed, rapid bursts of morphological innovation are generally associated with a high rate of microRNA accumulation.[106][108]

New microRNAs are created in multiple different ways. Novel microRNAs can originate from the random formation of hairpins in "non-coding" sections of DNA (i.e. introns or intergene regions), but also by the duplication and modification of existing microRNAs.[110] MicroRNAs can also form from inverted duplications of protein-coding sequences, which allows for the creation of a foldback hairpin structure.[111] The rate of evolution (i.e. nucleotide substitution) in recently originated microRNAs is comparable to that elsewhere in the non-coding DNA, implying evolution by neutral drift; however, older microRNAs have a much lower rate of change (often less than one substitution per hundred million years),[109] suggesting that once a microRNA gains a function, it undergoes extreme purifying selection.[110] Additionally, different regions within an miRNA gene seem to be under different evolutionary pressures, where regions that are vital for processing and function have much higher levels of conservation.[112] At this point, a microRNA is rarely lost from an animal's genome,[109] although microRNAs that are more recently derived (and thus presumably non-functional) are frequently lost.[110] In *Arabidopsis thaliana*, the net flux of miRNA genes has been predicted to be between 1.2 and 3.3 genes per million years.[113] This makes them a valuable phylogenetic marker, and they are being looked upon as a possible solution to such outstanding phylogenetic problems as the relationships of arthropods.[114] On the other hand, there are a number of cases where microRNAs correlate very poorly with phylogeny, and it is possible that their high phylogenetic concordance largely reflects the limited sampling of microRNAs currently available.[115]

MicroRNAs feature in the genomes of most eukaryotic organisms, from the brown algae[116] to the animals. However, the difference in how these microRNAs function and the way they are processed suggests that microRNAs arose independently in plants and animals.[117] Focusing on the animals, the genome of *Mnemiopsis leidyi* [118] appears to lack recognizable microRNAs, as well as the nuclear proteins Drosha and Pasha, which are critical to canonical microRNA biogenesis. It is the only animal thus far reported to be missing Drosha. MicroRNAs play a vital role in the regulation of gene expression in all non-ctenophore animals investigated thus far except for *Trichoplax adhaerens*, the only known member of the phylum Placozoa.[119]

Across all species, in excess of 5000 different miRNAs had been identified by March 2010.[120] Whilst short RNA sequences (50 – hundreds of base pairs) of a broadly comparable function occur in bacteria, bacteria lack true microRNAs.[121]

32.8   Experimental detection and manipulation of miRNA

While researchers have focused on the study of miRNA expression in physiological and pathological processes, various technical variables related to microRNA isolation have emerged. The stability of the stored miRNA samples has often been questioned.[122] MicroRNAs are degraded much more easily than mRNAs, partly due to their length, but also because of the ubiquitously present RNases. This makes it necessary to cool samples on ice and use RNase-free equipment whenever working with microRNAs.[123]

MicroRNA expression can be quantified in a two-step polymerase chain reaction process of modified RT-PCR followed by quantitative PCR. Variations of this method achieve absolute or relative quantification.[124] miRNAs can also be hybridized to microarrays, slides or chips with probes to hundreds or thousands of miRNA targets, so that relative levels of miRNAs can be determined in different samples.[125] MicroRNAs can be both discovered and profiled by high-throughput sequencing methods (MicroRNA Sequencing).[126] The activity of an miRNA can be experimentally inhibited using a locked nucleic acid (LNA) oligo, a Morpholino oligo[127][128] or a 2'-O-methyl RNA oligo.[129] Additionally, a specific miRNA can be silenced by a complementary antagomir. MicroRNA maturation can be inhibited at several points by steric-blocking oligos.[130] The miRNA target site of an mRNA transcript can also be blocked by a steric-blocking oligo.[131] For the "in situ" detection of miRNA, LNA[132] or Morpholino[133] probes can be used. The locked conformation of LNA results in enhanced hybridization properties and increases sensitivity and selectivity, making it ideal for detection of short miRNA.[134]

High-throughput quantification of miRNAs is often difficult and prone to errors, for the larger variance (compared to mRNAs) that comes with the methodological problems. mRNA-expression is therefore often analyzed as well to check for miRNA-effects in their levels (e. g. in [135][136]). To pair mRNA- and miRNA-data, databases can be used which predict miRNA-targets based on their base sequence.[137][138] While this is usually done after miRNAs of interest have been detected (e. g. because of high expression levels), ideas for analysis tools that integrate mRNA- and miRNA-expression information have been proposed.[139][140]

32.9   Disease

Just as miRNA is involved in the normal functioning of eukaryotic cells, so has dysregulation of miRNA been associated with disease.[141] A manually curated, publicly available database, miR2Disease, documents known relationships between miRNA dysregulation and human disease.[142]

32.9.1 Inherited diseases

A mutation in the seed region of **miR-96**, causes hereditary progressive hearing loss.[143]

A mutation in the seed region of **miR-184**, causes hereditary keratoconus with anterior polar cataract.[144]

Deletion of the **miR-17~92** cluster, causes skeletal and growth defects.[145]

32.9.2 Cancer

(A) miRNA functioning as a tumour suppressor gene

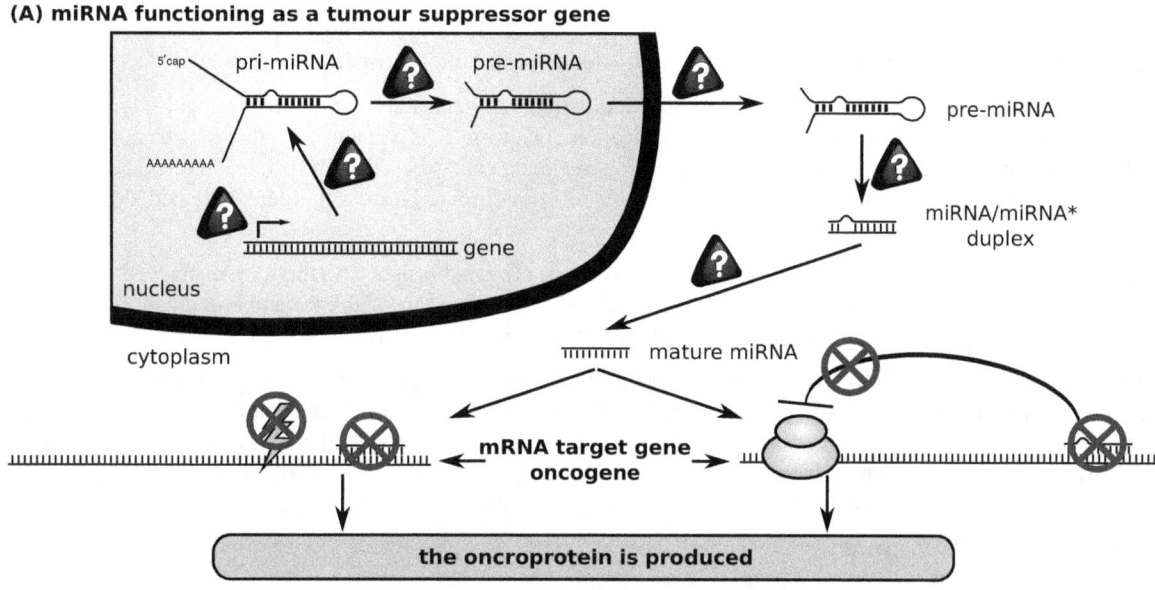

(B) miRNA functioning as an oncogene

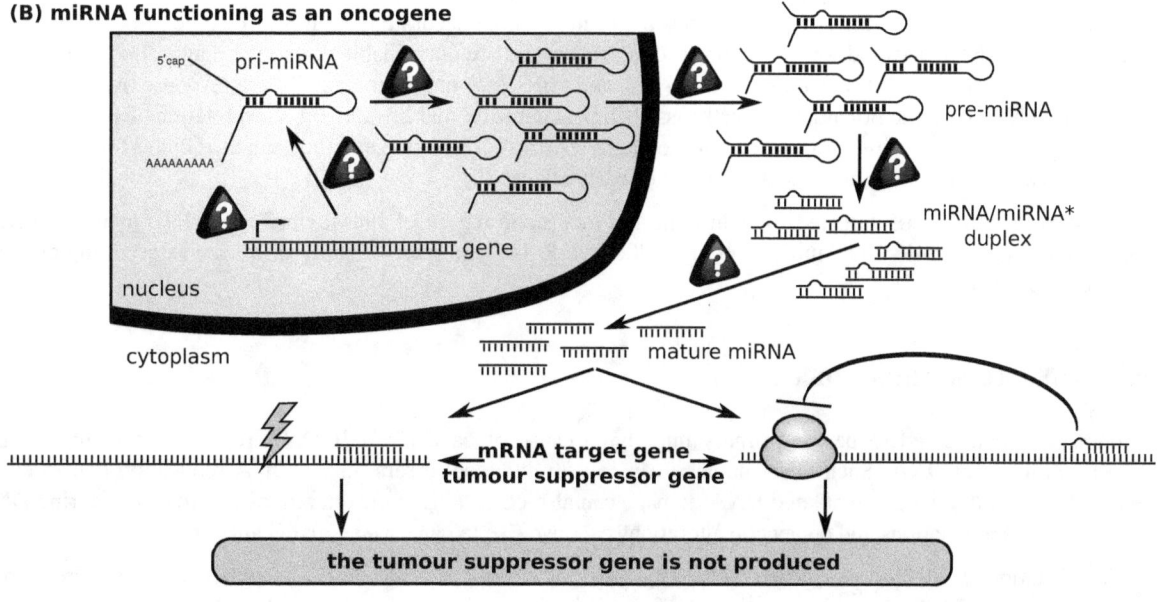

Role of miRNA in a cancer cell

The first human disease known to be associated with miRNA deregulation was chronic lymphocytic leukemia.[45] Many miRNAs have subsequently been found to have links with various types of cancer.[45] and accordingly are sometimes referred to as "oncomirs". In malignant B cells miRNAs participate in pathways fundamental to B cell development like B cell receptor (BCR) signalling, B cell migration/adhesion, cell-cell interactions in immune niches, and the production and class-switching of immunoglobulins. MiRNAs influence B cell maturation, generation of pre-, marginal zone, follicular, B1, plasma and memory B cells.[45]

A study of mice altered to produce excess c-Myc — a protein with mutated forms implicated in several cancers — shows that miRNA has an effect on the development of cancer. Mice that were engineered to produce a surplus of types of miRNA found in lymphoma cells developed the disease within 50 days and died two weeks later. In contrast, mice without the surplus miRNA lived over 100 days[45] Leukemia can be caused by the insertion of a viral genome next to the 17-92 array of microRNAs leading to increased expression of this microRNA[45]··

Another study found that two types of miRNA inhibit the E2F1 protein, which regulates cell proliferation. miRNA appears to bind to messenger RNA before it can be translated to proteins that switch genes on and off[45]

By measuring activity among 217 genes encoding miRNA, patterns of gene activity that can distinguish types of cancers can be discerned. miRNA signatures may enable classification of cancer. This will allow doctors to determine the original tissue type which spawned a cancer and to be able to target a treatment course based on the original tissue type.[146] miRNA profiling has already been able to determine whether patients with chronic lymphocytic leukemia had slow growing or aggressive forms of the cancer.[45]

Transgenic mice that over-express or lack specific miRNAs have provided insight into the role of small RNAs in various malignancies.[45] Much work has also been done on the role of microRNAs in establishing and maintaining cancer stem cells that are especially resistant to chemotherapy and often responsible for relapse.[147]

A novel miRNA-profiling-based screening assay for the detection of early-stage colorectal cancer has been developed and is currently in clinical trials. Early results showed that blood plasma samples collected from patients with early, resectable (Stage II) colorectal cancer could be distinguished from those of sex-and age-matched healthy volunteers. Sufficient selectivity and specificity could be achieved using small (less than 1 mL) samples of blood. The test has potential to be a cost-effective, non-invasive way to identify at-risk patients who should undergo colonoscopy.[148][149]

Another role for miRNA in cancers is to use their expression level as a prognostic. For example, one study on NSCLC samples found that low miR-324a levels could serve as a prognostic indicator of poor survival.[150] Either high miR-185 or low miR-133b levels may correlate with metastasis and poor survival in colorectal cancer.[151] Hepatocellular carcinoma cell proliferation may arise from miR-21 interaction with MAP2K3, a tumor repressor gene.[152] Optimal treatment for cancer involves accurately identifying patients for risk-stratified therapy. Those with a rapid response to initial treatment may benefit from truncated treatment regimens, thus the need for more accurate measures of disease response. Cell-free miRNA are highly stable in blood, are overexpressed in cancer, and are quantifiable within the diagnostic laboratory. In classical Hodgkin lymphoma, plasma miR-21, miR-494, and miR-1973 are promising disease response biomarkers.[153] Circulating miRNAs have the potential to greatly assist clinical decision making and aid interpretation of positron emission tomography combined with computerized tomography. A further advantage is they can also be performed at each consultation to assess disease response and detection of early relapse.

Recent studies have miR-205 targeted for inhibiting the metastatic nature of breast cancer.[154] Five members of the microRNA-200 family (miR-200a, miR-200b, miR-200c, miR-141 and miR-429) are down regulated in tumour progression of breast cancer.[155]

32.9.3   DNA repair and cancer

DNA damage is considered to be the primary underlying cause of cancer.[156] If DNA repair is deficient, damage tends to accumulate in DNA. Such DNA damage can cause mutational errors during DNA replication due to error-prone translesion synthesis. Accumulated DNA damage can also cause epigenetic alterations due to errors during DNA repair.[157][158] Such mutations and epigenetic alterations can give rise to cancer (see malignant neoplasms).

Germ line mutations in DNA repair genes cause only 2–5% of colon cancer cases.[159] However, altered expression of microRNAs, causing DNA repair deficiencies, are frequently associated with cancers and may be an important causal factor for these cancers.

Among 68 sporadic colon cancers with reduced expression of the DNA mismatch repair protein MLH1, most were found to be deficient due to epigenetic methylation of the CpG island of the MLH1 gene.[160] However, up to 15% of the MLH1-deficiencies in sporadic colon cancers appeared to be due to over-expression of the microRNA miR-155, which represses MLH1 expression.[161]

In 29–66%[162][163] of glioblastomas, DNA repair is deficient due to epigenetic methylation of the MGMT gene, which reduces protein expression of MGMT. However, for 28% of glioblastomas, the MGMT protein is deficient but the MGMT promoter is not methylated.[162] In the glioblastomas without methylated MGMT promoters, the level of microRNA miR-181d is inversely correlated with protein expression of MGMT and the direct target of miR-181d is the MGMT mRNA 3'UTR (the three prime untranslated region of MGMT mRNA).[162] Thus, in 28% of glioblastomas, increased expression of miR-181d and reduced expression of DNA repair enzyme MGMT may be a causal factor.

HMGA proteins (HMGA1a, HMGA1b and HMGA2) are implicated in cancer, and expression of these proteins is regulated by microRNAs. HMGA expression is almost undetectable in differentiated adult tissues but is elevated in many cancers. HMGA proteins are polypeptides of ~100 amino acid residues characterized by a modular sequence organization. These proteins have three highly positively charged regions, termed AT hooks, that bind the minor groove of AT-rich DNA stretches in specific regions of DNA. Human neoplasias, including thyroid, prostatic, cervical, colorectal, pancreatic and ovarian carcinoma, show a strong increase of HMGA1a and HMGA1b proteins.[164] Transgenic mice with HMGA1 targeted to lymphoid cells develop aggressive lymphoma, showing that high HMGA1 expression is not only associated with cancers, but that the HMGA1 gene can act as an oncogene to cause cancer.[165] Baldassarre et al.,[166] showed that HMGA1 protein binds to the promoter region of DNA repair gene BRCA1 and inhibits BRCA1 promoter activity. They also showed that while only 11% of breast tumors had hypermethylation of the BRCA1 gene, 82% of aggressive breast cancers have low BRCA1 protein expression, and most of these reductions were due to chromatin remodeling by high levels of HMGA1 protein.

HMGA2 protein specifically targets the promoter of ERCC1, thus reducing expression of this DNA repair gene.[167] ERCC1 protein expression was deficient in 100% of 47 evaluated colon cancers (though the extent to which HGMA2 was involved is not known).[168] Palmieri et al.[169] showed that in normal tissues, HGMA1 and HMGA2 genes are targeted (and thus strongly reduced in expression) by miR-15, miR-16, miR-26a, miR-196a2 and Let-7a. However, each of these HMGA-targeting miRNAs are drastically reduced in almost all human pituitary adenomas studied, when compared with the normal pituitary gland. Consistent with the down-regulation of these HMGA-targeting miRNAs, an increase in the HMGA1 and HMGA2-specific mRNAs was observed. Three of these microRNAs (miR-16, miR-196a and Let-7a)[170][171] have methylated promoters and therefore low expression in colon cancer. For two of these, miR-15 and miR-16, the coding regions are epigenetically silenced in cancer due to histone deacetylase activity.[172] When these microRNAs are expressed at a low level, then HMGA1 and HMGA2 proteins are expressed at a high level. HMGA1 and HMGA2 target (reduce expression of) BRCA1 and ERCC1 DNA repair [173] genes. Thus DNA repair can be reduced, likely contributing to cancer progression.[156]

In contrast to the previous example, where under-expression of miRNAs indirectly caused reduced expression of DNA repair genes, in some cases over-expression of certain miRNAs may directly reduce expression of specific DNA repair proteins. Wan et al.[174] referred to 6 DNA repair genes that are directly targeted by the miRNAs indicated: *ATM* (miR-421), *RAD52* (miR-210, miR-373), *RAD23B* (miR-373), *MSH2* (miR-21), *BRCA1* (miR-182) and *P53* (miR-504, miR-125b). More recently, Tessitore et al.[175] listed multiple DNA repair genes directly targeted by these additional miRNAs: *ATM* (miR-100, miR18a, miR-101), *DNA-PK* (miR-101), *ATR* (mir-185), *Wip1* (miR-16), *MLH1*, *MSH2*, *MSH6* (miR-155), *ERCC3*, *ERCC4* (miR-192) and *UNG2* (miR-16, miR-34c). Among these miRNAs, miR-16, miR-18a, miR-21, miR-34c, miR-101, miR-125b, miR-155, miR-182, miR-185, miR-192 and miR-373 were identified by Schnekenburger and Diederich[171] as over-expressed in colon cancer through epigenetic hypomethylation. Over expression of any one of these miRNAs can cause reduced expression of its target DNA repair gene.

32.9.4 Heart disease

The global role of miRNA function in the heart has been addressed by conditionally inhibiting miRNA maturation in the murine heart, and has revealed that miRNAs play an essential role during its development.[176][177] miRNA expression profiling studies demonstrate that expression levels of specific miRNAs change in diseased human hearts, pointing to their involvement in cardiomyopathies.[178][179][180] Furthermore, studies on specific miRNAs in animal models have identified distinct roles for miRNAs both during heart development and under pathological conditions, including the regulation of key factors important for cardiogenesis, the hypertrophic growth response, and cardiac conductance.[177][181][182][183][184][185] miRNA's in animal models have also been linked to cholesterol metabolism and regulation.[186]

miRNA-712

Murine microRNA-712 is a potential biomarker (i.e. predictor) for atherosclerosis, a cardiovascular disease of the arterial wall associated with lipid retention and inflammation.[187] Non-laminar blood flow also correlates with development of atherosclerosis as mechanosenors of epithelial cells respond to the sheer force of disturbed flow (d-flow).[173] A number of pro-atherogenic genes including matrix metalloproteinases (MMPs) are upregulated by d-flow ,[173] mediating pro-inflammatory and pro-angiogenic signals. These findings were observed in ligated carotid arteries of mice to mimic the effects of d-flow. Within 24 hours, pre-existing immature miR-712 formed mature miR-712 suggesting that miR-712 is flow-sensitive.[173] Coinciding with these results, miR-712 is also upregulated in endothelial cells exposed to naturally occurring d-flow in the greater curvature of the aortic arch.[173]

Gene Origin

Pre-mRNA sequence of miR-712 is generated from the murine ribosomal RN45s gene at the internal transcribed spacer region 2 (ITS2).[173] XRN1 is an exonuclease that degrades the ITS2 region during processing of RN45s.[173] Reduction of XRN1 under d-flow conditions therefore leads to the accumulation of miR-712.[173]

Mechanism

MiR-712 targets tissue inhibitor of metalloproteinases 3 (TIMP3).[173] TIMPs normally regulate activity of matrix metalloproteinases (MMPs) which degrade the extracellular matrix (ECM). Arterial ECM is mainly composed of collagen and elastin fibers, providing the structural support and recoil properties of arteries.[188] These fibers play a critical role in regulation of vascular inflammation and permeability, which are important in the development of atherosclerosis.[189] Expressed by endothelial cells, TIMP3 is the only ECM bound TIMP.[188] A decrease in TIMP3 expression results in an increase of ECM degradation in the presence of d-flow. Consistent with these findings, inhibition of pre-miR712 increases expression of TIMP3 in cells, even when exposed to turbulent flow.[173]

TIMP3 also decreases the expression of TNFα (a pro-inflammatory regulator) during turbulent flow.[173] Activity of TNFα in turbulent flow was measured by the expression of TNFα converting enzyme (TACE) in blood. TNFα decreased if miR-712 was inhibited or TIMP3 overexpressed,[173] suggesting that miR-712 and TIMP3 regulate TACE activity in turbulent flow conditions.

Anti-miR-712 effectively suppresses d-flow induced miR-712 expression and increases TIMP3 expression.[173] Anti-miR-712 also inhibits vascular hyperpermeability, thereby significantly reducing atherosclerosis lesion development and immune cell infiltration.[173]

Human Homolog microRNA-205

The human homolog of miR-712 was found on the RN45s homolog gene, which maintains similar miRNAs to mice.[173] MiR-205 of humans share similar sequences with miR-712 of mice and is conserved across most vertebrates.[173] MiR-205 and miR-712 also share more than 50% of the cell signaling targets, including TIMP3.[173]

When tested, d-flow decreased the expression of XRN1 in humans as it did in mice endothelial cells, indicating that there may be a common role of XRN1 in human.[173] While the human homolog has not been thoroughly studied, the discovery and function of miRNA-712 can give weight for future research on its potential as a biomarker in mice models of atherosclerosis.

32.9.5 Kidney disease

Targeted deletion of Dicer in the FoxD1-derived renal progenitor cells in a murine model resulted in a complex renal phenotype including expansion of nephron progenitors, fewer renin cells, smooth muscle arterioles, progressive mesangial loss and glomerular aneurysms.[190] High throughput whole transcriptome profiling of the FoxD1-Dicer knockout mouse model revealed ectopic upregulation of pro-apoptotic gene, Bcl2L11 (Bim) and dysregulation of the p53 pathway with increase in p53 effector genes including Bax, Trp53inp1, Jun, Cdkn1a, Mmp2, and Arid3a. Interestingly, p53 protein levels remained unchanged, suggesting that FoxD1 stromal miRNAs directly repress p53-effector genes. Using a lineage tracing approach followed by Fluorescent Activated Cell Sorting, miRNA profiling of the FoxD1-derived cells not only comprehensively defined the transcriptional landscape of miRNAs that are critical for vascular development, but also identified key miRNAs that are likely to modulate the renal phenotype in its absence. These miRNAs include miRs-10a, 18a, 19b, 24, 30c, 92a, 106a, 130a, 152, 181a, 214, 222, 302a, 370, and 381 that regulate Bcl2L11 (Bim) and miRs-15b, 18a, 21, 30c, 92a, 106a, 125b-5p, 145, 214, 222, 296-5p and 302a that regulate p53-effector genes. Consistent with the profiling results, ectopic apoptosis was observed in the cellular derivatives of the FoxD1 derived progenitor lineage and reiterates the importance of renal stromal miRNAs in cellular homeostasis.[190]

32.9.6 Nervous system

miRNAs appear to regulate the development and function of the nervous system.[191] Neural miRNAs are involved at various stages of synaptic development, including dendritogenesis (involving miR-132, miR-134 and miR-124), synapse formation [192] and synapse maturation (where miR-134 and miR-138 are thought to be involved).[193] Some studies find altered miRNA expression in schizophrenia, as well as bipolar disorder, and major depression and anxiety disorders.[194][195][196]

Alcoholism

The vital role of miRNAs in gene expression is significant to addiction, specifically alcoholism.[197] Chronic alcohol abuse results in persistent changes in brain function mediated in part by alterations in gene expression.[197] miRNA global regulation of many downstream genes deems significant regarding the reorganization or synaptic connections or long term neuroadaptions involving the behavioral change from alcohol consumption to withdrawal and/or dependence.[198] Up to 35 different miRNAs have been found to be altered in the alcoholic post-mortem brain, all of which target genes that include the regulation of the cell cycle, apoptosis, cell adhesion, nervous system development, and cell signaling.[197] Altered miRNA levels were also found in the medial prefrontal cortex of alcohol-dependent mice, suggesting the role of miRNA in orchestrating translational imbalances and the creation of differentially expressed proteins within an area of the brain where complex cognitive behavior and decision making likely originate.[199]

miRNAs can be either upregulated or downregulated in response to chronic alcohol use. miR-206 expression increased in the prefrontal cortex of alcohol dependent rats, targeting the transcription factor Brain-derived neurotrophic factor, or BDNF and ultimately reducing its expression. BDNF plays a critical role in the formation and maturation of new neurons and synapses, suggesting a possible implication in synapse growth/synaptic plasticity in alcohol abusers.[200] miR-155, important in regulating alcohol-induced neuroinflammation responses, was also found to be upregulated, suggesting the role of microglia and inflammatory cytokines in alcohol pathophysiology.[201] Downregulation of miR-382 was found in the nucleus accumbens, a structure in the basal forebrain significant in regulating feelings of reward that power motivational habits. miR-382 is the target for the dopamine receptor D1 (DRD1), and its overexpression results in the upregulation of DRD1 and delta fosB, a transcription factor which activates a series of transcription events in the nucleus accumbens that ultimately result in addictive behaviors.[202] Alternatively, overexpressing miR-382 resulted in attenuated drinking and the inhibition of DRD1 and delta fosB upregulation in rat models of alcoholism, demonstrating the possibility of using miRNA-targeted pharmaceuticals in the treatment of alcoholism.[202]

32.9.7 Obesity

miRNAs play crucial roles in the regulation of stem cell progenitors differentiating into adipocytes.[203] Studies to determine what role pluripotent stem cells play in adipogenesis, were examined in the immortalized human bone marrow-derived stromal cell line hMSC-Tert20.[204] Decreased expression of miR-155,miR-221,and miR-222, have been found during the adipogenic programming of both immortalized and primary hMSCs, suggesting that they act as negative regulators of differentiation. Conversely, ectopic expression of the miRNAs 155,221, and 222 significantly inhibited adipogenesis and repressed induction of the master regulators PPARγ and CCAAT/enhancer-binding protein alpha (CEBPA).[205] This paves the way for possible obesity treatments on the genetic level.

Another class of miRNAs that regulate insulin resistance, obesity, and diabetes, is the let-7 family. Let-7 is known to accumulate in human tissues during the course of aging. When let-7 was ectopically overexpressed to mimic accelerated aging, mice became insulin-resistant, and thus more prone to high fat diet-induced obesity and diabetes.[206] In contrast when let-7 was inhibited by injections of let-7-specific antagomirs, mice become more insulin-sensitive, and remarkably resistant to high fat diet-induced obesity and diabetes. Not only could let-7 inhibition prevent obesity and diabetes, it could also reverse and cure diabetes.[207] These experimental findings suggest that let-7 inhibition could represent a new therapy for obesity and type 2 diabetes.

32.10   Non-coding RNAs

When the human genome project mapped its first chromosome in 1999, it was predicted the genome would contain over 100,000 protein coding genes. However, only around 20,000 were eventually identified (International Human Genome Sequencing Consortium, 2004).[208] Since then, the advent of bioinformatics approaches combined with genome tiling studies examining the transcriptome,[209] systematic sequencing of full length cDNA libraries,[210] and experimental validation[211] (including the creation of miRNA derived antisense oligonucleotides called antagomirs) have revealed that many transcripts are non protein-coding RNA, including several snoRNAs and miRNAs.[212]

32.11   Viruses

The expression of transcription activators by *human herpesvirus-6* DNA, is believed to be regulated by viral miRNA.[213]

32.12   Target prediction

miRNAs can bind to target messenger RNA (mRNA) transcripts of protein-coding genes and negatively control their translation or cause mRNA degradation. It is of key importance to identify the miRNA targets accurately. A detailed review for the advances in the miRNA target identification methods and available resources has been published by Zheng et al.[214] A comparison of the predictive performance of eighteen *in silico* algorithms is available in Agarwal et al.[215]

- See also: List of miRNA target prediction tools

32.13   See also

- Gene expression

- List of miRNA gene prediction tools

- List of miRNA target prediction tools

- RNAi

- siRNA

- Small nucleolar RNA derived microRNA

- Anti-miRNA oligonucleotides

32.14 References

[1] Ambros, V (Sep 16, 2004). "The functions of animal microRNAs". *Nature* **431** (7006): 350–5. doi:10.1038/nature02871. PMID 15372042.

[2] Bartel, DP (Jan 23, 2004). "MicroRNAs: genomics, biogenesis, mechanism, and function". *Cell* **116** (2): 281–97. doi:10.1016/S0092-8674(04)00045-5. PMID 14744438.

[3] Bartel DP (January 2009). "MicroRNAs: target recognition and regulatory functions". *Cell* **136** (2): 215–33. doi:10.1016/j.cell.2009.01.002. PMC 3794896. PMID 19167326.

[4] Fabian, MR; Sonenberg, N; Filipowicz, W (2010). "Regulation of mRNA translation and stability by microRNAs". *Annual review of biochemistry* **79**: 351–79. doi:10.1146/annurev-biochem-060308-103103. PMID 20533884.

[5] *Homo sapiens* miRNAs in the miRBase at Manchester University

[6] Bentwich I, Avniel A, Karov Y, Aharonov R, Gilad S, Barad O, Barzilai A, Einat P, Einav U, Meiri E, Sharon E, Spector Y, Bentwich Z; Avniel; Karov; Aharonov; Gilad; Barad; Barzilai; Einat; Einav; Meiri; Sharon; Spector; Bentwich (July 2005). "Identification of hundreds of conserved and nonconserved human microRNAs". *Nat. Genet.* **37** (7): 766–70. doi:10.1038/ng1590. PMID 15965474.

[7] Lim LP, Lau NC, Weinstein EG, Abdelhakim A, Yekta S, Rhoades MW, Burge CB, Bartel DP; Lau; Weinstein; Abdelhakim; Yekta; Rhoades; Burge; Bartel (April 2003). "The microRNAs of Caenorhabditis elegans". *Genes Dev.* **17** (8): 991–1008. doi:10.1101/gad.1074403. PMC 196042. PMID 12672692.

[8] Lagos-Quintana M, Rauhut R, Yalcin A, Meyer J, Lendeckel W, Tuschl T; Rauhut; Yalcin; Meyer; Lendeckel; Tuschl (April 2002). "Identification of tissue-specific microRNAs from mouse". *Curr. Biol.* **12** (9): 735–9. doi:10.1016/S0960-9822(02)00809-6. PMID 12007417.

[9] Lewis BP, Burge CB, Bartel DP (2005). "Conserved seed pairing, often flanked by adenosines, indicates that thousands of human genes are microRNA targets". *Cell* **120** (1): 15–20. doi:10.1016/j.cell.2004.12.035. PMID 15652477.

[10] Friedman RC, Farh KK, Burge CB, Bartel DP; Farh; Burge; Bartel (January 2009). "Most mammalian mRNAs are conserved targets of microRNAs". *Genome Res.* **19** (1): 92–105. doi:10.1101/gr.082701.108. PMC 2612969. PMID 18955434.

[11] Axtell, MJ; Bartel, DP (Jun 2005). "Antiquity of microRNAs and their targets in land plants". *The Plant cell* **17** (6): 1658–73. doi:10.1105/tpc.105.032185. PMC 1143068. PMID 15849273.

[12] Tanzer A, Stadler PF; Stadler (May 2004). "Molecular evolution of a microRNA cluster". *J. Mol. Biol.* **339** (2): 327–35. doi:10.1016/j.jmb.2004.03.065. PMID 15136036.

[13] Chen, Kevin; Rajewsky, Nikolaus (2007). "The evolution of gene regulation by transcription factors and microRNAs". *Nature Reviews Genetics* **8** (2): 93–103. doi:10.1038/nrg1990. PMID 17230196.

[14] Lee CT, Risom T, Strauss WM; Risom; Strauss (April 2007). "Evolutionary conservation of microRNA regulatory circuits: an examination of microRNA gene complexity and conserved microRNA-target interactions through metazoan phylogeny". *DNA Cell Biol.* **26** (4): 209–18. doi:10.1089/dna.2006.0545. PMID 17465887.

[15] Peterson, KJ; Dietrich, MR; McPeek, MA (Jul 2009). "MicroRNAs and metazoan macroevolution: insights into canalization, complexity, and the Cambrian explosion". *BioEssays : news and reviews in molecular, cellular and developmental biology* **31** (7): 736–47. doi:10.1002/bies.200900033. PMID 19472371.

[16] Shabalina SA, Koonin EV; Koonin (October 2008). "Origins and evolution of eukaryotic RNA interference". *Trends in Ecology and Evolution.* **10** (10): 578–587. doi:10.1016/j.tree.2008.06.005. PMC 2695246. PMID 18715673.

[17] Axtell, MJ; Westholm, JO; Lai, EC (2011). "Vive la différence: biogenesis and evolution of microRNAs in plants and animals". *Genome Biology* **12** (4): 221. doi:10.1186/gb-2011-12-4-221. PMC 3218855. PMID 21554756.

[18] Jones-Rhoades, MW; Bartel, DP; Bartel, B (2006). "MicroRNAS and their regulatory roles in plants". *Annual review of plant biology* **57**: 19–53. doi:10.1146/annurev.arplant.57.032905.105218. PMID 16669754.

[19] Lewis BP, Shih IH, Jones-Rhoades M, Bartel DP, Burge CB (2003). "Prediction of Mammalian MicroRNA Targets". *Cell* **115** (7): 787–798. doi:10.1016/S0092-8674(03)01018-3. PMID 14697198.

[20] Ellwanger DC, Büttner FA, Mewes HW, Stümpflen V (2011). "The sufficient minimal set of miRNA seed types". *Bioinformatics* **27** (10): 1346–50. doi:10.1093/bioinformatics/btr149. PMC 3087955. PMID 21441577.

[21] Rajewsky, Nikolaus (2006). "microRNA target predictions in animals". *Nature Genetics* **38** (6s): S8–S13. doi:10.1038/ng1798.

[22] Krek, Azra; Grün, Dominic; Poy, Matthew N; Wolf, Rachel; Rosenberg, Lauren; Epstein, Eric J; MacMenamin, Philip; da Piedade, Isabelle; Gunsalus, Kristin C; Stoffel, Markus; Rajewsky, Nikolaus (May 2005). "Combinatorial microRNA target predictions". *Nature Genetics* **37** (5): 495–500. doi:10.1038/ng1536. PMID 15806104.

[23] Lee RC, Feinbaum RL, Ambros V; Feinbaum; Ambros (Dec 3, 1993). "The C. elegans heterochronic gene lin-4 encodes small RNAs with antisense complementarity to lin-14". *Cell* **75** (5): 843–54. doi:10.1016/0092-8674(93)90529-Y. PMID 8252621.

[24] Wightman B, Ha I, Ruvkun G (Dec 3, 1993). "Posttranscriptional regulation of the heterochronic gene lin-14 by lin-4 mediates temporal pattern formation in C. elegans". *Cell* **75** (5): 855–62. doi:10.1016/0092-8674(93)90530-4. PMID 8252622.

[25] Reinhart BJ, Slack FJ, Basson M, Pasquinelli AE, Bettinger JC, Rougvie AE, Horvitz HR, Ruvkun G; Slack; Basson; Pasquinelli; Bettinger; Rougvie; Horvitz; Ruvkun (February 2000). "The 21-nucleotide let-7 RNA regulates developmental timing in Caenorhabditis elegans". *Nature* **403** (6772): 901–6. Bibcode:2000Natur.403..901R. doi:10.1038/35002607. PMID 10706289.

[26] Pasquinelli AE, Reinhart BJ, Slack F, Martindale MQ, Kuroda MI, Maller B, Hayward DC, Ball EE, Degnan B, Müller P, Spring J, Srinivasan A, Fishman M, Finnerty J, Corbo J, Levine M, Leahy P, Davidson E, Ruvkun G; Reinhart; Slack; Martindale; Kuroda; Maller; Hayward; Ball; Degnan; Müller; Spring; Srinivasan; Fishman; Finnerty; Corbo; Levine; Leahy; Davidson; Ruvkun (November 2000). "Conservation of the sequence and temporal expression of let-7 heterochronic regulatory RNA". *Nature* **408** (6808): 86–9. doi:10.1038/35040556. PMID 11081512.

[27] Lagos-Quintana M, Rauhut R, Lendeckel W, Tuschl T; Rauhut; Lendeckel; Tuschl (October 2001). "Identification of novel genes coding for small expressed RNAs". *Science* **294** (5543): 853–8. Bibcode:2001Sci...294..853L. doi:10.1126/science.1064921. PMID 11679670.

[28] Lau NC, Lim LP, Weinstein EG, Bartel DP; Lim; Weinstein; Bartel (October 2001). "An abundant class of tiny RNAs with probable regulatory roles in Caenorhabditis elegans". *Science* **294** (5543): 858–62. Bibcode:2001Sci...294..858L. doi:10.1126/science.106506 PMID 11679671.

[29] Lee RC, Ambros V; Ambros (October 2001). "An extensive class of small RNAs in Caenorhabditis elegans". *Science* **294** (5543): 862–4. Bibcode:2001Sci...294..862L. doi:10.1126/science.1065329. PMID 11679672.

[30] Wienholds, E; Kloosterman, WP; Miska, E; Alvarez-Saavedra, E; Berezikov, E; de Bruijn, E; Horvitz, HR; Kauppinen, S; Plasterk, RH (Jul 8, 2005). "MicroRNA expression in zebrafish embryonic development". *Science* **309** (5732): 310–1. doi:10.1126/science.1114519. PMID 15919954.

[31] Brennecke J, Hipfner DR, Stark A, Russell RB, Cohen SM; Hipfner; Stark; Russell; Cohen (April 2003). "bantam encodes a developmentally regulated microRNA that controls cell proliferation and regulates the proapoptotic gene hid in Drosophila". *Cell* **113** (1): 25–36. doi:10.1016/S0092-8674(03)00231-9. PMID 12679032.

[32] Cuellar TL, McManus MT; McManus (December 2005). "MicroRNAs and endocrine biology". *J. Endocrinol.* **187** (3): 327–32. doi:10.1677/joe.1.06426. PMID 16423811.

[33] Poy MN, Eliasson L, Krutzfeldt J, Kuwajima S, Ma X, Macdonald PE, Pfeffer S, Tuschl T, Rajewsky N, Rorsman P, Stoffel M; Eliasson; Krutzfeldt; Kuwajima; Ma; MacDonald; Pfeffer; Tuschl; Rajewsky; Rorsman; Stoffel (November 2004). "A pancreatic islet-specific microRNA regulates insulin secretion". *Nature* **432** (7014): 226–30. Bibcode:2004Natur.432..226P. doi:10.1038/nature03076. PMID 15538371.

[34] Chen CZ, Li L, Lodish HF, Bartel DP; Li; Lodish; Bartel (January 2004). "MicroRNAs modulate hematopoietic lineage differentiation". *Science* **303** (5654): 83–6. Bibcode:2004Sci...303...83C. doi:10.1126/science.1091903. PMID 14657504.

[35] Wilfred BR, Wang WX, Nelson PT; Wang; Nelson (July 2007). "Energizing miRNA research: a review of the role of miRNAs in lipid metabolism, with a prediction that miR-103/107 regulates human metabolic pathways". *Mol. Genet. Metab.* **91** (3): 209–17. doi:10.1016/j.ymgme.2007.03.011. PMC 1978064. PMID 17521938.

[36] Harfe BD, McManus MT, Mansfield JH, Hornstein E, Tabin CJ; McManus; Mansfield; Hornstein; Tabin (August 2005). "The RNaseIII enzyme Dicer is required for morphogenesis but not patterning of the vertebrate limb". *Proc. Natl. Acad. Sci. U.S.A.* **102** (31): 10898–903. Bibcode:2005PNAS..10210898H. doi:10.1073/pnas.0504834102. PMC 1182454. PMID 16040801.

[37] Trang P, Weidhaas JB, Slack FJ; Weidhaas; Slack (December 2008). "MicroRNAs as potential cancer therapeutics". *Oncogene.* 27 Suppl 2: S52–7. doi:10.1038/onc.2009.353. PMID 19956180.

[38] Li C, Feng Y, Coukos G, Zhang L; Feng; Coukos; Zhang (December 2009). "Therapeutic microRNA strategies in human cancer". *AAPS J* **11** (4): 747–57. doi:10.1208/s12248-009-9145-9. PMC 2782079. PMID 19876744.

[39] Fasanaro P, Greco S, Ivan M, Capogrossi MC, Martelli F; Greco; Ivan; Capogrossi; Martelli (January 2010). "microRNA: emerging therapeutic targets in acute ischemic diseases". *Pharmacol. Ther.* **125** (1): 92–104. doi:10.1016/j.pharmthera.2009.10.003. PMID 19896977.

[40] Hydbring, Per; Badalian-Very, Gayane (August 2013). "Clinical applications of microRNAs". *F1000Research* **2**. doi:10.12688/f1000research.2-136.v2.

[41] Thomson DW, Bracken CP, Goodall GJ; Bracken; Goodall (September 2011). "Experimental strategies for microRNA target identification". *Nucleic Acids Res.* **39** (16): 6845–53. doi:10.1093/nar/gkr330. PMC 3167600. PMID 21652644.

[42] Lim LP, Lau NC, Garrett-Engele P, Grimson A, Schelter JM, Castle J, Bartel DP, Linsley PS, Johnson JM; Lau; Garrett-Engele; Grimson; Schelter; Castle; Bartel; Linsley; Johnson (February 2005). "Microarray analysis shows that some microRNAs downregulate large numbers of target mRNAs". *Nature* **433** (7027): 769–73. Bibcode:2005Natur.433..769L. doi:10.1038/nature03315. PMID 15685193.

[43] Selbach M, Schwanhäusser B, Thierfelder N, Fang Z, Khanin R, Rajewsky N; Schwanhäusser; Thierfelder; Fang; Khanin; Rajewsky (September 2008). "Widespread changes in protein synthesis induced by microRNAs". *Nature* **455** (7209): 58–63. doi:10.1038/nature07228. PMID 18668040.

[44] Baek D, Villén J, Shin C, Camargo FD, Gygi SP, Bartel DP; Villén; Shin; Camargo; Gygi; Bartel (September 2008). "The impact of microRNAs on protein output". *Nature* **455** (7209): 64–71. doi:10.1038/nature07242. PMC 2745094. PMID 18668037.

[45] Musilova K, Mraz M; Mraz (2014). "MicroRNAs in B cell lymphomas: How a complex biology gets more complex". *Leukemia.* doi:10.1038/leu.2014.351. PMID 25541152.

[46] Ambros V, Bartel B, Bartel DP, Burge CB, Carrington JC, Chen X, Dreyfuss G, Eddy SR, Griffiths-Jones S, Marshall M, Matzke M, Ruvkun G, Tuschl T; Bartel; Bartel; Burge; Carrington; Chen; Dreyfuss; Eddy; Griffiths-Jones; Marshall; Matzke; Ruvkun; Tuschl (March 2003). "A uniform system for microRNA annotation". *RNA* **9** (3): 277–9. doi:10.1261/rna.2183803. PMC 1370393. PMID 12592000.

[47] Griffiths-Jones S, Grocock RJ, van Dongen S, Bateman A, Enright AJ; Grocock; Van Dongen; Bateman; Enright (January 2006). "miRBase: microRNA sequences, targets and gene nomenclature". *Nucleic Acids Res.* **34** (Database issue): D140–4. doi:10.1093/nar/gkj112. PMC 1347474. PMID 16381832.

[48] Wright, MW; Bruford, EA (Jan 2011). "Naming 'junk': human non-protein coding RNA (ncRNA) gene nomenclature". *Human genomics* **5** (2): 90–8. doi:10.1186/1479-7364-5-2-90. PMC 3051107. PMID 21296742.

[49] Lee Y, Kim M, Han J, Yeom KH, Lee S, Baek SH, Kim VN; Kim; Han; Yeom; Lee; Baek; Kim (October 2004). "MicroRNA genes are transcribed by RNA polymerase II". *EMBO J.* **23** (20): 4051–60. doi:10.1038/sj.emboj.7600385. PMC 524334. PMID 15372072.

[50] Mraz M, Dolezalova D, Plevova K, Stano Kozubik K, Mayerova V, Cerna K, Musilova K, Tichy B, Pavlova S, Borsky M, Verner J, Doubek M, Brychtova Y, Trbusek M, Hampl A, Mayer J, Pospisilova S (March 2012). "MicroRNA-650 expression is influenced by immunoglobulin gene rearrangement and affects the biology of chronic lymphocytic leukemia". *Blood* **119** (9): 2110–2113. doi:10.1182/blood-2011-11-394874. PMID 22234685.

[51] Lisse TS, Chun RF, Rieger S, Adams JS, Hewison M (June 2013). "Vitamin D activation of functionally distinct regulatory miRNAs in primary human osteoblasts". *J Bone Miner Res.* **28** (6): 1478–14788. doi:10.1002/jbmr.1882. PMID 23362149.

[52] Rodriguez A, Griffiths-Jones S, Ashurst JL, Bradley A; Griffiths-Jones; Ashurst; Bradley (October 2004). "Identification of mammalian microRNA host genes and transcription units". *Genome Res.* **14** (10A): 1902–10. doi:10.1101/gr.2722704. PMC 524413. PMID 15364901.

[53] Cai X, Hagedorn CH, Cullen BR; Hagedorn; Cullen (December 2004). "Human microRNAs are processed from capped, polyadenylated transcripts that can also function as mRNAs". *RNA* **10** (12): 1957–66. doi:10.1261/rna.7135204. PMC 1370684. PMID 15525708.

[54] Weber MJ (January 2005). "New human and mouse microRNA genes found by homology search". *FEBS J.* **272** (1): 59–73. doi:10.1111/j.1432-1033.2004.04389.x. PMID 15634332.

[55] Kim YK, Kim VN; Kim (February 2007). "Processing of intronic microRNAs". *EMBO J.* **26** (3): 775–83. doi:10.1038/sj.emboj.7601512. PMC 1794378. PMID 17255951.

[56] Baskerville S, Bartel DP; Bartel (March 2005). "Microarray profiling of microRNAs reveals frequent coexpression with neighboring miRNAs and host genes". *RNA* **11** (3): 241–7. doi:10.1261/rna.7240905. PMC 1370713. PMID 15701730.

[57] Altuvia Y, Landgraf P, Lithwick G, Elefant N, Pfeffer S, Aravin A, Brownstein MJ, Tuschl T, Margalit H; Landgraf; Lithwick; Elefant; Pfeffer; Aravin; Brownstein; Tuschl; Margalit (2005). "Clustering and conservation patterns of human microRNAs". *Nucleic Acids Res.* **33** (8): 2697–706. doi:10.1093/nar/gki567. PMC 1110742. PMID 15891114.

[58] Zhou X, Ruan J, Wang G, Zhang W; Ruan; Wang; Zhang (March 2007). "Characterization and identification of microRNA core promoters in four model species". *PLoS Comput. Biol.* **3** (3): e37. Bibcode:2007PLSCB...3...37Z. doi:10.1371/journal.pcbi.0030037. PMC 1817659. PMID 17352530.

[59] Faller M, Guo F; Guo (November 2008). "MicroRNA biogenesis: there's more than one way to skin a cat". *Biochim. Biophys. Acta* **1779** (11): 663–7. doi:10.1016/j.bbagrm.2008.08.005. PMC 2633599. PMID 18778799.

[60] Lee, Y; Ahn, C; Han, J; Choi, H; Kim, J; Yim, J; Lee, J; Provost, P; Rådmark, O; Kim, S; Kim, VN (25 September 2003). "The nuclear RNase III Drosha initiates microRNA processing.". *Nature* **425** (6956): 415–9. doi:10.1038/nature01957. PMID 14508493.

[61] Gregory RI, Chendrimada TP, Shiekhattar R; Chendrimada; Shiekhattar (2006). "MicroRNA biogenesis: isolation and characterization of the microprocessor complex". *Methods Mol. Biol.* **342**: 33–47. doi:10.1385/1-59745-123-1:33. ISBN 1-59745-123-1. PMID 16957365.

[62] Han, J; Lee, Y; Yeom, KH; Kim, YK; Jin, H; Kim, VN (15 December 2004). "The Drosha-DGCR8 complex in primary microRNA processing.". *Genes & Development* **18** (24): 3016–27. doi:10.1101/gad.1262504. PMC 535913. PMID 15574589.

[63] Han, J; Lee, Y; Yeom, KH; Nam, JW; Heo, I; Rhee, JK; Sohn, SY; Cho, Y; Zhang, BT; Kim, VN (2 June 2006). "Molecular basis for the recognition of primary microRNAs by the Drosha-DGCR8 complex.". *Cell* **125** (5): 887–901. doi:10.1016/j.cell.2006.03.043. PMID 16751099.

[64] Conrad, Thomas; Annalisa, Marsico; Gehre, Maja; Ørom, Ulf (Oct 23, 2014). "Microprocessor activity controls differential miRNA biogenesis In Vivo.". *Cell Reports* **9** (2): 542–554. doi:10.1016/j.celrep.2014.09.007. PMID 25310978.

[65] Auyeung, Vincent; Igor, Ulitsky; McGeary, SE; Bartel, DP (Feb 14, 2013). "Beyond secondary structure: primary-sequence determinants license pri-miRNA hairpins for processing.". *Cell* **152** (4): 844–858. doi:10.1016/j.cell.2013.01.031. PMID 23415231.

[66] Ali PS, Ghoshdastider U, Hoffmann J, Brutschy B, Filipek S (2012). "Recognition of the let-7g miRNA precursor by human Lin28B". *FEBS Letters* **586** (22): 3986–90. doi:10.1016/j.febslet.2012.09.034. PMID 23063642.

[67] Berezikov E, Chung WJ, Willis J, Cuppen E, Lai EC; Chung; Willis; Cuppen; Lai (October 2007). "Mammalian mirtron genes". *Mol. Cell* **28** (2): 328–36. doi:10.1016/j.molcel.2007.09.028. PMC 2763384. PMID 17964270.

[68] Kawahara Y, Megraw M, Kreider E, Iizasa H, Valente L, Hatzigeorgiou AG, Nishikura K; Megraw; Kreider; Iizasa; Valente; Hatzigeorgiou; Nishikura (September 2008). "Frequency and fate of microRNA editing in human brain". *Nucleic Acids Res.* **36** (16): 5270–80. doi:10.1093/nar/gkn479. PMC 2532740. PMID 18684997.

[69] Winter J, Jung S, Keller S, Gregory RI, Diederichs S; Jung; Keller; Gregory; Diederichs (March 2009). "Many roads to maturity: microRNA biogenesis pathways and their regulation". *Nat. Cell Biol.* **11** (3): 228–34. doi:10.1038/ncb0309-228. PMID 19255566.

[70] Ohman M (October 2007). "A-to-I editing challenger or ally to the microRNA process". *Biochimie* **89** (10): 1171–6. doi:10.1016/j.biochi.2007 PMID 17628290.

[71] Murchison EP, Hannon GJ; Hannon (June 2004). "miRNAs on the move: miRNA biogenesis and the RNAi machinery". *Curr. Opin. Cell Biol.* **16** (3): 223–9. doi:10.1016/j.ceb.2004.04.003. PMID 15145345.

[72] Lund E, Dahlberg JE; Dahlberg (2006). "Substrate selectivity of exportin 5 and Dicer in the biogenesis of microRNAs". *Cold Spring Harb. Symp. Quant. Biol.* **71**: 59–66. doi:10.1101/sqb.2006.71.050. PMID 17381281.

[73] Park, JE; Heo, I; Tian, Y; Simanshu, DK; Chang, H; Jee, D; Patel, DJ; Kim, VN (13 July 2011). "Dicer recognizes the 5' end of RNA for efficient and accurate processing.". *Nature* **475** (7355): 201–5. doi:10.1038/nature10198. PMID 21753850.

[74] Ji X (2008). "The mechanism of RNase III action: how dicer dices". *Curr. Top. Microbiol. Immunol.* Current Topics in Microbiology and Immunology **320**: 99–116. doi:10.1007/978-3-540-75157-1_5. ISBN 978-3-540-75156-4. PMID 18268841.

[75] Mirihana Arachchilage G, Dassanayake AC, Basu S (2015). "A Potassium Ion-Dependent RNA Structural Switch Regulates Human Pre-miRNA 92b Maturation". *Chem. Biol.* **22**: 262–272. doi:10.1016/j.chembiol.2014.12.013. PMID 25641166.

[76] Lelandais-Brière C, Sorin C, Declerck M, Benslimane A, Crespi M, Hartmann C; Sorin; Declerck; Benslimane; Crespi; Hartmann (March 2010). "Small RNA diversity in plants and its impact in development". *Current Genomics* **11** (1): 14–23. doi:10.2174/138920210790217918. PMC 2851111. PMID 20808519.

[77] Rana TM (January 2007). "Illuminating the silence: understanding the structure and function of small RNAs". *Nat. Rev. Mol. Cell Biol.* **8** (1): 23–36. doi:10.1038/nrm2085. PMID 17183358.

[78] Schwarz DS, Zamore PD; Zamore (May 2002). "Why do miRNAs live in the miRNP?". *Genes Dev.* **16** (9): 1025–31. doi:10.1101/gad.992502. PMID 12000786.

[79] Krol J, Sobczak K, Wilczynska U, Drath M, Jasinska A, Kaczynska D, Krzyzosiak WJ; Sobczak; Wilczynska; Drath; Jasinska; Kaczynska; Krzyzosiak (2004). "Structural features of microRNA (miRNA) precursors and their relevance to miRNA biogenesis and small interfering RNA/short hairpin RNA design". *J Biol Chem* **279** (40): 42230–9. doi:10.1074/jbc.M404931200. PMID 15292246.

[80] Khvorova A, Reynolds A, Jayasena SD; Reynolds; Jayasena (2003). "Functional siRNAs and miRNAs exhibit strand bias". *Cell* **115** (2): 209–16. doi:10.1016/S0092-8674(03)00801-8. PMID 14567918.

[81] Schwarz DS, Hutvágner G, Du T, Xu Z, Aronin N, Zamore PD; Hutvágner; Du; Xu; Aronin; Zamore (2003). "Asymmetry in the assembly of the RNAi enzyme complex". *Cell* **115** (2): 199–208. doi:10.1016/S0092-8674(03)00759-1. PMID 14567917.

[82] Lin SL, Chang D, Ying SY; Chang; Ying (2005). "Asymmetry of intronic pre-miRNA structures in functional RISC assembly". *Gene* **356**: 32–8. doi:10.1016/j.gene.2005.04.036. PMC 1788082. PMID 16005165.

[83] Okamura K, Chung WJ, Lai EC; Chung; Lai (2008). "The long and short of inverted repeat genes in animals: microRNAs, mirtrons and hairpin RNAs". *Cell Cycle* **7** (18): 2840–5. doi:10.4161/cc.7.18.6734. PMC 2697033. PMID 18769156.

[84] Pratt AJ, MacRae IJ; MacRae (July 2009). "The RNA-induced silencing complex: a versatile gene-silencing machine". *J. Biol. Chem.* **284** (27): 17897–901. doi:10.1074/jbc.R900012200. PMC 2709356. PMID 19342379.

[85] MacRae IJ, Ma E, Zhou M, Robinson CV, Doudna JA; Ma; Zhou; Robinson; Doudna (January 2008). "In vitro reconstitution of the human RISC-loading complex". *Proc. Natl. Acad. Sci. U.S.A.* **105** (2): 512–7. Bibcode:2008PNAS..105..512M. doi:10.1073/pnas.0710869105. PMC 2206567. PMID 18178619.

[86] Mourelatos Z, Dostie J, Paushkin S, Sharma A, Charroux B, Abel L, Rappsilber J, Mann M, Dreyfuss G; Dostie; Paushkin; Sharma; Charroux; Abel; Rappsilber; Mann; Dreyfuss (March 2002). "miRNPs: a novel class of ribonucleoproteins containing numerous microRNAs". *Genes Dev.* **16** (6): 720–8. doi:10.1101/gad.974702. PMC 155365. PMID 11914277.

[87] Meister G, Landthaler M, Peters L, Chen P, Urlaub H, Lurhmann R, Tuschl T; Landthaler; Peters; Chen; Urlaub; Lührmann; Tuschl (December 2005). "Identification of Novel Argonaute-Associated Proteins". *Current Biology* **15** (23): 2149–55. doi:10.1016/j.cub.2005.10.048. PMID 16289642.

[88] Jing Q, Huang S, Guth S, Zarubin T, Motoyama A, Chen J, Di Padova F, Lin SC, Gram H, Han J (2005). "Involvement of microRNA in AU-rich element-mediated mRNA instability". *Cell* **120** (5): 623–34. doi:10.1016/j.cell.2004.12.038. PMID 15766526.

[89] Kai ZS, Pasquinelli AE; Pasquinelli (January 2010). "MicroRNA assassins: factors that regulate the disappearance of miRNAs". *Nat. Struct. Mol. Biol.* **17** (1): 5–10. doi:10.1038/nsmb.1762. PMID 20051982.

[90] Chatterjee S, Großhans H; Grosshans (September 2009). "Active turnover modulates mature microRNA activity in *Caenorhabditis elegans*". *Nature* **461** (7263): 546–459. Bibcode:2009Natur.461..546C. doi:10.1038/nature08349. PMID 19734881.

[91] Morozova N, Zinovyev A, Nonne N, Pritchard LL, Gorban AN, Harel-Bellan A (September 2012). "Kinetic signatures of microRNA modes of action". *RNA* **18** (9): 1635–55. doi:10.1261/rna.032284.112. PMC 3425779. PMID 22850425.

[92] Wang XJ, Reyes JL, Chua NH, Gaasterland T; Reyes; Chua; Gaasterland (2004). "Prediction and identification of Arabidopsis thaliana microRNAs and their mRNA targets". *Genome Biol.* **5** (9): R65. doi:10.1186/gb-2004-5-9-r65. PMC 522872. PMID 15345049.

[93] Kawasaki H, Taira K; Taira (2004). "MicroRNA-196 inhibits HOXB8 expression in myeloid differentiation of HL60 cells". *Nucleic Acids Symp Ser* **48** (1): 211–2. doi:10.1093/nass/48.1.211. PMID 17150553.

[94] Moxon S, Jing R, Szittya G, Schwach F, Rusholme Pilcher RL, Moulton V, Dalmay T; Jing; Szittya; Schwach; Rusholme Pilcher; Moulton; Dalmay (October 2008). "Deep sequencing of tomato short RNAs identifies microRNAs targeting genes involved in fruit ripening". *Genome Res.* **18** (10): 1602–9. doi:10.1101/gr.080127.108. PMC 2556272. PMID 18653800.

[95] Mazière P, Enright AJ; Enright (June 2007). "Prediction of microRNA targets". *Drug Discov. Today* **12** (11–12): 452–8. doi:10.1016/j.drudis.2007.04.002. PMID 17532529.

[96] Williams AE (February 2008). "Functional aspects of animal microRNAs". *Cell. Mol. Life Sci.* **65** (4): 545–62. doi:10.1007/s00018-007-7355-9. PMID 17965831.

[97] Eulalio A, Huntzinger E, Nishihara T, Rehwinkel J, Fauser M, Izaurralde E; Huntzinger; Nishihara; Rehwinkel; Fauser; Izaurralde (January 2009). "Deadenylation is a widespread effect of miRNA regulation". *RNA* **15** (1): 21–32. doi:10.1261/rna.1399509. PMC 2612776. PMID 19029310.

[98] Bazzini AA, Lee MT, Giraldez AJ; Lee; Giraldez (April 2012). "Ribosome profiling shows that miR-430 reduces translation before causing mRNA decay in zebrafish". *Science* **336** (6078): 233–7. Bibcode:2012Sci...336..233B. doi:10.1126/science.1215704. PMC 3547538. PMID 22422859.

[99] Djuranovic S, Nahvi A, Green R; Nahvi; Green (April 2012). "miRNA-mediated gene silencing by translational repression followed by mRNA deadenylation and decay". *Science* **336** (6078): 237–40. Bibcode:2012Sci...336..237B. doi:10.1126/science.1215691. PMC 3971879. PMID 22499947.

[100] Tan Y, Zhang B, Wu T, Skogerbø G, Zhu X, Guo X, He S, Chen R; Zhang; Wu; Skogerbø; Zhu; Guo; He; Chen (2009). "Transcriptional inhibiton of Hoxd4 expression by miRNA-10a in human breast cancer cells". *BMC Mol. Biol.* **10** (1): 12. doi:10.1186/1471-2199-10-12. PMC 2680403. PMID 19232136.

[101] Hawkins PG, Morris KV; Morris (March 2008). "RNA and transcriptional modulation of gene expression". *Cell Cycle* **7** (5): 602–7. doi:10.4161/cc.7.5.5522. PMC 2877389. PMID 18256543.

[102] Stark A, Brennecke J, Bushati N, Russell RB, Cohen SM; Brennecke; Bushati; Russell; Cohen (2005). "Animal MicroRNAs confer robustness to gene expression and have a significant impact on 3'UTR evolution". *Cell* **123** (6): 1133–46. doi:10.1016/j.cell.2005.11.023. PMID 16337999.

[103] Li LC (2008). "Small RNA-Mediated Gene Activation". *RNA and the Regulation of Gene Expression: A Hidden Layer of Complexity*. Caister Academic Press. ISBN 978-1-904455-25-7. http://www.horizonpress.com/rnareg.

[104] Place RF, Li LC, Pookot D, Noonan EJ, Dahiya R (2008). "MicroRNA-373 induces expression of genes with complementary promoter sequences". *Proc. Natl. Acad. Sci. U.S.A.* **105** (5): 1608–13. Bibcode:2008PNAS..105.1608P. doi:10.1073/pnas.0707594105. PMC 2234192. PMID 18227514.

[105] Salmena L, Poliseno L, Tay Y, Kats L, Pandolfi PP (August 2011). "A ceRNA hypothesis: the Rosetta Stone of a hidden RNA language?". *Cell* **146** (3): 353–8. doi:10.1016/j.cell.2011.07.014. PMC 3235919. PMID 21802130.

[106] Wheeler BM, Heimberg AM, Moy VN, Sperling EA, Holstein TW, Heber S, Peterson KJ (2009). "The deep evolution of metazoan microRNAs". *Evol. Dev.* **11** (1): 50–68. doi:10.1111/j.1525-142X.2008.00302.x. PMID 19196333.

[107] Pashkovskiy, P. P.; Ryazansky, S. S. (2013). "Biogenesis, evolution, and functions of plant microRNAs". *Biochemistry-Moscow* **78** (6): 627–637. doi:10.1134/S0006297913060084. PMID 23980889.

[108] Heimberg AM, Sempere LF, Moy VN, Donoghue PC, Peterson KJ (February 2008). "MicroRNAs and the advent of vertebrate morphological complexity". *Proc. Natl. Acad. Sci. U.S.A.* **105** (8): 2946–50. Bibcode:2008PNAS..105.2946H. doi:10.1073/pnas.0712259105. PMC 2268565. PMID 18287013.

[109] Peterson KJ, Dietrich MR, McPeek MA (July 2009). "MicroRNAs and metazoan macroevolution: insights into canalization, complexity, and the Cambrian explosion". *BioEssays* **31** (7): 736–47. doi:10.1002/bies.200900033. PMID 19472371.

[110] Nozawa M, Miura S, Nei M (2010). "Origins and evolution of microRNA genes in Drosophila species". *Genome Biol Evol* **2**: 180–9. doi:10.1093/gbe/evq009. PMC 2942034. PMID 20624724.

[111] Allen, E.; Z. X. Xie; A. M. Gustafson; G. H. Sung; J. W. Spatafora; J. C. Carrington (2004). "Evolution of microRNA genes by inverted duplication of target gene sequences in Arabidopsis thaliana". *Nature Genetics* **36** (12): 1282–1290. doi:10.1038/ng1478. PMID 15565108.

[112] Warthmann, N.; S. Das; C. Lanz; D. Weigel (2008). "Comparative analysis of the MIR319a MicroRNA locus in Arabidopsis and related Brassicaceae". *Molecular Biology and Evolution* **25** (5): 892–902. doi:10.1093/molbev/msn029. PMID 18296705.

[113] Fahlgren, N.; S. Jogdeo; K. D. Kasschau; C. M. Sullivan; E. J. Chapman; S. Laubinger; L. M. Smith; M. Dasenko; S. A. Givan; D. Weigel; J. C. Carrington (2010). "MicroRNA gene evolution in Arabidopsis lyrata and Arabidopsis thaliana". *Plant Cell* **22** (4): 1074–1089. doi:10.1105/tpc.110.073999.

[114] Caravas J, Friedrich M (June 2010). "Of mites and millipedes: recent progress in resolving the base of the arthropod tree". *BioEssays* **32** (6): 488–95. doi:10.1002/bies.201000005. PMID 20486135.

[115] Kenny NJ, Namigai EK, Marlétaz F, Hui JH, Shimeld SM (December 2015). "Draft genome assemblies and predicted microRNA complements of the intertidal lophotrochozoans Patella vulgata (Mollusca, Patellogastropoda) and Spirobranchus (Pomatoceros) lamarcki (Annelida, Serpulida)". *Marine Genomics* **24** (2): 139–146. doi:10.1016/j.margen.2015.07.004.

[116] Cock JM, Sterck L, Rouzé P, Scornet D, Allen AE, Amoutzias G, Anthouard V, Artiguenave F, Aury JM, Badger JH; et al. (June 2010). "The Ectocarpus genome and the independent evolution of multicellularity in brown algae". *Nature* **465** (7298): 617–21. Bibcode:2010Natur.465..617C. doi:10.1038/nature09016. PMID 20520714.

[117] Cuperus, J. T.; N. Fahlgren,; J. C. Carrington (2011). "Evolution and functional diversification of MIRNA genes". *Plant Cell* **23** (2): 431–442. doi:10.1105/tpc.110.082784. PMID 21317375.

[118] Ryan, J. F.; Pang, K.; Schnitzler, C. E.; Nguyen, A.-D.; Moreland, R. T.; Simmons, D. K.; Koch, B. J.; Francis, W. R.; Havlak, P.; Smith, S. A.; Putnam, N. H.; Haddock, S. H. D.; Dunn, C. W.; Wolfsberg, T. G.; Mullikin, J. C.; Martindale, M. Q.; Baxevanis, A. D. (2013). "The Genome of the Ctenophore Mnemiopsis leidyi and Its Implications for Cell Type Evolution". *Science* **342** (6164): 1242592. doi:10.1126/science.1242592. PMC 3920664. PMID 24337300.

[119] Maxwell, E.K.; Ryan, J.F.; Schnitzler, C.E.; Browne, W.E.; Baxevanis, A.D. (December 2012). "MicroRNAs and essential components of the microRNA processing machinery are not encoded in the genome of the ctenophore *Mnemiopsis leidyi*". *BMC Genomics* **13** (1): 714. doi:10.1186/1471-2164-13-714. PMC 3563456. PMID 23256903.

[120] Dimond PF (15 March 2010). "miRNAs' Therapeutic Potential". *Genetic Engineering & Biotechnology News* **30** (6). p. 1. Archived from the original on 10 July 2010. Retrieved 10 July 2010.

[121] Tjaden B, Goodwin SS, Opdyke JA, Guillier M, Fu DX, Gottesman S, Storz G (2006). "Target prediction for small, noncoding RNAs in bacteria". *Nucleic Acids Res.* **34** (9): 2791–802. doi:10.1093/nar/gkl356. PMC 1464411. PMID 16717284.

[122] Mraz M, Malinova K, Mayer J, Pospisilova S (December 2009). "MicroRNA isolation and stability in stored RNA samples". *Biochem. Biophys. Res. Commun.* **390** (1): 1–4. doi:10.1016/j.bbrc.2009.09.061. PMID 19769940.

[123] Liu CG, Calin GA, Volinia S, Croce CM (2008). "MicroRNA expression profiling using microarrays". *Nat Protoc* **3** (4): 563–78. doi:10.1038/nprot.2008.14. PMID 18388938.

[124] Chen C, Ridzon DA, Broomer AJ, Zhou Z, Lee DH, Nguyen JT, Barbisin M, Xu NL, Mahuvakar VR, Andersen MR, Lao KQ, Livak KJ, Guegler KJ (2005). "Real-time quantification of microRNAs by stem-loop RT-PCR". *Nucleic Acids Res.* **33** (20): e179. doi:10.1093/nar/gni178. PMC 1292995. PMID 16314309.

[125] Shingara J, Keiger K, Shelton J, Laosinchai-Wolf W, Powers P, Conrad R, Brown D, Labourier E (September 2005). "An optimized isolation and labeling platform for accurate microRNA expression profiling". *RNA* **11** (9): 1461–70. doi:10.1261/rna.2610405. PMC 1370829. PMID 16043497.

[126] Buermans HP, Ariyurek Y, van Ommen G, den Dunnen JT, 't Hoen PA. (December 2010). "New methods for next generation sequencing based microRNA expression profiling". *BMC Genomics* **11**: 716. doi:10.1186/1471-2164-11-716. PMC 3022920. PMID 21171994.

[127] Kloosterman WP, Wienholds E, Ketting RF, Plasterk RH (2004). "Substrate requirements for let-7 function in the developing zebrafish embryo". *Nucleic Acids Res.* **32** (21): 6284–91. doi:10.1093/nar/gkh968. PMC 535676. PMID 15585662.

[128] Flynt AS, Li N, Thatcher EJ, Solnica-Krezel L, Patton JG (February 2007). "Zebrafish miR-214 modulates Hedgehog signaling to specify muscle cell fate". *Nat. Genet.* **39** (2): 259–63. doi:10.1038/ng1953. PMID 17220889.

[129] Meister G, Landthaler M, Dorsett Y, Tuschl T (March 2004). "Sequence-specific inhibition of microRNA- and siRNA-induced RNA silencing". *RNA* **10** (3): 544–50. doi:10.1261/rna.5235104. PMC 1370948. PMID 14970398.

[130] Kloosterman WP, Lagendijk AK, Ketting RF, Moulton JD, Plasterk RH (August 2007). "Targeted inhibition of miRNA maturation with morpholinos reveals a role for miR-375 in pancreatic islet development". *PLoS Biol.* **5** (8): e203. doi:10.1371/journal.pbio.0050203. PMC 1925136. PMID 17676975.

[131] Choi WY, Giraldez AJ, Schier AF (October 2007). "Target protectors reveal dampening and balancing of Nodal agonist and antagonist by miR-430". *Science* **318** (5848): 271–4. Bibcode:2007Sci...318..271C. doi:10.1126/science.1147535. PMID 17761850.

[132] You Y, Moreira BG, Behlke MA, Owczarzy R (2006). "Design of LNA probes that improve mismatch discrimination". *Nucleic Acids Res* **34** (8): e60. doi:10.1093/nar/gkl175. PMC 1456327. PMID 16670427.

[133] Lagendijk AK, Moulton JD, Bakkers J (2012). "Revealing details: whole mount microRNA in situ hybridization protocol for zebrafish embryos and adult tissues". *Bio Open* **1** (6): 566–569. doi:10.1242/bio.2012810.

[134] Kaur H, Arora A, Wengel J, Maiti S, Arora A, Wengel J, Maiti S (2006). "Thermodynamic, Counterion, and Hydration Effects for the Incorporation of Locked Nucleic Acid Nucleotides into DNA Duplexes". *Biochemistry* **45** (23): 7347–55. doi:10.1021/bi060307w. PMID 16752924.

[135] Nielsen JA, Lau P, Maric D, Barker JL, Hudson LD (2009). "Integrating microRNA and mRNA expression profiles of neuronal progenitors to identify regulatory networks underlying the onset of cortical neurogenesis". *BMC Neurosci* **10**: 98. doi:10.1186/1471-2202-10-98. PMC 2736963. PMID 19689821.

[136] Gupta A, Nagilla P, Le HS, Bunney C, Zych C, Thalamuthu A, Bar-Joseph Z, Mathavan S, Ayyavoo V (2011). Mammano, Fabrizio, ed. "Comparative expression profile of miRNA and mRNA in primary peripheral blood mononuclear cells infected with human immunodeficiency virus (HIV-1)". *PLoS ONE* **6** (7): e22730. doi:10.1371/journal.pone.0022730. PMC 3145673. PMID 21829495.

[137] Grimson A, Farh KK, Johnston WK, Garrett-Engele P, Lim LP, Bartel DP (July 2007). "MicroRNA targeting specificity in mammals: determinants beyond seed pairing". *Mol. Cell* **27** (1): 91–105. doi:10.1016/j.molcel.2007.06.017. PMC 3800283. PMID 17612493.

[138] Griffiths-Jones S, Saini HK, van Dongen S, Enright AJ (January 2008). "miRBase: tools for microRNA genomics". *Nucleic Acids Res.* **36** (Database issue): D154–8. doi:10.1093/nar/gkm952. PMC 2238936. PMID 17991681.

[139] Nam S, Li M, Choi K, Balch C, Kim S, Nephew KP (July 2009). "MicroRNA and mRNA integrated analysis (MMIA): a web tool for examining biological functions of microRNA expression". *Nucleic Acids Res.* **37** (Web Server issue): W356–62. doi:10.1093/nar/gkp294. PMC 2703907. PMID 19420067.

[140] Artmann S, Jung K, Bleckmann A, Beissbarth T (2012). Provero, Paolo, ed. "Detection of simultaneous group effects in microRNA expression and related target gene sets". *PLoS ONE* **7** (6): e38365. Bibcode:2012PLoSO...738365A. doi:10.1371/journal.pone.00383. PMC 3378551. PMID 22723856.

[141] Mraz M, Pospisilova S (December 2012). "MicroRNAs in chronic lymphocytic leukemia: from causality to associations and back". *Expert Review of Hematology* **5** (6): 579–81. doi:10.1586/ehm.12.54. PMID 23216588.

[142] Jiang Q, Wang Y, Hao Y, Juan L, Teng M, Zhang X, Li M, Wang G, Liu Y. (January 2009). "miR2Disease: a manually curated database for microRNA deregulation in human disease". *Nucleic Acids Research*. 37. (Database issue) (Database issue): D98–104. doi:10.1093/nar/gkn714. PMC 2686559. PMID 18927107.

[143] Mencía A, Modamio-Høybjør S, Redshaw N, Morín M, Mayo-Merino F, Olavarrieta L, Aguirre LA, del Castillo I, Steel KP, Dalmay T, Moreno F, Moreno-Pelayo MA (May 2009). "Mutations in the seed region of human miR-96 are responsible for nonsyndromic progressive hearing loss". *Nat. Genet.* **41** (5): 609–13. doi:10.1038/ng.355. PMID 19363479.

[144] Hughes AE, Bradley DT, Campbell M, Lechner J, Dash DP, Simpson DA, Willoughby CE (2011). "Mutation Altering the miR-184 Seed Region Causes Familial Keratoconus with Cataract". *The American Journal of Human Genetics* **89** (5): 628–633. doi:10.1016/j.ajhg.2011.09.014. PMC 3213395. PMID 21996275.

[145] de Pontual L, Yao E, Callier P, Faivre L, Drouin V, Cariou S, Van Haeringen A, Geneviève D, Goldenberg A, Oufadem M, Manouvrier S, Munnich A, Vidigal JA, Vekemans M, Lyonnet S, Henrion-Caude A, Ventura A, Amiel J (October 2011). "Germline deletion of the miR-17~92 cluster causes skeletal and growth defects in humans". *Nat. Genet.* **43** (10): 1026–30. doi:10.1038/ng.915. PMC 3184212. PMID 21892160.

[146] Lu J, Getz G, Miska EA, Alvarez-Saavedra E, Lamb J, Peck D, Sweet-Cordero A, Ebert BL, Mak RH, Ferrando AA, Downing JR, Jacks T, Horvitz HR, Golub TR (June 2005). "MicroRNA expression profiles classify human cancers". *Nature* **435** (7043): 834–8. Bibcode:2005Natur.435..834L. doi:10.1038/nature03702. PMID 15944708.

[147] Jun Qian, Vinayakumar Siragam, Jiang Lin, Jichun Ma, Zhaoqun Deng (2011). "The role of microRNAs in the formation of cancer stem cells: Future directions for miRNAs". *Hypothesis* **9** (1): e10. doi:10.5779/hypothesis.v9i1.224.

[148] "Screening Tool Can Detect Colorectal Cancer from a Small Blood Sample" (Press release). American Association for Cancer Research. 29 September 2010. Retrieved 29 November 2010.

[149] Nielsen BS, Jørgensen S, Fog JU, Søkilde R, Christensen IJ, Hansen U, Brünner N, Baker A, Møller S, Nielsen HJ (October 2010). "High levels of microRNA-21 in the stroma of colorectal cancers predict short disease-free survival in stage II colon cancer patients". *Clin Exp Metastasis* **28** (1): 27–38. doi:10.1007/s10585-010-9355-7. PMC 2998639. PMID 21069438.

[150] Võsa U, Vooder T, Kolde R, Fischer K, Välk K, Tõnisson N, Roosipuu R, Vilo J, Metspalu A, Annilo T (October 2011). "Identification of miR-374a as a prognostic marker for survival in patients with early-stage nonsmall cell lung cancer". *Genes Chromosomes Cancer* **50** (10): 812–22. doi:10.1002/gcc.20902. PMID 21748820.

[151] Akçakaya P, Ekelund S, Kolosenko I, Caramuta S, Ozata DM, Xie H, Lindforss U, Olivecrona H, Lui WO (August 2011). "miR-185 and miR-133b deregulation is associated with overall survival and metastasis in colorectal cancer". *Int. J. Oncol.* **39** (2): 311–8. doi:10.3892/ijo.2011.1043. PMID 21573504.

[152] MicroRNA-21 promotes hepatocellular carcinoma HepG2 cell proliferation through repression of mitogen-activated protein kinase-kinase 3. Guangxian Xu et al, 2013

[153] Jones, K; Nourse JP; Keane C; Bhatnagar A; Gandhi MK. (Jan 2014). "Plasma MicroRNA Are Disease Response Biomarkers in Classical Hodgkin Lymphoma". *Clin Can Res* **20** (1): 253–64. doi:10.1158/1078-0432.CCR-13-1024. PMID 24222179.

[154] Wu H, Mo YY (December 2009). "Targeting miR-205 in breast cancer". *Expert Opin. Ther. Targets* **13** (12): 1439–48. doi:10.1517/14728220903338777. PMID 19839716.

[155] Gregory PA, Bert AG, Paterson EL, Barry SC, Tsykin A, Farshid G, Vadas MA, Khew-Goodall Y, Goodall GJ (May 2008). "The miR-200 family and miR-205 regulate epithelial to mesenchymal transition by targeting ZEB1 and SIP1". *Nat. Cell Biol.* **10** (5): 593–601. doi:10.1038/ncb1722. PMID 18376396.

[156] Bernstein C, Prasad AR Nfonsam V, Bernstei H (2013). "DNA Damage, DNA Repair and Cancer". *New Research Directions in DNA Repair.* pp. 413–65. doi:10.5772/53919. ISBN 978-953-51-1114-6.

[157] O'Hagan, HM; Mohammad, HP; Baylin, SB (2008). "Double strand breaks can initiate gene silencing and SIRT1-dependent onset of DNA methylation in an exogenous promoter CpG island". *PLoS Genet* **4** (8): e1000155. doi:10.1371/journal.pgen.1000155. PMC 2491723. PMID 18704159.

[158] Cuozzo, C; Porcellini, A; Angrisano, T; Morano, A; Lee, B; Di Pardo, A; Messina, S; Iuliano, R; Fusco, A; Santillo, MR; Muller, MT; Chiariotti, L; Gottesman, ME; Avvedimento, EV (2007). "DNA damage, homology-directed repair, and DNA methylation". *PLoS Genet* **3** (7): e110. doi:10.1371/journal.pgen.0030110. PMC 1913100. PMID 17616978.

[159] Jasperson, KW; Tuohy, TM; Neklason, DW; Burt, RW (2010). "Hereditary and familial colon cancer". *Gastroenterology* **138** (6): 2044–2058. doi:10.1053/j.gastro.2010.01.054. PMID 20420945.

[160] Truninger K, Menigatti M, Luz J, et al. (May 2005). "Immunohistochemical analysis reveals high frequency of PMS2 defects in colorectal cancer". *Gastroenterology* **128** (5): 1160–71. doi:10.1053/j.gastro.2005.01.056. PMID 15887099.

[161] Valeri N, Gasparini P, Fabbri M, et al. (April 2010). "Modulation of mismatch repair and genomic stability by miR-155". *Proc. Natl. Acad. Sci. U.S.A.* **107** (15): 6982–7. Bibcode:2010PNAS..107.6982V. doi:10.1073/pnas.1002472107. PMC 2872463. PMID 20351277.

[162] Zhang W, Zhang J, Hoadley K, et al. (June 2012). "miR-181d: a predictive glioblastoma biomarker that downregulates MGMT expression". *Neuro-oncology* **14** (6): 712–9. doi:10.1093/neuonc/nos089. PMC 3367855. PMID 22570426.

[163] Spiegl-Kreinecker S, Pirker C, Filipits M, et al. (January 2010). "O6-Methylguanine DNA methyltransferase protein expression in tumor cells predicts outcome of temozolomide therapy in glioblastoma patients". *Neuro-oncology* **12** (1): 28–36. doi:10.1093/neuonc/nop003. PMC 2940563. PMID 20150365.

[164] Sgarra R, Rustighi A, Tessari MA, et al. (September 2004). "Nuclear phosphoproteins HMGA and their relationship with chromatin structure and cancer". *FEBS Lett.* **574** (1–3): 1–8. doi:10.1016/j.febslet.2004.08.013. PMID 15358530.

[165] Xu Y, Sumter TF, Bhattacharya R, et al. (May 2004). "The HMG-I oncogene causes highly penetrant, aggressive lymphoid malignancy in transgenic mice and is overexpressed in human leukemia". *Cancer Res.* **64** (10): 3371–5. doi:10.1158/0008-5472.CAN-04-0044. PMID 15150086.

[166] Baldassarre G, Battista S, Belletti B, et al. (April 2003). "Negative regulation of BRCA1 gene expression by HMGA1 proteins accounts for the reduced BRCA1 protein levels in sporadic breast carcinoma". *Mol. Cell. Biol.* **23** (7): 2225–38. doi:10.1128/MCB.23.7.2225-2238.2003. PMC 150734. PMID 12640109.

[167] Borrmann L, Schwanbeck R, Heyduk T, et al. (December 2003). "High mobility group A2 protein and its derivatives bind a specific region of the promoter of DNA repair gene ERCC1 and modulate its activity". *Nucleic Acids Res.* **31** (23): 6841–51. doi:10.1093/nar/gkg884. PMC 290254. PMID 14627817.

[168] Facista A, Nguyen H, Lewis C, Prasad AR, Ramsey L, Zaitlin B, Nfonsam V, Krouse RS, Bernstein H, Payne CM, Stern S, Oatman N, Banerjee B, Bernstein C (2012). "Deficient expression of DNA repair enzymes in early progression to sporadic colon cancer". *Genome Integrity* **3** (1): 3. doi:10.1186/2041-9414-3-3. PMC 3351028. PMID 22494821.

[169] Palmieri D, D'Angelo D, Valentino T, et al. (August 2012). "Downregulation of HMGA-targeting microRNAs has a critical role in human pituitary tumorigenesis". *Oncogene* **31** (34): 3857–65. doi:10.1038/onc.2011.557. PMID 22139073.

[170] Malumbres M (2013). "miRNAs and cancer: an epigenetics view". *Mol. Aspects Med.* **34** (4): 863–74. doi:10.1016/j.mam.2012.06.005. PMID 22771542.

[171] Schnekenburger M, Diederich M (March 2012). "Epigenetics Offer New Horizons for Colorectal Cancer Prevention". *Curr Colorectal Cancer Rep* **8** (1): 66–81. doi:10.1007/s11888-011-0116-z. PMC 3277709. PMID 22389639.

[172] Sampath D, Liu C, Vasan K, et al. (February 2012). "Histone deacetylases mediate the silencing of miR-15a, miR-16, and miR-29b in chronic lymphocytic leukemia". *Blood* **119** (5): 1162–72. doi:10.1182/blood-2011-05-351510. PMC 3277352. PMID 22096249.

[173] Son DJ, Kumar S, Takabe W, Kim CW, Ni CW, Alberts-Grill N, Jang IH, Kim S, Kim W, Won Kang S, Baker AH, Woong Seo J, Ferrara KW, Jo H (2013). "The atypical mechanosensitive microRNA-712 derived from pre-ribosomal RNA induces endothelial inflammation and atherosclerosis". *Nature Communications* **4**: 3000. doi:10.1038/ncomms4000. PMC 3923891. PMID 24346612.

[174] Wan G, Mathur R, Hu X, Zhang X, Lu X (September 2011). "miRNA response to DNA damage". *Trends Biochem. Sci.* **36** (9): 478–84. doi:10.1016/j.tibs.2011.06.002. PMC 3532742. PMID 21741842.

[175] Tessitore A, Cicciarelli G, Del Vecchio F, Gaggiano A, Verzella D, Fischietti M, Vecchiotti D, Capece D, Zazzeroni F, Alesse E (2014). "MicroRNAs in the DNA Damage/Repair Network and Cancer". *Int J Genomics* **2014**: 820248. doi:10.1155/2014/820248. PMC 3926391. PMID 24616890.

[176] Chen JF, Murchison EP, Tang R, Callis TE, Tatsuguchi M, Deng Z, Rojas M, Hammond SM, Schneider MD, Selzman CH, Meissner G, Patterson C, Hannon GJ, Wang DZ (February 2008). "Targeted deletion of Dicer in the heart leads to dilated cardiomyopathy and heart failure". *Proc. Natl. Acad. Sci. U.S.A.* **105** (6): 2111–6. Bibcode:2008PNAS..105.2111C. doi:10.1073/pnas.0710228105. PMC 2542870. PMID 18256189.

[177] Zhao Y, Ransom JF, Li A, Vedantham V, von Drehle M, Muth AN, Tsuchihashi T, McManus MT, Schwartz RJ, Srivastava D (April 2007). "Dysregulation of cardiogenesis, cardiac conduction, and cell cycle in mice lacking miRNA-1-2". *Cell* **129** (2): 303–17. doi:10.1016/j.cell.2007.03.030. PMID 17397913.

[178] Thum T, Galuppo P, Wolf C, Fiedler J, Kneitz S, van Laake LW, Doevendans PA, Mummery CL, Borlak J, Haverich A, Gross C, Engelhardt S, Ertl G, Bauersachs J (July 2007). "MicroRNAs in the human heart: a clue to fetal gene reprogramming in heart failure". *Circulation* **116** (3): 258–67. doi:10.1161/CIRCULATIONAHA.107.687947. PMID 17606841.

[179] van Rooij E, Sutherland LB, Liu N, Williams AH, McAnally J, Gerard RD, Richardson JA, Olson EN (November 2006). "A signature pattern of stress-responsive microRNAs that can evoke cardiac hypertrophy and heart failure". *Proc. Natl. Acad. Sci. U.S.A.* **103** (48): 18255–60. Bibcode:2006PNAS..10318255V. doi:10.1073/pnas.0608791103. PMC 1838739. PMID 17108080.

[180] Tatsuguchi M, Seok HY, Callis TE, Thomson JM, Chen JF, Newman M, Rojas M, Hammond SM, Wang DZ (June 2007). "Expression of microRNAs is dynamically regulated during cardiomyocyte hypertrophy". *J. Mol. Cell. Cardiol.* **42** (6): 1137–41. doi:10.1016/j.yjmcc.2007.04.004. PMC 1934409. PMID 17498736.

[181] Zhao Y, Samal E, Srivastava D (July 2005). "Serum response factor regulates a muscle-specific microRNA that targets Hand2 during cardiogenesis". *Nature* **436** (7048): 214–20. Bibcode:2005Natur.436..214Z. doi:10.1038/nature03817. PMID 15951802.

[182] Xiao J, Luo X, Lin H, Zhang Y, Lu Y, Wang N, Zhang Y, Yang B, Wang Z (April 2007). "MicroRNA miR-133 represses HERG K+ channel expression contributing to QT prolongation in diabetic hearts". *J. Biol. Chem.* **282** (17): 12363–7. doi:10.1074/jbc.C700015200. PMID 17344217.

[183] Yang B, Lin H, Xiao J, Lu Y, Luo X, Li B, Zhang Y, Xu C, Bai Y, Wang H, Chen G, Wang Z (April 2007). "The muscle-specific microRNA miR-1 regulates cardiac arrhythmogenic potential by targeting GJA1 and KCNJ2". *Nat. Med.* **13** (4): 486–91. doi:10.1038/nm1569. PMID 17401374.

[184] Carè A, Catalucci D, Felicetti F, Bonci D, Addario A, Gallo P, Bang ML, Segnalini P, Gu Y, Dalton ND, Elia L, Latronico MV, Høydal M, Autore C, Russo MA, Dorn GW, Ellingsen O, Ruiz-Lozano P, Peterson KL, Croce CM, Peschle C, Condorelli G (May 2007). "MicroRNA-133 controls cardiac hypertrophy". *Nat. Med.* **13** (5): 613–8. doi:10.1038/nm1582. PMID 17468766.

[185] van Rooij E, Sutherland LB, Qi X, Richardson JA, Hill J, Olson EN (April 2007). "Control of stress-dependent cardiac growth and gene expression by a microRNA". *Science* **316** (5824): 575–9. Bibcode:2007Sci...316..575V. doi:10.1126/science.1139089. PMID 17379774.

[186] Wagschal, Alexandre (2 September 2015). "Genome-wide identification of microRNAs regulating cholesterol and triglyceride homeostasis". *Nature Medicine* **21**: 1290–1297. doi:10.1038/nm.3980. Retrieved 27 October 2015.

[187] Insull W (January 2009). "The pathology of atherosclerosis: plaque development and plaque responses to medical treatment". *The American Journal of Medicine* **122** (1 Suppl): S3–S14. doi:10.1016/j.amjmed.2008.10.013. PMID 19110086.

[188] Basu R, Fan D, Kandalam V, Lee J, Das SK, Wang X, Baldwin TA, Oudit GY, Kassiri Z (December 2012). "Loss of Timp3 gene leads to abdominal aortic aneurysm formation in response to angiotensin II". *The Journal of Biological Chemistry* **287** (53): 44083–96. doi:10.1074/jbc.M112.425652. PMID 23144462.

[189] Libby P (2002). "Inflammation in atherosclerosis". *Nature* **420** (6917): 868–74. doi:10.1038/nature01323. PMID 12490960.

[190] Phua, YL; Chu, JY; Marrone, AK; Bodnar, AJ; Sims-Lucas, S; Ho, J (October 2015). "Renal stromal miRNAs are required for normal nephrogenesis and glomerular mesangial survival.". *Physiological reports* **3** (10): e12537. doi:10.14814/phy2.12537. PMID 26438731.

[191] Maes OC, Chertkow HM, Wang E, Schipper HM (May 2009). "MicroRNA: Implications for Alzheimer Disease and other Human CNS Disorders". *Current Genomics* **10** (3): 154–68. doi:10.2174/138920209788185252. PMC 2705849. PMID 19881909.

[192] Amin, ND; Bai, G; Klug, JR; Bonanomi, D; Pankratz, MT; Gifford, WD; Hinckley, CA; Sternfeld, MJ; Driscoll, SP; Dominguez, B; Lee, K; Jin, X; Pfaff, SL (Dec 18, 2015). "Loss of motoneuron-specific microRNA-218 causes systemic neuromuscular failure.". *Science* **350** (6267): 1525–29. doi:10.1126/science.aad2509. PMID 26680198.

[193] Schratt G (December 2009). "microRNAs at the synapse". *Nat. Rev. Neurosci.* **10** (12): 842–9. doi:10.1038/nrn2763. PMID 19888283.

[194] Hommers LG, Domschke K, Deckert J (2015). "Heterogeneity and Individuality: microRNAs in Mental Disorders". *J Neural Transm.* **122** (1): 79–97. doi:10.1007/s00702-014-1338-4. PMID 25395183.

[195] Feng J, Sun G, Yan J, Noltner K, Li W, Buzin CH, Longmate J, Heston LL, Rossi J, Sommer SS (2009). Reif, Andreas, ed. "Evidence for X-chromosomal schizophrenia associated with microRNA alterations". *PLoS ONE* **4** (7): e6121. Bibcode:2009PLoSO...4.6121F. doi:10.1371/journal.pone.0006121. PMC 2699475. PMID 19568434.

[196] Beveridge NJ, Gardiner E, Carroll AP, Tooney PA, Cairns MJ (September 2009). "Schizophrenia is associated with an increase in cortical microRNA biogenesis". *Mol. Psychiatry* **15** (12): 1176–89. doi:10.1038/mp.2009.84. PMC 2990188. PMID 19721432.

[197] Lewohl JM, Nunez YO, Dodd PR, Tiwari GR, Harris RA, Mayfield RD (2011). "Up-regulation of microRNAs in brain of human alcoholics". *Alcohol Clin Exp Res.* **35** (11): 1928–37. doi:10.1111/j.1530-0277.2011.01544.x. PMC 3170679. PMID 21651580.

[198] Tapocik JD, Solomon M, Flanigan M, Meinhardt M, Barbier E, Schank JR, Schwandt M, Sommer WH, Heilig M (2013). "Coordinated dysregulation of mRNAs and microRNAs in the rat medial prefrontal cortex following a history of alcohol dependence". *Pharmacogenomics J* **13** (3): 286–96. doi:10.1038/tpj.2012.17. PMC 3546132. PMID 22614244.

[199] Gorini G, Nunez YO, Mayfield RD (2013). "Integration of miRNA and protein profiling reveals coordinated neuroadaptations in the alcohol-dependent mouse brain". *PLOS ONE* **8** (12): e82565. doi:10.1371/journal.pone.0082565. PMC 3865091. PMID 24358208.

[200] Tapocik JD, Barbier E, Flanigan M, Solomon M, Pincus A, Pilling A, Sun H, Schank JR, King C, Heilig M (2014). "microRNA-206 in rat medial prefrontal cortex regulates BDNF expression and alcohol drinking". *J Neurosci.* **34** (13): 4581–4588. doi:10.1523/JNEUROSCI.0445-14.2014. PMC 3965783. PMID 24672003.

[201] Lippai D, Bala S, Csak T, Kurt-Jones EA, Szabo G (2013). "Chronic alcohol-induced microRNA-155 contributes to neuroinflammation in a TLR4-dependent manner in mice". *PLOS ONE* **8** (8): e70945. doi:10.1371/journal.pone.0070945. PMC 3739772. PMID 23951048.

[202] Li J, Li J, Liu X, Qin S, Guan Y, Liu Y, Cheng Y, Chen X, Li W, Wang S, Xiong M, Kuzhikandathil EV, Ye JH, Zhang C (2013). "MicroRNA expression profile and functional analysis reveal that miR-382 is a critical novel gene of alcohol addiction". *EMBO Mol Med.* **5** (9): 1402–14. doi:10.1002/emmm.201201900. PMC 3799494. PMID 23873704.

[203] Romao JM, Jin W, Dodson MV, Hausman GJ, Moore SS, Guan LL (September 2011). "MicroRNA regulation in mammalian adipogenesis". *Exp. Biol. Med. (Maywood)* **236** (9): 997–1004. doi:10.1258/ebm.2011.011101. PMID 21844119.

[204] Skårn M, Namløs HM, Noordhuis P, Wang MY, Meza-Zepeda LA, Myklebost O (April 2012). "Adipocyte differentiation of human bone marrow-derived stromal cells is modulated by microRNA-155, microRNA-221, and microRNA-222". *Stem Cells Dev.* **21** (6): 873–83. doi:10.1089/scd.2010.0503. PMID 21756067.

[205] Zuo Y, Qiang L, Farmer SR (March 2006). "Activation of CCAAT/enhancer-binding protein (C/EBP) alpha expression by C/EBP beta during adipogenesis requires a peroxisome proliferator-activated receptor-gamma-associated repression of HDAC1 at the C/ebp alpha gene promoter". *J. Biol. Chem.* **281** (12): 7960–7. doi:10.1074/jbc.M510682200. PMID 16431920.

[206] Zhu H, Shyh-Chang N, Segrè AV, Shinoda G, Shah SP, Einhorn WS, Takeuchi A, Engreitz JM, Hagan JP, Kharas MG, Urbach A, Thornton JE, Triboulet R, Gregory RI; DIAGRAM Consortium; MAGIC Investigators, Altshuler D, Daley GQ (September 2011). "The Lin28/let-7 axis regulates glucose metabolism". *Cell* **147** (1): 81–94. doi:10.1016/j.cell.2011.08.033. PMC 3353524. PMID 21962509.

[207] Frost RJ, Olson EN. (December 2011). "Control of glucose homeostasis and insulin sensitivity by the Let-7 family of microRNAs". *Proc Natl Acad Sci U S A.* **108** (52): 21075–80. Bibcode:2011PNAS..10821075F. doi:10.1073/pnas.1118922109. PMC 3248488. PMID 22160727.

[208] Pheasant M, Mattick JS (September 2007). "Raising the estimate of functional human sequences". *Genome Res.* **17** (9): 1245–53. doi:10.1101/gr.6406307. PMID 17690206.

[209] Bertone P, Stolc V, Royce TE, Rozowsky JS, Urban AE, Zhu X, Rinn JL, Tongprasit W, Samanta M, Weissman S, Gerstein M, Snyder M (December 2004). "Global identification of human transcribed sequences with genome tiling arrays". *Science* **306** (5705): 2242–6. Bibcode:2004Sci...306.2242B. doi:10.1126/science.1103388. PMID 15539566.

[210] Ota T, Suzuki Y, Nishikawa T, Otsuki T, Sugiyama T, Irie R, Wakamatsu A, Hayashi K, Sato H, Nagai K, Kimura K, Makita H, Sekine M, Obayashi M, Nishi T, Shibahara T, Tanaka T, Ishii S, Yamamoto J, Saito K, Kawai Y, Isono Y, Nakamura Y, Nagahari K, Murakami K, Yasuda T, Iwayanagi T, Wagatsuma M; et al. (January 2004). "Complete sequencing and characterization of 21,243 full-length human cDNAs". *Nat. Genet.* **36** (1): 40–5. doi:10.1038/ng1285. PMID 14702039.

[211] Kuhn DE, Martin MM, Feldman DS, Terry AV, Nuovo GJ, Elton TS (January 2008). "Experimental validation of miRNA targets". *Methods* **44** (1): 47–54. doi:10.1016/j.ymeth.2007.09.005. PMC 2237914. PMID 18158132.

[212] Hüttenhofer A, Schattner P, Polacek N (May 2005). "Non-coding RNAs: hope or hype?". *Trends Genet.* **21** (5): 289–97. doi:10.1016/j.tig.2005.03.007. PMID 15851066.

[213] Tuddenham L, Jung JS, Chane-Woon-Ming B, Dölken L, Pfeffer S (February 2012). "Small RNA deep sequencing identifies microRNAs and other small noncoding RNAs from human herpesvirus 6B". *J. Virol.* **86** (3): 1638–49. doi:10.1128/JVI.05911-11. PMC 3264354. PMID 22114334.

[214] Zheng H, Fu R, Wang JT, Liu Q, Chen H, Jiang SW (Apr 2013). "Advances in the Techniques for the Prediction of microRNA Targets". *Int J Mol Sci.* **14** (4): 8179–87. doi:10.3390/ijms14048179. PMC 3645737. PMID 23591837.

[215] Agarwal, Vikram; Bell, George W.; Nam, Jin-Wu; Bartel, David P. (2015-08-12). "Predicting effective microRNA target sites in mammalian mRNAs". *eLife* **4**: e05005. doi:10.7554/eLife.05005. ISSN 2050-084X. PMC 4532895. PMID 26267216.

32.15   Bibliography

- *miRNA definition and classification:* Ambros V, Bartel B, Bartel DP, Burge CB, Carrington JC, Chen X, Dreyfuss G, Eddy SR, Griffiths-Jones S, Marshall M, Matzke M, Ruvkun G, Tuschl T (2003). "A uniform system for microRNA annotation". *RNA* **9** (3): 277–279. doi:10.1261/rna.2183803. PMC 1370393. PMID 12592000.

- *Science* review of small RNA: *Baulcombe D (2002). "DNA events. An RNA microcosm".* Science **297** *(5589): 2002–2003. doi:10.1126/science.1077906. PMID 12242426.*

- *Discovery of* lin-4*, the first miRNA to be discovered:* Lee RC, Feinbaum RL, Ambros V (1993). "The *C. elegans* heterochronic gene lin-4 encodes small RNAs with antisense complementarity to lin-14". *Cell* **75** (5): 843–854. doi:10.1016/0092-8674(93)90529-Y. PMID 8252621.

32.16   External links

- The miRBase database

- The miRNA Blog

- miR2Disease, a manually curated database documenting known relationships between miRNA dysregulation and human disease.

- semirna, Web application to search for microRNAs in a plant genome.

- miRandola: Extracellular Circulating microRNAs Database.

- MirOB: MicroRNA targets database and data analysis and dataviz tool.

Chapter 33

Small nuclear RNA

Small nuclear ribonucleic acid (snRNA), also commonly referred to as **U-RNA**, is a class of small RNA molecules that are found within the splicing speckles and Cajal bodies of the cell nucleus in eukaryotic cells. The length of an average snRNA is approximately 150 nucleotides. They are transcribed by either RNA polymerase II or RNA polymerase III, and studies have shown that their primary function is in the processing of pre-messenger RNA (hnRNA) in the nucleus. They have also been shown to aid in the regulation of transcription factors (7SK RNA) or RNA polymerase II (B2 RNA), and maintaining the telomeres.

snRNA are always associated with a set of specific proteins, and the complexes are referred to as small nuclear ribonucleoproteins (snRNP, often pronounced "snurps"). Each snRNP particle is composed of several Sm proteins, the snRNA component, and snRNP-specific proteins. The most common snRNA components of these complexes are known, respectively, as: U1 spliceosomal RNA, U2 spliceosomal RNA, U4 spliceosomal RNA, U5 spliceosomal RNA, and U6 spliceosomal RNA. Their nomenclature derives from their high uridine content.

snRNAs were discovered by accident during a gel electrophoresis experiment in 1966.[1] An unexpected type of RNA was found in the gel and investigated. Later analysis has shown that these RNA were high in uridylate and were established in the nucleus.

A large group of snRNAs are known as small nucleolar RNAs (snoRNAs). These are small RNA molecules that play an essential role in RNA biogenesis and guide chemical modifications of ribosomal RNAs (rRNAs) and other RNA genes (tRNA and snRNAs). They are located in the nucleolus and the Cajal bodies of eukaryotic cells (the major sites of RNA synthesis), where they are called scaRNAs (small Cajal body-specific RNAs).

33.1   Classes of snRNA

snRNA are often divided into two classes based upon both common sequence features as well as associated protein factors such as the RNA-binding LSm proteins.[2]

The first class, known as **Sm-class snRNA**, is more widely studied and consists of U1, U2, U4, U4atac, U5, U7, U11, and U12. Sm-class snRNA are transcribed by RNA polymerase II. The pre-snRNA are transcribed and receive the usual 7-methylguanosine five-prime cap in the nucleus. They are then exported to the cytoplasm through nuclear pores for further processing. In the cytoplasm, the snRNA receive 3' trimming to form a 3' stem-loop structure, as well as hypermethylation of the 5' cap to form trimethylguanosine.[3] The 3' stem structure is necessary for recognition by the survival of motor neuron (SMN) protein.[4] This complex assembles the snRNA into stable ribonucleoproteins (RNPs). The modified 5' cap is then required to import the snRNP back into the nucleus. All of these uridine-rich snRNA, with the exception of U7, form the core of the spliceosome. Splicing, or the removal of introns, is a major aspect of post-transcriptional modification, and takes place only in the nucleus of eukaryotes. U7 snRNA has been found to function in histone pre-mRNA processing.

The second class, known as **Lsm-class snRNA**, consists of U6 and U6atac. Lsm-class snRNAs are transcribed by

RNA polymerase III and never leave the nucleus, in contrast to Sm-class snRNA. Lsm-class snRNAs contain a 5'-γ-monomethylphosphate cap[5] and a 3' stem–loop, terminating in a stretch of uridines that form the binding site for a distinct heteroheptameric ring of Lsm proteins.[6]

33.2 snRNA in the spliceosome

Spliceosomes are a major component of an integral step in eukaryotic precursor messenger RNA maturation. A mistake in even a single nucleotide can be devastating to the cell, and a reliable, repeatable method of RNA processing is necessary to ensure cell survival. The spliceosome is a large, protein-RNA complex that consists of five small nuclear RNAs (U1, U2, U4, U5, and U6) and over 150 proteins. The snRNAs, along with their associated proteins, form ribonucleoprotein complexes (snRNPs), which bind to specific sequences on the pre-mRNA substrate.[7] This intricate process results in two sequential transesterification reactions. These reactions will produce a free lariat intron and ligate two exons to form a mature mRNA. There are two separate classes of spliceosomes. The major class, which is far more abundant in eukaryotic cells, splices primarily U2-type introns. The initial step of splicing is the bonding of the U1 snRNP and its associated proteins to the 5' splice end to the hnRNA. This creates the commitment complex which will constrain the hnRNA to the splicing pathway.[8] Then, U2 snRNP is recruited to the spliceosome binding site and forms complex A.[9] U2 snRNP changes the conformation of the hnRNA-snRNP complex, exposing the nucleotide favorably for splicing. Following the conformation change, the U4/U5/U6 tri-snRNP complex binds to complex A to form the structure known as complex B. After rearrangement, complex C is formed, and the spliceosome is active for catalysis.[10]

In addition to this main spliceosome complex, there exists a much less common (~1%) minor spliceosome. This complex comprises U11, U12, U4atac, U6atac and U5 snRNPs. These snRNPs are functional analogs of the snRNPs used in the major spliceosome. The minor spliceosome splices U-12 type introns. The two types of introns mainly differ in their splicing sites: U2-type introns have GT-AG 5' and 3' splice sites while U12-type introns have AT-AC at their 5' and 3' ends. The minor spliceosome carries out its function through a different pathway from the major spliceosome.

33.3 U1 snRNA

U1 snRNP is the initiator of spliceosomal activity in the cell by base pairing with the hnRNA. In the major spliceosome, experimental data has shown that the U1 snRNP is present in equal stoichiometry with U2, U4, U5, and U6 snRNP. However, U1 snRNP's abundance in human cells is far greater than that of the other snRNPs.[11] Through U1 snRNA gene knockdown in HeLa cells, studies have shown the U1 snRNA holds great importance for cellular function. When U1 snRNA genes were knocked out, genomic microarrays showed an increased accumulation of unspliced pre-mRNA.[12] In addition, the knockout was shown to cause premature cleavage and polyadenylation primarily in introns located near the beginning of the transcript. When other uridine based snRNAs were knocked out, this effect was not seen. Thus, U1 snRNA–pre-mRNA base pairing was shown to protect pre-mRNA from polyadenylation as well as premature cleavage. This special protection may explain the overabundance of U1 snRNA in the cell.

33.4 snRNPs and human disease

Through the study of small nuclear ribonucleoproteins (snRNPs) and small nucleolar (sno)RNPs we have been able to better understand many important diseases.

Spinal muscular atrophy - Mutations in the survival motor neuron-1 (SMN1) gene result in the degeneration of spinal motor neurons and severe muscle wasting. The SMN protein assembles Sm-class snRNPs, and probably also snoRNPs and other RNPs.[13] Spinal muscular atrophy affects up to 1 in 6,000 people and is the second leading cause of neuromuscular disease, after Duchenne muscular dystrophy.[14]

Dyskeratosis congenital – Mutations in the assembled snRNPs are also found to be a cause of dyskeratosis congenital, a rare syndrome that presents by abnormal changes in the skin, nails and mucous membrane. Some ultimate effects of this disease include bone-marrow failure as well as cancer. This syndrome has been shown to arise from mutations in multiple genes, including dyskerin, telomerase RNA and telomerase reverse transcriptase.[15]

Prader–Willi syndrome - This syndrome affects as many as 1 in 12,000 people and has a presentation of extreme hunger, cognitive and behavioural problems, poor muscle tone and short stature.[16] The syndrome has been linked to the deletion of a region of paternal chromosome 15 that is not expressed on the maternal chromosome. This region includes a brain-specific snRNA that targets the serotonin−2C receptor mRNA.

33.5   See also

- MicroRNA

33.6   References

[1] Hadjiolov, A.A.; Venkov, P.V.; Tsanev, R.G. (November 1966). "Ribonucleic acids fractionation by density-gradient centrifugation and by agar gel electrophoresis: A comparison". *Analytical Biochemistry* **17** (2): 263–267. doi:10.1016/0003-2697(66)90204-1. Retrieved 12 December 2014.

[2] Matera, A. Gregory; Terns, Rebecca M.; Terns, Michael P. (March 2007). "Non-coding RNAs: lessons from the small nuclear and small nucleolar RNAs". *Nature Reviews Molecular Cell Biology* **8** (3): 209–220. doi:10.1038/nrm2124. PMID 17318225. Retrieved 12 December 2014.

[3] Hamm, Jörg; Darzynkiewicz, Edward; Tahara, Stanley M.; Mattaj, Iain W. (August 1990). "The trimethylguanosine cap structure of U1 snRNA is a component of a bipartite nuclear targeting signal". *Cell* **62** (3): 569–577. doi:10.1016/0092-8674(90)90021-6. Retrieved 12 December 2014.

[4] Sattler, Michael; Selenko, Philipp; Sprangers, Remco; Stier, Gunter; Bühler, Dirk; Fischer, Utz (1 January 2001). "SMN Tudor domain structure and its interaction with the Sm proteins". *Nature Structural Biology* **8** (1): 27–31. doi:10.1038/83014. PMID 11135666. Retrieved 12 December 2014.

[5] Singh, R; Reddy, R (November 1989). "Gamma-monomethyl phosphate: a cap structure in spliceosomal U6 small nuclear RNA.". *Proceedings of the National Academy of Sciences of the United States of America* **86** (21): 8280–3. doi:10.1073/pnas.86.21.8280. PMID 2813391. Retrieved 12 December 2014.

[6] Kiss, Tamás (1 December 2004). "Biogenesis of small nuclear RNPs". *Journal of Cell Science* **117** (25): 5949–5951. doi:10.1242/jcs.01487. PMID 15564372. Retrieved 12 December 2014.

[7] Guo, Zhuojun; Karunatilaka, Krishanthi S; Rueda, David (1 November 2009). "Single-molecule analysis of protein-free U2–U6 snRNAs". *Nature Structural & Molecular Biology* **16** (11): 1154–1159. doi:10.1038/nsmb.1672. Retrieved 12 December 2014.

[8] Legrain, P; Seraphin, B; Rosbash, M (September 1988). "Early Commitment of Yeast Pre-mRNA to the Spliceosome Pathway". *Molecular and Cellular Biology* **8** (9): 3755–3760. doi:10.1128/MCB.8.9.3755. Retrieved 12 December 2014.

[9] Newby, Meredith I.; Greenbaum, Nancy L. (11 November 2002). "Sculpting of the spliceosomal branch site recognition motif by a conserved pseudouridine". *Nature Structural Biology* **9** (12): 958–965. doi:10.1038/nsb873. PMID 12426583. Retrieved 12 December 2014.

[10] Burge, Christopher B; Tuschl, Thomas; Sharp, Phillip A (1999). "Splicing of Precursors to mRNAs by the Spliceosomes". *CSH Monographs Volume 37 (1999): The RNA World, 2nd Ed.: The Nature of Modern RNA Suggests a Prebiotic RNA World*: 525–560. doi:10.1101/087969589.37.525. Retrieved 12 December 2014.

[11] Baserga, Susan J; Steitz, Joan A (1993). "The Diverse World of Small Ribonucleoproteins". *CSH Monographs Volume 24 (1993): The RNA World*: 359–381. doi:10.1101/087969380.24.359. Retrieved 12 December 2014.

[12] Kaida, Daisuke; Berg, Michael G.; Younis, Ihab; Kasim, Mumtaz; Singh, Larry N.; Wan, Lili; Dreyfuss, Gideon (29 September 2010). "U1 snRNP protects pre-mRNAs from premature cleavage and polyadenylation". *Nature* **468** (7324): 664–668. doi:10.1038/nature09479. Retrieved 12 December 2014.

[13] Matera, A Gregory; Shpargel, Karl B (June 2006). "Pumping RNA: nuclear bodybuilding along the RNP pipeline". *Current Opinion in Cell Biology* **18** (3): 317–324. doi:10.1016/j.ceb.2006.03.005. Retrieved 12 December 2014.

[14] (Sarnat HB. Spinal muscular atrophies. In: Kliegman RM, Behrman RE, Jenson HB, Stanton BF. Nelson Textbook of Pediatrics. 19th ed. Philadelphia, Pa: Elsevier; 2011:chap 604.2.)

[15] (Wattendorf, D. J. & Muenke, M. Prader–Willi syndrome. Am. Fam. Physician 72, 827–830 (2005).)

[16] (Cooke DW, Divall SA, Radovick S. Normal and aberrant growth. In: Melmed S, ed. Williams Textbook of Endocrinology. 12th ed. Philadelphia, Pa: Saunders Elsevier; 2011:chap 24.)

33.7 External links

- Small Nuclear RNA at the US National Library of Medicine Medical Subject Headings (MeSH)

- Small Nucleolar RNA at the US National Library of Medicine Medical Subject Headings (MeSH)

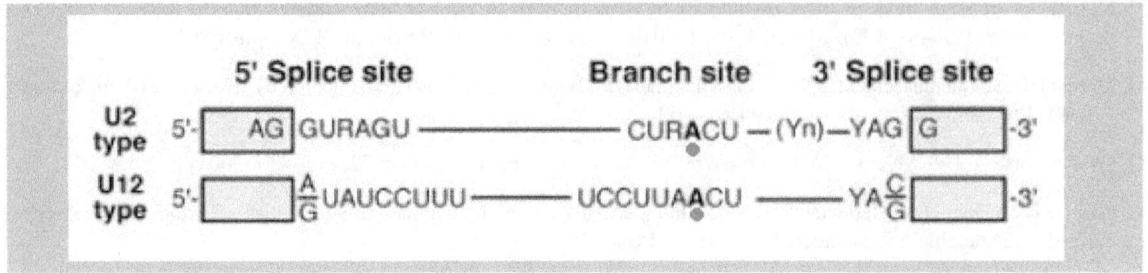

U2 dependent spliceosome   U12 dependent spliceosome

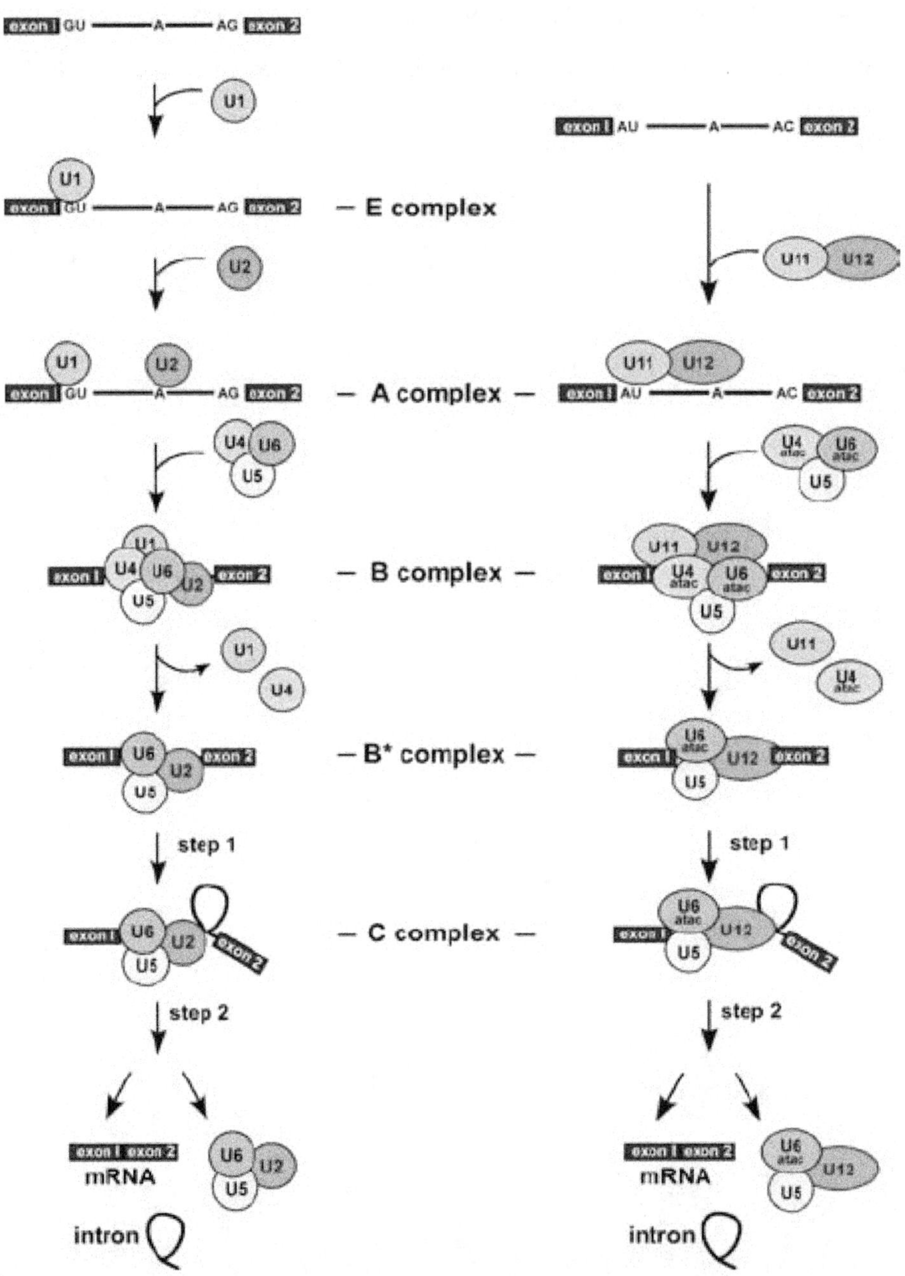

U5 is believed to be the only common component between major and minor spliceosomes.

Reference: Will CL, Luhrmann R.Biol Chem. 2005 Aug;386(8):713–24.

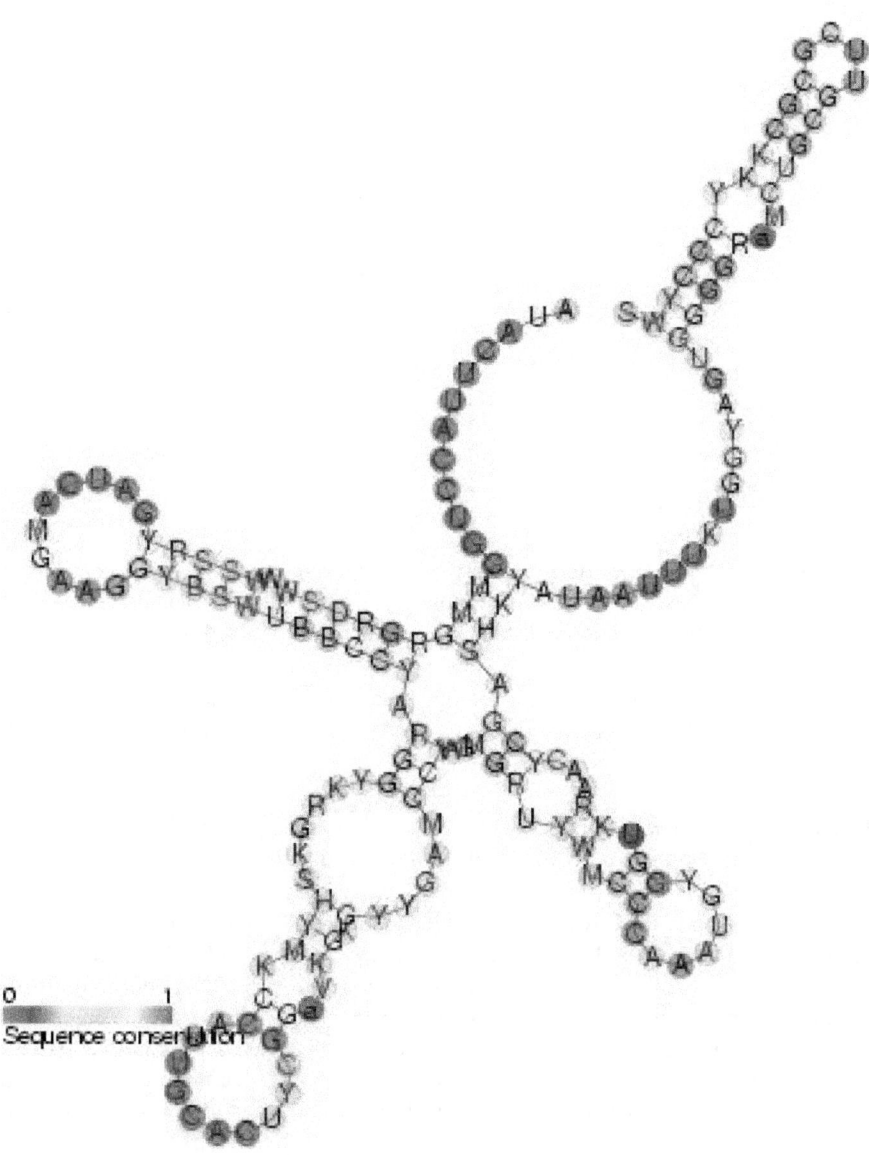

Predicted secondary structure and sequence conservation of U1 snRNA

Chapter 34

Small nucleolar RNA

Small nucleolar RNAs (**snoRNAs**) are a class of small RNA molecules that primarily guide chemical modifications of other RNAs, mainly ribosomal RNAs, transfer RNAs and small nuclear RNAs. There are two main classes of snoRNA, the C/D box snoRNAs, which are associated with methylation, and the H/ACA box snoRNAs, which are associated with pseudouridylation. SnoRNAs are commonly referred to as guide RNAs but should not be confused with the guide RNAs that direct RNA editing in trypanosomes.

34.1   snoRNA guided modifications

After transcription, nascent rRNA molecules (termed pre-rRNA) undergo a series of processing steps to generate the mature rRNA molecule. Prior to cleavage by exo- and endonucleases, the pre-rRNA undergoes a complex pattern of nucleoside modifications. These include methylations and pseudouridylations, guided by snoRNAs.

- Methylation is the attachment or substitution of a methyl group onto various substrates. The rRNA of humans contain approximately 115 methyl group modifications. The majority of these are 2'O-ribose-methylations (where the methyl group is attached to the ribose group).[1]

- Pseudouridylation is the conversion (isomerisation) of the nucleoside uridine to a different isomeric form pseudouridine (Ψ). Mature human rRNAs contain approximately 95 Ψ modifications.[1]

Each snoRNA molecule acts as a guide for only one (or two) individual modifications in a target RNA. In order to carry out modification, each snoRNA associates with at least four protein molecules in an RNA/protein complex referred to as a small nucleolar ribonucleoprotein (snoRNP). The proteins associated with each RNA depend on the type of snoRNA molecule (see snoRNA guide families below). The snoRNA molecule contains an antisense element (a stretch of 10-20 nucleotides), which are base complementary to the sequence surrounding the base (nucleotide) targeted for modification in the pre-RNA molecule. This enables the snoRNP to recognise and bind to the target RNA. Once the snoRNP has bound to the target site, the associated proteins are in the correct physical location to catalyse the chemical modification of the target base.

34.2   snoRNA guide families

The two different types of rRNA modification (methylation and pseudouridylation) are directed by two different families of snoRNPs. These families of snoRNAs are referred to as antisense C/D box and H/ACA box snoRNAs based on the presence of conserved sequence motifs in the snoRNA. There are exceptions, but as a general rule C/D box members guide methylation and H/ACA members guide pseudouridylation. The members of each family may vary in biogenesis, structure, and function, but each family is classified by the following generalised characteristics. For more detail, see review.[2] SnoRNAs are classified under small nuclear RNA in MeSH. The HGNC, in collaboration with snoRNABase and experts in the field, has approved unique names for human genes that encode snoRNAs.[3]

34.2.1   C/D box

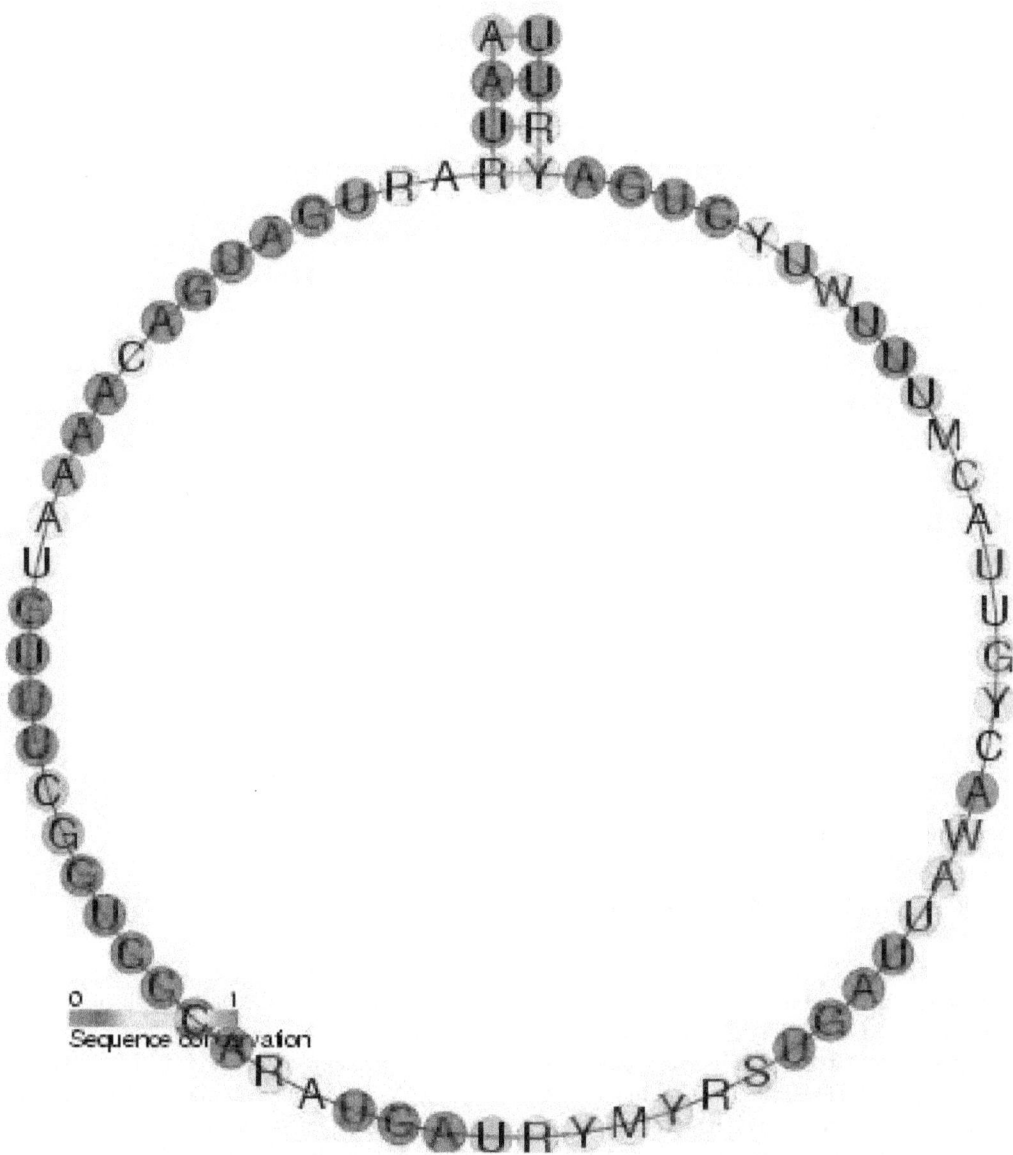

Example of a C/D box snoRNA secondary structure taken from the Rfam database. This example is SNORD73 (RF00071).

C/D box snoRNAs contain two short conserved sequence motifs, C (RUGAUGA) and D (CUGA), located near the 5' and 3' ends of the snoRNA, respectively. Short regions (~ 5 nucleotides) located upstream of the C box and downstream of the D box are usually base complementary and form a stem-box structure, which brings the C and D box motifs into close proximity. This stem-box structure has been shown to be essential for correct snoRNA synthesis and nucleolar localization.[4] Many C/D box snoRNA also contain an additional less-well-conserved copy of the C and D motifs (referred to as C' and D') located in the central portion of the snoRNA molecule. A conserved region of 10-21 nucleotides upstream of the D box is complementary to the methylation site of the target RNA and enables the snoRNA to form an RNA duplex with the RNA.[5] The nucleotide to be modified in the target RNA is usually located at the 5th position upstream from the

D box (or D' box).[6][7] C/D box snoRNAs associate with four evolutionary conserved and essential proteins—fibrillarin (Nop1p), NOP56, NOP58, and Snu13 (15.5-kD protein in eukaryotes; its archaeal homolog is L7Ae)—which make up the core C/D box snoRNP.[2]

There exists a eukaryotic C/D box snoRNA (snoRNA U3) that has not been shown to guide 2'-O-methylation. Instead, it functions in rRNA processing by directing pre-rRNA cleavage.

34.2.2   H/ACA box

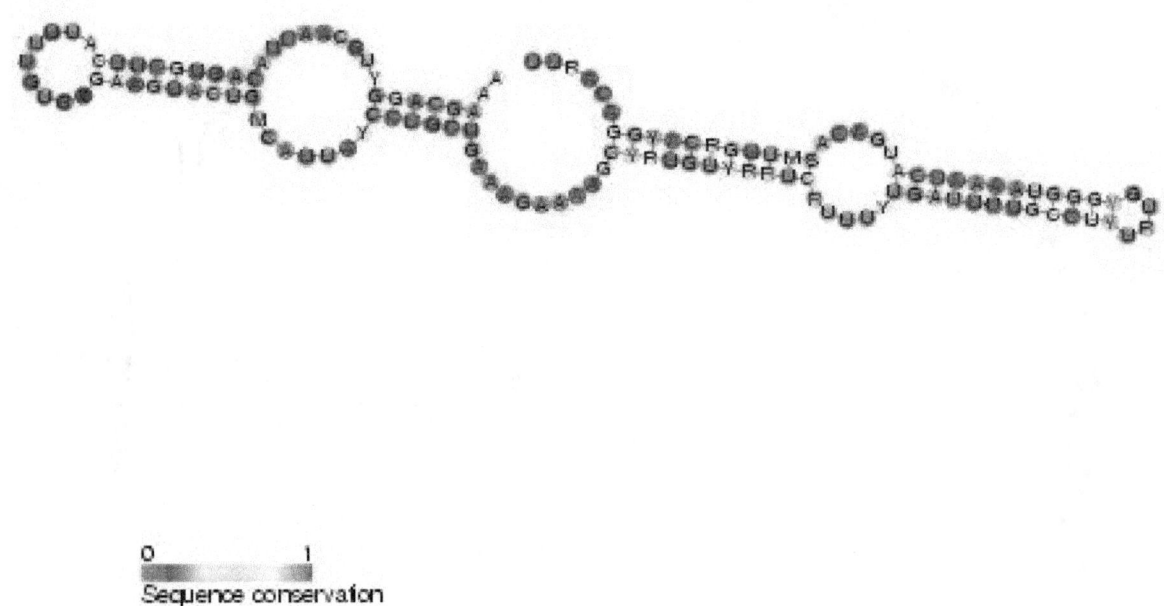

0   1
Sequence conservation

Example of a H/ACA box snoRNA secondary structure taken from the Rfam database. This example is SNORA69 (RF00265).

H/ACA box snoRNAs have a common secondary structure consisting of a two hairpins and two single-stranded regions termed a hairpin-hinge-hairpin-tail structure.[2] H/ACA snoRNAs also contain conserved sequence motifs known as H box (consensus ANANNA) and the ACA box (ACA). Both motifs are usually located in the single-stranded regions of the secondary structure. The H motif is located in the hinge and the ACA motif is located in the tail region; 3 nucleotides from the 3' end of the sequence.[8] The hairpin regions contain internal bulges known as recognition loops in which the antisense guide sequences (bases complementary to the target sequence) are located. This recognition sequence is bipartite (constructed from the two different arms of the loop region) and forms complex pseudo-knots with the target RNA. H/ACA box snoRNAs associate with four evolutionary conserved and essential proteins—dyskerin (Cbf5p), GAR1, NHP2, and NOP10—which make up the core of the H/ACA box snoRNP.[2] However, in lower eukaryotic cells such as trypanosomes, similar RNAs exist in the form of single hairpin structure and an AGA box instead of ACA box at the 3' end of the RNA.[9]

The RNA component of human telomerase (hTERC) contains an H/ACA domain for pre-RNP formation and nucleolar localization of the telomerase RNP itself.[10] The H/ACA snoRNP has been implicated in the rare genetic disease dyskeratosis congenita (DKC) due to its affiliation with human telomerase. Mutations in the protein component of the H/ACA snoRNP result in a reduction in physiological TERC levels. This has been strongly correlated with the pathology behind DKC, which seems to be primarily a disease of poor telomere maintenance.

34.2.3 Composite H/ACA and C/D box

An unusual guide snoRNA U85 that functions in both 2'-O-ribose methylation and pseudouridylation of small nuclear RNA (snRNA) U5 has been identified.[11] This composite snoRNA contains both C/D and H/ACA box domains and associates with the proteins specific to each class of snoRNA (fibrillarin and Gar1p, respectively). More composite snoRNAs have now been characterised.[12]

These composite snoRNAs have been found to accumulate in a subnuclear organelle called the Cajal body and are referred to as small Cajal body-specific RNAs. This is in contrast to the majority of C/D box or H/ACA box snoRNAs, which localise to the nucleolus. These Cajal body specific RNAs are proposed to be involved in the modification of RNA polymerase II transcribed spliceosomal RNAs U1, U2, U4, U5 and U12.[12] Not all snoRNAs that have been localised to Cajal bodies are composite C/D and H/ACA box snoRNAs.

34.2.4 Orphan snoRNAs

The targets for newly identified snoRNAs are predicted on the basis of sequence complementarity between putative target RNAs and the antisense elements or recognition loops in the snoRNA sequence. However, there are increasing numbers of 'orphan' guides without any known RNA targets, which suggests that there might be more proteins or transcripts involved in rRNA than previously and/or that some snoRNAs have different functions not concerning rRNA.[13][14] There is evidence that some of these orphan snoRNAs regulate alternatively spliced transcripts.[15] For example, it appears that the C/D box snoRNA SNORD115 regulates the alternative splicing of the serotonin 2C receptor mRNA via a conserved region of complementarity.[16][17] Another C/D box snoRNA, SNORD116, that resides in the same cluster as SNORD115 has been predicted to have 23 possible targets within protein coding genes using a bioinformatic approach. Of these, a large fraction were found to be alternatively spliced, suggesting a role of SNORD116 in the regulation of alternative splicing.[18]

34.3 Target modifications

The precise effect of the methylation and pseudouridylation modifications on the function of the mature RNAs is not yet known. The modifications do not appear to be essential but are known to subtly enhance the RNA folding and interaction with ribosomal proteins. In support of their importance, target site modifications are exclusively located within conserved and functionally important domains of the mature RNA and are commonly conserved among distant eukaryotes.[2]

1. 2'-O-methylated ribose causes an increase in the 3'-endo conformation

2. Pseudouridine (psi/Ψ) adds another option for H-bonding.

3. Heavily methylated RNA is protected from hydrolysis. rRNA acts as a ribozyme by catalyzing its own hydrolysis and splicing.

34.4 Genomic organisation

SnoRNAs are located diversely in the genome, the majority of vertebrate snoRNA genes are encoded in the introns of genes encoding proteins involved in ribosome synthesis or translation, and are synthesized by RNA polymerase II, they are also shown to be located in intergenic regions, ORFs of protein coding genes, and UTRs.[19] SnoRNAs can also be transcribed from their own promoters by RNA polymerase II or III.

34.4.1 Imprinted loci

In the human genome, there are at least two examples where C/D box snoRNAs are found in tandem repeats within imprinted loci. These two loci (14q32 on chromosome 14 and 15q11q13 on chromosome 15) have been extensively characterised, and in both regions multiple snoRNAs have been found located within introns in clusters of closely related copies.

In 15q11q13, five different snoRNAs have been identified (SNORD64, SNORD107, SNORD108, SNORD109 (two copies), SNORD116 (29 copies) and SNORD115 (48 copies). Loss of the 29 copies of SNORD116 (HBII-85) from this region has been identified as a cause of Prader-Willi syndrome[20][21][22][23] whereas gain of additional copies of SNORD115 has been linked to autism.[24][25] [26]

Region 14q32 contains repeats of two snoRNAs SNORD113 (9 copies) and SNORD114 (31 copies) within the introns of a tissue-specific ncRNA transcript (MEG8). The 14q32 domain has been shown to share common genomic features with the imprinted 15q11-q13 loci and a possible role for tandem repeats of C/D box snoRNAs in the evolution or mechanism of imprinted loci has been suggested.[27][28]

34.5   Other functions

snoRNAs can function as miRNAs. It has been shown that human ACA45 is a *bona fide* snoRNA that can be processed into a 21-nucleotides-long mature miRNA by the RNAse III family endoribonuclease dicer.[29] This snoRNA product has previously been identified as mmu-miR-1839 and was shown to be processed independently from the other miRNA-generating endoribonuclease drosha.[30] Bioinformatical analyses have revealed that putatively snoRNA-derived, miRNA-like fragments occur in different organisms.[31]

Recently, it has been found that snoRNAs can have functions not related to rRNA. One such function is the regulation of alternative splicing of the *trans* gene transcript, which is done by the snoRNA HBII-52, which is also known as SNORD115.[32]

Recent discoveries also show the existence of snoRNA, microRNA, piRNA characteristics in a novel non-coding RNA: x-ncRNA and its biological implication in *Homo sapiens*.[33]

In November 2012, Schubert et al. reveal that specific RNAs control chromatin compaction and accessibility in Drosophila cells.[34]

34.6   References

[1] Maden BE, Hughes JM (1997). "Eukaryotic ribosomal RNA: the recent excitement in the nucleotide modification problem". *Chromosoma* **105** (7–8): 391–400. doi:10.1007/BF02510475. PMID 9211966.

[2] Bachellerie, Jean-Pierre; Cavaillé, Jérôme; Hüttenhofer, Alexander (August 2002). "The expanding snoRNA world". *Biochimie* **84** (8): 775–790. doi:10.1016/S0300-9084(02)01402-5. PMID 12457565.

[3] Wright, MW; Bruford, EA (Jan 2011). "Naming 'junk': human non-protein coding RNA (ncRNA) gene nomenclature.". *Human genomics* **5** (2): 90–8. doi:10.1186/1479-7364-5-2-90. PMC 3051107. PMID 21296742.

[4] Samarsky, DA; Fournier MJ; Singer RH; Bertrand E (1998). "The snoRNA box C/D motif directs nucleolar targeting and also couples snoRNA synthesis and localization". *EMBO* **17** (13): 3747–57. doi:10.1093/emboj/17.13.3747. PMC 1170710. PMID 9649444.

[5] Kiss-László Z, Henry Y, Kiss T (1998). "Sequence and structural elements of methylation guide snoRNAs essential for site-specific ribose methylation of pre-rRNA". *EMBO J.* **17** (3): 797–807. doi:10.1093/emboj/17.3.797. PMC 1170428. PMID 9451004.

[6] Cavaillé J, Nicoloso M, Bachellerie JP (1996). "Targeted ribose methylation of RNA in vivo directed by tailored antisense RNA guides". *Nature* **383** (6602): 732–5. Bibcode:1996Natur.383..732C. doi:10.1038/383732a0. PMID 8878486.

[7] Kiss-László, Zsuzsanna; Henry, Yves; Bachellerie, Jean-Pierre; Caizergues-Ferrer, Michèle; Kiss, Tamás (June 1996). "Site-specific ribose methylation of preribosomal RNA: a novel function for small nucleolar RNAs". *Cell* **85** (7): 1077–1088. doi:10.1016/S0092-8674(00)81308-2. PMID 8674114.

[8] Ganot, Philippe; Caizergues-Ferrer, Michèle; Kiss, Tamás (1 April 1997). "The family of box ACA small nucleolar RNAs is defined by an evolutionarily conserved secondary structure and ubiquitous sequence elements essential for RNA accumulation". *Genes & Development* **11** (7): 941–956. doi:10.1101/gad.11.7.941. PMID 9106664.

[9] Liang XH, Liu L, Michaeli S (2001). "Identification of the first trypanosome H/ACA RNA that guides pseudouridine formation on rRNA". *JBC.* **276** (43): 40313–8. doi:10.1074/jbc.M104488200. PMID 11483606.

[10] Trahan C, Dragon F (February 2009). "Dyskeratosis congenita mutations in the H/ACA domain of human telomerase RNA affect its assembly into a pre-RNP". *RNA* **15** (2): 235–43. doi:10.1261/rna.1354009. PMC 2648702. PMID 19095616.

[11] Jády BE, Kiss T (2001). "A small nucleolar guide RNA functions both in 2'-O-ribose methylation and pseudouridylation of the U5 spliceosomal RNA". *EMBO J.* **20** (3): 541–51. doi:10.1093/emboj/20.3.541. PMC 133463. PMID 11157760.

[12] Darzacq, Xavier; Jády, Beáta E.; Verheggen, Céline; Kiss, Arnold M.; Bertrand, Edouard; Kiss, Tamás (2002). "Cajal body-specific small nuclear RNAs: A novel class of 2'-O-methylation and pseudouridylation guide RNAs". *The EMBO Journal* **21** (11): 2746–2756. doi:10.1093/emboj/21.11.2746. PMC 126017. PMID 12032087.

[13] Jady, B. K.; Kiss, T. (Mar 2000). "Characterisation of the U83 and U84 small nucleolar RNAs: two novel 2'-O-ribose methylation guide RNAs that lack complementarities to ribosomal RNAs" (Free full text). *Nucleic Acids Research* **28** (6): 1348–1354. doi:10.1093/nar/28.6.1348. ISSN 0305-1048. PMC 111033. PMID 10684929.

[14] Li, G. .; Zhou, H. .; Luo, P. .; Zhang, P. .; Qu, H. . (Apr 2005). "Identification and functional analysis of 20 Box H/ACA small nucleolar RNAs (snoRNAs) from Schizosaccharomyces pombe" (Free full text). *The Journal of Biological Chemistry* **280** (16): 16446–16455. doi:10.1074/jbc.M500326200. ISSN 0021-9258. PMID 15716270.

[15] Kishore, S. .; Stamm, S. . (2006). "Regulation of alternative splicing by snoRNAs". *Cold Spring Harbor symposia on quantitative biology* **71**: 329–334. doi:10.1101/sqb.2006.71.024. ISSN 0091-7451. PMID 17381313.

[16] Kishore S, Stamm S (January 2006). "The snoRNA HBII-52 regulates alternative splicing of the serotonin receptor 2C". *Science* **311** (5758): 230–2. Bibcode:2006Sci...311..230K. doi:10.1126/science.1118265. PMID 16357227.

[17] Doe, M. .; Relkovic, D. .; Garfield, S. .; Dalley, W. .; Theobald, E. .; Humby, T. .; Wilkinson, S. .; Isles, R. . (Jun 2009). "Loss of the imprinted snoRNA mbii-52 leads to increased 5htr2c pre-RNA editing and altered 5HT2CR-mediated behaviour". *Human Molecular Genetics* **18** (12): 2140–2148. doi:10.1093/hmg/ddp137. ISSN 0964-6906. PMC 2685753. PMID 19304781.

[18] Bazeley PS, Shepelev V, Talebizadeh Z, Butler MG, Fedorova L, Filatov V, Fedorov A (2008). "snoTARGET shows that human orphan snoRNA targets locate close to alternative splice junctions". *Gene* **408** (1–2): 172–9. doi:10.1016/j.gene.2007.10.037. PMID 18160232.

[19] Kaur D, Gupta AK, Kumari V, Sharma R, Bhattacharya A, Bhattacharya S (2012). "Computational prediction and validation of C/D, H/ACA and Eh_U3 snoRNAs of Entamoeba histolytica". BMC Genomics. 14;13:390. doi: 10.1186/1471-2164-13-390

[20] Skryabin BV, Gubar LV, Seeger B, et al. (2007). "Deletion of the MBII-85 snoRNA gene cluster in mice results in postnatal growth retardation". *PLoS Genet.* **3** (12): e235. doi:10.1371/journal.pgen.0030235. PMC 2323313. PMID 18166085.

[21] Sahoo T, del Gaudio D, German JR, Shinawi M, Peters SU, Person RE, Garnica A, Cheung SW, Beaudet AL (2008). "Prader-Willi phenotype caused by paternal deficiency for the HBII-85 C/D box small nucleolar RNA cluster". *Nat Genet* **40** (6): 719–21. doi:10.1038/ng.158. PMC 2705197. PMID 18500341.

[22] Ding F, Li HH, Zhang S, Solomon NM, Camper SA, Cohen P, Francke U (2008). Akbarian, Schahram, ed. "SnoRNA Snord116 (Pwcr1/MBII-85) deletion causes growth deficiency and hyperphagia in mice". *PLoS ONE* **3** (3): e1709. Bibcode:2008PLoSO...3.1709D. doi:10.1371/journal.pone.0001709. PMC 2248623. PMID 18320030.

[23] Ding F, Prints Y, Dhar MS, Johnson DK, Garnacho-Montero C, Nicholls RD, Francke U (2005). "Lack of Pwcr1/MBII-85 snoRNA is critical for neonatal lethality in Prader-Willi syndrome mouse models". *Mamm Genome* **16** (6): 424–31. doi:10.1007/s00335-005-2460-2. PMID 16075369.

[24] Nakatani J, Tamada K, Hatanaka F, et al. (June 2009). "Abnormal behavior in a chromosome-engineered mouse model for human 15q11-13 duplication seen in autism". *Cell* **137** (7): 1235–46. doi:10.1016/j.cell.2009.04.024. PMC 3710970. PMID 19563756.

[25] Bolton PF, Veltman MW, Weisblatt E, et al. (September 2004). "Chromosome 15q11-13 abnormalities and other medical conditions in individuals with autism spectrum disorders". *Psychiatr. Genet.* **14** (3): 131–7. doi:10.1097/00041444-200409000-00002. PMID 15318025.

[26] Cook EH, Scherer SW (October 2008). "Copy-number variations associated with neuropsychiatric conditions". *Nature* **455** (7215): 919–23. Bibcode:2008Natur.455..919C. doi:10.1038/nature07458. PMID 18923514.

[27] Cavaillé J, Seitz H, Paulsen M, Ferguson-Smith AC, Bachellerie JP (2002). "Identification of tandemly-repeated C/D snoRNA genes at the imprinted human 14q32 domain reminiscent of those at the Prader-Willi/Angelman syndrome region". *Hum. Mol. Genet.* **11** (13): 1527–38. doi:10.1093/hmg/11.13.1527. PMID 12045206.

[28] Labialle S, Cavaillé J. (2011). "Do repeated arrays of regulatory small-RNA genes elicit genomic imprinting?". *BioEssays* **33** (8): 565–73. doi:10.1002/bies.201100032. PMID 21618561.

[29] Ender, C.; Krek, A.; Friedländer, M.; Beitzinger, M.; Weinmann, L.; Chen, W.; Pfeffer, S.; Rajewsky, N.; Meister, G. (2008). "A human snoRNA with microRNA-like functions". *Molecular Cell* **32** (4): 519–528. doi:10.1016/j.molcel.2008.10.017. PMID 19026782.

[30] Babiarz, J.; Ruby, J.; Wang, Y.; Bartel, D.; Blelloch, R. (2008). "Mouse ES cells express endogenous shRNAs, siRNAs, and other Microprocessor-independent, Dicer-dependent small RNAs". *Genes & Development* **22** (20): 2773–2785. doi:10.1101/gad.1705308. PMC 2569885. PMID 18923076.

[31] Taft, R.; Glazov, E.; Lassmann, T.; Hayashizaki, Y.; Carninci, P.; Mattick, J. (2009). "Small RNAs derived from snoRNAs". *RNA* **15** (7): 1233–1240. doi:10.1261/rna.1528909. PMC 2704076. PMID 19474147.

[32] Kishore S, Stamm S (2006). "The snoRNA HBII-52 regulates alternative splicing of the serotonin receptor 2C". *Science* **311** (5758): 230–1. Bibcode:2006Sci...311..230K. doi:10.1126/science.1118265. PMID 16357227.

[33] Kandhavelu M, Lammi C, Buccioni M, Dal Ben D, Volpini R, Marucci G (2009). "Existence of snoRNA, microRNA, piRNA characteristics in a novel non-coding RNA: x-ncRNA and its biological implication in *Homo sapiens*". *Journal of Bioinformatics and Sequence Analysis* **1** (2): 031–040.

[34] Thomas Schubert, Miriam Caroline Pusch, Sarah Diermeier, Vladimir Benes, Elisabeth Kremmer, Axel Imhof, Gernot Langst (2012). "Df31 Protein and snoRNAs Maintain Accessible Higher-Order Structures of Chromatin". *Molecular Cell* (48): 031–040. doi:10.1016/j.molcel.2012.08.021.

34.7   External links

- Small Nucleolar RNA at the US National Library of Medicine Medical Subject Headings (MeSH)

- Small Nuclear RNA at the US National Library of Medicine Medical Subject Headings (MeSH)

- plant snoRNA database

- snoRNAbase: human H/ACA and C/D box snoRNA database

- snoRNP Database

- The yeast snoRNA database

- Rfam page for C/D box snoRNAs

- Rfam page for H/ACA box snoRNAs

- Rfam page for scaRNA snoRNAs

Chapter 35

Long non-coding RNA

Long non-coding RNAs (long ncRNAs, lncRNA) are non-protein coding transcripts longer than 200 nucleotides.[1] This somewhat arbitrary limit distinguishes long ncRNAs from small regulatory RNAs such as microRNAs (miRNAs), short interfering RNAs (siRNAs), Piwi-interacting RNAs (piRNAs), small nucleolar RNAs (snoRNAs), and other short RNAs.[2]

35.1 Abundance of long ncRNAs

A recent study found only one-fifth of transcription across the human genome is associated with protein-coding genes,[3] indicating at least four times more long non-coding than coding RNA sequences. However, it is large-scale complementary DNA (cDNA) sequencing projects such as FANTOM (Functional Annotation of Mammalian cDNA) that reveal the complexity of this transcription.[4] The FANTOM3 project identified ~35,000 non-coding transcripts from ~10,000 distinct loci that bear many signatures of mRNAs, including 5' capping, splicing, and poly-adenylation, but have little or no open reading frame (ORF).[5] While the abundance of long ncRNAs was unanticipated, this number, nevertheless, represents a conservative lower estimate, since it omitted many singleton transcripts and non-polyadenylated transcripts (tiling array data shows more than 40% of transcripts are non-polyadenylated).[6] However, unambiguously identifying ncRNAs within these cDNA libraries is challenging, since it can be difficult to distinguish protein-coding transcripts from non-coding transcripts. It has been suggested through multiple studies that the brain and central nervous system express the greatest amount of long non-coding RNAs of any tissue type.[7]

35.2 Genomic organization of long ncRNAs

The current landscape of the mammalian genome is described as numerous 'foci' of transcription that are separated by long stretches of intergenic space.[8] While long ncRNAs are located and transcribed within the intergenic stretches, the majority are transcribed as complex, interlaced networks of overlapping sense and antisense transcripts that often includes protein-coding genes.[9] Genomic sequences within these transcriptional foci are often shared within a number of different coding and non-coding transcripts in the sense and antisense directions[10] giving rise to a complex hierarchy of overlapping isoforms. For example, 3012 out of 8961 cDNAs previously annotated as truncated coding sequences within FANTOM2 were later designated as genuine ncRNA variants of protein-coding cDNAs.[11] While the abundance and conservation of these interleaved arrangements suggest they have biological relevance, the complexity of these foci frustrates easy evaluation.

The GENCODE consortium has collated and analysed a comprehensive set of human lncRNA annotations and their genomic organisation, modifications, cellular locations and tissue expression profiles.[12] Their analysis indicates human lncRNAs show a bias toward two-exon transcripts.[13]

35.3   Conservation of long ncRNAs

Many small RNAs, such as microRNAs or snoRNAs, exhibit strong conservation across diverse species.[14] In contrast, in general long ncRNAs lack strong conservation, which is often cited as evidence of non-functionality.[15][16] However, many well-described long ncRNAs, such as Air and Xist, are poorly conserved,[17] suggesting that ncRNAs may be subject to different selection pressures.[18] Unlike mRNAs, which have to conserve the codon usage and prevent frameshift mutations in a single long ORF, selection may conserve only short regions of long ncRNAs that are constrained by structure or sequence-specific interactions. Therefore, we may see selection act only over small regions of the long ncRNA transcript. Nevertheless, despite low conservation of long ncRNAs in general, it should be noted that many long ncRNAs still contain strongly conserved elements. For example, 19% of highly conserved phastCons elements occur in known introns, and another 32% in unannotated regions.[19] Furthermore, a representative set of human long ncRNAs exhibit small, yet significant, reductions in substitution and insertion/deletion rates indicative of purifying selection that conserve the integrity of the transcript at the levels of sequence, promoter and splicing.[20]

The poor conservation of ncRNAs may be the result of recent and rapid adaptive selection. For instance, ncRNAs may be more pliant to evolutionary pressures than protein-coding genes, as evidenced by the existence of many lineage specific ncRNAs, such as Xist or Air.[21] Indeed, those conserved regions of the human genome that are subject to recent evolutionary change relative to the chimpanzee genome occurs mainly in non-coding regions, many of which are transcribed.[22][23] This includes a ncRNA, HAR1F, which has undergone rapid evolutionary change in humans and is specifically expressed in the Cajal-Retzius cells in the human neocortex.[24] The observation that many functionally validated RNAs are evolving quickly,[25][26] may result from these sequences having more plastic structure-function constraints, and we may expect a great deal of evolutionary innovation to occur in such sequences. This is supported by the existence of thousands of sequences in the mammalian genome that show poor conservation at the primary sequence level but have evidence of conserved RNA secondary structures.[27]

35.4   Long ncRNA functions

Large-scale sequencing of cDNA libraries and more recently transcriptomic sequencing by next generation sequencing indicate that long noncoding RNAs number in the order of tens of thousands in mammals. However, despite accumulating evidence suggesting that the majority of these are likely to be functional,[28][29] only a relatively small proportion has been demonstrated to be biologically relevant. As of June 2014, 197 LncRNAs have been functionally annotated in LncRNAdb (a database of literature described LncRNAs),[30] thereby, the majority (118 genes) were described in human. A further large-scale sequencing study provides evidence that many transcripts thought to be LncRNAs may, in fact, be translated into proteins.[31]

35.4.1   Long ncRNAs in the regulation of gene transcription

Long ncRNAs in gene-specific transcription

In eukaryotes, RNA transcription is a tightly regulated process. NcRNAs can target different aspects of this process, targeting transcriptional activators or repressors, different components of the transcription reaction including RNA polymerase (RNAP) II and even the DNA duplex to regulate gene transcription and expression (Goodrich 2006). In combination these ncRNAs may comprise a regulatory network that, including transcription factors, finely control gene expression in complex eukaryotes.

NcRNAs modulate the function of transcription factors by several different mechanisms, including functioning themselves as co-regulators, modifying transcription factor activity, or regulating the association and activity of co-regulators. For example, the ncRNA Evf-2 functions as a co-activator for the homeobox transcription factor Dlx2, which plays important roles in forebrain development and neurogenesis (Feng 2006; Panganiban 2002). Sonic hedgehog induces transcription of Evf-2 from an ultra-conserved element located between the Dlx5 and Dlx6 genes during forebrain development (Feng 2006). Evf-2 then recruits the Dlx2 transcription factor to the same ultra-conserved element whereby Dlx2 subsequently induces expression of Dlx5. The existence of other similar ultra- or highly conserved elements within the mammalian genome that are both transcribed and fulfil enhancer functions suggest Evf-2 may be illustrative of a generalised mechanism that tightly regulates important developmental genes with complex expression patterns during vertebrate growth (Pennacchio 2006; Visel 2008). Indeed, the transcription and expression of similar non-coding ultraconserved elements was recently shown to be abnormal in human leukaemia and to contribute to apoptosis in colon cancer cells, suggesting their involvement in tumorogenesis (Calin 2007).

Local ncRNAs can also recruit transcriptional programmes to regulate adjacent protein-coding gene expression. The RNA binding protein TLS, binds and inhibits the CREB binding protein and p300 histone acetyltransferease activities on a repressed gene target, cyclin D1. The recruitment of TLS to the promoter of cyclin D1 is directed by long ncRNAs expressed at low levels and tethered to 5' regulatory regions in response to DNA damage signals (Wang 2008). Moreover, these local ncRNAs act cooperatively as ligands to modulate the activities of TLS. In the broad sense, this mechanism allows the cell to harness RNA-binding proteins, which make up one of the largest classes within the mammalian proteome, and integrate their function in transcriptional programs. A recent study found that a lncRNA in the antisense direction of the Apolipoprotein A1 (APOA1) regulates the transcription of APOA1 through epigenetic modifications.[32]

Recent evidence has raised the possibility that transcription of genes that escape from X-inactivation might be mediated by expression of long non-coding RNA within the escaping chromosomal domains (Reinius 2010).

Long ncRNAs regulating basal transcription machinery

NcRNAs also target general transcription factors required for the RNAP II transcription of all genes (Goodrich 2006). These general factors include components of the initiation complex that assemble on promoters or involved in transcription elongation. A ncRNA transcribed from an upstream minor promoter of the dihydrofolate reductase (DHFR) gene forms a stable RNA-DNA triplex within the major promoter of DHFR to prevent the binding of the transcriptional co-factor TFIIB (Martianov 2007). This novel mechanism of regulating gene expression may in fact represent a widespread method of controlling promoter usage given that thousands of such triplexes exist in eukaryotic chromosome (Lee 1987). The U1 ncRNA can induce transcription initiation by specifically binding to and stimulating TFIIH to phosphorylate the C-terminal domain of RNAP II (Kwek 2002). In contrast the ncRNA 7SK, is able to repress transcription elongation by, in combination with HEXIM1/2, forming an inactive complex that prevents the PTEFb general transcription factor from phosphorylating the C-terminal domain of RNAP II (Kwek 2002; Yang 2001; Yik 2003), thereby repressing global elongation under stressful conditions. These examples, which bypass specific modes of regulation at individual promoters to mediate changes directly at the level of initiation and elongation transcriptional machinery, provide a means of quickly affecting global changes in gene expression.

The ability to quickly mediate global changes is also apparent in the rapid expression of non-coding repetitive sequences. The short interspersed nuclear (SINE) Alu elements in humans and analogous B1 and B2 elements in mice have succeeded in becoming the most abundant mobile elements within the genomes, comprising ~10% of the human and ~6% of the mouse genome, respectively (Lander 2001; Waterston 2002). These elements are transcribed as ncRNAs by RNAP III in response to environmental stresses such as heat shock (Liu 1995), where they then bind to RNAP II with high affinity and prevent the formation of active pre-initiation complexes (Allen 2004; Espinoza 2004; Espinoza 2007; Mariner & Walters 2008). This allows for the broad and rapid repression of gene expression in response to stress (Allen 2004; Mariner & Walters 2008).

A dissection of the functional sequences within Alu RNA transcripts has drafted a modular structure analogous to the organization of domains in protein transcription factors (Shamovsky 2008). The Alu RNA contains two 'arms', each of which may bind one RNAP II molecule, as well as two regulatory domains that are responsible for RNAP II transcriptional repression in vitro (Mariner 2008). These two loosely structured domains may even be concatenated to other ncRNAs such as B1 elements to impart their repressive role (Mariner & Walters 2008). The abundance and distribution of Alu elements and similar repetitive elements throughout the mammalian genome may be partly due to these functional domains being co-opted into other long ncRNAs during evolution, with the presence of functional repeat sequence domains being a common characteristic of several known long ncRNAs including Kcnq1ot1, Xlsirt and Xist (Mattick 2003; Mohammad 2008; Wutz 2002; Zearfoss 2003).

In addition to heat shock, the expression of SINE elements (including Alu, B1, and B2 RNAs) increases during cellular stress such as viral infection (Singh 1985) in some cancer cells (Tang 2005) where they may similarly regulate global changes to gene expression. The ability of Alu and B2 RNA to bind directly to RNAP II provides a broad mechanism to repress transcription (Espinoza 2004; Mariner & Walters 2008). Nevertheless, there are specific exceptions to this global response where Alu or B2 RNAs are not found at activated promoters of genes undergoing induction, such as the heat shock genes (Mariner & Walters 2008). This additional hierarchy of regulation that exempts individual genes from the generalised repression also involves a long ncRNA, heat shock RNA-1 (HSR-1). It was argued that HSR-1 is present in mammalian cells in an inactive state, but upon stress is activated to induce the expression of heat shock genes (Shamovsky 2006). The authors found that this activation involves a conformational alteration to the structure of HSR-1 in response to rising temperatures, thereby permitting its interaction with the transcriptional activator HSF-1 that subsequently undergoes trimerisation and induces the expression of heat shock genes (Shamovsky 2006). In the broad sense, these examples illustrate a regulatory circuit nested within ncRNAs whereby Alu or B2 RNAs repress general gene expression, while other ncRNAs activate the expression of specific genes.

Long ncRNA transcribed by RNA polymerase III

Many of the ncRNAs that interact with general transcription factors or RNAP II itself (including 7SK, Alu and B1 and B2 RNAs) are transcribed by RNAP III,[33] thereby uncoupling the expression of these ncRNAs from the RNAP II transcriptional reaction they regulate. RNAP III also transcribes a number of additional novel ncRNAs, such as BC2, BC200 and some microRNAs and snoRNAs, in addition to the highly expressed infrastructural 'housekeeping' ncRNA genes such as tRNAs, 5S rRNAs and snRNAs.[34] The existence of an RNAP III-dependent ncRNA transcriptome that regulates its RNAP II-dependent counterpart was supported by a recent study that described a novel set of ncRNAs transcribed by RNAP III with sequence homology to protein-coding genes. This prompted the authors to posit a 'cogene/gene' functional regulatory network,[35] showing that one of these ncRNAs, 21A, regulates the expression its antisense partner gene, CENP-F in trans.

35.4.2   Long non-coding RNAs in post-transcriptional regulation

In addition to regulating transcription, ncRNAs also control various aspects of post-transcriptional mRNA processing. Similar to small regulatory RNAs such as microRNAs and snoRNAs, these functions often involve complementary base pairing with the target mRNA. The formation of RNA duplexes between complementary ncRNA and mRNA may mask key elements within the mRNA required to bind trans-acting factors, potentially affecting any step in post-transcriptional gene expression including pre-mRNA processing and splicing, transport, translation, and degradation.

Long ncRNAs in splicing

The splicing of mRNA can induce its translation and functionally diversify the repertoire of proteins it encodes. The Zeb2 mRNA, which has a particularly long 5'UTR, requires the retention of a 5'UTR intron that contains an internal ribosome entry site for efficient translation.[36] However, retention of the intron is dependent on the expression of an antisense transcript that complements the intronic 5' splice site.[37] Therefore, the ectopic expression of the antisense transcript represses splicing and induces translation of the Zeb2 mRNA during mesenchymal development. Likewise, the expression of an overlapping antisense Rev-ErbAα2 transcript controls the alternative splicing of the thyroid hormone receptor ErbAα2 mRNA to form two antagonistic isoforms.[38]

Long ncRNAs in translation

NcRNA may also apply additional regulatory pressures during translation, a property particularly exploited in neurons where the dendritic or axonal translation of mRNA in response to synaptic activity contributes to changes in synaptic plasticity and the remodelling of neuronal networks. The RNAP III transcribed BC1 and BC200 ncRNAs, that previously derived from tRNAs, are expressed in the mouse and human central nervous system, respectively.[39][40] BC1 expression is induced in response to synaptic activity and synaptogenesis and is specifically targeted to dendrites in neurons.[41] Sequence complementarity between BC1 and regions of various neuron-specific mRNAs also suggest a role for BC1 in targeted translational repression.[42] Indeed, it was recently shown that BC1 is associated with translational repression in dendrites to control the efficiency of dopamine D2 receptor-mediated transmission in the striatum[43] and BC1 RNA-deleted mice exhibit behavioural changes with reduced exploration and increased anxiety.[44]

Long ncRNAs in siRNA-directed gene regulation

In addition to masking key elements within single-stranded RNA, the formation of double-stranded RNA duplexes can also provide a substrate for the generation of endogenous siRNAs (endo-siRNAs) in Drosophila and mouse oocytes.[45] The annealing of complementary sequences, such as antisense or repetitive regions between transcripts, forms an RNA duplex that may be processed by Dicer-2 into endo-siRNAs. Also, long ncRNAs that form extended intramolecular hairpins may be processed into siRNAs, compellingly illustrated by the esi-1 and esi-2 transcripts.[46] Endo-siRNAs generated from these transcripts seem particularly useful in suppressing the spread of mobile transposon elements within the genome in the germline. However, the generation of endo-siRNAs from antisense transcripts or pseudogenes may also silence the expression of their functional counterparts via RISC effector complexes, acting as an important node that integrates various modes of long and short RNA regulation, as exemplified by the Xist and Tsix (see above).[47]

35.4.3 Long ncRNAs in epigenetic regulation

Epigenetic modifications, including histone and DNA methylation, histone acetylation and sumoylation, affect many aspects of chromosomal biology, primarily including regulation of large numbers of genes by remodeling broad chromatin domains (Kiefer 2007; Mikkelsen 2007). While it has been known for some time that RNA is an integral component of chromatin (Nickerson 1989; Rodriguez-Campos 2007), it is only recently that we are beginning to appreciate the means by which RNA is involved in pathways of chromatin modification (Chen 2008; Rinn 2007; Sanchez-Elsner 2006).

In Drosophila, long ncRNAs induce the expression of the homeotic gene, Ubx, by recruiting and directing the chromatin modifying functions of the trithorax protein Ash1 to Hox regulatory elements (Sanchez-Elsner 2006). Similar models have been proposed in mammals, where strong epigenetic mechanisms are thought to underlie the embryonic expression profiles of the Hox genes that persist throughout human development (Mazo 2007; Rinn 2007). Indeed, the human Hox genes are associated with hundreds of ncRNAs that are sequentially expressed along both the spatial and temporal axes of human development and define chromatin domains of differential histone methylation and RNA polymerase accessibility (Rinn 2007). One ncRNA, termed HOTAIR, that originates from the HOXC locus represses transcription across 40 kb of the HOXD locus by altering chromatin trimethylation state. HOTAIR is thought to achieve this by directing the action of Polycomb chromatin remodeling complexes in trans to govern the cells' epigenetic state and subsequent gene expression. Components of the Polycomb complex, including Suz12, EZH2 and EED, contain RNA binding domains that may potentially bind HOTAIR and probably other similar ncRNAs (Denisenko 1998; Katayama 2005). This example nicely illustrates a broader theme whereby ncRNAs recruit the function of a generic suite of chromatin modifying proteins to specific genomic loci, underscoring the complexity of recently published genomic maps (Mikkelsen 2007). Indeed, the prevalence of long ncRNAs associated with protein coding genes may contribute to localised patterns of chromatin modifications that regulate gene expression during development. For example, the majority of protein-coding genes have antisense partners, including many tumour suppressor genes that are frequently silenced by epigenetic mechanisms in cancer (Yu 2008). A recent study observed an inverse expression profile of the p15 gene and an antisense ncRNA in leukaemia (Yu 2008). A detailed analysis showed the p15 antisense ncRNA (CDKN2BAS) was able to induce changes to heterochromatin and DNA methylation status of p15 by an unknown mechanism, thereby regulating p15 expression (Yu 2008). Therefore, misexpression of the associated antisense ncRNAs may subsequently silence the tumour suppressor gene contributing towards oncogenesis.

Imprinting

Many emergent themes of ncRNA-directed chromatin modification were first apparent within the phenomenon of imprinting, whereby only one allele of a gene is expressed from either the maternal or the paternal chromosome. In general, imprinted genes are clustered together on chromosomes, suggesting the imprinting mechanism acts upon local chromosome domains rather than individual genes. These clusters are also often associated with long ncRNAs whose expression is correlated with the repression of the linked protein-coding gene on the same allele (Pauler 2007). Indeed, detailed analysis has revealed a crucial role for the ncRNAs Kcnqot1 and Igf2r/Air in directing imprinting (Braidotti 2004).

Almost all the genes at the Kcnq1 loci are maternally inherited, except the paternally expressed antisense ncRNA Kcnqot1 (Mitsuya 1999). Transgenic mice with truncated Kcnq1ot fail to silence the adjacent genes, suggesting that Kcnqot1 is crucial to the imprinting of genes on the paternal chromosome (Mancini-Dinardo 2006). It appears that Kcnqot1 is able to direct the trimethylation of lysine 9 (H3K9me3) and 27 of histone 3 (H3K27me3) to an imprinting centre that overlaps the Kcnqot1 promoter and actually resides within a Kcnq1 sense exon (Umlauf 2004). Similar to HOTAIR (see above), Eed-Ezh2 Polycomb complexes are recruited to the Kcnq1 loci paternal chromosome, possibly by Kcnqot1, where they may mediate gene silencing through repressive histone methylation (Umlauf 2004). A differentially methylated imprinting centre also overlaps the promoter of a long antisense ncRNA Air that is responsible for the silencing of neighbouring genes at the Igf2r locus on the paternal chromosome (Sleutels 2002; Zwart 2001). The presence of allele-specific histone methylation at the Igf2r locus suggests Air also mediates silencing via chromatin modification (Fournier 2002).

Xist and X-chromosome inactivation

The inactivation of a X-chromosome in female placental mammals is directed by one of the earliest and best character-ized long ncRNAs, Xist (Wutz 2007). The expression of Xist from the future inactive X-chromosome, and its subsequent coating of the inactive X-chromosome, occurs during early embryonic stem cell differentiation. Xist expression is fol-lowed by irreversible layers of chromatin modifications that include the loss of the histone (H3K9) acetylation and H3K4 methylation that are associated with active chromatin, and the induction of repressive chromatin modifications including H4 hypoacetylation, H3K27 trimethylation (Wutz 2007), H3K9 hypermethylation and H4K20 monomethylation as well as H2AK119 monoubiquitylation. These modifications coincide with the transcriptional silencing of the X-linked genes (Morey 2004). Xist RNA also localises the histone variant macroH2A to the inactive X–chromosome (Costanzi 1998). There are additional ncRNAs that are also present at the Xist loci, including an antisense transcript Tsix, which is ex-pressed from the future active chromosome and able to repress Xist expression by the generation of endogenous siRNA (Ogawa 2008). Together these ncRNAs ensure that only one X-chromosome is active in female mammals.

Telomeric non-coding RNAs

Telomeres form the terminal region of mammalian chromosomes and are essential for stability and aging and play central roles in diseases such as cancer.[48] Telomeres have been long considered transcriptionally inert DNA-protein complexes until it was recently shown that telomeric repeats may be transcribed as telomeric RNAs (TelRNAs)[49] or telomeric repeat-containing RNAs.[50] These ncRNAs are heterogeneous in length, transcribed from several sub-telomeric loci and physically localise to telomeres. Their association with chromatin, which suggests an involvement in regulating telom-ere specific heterochromatin modifications, is repressed by SMG proteins that protect chromosome ends from telomere loss.[51] In addition, TelRNAs block telomerase activity in vitro and may therefore regulate telomerase activity.[52] Al-though early, these studies suggest an involvement for telomeric ncRNAs in various aspects of telomere biology.

35.5   Long non-coding RNAs in aging and disease

Recent recognition that long ncRNAs function in various aspects of cell biology has focused increasing attention on their potential to contribute towards disease etiology. A handful of studies have implicated long ncRNAs in a variety of disease states and support an involvement and co-operation in neurological disease and oncogenesis.

The first published report of an alteration in lncRNA abundance in aging and human neurological disease was provided by Lukiw et al.[53] in a study using short post-mortem interval Alzheimer's disease and non-Alzheimer's dementia (NAD) tissues; this early work was based on the prior identification of a primate brain-specific cytoplasmic transcript of the Alu repeat family by Watson and Sutcliffe in 1987 known as BC200 (brain, cytoplasmic, 200 nucleotide).[54]

While many association studies have identified long ncRNAs that are aberrantly expressed in disease states, we have little understanding of their contribution within disease etiology. Expression analyses that compare tumor cells and normal cells have revealed changes in the expression of ncRNAs in several forms of cancer. For example, in prostate tumours, one of two overexpressed ncRNAs, PCGEM1, is correlated with increased proliferation and colony formation suggesting an involvement in regulating cell growth.[55] MALAT1 (also known as NEAT2) was originally identified as an abundantly expressed ncRNA that is upregulated during metastasis of early-stage non-small cell lung cancer and its overexpression is an early prognostic marker for poor patient survival rates.[56] More recently, the highly conserved mouse homologue of MALAT1 was found to be highly expressed in hepatocellular carcinoma.[57] Intronic antisense ncRNAs with expression correlated to the degree of tumor differentiation in prostate cancer samples have also been reported.[58] Despite a number of long ncRNAs having aberrant expression in cancer, their function and potential role in tumourogenesis is relatively unknown. For example, the ncRNAs HIS-1 and BIC have been implicated in oncogenesis and growth control, but their function in normal cells is unknown.[59][60] In addition to cancer, ncRNAs also exhibit aberrant expression in other disease states. Overexpression of PRINS is associated with psoriasis susceptibility, with PRINS expression being elevated in the uninvolved epidermis of psoriatic patients compared with both psoriatic lesions and healthy epidermis.[61]

Genome-wide profiling revealed that many transcribed non-coding ultraconserved regions exhibit distinct profiles in various human cancer states.[62] An analysis of chronic lymphocytic leukaemia, colorectal carcinoma and hepatocellular carcinoma found that all three cancers exhibited aberrant expression profiles for ultraconserved ncRNAs relative to normal cells. Further analysis of one ultraconserved ncRNA suggested it behaved like an oncogene by mitigating apoptosis and subsequently expanding the number of malignant cells in colorectal cancers.[63] Many of these transcribed ultraconserved sites that exhibit distinct signatures in cancer are found at fragile sites and genomic regions associated with cancer. It seems likely that the aberrant expression of these ultraconserved ncRNAs within malignant processes results from important functions they fulfil in normal human development.

Recently, a number of association studies examining single nucleotide polymorphisms (SNPs) associated with disease states have been mapped to long ncRNAs. For example, SNPs that identified a susceptibility locus for myocardial infarction mapped to a long ncRNA, MIAT (myocardial infarction associated transcript).[64] Likewise, genome-wide association studies identified a region associated with coronary artery disease[65] that encompassed a long ncRNA, ANRIL.[66] ANRIL is expressed in tissues and cell types affected by atherosclerosis[67][68] and its altered expression is associated with a high-risk haplotype for coronary artery disease.[69][70]

The complexity of the transcriptome, and our evolving understanding of its structure may inform a reinterpretation of the functional basis for many natural polymorphisms associated with disease states. Many SNPs associated with certain disease conditions are found within non-coding regions and the complex networks of non-coding transcription within these regions make it particularly difficult to elucidate the functional effects of polymorphisms. For example, a SNP both within the truncated form of ZFAT and the promoter of an antisense transcript increases the expression of ZFAT not through increasing the mRNA stability, but rather by repressing the expression of the antisense transcript.[71]

The ability of long ncRNAs to regulate associated protein-coding genes may contribute to disease if misexpression of a long ncRNA deregulates a protein coding gene with clinical significance. In similar manner, an antisense long ncRNA that regulates the expression of the sense BACE1 gene, a crucial enzyme in Alzheimer's disease etiology, exhibits elevated expression in several regions of the brain in individuals with Alzheimer's disease[72] Alteration of the expression of ncRNAs may also mediate changes at an epigenetic level to affect gene expression and contribute to disease aetiology. For example, the induction of an antisense transcript by a genetic mutation led to DNA methylation and silencing of sense genes, causing β-thalassemia in a patient.[73]

35.6   Long intergenic non-coding RNAs (lincRNA)

"Intergenic" refers to long non-coding RNAs that are transcribed from non-coding DNA sequences between protein-coding genes.[74][75] A 2013 study identified tens of thousands of human lincRNAs.[76] Some lincRNAs attach to messenger RNA to block protein production. At least 26 different lincRNAs are needed to prevent an embryonic stem cell from differentiating. Additionally, it was proposed to classify intergenic RNA domains of at least 50 kb in length as "very long intergenic non-coding" (vlincRNAs) regions.[77]

35.7   See also

- Ribozyme

- Group II intron

- Ribosomal RNA

- OLE RNA

- LncRNAdb

35.8 References

[1] Perkel 2013

[2] Ma 2013

[3] Kapranov 2007

[4] Carninci 2005

[5] Carninci 2005

[6] Cheng 2005

[7] Derrien 2012

[8] Carninci 2005

[9] Kapranov 2007

[10] Birney 2007

[11] Carninci 2005

[12] Derrien 2012

[13] Derrien 2012

[14] Bentwich 2005

[15] Brosius 2005

[16] Struhl 2007

[17] Nesterova 2001

[18] Pang 2006

[19] Siepel 2005

[20] Ponjavic 2007

[21] Pang 2006

[22] Pollard 2006

[23] Pollard 2006

[24] Pollard 2006

[25] Pang 2006

[26] Smith 2004

[27] Torarinsson 2006

[28] Mercer, T. R.; Dinger, M. E.; Mattick, J. S. (2009). ""Long non-coding RNAs: Insights into functions"". *Nature Reviews Genetics* **10** (3): 155–159. doi:10.1038/nrg2521. PMID 19188922.

[29] Dinger, M. E.; Amaral, P. P.; Mercer, T. R.; Mattick, J. S. (2009). ""Pervasive transcription of the eukaryotic genome: Functional indices and conceptual implications"". *Briefings in Functional Genomics and Proteomics* **8** (6): 407–423. doi:10.1093/bfgp/elp038. PMID 19770204.

[30] Amaral, P. P.; Clark, M. B.; Gascoigne, D. K.; Dinger, M. E.; Mattick, J. S. (2010). ""LncRNAdb: A reference database for long noncoding RNAs"". *Nucleic Acids Research* **39** ((Database issue)): D146–D151. doi:10.1093/nar/gkq1138. PMC 3013714. PMID 21112873.

[31] Smith, JE; Alvarez-Dominguez, JR; Kline, N; Huynh, NJ; Geisler, S; Hu, W; Coller, J; Baker, KE (Jun 26, 2014). "Translation of Small Open Reading Frames within Unannotated RNA Transcripts in Saccharomyces cerevisiae.". *Cell reports* **7** (6): 1858–66. doi:10.1016/j.celrep.2014.05.023. PMID 24931603.

[32] Halley, Paul; Kadakkuzha, Beena (2014). "Regulation of the apolipoprotein gene cluster by a long noncoding RNA.". *Cell Reports* **6** (1): 222–30. doi:10.1016/j.celrep.2013.12.015. PMID 24388749.

[33] (Dieci 2007

[34] Dieci 2007

[35] Pagano 2007

[36] Beltran 2008

[37] Beltran 2008

[38] (Munroe 1991

[39] Tiedge 1993

[40] Tiedge 1991)

[41] Muslimov 1998

[42] Wang 2005

[43] Centonze 2007

[44] Lewejohann 2004

[45] Golden 2008

[46] Czech 2008

[47] (Ogawa 2008

[48] Blasco 2007

[49] Schoeftner 2008

[50] Azzalin 2007

[51] Azzalin 2007

[52] Schoeftner 2008

[53] Lukiw WJ, Handley P, Wong L, Crapper McLachlan DR (Jun 1992). "BC200 RNA in normal human neocortex, non-Alzheimer dementia (NAD), and senile dementia of the Alzheimer type (AD)". *Neurochem Res.* **17** (6): 591–7. doi:10.1007/bf00968788. PMID 1603265.

[54] Watson JB, Sutcliffe JG (Sep 1987). "Primate brain-specific cytoplasmic transcript of the Alu repeat family". *Mol Cell Biol* **7** (9): 3324–7. PMID 2444875.

[55] Fu 2006

[56] Fu 2006

[57] Lin 2007

[58] Reis 2004

[59] Eis 2005

[60] Li 1997)

[61] Sonkoly 2005

[62] Calin 2007

[63] Calin 2007

[64] Ishii 2006

[65] McPherson 2007

[66] Pasmant 2007

[67] Broadbend 2008

[68] Jarinova 2009

[69] Jarinova 2009

[70] Liu 2009

[71] Shirasawa 2004

[72] Faghihi 2008

[73] Tufarelli 2003

[74] Hesman Saey 2011

[75] Rinn Lab lincRNA homepage

[76] Hangauer, Matthew J.; Vaughn, Ian W.; McManus, Michael T.; Rinn, John L. (20 June 2013). "Pervasive Transcription of the Human Genome Produces Thousands of Previously Unidentified Long Intergenic Noncoding RNAs". *PLoS Genetics* **9** (6): e1003569. doi:10.1371/journal.pgen.1003569.

[77] Laurent 2010

35.9 Bibliography

• Allen E, Xie Z, Gustafson AM, Sung GH, Spatafora JW, Carrington JC (December 2004). "Evolution of microRNA genes by inverted duplication of target gene sequences in Arabidopsis thaliana". *Nature Genetics* **36** (12): 1282–90. doi:10.1038/ng1478. PMID 15565108.

• Amaral, P. P.; Clark, M. B.; Gascoigne, D. K.; Dinger, M. E.; Mattick, J. S. (2010). "LncRNAdb: A reference database for long noncoding RNAs". *Nucleic Acids Research* **39** (Database issue): D146–D151. doi:10.1093/nar/gkq1138. PMC 3013714. PMID 21112873.

• Azzalin CM, Reichenbach P, Khoriauli L, Giulotto E, Lingner J (November 2007). "Telomeric repeat containing RNA and RNA surveillance factors at mammalian chromosome ends". *Science* **318** (5851): 798–801. Bibcode:2007Sci...318..798A. doi:10.1126/science.1147182. PMID 17916692.

• Beltran M, Puig I, Peña C, et al. (March 2008). "A natural antisense transcript regulates Zeb2/Sip1 gene expression during Snail1-induced epithelial-mesenchymal transition". *Genes & Development* **22** (6): 756–69. doi:10.1101/gad.455708. PMC 2275429. PMID 18347095.

• Bentwich I, Avniel A, Karov Y, et al. (July 2005). "Identification of hundreds of conserved and nonconserved human microRNAs". *Nature Genetics* **37** (7): 766–70. doi:10.1038/ng1590. PMID 15965474.

• Birney E, Stamatoyannopoulos JA, Dutta A, et al. (June 2007). "Identification and analysis of functional elements in 1% of the human genome by the ENCODE pilot project". *Nature* **447** (7146): 799–816. Bibcode:2007Natur.447..799B. doi:10.1038/nature05874. PMC 2212820. PMID 17571346.

- Blasco MA (October 2007). "Telomere length, stem cells and aging". *Nature Chemical Biology* **3** (10): 640–9. doi:10.1038/nchembio.2007.38. PMID 17876321.

- Braidotti G, Baubec T, Pauler F, et al. (2004). "The Air noncoding RNA: an imprinted cis-silencing transcript". *Cold Spring Harbor Symposia on Quantitative Biology* **69**: 55–66. doi:10.1101/sqb.2004.69.55. PMC 2847179. PMID 16117633.

- Broadbent HM, Peden JF, Lorkowski S, et al. (2008). "Susceptibility to coronary artery disease and diabetes is encoded by distinct, tightly linked SNPs in the ANRIL locus on chromosome 9p". *Human Molecular Genetics* **17** (6): 806–14. doi:10.1093/hmg/ddm352. PMID 18048406.

- Brosius J (May 2005). "Waste not, want not--transcript excess in multicellular eukaryotes". *Trends in Genetics* **21** (5): 287–8. doi:10.1016/j.tig.2005.02.014. PMID 15851065.

- Calin GA, Liu CG, Ferracin M, et al. (September 2007). "Ultraconserved regions encoding ncRNAs are altered in human leukemias and carcinomas". *Cancer Cell* **12** (3): 215–29. doi:10.1016/j.ccr.2007.07.027. PMID 17785203.

- Carninci P, Kasukawa T, Katayama S, et al. (September 2005). "The transcriptional landscape of the mammalian genome". *Science* **309** (5740): 1559–63. Bibcode:2005Sci...309.1559F. doi:10.1126/science.1112014. PMID 16141072.

- Centonze D, Rossi S, Napoli I, et al. (August 2007). "The brain cytoplasmic RNA BC1 regulates dopamine D2 receptor-mediated transmission in the striatum". *The Journal of Neuroscience* **27** (33): 8885–92. doi:10.1523/JNEUROSCI.0548-07.2007. PMID 17699670.

- Chen X, Xu H, Yuan P, et al. (June 2008). "Integration of external signaling pathways with the core transcriptional network in embryonic stem cells". *Cell* **133** (6): 1106–17. doi:10.1016/j.cell.2008.04.043. PMID 18555785.

- Cheng J, Kapranov P, Drenkow J, et al. (May 2005). "Transcriptional maps of 10 human chromosomes at 5-nucleotide resolution". *Science* **308** (5725): 1149–54. Bibcode:2005Sci...308.1149C. doi:10.1126/science.1108625. PMID 15790807.

- Costanzi C, Pehrson JR (June 1998). "Histone macroH2A1 is concentrated in the inactive X chromosome of female mammals". *Nature* **393** (6685): 599–601. Bibcode:1998Natur.393..599C. doi:10.1038/31275. PMID 9634239.

- Czech B, Malone CD, Zhou R, et al. (June 2008). "An endogenous small interfering RNA pathway in Drosophila". *Nature* **453** (7196): 798–802. Bibcode:2008Natur.453..798C. doi:10.1038/nature07007. PMC 2895258. PMID 18463631.

- Denisenko O, Shnyreva M, Suzuki H, Bomsztyk K (1 October 1998). "Point mutations in the WD40 domain of Eed block its interaction with Ezh2". *Molecular and Cellular Biology* **18** (10): 5634–42. PMC 109149. PMID 9742080.

- Derrien, T.; Johnson, R.; Bussotti, G.; Tanzer, A.; Djebali, S.; Tilgner, H.; Guernec, G.; Martin, D.; Merkel, A.; Knowles, D. G.; Lagarde, J.; Veeravalli, L.; Ruan, X.; Ruan, Y.; Lassmann, T.; Carninci, P.; Brown, J. B.; Lipovich, L.; Gonzalez, J. M.; Thomas, M.; Davis, C. A.; Shiekhattar, R.; Gingeras, T. R.; Hubbard, T. J.; Notredame, C.; Harrow, J.; Guigó, R. (2012). "The GENCODE v7 catalog of human long noncoding RNAs: Analysis of their gene structure, evolution, and expression". *Genome Research* **22** (9): 1775–1789. doi:10.1101/gr.132159.111. PMC 3431493. PMID 22955988.

- Dieci G, Fiorino G, Castelnuovo M, Teichmann M, Pagano A (December 2007). "The expanding RNA polymerase III transcriptome". *Trends in Genetics* **23** (12): 614–22. doi:10.1016/j.tig.2007.09.001. PMID 17977614.

- Dinger, M. E.; Amaral, P. P.; Mercer, T. R.; Mattick, J. S. (2009). "Pervasive transcription of the eukaryotic genome: Functional indices and conceptual implications". *Briefings in Functional Genomics and Proteomics* **8** (6): 407–423. doi:10.1093/bfgp/elp038. PMID 19770204.

• Eis PS, Tam W, Sun L, et al. (March 2005). "Accumulation of miR-155 and BIC RNA in human B cell lymphomas". *Proceedings of the National Academy of Sciences of the United States of America* **102** (10): 3627–32. Bibcode:2005PNAS..102.3627E. doi:10.1073/pnas.0500613102. PMC 552785. PMID 15738415.

• Espinoza CA, Allen TA, Hieb AR, Kugel JF, Goodrich JA (September 2004). "B2 RNA binds directly to RNA polymerase II to repress transcript synthesis". *Nature Structural & Molecular Biology* **11** (9): 822–9. doi:10.1038/nsmb812. PMID 15300239.

• Espinoza CA, Goodrich JA, Kugel JF (April 2007). "Characterization of the structure, function, and mechanism of B2 RNA, an ncRNA repressor of RNA polymerase II transcription". *RNA* **13** (4): 583–96. doi:10.1261/rna.310307. PMC 1831867. PMID 17307818.

• Faghihi MA, Modarresi F, Khalil AM, et al. (July 2008). "Expression of a noncoding RNA is elevated in Alzheimer's disease and drives rapid feed-forward regulation of beta-secretase". *Nature Medicine* **14** (7): 723–30. doi:10.1038/nm1784. PMC 2826895. PMID 18587408.

• Feng J, Bi C, Clark BS, Mady R, Shah P, Kohtz JD (June 2006). "The Evf-2 noncoding RNA is transcribed from the Dlx-5/6 ultraconserved region and functions as a Dlx-2 transcriptional coactivator". *Genes & Development* **20** (11): 1470–84. doi:10.1101/gad.1416106. PMC 1475760. PMID 16705037.

• Fournier C, Goto Y, Ballestar E, et al. (December 2002). "Allele-specific histone lysine methylation marks regulatory regions at imprinted mouse genes". *The EMBO Journal* **21** (23): 6560–70. doi:10.1093/emboj/cdf655. PMC 136958. PMID 12456662.

• Fu X, Ravindranath L, Tran N, Petrovics G, Srivastava S (March 2006). "Regulation of apoptosis by a prostate-specific and prostate cancer-associated noncoding gene, PCGEM1". *DNA and Cell Biology* **25** (3): 135–41. doi:10.1089/dna.2006.25.135. PMID 16569192.

• Golden DE, Gerbasi VR, Sontheimer EJ (August 2008). "An inside job for siRNAs". *Molecular Cell* **31** (3): 309–12. doi:10.1016/j.molcel.2008.07.008. PMC 2675693. PMID 18691963.

• Goodrich JA, Kugel JF (August 2006). "Non-coding-RNA regulators of RNA polymerase II transcription". *Nature Reviews Molecular Cell Biology* **7** (8): 612–6. doi:10.1038/nrm1946. PMID 16723972.

• Hesman Saey, Tina (17 December 2011). "Missing Lincs". *Science News* **180** (13): 22–25. doi:10.1002/scin.5591801327.

• Ishii N, Ozaki K, Sato H, et al. (2006). "Identification of a novel non-coding RNA, MIAT, that confers risk of myocardial infarction". *Journal of Human Genetics* **51** (12): 1087–99. doi:10.1007/s10038-006-0070-9. PMID 17066261.

• Jarinova O, Stewart AF, Roberts R, et al. (October 2009). "Functional analysis of the chromosome 9p21.3 coronary artery disease risk locus". *Arteriosclerosis, Thrombosis, and Vascular Biology* **29** (10): 1671–77. doi:10.1161/ATVBAHA.109.1 PMID 19592466.

• Kapranov P, Cheng J, Dike S, et al. (June 2007). "RNA maps reveal new RNA classes and a possible function for pervasive transcription". *Science* **316** (5830): 1484–8. Bibcode:2007Sci...316.1484K. doi:10.1126/science.1138341. PMID 17510325.

• Kapranov P, Willingham AT, Gingeras TR (June 2007). "Genome-wide transcription and the implications for genomic organization". *Nature Reviews. Genetics* **8** (6): 413–23. doi:10.1038/nrg2083. PMID 17486121.

• Kapranov P, St Laurent G, Raz T, et al. (2010). "The majority of total nuclear-encoded non-ribosomal RNA in a human cell is 'dark matter' un-annotated RNA". *BMC Biol.* **8**: 149. doi:10.1186/1741-7007-8-149. PMC 3022773. PMID 21176148. Erratum in: BMC Biol. 2011;9:86.

• Katayama S, Tomaru Y, Kasukawa T, et al. (September 2005). "Antisense transcription in the mammalian transcriptome". *Science* **309** (5740): 1564–6. Bibcode:2005Sci...309.1564R. doi:10.1126/science.1112009. PMID 16141073.

- Kiefer JC (April 2007). "Epigenetics in development". *Developmental Dynamics* **236** (4): 1144–56. doi:10.1002/dvdy.21094. PMID 17304537.

- Kim (February 2010). "Evidence for bacterial origin of heat shock RNA-1". *RNA* **16** (2): 274–279. doi:10.1261/rna.1879610. PMC 2811656. PMID 20040589.

- Kwek KY, Murphy S, Furger A, et al. (November 2002). "U1 snRNA associates with TFIIH and regulates transcriptional initiation". *Nature Structural Biology* **9** (11): 800–5. doi:10.1038/nsb862. PMID 12389039.

- Lander ES, Linton LM, Birren B, et al. (February 2001). "Initial sequencing and analysis of the human genome". *Nature* **409** (6822): 860–921. doi:10.1038/35057062. PMID 11237011.

- Lee JS, Burkholder GD, Latimer LJ, Haug BL, Braun RP (February 1987). "A monoclonal antibody to triplex DNA binds to eucaryotic chromosomes". *Nucleic Acids Research* **15** (3): 1047–61. doi:10.1093/nar/15.3.1047. PMC 340507. PMID 2434928.

- Lewejohann L, Skryabin BV, Sachser N, et al. (September 2004). "Role of a neuronal small non-messenger RNA: behavioural alterations in BC1 RNA-deleted mice". *Behavioural Brain Research* **154** (1): 273–89. doi:10.1016/j.bbr.2004.02.015. PMID 15302134.

- Li J, Witte DP, Van Dyke T, Askew DS (April 1997). "Expression of the putative proto-oncogene His-1 in normal and neoplastic tissues". *The American Journal of Pathology* **150** (4): 1297–305. PMC 1858164. PMID 9094986.

- Lin R, Maeda S, Liu C, Karin M, Edgington TS (February 2007). "A large noncoding RNA is a marker for murine hepatocellular carcinomas and a spectrum of human carcinomas". *Oncogene* **26** (6): 851–8. doi:10.1038/sj.onc.1209846. PMID 16878148.

- lincRNA homepage of the Rinn Lab

- Liu WM, Chu WM, Choudary PV, Schmid CW (May 1995). "Cell stress and translational inhibitors transiently increase the abundance of mammalian SINE transcripts". *Nucleic Acids Research* **23** (10): 1758–65. doi:10.1093/nar/23.10.1758. PMC 306933. PMID 7784180.

- Liu Y, Sanoff HK, Cho H, et al. (April 2009). "INK4/ARF transcript expression is associated with chromosome 9p21 variants linked to atherosclerosis". *PloS One* **4** (4): e5027. Bibcode:2009PLoSO...4.5027L. doi:10.1371/journal.pone.0005027. PMC 2660422. PMID 19343170.

- Mancini-Dinardo D, Steele SJ, Levorse JM, Ingram RS, Tilghman SM (May 2006). "Elongation of the Kcnq1ot1 transcript is required for genomic imprinting of neighboring genes". *Genes & Development* **20** (10): 1268–82. doi:10.1101/gad.1416906. PMC 1472902. PMID 16702402.

- Mariner PD, Walters RD, Espinoza CA, et al. (February 2008). "Human Alu RNA is a modular transacting repressor of mRNA transcription during heat shock". *Molecular Cell* **29** (4): 499–509. doi:10.1016/j.molcel.2007.12.013. PMID 18313387.

- Martianov I, Ramadass A, Serra Barros A, Chow N, Akoulitchev A (February 2007). "Repression of the human dihydrofolate reductase gene by a non-coding interfering transcript". *Nature* **445** (7128): 666–70. doi:10.1038/nature05519. PMID 17237763.

- Mattick JS (October 2003). "Challenging the dogma: the hidden layer of non-protein-coding RNAs in complex organisms". *BioEssays* **25** (10): 930–9. doi:10.1002/bies.10332. PMID 14505360.

- Mazo A, Hodgson JW, Petruk S, Sedkov Y, Brock HW (August 2007). "Transcriptional interference: an unexpected layer of complexity in gene regulation". *Journal of Cell Science* **120** (Pt 16): 2755–61. doi:10.1242/jcs.007633. PMID 17690303.

- McPherson R, Pertsemlidis A, Kavaslar N, et al. (May 2007). "A Common Allele on Chromosome 9 Associated with Coronary Heart Disease". *Science* **316** (5830): 1488–91. Bibcode:2007Sci...316.1488M. doi:10.1126/science.1142447. PMC 2711874. PMID 17478681.

- Mercer, T. R.; Dinger, M. E.; Mattick, J. S. (2009). "Long non-coding RNAs: Insights into functions". *Nature Reviews Genetics* **10** (3): 155–159. doi:10.1038/nrg2521. PMID 19188922.

- Mikkelsen TS, Ku M, Jaffe DB, et al. (August 2007). "Genome-wide maps of chromatin state in pluripotent and lineage-committed cells". *Nature* **448** (7153): 553–60. Bibcode:2007Natur.448..553M. doi:10.1038/nature06008. PMC 2921165. PMID 17603471.

- Mitsuya K, Meguro M, Lee MP, et al. (July 1999). "LIT1, an imprinted antisense RNA in the human KvLQT1 locus identified by screening for differentially expressed transcripts using monochromosomal hybrids". *Human Molecular Genetics* **8** (7): 1209–17. doi:10.1093/hmg/8.7.1209. PMID 10369866.

- Mohammad F, Pandey RR, Nagano T, et al. (June 2008). "Kcnq1ot1/Lit1 noncoding RNA mediates transcriptional silencing by targeting to the perinucleolar region". *Molecular and Cellular Biology* **28** (11): 3713–28. doi:10.1128/MCB.02263-07. PMC 2423283. PMID 18299392.

- Morey C, Navarro P, Debrand E, Avner P, Rougeulle C, Clerc P (February 2004). "The region 3' to Xist mediates X chromosome counting and H3 Lys-4 dimethylation within the Xist gene". *The EMBO Journal* **23** (3): 594–604. doi:10.1038/sj.emboj.7600071. PMC 1271805. PMID 14749728.

- Munroe SH, Lazar MA (25 November 1991). "Inhibition of c-erbA mRNA splicing by a naturally occurring antisense RNA". *The Journal of Biological Chemistry* **266** (33): 22083–6. PMID 1657988.

- Muslimov IA, Banker G, Brosius J, Tiedge H (June 1998). "Activity-dependent regulation of dendritic BC1 RNA in hippocampal neurons in culture". *The Journal of Cell Biology* **141** (7): 1601–11. doi:10.1083/jcb.141.7.1601. PMC 1828539. PMID 9647652.

- Nesterova TB, Barton SC, Surani MA, Brockdorff N (July 2001). "Loss of Xist imprinting in diploid parthenogenetic preimplantation embryos". *Developmental Biology* **235** (2): 343–50. doi:10.1006/dbio.2001.0295. PMID 11437441.

- Nickerson JA, Krochmalnic G, Wan KM, Penman S (January 1989). "Chromatin architecture and nuclear RNA". *Proceedings of the National Academy of Sciences of the United States of America* **86** (1): 177–81. Bibcode:1989PNAS...86..177N. doi:10.1073/pnas.86.1.177. PMC 286427. PMID 2911567.

- Ogawa Y, Sun BK, Lee JT (June 2008). "Intersection of the RNA interference and X-inactivation pathways". *Science* **320** (5881): 1336–41. Bibcode:2008Sci...320.1336O. doi:10.1126/science.1157676. PMC 2584363. PMID 18535243.

- Pagano JM, Farley BM, McCoig LM, Ryder SP (March 2007). "Molecular basis of RNA recognition by the embryonic polarity determinant MEX-5". *The Journal of Biological Chemistry* **282** (12): 8883–94. doi:10.1074/jbc.M700079200. PMID 17264081.

- Pang KC, Frith MC, Mattick JS (January 2006). "Rapid evolution of noncoding RNAs: lack of conservation does not mean lack of function". *Trends in Genetics* **22** (1): 1–5. doi:10.1016/j.tig.2005.10.003. PMID 16290135.

- Panganiban G, Rubenstein JL (1 October 2002). "Developmental functions of the Distal-less/Dlx homeobox genes". *Development* **129** (19): 4371–86. PMID 12223397.

- Pasmant E, Laurendeau I, Héron D, Vidaud M, Vidaud D, Bièche I (April 2007). "Characterization of a germline deletion, including the entire INK4/ARF locus, in a melanoma-neural system tumor family: identification of ANRIL, an antisense noncoding RNA whose expression coclusters with ARF". *Cancer Research* **67** (8): 3963–9. doi:10.1158/0008-5472.CAN-06-2004. PMID 17440112.

- Pauler FM, Koerner MV, Barlow DP (June 2007). "Silencing by imprinted noncoding RNAs: is transcription the answer?". *Trends in Genetics* **23** (6): 284–92. doi:10.1016/j.tig.2007.03.018. PMC 2847181. PMID 17445943.

- Pennacchio LA, Ahituv N, Moses AM; et al. (November 2006). "In vivo enhancer analysis of human conserved non-coding sequences". *Nature* **444** (7118): 499–502. Bibcode:2006Natur.444..499P. doi:10.1038/nature05295. PMID 17086198.

- Perkel, Jeffrey M. (2013). "Visiting "Noncodarnia"". *BioTechniques* (paper) **54** (6): 301–304. doi:10.2144/000114037. PMID 23750541. "We're calling long noncoding RNAs a class, when actually the only definition is that they are longer than 200 bp," says Ana Marques, a Research Fellow at the University of Oxford who uses evolutionary approaches to understand lncRNA function.

- Pibouin L, Villaudy J, Ferbus D, et al. (February 2002). "Cloning of the mRNA of overexpression in colon carcinoma-1: a sequence overexpressed in a subset of colon carcinomas". *Cancer Genetics and Cytogenetics* **133** (1): 55–60. doi:10.1016/S0165-4608(01)00634-3. PMID 11890990.

- Pollard KS, Salama SR, King B, et al. (October 2006). "Forces shaping the fastest evolving regions in the human genome". *PLoS Genetics* **2** (10): e168. doi:10.1371/journal.pgen.0020168. PMC 1599772. PMID 17040131.

- Pollard KS, Salama SR, Lambert N, et al. (September 2006). "An RNA gene expressed during cortical development evolved rapidly in humans". *Nature* **443** (7108): 167–72. Bibcode:2006Natur.443..167P. doi:10.1038/nature05113. PMID 16915236.

- Ponjavic J, Ponting CP, Lunter G (May 2007). "Functionality or transcriptional noise? Evidence for selection within long noncoding RNAs". *Genome Research* **17** (5): 556–65. doi:10.1101/gr.6036807. PMC 1855172. PMID 17387145.

- Reinius, B.; Shi, C.; Hengshuo, L.; Sandhu, K.; Radomska, K. J.; Rosen, G. D.; Lu, L.; Kullander, K.; Williams, R. W.; Jazin, E. (2010). "Female-biased expression of long non-coding RNAs in domains that escape X-inactivation in mouse". *BMC Genomics* **11**: 614. doi:10.1186/1471-2164-11-614. PMC 3091755. PMID 21047393.

- Reis EM, Nakaya HI, Louro R, et al. (August 2004). "Antisense intronic non-coding RNA levels correlate to the degree of tumor differentiation in prostate cancer". *Oncogene* **23** (39): 6684–92. doi:10.1038/sj.onc.1207880. PMID 15221013.

- Rinn JL, Kertesz M, Wang JK, et al. (June 2007). "Functional demarcation of active and silent chromatin domains in human HOX loci by noncoding RNAs". *Cell* **129** (7): 1311–23. doi:10.1016/j.cell.2007.05.022. PMC 2084369. PMID 17604720.

- Rodríguez-Campos A, Azorín F (2007). "RNA is an integral component of chromatin that contributes to its structural organization". *PLoS ONE* **2** (11): e1182. Bibcode:2007PLoSO...2.1182R. doi:10.1371/journal.pone.0001182. PMC 2063516. PMID 18000552.

- Sanchez-Elsner T, Gou D, Kremmer E, Sauer F (February 2006). "Noncoding RNAs of trithorax response elements recruit Drosophila Ash1 to Ultrabithorax". *Science* **311** (5764): 1118–23. Bibcode:2006Sci...311.1118S. doi:10.1126/science.1117705. PMID 16497925.

- Schoeftner S, Blasco MA (February 2008). "Developmentally regulated transcription of mammalian telomeres by DNA-dependent RNA polymerase II". *Nature Cell Biology* **10** (2): 228–36. doi:10.1038/ncb1685. PMID 18157120.

- Shamovsky I, Nudler E (October 2006). "Gene control by large noncoding RNAs". *Science's STKE : Signal Transduction Knowledge Environment* **2006** (355): pe40. doi:10.1126/stke.3552006pe40. PMID 17018852.

- Shamovsky I, Nudler E (February 2008). "Modular RNA heats up". *Molecular Cell* **29** (4): 415–7. doi:10.1016/j.molcel.2008.02.001. PMID 18313380.

- Shirasawa S, Harada H, Furugaki K, et al. (October 2004). "SNPs in the promoter of a B cell-specific antisense transcript, SAS-ZFAT, determine susceptibility to autoimmune thyroid disease". *Human Molecular Genetics* **13** (19): 2221–31. doi:10.1093/hmg/ddh245. PMID 15294872.

- Siepel A, Bejerano G, Pedersen JS, et al. (August 2005). "Evolutionarily conserved elements in vertebrate, insect, worm, and yeast genomes". *Genome Research* **15** (8): 1034–50. doi:10.1101/gr.3715005. PMC 1182216. PMID 16024819.

- Singh K, Carey M, Saragosti S, Botchan M (1985). "Expression of enhanced levels of small RNA polymerase III transcripts encoded by the B2 repeats in simian virus 40-transformed mouse cells". *Nature* **314** (6011): 553–6. Bibcode:1985Natur.314..553S. doi:10.1038/314553a0. PMID 2581137.

- Sleutels F, Zwart R, Barlow DP (February 2002). "The non-coding Air RNA is required for silencing autosomal imprinted genes". *Nature* **415** (6873): 810–3. doi:10.1038/415810a. PMID 11845212.

- Smith NG, Brandström M, Ellegren H (November 2004). "Evidence for turnover of functional noncoding DNA in mammalian genome evolution". *Genomics* **84** (5): 806–13. doi:10.1016/j.ygeno.2004.07.012. PMID 15475259.

- Sonkoly E, Bata-Csorgo Z, Pivarcsi A, et al. (June 2005). "Identification and characterization of a novel, psoriasis susceptibility-related noncoding RNA gene, PRINS". *The Journal of Biological Chemistry* **280** (25): 24159–67. doi:10.1074/jbc.M501704200. PMID 15855153.

- Struhl K (February 2007). "Transcriptional noise and the fidelity of initiation by RNA polymerase II". *Nature Structural & Molecular Biology* **14** (2): 103–5. doi:10.1038/nsmb0207-103. PMID 17277804.

- Tang RB, Wang HY, Lu HY, et al. (February 2005). "Increased level of polymerase III transcribed Alu RNA in hepatocellular carcinoma tissue". *Molecular Carcinogenesis* **42** (2): 93–6. doi:10.1002/mc.20057. PMID 15593371.

- Tiedge H, Chen W, Brosius J (1 June 1993). "Primary structure, neural-specific expression, and dendritic location of human BC200 RNA". *Journal of Neuroscience* **13** (6): 2382–90. PMID 7684772.

- Tiedge H, Fremeau RT, Weinstock PH, Arancio O, Brosius J (March 1991). "Dendritic location of neural BC1 RNA". *Proceedings of the National Academy of Sciences of the United States of America* **88** (6): 2093–7. Bibcode:1991PNAS...88 doi:10.1073/pnas.88.6.2093. PMC 51175. PMID 1706516.

- Torarinsson E, Sawera M, Havgaard JH, Fredholm M, Gorodkin J (July 2006). "Thousands of corresponding human and mouse genomic regions unalignable in primary sequence contain common RNA structure". *Genome Research* **16** (7): 885–9. doi:10.1101/gr.5226606. PMC 1484455. PMID 16751343.

- Tufarelli C, Stanley JA, Garrick D, et al. (June 2003). "Transcription of antisense RNA leading to gene silencing and methylation as a novel cause of human genetic disease". *Nature Genetics* **34** (2): 157–65. doi:10.1038/ng1157. PMID 12730694.

- Umlauf D, Goto Y, Cao R, et al. (December 2004). "Imprinting along the Kcnq1 domain on mouse chromosome 7 involves repressive histone methylation and recruitment of Polycomb group complexes". *Nature Genetics* **36** (12): 1296–300. doi:10.1038/ng1467. PMID 15516932.

- Visel A, Prabhakar S, Akiyama JA, et al. (February 2008). "Ultraconservation identifies a small subset of extremely constrained developmental enhancers". *Nature Genetics* **40** (2): 158–60. doi:10.1038/ng.2007.55. PMC 2647775. PMID 18176564.

- Wang H, Iacoangeli A, Lin D, et al. (December 2005). "Dendritic BC1 RNA in translational control mechanisms". *The Journal of Cell Biology* **171** (5): 811–21. doi:10.1083/jcb.200506006. PMC 1828541. PMID 16330711.

- Wang X, Arai S, Song X, et al. (July 2008). "Induced ncRNAs allosterically modify RNA-binding proteins in cis to inhibit transcription". *Nature* **454** (7200): 126–30. Bibcode:2008Natur.454..126W. doi:10.1038/nature06992. PMC 2823488. PMID 18509338.

- Waterston RH, Lindblad-Toh K, Birney E, et al. (December 2002). "Initial sequencing and comparative analysis of the mouse genome". *Nature* **420** (6915): 520–62. Bibcode:2002Natur.420..520W. doi:10.1038/nature01262. PMID 12466850.

- Wutz A, Gribnau J (October 2007). "X inactivation Xplained". *Current Opinion in Genetics & Development* **17** (5): 387–93. doi:10.1016/j.gde.2007.08.001. PMID 17869504.

- Wutz A, Rasmussen TP, Jaenisch R (February 2002). "Chromosomal silencing and localization are mediated by different domains of Xist RNA". *Nature Genetics* **30** (2): 167–74. doi:10.1038/ng820. PMID 11780141.

- Yang S, Tutton S, Pierce E, Yoon K (November 2001). "Specific double-stranded RNA interference in undifferentiated mouse embryonic stem cells". *Molecular and Cellular Biology* **21** (22): 7807–16. doi:10.1128/MCB.21.22.7807-7816.2001. PMC 99950. PMID 11604515.

- Yik JH, Chen R, Nishimura R, Jennings JL, Link AJ, Zhou Q (October 2003). "Inhibition of P-TEFb (CDK9/Cyclin T) kinase and RNA polymerase II transcription by the coordinated actions of HEXIM1 and 7SK snRNA". *Molecular Cell* **12** (4): 971–82. doi:10.1016/S1097-2765(03)00388-5. PMID 14580347.

- Yu W, Gius D, Onyango P, et al. (January 2008). "Epigenetic silencing of tumour suppressor gene p15 by its antisense RNA". *Nature* **451** (7175): 202–6. Bibcode:2008Natur.451..202Y. doi:10.1038/nature06468. PMC 2743558. PMID 18185590.

- Zearfoss NR, Chan AP, Kloc M, Allen LH, Etkin LD (April 2003). "Identification of new Xlsirt family members in the Xenopus laevis oocyte". *Mechanisms of Development* **120** (4): 503–9. doi:10.1016/S0925-4773(02)00459-8. PMID 12676327.

- Zwart R, Sleutels F, Wutz A, Schinkel AH, Barlow DP (September 2001). "Bidirectional action of the Igf2r imprint control element on upstream and downstream imprinted genes". *Genes & Development* **15** (18): 2361–6. doi:10.1101/gad.206201. PMC 312779. PMID 11562346.

35.10 External links

- **LncRNABase database**: is designed for decoding miRNA-lncRNA(lncRNAs, pseudogenes, circRNAs) and miRNA-ceRNA interaction networks from 108 CLIP-Seq datasets and provides **Pan-Cancer(14 cancer types with >6000 tumor samples) interaction networks** of lncRNAs, miRNAs, ceRNAs, mRNAs and RNA-Binding Proteins.

- **ChIPBase database**: decodes the transcriptional regulation of lncRNAs from **ChIP-Seq** and provides **expression profiles** of lncRNAs (lincRNAs) from **RNA-Seq data across 22 tissues**.

- **lncRNAdb database**: providing comprehensive annotations of functional long non-coding RNAs (lncRNAs).

- **Long RNA Synthesis**: chemical long RNA synthesis and long RNA synthesis by transcription.

- **MONOCLdb**: provides the annotations and expression profiles of several thousands mouse long non-coding RNAs (lncRNAs) involved in Influenza and SARS-CoV infections.

- **HGNC**: provides approved gene names for human long non-coding RNAs (lncRNAs) - Wright, MW (Apr 9, 2014). "A short guide to long non-coding RNA gene nomenclature.". *Human genomics* **8**: 7. doi:10.1186/1479-7364-8-7. PMC 4021045. PMID 24716852.

- **LncRNAWiki**: a wiki-based database for management of human long non-coding RNAs (lncRNAs).

- **LNCipedia**: a database for annotated human lncRNA sequences.

- **LncRNA2Target database**: providing the manually curated target genes of long non-coding RNAs (lncRNAs).

Chapter 36

Alternative splicing

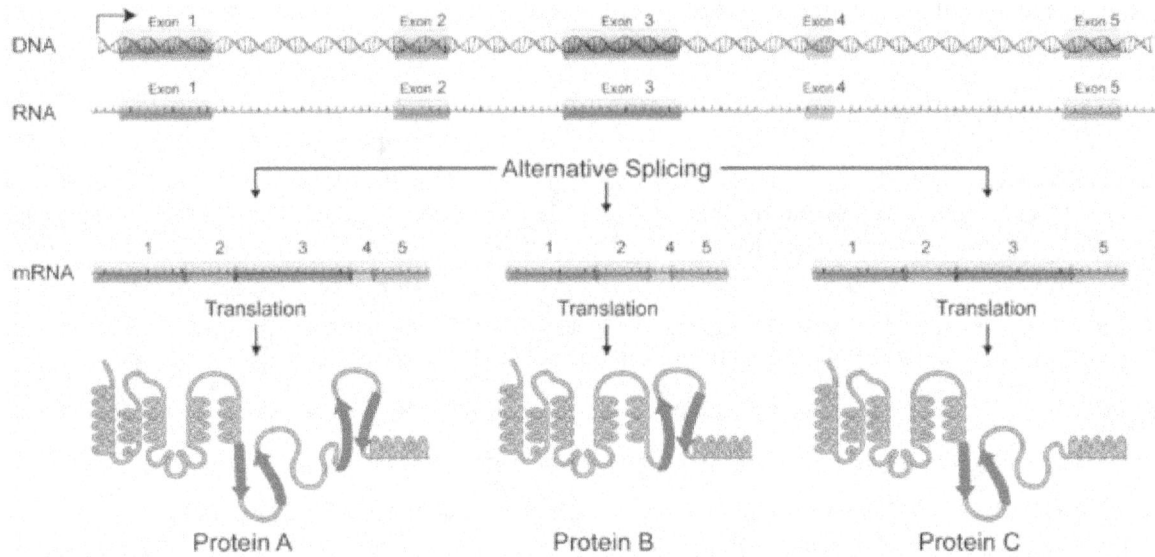

Alternative splicing produces three protein isoforms.

Alternative splicing is a regulated process during gene expression that results in a single gene coding for multiple proteins. In this process, particular exons of a gene may be included within or excluded from the final, processed messenger RNA (mRNA) produced from that gene.[1] Consequently the proteins translated from alternatively spliced mRNAs will contain differences in their amino acid sequence and, often, in their biological functions (see Figure). Notably, alternative splicing allows the human genome to direct the synthesis of many more proteins than would be expected from its 20,000 protein-coding genes. Alternative splicing is sometimes termed *differential splicing*.

Alternative splicing occurs as a normal phenomenon in eukaryotes, where it greatly increases the biodiversity of proteins that can be encoded by the genome;[1] in humans, ~95% of multi-exonic genes are alternatively spliced.[2] There are numerous modes of alternative splicing observed, of which the most common is exon skipping. In this mode, a particular exon may be included in mRNAs under some conditions or in particular tissues, and omitted from the mRNA in others.[1]

The production of alternatively spliced mRNAs is regulated by a system of trans-acting proteins that bind to cis-acting sites on the primary transcript itself. Such proteins include splicing activators that promote the usage of a particular splice site, and splicing repressors that reduce the usage of a particular site. Mechanisms of alternative splicing are highly variable, and new examples are constantly being found, particularly through the use of high-throughput techniques. Researchers hope to fully elucidate the regulatory systems involved in splicing, so that alternative splicing products from a given gene under particular conditions could be predicted by a "splicing code".[3][4]

Abnormal variations in splicing are also implicated in disease; a large proportion of human genetic disorders result from splicing variants.[3] Abnormal splicing variants are also thought to contribute to the development of cancer.[5][6][7]

36.1 Discovery

Alternative splicing was first observed in 1977.[8][9] The Adenovirus produces five primary transcripts early in its infectious cycle, prior to viral DNA replication, and an additional one later, after DNA replication begins. The early primary transcripts continue to be produced after DNA replication begins. The additional primary transcript produced late in infection is large and comes from 5/6 of the 32kb adenovirus genome. This is much larger than any of the individual adenovirus mRNAs present in infected cells. Researchers found that the primary RNA transcript produced by adenovirus type 2 in the late phase was spliced in many different ways, resulting in mRNAs encoding different viral proteins. In addition, the primary transcript contained multiple polyadenylation sites, giving different 3' ends for the processed mRNAs.[10][11][12]

In 1981, the first example of alternative splicing in a transcript from a normal, endogenous gene was characterized.[10] The gene encoding the thyroid hormone calcitonin was found to be alternatively spliced in mammalian cells. The primary transcript from this gene contains 6 exons; the calcitonin mRNA contains exons 1–4, and terminates after a polyadenylation site in exon 4. Another mRNA is produced from this pre-mRNA by skipping exon 4, and includes exons 1–3, 5, and 6. It encodes a protein known as CGRP (calcitonin gene related peptide).[13][14] Examples of alternative splicing in immunoglobin gene transcripts in mammals were also observed in the early 1980s.[10][15]

Since then, alternative splicing has been found to be ubiquitous in eukaryotes.[1] The "record-holder" for alternative splicing is a *D. melanogaster* gene called Dscam, which could potentially have 38,016 splice variants.[16]

36.2 Modes

Five basic modes of alternative splicing are generally recognized.[1][2][3][17]

- **Exon skipping** or **cassette exon**: in this case, an exon may be spliced out of the primary transcript or retained. This is the most common mode in mammalian pre-mRNAs.[17]

- **Mutually exclusive exons**: One of two exons is retained in mRNAs after splicing, but not both.

- **Alternative donor site**: An alternative 5' splice junction (donor site) is used, changing the 3' boundary of the upstream exon.

- **Alternative acceptor site**: An alternative 3' splice junction (acceptor site) is used, changing the 5' boundary of the downstream exon.

- **Intron retention**: A sequence may be spliced out as an intron or simply retained. This is distinguished from exon skipping because the retained sequence is not flanked by introns. If the retained intron is in the coding region, the intron must encode amino acids in frame with the neighboring exons, or a stop codon or a shift in the reading frame will cause the protein to be non-functional. This is the rarest mode in mammals.[17]

In addition to these primary modes of alternative splicing, there are two other main mechanisms by which different mRNAs may be generated from the same gene; multiple promoters and multiple polyadenylation sites. Use of multiple promoters is properly described as a transcriptional regulation mechanism rather than alternative splicing; by starting transcription at different points, transcripts with different 5'-most exons can be generated. At the other end, multiple polyadenylation sites provide different 3' end points for the transcript. Both of these mechanisms are found in combination with alternative splicing and provide additional variety in mRNAs derived from a gene.[1][3]

These modes describe basic splicing mechanisms, but may be inadequate to describe complex splicing events. For instance, the figure to the right shows 3 spliceforms from the mouse hyaluronidase 3 gene. Comparing the exonic structure shown in the first line (green) with the one in the second line (yellow) shows intron retention, whereas the comparison between the second and the third spliceform (yellow vs. blue) exhibits exon skipping. A model nomenclature to uniquely designate all possible splicing patterns has recently been proposed.[17]

36.3 Alternative splicing mechanisms

36.3.1 General splicing mechanism

Main article: RNA splicing

 When the pre-mRNA has been transcribed from the DNA, it includes several introns and exons. (In nematodes, the mean is 4–5 exons and introns; in the fruit fly *Drosophila* there can be more than 100 introns and exons in one transcribed pre-mRNA.) The exons to be retained in the mRNA are determined during the splicing process. The regulation and selection of splice sites are done by trans-acting splicing activator and splicing repressor proteins as well as cis-acting elements within the pre-mRNA itself such as exonic splicing enhancers and exonic splicing silencers.

The typical eukaryotic nuclear intron has consensus sequences defining important regions. Each intron has GU at its 5' end. Near the 3' end there is a branch site. The nucleotide at the branchpoint is always an A; the consensus around this sequence varies somewhat. In humans the branch site consensus sequence is yUnAy.[18] The branch site is followed by a series of pyrimidines - the polypyrimidine tract - then by AG at the 3' end.[3]

Splicing of mRNA is performed by an RNA and protein complex known as the spliceosome, containing snRNPs designated U1, U2, U4, U5, and U6 (U3 is not involved in mRNA splicing).[19] U1 binds to the 5' GU and U2, with the assistance of the U2AF protein factors, binds to the branchpoint A within the branch site. The complex at this stage is known as the spliceosome A complex. Formation of the A complex is usually the key step in determining the ends of the intron to be spliced out, and defining the ends of the exon to be retained.[3] (The U nomenclature derives from their high uridine content).

The U4,U5,U6 complex binds, and U6 replaces the U1 position. U1 and U4 leave. The remaining complex then performs two transesterification reactions. In the first transesterification, 5' end of the intron is cleaved from the upstream exon and joined to the branch site A by a 2',5'-phosphodiester linkage. In the second transesterification, the 3' end of the intron is cleaved from the downstream exon, and the two exons are joined by a phosphodiester bond. The intron is then released in lariat form and degraded.[1]

36.3.2 Regulatory elements and proteins

Splicing is regulated by trans-acting proteins (repressors and activators) and corresponding cis-acting regulatory sites (silencers and enhancers) on the pre-mRNA. However, as part of the complexity of alternative splicing, it is noted that the effects of a splicing factor are frequently position-dependent. That is, a splicing factor that serves as a splicing activator when bound to an intronic enhancer element may serve as a repressor when bound to its splicing element in the context of an exon, and vice versa.[20] The secondary structure of the pre-mRNA transcript also plays a role in regulating splicing, such as by bringing together splicing elements or by masking a sequence that would otherwise serve as a binding element for a splicing factor.[21][22] Together, these elements form a "splicing code" that governs how splicing will occur under different cellular conditions.[23][24]

There are two major types of cis-acting RNA sequence elements present in pre-mRNAs and they have corresponding trans-acting RNA-binding proteins. Splicing *silencers* are sites to which splicing repressor proteins bind, reducing the probability that a nearby site will be used as a splice junction. These can be located in the intron itself (intronic splicing silencers, ISS) or in a neighboring exon (exonic splicing silencers, ESS). They vary in sequence, as well as in the types of proteins that bind to them. The majority of splicing repressors are heterogeneous nuclear ribonucleoproteins (hnRNPs) such as hnRNPA1 and polypyrimidine tract binding protein (PTB).[3][23] Splicing *enhancers* are sites to which splicing activator proteins bind, increasing the probability that a nearby site will be used as a splice junction. These also may occur in the intron (intronic splicing enhancers, ISE) or exon (exonic splicing enhancers, ESE). Most of the activator proteins that bind to ISEs and ESEs are members of the SR protein family. Such proteins contain RNA recognition motifs and arginine and serine-rich (RS) domains.[3][23]

In general, the determinants of splicing work in an inter-dependent manner that depends on context, so that the rules governing how splicing is regulated from a splicing code.[24] The presence of a particular cis-acting RNA sequence element may increase the probability that a nearby site will be spliced in some cases, but decrease the probability in other cases, depending on context. The context within which regulatory elements act includes cis-acting context that is established by the presence of other RNA sequence features, and trans-acting context that is established by cellular conditions. For example, some cis-acting RNA sequence elements influence splicing only if multiple elements are present in the same region so as to establish context. As another example, a cis-acting element can have opposite effects on splicing, depending on which proteins are expressed in the cell (e.g., neuronal versus non-neuronal PTB). The adaptive significance of splicing silencers and enhancers is attested by studies showing that there is strong selection in human genes against mutations that produce new silencers or disrupt existing enhancers.[25][26]

36.3.3 Examples

Exon skipping: *Drosophila dsx*

Pre-mRNAs from the *D. melanogaster* gene *dsx* contain 6 exons. In males, exons 1,2,3,5,and 6 are joined to form the mRNA, which encodes a transcriptional regulatory protein required for male development. In females, exons 1,2,3, and 4 are joined, and a polyadenylation signal in exon 4 causes cleavage of the mRNA at that point. The resulting mRNA is a transcriptional regulatory protein required for female development.[27]

This is an example of exon skipping. The intron upstream from exon 4 has a polypyrimidine tract that doesn't match the consensus sequence well, so that U2AF proteins bind poorly to it without assistance from splicing activators. This 3' splice acceptor site is therefore not used in males. Females, however, produce the splicing activator Transformer (Tra) (see below). The SR protein Tra2 is produced in both sexes and binds to an ESE in exon 4; if Tra is present, it binds to Tra2 and, along with another SR protein, forms a complex that assists U2AF proteins in binding to the weak polypyrimidine tract. U2 is recruited to the associated branchpoint, and this leads to inclusion of exon 4 in the mRNA.[27][28]

Alternative acceptor sites: *Drosophila Transformer*

Pre-mRNAs of the *Transformer* (Tra) gene of *Drosophila melanogaster* undergo alternative splicing via the alternative acceptor site mode. The gene Tra encodes a protein that is expressed only in females. The primary transcript of this gene contains an intron with two possible acceptor sites. In males, the upstream acceptor site is used. This causes a longer version of exon 2 to be included in the processed transcript, including an early stop codon. The resulting mRNA encodes a truncated protein product that is inactive. Females produce the master sex determination protein Sex lethal (Sxl). The Sxl protein is a splicing repressor that binds to an ISS in the RNA of the Tra transcript near the upstream acceptor site, preventing U2AF protein from binding to the polypyrimidine tract. This prevents the use of this junction, shifting the spliceosome binding to the downstream acceptor site. Splicing at this point bypasses the stop codon, which is excised as part of the intron. The resulting mRNA encodes an active Tra protein, which itself is a regulator of alternative splicing of other sex-related genes (see *dsx* above).[1]

Exon definition: Fas receptor

Multiple isoforms of the Fas receptor protein are produced by alternative splicing. Two normally occurring isoforms in humans are produced by an exon-skipping mechanism. An mRNA including exon 6 encodes the membrane-bound form of the Fas receptor, which promotes apoptosis, or programmed cell death. Increased expression of Fas receptor in skin cells chronically exposed to the sun, and absence of expression in skin cancer cells, suggests that this mechanism may be important in elimination of pre-cancerous cells in humans.[29] If exon 6 is skipped, the resulting mRNA encodes a soluble Fas protein that does not promote apoptosis. The inclusion or skipping of the exon depends on two antagonistic proteins, TIA-1 and polypyrimidine tract-binding protein (PTB).

- The 5' donor site in the intron downstream from exon 6 in the pre-mRNA has a weak agreement with the consensus sequence, and is not bound usually by the U1 snRNP. If U1 does not bind, the exon is skipped (see "a" in accompanying figure).

- Binding of TIA-1 protein to an intronic splicing enhancer site stabilizes binding of the U1 snRNP.[3] The resulting 5' donor site complex assists in binding of the splicing factor U2AF to the 3' splice site upstream of the exon, through a mechanism that is not yet known (see b).[30]

- Exon 6 contains a pyrimidine-rich exonic splicing silencer, *ure6*, where PTB can bind. If PTB binds, it inhibits the effect of the 5' donor complex on the binding of U2AF to the acceptor site, resulting in exon skipping (see c).

This mechanism is an example of exon definition in splicing. A spliceosome assembles on an intron, and the snRNP subunits fold the RNA so that the 5' and 3' ends of the intron are joined. However, recently studied examples such as this one show that there are also interactions between the ends of the exon. In this particular case, these exon definition interactions are necessary to allow the binding of core splicing factors prior to assembly of the spliceosomes on the two flanking introns.[30]

Repressor-activator competition: HIV-1 *tat* exon 2

HIV, the retrovirus that causes AIDS in humans, produces a single primary RNA transcript, which is alternatively spliced in multiple ways to produce over 40 different mRNAs.[31] Equilibrium among differentially spliced transcripts provides multiple mRNAs encoding different products that are required for viral multiplication.[32] One of the differentially spliced transcripts contains the *tat* gene, in which exon 2 is a cassette exon that may be skipped or included. The inclusion of tat exon 2 in the RNA is regulated by competition between the splicing repressor hnRNP A1 and the SR protein SC35. Within exon 2 an exonic splicing silencer sequence (ESS) and an exonic splicing enhancer sequence (ESE) overlap. If A1 repressor protein binds to the ESS, it initiates cooperative binding of multiple A1 molecules, extending into the 5' donor site upstream of exon 2 and preventing the binding of the core splicing factor U2AF35 to the polypyrimidine tract. If SC35 binds to the ESE, it prevents A1 binding and maintains the 5' donor site in an accessible state for assembly of the spliceosome. Competition between the activator and repressor ensures that both mRNA types (with and without exon 2) are produced.[31]

36.4   Adaptive significance

Alternative splicing is one of several exceptions to the original idea that one DNA sequence codes for one polypeptide (the One gene-one enzyme hypothesis). It might be more correct now to say "One gene – many polypeptides".[33] External information is needed in order to decide which polypeptide is produced, given a DNA sequence and pre-mRNA. Since the methods of regulation are inherited, this provides novel ways for mutations to affect gene expression.[7]

It has been proposed that for eukaryotes alternative splicing was a very important step towards higher efficiency, because information can be stored much more economically. Several proteins can be encoded by a single gene, rather than requiring a separate gene for each, and thus allowing a more varied proteome from a genome of limited size.[1] It also provides evolutionary flexibility. A single point mutation may cause a given exon to be occasionally excluded or included from a transcript during splicing, allowing production of a new protein isoform without loss of the original protein.[1] Studies have identified intrinsically disordered regions (see Intrinsically unstructured proteins) as enriched in the non-constitutive exons[34] suggesting that protein isoforms may display functional diversity due to the alteration of functional modules within these regions. Such functional diversity achieved by isoforms is reflected by the their expression patterns and can be predicted by machine learning approaches.[35][36] Comparative studies indicate that alternative splicing preceded multicellularity in evolution, and suggest that this mechanism might have been co-opted to assist in the development of multicellular organisms.[37]

Research based on the Human Genome Project and other genome sequencing has shown that humans have only about 30% more genes than the roundworm *Caenorhabditis elegans*, and only about twice as many as the fly *Drosophila melanogaster*. This finding led to speculation that the perceived greater complexity of humans, or vertebrates generally, might be due to higher rates of alternative splicing in humans than are found in invertebrates.[38][39] However, a study on samples of 100,000 ESTs each from human, mouse, rat, cow, fly (*D. melanogaster*), worm (*C. elegans*), and the plant *Arabidopsis thaliana* found no large differences in frequency of alternatively spliced genes among humans and any of the other animals tested.[40] Another study, however, proposed that these results were an artifact of the different numbers of ESTs available for the various organisms. When they compared alternative splicing frequencies in random subsets of genes from each organism, the authors concluded that vertebrates do have higher rates of alternative splicing than invertebrates.[41]

36.5   Alternative splicing and disease

Changes in the RNA processing machinery may lead to mis-splicing of multiple transcripts, while single-nucleotide alterations in splice sites or cis-acting splicing regulatory sites may lead to differences in splicing of a single gene, and thus in the mRNA produced from a mutant gene's transcripts. A study in 2005 involving probabilistic analyses indicated that greater than 60% of human disease-causing mutations affect splicing rather than directly affecting coding sequences.[42] A more recent study indicates that one-third of all hereditary diseases are likely to have a splicing component.[20] Regardless of exact percentage, a number of splicing-related diseases do exist.[43] As described below, a prominent example of splicing-related diseases is cancer.

Abnormally spliced mRNAs are also found in a high proportion of cancerous cells.[5][6] Combined RNA-Seq and proteomics analyses have revealed striking differential expression of splice isoforms of key proteins in important cancer pathways.[44] Until recently, it was unclear whether such aberrant patterns of splicing played a role in causing cancerous growth, or were merely a consequence of cellular abnormalities associated with cancer. It has been shown that there is actually a reduction of alternative splicing in cancerous cells compared to normal ones, and the types of splicing differ; for instance, cancerous cells show higher levels of intron retention than normal cells, but lower levels of exon skipping.[45] Some of the differences in splicing in cancerous cells may result from changes in phosphorylation of trans-acting splicing factors.[7] Others may be produced by changes in the relative amounts of splicing factors produced; for instance, breast cancer cells have been shown to have increased levels of the splicing factor SF2/ASF.[46] One study found that a relatively small percentage (383 out of over 26000) of alternative splicing variants were significantly higher in frequency in tumor cells than normal cells, suggesting that there is a limited set of genes which, when mis-spliced, contribute to tumor development.[47] It is believed however that the deleterious effects of mis-spliced transcripts are usually safeguarded and eliminated by a cellular posttranscriptional quality control mechanism termed Nonsense-mediated mRNA decay [NMD].[48]

One example of a specific splicing variant associated with cancers is in one of the human DNMT genes. Three DNMT genes encode enzymes that add methyl groups to DNA, a modification that often has regulatory effects. Several abnormally spliced DNMT3B mRNAs are found in tumors and cancer cell lines. In two separate studies, expression of two of these abnormally spliced mRNAs in mammalian cells caused changes in the DNA methylation patterns in those cells. Cells with one of the abnormal mRNAs also grew twice as fast as control cells, indicating a direct contribution to tumor development by this product.[7]

Another example is the *Ron (MST1R)* proto-oncogene. An important property of cancerous cells is their ability to move and invade normal tissue. Production of an abnormally spliced transcript of *Ron* has been found to be associated with increased levels of the SF2/ASF in breast cancer cells. The abnormal isoform of the Ron protein encoded by this mRNA leads to cell motility.[46]

Overexpression of a truncated splice variant of the FOSB gene – ΔFosB – in a specific population of neurons in the nucleus accumbens has been identified as the causal mechanism involved in the induction and maintenance of an addiction to drugs and natural rewards.[49][50][51][52]

Recent provocative studies point to a key function of chromatin structure and histone modifications in alternative splicing regulation. These insights suggest that epigenetic regulation determines not only what parts of the genome are expressed but also how they are spliced.[53]

36.6   Genome-wide analysis of alternative splicing

Genome-wide analysis of alternative splicing is a challenging task. Typically, alternatively spliced transcripts have been found by comparing EST sequences, but this requires sequencing of very large numbers of ESTs. Most EST libraries come from a very limited number of tissues, so tissue-specific splice variants are likely to be missed in any case. High-throughput approaches to investigate splicing have, however, been developed, such as: DNA microarray-based analyses,

RNA-binding assays, and deep sequencing. These methods can be used to screen for polymorphisms or mutations in or around splicing elements that affect protein binding. When combined with splicing assays, including *in vivo* reporter gene assays, the functional effects of polymorphisms or mutations on the splicing of pre-mRNA transcripts can then be analyzed.[20][23][54]

In microarray analysis, arrays of DNA fragments representing individual exons (*e.g.* Affymetrix exon microarray) or exon/exon boundaries (*e.g.* arrays from ExonHit or Jivan) have been used. The array is then probed with labeled cDNA from tissues of interest. The probe cDNAs bind to DNA from the exons that are included in mRNAs in their tissue of origin, or to DNA from the boundary where two exons have been joined. This can reveal the presence of particular alternatively spliced mRNAs.[2]

CLIP (Cross-linking and immunoprecipitation) uses UV radiation to link proteins to RNA molecules in a tissue during splicing. A trans-acting splicing regulatory protein of interest is then precipitated using specific antibodies. When the RNA attached to that protein is isolated and cloned, it reveals the target sequences for that protein.[4] Another method for identifying RNA-binding proteins and mapping their binding to pre-mRNA transcripts is "Microarray Evaluation of Genomic Aptamers by shift (MEGAshift)".[55][56] This method involves an adaptation of the "Systematic Evolution of Ligands by Exponential Enrichment (SELEX)" method[57] together with a microarray-based readout. Use of the MEGAshift method has provided insights into the regulation of alternative splicing by allowing for the identification of sequences in pre-mRNA transcripts surrounding alternatively spliced exons that mediate binding to different splicing factors, such as ASF/SF2 and PTB.[58] This approach has also been used to aid in determining the relationship between RNA secondary structure and the binding of splicing factors.[22]

Deep sequencing technologies have been used to conduct genome-wide analyses of mRNAs - unprocessed and processed - thus providing insights into alternative splicing. For example, results from use of deep sequencing indicate that, in humans, an estimated 95% of transcripts from multiexon genes undergo alternative splicing, with a number of pre-mRNA transcripts spliced in a tissue-specific manner.[2] Functional genomics and computational approaches based on multiple instance learning have also been developed to integrate RNA-seq data to predict functions for alternatively spliced isoforms.[36] Deep sequencing has also aided in the *in vivo* detection of the transient lariats that are released during splicing, the determination of branch site sequences, and the large-scale mapping of branchpoints in human pre-mRNA transcripts.[59]

Use of reporter assays makes it possible to find the splicing proteins involved in a specific alternative splicing event by constructing reporter genes that will express one of two different fluorescent proteins depending on the splicing reaction that occurs. This method has been used to isolate mutants affecting splicing and thus to identify novel splicing regulatory proteins inactivated in those mutants.[4]

36.7   See also

- AspicDB database

- Exitron

- Polyadenylation § Alternative polyadenylation

36.8   References

[1] Black, Douglas L. (2003). "Mechanisms of alternative pre-messenger RNA splicing". *Annual Reviews of Biochemistry* **72** (1): 291–336. doi:10.1146/annurev.biochem.72.121801.161720. PMID 12626338.

[2] Pan, Q; Shai O; Lee LJ; Frey BJ; Blencowe BJ (Dec 2008). "Deep surveying of alternative splicing complexity in the human transcriptome by high-throughput sequencing". *Nature Genetics* **40** (12): 1413–1415. doi:10.1038/ng.259. PMID 18978789.

[3] Matlin, AJ; Clark, F; Smith, CWJ (May 2005). "Understanding alternative splicing: towards a cellular code". *Nature Reviews* **6** (5): 386–398. doi:10.1038/nrm1645. PMID 15956978.

[4] David, C. J.; Manley, J. L. (2008). "The search for alternative splicing regulators: new approaches offer a path to a splicing code". *Genes & Development* **22** (3): 279–85. doi:10.1101/gad.1643108. PMC 2731647. PMID 18245441.

[5] Skotheim, R I and Nees, M (2007). "Alternative splicing in cancer: noise, functional, or systematic?". *The international journal of biochemistry & cell biology* **39** (7-8): 1432–49. doi:10.1016/j.biocel.2007.02.016. PMID 17416541.

[6] Bauer, Joseph Alan; He, Chunjiang; Zhou, Fang; Zuo, Zhixiang; Cheng, Hanhua; Zhou, Rongjia (2009). Bauer, Joseph Alan, ed. "A Global View of Cancer-Specific Transcript Variants by Subtractive Transcriptome-Wide Analysis". *PLoS ONE* **4** (3): e4732. Bibcode:2009PLoSO...4.4732H. doi:10.1371/journal.pone.0004732. PMC 2648985. PMID 19266097.

[7] Fackenthal, Jd; Godley, La (2008). "Aberrant RNA splicing and its functional consequences in cancer cells" (Free full text). *Disease models & mechanisms* **1** (1): 37–42. doi:10.1242/dmm.000331. PMC 2561970. PMID 19048051.

[8] Chow LT, Gelinas RE, Broker TR, Roberts RJ (1977). "An amazing sequence arrangement at the 5' ends of adenovirus 2 messenger RNA". *Cell* **12** (1): 1–8. doi:10.1016/0092-8674(77)90180-5. PMID 902310.

[9] Berget SM, Moore C, Sharp PA (1977). "Spliced segments at the 5' terminus of adenovirus 2 late mRNA". *Proc. Natl. Acad. Sci. U.S.A.* **74** (8): 3171–5. Bibcode:1977PNAS...74.3171B. doi:10.1073/pnas.74.8.3171. PMC 431482. PMID 269380.

[10] Leff SE, Rosenfeld MG, Evans RM (1986). "Complex transcriptional units: diversity in gene expression by alternative RNA processing". *Annu. Rev. Biochem.* **55** (1): 1091–117. doi:10.1146/annurev.bi.55.070186.005303. PMID 3017190.

[11] Chow LT, Broker TR (1978). "The spliced structures of adenovirus 2 fiber message and the other late mRNAs". *Cell* **15** (2): 497–510. doi:10.1016/0092-8674(78)90019-3. PMID 719751.

[12] Nevins JR, Darnell JE (1978). "Steps in the processing of Ad2 mRNA: poly(A)+ nuclear sequences are conserved and poly(A) addition precedes splicing". *Cell* **15** (4): 1477–93. doi:10.1016/0092-8674(78)90071-5. PMID 729004.

[13] Rosenfeld MG, Amara SG, Roos BA, Ong ES, Evans RM (1981). "Altered expression of the calcitonin gene associated with RNA polymorphism". *Nature* **290** (5801): 63–5. Bibcode:1981Natur.290...63R. doi:10.1038/290063a0. PMID 7207587.

[14] Rosenfeld MG, Lin CR, Amara SG; et al. (1982). "Calcitonin mRNA polymorphism: peptide switching associated with alternative RNA splicing events". *Proc. Natl. Acad. Sci. U.S.A.* **79** (6): 1717–21. Bibcode:1982PNAS...79.1717R. doi:10.1073/pnas.79.6.1717. PMC 346051. PMID 6952224.

[15] Maki R, Roeder W, Traunecker A; et al. (1981). "The role of DNA rearrangement and alternative RNA processing in the expression of immunoglobulin delta genes". *Cell* **24** (2): 353–65. doi:10.1016/0092-8674(81)90325-1. PMID 6786756.

[16] Schmucker D, Clemens JC, Shu H, Worby CA, Xiao J, Muda M, Dixon JE, Zipursky SL (2000). "Drosophila Dscam is an axon guidance receptor exhibiting extraordinary molecular diversity". *Cell* **101** (6): 671–84. doi:10.1016/S0092-8674(00)80878-8. PMID 10892653.

[17] Michael Sammeth; Sylvain Foissac; Roderic Guigó (2008). Brent, Michael R., ed. "A general definition and nomenclature for alternative splicing events". *PLoS Comput Biol.* **4** (8): e1000147. Bibcode:2008PLSCB...4E0147S. doi:10.1371/journal.pcbi.1000147. PMC 2467475. PMID 18688268.

[18] Gao, K.; Masuda, A.; Matsuura, T.; Ohno, K. (2008). "Human branch point consensus sequence is yUnAy". *Nucleic Acids Research* **36** (7): 2257–67. doi:10.1093/nar/gkn073. PMC 2367711. PMID 18285363.

[19] Clark, David (2005). *Molecular biology*. Amsterdam: Elsevier Academic Press. ISBN 0-12-175551-7.

[20] Lim, KH; Ferraris, L; Filloux, ME; Raphael, BJ; Fairbrother, WG (2011). "Using positional distribution to identify splicing elements and predict pre-mRNA processing defects in human genes". *Proc. Natl. Acad. Sci. USA* **108** (27): 11093–11098. doi:10.1073/pnas.1101135108. PMC 3131313. PMID 21685335.

[21] Warf, MB; Berglund, JA (2010). "Role of RNA structure in regulating pre-mRNA splicing". *Trends Biochem. Sci.* **35** (3): 169–178. doi:10.1016/j.tibs.2009.10.004. PMC 2834840. PMID 19959365.

[22] Reid, DC; Chang, BL; Gunderson, SI; Alpert, L; Thompson, WA; Fairbrother, WG (2009). "Next-generation SELEX identifies sequence and structural determinants of splicing factor binding in human pre-mRNA sequence". *RNA* **15** (12): 2385–2397. doi:10.1261/rna.1821809. PMC 2779669. PMID 19861426.

[23] Wang, Z; Burge, Cb (2008). "Splicing regulation: from a parts list of regulatory elements to an integrated splicing code" (Free full text). *RNA* **14** (5): 802–13. doi:10.1261/rna.876308. PMC 2327353. PMID 18369186.

[24] Barash, Y; et al. (2010). "Deciphering the splicing code". *Nature* **465** (7294): 53–59. Bibcode:2010Natur.465...53B. doi:10.1038/nature09000. PMID 20445623.

[25] Ke S, Zhang XH, Chasin LA (2008). "Positive selection acting on splicing motifs reflects compensatory evolution". *Genome Res.* **18** (4): 533–43. doi:10.1101/gr.070268.107. PMC 2279241. PMID 18204002.

[26] Fairbrother, WG; Holste, D; Burge, CB; Sharp, PA (2004). "Single nucleotide polymorphism–based validation of exonic splicing enhancers". *PLoS Biol* **2** (9): e268. doi:10.1371/journal.pbio.0020268. PMC 514884. PMID 15340491.

[27] Lynch KW, Maniatis T (1996). "Assembly of specific SR protein complexes on distinct regulatory elements of the Drosophila doublesex splicing enhancer". *Genes Dev.* **10** (16): 2089–101. doi:10.1101/gad.10.16.2089. PMID 8769651.

[28] Graveley BR, Hertel KJ, Maniatis T (2001). "The role of U2AF35 and U2AF65 in enhancer-dependent splicing". *RNA* **7** (6): 806–18. doi:10.1017/S1355838201010317. PMC 1370132. PMID 11421359.

[29] Filipowicz, Ewa; Adegboyega, P.; Sanchez, R. L.; Gatalica, Zoran (2002). "Expression of CD95 (Fas) in sun-exposed human skin and cutaneous carcinomas". *Cancer* **94** (3): 814–9. doi:10.1002/cncr.10277. PMID 11857317.

[30] Izquierdo JM, Majós N, Bonnal S; et al. (2005). "Regulation of Fas alternative splicing by antagonistic effects of TIA-1 and PTB on exon definition". *Mol. Cell* **19** (4): 475–84. doi:10.1016/j.molcel.2005.06.015. PMID 16109372.

[31] Zahler, A. M.; Damgaard, CK; Kjems, J; Caputi, M (2003). "SC35 and Heterogeneous Nuclear Ribonucleoprotein A/B Proteins Bind to a Juxtaposed Exonic Splicing Enhancer/Exonic Splicing Silencer Element to Regulate HIV-1 tat Exon 2 Splicing". *Journal of Biological Chemistry* **279** (11): 10077–84. doi:10.1074/jbc.M312743200. PMID 14703516.

[32] Jacquenet, S.; Méreau, A; Bilodeau, PS; Damier, L; Stoltzfus, CM; Branlant, C (2001). "A Second Exon Splicing Silencer within Human Immunodeficiency Virus Type 1 tat Exon 2 Represses Splicing of Tat mRNA and Binds Protein hnRNP H". *Journal of Biological Chemistry* **276** (44): 40464–75. doi:10.1074/jbc.M104070200. PMID 11526107.

[33] "HHMI Bulletin September 2005: Alternative Splicing". www.hhmi.org. Archived from the original on 22 June 2009. Retrieved 2009-05-26.

[34] Romero, Pedro; Zaidi, Saima; Fang, Ya Yin; Uversky, Vladimir; Dunker, Keith (2006). "Alternative splicing in concert with protein intrinsic disorder enables increased functional diversity in multicellular organisms.". *Proc Natl Acad Sci U S A* **103**: 8390–8395. Bibcode:2006PNAS..103.8390R. doi:10.1073/pnas.0507916103. PMC 1482503. PMID 16717195.

[35] Li, HD; Menon, R; Omenn, GS; Guan, Y (Jun 17, 2014). "The emerging era of genomic data integration for analyzing splice isoform function.". *Trends in genetics : TIG* **30** (8): 340–347. doi:10.1016/j.tig.2014.05.005. PMID 24951248.

[36] Eksi, R; Li, HD; Menon, R; Wen, Y; Omenn, GS; Kretzler, M; Guan, Y (Nov 2013). "Systematically differentiating functions for alternatively spliced isoforms through integrating RNA-seq data.". *PLoS computational biology* **9** (11): e1003314. Bibcode:2013PLSCB...9E3314E. doi:10.1371/journal.pcbi.1003314. PMC 3820534. PMID 24244129.

[37] Irimia, Manuel; Rukov, Jakob; Penny, David; Roy, Scott (2007). "Functional and evolutionary analysis of alternatively spliced genes is consistent with an early eukaryotic origin of alternative splicing". *BMC Evolutionary Biology* **7**: 188. doi:10.1186/1471-2148-7-188. PMC 2082043. PMID 17916237.

[38] Ewing, B; Green P (June 2000). "Analysis of expressed sequence tags indicates 35,000 human genes". *Nature Genetics* **25** (2): 232–234. doi:10.1038/76115. PMID 10835644.

[39] Crollius, HR; et al. (2000). "Estimate of human gene number provided by genome-wide analysis using *Tetraodon nigroviridis* DNA sequence". *Nature Genetics* **25** (2): 235–238. doi:10.1038/76118. PMID 10835645.

[40] David Brett; Heike Pospisil; Juan Valcárcel; Jens Reich; Peer Bork (2002). "Alternative splicing and genome complexity". *Nature Genetics* **30** (1): 29–30. doi:10.1038/ng803. PMID 11743582.

[41] Kim, E.; Magen, A.; Ast, G. (2006). "Different levels of alternative splicing among eukaryotes". *Nucleic Acids Research* **35** (1): 125–31. doi:10.1093/nar/gkl924. PMC 1802581. PMID 17158149.

[42] López-Bigas, Núria; Audit, Benjamin; Ouzounis, Christos; Parra, Genís; Guigó, Roderic (2005). "Are splicing mutations the most frequent cause of hereditary disease?". *FEBS Letters* **579** (9): 1900–3. doi:10.1016/j.febslet.2005.02.047. PMID 15792793.

[43] Ward, AJ; Cooper, TA (2010). "The pathobiology of splicing". *J. Pathol* **220** (2): 152–163. doi:10.1002/path.2649. PMC 2855871. PMID 19918805.

[44] Omenn, GS; Guan, Y; Menon, R (May 3, 2014). "A New Class of Protein Cancer Biomarker Candidates: Differentially-Expressed Splice Variants of ERBB2 (HER2/neu) and ERBB1 (EGFR) in Breast Cancer Cell Lines.". *Journal of proteomics* **107C**: 103–112. doi:10.1016/j.jprot.2014.04.012. PMID 24802673.

[45] Kim E, Goren A, Ast G (2008). "Insights into the connection between cancer and alternative splicing". *Trends Genet.* **24** (1): 7–10. doi:10.1016/j.tig.2007.10.001. PMID 18054115.

[46] Ghigna C, Giordano S, Shen H; et al. (2005). "Cell motility is controlled by SF2/ASF through alternative splicing of the *Ron proto-oncogene*". Mol. Cell *20 (6): 881–90. doi:10.1016/j.molcel.2005.10.026. PMID 16364913.*

[47] Hui L, Zhang X, Wu X; et al. (2004). "Identification of alternatively spliced mRNA variants related to cancers by genome-wide ESTs alignment". *Oncogene* **23** (17): 3013–23. doi:10.1038/sj.onc.1207362. PMID 15048092.

[48] Danckwardt S, Neu-Yilik G, Thermann R, Frede U, Hentze MW, Kulozik AE (2002). "Abnormally spliced beta-globin mR-NAs: a single point mutation generates transcripts sensitive and insensitive to nonsense-mediated mRNA decay". *Blood* **99** (5): 1811–6. doi:10.1182/blood.V99.5.1811. PMID 11861299.

[49] Nestler EJ (December 2013). "Cellular basis of memory for addiction". *Dialogues Clin. Neurosci.* **15** (4): 431–443. PMC 3898681. PMID 24459410. DESPITE THE IMPORTANCE OF NUMEROUS PSYCHOSOCIAL FACTORS, AT ITS CORE, DRUG ADDICTION INVOLVES A BIOLOGICAL PROCESS: the ability of repeated exposure to a drug of abuse to induce changes in a vulnerable brain that drive the compulsive seeking and taking of drugs, and loss of control over drug use, that define a state of addiction. ... A large body of literature has demonstrated that such ΔFosB induction in D1-type NAc neurons increases an animal's sensitivity to drug as well as natural rewards and promotes drug self-administration, presumably through a process of positive reinforcement

[50] Ruffle JK (November 2014). "Molecular neurobiology of addiction: what's all the (Δ)FosB about?". *Am J Drug Alcohol Abuse* **40** (6): 428–437. doi:10.3109/00952990.2014.933840. PMID 25083822. ΔFosB is an essential transcription factor implicated in the molecular and behavioral pathways of addiction following repeated drug exposure. The formation of ΔFosB in multiple brain regions, and the molecular pathway leading to the formation of AP-1 complexes is well understood. The establishment of a functional purpose for ΔFosB has allowed further determination as to some of the key aspects of its molecular cascades, involving effectors such as GluR2 (87,88), Cdk5 (93) and NFkB (100). Moreover, many of these molecular changes identified are now directly linked to the structural, physiological and behavioral changes observed following chronic drug exposure (60,95,97,102). New frontiers of research investigating the molecular roles of ΔFosB have been opened by epigenetic studies, and recent advances have illustrated the role of ΔFosB acting on DNA and histones, truly as a "molecular switch" (34). As a consequence of our improved understanding of ΔFosB in addiction, it is possible to evaluate the addictive potential of current medications (119), as well as use it as a biomarker for assessing the efficacy of therapeutic interventions (121,122,124). Some of these proposed interventions have limitations (125) or are in their infancy (75). However, it is hoped that some of these preliminary findings may lead to innovative treatments, which are much needed in addiction.

[51] Biliński P, Wojtyła A, Kapka-Skrzypczak L, Chwedorowicz R, Cyranka M, Studziński T (2012). "Epigenetic regulation in drug addiction". *Ann. Agric. Environ. Med.* **19** (3): 491–496. PMID 23020045. For these reasons, ΔFosB is considered a primary and causative transcription factor in creating new neural connections in the reward centre, prefrontal cortex, and other regions of the limbic system. This is reflected in the increased, stable and long-lasting level of sensitivity to cocaine and other drugs, and tendency to relapse even after long periods of abstinence. These newly constructed networks function very efficiently via new pathways as soon as drugs of abuse are further taken

[52] Olsen CM (December 2011). "Natural rewards, neuroplasticity, and non-drug addictions". *Neuropharmacology* **61** (7): 1109–1122. doi:10.1016/j.neuropharm.2011.03.010. PMC 3139704. PMID 21459101.

[53] Luco, RF; Allo, M; Schor, IE; Kornblihtt, AR; Misteli, T. (2011). "Epigenetics in alternative pre-mRNA splicing". *Cell* **144** (1): 16–26. doi:10.1016/j.cell.2010.11.056. PMC 3038581. PMID 21215366.

[54] Fairbrother, WG; Yeh, RF; Sharp, PA; Burge, CB (2002). "Predictive identification of exonic splicing enhancers in human genes". *Science* **297** (5583): 1007–1013. Bibcode:2002Sci...297.1007F. doi:10.1126/science.1073774. PMID 12114529.

[55] Tantin, D; Gemberling, M; Callister, C; Fairbrother, WG (2008). "High-throughput biochemical analysis of in vivo location data reveals novel distinct classes of POU5F1(Oct4)/DNA complexes". *Genome Res.* **18** (4): 631–639. doi:10.1101/gr.072942.107. PMID 18212089.

[56] Watkins, KH; Stewart, A; Fairbrother, WG (2009). "A rapid high-throughput method for mapping ribonucleoproteins (RNPs) on human pre-mRNA". *J. Vis. Exp.* **34**: 1622. doi:10.3791/1622.

[57] Tuerk, C; Gold, L (1990). "Systematic evolution of ligands by exponential enrichment: RNA ligands to bacteriophage T4 DNA polymerase". *Science* **249** (4968): 505–510. Bibcode:1990Sci...249..505T. doi:10.1126/science.2200121. PMID 2200121.

[58] Chang, B; Levin, J; Thompson, WA; Fairbrother, WG (2010). "High-throughput binding analysis determines the binding specificity of ASF/SF2 on alternatively spliced human pre-mRNAs". *Comb. Chem. High Throughput Screen* **13** (3): 242–252. doi:10.2174/138620710790980522. PMID 20015017.

[59] Taggart, AJ; DeSimone, AM; Shih, JS; Filloux, ME; Fairbrother, WG (2012). "Large-scale mapping of branchpoints in human pre-mRNA transcripts in vivo". *Nat. Struct. Mol. Biol.* **19** (7): 719–721. doi:10.1038/nsmb.2327. PMC 3465671. PMID 22705790.

36.9   External links

- A General Definition and Nomenclature for Alternative Splicing Events at SciVee

- AStalavista (Alternative Splicing landscape visualization tool), a method for the computationally exhaustive classification of Alternative Splicing Structures

- IsoPred: computationally predicted isoform functions

- Stamms-lab.net: Research Group dealing with alternative Splicing issues and mis-splicing in human diseases

- Alternative Splicing of ion channels in the brain, connected to mental and neurological diseases

- BIPASS: Web Services in Alternative Splicing

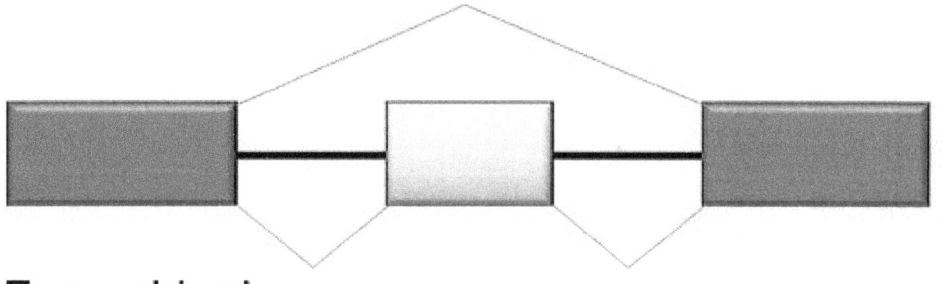

Exon skipping

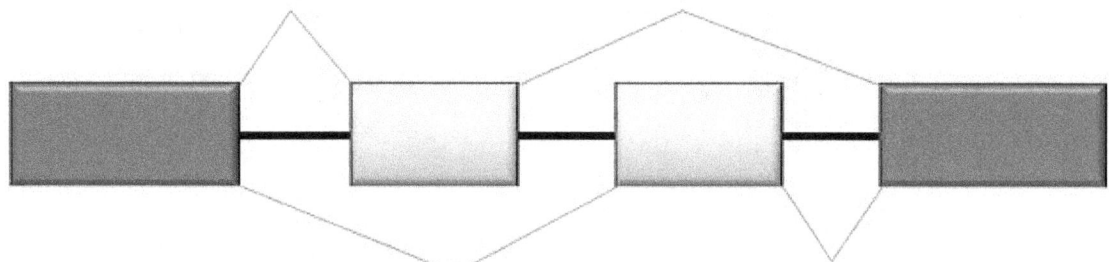

Mutually exclusive exons

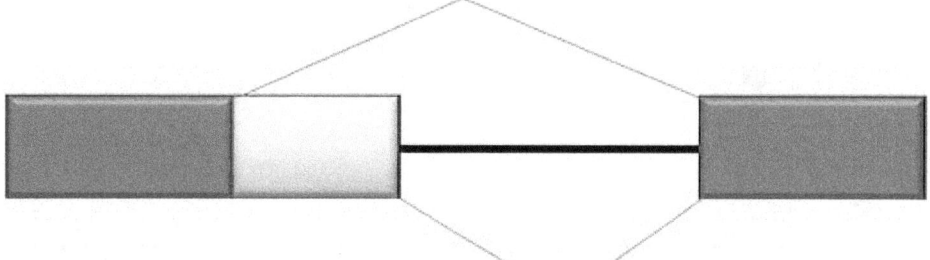

Alternative 5' donor sites

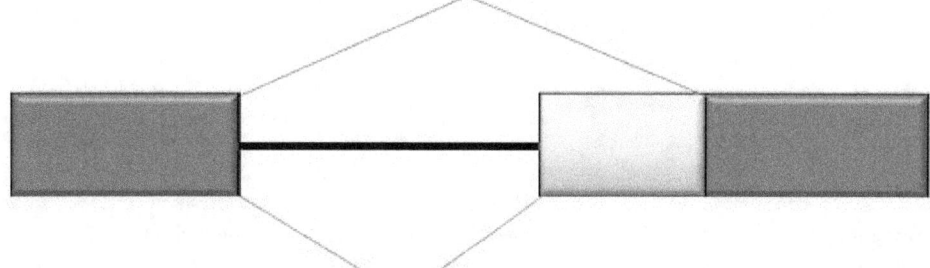

Alternative 3' acceptor sites

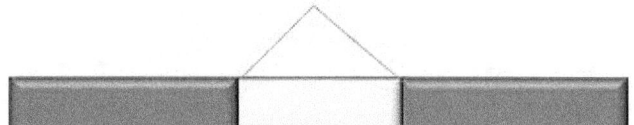

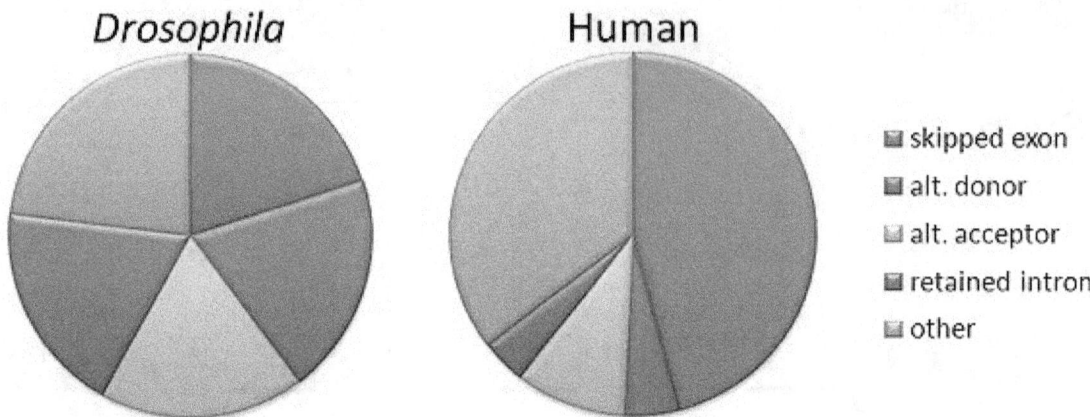

Relative frequencies of types of alternative splicing events differ between vertebrates and invertebrates.[17]

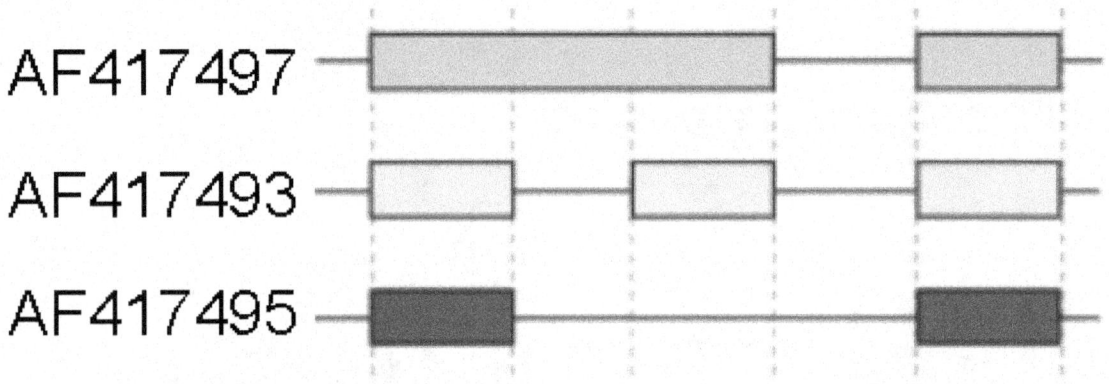

Schematic cutoff from 3 splicing structures in the murine hyaluronidase gene. Directionality of transcription from 5' to 3' is shown from left to right. Exons and introns are not drawn to scale.

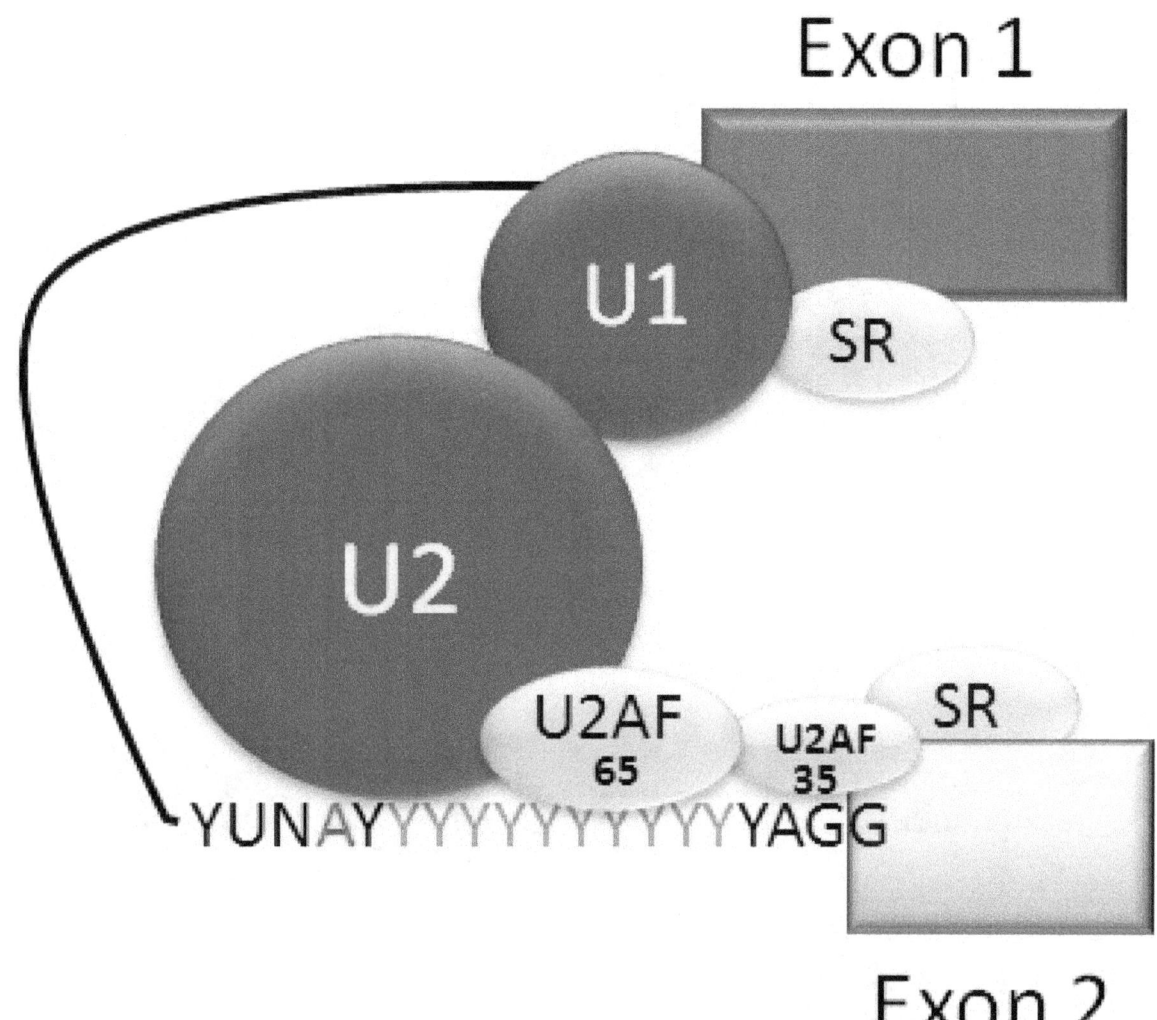

Spliceosome A complex defines the 5' and 3' ends of the intron before removal[3]

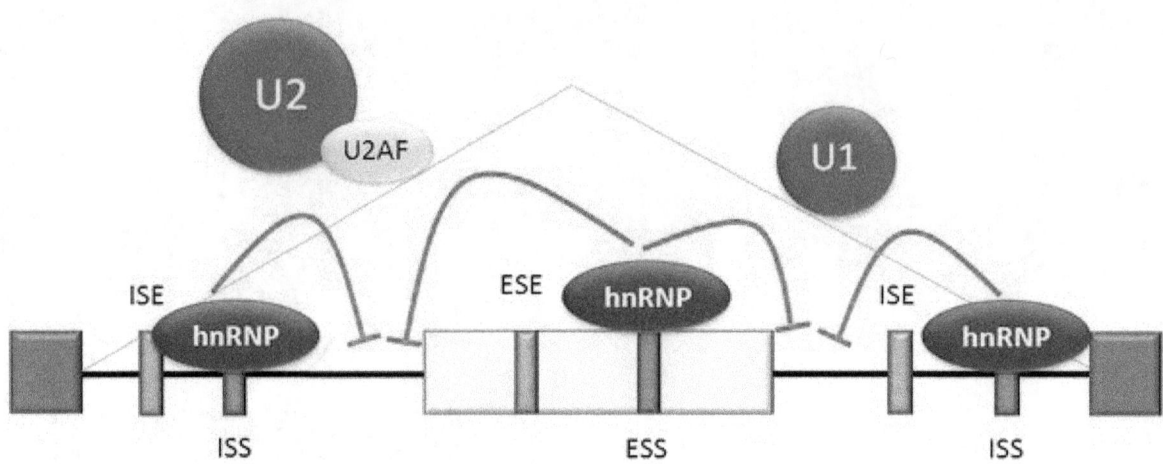

Splicing repression

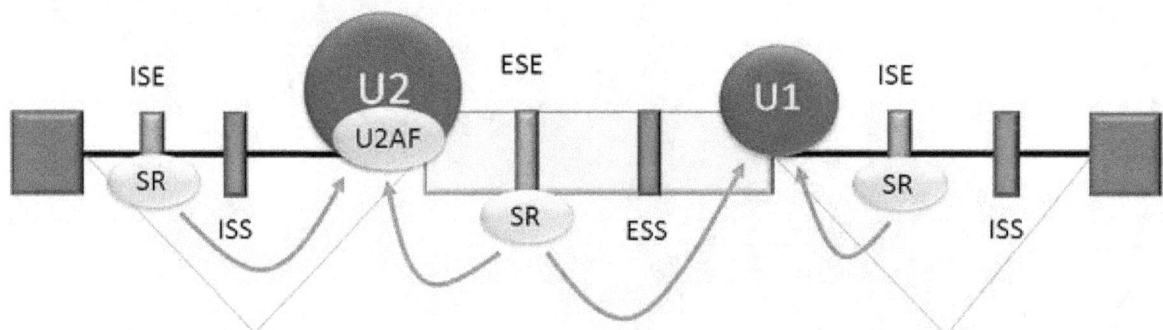

Splicing activation

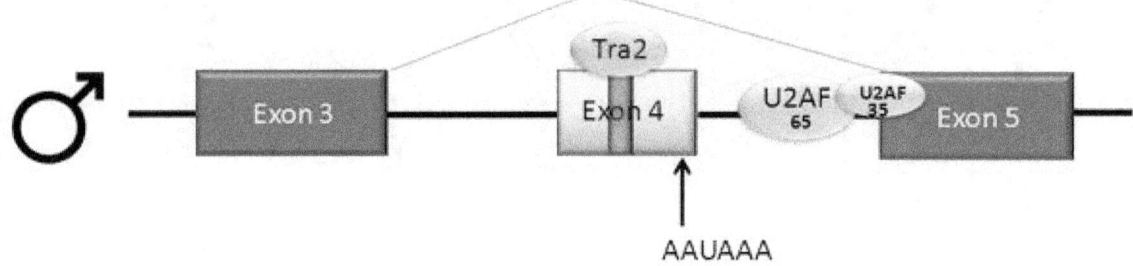

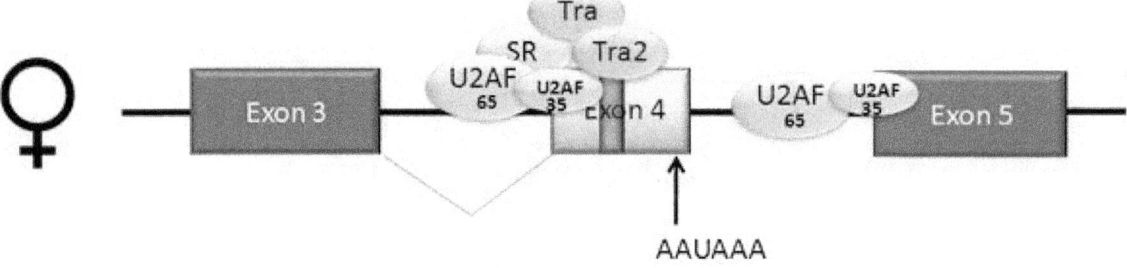

Alternative splicing of dsx *pre-mRNA*

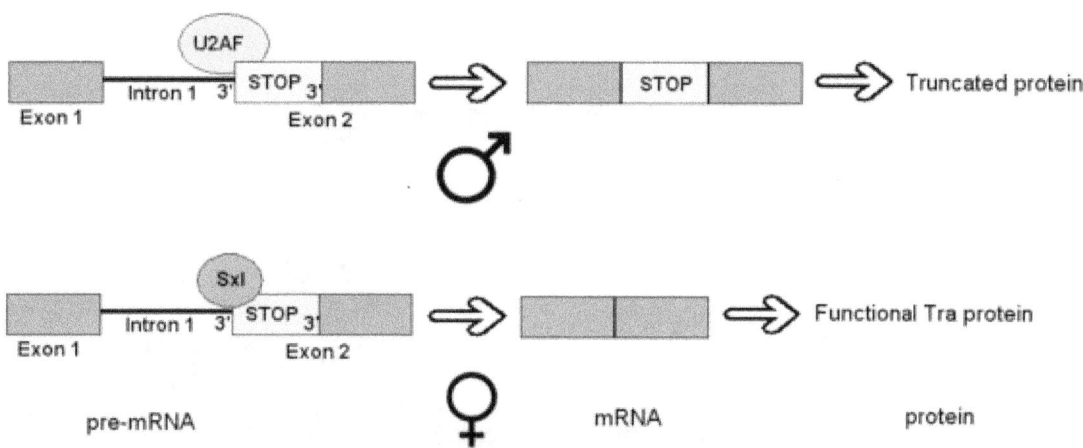

Alternative splicing of the Drosophila Transformer *gene product.*

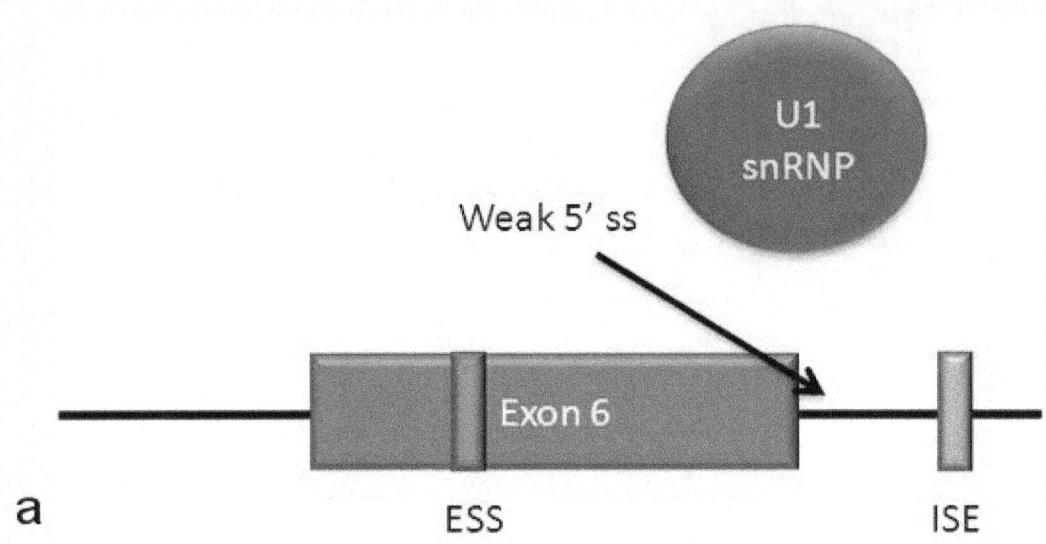

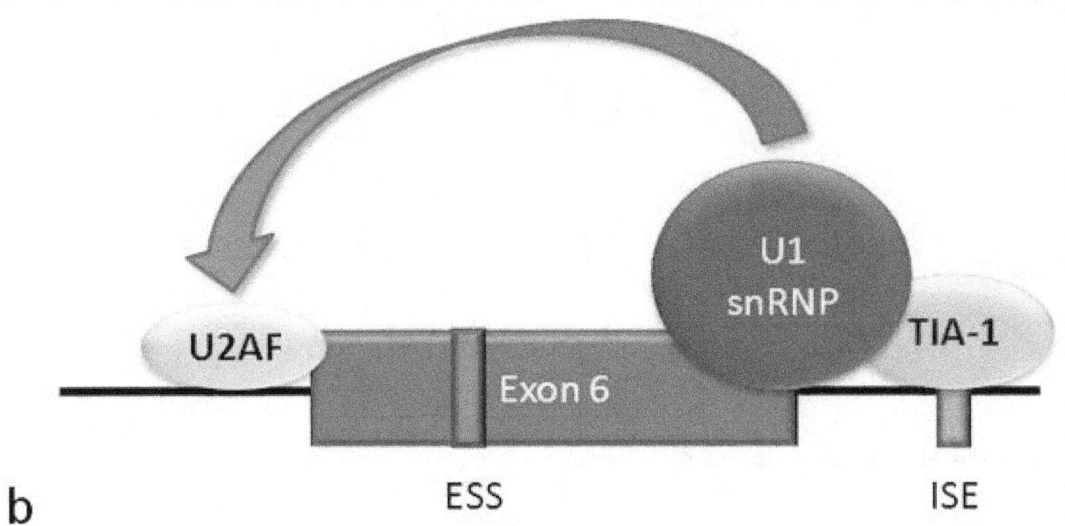

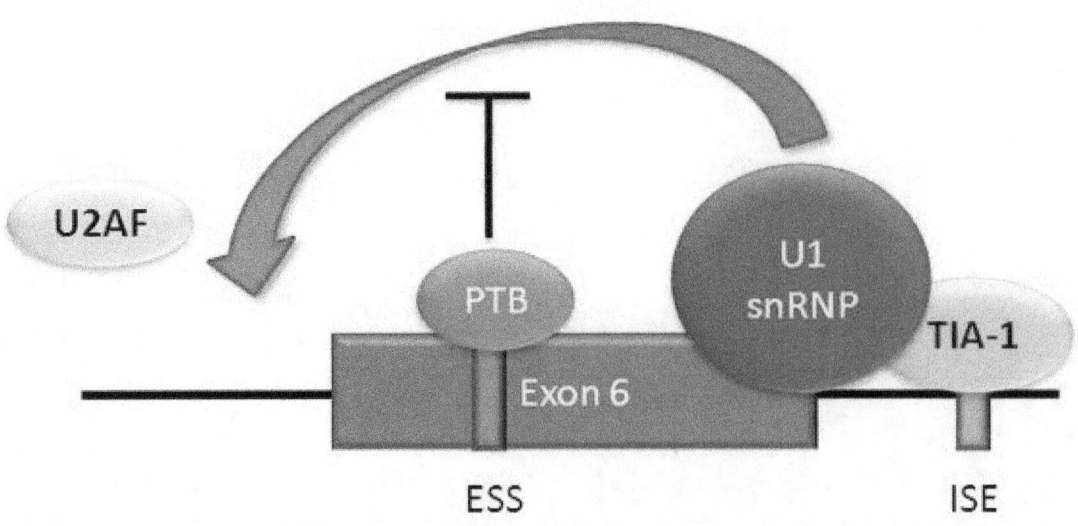

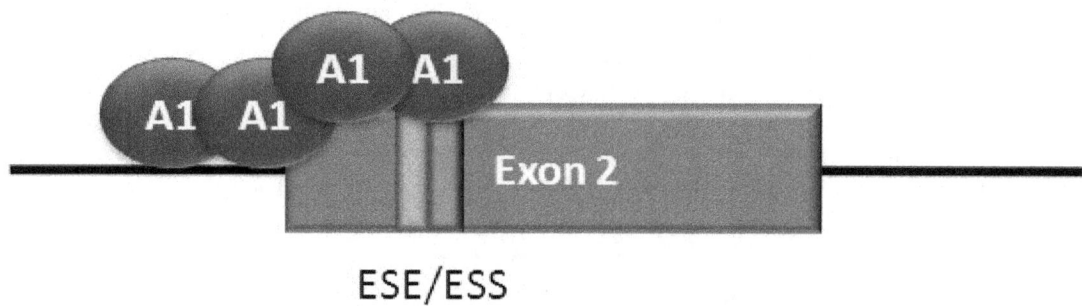

ESE/ESS

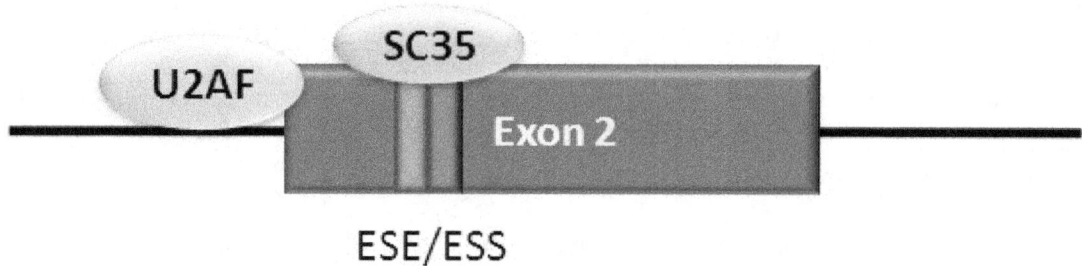

ESE/ESS

Alternative splicing of HIV-1 tat exon 2

Chapter 37

Translation (biology)

In molecular biology and genetics, **translation** is the process in which cellular ribosomes create proteins.

In translation, messenger RNA (mRNA)—produced by transcription from DNA—is decoded by a ribosome to produce a specific amino acid chain, or polypeptide. The polypeptide later folds into an active protein and performs its functions in the cell. The ribosome facilitates decoding by inducing the binding of complementary tRNA anticodon sequences to mRNA codons. The tRNAs carry specific amino acids that are chained together into a polypeptide as the mRNA passes through and is "read" by the ribosome. The entire process is a part of gene expression.

In brief, translation proceeds in four phases:

1. **Initiation**: The ribosome assembles around the target mRNA. The first tRNA is attached at the start codon.

2. **Elongation**: The tRNA transfers an amino acid to the tRNA corresponding to the next codon.

3. **Translocation**: The ribosome then moves (*translocates*) to the next mRNA codon to continue the process, creating an amino acid chain.

4. **Termination**: When a stop codon is reached, the ribosome releases the polypeptide.

In bacteria, translation occurs in the cell's cytoplasm, where the large and small subunits of the ribosome bind to the mRNA. In eukaryotes, translation occurs in the cytosol or across the membrane of the endoplasmic reticulum in a process called vectorial synthesis. In many instances, the entire ribosome/mRNA complex binds to the outer membrane of the rough endoplasmic reticulum (ER); the newly created polypeptide is stored inside the ER for later vesicle transport and secretion outside of the cell.

Many of transcribed RNA, such as transfer RNA, ribosomal RNA, and small nuclear RNA, do not undergo translation into proteins.

A number of antibiotics act by inhibiting translation. These include anisomycin, cycloheximide, chloramphenicol, tetracycline, streptomycin, erythromycin, and puromycin. Prokaryotic ribosomes have a different structure from that of eukaryotic ribosomes, and thus antibiotics can specifically target bacterial infections without any harm to a eukaryotic host's cells.

37.1   Basic mechanisms

Further information: Prokaryotic translation and Eukaryotic translation

The basic process of protein production is addition of one amino acid at a time to the end of a protein. This operation is performed by a ribosome. The choice of amino acid type to add is determined by an mRNA molecule. Each amino acid added is matched to a three nucleotide subsequence of the mRNA. For each such triplet possible, the corresponding amino acid is accepted. The successive amino acids added to the chain are matched to successive nucleotide triplets in the mRNA. In this way the sequence of nucleotides in the template mRNA chain determines the sequence of amino acids in

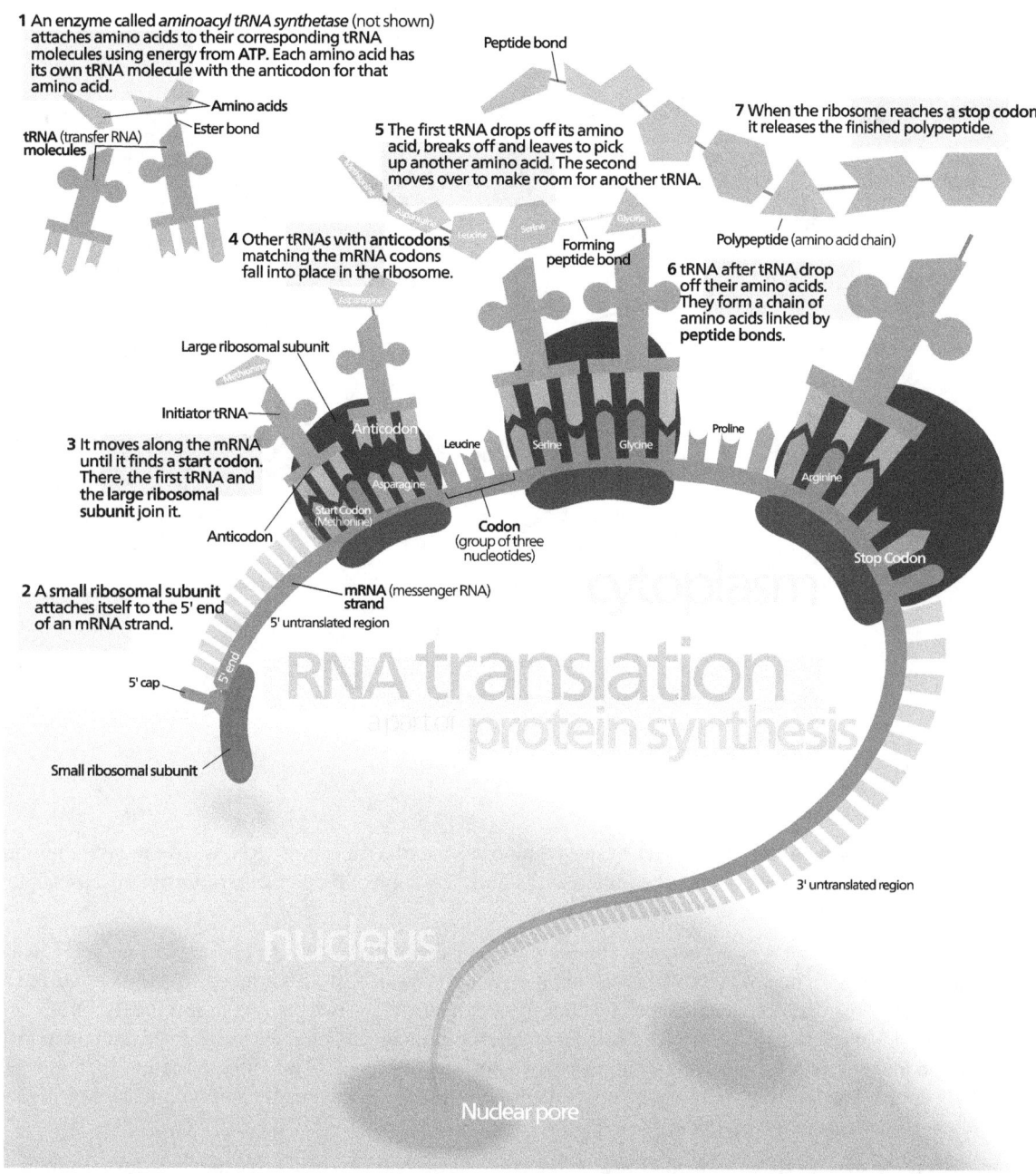

Overview of the translation of eukaryotic messenger RNA

the generated amino acid chain.[1] Addition of an amino acid occurs at the C-terminus of the peptide and thus translation is said to be amino-to-carboxyl directed.[2]

The mRNA carries genetic information encoded as a ribonucleotide sequence from the chromosomes to the ribosomes. The ribonucleotides are "read" by translational machinery in a sequence of nucleotide triplets called codons. Each of those triplets codes for a specific amino acid.

The ribosome molecules translate this code to a specific sequence of amino acids. The ribosome is a multisubunit structure containing rRNA and proteins. It is the "factory" where amino acids are assembled into proteins. tRNAs are small

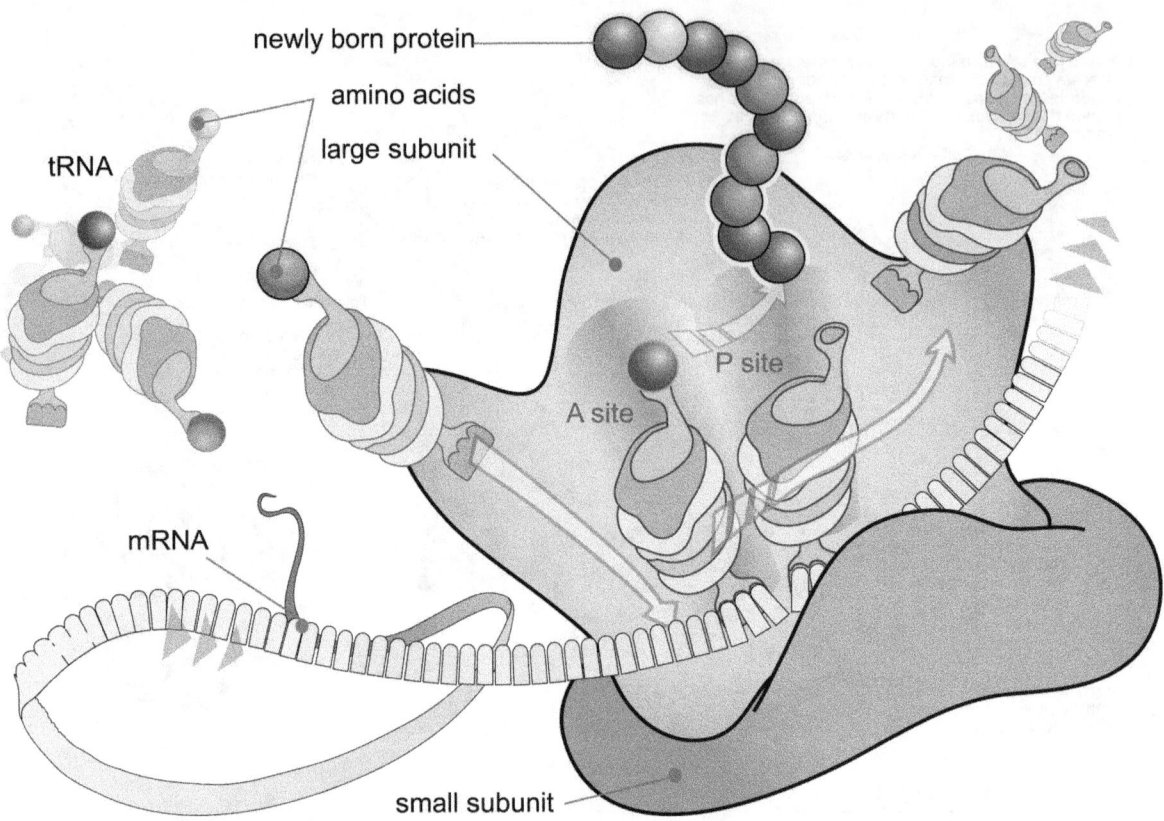

Diagram showing the translation of mRNA and the synthesis of proteins by a ribosome

noncoding RNA chains (74-93 nucleotides) that transport amino acids to the ribosome. tRNAs have a site for amino acid attachment, and a site called an anticodon. The anticodon is an RNA triplet complementary to the mRNA triplet that codes for their cargo amino acid.

Aminoacyl tRNA synthetases (enzymes) catalyze the bonding between specific tRNAs and the amino acids that their anticodon sequences call for. The product of this reaction is an aminoacyl-tRNA. This aminoacyl-tRNA is carried to the ribosome by EF-Tu, where mRNA codons are matched through complementary base pairing to specific tRNA anticodons. Aminoacyl-tRNA synthetases that mispair tRNAs with the wrong amino acids can produce mischarged aminoacyl-tRNAs, which can result in inappropriate amino acids at the respective position in protein. This "mistranslation"[3] of the genetic code naturally occurs at low levels in most organisms, but certain cellular environments cause an increase in permissive mRNA decoding, sometimes to the benefit of the cell.

The ribosome has three sites for tRNA to bind. They are the aminoacyl site (abbreviated A), the peptidyl site (abbreviated P) and the exit site (abbreviated E). With respect to the mRNA, the three sites are oriented 5' to 3' E-P-A, because ribosomes move toward the 3' end of mRNA. The A site binds the incoming tRNA with the complementary codon on the mRNA. The P site holds the tRNA with the growing polypeptide chain. The E site holds the tRNA without its amino acid. When an aminoacyl-tRNA initially binds to its corresponding codon on the mRNA, it is in the A site. Then, a peptide bond forms between the amino acid of the tRNA in the A site and the amino acid of the charged tRNA in the P site. The growing polypeptide chain is transferred to the tRNA in the A site. Translocation occurs, moving the tRNA in the P site, now without an amino acid, to the E site; the tRNA that was in the A site, now charged with the polypeptide chain, is moved to the P site. The tRNA in the E site leaves and another aminoacyl-tRNA enters the A site to repeat the process.[4]

After the new amino acid is added to the chain, and after the mRNA is released out of the nucleus and into the ribosome's core, the energy provided by the hydrolysis of a GTP bound to the translocase EF-G (in prokaryotes) and eEF-2 (in eukaryotes) moves the ribosome down one codon towards the 3' end. The energy required for translation of proteins is

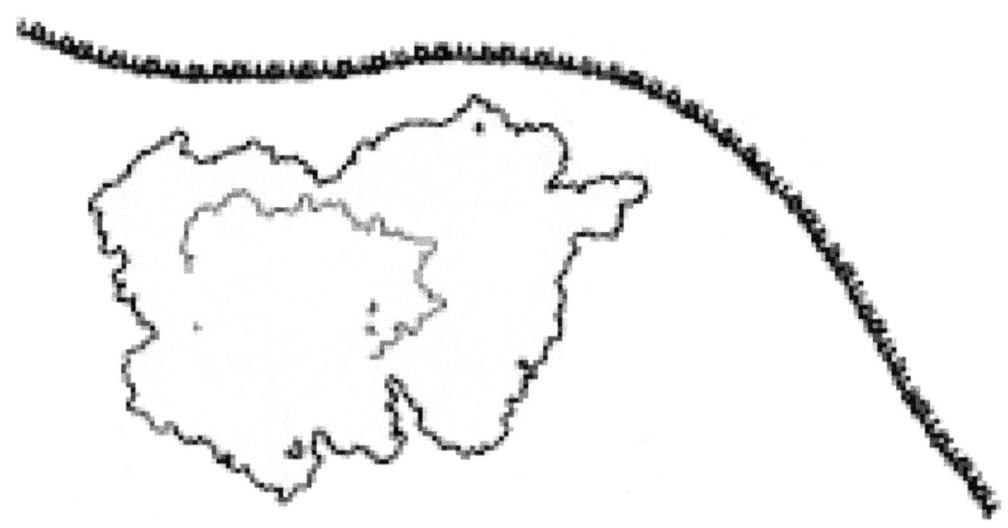

A ribosome translating a protein that is secreted into the endoplasmic reticulum. tRNAs are colored dark blue.

significant. For a protein containing n amino acids, the number of high-energy phosphate bonds required to translate it is $4n-1$. The rate of translation varies; it is significantly higher in prokaryotic cells (up to 17-21 amino acid residues per second) than in eukaryotic cells (up to 6-9 amino acid residues per second).[5]

In activation, the correct amino acid is covalently bonded to the correct transfer RNA (tRNA). The amino acid is joined by its carboxyl group to the 3' OH of the tRNA by an ester bond. When the tRNA has an amino acid linked to it, it is termed "charged". Initiation involves the small subunit of the ribosome binding to the 5' end of mRNA with the help of initiation factors (IF). Termination of the polypeptide happens when the A site of the ribosome faces a stop codon (UAA, UAG, or UGA). No tRNA can recognize or bind to this codon. Instead, the stop codon induces the binding of a release factor protein that prompts the disassembly of the entire ribosome/mRNA complex.

The process of translation is highly regulated in both eukaryotic and prokaryotic organisms. Regulation of translation can impact the global rate of protein synthesis which is closely coupled to the metabolic and proliferative state of a cell. In addition, recent work has revealed that genetic differences and their subsequent expression as mRNAs can also impact

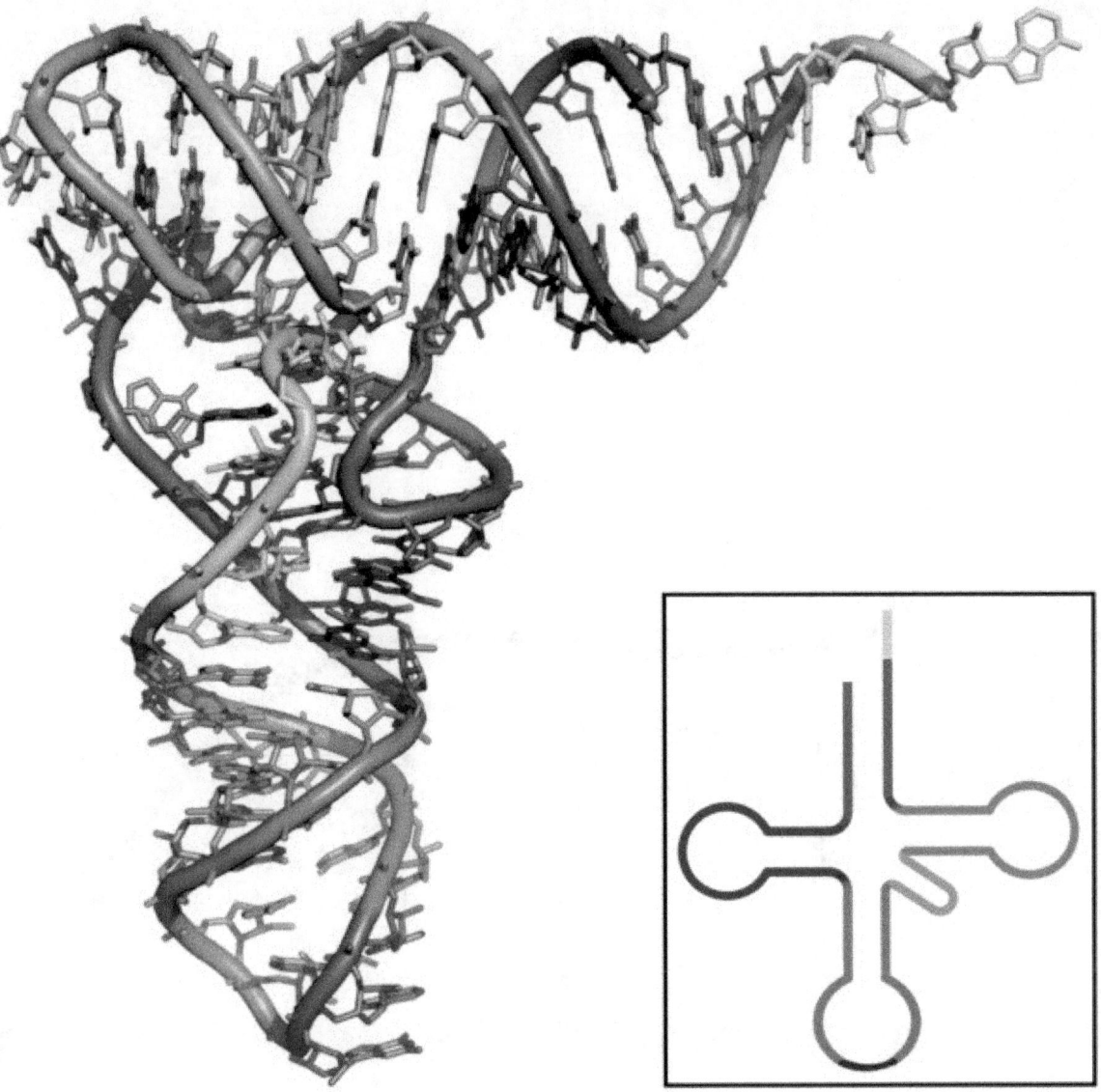

Tertiary structure of tRNA. CCA tail *in orange,* Acceptor stem *in purple,* D arm *in red,* Anticodon arm *in blue with* Anticodon *in black,* T arm *in green.*

translation rate in an RNA-specific manner.[6]

37.2   Genetic code

Main article: Genetic code

Whereas other aspects such as the 3D structure, called tertiary structure, of protein can only be predicted using sophisticated algorithms, the amino acid sequence, called primary structure, can be determined solely from the nucleic acid sequence with the aid of a translation table.

This approach may not give the correct amino acid composition of the protein, in particular if unconventional amino acids such as selenocysteine are incorporated into the protein, which is coded for by a conventional stop codon in combination with a downstream hairpin (SElenoCysteine Insertion Sequence, or SECIS).

There are many computer programs capable of translating a DNA/RNA sequence into a protein sequence. Normally this is performed using the Standard Genetic Code; many bioinformaticians have written at least one such program at some point in their education. However, few programs can handle all the "special" cases, such as the use of the alternative initiation codons. For instance, the rare alternative start codon CTG codes for Methionine when used as a start codon, and for Leucine in all other positions.

Example: Condensed translation table for the Standard Genetic Code (from the NCBI Taxonomy webpage).

AAs = FFLLSSSSYY**CC*WLLLLPPPPHHQQRRRRIIIMTTTTNNKKSSRRVVVVAAAADDEEGGGG Starts = ---M--------------M--------------M-------------------------- Base1 = TTTTTTTTTTTTTTTTCCCCCCCCCCCCCCCC-CAAAAAAAAAAAAAAAAGGGGGGGGGGGGGGGG Base2 = TTTTCCCCAAAAGGGGTTTTCCCCAAAAGGGGTTTTC-CCCAAAAGGGGTTTTCCCCAAAAGGGG Base3 = TCAGTCAGTCAGTCAGTCAGTCAGTCAGTCAGTCAGTCAGTCAGTCAGTCAGTCAGTCAGTCAG

37.2.1 Translation tables

Even when working with ordinary eukaryotic sequences such as the Yeast genome, it is often desired to be able to use alternative translation tables—namely for translation of the mitochondrial genes. Currently the following translation tables are defined by the NCBI Taxonomy Group for the translation of the sequences in GenBank:

1: The Standard 2: The Vertebrate Mitochondrial Code 3: The Yeast Mitochondrial Code 4: The Mold, Protozoan, and Coelenterate Mitochondrial Code and the Mycoplasma/Spiroplasma Code 5: The Invertebrate Mitochondrial Code 6: The Ciliate, Dasycladacean and Hexamita Nuclear Code 9: The Echinoderm and Flatworm Mitochondrial Code 10: The Euplotid Nuclear Code 11: The Bacterial and Plant Plastid Code 12: The Alternative Yeast Nuclear Code 13: The Ascidian Mitochondrial Code 14: The Alternative Flatworm Mitochondrial Code 15: Blepharisma Nuclear Code 16: Chlorophycean Mitochondrial Code 21: Trematode Mitochondrial Code 22: Scenedesmus obliquus mitochondrial Code 23: Thraustochytrium Mitochondrial Code

37.3 See also

- DNA codon table

- Expanded genetic code

- Protein methods

- Start codon

37.4 References

[1] Neill, Campbell (1996). *Biology; Fourth edition*. The Benjamin/Cummings Publishing Company. p. 309,310. ISBN 0-8053-1940-9.

[2] Stryer, Lubert (2002). *Biochemistry; Fifth edition*. W. H. Freeman and Company. p. 826. ISBN 0-7167-4684-0.

[3] Moghal, A., Mohler, K., and Ibba, M. (September 2014). "Mistranslation of the genetic code.". *FEBS Letters* **588**: 4305–10. doi:10.1016/j.febslet.2014.08.035. PMID 25220850.

[4] Griffiths, Anthony (2008). "9". *Introduction to Genetic Analysis* (9th ed.). New York: W.H. Freeman and Company. pp. 335–339. ISBN 978-0-7167-6887-6.

[5] Ross JF, Orlowski M (February 1982). "Growth-rate-dependent adjustment of ribosome function in chemostat-grown cells of the fungus *Mucor racemosus*". *J. Bacteriol.* **149** (2): 650–3. PMC 216554. PMID 6799491.

[6] Cenik C, Cenik ES, Byeon GW, Grubert F, Candille SI, Spacek D, Alsallakh B, Tilgner H, Araya CL, Tang H, Ricci E, Snyder MP (2015). "Integrative analysis of RNA, translation, and protein levels reveals distinct regulatory variation across humans". *Genome Res.* **25**: 1610–21. doi:10.1101/gr.193342.115. PMID 26297486.

37.5   Further reading

- Champe, Pamela C; Harvey, Richard A; Ferrier, Denise R (2004). *Lippincott's Illustrated Reviews: Biochemistry* (3rd ed.). Hagerstwon, MD: Lippincott Williams & Wilkins. ISBN 0-7817-2265-9.

- Cox, Michael; Nelson, David R.; Lehninger, Albert L (2005). *Lehninger principles of biochemistry* (4th ed.). San Francisco...: W.H. Freeman. ISBN 0-7167-4339-6.

- Malys N, McCarthy JEG (2010). "Translation initiation: variations in the mechanism can be anticipated". *Cellular and Molecular Life Sciences* **68** (6): 991–1003. doi:10.1007/s00018-010-0588-z. PMID 21076851.

37.6   External links

- Virtual Cell Animation Collection: Introducing Translation

- Translate tool (from DNA or RNA sequence)

Chapter 38

Euchromatin

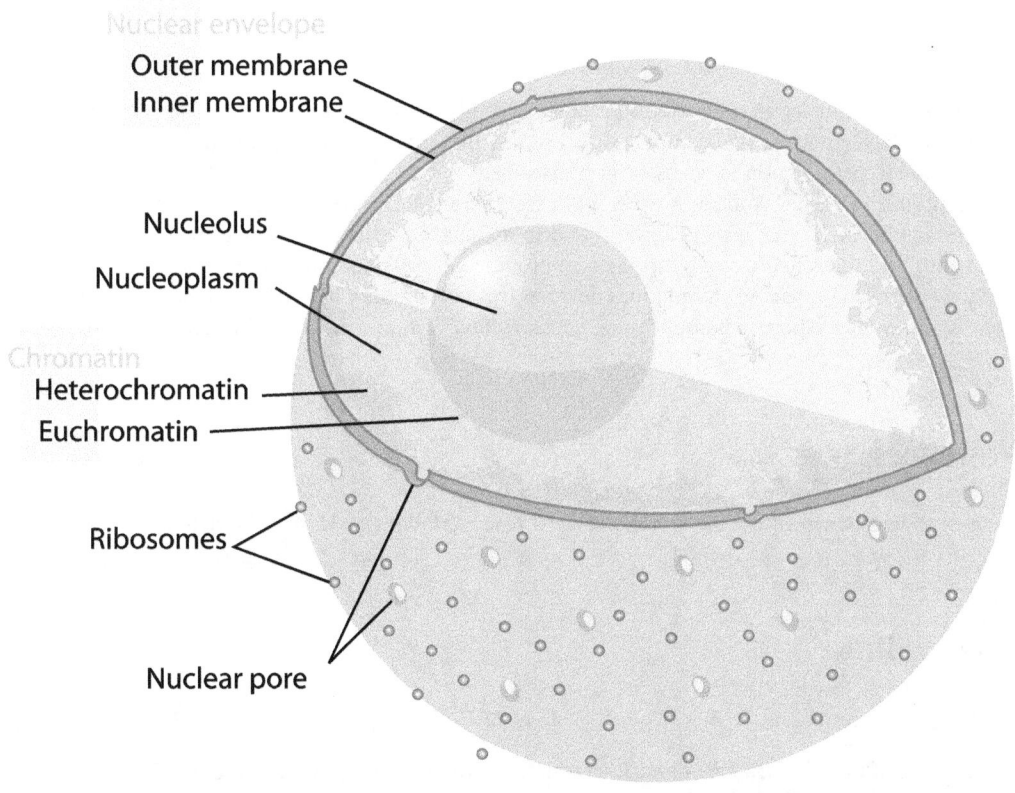

The nucleus of a human cell showing the location of euchromatin

Euchromatin is a lightly packed form of chromatin (DNA, RNA and protein) that is enriched in genes, and is often (but not always) under active transcription. Euchromatin comprises the most active portion of the genome within the cell nucleus. 92% of the human genome is euchromatic.[1] The remainder is called heterochromatin.

38.1   Structure

The structure of euchromatin is reminiscent of an unfolded set of beads along a string, wherein those beads represent nucleosomes. Nucleosomes consist of eight proteins known as histones, with approximately 147 base pairs of DNA wound around them; in euchromatin, this wrapping is loose so that the raw DNA may be accessed. Each core histone possesses a `tail' structure, which can vary in several ways; it is thought that these variations act as "master control switches," which determine the overall arrangement of the chromatin. In particular, it is believed that the presence of methylated lysine 4 on the histone tails acts as a general marker for euchromatin.

38.2   Appearance

In general, euchromatin appears as light-colored bands when stained in G banding and observed under an optical microscope, in contrast to heterochromatin, which stains darkly. This lighter staining is due to the less compact structure of euchromatin. The basic structure of euchromatin is an elongated, open, 10 nm microfibril, as noted by electron microscopy. In prokaryotes, euchromatin is the *only* form of chromatin present; this indicates that the heterochromatin structure evolved later along with the nucleus, possibly as a mechanism to handle increasing genome size.

38.3   Function

Euchromatin participates in the active transcription of DNA to mRNA products. The unfolded structure allows gene regulatory proteins and RNA polymerase complexes to bind to the DNA sequence, which can then initiate the transcription process. Not all euchromatin is necessarily transcribed, but in general that which is not is transformed into heterochromatin to protect the genes while they are not in use. There is therefore a direct link to how actively productive a cell is and the amount of euchromatin that can be found in its nucleus. It is thought that the cell uses transformation from euchromatin into heterochromatin as a method of controlling gene expression and replication, since such processes behave differently on densely compacted chromatin, known as the `accessibility hypothesis'. One example of constitutive euchromatin that is 'always turned on' is housekeeping genes, which code for the proteins needed for basic functions of cell survival.

38.4   References

[1] "Finishing the euchromatic sequence of the human genome". *Nature* **431** (7011): 931–45. 21 October 2004. doi:10.1038/nature03001. PMID 15496913.

38.5   External links

- Research news in Euchromatin

- Zheng C, Hayes J (2003). "Structures and interactions of the core histone tail domains.". *Biopolymers* **68** (4): 539–46. doi:10.1002/bip.10303. PMID 12666178.

- Muegge K (2003). "Modifications of histone cores and tails in V(D)J recombination". *Genome Biol* **4** (4): 211. doi:10.1186/gb-2003-4-4-211. PMC 154571. PMID 12702201. Article

- Histology image: 20102loa – Histology Learning System at Boston University

Chapter 39

Heterochromatin

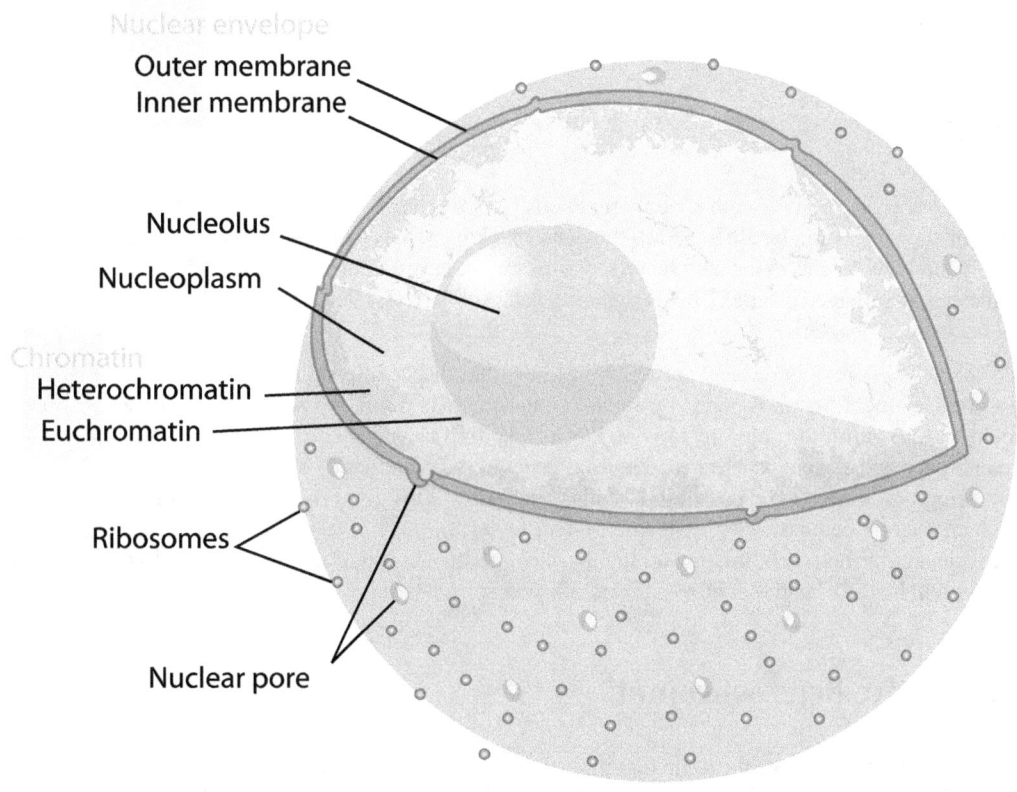

Nuclear envelope
Outer membrane
Inner membrane

Nucleolus
Nucleoplasm
Chromatin
Heterochromatin
Euchromatin

Ribosomes

Nuclear pore

The nucleus of a human cell showing the location of heterochromatin

Heterochromatin is a tightly packed form of DNA, which comes in multiple varieties. These varieties lie on a continuum between the two extremes of **constitutive** and **facultative heterochromatin**. Both play a role in the expression of genes.

Constitutive heterochromatin can affect the genes near them (position-effect variegation). It is usually repetitive and forms structural functions such as centromeres or telomeres, in addition to acting as an attractor for other gene-expression or

447

repression signals.

Facultative heterochromatin is the result of genes that are silenced through a mechanism such as histone deacetylation or Piwi-interacting RNA (piRNA) through RNAi. It is not repetitive and shares the compact structure of constitutive heterochromatin. However, under specific developmental or environmental signaling cues, it can lose its condensed structure and become transcriptionally active.[1]

Heterochromatin has been associated with the di- and tri-methylation of H3K9 in certain portions of the genome.[2]

39.1   Structure

Chromatin is found in two varieties: euchromatin and heterochromatin.[3] Originally, the two forms were distinguished cytologically by how intensely they stained – the euchromatin is less intense, while heterochromatin stains intensely, indicating tighter packing. Heterochromatin is usually localized to the periphery of the nucleus. Despite this early dichotomy, recent evidence in both animals[4] and plants[5] has suggested that there are more than two distinct heterochromatin states, and it may in fact exist in four or five 'states', each marked by different combinations of epigenetic marks.

Heterochromatin mainly consists of genetically inactive satellite sequences,[6] and many genes are repressed to various extents, although some cannot be expressed in euchromatin at all.[7] Both centromeres and telomeres are heterochromatic, as is the Barr body of the second, inactivated X-chromosome in a female.

39.2   Function

Heterochromatin has been associated with several functions, from gene regulation to the protection of chromosome integrity;[8] some of these roles can be attributed to the dense packing of DNA, which makes it less accessible to protein factors that usually bind DNA or its associated factors. For example, naked double-stranded DNA ends would usually be interpreted by the cell as damaged or viral DNA, triggering cell cycle arrest, DNA repair or destruction of the fragment, such as by endonucleases in bacteria.

Some regions of chromatin are very densely packed with fibers that display a condition comparable to that of the chromosome at mitosis. Heterochromatin is generally clonally inherited; when a cell divides, the two daughter cells typically contain heterochromatin within the same regions of DNA, resulting in epigenetic inheritance. Variations cause heterochromatin to encroach on adjacent genes or recede from genes at the extremes of domains. Transcribable material may be repressed by being positioned (in *cis*) at these boundary domains. This gives rise to expression levels that vary from cell to cell,[9] which may be demonstrated by position-effect variegation.[10] Insulator sequences may act as a barrier in rare cases where constitutive heterochromatin and highly active genes are juxtaposed (e.g. the 5'HS4 insulator upstream of the chicken β-globin locus,[11] and loci in two *Saccharomyces* spp.[12][13]).

39.3   Constitutive heterochromatin

All cells of a given species that package the same regions of DNA in constitutive heterochromatin, and thus in all cells any genes contained within the constitutive heterochromatin, will be poorly expressed. For example, all human chromosomes 1, 9, 16, and the Y-chromosome contain large regions of constitutive heterochromatin. In most organisms, constitutive heterochromatin occurs around the chromosome centromere and near telomeres.

39.4   Facultative heterochromatin

The regions of DNA packaged in facultative heterochromatin will not be consistent between the cell types within a species, and thus a sequence in one cell that is packaged in facultative heterochromatin (and the genes within are poorly expressed) may be packaged in euchromatin in another cell (and the genes within are no longer silenced). However, the formation of facultative heterochromatin is regulated, and is often associated with morphogenesis or differentiation. An example of facultative heterochromatin is X chromosome inactivation in female mammals: one X chromosome is packaged as facultative heterochromatin and silenced, while the other X chromosome is packaged as euchromatin and expressed.

Among the molecular components that appear to regulate the spreading of heterochromatin are the Polycomb-group proteins and non-coding genes such as Xist. The mechanism for such spreading is still a matter of controversy.[14]

39.5 Yeast heterochromatin

Saccharomyces cerevisiae, or budding yeast, is a model eukaryote and its heterochromatin has been defined thoroughly. Although most of its genome can be characterized as euchromatin, *S. cerevisiae* has regions of DNA that are transcribed very poorly. These loci are the so-called silent mating type loci (HML and HMR), the rDNA (encoding ribosomal RNA), and the sub-telomeric regions. Fission yeast (*Schizosaccharomyces pombe*) uses another mechanism for heterochromatin formation at its centromeres. Gene silencing at this location depends on components of the RNAi pathway. Double-stranded RNA is believed to result in silencing of the region through a series of steps.

In the fission yeast *Schizosaccharomyces pombe*, two RNAi complexes, the RNAi-induced transcriptional gene silencing (RITS) complex and the RNA-directed RNA polymerase complex (RDRC), are part of an RNAi machinery involved in the initiation, propagation and maintenance of heterochromatin assembly. These two complexes localize in a siRNA-dependent manner on chromosomes, at the site of heterochromatin assembly. RNA polymerase II synthesizes a transcript that serves as a platform to recruit RITS, RDRC and possibly other complexes required for heterochromatin assembly.[15][16] Both RNAi and an exosome-dependent RNA degradation process contribute to heterochromatic gene silencing. These mechanisms of *Schizosaccharomyces pombe* may occur in other eukaryotes.[17] A large RNA structure called RevCen has also been implicated in the production of siRNAs to mediate heterochromatin formation in some fission yeast.[18]

39.6 See also

- Centric heterochromatin

39.7 References

[1] Oberdoerffer, P; Sinclair, D (2007). "The role of nuclear architecture in genomic instability and ageing". *Nature Reviews Molecular Cell Biology* **8**: 692–702. doi:10.1038/nrm2238.

[2] Rosenfeld, Jeffrey A; Wang, Zhibin; Schones, Dustin; Zhao, Keji; Desalle, Rob; Zhang, Michael Q (31 March 2009). "Determination of enriched histone modifications in non-genic portions of the human genome". *BMC Genomics* **10** (1): 143. doi:10.1186/1471-2164-10-143. PMC 2667539. PMID 19335899.

[3] Elgin, S.C. (1996). "Heterochromatin and gene regulation in *Drosophila*". *Current Opinion in Genetics & Development* **6** (2): 193–202. doi:10.1016/S0959-437X(96)80050-5. ISSN 0959-437X.

[4] van Steensel, B. (2011). "Chromatin: constructing the big picture". *The EMBO Journal* **30** (10): 1885–95. doi:10.1038/emboj.2011.135. PMC 3098493. PMID 21527910.

[5] Roudier, François; et al. (2011). "Integrative epigenomic mapping defines four main chromatin states in Arabidopsis". *The EMBO Journal* **30** (10): 1928–1938. doi:10.1038/emboj.2011.103. PMC 3098477. PMID 21487388.

[6] Lohe, A.R.; et al. (August 1, 1993). "Mapping Simple Repeated DNA Sequences in Heterochromatin of Drosophila Melanogaster". *Genetics* **134** (4): 1149–74. ISSN 0016-6731. PMC 1205583. PMID 8375654.

[7] Lu, B.Y.; et al. (June 1, 2000). "Heterochromatin protein 1 is required for the normal expression of two heterochromatin genes in Drosophila". *Genetics* **155** (2): 699–708. ISSN 0016-6731. PMC 1461102. PMID 10835392.

[8] Grewal SIS & Jia S (January 2007). "Heterochromatin revisited". *Nature Reviews Genetics* **8** (1): 35–. doi:10.1038/nrg2008. PMID 17173056. Retrieved 18 September 2013. An up-to-date account of the current understanding of repetitive DNA, which usually doesn't contain genetic information. If evolution makes sense only in the context of the regulatory control of genes, we propose that heterochromatin, which is the main form of chromatin in higher eukaryotes, is positioned to be a deeply effective target for evolutionary change. Future investigations into assembly, maintenance and the many other functions of heterochromatin will shed light on the processes of gene and chromosome regulation.

[9] Fisher, Amanda G.; Matthias Merkenschlager (April 2002). "Gene silencing, cell fate and nuclear organisation". *Current Opinion in Genetics & Development* **12** (2): 193–7. doi:10.1016/S0959-437X(02)00286-1. ISSN 0959-437X. PMID 11893493.

[10] Zhimulev, I.F.; et al. (December 1986). "Cytogenetic and molecular aspects of position effect variegation in Drosophila melanogaster". *Chromosoma* **94** (6): 492–504. doi:10.1007/BF00292759. ISSN 1432-0886.

[11] Burgess-Beusse, B; et al. (December 2002). "The insulation of genes from external enhancers and silencing chromatin". *Proc. Natl Acad. Sci. USA* **9** (Suppl 4): 16433–7. doi:10.1073/pnas.162342499. PMC 139905. PMID 12154228.

[12] Noma, K.; et al. (August 2001). "transitions in distinct histone H3 methylation patterns at the heterochromatin domain boundaries". *Science* **293** (5532): 1150–5. doi:10.1126/science.1064150. PMID 11498594.

[13] Donze, D. & R.T. Kamakaka (2000). "RNA polymerase III and RNA polymerase II promoter complexes are heterochromatin barriers in Saccharomyces cerevisiae". *The EMBO Journal* **20** (3): 520–31. doi:10.1093/emboj/20.3.520. PMC 133458. PMID 11157758.

[14] Talbert PB, Henikoff S (October 2006). "Spreading of silent chromatin: inaction at a distance". *Nature Reviews Genetics* **7** (10): 793–803. doi:10.1038/nrg1920. PMID 16983375.

[15] Kato, H.; et al. (2005). "RNA Polymerase II Is Required for RNAi-Dependent Heterochromatin Assembly". *Science* **309**: 467–469. doi:10.1126/science.1114955.

[16] Djupedal, I.; et al. (2005). "RNA Pol II subunit Rpb7 promotes centromeric transcription and RNAi-directed chromatin silencing". *Genes & Development* **19**: 2301–2306. doi:10.1101/gad.344205.

[17] Vavasseur; et al. (2008). "Heterochromatin Assembly and Transcriptional Gene Silencing under the Control of Nuclear RNAi: Lessons from Fission Yeast". *RNA and the Regulation of Gene Expression: A Hidden Layer of Complexity*. Caister Academic Press. ISBN 978-1-904455-25-7.

[18] Djupedal I, Kos-Braun IC, Mosher RA, et al. (December 2009). "Analysis of small RNA in fission yeast; centromeric siRNAs are potentially generated through a structured RNA". *EMBO J.* **28** (24): 3832–44. doi:10.1038/emboj.2009.351. PMC 2797062. PMID 19942857.

39.8 External links

- Histology image: 20102loa – Histology Learning System at Boston University

- Avramova, Z (May 2002). "Heterochromatin in Animals and Plants. Similarities and Differences". *Plant Physiology* **129** (1): 40–9. doi:10.1104/pp.010981. PMC 1540225. PMID 12011336.

- Caron H, et al. (2001). "The Human Transcriptome Map: Clustering of Highly Expressed Genes in Chromosomal Domains". *Science* **291** (5507): 1289–92. doi:10.1126/science.1056794. PMID 11181992.

- Cha, Ariana Eunjung; Bernstein, Lenny (April 30, 2015). "Scientists discover an important new driver of aging". New York Times. Retrieved May 2015.

Chapter 40

Coding region

The **coding region** of a gene, also known as the **coding sequence** or **CDS** (from **coding DNA sequence**), is that portion of a gene's DNA or RNA, composed of exons, that codes for protein. The region is bounded nearer the 5' end by a start codon and nearer the 3' end with a stop codon. The coding region in mRNA is bounded by the five prime untranslated region and the three prime untranslated region, which are also parts of the exons.[1] The CDS is that portion of an mRNA transcript that is translated by a ribosome.

The coding region of an organism is the sum total of the organism's genome that is composed of gene coding regions.[2]

40.1 Coding sequence annotation

While identification of open reading frames within a DNA sequence is straightforward, identifying coding sequences is not, because the cell translates only a subset of all open reading frames to proteins.[3] Currently CDS prediction uses sampling and sequencing of mRNA from cells, although there is still the problem of determining which parts of a given mRNA are actually translated to protein. CDS prediction is a subset of gene prediction, the latter also including prediction of DNA sequences that code not only for protein but also for other functional elements such as RNA genes and regulatory sequences.

40.2 References

[1] Twyman, Richard (1 August 2003). "Gene Structure". The Wellcome Trust. Retrieved 6 April 2003.

[2] Goto, Mami; et al. (April 8, 2000). *Analysis of CpG Dinucleotide Frequency in Bacterial Genomes with Respect to Genomic Regions and Codon* (PDF). The Fourth Annual International Conference on Computational Molecular Biology, Tokyo, Japan. Retrieved 6 April 2009.

[3] Furuno, Masaaki; Kasukawa, Takeya; Saito, Rintaro; Adachi, Jun; Suzuki, Harukazu; Baldarelli, Richard; Hayashizaki, Yoshihide; Okazaki, Yasushi (September 2011). "CDS Annotation in Full-Length cDNA Sequence" (PDF). *Genome Research* (Cold Spring Harbor Laboratory Press) **21** (9): 1478–1487. doi:10.1101/gr.1060303. Retrieved 18 September 2011

Chapter 41

ENCODE

The **Encyclopedia of DNA Elements** (**ENCODE**) is a public research project launched by the US National Human Genome Research Institute (NHGRI) in September 2003.[1][2][3][4][5]

Intended as a follow-up to the Human Genome Project (Genomic Research), the ENCODE project aims to identify all functional elements in the human genome.

The project involves a worldwide consortium of research groups, and data generated from this project can be accessed through public databases.

41.1   Motivation and significance

Humans are estimated to have approximately 20,000 protein-coding genes (collectively known as the exome), which account for only about 1.5% of DNA in the human genome. The primary goal of the ENCODE project is to determine the role of the remaining component of the genome, much of which was traditionally regarded as "junk" (i.e. DNA that is not transcribed).

Approximately 90% of single-nucleotide polymorphisms in the human genome (that have been linked to various diseases by genome-wide association studies) are found outside of protein-coding regions.[6]

The activity and expression of protein-coding genes can be modulated by the regulome - a variety of DNA elements, such as promoter, transcriptional regulatory sequences and regions of chromatin structure and histone modification. It is thought that changes in the regulation of gene activity can disrupt protein production and cell processes and result in disease (ENCODE Project Background). Determining the location of these regulatory elements and how they influence gene transcription could reveal links between variations in the expression of certain genes and the development of disease.[7]

ENCODE is intended as a comprehensive resource to allow the scientific community to better understand how the genome can affect human health, and to "stimulate the development of new therapies to prevent and treat these diseases".[2]

To date, the project has facilitated the identification of novel DNA regulatory elements, providing new insights into the organization and regulation of our genes and genome, and how differences in DNA sequence could influence disease.[6] One main accomplishment described by the Consortium has been that 80% of the human genome is now "associated with at least one biochemical function".[8][9] Much of this functional non-coding DNA is involved in the regulation of the expression of coding genes.[8] Furthermore the expression of each coding gene is controlled by multiple regulatory sites located both near and distant from the gene. These results demonstrate that gene regulation is far more complex than was previously believed.[10]

41.2   The ENCODE Project

ENCODE is implemented in three phases: the pilot phase, the technology development phase and the production phase.

Along the pilot phase, the ENCODE Consortium evaluated strategies for identifying various types of genomic elements. The goal of the pilot phase was to identify a set of procedures that, in combination, could be applied cost-effectively and at high-throughput to accurately and comprehensively characterize large regions of the human genome. The pilot phase had to reveal gaps in the current set of tools for detecting functional sequences, and was also thought to reveal whether some methods used by that time were inefficient or unsuitable for large-scale utilization. Some of these problems had to be addressed in the ENCODE technology development phase (being executed concurrently with the pilot phase), which aimed to devise new laboratory and computational methods that would improve our ability to identify known functional sequences or to discover new functional genomic elements. The results of the first two phases determined the best path forward for analysing the remaining 99% of the human genome in a cost-effective and comprehensive production phase.[2]

41.2.1   The ENCODE Phase I Project: The Pilot Project

The pilot phase tested and compared existing methods to rigorously analyze a defined portion of the human genome sequence. It was organized as an open consortium and brought together investigators with diverse backgrounds and expertise to evaluate the relative merits of each of a diverse set of techniques, technologies and strategies. The concurrent technology development phase of the project aimed to develop new high throughput methods to identify functional elements. The goal of these efforts was to identify a suite of approaches that would allow the comprehensive identification of all the functional elements in the human genome. Through the ENCODE pilot project, National Human Genome Research Institute (NHGRI) assessed the abilities of different approaches to be scaled up for an effort to analyse the entire human genome and to find gaps in the ability to identify functional elements in genomic sequence.

The ENCODE pilot project process involved close interactions between computational and experimental scientists to evaluate a number of methods for annotating the human genome. A set of regions representing approximately 1% (30 Mb) of the human genome was selected as the target for the pilot project and was analyzed by all ENCODE pilot project investigators. All data generated by ENCODE participants on these regions was rapidly released into public databases.[4][11]

Target Selection

For use in the ENCODE pilot project, defined regions of the human genome - corresponding to 30Mb, roughly 1% of the total human genome - were selected. These regions served as the foundation on which to test and evaluate the effectiveness and efficiency of a diverse set of methods and technologies for finding various functional elements in human DNA.

Prior to embarking upon the target selection, it was decided that 50% of the 30Mb of sequence would be selected manually while the remaining sequence would be selected randomly. The two main criteria for manually selected regions were: 1) the presence of well-studied genes or other known sequence elements, and 2) the existence of a substantial amount of comparative sequence data. A total of 14.82Mb of sequence was manually selected using this approach, consisting of 14 targets that range in size from 500kb to 2Mb.

The remaining 50% of the 30Mb of sequence were composed of thirty, 500kb regions selected according to a stratified random-sampling strategy based on gene density and level of non-exonic conservation. The decision to use these particular criteria was made in order to ensure a good sampling of genomic regions varying widely in their content of genes and other functional elements. The human genome was divided into three parts - top 20%, middle 30%, and bottom 50% - along each of two axes: 1) gene density and 2) level of non-exonic conservation with respect to the orthologous mouse genomic sequence (see below), for a total of nine strata. From each stratum, three random regions were chosen for the pilot project. For those strata underrepresented by the manual picks, a fourth region was chosen, resulting in a total of 30 regions. For all strata, a "backup" region was designated for use in the event of unforeseen technical problems.

In greater detail, the stratification criteria were as follows:

- Gene density: The gene density score of a region was the percentage of bases covered either by genes in the Ensembl database, or by human mRNA best BLAT (BLAST-like alignment tool) alignments in the UCSC Genome Browser database.

- Non-exonic conservation: The region was divided into non-overlapping subwindows of 125 bases. Subwindows that showed less than 75% base alignment with mouse sequence were discarded. For the remaining subwindows, the percentage with at least 80% base identity to mouse, and which did not correspond to Ensembl genes, GenBank mRNA BLASTZ alignments, Fgenesh++ gene predictions, TwinScan gene predictions, spliced EST alignments, or repeated sequences (DNA), was used as the non-exonic conservation score.

The above scores were computed within non-overlapping 500 kb windows of finished sequence across the genome, and used to assign each window to a stratum.[12]

Pilot Phase Results

The pilot phase was successfully finished and the results were published in June 2007 in *Nature*[4] and in a special issue of *Genome Research*;[13] the results published in the first paper mentioned advanced the collective knowledge about human genome function in several major areas, included in the following highlights:[4]

- The human genome is pervasively transcribed, such that the majority of its bases are associated with at least one primary transcript and many transcripts link distal regions to established protein-coding loci.

- Many novel non-protein-coding transcripts have been identified, with many of these overlapping protein-coding loci and others located in regions of the genome previously thought to be transcriptionally silent.

- Numerous previously unrecognized transcription start sites have been identified, many of which show chromatin structure and sequence-specific protein-binding properties similar to well-understood promoters.

- Regulatory sequences that surround transcription start sites are symmetrically distributed, with no bias towards upstream regions.

- chromatin accessibility and histone modification patterns are highly predictive of both the presence and activity of transcription start sites.

- Distal DNaseI hypersensitive sites have characteristic histone modification patterns that reliably distinguish them from promoters; some of these distal sites show marks consistent with insulator function.

- DNA replication timing is correlated with chromatin structure.

- A total of 5% of the bases in the genome can be confidently identified as being under evolutionary constraint in mammals; for approximately 60% of these constrained bases, there is evidence of function on the basis of the results of the experimental assays performed to date.

- Although there is general overlap between genomic regions identified as functional by experimental assays and those under evolutionary constraint, not all bases within these experimentally defined regions show evidence of constraint.

- Different functional elements vary greatly in their sequence variability across the human population and in their likelihood of residing within a structurally variable region of the genome.

- Surprisingly, many functional elements are seemingly unconstrained across mammalian evolution. This suggests the possibility of a large pool of neutral elements that are biochemically active but provide no specific benefit to the organism. This pool may serve as a 'warehouse' for natural selection, potentially acting as the source of lineage-specific elements and functionally conserved but non-orthologous elements between species.

41.2.2 The ENCODE Phase II Project: The Production Phase Project

In September 2007, National Human Genome Research Institute (NHGRI) began funding the production phase of the ENCODE project. In this phase, the goal was to analyze the entire genome and to conduct "additional pilot-scale studies".[14]

As in the pilot project, the production effort is organized as an open consortium. In October 2007, NHGRI awarded grants totaling more than $80 million over four years.[15] The production phase also includes a Data Coordination Center, a Data

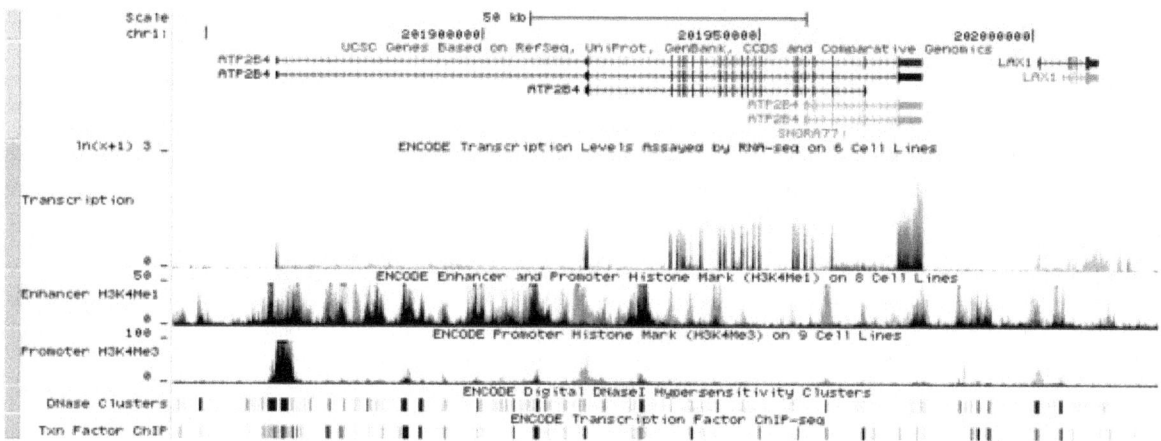

Image of ENCODE data in the UCSC Genome Browser. This shows several tracks containing information on gene regulation. The gene on the left (ATP2B4) is transcribed in a wide variety of cells. The gene on the right is only transcribed in a few types of cells, including embryonic stem cells.

Analysis Center, and a Technology Development Effort.[16] At that time the project evolved into a truly global enterprise, involving 440 scientists from 32 laboratories worldwide. Once the pilot phase was completed, the project "scaled up" in 2007, profiting immensely from new-generation sequencing machines. And the data was, indeed, big; researchers generated around 15 terabytes of raw data.

By 2010, over 1,000 genome-wide data sets had been produced by the ENCODE project. Taken together, these data sets show which regions are transcribed into RNA, which regions are likely to control the genes that are used in a particular type of cell, and which regions are associated with a wide variety of proteins. The primary assays used in ENCODE are ChIP-seq, DNase I Hypersensitivity, RNA-seq, and assays of DNA methylation.

Production Phase Results

In September 2012, the project released a much more extensive set of results, in 30 papers published simultaneously in several journals, including six in *Nature*, six in *Genome Biology* and a special issue with 18 publications of *Genome Research*.[17]

The authors described the production and the initial analysis of 1,640 data sets designed to annotate functional elements in the entire human genome, integrating results from diverse experiments within cell types, related experiments involving 147 different cell types, and all ENCODE data with other resources, such as candidate regions from genome-wide association studies (GWAS) and evolutionary constrained regions. Together, these efforts revealed important features about the organization and function of the human genome, which were summarized in an overview paper as follows:[8]

1. The vast majority (80.4%) of the human genome participates in at least one biochemical RNA and/or chromatin associated event in at least one cell type. Much of the genome lies close to a regulatory event: 95% of the genome lies within 8kb of a DNA-protein interaction (as assayed by bound ChIP-seq motifs or DNaseI footprints), and 99% is within 1.7kb of at least one of the biochemical events measured by ENCODE.

2. Primate-specific elements as well as elements without detectable mammalian constraint show, in aggregate, evidence of negative selection; thus some of them are expected to be functional.

3. Classifying the genome into seven chromatin states suggests an initial set of 399,124 regions with enhancer-like features and 70,292 regions with promoters-like features, as well hundreds of thousands of quiescent regions. High-resolution analyses further subdivide the genome into thousands of narrow states with distinct functional properties.

4. It is possible to quantitatively correlate RNA sequence production and processing with both chromatin marks and transcription factor (TF) binding at promoters, indicating that promoter functionality can explain the majority of RNA expression variation.

5. Many non-coding variants in individual genome sequences lie in ENCODE- annotated functional regions; this number is at least as large as those that lie in protein coding genes.

6. SNPs associated with disease by GWAS are enriched within non-coding functional elements, with a majority residing in or near ENCODE-defined regions that are outside of protein coding genes. In many cases, the disease phenotypes can be associated with a specific cell type or TF.

The most striking finding was that the fraction of human DNA that is biologically active is considerably higher than even the most optimistic previous estimates. In an overview paper, the ENCODE Consortium reported that its members were able to assign biochemical functions to over 80% of the genome.[8] Much of this was found to be involved in controlling the expression levels of coding DNA, which makes up less than 1% of the genome.

The most important new elements of the "encyclopedia" include:

- A comprehensive map of DNase 1 hypersensitive sites, which are markers for regulatory DNA that is typically located adjacent to genes and allows chemical factors to influence their expression. The map identified nearly 3 million sites of this type, including nearly all that were previously known and many that are novel.[18]

- A lexicon of short DNA sequences that form recognition motifs for DNA-binding proteins. Approximately 8.4 million such sequences were found, comprising a fraction of the total DNA roughly twice the size of the exome. Thousands of transcription promoters were found to make use of a single stereotyped 50-base-pair footprint.[19]

- A preliminary sketch of the architecture of the network of human transcription factors, that is, factors that bind to DNA in order to promote or inhibit the expression of genes. The network was found to be quite complex, with factors that operate at different levels as well as numerous feedback loops of various types.[20]

- A measurement of the fraction of the human genome that is capable of being transcribed into RNA. This fraction was estimated to add up to more than 75% of the total DNA, a much higher value than previous estimates. The project also began to characterize the types of RNA transcripts that are generated at various locations.[21]

41.2.3   Data Management and Analysis

Capturing, storing, integrating, and displaying the diverse data generated is challenging. The ENCODE Data Coordination Center (DCC) organizes and displays the data generated by the labs in the consortium, and ensures that the data meets specific quality standards when it is released to the public. Before a lab submits any data, the DCC and the lab draft a data agreement that defines the experimental parameters and associated metadata. The DCC validates incoming data to ensure consistency with the agreement. It then loads the data onto a test server for preliminary inspection, and coordinates with the labs to organize the data into a consistent set of tracks. When the tracks are ready, the DCC Quality Assurance team performs a series of integrity checks, verifies that the data is presented in a manner consistent with other browser data, and perhaps most importantly, verifies that the metadata and accompanying descriptive text are presented in a way that is useful to our users. The data is released on the public UCSC Genome Browser website only after all of these checks have been satisfied. In parallel, data is analyzed by the ENCODE Data Analysis Center, a consortium of analysis teams from the various production labs plus other researchers. These teams develop standardized protocols to analyze data from novel assays, determine best practices, and produce a consistent set of analytic methods such as standardized peak callers and signal generation from alignment pile-ups.[22]

The National Human Genome Research Institute (NHGRI) has identified ENCODE as a "community resource project". This important concept was defined at an international meeting held in Ft. Lauderdale in January 2003 as a research project specifically devised and implemented to create a set of data, reagents, or other material whose primary utility will be as a resource for the broad scientific community. Accordingly, the ENCODE data release policy stipulates that data, once verified, will be deposited into public databases and made available for all to use without restriction.[22]

41.2.4 Future Perspectives

To date, ENCODE has sampled 119 of 1,800 known TFs and general components of the transcriptional machinery on a limited number of cell types and 13 of more than 60 currently known histone or DNA modifications across 147 cell types. DNaseI, FAIRE and extensive RNA assays across subcellular fractionations have been undertaken on many cell types, but overall these data reflect a minor fraction of the potential functional information encoded in the human genome. An important future goal will be to enlarge this dataset to additional factors, modifications and cell types, complementing the other related projects in this area (e.g., Roadmap Epigenomics Project and International Human Epigenome (HEP) Consortium). These projects will constitute foundational resources for human genomics, allowing a deeper interpretation of the organization of gene and regulatory information and the mechanisms of regulation and thereby provide important insights in human health and disease.[8]

41.3 The ENCODE Consortium

The ENCODE Consortium is composed primarily of scientists who were funded by US National Human Genome Research Institute (NHGRI). Other participants contributing to the project are brought up into the Consortium or Analysis Working Group.

The pilot phase consisted of eight research groups and twelve groups participating in the ENCODE Technology Development Phase (ENCODE Pilot Project: Participants and Projects). After 2007, the number of participants grew up to 440 scientists from 32 laboratories worldwide as the pilot phase was officially over. At the moment the consortium consists of different centers which perform different tasks (ENCODE Participants and Projects):

1. ENCODE Production Centers

2. ENCODE Data Coordination Center

3. ENCODE Data Analysis Center

4. ENCODE Computational Analysis Awards

5. ENCODE Technology Development Effort

41.4 Controversy

Although the consortium claims they are far from finished with the ENCODE project, many reactions to the published papers and the news coverage that accompanied the release were favorable. The Nature editors and ENCODE authors "... collaborated over many months to make the biggest splash possible and capture the attention of not only the research community but also of the public at large".[23] The ENCODE project's claim that 80% of the human genome has biochemical function[8] was rapidly picked up by the popular press who described the results of the project as leading to the death of junk DNA.[24][25]

However the conclusion that most of the genome is "functional" has been criticized on the grounds that ENCODE project used a liberal definition of "functional", namely anything that is transcribed must be functional. This conclusion was arrived at despite the widely accepted view, based on genomic conservation estimates from comparative genomics, that many DNA elements such as pseudogenes that are transcribed are nevertheless non-functional . Furthermore the ENCODE project has emphasized sensitivity over specificity leading possibly to the detection of many false positives.[26][27][28] Somewhat arbitrary choice of cell lines and transcription factors as well as lack of appropriate control experiments were additional major criticisms of ENCODE as random DNA mimics ENCODE-like 'functional' behavior.[29]

In response to some of the criticisms, other scientists argued that the wide spread transcription and splicing that is observed in the human genome directly by biochemcial testing is a more accurate indicator of genetic function than genomic conservation estimates because conservation estimates are all relative and difficult to align due to incredible variations in genome sizes of even closely related species, it is partially tautological, and these estimates are not based on direct testing for functionality on the genome.[30][31] Conservation estimates may be used to provide clues to identify possible functional elements in the genome, but it does not limit or cap the total amount of functional elements that could possibly exist in the genome.[31] Furthermore, much of the genome that is being disputed by critics seems to be involved in epigenetic regulation such as gene expression and appears to be necessary for the development of complex organisms.[30][32] The ENCODE results were not necessarily unexpected since increases in attributions of functionality were foreshadowed by previous decades of research.[30][32] Additionally, others have noted that the ENCODE project from the very beginning had a scope that was based on seeking biomedically relevant functional elements in the genome not evolutionary functional elements, which are not necessarily the same thing since evolutionary selection is neither sufficient nor necessary to establish a function. It is a very useful proxy to relevant functions, but an imperfect one and not the only one.[33]

In response to the complaints about the definition of the word "function" some have noted that ENCODE did define what it meant and since the scope of ENCODE was seeking biomedically relevant functional elements in the genome, then the conclusion of the project should be interpreted *"as saying that 80 % of the genome is engaging in relevant biochemical activities that are very likely to have causal roles in phenomena deemed relevant to biomedical research."* [33] The issue of function is more about definitional differences than about the strength of the project, which was in providing data for further research on biochemical activity of non-protein coding parts of DNA. Though definitions are important and science is bounded by the limits of language, it seems that ENCODE has been well received for its purpose since there are now more research papers using ENCODE data than there are papers arguing over the definition of function, as of March 2013.[34] Ewan Birney, one of the ENCODE researchers, commented that "function" was used pragmatically to mean "specific biochemical activity" which included different classes of assays: RNA, "broad" histone modifications, "narrow" histone modifications, DNaseI hypersensitive sites, Transcription Factor ChIP-seq peaks, DNaseI Footprints, Transcription Factor bound motifs, and Exons.[35]

In 2014, ENCODE researchers noted that in the literature, functional parts of the genome have been identified differently in previous studies depending on the approaches used. There have been three general approaches used to identify functional parts of the human genome: genetic approaches (which rely on changes in phenotype), evolutionary approaches (which rely on conservation) and biochemical approaches (which rely on biochemical testing and was used by ENCODE). All three have limitations: genetic approaches may miss functional elements that do not manifest physically on the organism, evolutionary approaches have difficulties using accurate multispecies sequence alignments since genomes of even closely related species vary considerably, and with biochemical approaches, though having high reproducibility, the biochemical signatures do not always automatically signify a function. They concluded that in contrast to evolutionary and genetic evidence, biochemical data offer clues about both the molecular function served by underlying DNA elements and the cell types in which they act and ultimately all three approaches can be used in a complementary way to identify regions that may be functional in human biology and disease. Furthermore, they noted that the biochemical maps provided by ENCODE were the most valuable things from the project since they provide a starting point for testing how these signatures relate to molecular, cellular, and organismal function.[31]

The project has also been criticized for its high cost (~$400 million in total) and favoring big science which takes money away from highly productive investigator-initiated research.[36] The pilot ENCODE project cost an estimated $55 million; the scale-up was about $130 million and the US National Human Genome Research Institute NHGRI could award up to $123 million for the next phase. Some researchers argue that a solid return on that investment has yet to be seen. There have been attempts to scour the literature for the papers in which ENCODE plays a significant part and since 2012 there have been 300 papers, 110 of which come from labs without ENCODE funding. An additional problem is that ENCODE is not a unique name dedicated to the ENCODE project exclusively, so the word 'encode' comes up in many genetics and genomics literature.[6]

Another major critique is that the results do not justify the amount of time spent on the project and that the project itself is essentially unfinishable. Although often compared to Human Genome Project (HGP) and even termed as the HGP next step, the HGP had a clear endpoint which ENCODE currently lacks.

The authors seem to sympathize with the scientific concerns and at the same time try to justify their efforts by giving interviews and explaining ENCODE details not just to the scientific public, but also to mass media. They also claim that it took more than half a century from the realization that DNA is the hereditary material of life to the human genome sequence, so that their plan for the next century would be to really understand the sequence itself.[6]

41.5 modENCODE project

The Model Organism ENCyclopedia Of DNA Elements (modENCODE) project is a continuation of the original EN-CODE project targeting the identification of functional elements in selected model organism genomes, specifically, *Drosophila melanogaster* and *Caenorhabditis elegans*.[37] The extension to model organisms permits biological validation of the computational and experimental findings of the ENCODE project, something that is difficult or impossible to do in humans.[37]

Funding for the modENCODE project was announced by the National Institutes of Health (NIH) in 2007 and included several different research institutions in the US.[38][39]

In late 2010, the modENCODE consortium unveiled its first set of results with publications on annotation and integrative analysis of the worm and fly genomes in *Science*.[40][41] Data from these publications is available from the modENCODE web site.[42]

At the moment, modENCODE is run as a Research Network and the consortium is formed by 11 primary projects, divided between worm and fly. The projects spans the following:

- Gene structure

- mRNA and ncRNA expression profiling

- Transcription factor binding sites

- Histone modifications and replacement

- Chromatin structure

- DNA replication initiation and timing

- Copy number variation.[43]

41.6 FactorBook

The analysis of transcription factor binding data generated by the ENCODE project is currently available in the web-accessible repository FactorBook.[44] Essentially, Factorbook.org is a Wiki-based database for transcription factor-binding data generated by the ENCODE consortium. In the first release, Factorbook contains:

- 457 ChIP-seq datasets on 119 TFs in a number of human cell lines

- The average profiles of histone modifications and nucleosome positioning around the TF-binding regions

- Sequence motifs enriched in the regions and the distance and orientation preferences between motif sites.[45]

41.7 See also

- GENCODE

- SIMAP

- Functional genomics

- Human Genome Project

- 1000 Genomes Project

- International HapMap Project

- List of biological databases

41.8   References

[1] Raney BJ, Cline MS, Rosenbloom KR, Dreszer TR, Learned K, Barber GP, Meyer LR, Sloan CA, Malladi VS, Roskin KM, Suh BB, Hinrichs AS, Clawson H, Zweig AS, Kirkup V, Fujita PA, Rhead B, Smith KE, Pohl A, Kuhn RM, Karolchik D, Haussler D, Kent, WJ (January 2011). "ENCODE whole-genome data in the UCSC genome browser (2011 update)". *Nucleic Acids Res.* **39** (Database issue): D871–5. doi:10.1093/nar/gkq1017. PMC 3013645. PMID 21037257.

[2] The ENCODE Project Consortium (2004). "The ENCODE (ENCyclopedia Of DNA Elements) Project". Science.

[3] ENCODE Project Consortium (2011). Becker PB, ed. "A User's Guide to the Encyclopedia of DNA Elements (ENCODE)". *PLOS Biology* **9** (4): e1001046. doi:10.1371/journal.pbio.1001046. PMC 3079585. PMID 21526222.

[4] ENCODE Project Consortium, Birney E, Stamatoyannopoulos JA, Dutta A, Guigó R, Gingeras TR, Margulies EH, Weng Z, Snyder M, Dermitzakis ET; et al. (2007). "Identification and analysis of functional elements in 1% of the human genome by the ENCODE pilot project". *Nature* **447** (7146): 799–816. Bibcode:2007Natur.447..799B. doi:10.1038/nature05874. PMC 2212820. PMID 17571346.

[5] Guigó R, Flicek P, Abril JF, Reymond A, Lagarde J, Denoeud F, Antonarakis S, Ashburner M, Bajic VB, Birney E, Castelo R, Eyras E, Ucla C, Gingeras TR, Harrow J, Hubbard T, Lewis SE, Reese MG (2006). "EGASP: The human ENCODE Genome Annotation Assessment Project". *Genome Biology* **7**: S2. doi:10.1186/gb-2006-7-s1-s2. PMC 1810551. PMID 16925836.

[6] Maher B (September 2012). "ENCODE: The human encyclopaedia". *Nature* **489** (7414): 46–8. doi:10.1038/489046a. PMID 22962707.

[7] Saey, Tina Hesman (6 October 2012). "Team releases sequel to the human genome". Society for Science & the Public. Retrieved 18 October 2012.

[8] Bernstein BE, Birney E, Dunham I, Green ED, Gunter C, Snyder M (September 2012). "An integrated encyclopedia of DNA elements in the human genome". *Nature* **489** (7414): 57–74. Bibcode:2012Natur.489...57T. doi:10.1038/nature11247. PMC 3439153. PMID 22955616.

[9] Timmer J (2012-09-10). "Most of what you read was wrong: how press releases rewrote scientific history". *Staff / From the Minds of Ars*. Ars Technica. Retrieved 2012-09-10.

[10] Pennisi E (September 2012). "Genomics. ENCODE project writes eulogy for junk DNA". *Science* **337** (6099): 1159, 1161. doi:10.1126/science.337.6099.1159. PMID 22955811.

[11] ENCODE Program Staff (2012-10-18). "ENCODE: Pilot Project: overview". National Human Genome Research Institute.

[12] ENCODE Program Staff (2012-02-19). "ENCODE: Pilot Project: Target Selection". National Human Genome Research Institute.

[13] Weinstock GM (2007). "ENCODE: More genomic empowerment". *Genome Research* **17** (6): 667–668. doi:10.1101/gr.6534207. PMID 17567987.

[14] "Genome.gov | ENCODE and modENCODE Projects". *The ENCODE Project: ENCyclopedia Of DNA Elements*. United States National Human Genome Research Institute. 2011-08-01. Retrieved 2011-08-05.

[15] "National Human Genome Research Institute - Organization". *The NIH Almanac*. United States National Institutes of Health. Retrieved 2011-08-05.

[16] "Genome.gov | ENCODE Participants and Projects". *The ENCODE Project: ENCyclopedia Of DNA Elements*. United States National Human Genome Research Institute. 2011-08-01. Retrieved 2011-08-05.

[17] Ecker JR, Bickmore WA, Barroso I, Pritchard JK, Gilad Y, Segal E (September 2012). "Genomics: ENCODE explained". *Nature* **489** (7414): 52–5. Bibcode:2012Natur.489...52E. doi:10.1038/489052a. PMID 22955614.

[18] Thurman RE, Rynes E, Humbert R, Vierstra J, Maurano MT, Haugen E, Sheffield NC, Stergachis AB, Wang H; et al. (September 2012). "The accessible chromatin landscape of the human genome". *Nature* **489** (7414): 75–82. Bibcode:2012Natur.489...75T. doi:10.1038/nature11232. PMC 3721348. PMID 22955617.

[19] Neph S, Vierstra J, Stergachis AB, Reynolds AP, Haugen E, Vernot B, Thurman RE, John S, Sandstrom R; et al. (September 2012). "An expansive human regulatory lexicon encoded in transcription factor footprints". *Nature* **489** (7414): 83–90. Bibcode:2012Natur.489...83N. doi:10.1038/nature11212. PMID 22955618.

[20] Gerstein MB, Kundaje A, Hariharan M, Landt SG, Yan KK, Cheng C, Mu XJ, Khurana E, Rozowsky J; et al. (September 2012). "Architecture of the human regulatory network derived from ENCODE data". *Nature* **489** (7414): 91–100. Bibcode:2012Natur.489...91G. doi:10.1038/nature11245. PMID 22955619.

[21] Djebali S, Davis CA, Merkel A, Dobin A, Lassmann T, Mortazavi A, Tanzer A, Lagarde J, Lin W; et al. (September 2012). "Landscape of transcription in human cells". *Nature* **489** (7414): 101–8. Bibcode:2012Natur.489..101D. doi:10.1038/nature11233. PMID 22955620.

[22] Brian J. Raney; et al. (2010-10-30). "ENCODE whole-genome data in the UCSC genome browser (2011 update)". Nucleic Acids Research.

[23] Maher B (2012-09-06). "Fighting about ENCODE and junk". *News Blog*. Nature Publishing Group.

[24] Kolata G (2012-09-05). "Far From 'Junk,' DNA Dark Matter Proves Crucial to Health". The New York Times.

[25] Gregory TR (2012-09-06). "The ENCODE media hype machine". Genomicron.

[26] Graur D, Zheng Y, Price N, Azevedo RB, Zufall RA, Elhaik E (2013). "On the immortality of television sets: "function" in the human genome according to the evolution-free gospel of ENCODE". *Genome Biol Evol* **5** (3): 578–90. doi:10.1093/gbe/evt028. PMC 3622293. PMID 23431001.

[27] Moran LA (2013-03-15). "Sandwalk: On the Meaning of the Word "Function"". Sandwalk.

[28] Gregory TR (2013-04-11). "Critiques of ENCODE in peer-reviewed journals. « Genomicron". Genomicron.

[29] White MA, Myers CA, Corbo JC, Cohen BA (July 2013). "Massively parallel in vivo enhancer assay reveals that highly local features determine the cis-regulatory function of ChIP-seq peaks". *Proc. Natl. Acad. Sci. U.S.A.* **110** (29): 11952–7. Bibcode:2013PNAS..11011952W. doi:10.1073/pnas.1307449110. PMID 23818646. Lay summary – *thefinchandpea.com*.

[30] Mattick JS, Dinger ME (2013). "The extent of functionality in the human genome". *The HUGO Journal* **7** (1): 2. doi:10.1186/1877-6566-7-2.

[31] Kellis, M.; et al. (2014). "Defining functional DNA elements in the human genome". *PNAS* **111** (17): 6131–6138. Bibcode:2014PNAS..111.6131K. doi:10.1073/pnas.1318948111. PMC 4035993. PMID 24753594.

[32] Carey, Nessa (2015). *Junk DNA: A Journey Through the Dark Matter of the Genome*. Columbia University Press. ISBN 9780231170840.

[33] Germain, Pierre-Luc; Ratti, Emanuele; Boem, Federico (November 2014). "Junk or Functional DNA? ENCODE and the Function Controversy". *Biology & Philosophy* **29** (6): 807–831. doi:10.1007/s10539-014-9441-3.

[34] Nature Editorial (March 14, 2013). "Form and function". *Nature* **495**: 141–142. doi:10.1038/495141b.

[35] Birney, Ewan (September 5, 2012). "ENCODE: My own thoughts". *Ewan's Blog: Bioinformatician at large*.

[36] Timpson T (2013-03-05). "Debating ENCODE: Dan Graur, Michael Eisen". Mendelspod.

[37] "The modENCODE Project: Model Organism ENCyclopedia Of DNA Elements (modENCODE)". *NHGRI website*. Retrieved 2008-11-13.

[38] "modENCODE Participants and Projects". *NHGRI website*. Retrieved 2008-11-13.

[39] "Berkeley Lab Life Sciences Awarded NIH Grants for Fruit Fly, Nematode Studies". *Lawrence Berkeley National Laboratory website*. 2007-05-14. Retrieved 2008-11-13.

[40] Gerstein MB, Lu ZJ, Van Nostrand EL, Cheng C, Arshinoff BI, Liu T, Yip KY, Robilotto R, Rechtsteiner A; et al. (2010). "Integrative Analysis of the Caenorhabditis elegans Genome by the modENCODE Project". *Science* **330** (6012): 1775–1787. Bibcode:2010Sci...330.1775G. doi:10.1126/science.1196914. PMC 3142569. PMID 21177976.

[41] modENCODE Consortium, Roy S, Ernst J, Kharchenko PV, Kheradpour P, Negre N, Eaton ML, Landolin JM, Bristow CA, Ma L; et al. (2010). "Identification of Functional Elements and Regulatory Circuits by Drosophila modENCODE". *Science* **330** (6012): 1787–1797. Bibcode:2010Sci...330.1787R. doi:10.1126/science.1198374. PMC 3192495. PMID 21177974.

[42] "modENCODE". The National Human Genome Research Institute.

[43] Celniker S (2009-06-11). "Unlocking the secrets of the genome". Nature.

[44] FactorBook

[45] Wang J (2012-11-29). "Factorbook.org: a Wiki-based database for transcription factor-binding data generated by the ENCODE consortium". Nucleic Acid Research.

41.9 External links

- Official website

- ENCODE project at the National Human Genome Research Institute

- Encyclopedia of DNA Elements at the UCSC Genome Browser

- ENCODE/GENCODE project at the Wellcome Trust Sanger Institute

- ENCODE-sponsored introductory tutorial

- FactorBook

- modENCODE

- ENCODE threads Explorer at the Nature (journal)

Chapter 42

V(D)J recombination

V(D)J recombination, less commonly known as **somatic recombination**, is the unique mechanism of genetic recombination that occurs only in developing lymphocytes during the early stages of T and B cell maturation. The process results in the highly diverse repertoire of antibodies/immunoglobulins (Igs) and T cell receptors (TCRs) found on B cells and T cells, respectively. The process is a defining feature of the adaptive immune system and its development was a key event in the evolution of jawed vertebrates.

V(D)J recombination occurs in the primary lymphoid organs (bone marrow for B cells and thymus for T cells) and in a nearly random fashion rearranges variable (V), joining (J), and in some cases, diversity (D) gene segments. The process ultimately results in novel amino acid sequences in the antigen-binding regions of Igs and TCRs that allow for the recognition of antigens from nearly all pathogens including bacteria, viruses, parasites, and worms as well as "altered self cells" as seen in cancer. The recognition can also be allergic in nature (*e.g.*, to pollen or other allergens) or may be "autoreactive" and lead to autoimmunity.

In 1987, Susumu Tonegawa was awarded the Nobel Prize in Physiology or Medicine[1] "for his discovery of the genetic principle for generation of antibody diversity."[2]

42.1   Background

Human antibody molecules (and B cell receptors) are composed of heavy and light chains with both *constant* (C) and *variable* (V) regions that are encoded by genes on three loci.

1. Immunoglobulin heavy locus (IGH@) on chromosome 14, containing gene segments for the immunoglobulin heavy chain

2. Immunoglobulin kappa (κ) locus (IGK@) on chromosome 2, containing gene segments for the immunoglobulin light chain

3. Immunoglobulin lambda (λ) locus (IGL@) on chromosome 22, containing gene segments for the immunoglobulin light chain

Each heavy chain and light chain gene contains multiple copies of three different types of gene segments for the variable regions of the antibody proteins. For example, the human immunoglobulin heavy chain region contains 2 Constant (Cμ and Cδ) gene segments, 44 Variable (V) gene segments[3] plus 27 Diversity (D) gene segments and 6 Joining (J) gene segments.[4] The light chains also possess 2 Constant (Cμ and Cδ) gene segments and numerous V and J gene segments, but do not have D gene segments. DNA rearrangement causes one copy of each type of gene segment to go in any given lymphocyte, generating an enormous antibody repertoire; roughly 3×10^{11} combinations are possible, although some are removed due to self reactivity.

Most T cell receptors are composed of an alpha chain and a beta chain. The T cell receptor genes are similar to immunoglobulin genes in that they too contain multiple V, D and J gene segments in their beta chains (and V and J gene segments in their alpha chains) that are rearranged during the development of the lymphocyte to provide that cell with a unique antigen receptor. The T cell receptor in this sense is the topological equivalent to an antigen-binding fragment of the antibody, both being part of the immunoglobulin superfamily.

Failure of the cell to create a successful product that does not self-react leads to apoptosis. Autoimmunity is prevented by eliminating lymphocytes that self-react in the thymus by testing them against an array of self antigens expressed through the function of Aire.

42.2   In immunoglobulins

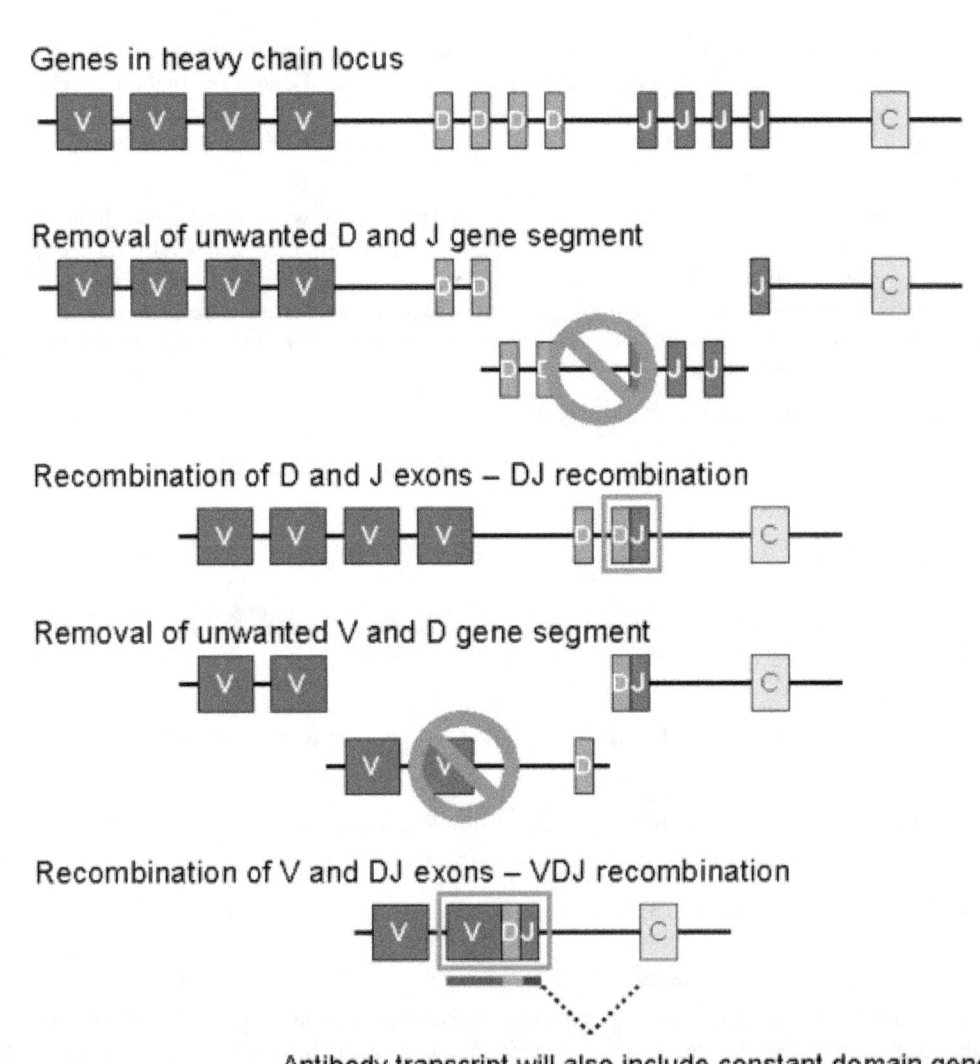

Simplistic overview of V(D)J recombination of immunoglobulin heavy chains

42.2.1 Heavy chain

In the developing B cell, the first recombination event to occur is between one D and one J gene segment of the heavy chain locus. Any DNA between these two gene segments is deleted. This D-J recombination is followed by the joining of one V gene segment, from a region upstream of the newly formed DJ complex, forming a rearranged VDJ gene segment. All other gene segments between V and D segments are now deleted from the cell's genome. Primary transcript (unspliced RNA) is generated containing the VDJ region of the heavy chain and both the constant *mu* and *delta* chains (Cµ and Cδ). (i.e. the primary transcript contains the segments: V-D-J-Cµ-Cδ). The primary RNA is processed to add a polyadenylated (poly-A) tail after the Cµ chain and to remove sequence between the VDJ segment and this constant gene segment. Translation of this mRNA leads to the production of the Ig µ heavy chain protein.

42.2.2 Light chain

The kappa (κ) and lambda (λ) chains of the immunoglobulin light chain loci rearrange in a very similar way, except the light chains lack a D segment. In other words, the first step of recombination for the light chains involves the joining of the V and J chains to give a VJ complex before the addition of the constant chain gene during primary transcription. Translation of the spliced mRNA for either the kappa or lambda chains results in formation of the Ig κ or Ig λ light chain protein. Interestingly, the immunoglobulin lambda light chain locus contains protein-coding genes that can be lost with its rearrangement.[5] This is based on a physiological mechanism and is not pathogenetic for leukemias or lymphomas.[6] However, the rearrangement of several lambda variable subgenes can activate expression of an overlapping miRNA gene, which has consequences for gene expression regulation.[7]

Assembly of the Ig µ heavy chain and one of the light chains results in the formation of membrane bound form of the immunoglobulin IgM that is expressed on the surface of the immature B cell.

42.3 In T cell receptors

During thymocyte development, the T cell receptor (TCR) chains undergo essentially the same sequence of ordered recombination events as that described for immunoglobulins. D-to-J recombination occurs first in the β chain of the TCR. This process can involve either the joining of the $D_\beta1$ gene segment to one of six $J_\beta1$ segments or the joining of the $D_\beta2$ gene segment to one of seven $J_\beta2$ segments. DJ recombination is followed (as above) with V_β-to-$D_\beta J_\beta$ rearrangements. All gene segments between the V_β-D_β-J_β gene segments in the newly formed complex are deleted and the primary transcript is synthesized that incorporates the constant domain gene (V_β-D_β-J_β-C_β). mRNA transcription splices out any intervening sequence and allows translation of the full length protein for the TCR C_β chain.

The rearrangement of the alpha (α) chain of the TCR follows β chain rearrangement, and resembles V-to-J rearrangement described for Ig light chains (see above). The assembly of the β- and α- chains results in formation of the αβ-TCR that is expressed on a majority of T cells.

42.4 Mechanism

42.4.1 Key Enzymes and Components

The process of V(D)J recombination is mediated by VDJ recombinase, which is a diverse collection of enzymes. The key enzymes involved are recombination activating genes 1 and 2 (RAG), terminal deoxynucleotidyl transferase (TdT), and Artemis nuclease, a member of the ubiquitous non-homologous end joining (NHEJ) pathway for DNA repair.[8] Several other enzymes are known to be involved in the process and include DNA-dependent protein kinase (DNA-PK), X-ray repair cross-complementing protein 4 (XRCC4), DNA ligase IV, non-homologous end-joining factor 1 (NHEJ1; also known as Cernunnos or XRCC4-like factor [XLF]), and DNA polymerases λ and µ.[9] Some enzymes involved are specific to lymphocytes (*e.g.*, RAG, TdT), while others are found in other cell types and even ubiquitously (*e.g.*, NHEJ components).

To maintain the specificity of recombination, V(D)J recombinase recognizes and binds to Recombination Signal Sequences (RSSs) flanking the variable (V), diversity (D), and joining (J) genes segments. RSSs are composed of three elements: a heptamer of seven conserved nucleotides, a spacer region of 12 or 23 basepairs in length, and a nonamer of nine conserved nucleotides. While the majority of RSSs vary in sequence, the consensus heptamer and nonamer sequences are CACAGTG and ACAAAAACC, respectively; and although the sequence of the spacer region is poorly conserved, the length is highly conserved.[10][11] The length of the spacer region corresponds to approximately one (12 basepairs) or two turns (23 basepairs) of the DNA helix. Following what is known as the 12/23 Rule, gene segments to be recombined are usually adjacent to RSSs of different spacer lengths (*i.e.*, one has a "12RSS" and one has a "23RSS").[12] This is an important feature in the regulation of V(D)J recombination.[13]

42.4.2   Process

V(D)J recombination begins when V(D)J recombinase (through the activity of RAG1) binds an RSS flanking a coding gene segment (V, D, or J) and creates a single-strand nick in the DNA between the first base of the RSS (just before the heptamer) and the coding segment. This is essentially energetically neutral (no need for ATP hydrolysis) and results in the formation of a free 3' hydroxyl group and a 5' phosphate group on the same strand. The reactive hydroxyl group is positioned by the recombinase to attack the phosphodiester bond of opposite strand, forming two DNA ends: a hairpin (stem-loop) on the coding segment and a blunt end on the signal segment.[14] The current model is that DNA nicking and hairpin formation occurs on both strands simultaneously (or nearly so) in a complex known as a *recombination center*.[15][16][17][18]

The blunt signal ends are flushly ligated together to form a circular piece of DNA containing all of the intervening sequences between the coding segments known as a signal joint (although circular in nature, this is not to be confused with a plasmid). While originally thought to be lost during successive cell divisions, there is evidence that signal joints may re-enter the genome and lead to pathologies by activating oncogenes or interrupting tumor suppressor gene function(s).

The coding ends are processed further prior to their ligation by several events that ultimately lead to junctional diversity.[19] Processing begins when DNA-PK binds to each broken DNA end and recruits several other proteins including Artemis, XRCC4, DNA ligase IV, Cernunnos, and several DNA polymerases.[20] DNA-PK forms a complex that leads to its autophosphorylation, resulting in activation of Artemis. The coding end hairpins are opened by the activity of Artemis.[21] If they are opened at the center, a blunt DNA end will result; however in many cases, the opening is "off-center" and results in extra bases remaining on one strand (an overhang). These are known as palindromic (P) nucleotides due to the palindromic nature of the sequence produced when DNA repair enzymes resolve the overhang.[22] The process of hairpin opening by Artemis is a crucial step of V(D)J recombination and is defective in the severe combined immunodeficiency (scid) mouse model.

Next, XRCC4, Cernunnos, and DNA-PK align the DNA ends and recruit terminal deoxynucleotidyl transferase (TdT), a template-independent DNA polymerase that adds non-templated (N) nucleotides to the coding end. The addition is mostly random, but TdT does exhibit a preference for G/C nucleotides.[23] As with all known DNA polymerases, the TdT adds nucleotides to one strand in a 5' to 3' direction.[24]

Lastly, exonucleases can remove bases from the coding ends (including any P or N nucleotides that may have formed). DNA polymerases λ and μ then insert additional nucleotides as needed to make the two ends compatible for joining. This is a stochastic process, therefore any combination of the addition of P and N nucleotides and exonucleolytic removal can occur (or none at all). Finally, the processed coding ends are ligated together by DNA ligase IV.[25]

All of these processing events result in an antigen-binding region that is highly variable, even when the same gene segments are recombined. V(D)J recombination allows for the generation of immunoglobulins and T cell receptors to antigens that neither the organism nor its ancestor(s) need to have previously encountered, allowing for an adaptive immune response to novel pathogens that develop or to those that frequently change (*e.g.*, seasonal influenza). However, a major caveat to this process is that the DNA sequence must remain in-frame in order to maintain the correct amino acid sequence in the final protein product. If the resulting sequence is out-of-frame, the development of the cell will be arrested, and the cell will not survive to maturity. V(D)J recombination is therefore a very costly process that must be (and is) strictly regulated and controlled.

42.5 See also

- Antibody

- B cell

- B cell receptor

- T cell

- T cell receptor

- Recombination-activating gene

- NKT cell

42.6 References

[1] "The Nobel Prize in Physiology or Medicine 1987". *nobelprize.org*. Retrieved 26 December 2014.

[2] Tonegawa, Susumu (1983). "Somatic Generation of Antibody Diversity". *Nature* **302** (5909): 575–581. doi:10.1038/302575a0. PMID 6300689.

[3] Matsuda, F; Ishii, K; Bourvagnet, P; Kuma, K; Hayashida, H; Miyata, T; Honjo, T (1998). "The complete nucleotide sequence of the human immunoglobulin heavy chain variable region locus". *The Journal of experimental medicine* **188** (11): 2151–62. doi:10.1084/jem.188.11.2151. PMC 2212390. PMID 9841928.

[4] Li A, Rue M, Zhou J, et al. (June 2004). "Utilization of Ig heavy chain variable, diversity, and joining gene segments in children with B-lineage acute lymphoblastic leukemia: implications for the mechanisms of VDJ recombination and for pathogenesis". *Blood* **103** (12): 4602–9. doi:10.1182/blood-2003-11-3857. PMID 15010366.

[5] Mraz, M.; Stano Kozubik, K.; Plevova, K.; Musilova, K.; Tichy, B.; Borsky, M.; Kuglik, P.; Doubek, M.; Brychtova, Y.; Mayer, J.; Pospisilova, S. (2013). "The origin of deletion 22q11 in chronic lymphocytic leukemia is related to the rearrangement of immunoglobulin lambda light chain locus". *Leukemia Research* **37** (7): 802–808. doi:10.1016/j.leukres.2013.03.018. PMID 23608880.

[6] Mraz, M.; Stano Kozubik, K.; Plevova, K.; Musilova, K.; Tichy, B.; Borsky, M.; Kuglik, P.; Doubek, M.; Brychtova, Y.; Mayer, J.; Pospisilova, S. (2013). "The origin of deletion 22q11 in chronic lymphocytic leukemia is related to the rearrangement of immunoglobulin lambda light chain locus". *Leukemia Research* **37** (7): 802–808. doi:10.1016/j.leukres.2013.03.018. PMID 23608880.

[7] Mraz, M.; Dolezalova, D.; Plevova, K.; Stano Kozubik, K.; Mayerova, V.; Cerna, K.; Musilova, K.; Tichy, B.; Pavlova, S.; Borsky, M.; Verner, J.; Doubek, M.; Brychtova, Y.; Trbusek, M.; Hampl, A.; Mayer, J.; Pospisilova, S. (2012). "MicroRNA-650 expression is influenced by immunoglobulin gene rearrangement and affects the biology of chronic lymphocytic leukemia". *Blood* **119** (9): 2110–2113. doi:10.1182/blood-2011-11-394874. PMID 22234685.

[8] Ma, Yunmei; Lu, Haihui; Schwarz, Klaus; Lieber, Michael (September 2005). "Repair of Double-Strand DNA Breaks by the Human Nonhomologous DNA End Joining Pathway: the Iterative Processing Model". *Cell Cycle* **4** (9): 1193–1200. doi:10.4161/cc.4.9.1977. PMID 16082219.

[9] Malu, Shruti; Malshetty, Vidyasagar; Francis, Dailia; Cortes, Patricia (2012). "Role of non-homologous end joining in V(D)J recombination.". *Immunologic Research* **54** (1-3): 233–246. doi:10.1007/s12026-012-8329-z. PMID 22569912.

[10] Ramsden, Dale; Baetz, Kristin; Wu, Gillian (1994). "Conservation of Sequence in Recombination Signal Sequence Spacers". *Nucleic Acids Research* **22** (10): 1785–1796. doi:10.1093/nar/22.10.1785. PMID 8208601.

[11] Cowell, Lindsay; Davila, Marco; Ramsden, Dale; Kelsoe, Garnett (2004). "Computational tools for understanding sequence variability in recombination signals.". *Immunological Reviews* **200**: 57–69. doi:10.1111/j.0105-2896.2004.00171.x. PMID 15242396.

[12] van Gent, Dik; Ramsden, Dale; Gellert, Martin (1996). "The RAG1 and RAG2 Proteins Establish the 12/23 Rule in V(D)J Recombination". *Cell* **85** (1): 107–13. doi:10.1016/s0092-8674(00)81086-7. PMID 8620529.

[13] Hiom, Kevin; Gellert, Martin (1998). "Assembly of a 12/23 Paired Signal Complex: a Critical Control Point in V(D)J Recombination". *Molecular Cell* **1** (7): 1011–1019. doi:10.1016/s1097-2765(00)80101-x. PMID 9651584.

[14] Schatz, David; Swanson, Patrick (2011). "V(D)J Recombination: Mechanisms of Initiation". *Annual Review of Genetics* **45**: 167–202. doi:10.1146/annurev-genet-110410-132552. PMID 21854230.

[15] Schatz, David; Ji, Yanhong (2011). "Recombination Centres and the Orchestration of V(D)J Recombination". *Nature Reviews Immunology* **11** (4): 251–263. doi:10.1038/nri2941. PMID 21394103.

[16] Curry, John; Geier, Jamie; Schlissel, Mark (2005). "Single-Strand Recombination Signal Sequence Nicks in vivo: Evidence for a Capture Model of Synapsis". *Nature Immunology* **6** (12): 1272–1279. doi:10.1038/ni1270. PMID 16286921.

[17] Agrawal, Alka; Schatz, David (1997). "RAG1 and RAG2 Form a Stable Postcleavage Synaptic Complex with DNA Containing Signal Ends in V(D)J Recombination". *Cell* **89** (1): 43–53. doi:10.1016/s0092-8674(00)80181-6. PMID 9094713.

[18] Fugmann, Sebastian; Lee, Alfred; Shockett, Penny; Villey, Isabelle; Schatz, David (2000). "The RAG Proteins and V(D)J Recombination: Complexes, Ends, and Transposition". *Annual Review of Immunology* **18**: 495–527. doi:10.1146/annurev.immunol.18.1.495. PMID 10837067.

[19] Lewis, Susanna (1994). "The Mechanism of V(D)J Joining: Lessons from Molecular, Immunological, and Comparative Analyses". *Advances in Immunology* **56**: 27–150. doi:10.1016/s0065-2776(08)60450-2. PMID 8073949.

[20] Helmink, Beth; Sleckman, Barry (2012). "The response to and repair of RAG-mediated DNA double-strand breaks.". *Annual Review of Immunology* **30**: 175–202. doi:10.1146/annurev-immunol-030409-101320. PMID 22224778.

[21] Ma, Yunmei; Schwarz, Klaus; Lieber, Michael (2005). "The Artemis:DNA-PKcs Endonuclease Cleaves DNA Loops, Flaps, and Gaps". *DNA Repair* **4** (7): 845–851. doi:10.1016/j.dnarep.2005.04.013. PMID 15936993.

[22] Lu, Haihui; Schwarz, Klaus; Lieber, Michael (2007). "Extent to Which Hairpin Opening by the Artemis:DNA-PKcs Complex can Contribute to Junctional Diversity in V(D)J Recombination". *Nucleic Acids Research* **35** (20): 6917–6923. doi:10.1093/nar/gkm823. PMC 2175297. PMID 17932067.

[23] Gauss, George; Lieber, Michael (1996). "Mechanistic Constraints on Diversity in Human V(D)J Recombination". *Molecular and Cellular Biology* **16** (1): 258–269. PMID 8524303.

[24] Benedict, Cindy; Gilfillan, Susan; Thai, To-Ha; Kearney, John (2000). "Terminal Deoxynucleotidyl Transferase and Repertoire Development.". *Immunological Reviews* **175**: 150–157. doi:10.1111/j.1600-065x.2000.imr017518.x. PMID 10933600.

[25] D.C. van Gent and M. van der Burg (2007). "Non-Homologous End-Joining, a Sticky Affair". *Oncogene* **26** (56): 7731–7740. doi:10.1038/sj.onc.1210871. PMID 18066085.

42.7   Further reading

- Hartwell LH, Hood L, Goldberg ML, Reynolds AE, Silver LM, Veres RC (2000). *Chapter 24, Evolution at the molecular level. In: Genetics*. New York: McGraw-Hill. pp. 805–807. ISBN 0-07-299587-4.

- V(D)J Recombination. Series: Advances in Experimental Medicine and Biology, Vol. 650 Ferrier, Pierre (Ed.) Landes Bioscience 2009, XII, 199 p. ISBN 978-1-4419-0295-5

Chapter 43

Pseudogene

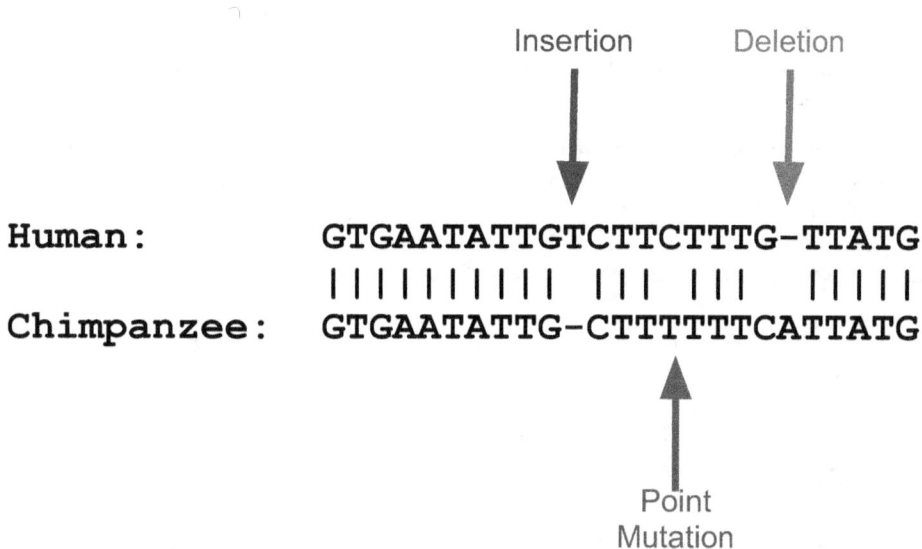

An illustration of the mutations that can cause pseudogenes. The human sequence is of a pseudogene in the olfactory gene family. The chimpanzee sequence is the functional ortholog. Key differences are highlighted.

Pseudogenes are functionless relatives of genes that have lost their gene expression in the cell or their ability to code protein.[1] Pseudogenes often result from the accumulation of multiple mutations within a gene whose product is not required for the survival of the organism. Although not protein-coding, the DNA of pseudogenes may be functional,[2] similar to other kinds of non-coding DNA which can have a regulatory role.

Although some do not have introns or promotor (these pseudogenes are copied from mRNA and incorporated into the chromosome and are called processed pseudogenes),[3] most have some gene-like features such as promoters, CpG islands, and splice sites. They are different from normal genes due to a lack of protein-coding ability resulting from a variety of disabling mutations (e.g. premature stop codons or frameshifts), a lack of transcription, or their inability to encode RNA (such as with rRNA pseudogenes). The term was coined in 1977 by Jacq et al.[4]

Because pseudogenes are generally thought of as the last stop for genomic material that is to be removed from the genome,[5] they are often labeled as junk DNA. We can define a pseudogene operationally as a fragment of nucleotide sequence that resembles a known protein's domains but with stop codons or frameshifts mid-domain. Nonetheless, pseudogenes contain fascinating biological and evolutionary histories within their sequences. This is due to a pseudogene's shared ancestry with a functional gene: in the same way that Darwin thought of two species as possibly having a shared common ancestry followed by millions of years of evolutionary divergence (see speciation), a pseudogene and its associated functional gene also share a common ancestor and have diverged as separate genetic entities over millions of years.

43.1   Properties

Pseudogenes are characterized by a combination of homology to a known gene and nonfunctionality. That is, although every pseudogene has a DNA sequence that is similar to some functional gene, they are nonetheless unable to produce functional final protein products.[6] Pseudogenes are sometimes difficult to identify and characterize in genomes, because the two requirements of homology and nonfunctionality are usually implied through sequence alignments rather than biologically proven.

1. Homology is implied by sequence identity between the DNA sequences of the pseudogene and parent gene. After aligning the two sequences, the percentage of identical base pairs is computed. A high sequence identity (usually between 40% and 100%) means that it is highly likely that these two sequences diverged from a common ancestral sequence (are homologous), and highly unlikely that these two sequences have evolved independently (see Convergent evolution).

2. Nonfunctionality can manifest itself in many ways. Normally, a gene must go through several steps to a fully functional protein: Transcription, pre-mRNA processing, translation, and protein folding are all required parts of this process. If any of these steps fails, then the sequence may be considered nonfunctional. In high-throughput pseudogene identification, the most commonly identified disablements are premature stop codons and frameshifts, which almost universally prevent the translation of a functional protein product.

Pseudogenes for RNA genes are usually more difficult to discover as they do not need to be translated and thus do not have "reading frames".

43.2   Types and origin

There are three main types of pseudogenes, all with distinct mechanisms of origin and characteristic features. The classifications of pseudogenes are as follows:

43.2.1   Processed

Processed (or retrotransposed) pseudogenes. In higher eukaryotes, particularly mammals, retrotransposition is a fairly common event that has had a huge impact on the composition of the genome. For example, somewhere between 30% - 44% of the human genome consists of repetitive elements such as SINEs and LINEs (see retrotransposons).[7][8] In the process of retrotransposition, a portion of the mRNA transcript of a gene is spontaneously reverse transcribed back into DNA and inserted into chromosomal DNA. Although retrotransposons usually create copies of themselves, it has been shown in an *in vitro* system that they can create retrotransposed copies of random genes, too.[9] Once these pseudogenes are inserted back into the genome, they usually contain a poly-A tail, and usually have had their introns spliced out; these are both hallmark features of cDNAs. However, because they are derived from a mature mRNA product, processed pseudogenes also lack the upstream promoters of normal genes; thus, they are considered "dead on arrival", becoming non-functional pseudogenes immediately upon the retrotransposition event.[10] However, these insertions occasionally contribute exons to existing genes, usually via alternatively spliced transcripts.[11] A further characteristic of processed pseudogenes is common truncation of the 5' end relative to the parent sequence, which is a result of the relatively non-processive retrotransposition mechanism that creates processed pseudogenes.[12]

43.2.2 Non-processed

Non-processed (or duplicated) pseudogenes. Gene duplication is another common and important process in the evolution of genomes. A copy of a functional gene may arise as a result of a gene duplication event and subsequently acquire mutations that cause it to become nonfunctional. Duplicated pseudogenes usually have all the same characteristics as genes, including an intact exon-intron structure and promoter sequences. The loss of a duplicated gene's functionality usually has little effect on an organism's fitness, since an intact functional copy still exists. According to some evolutionary models, shared duplicated pseudogenes indicate the evolutionary relatedness of humans and the other primates.[13] If pseudogenization is due to gene duplication, it usually occurs in the first few million years after the gene duplication, provided the gene has not been subjected to any selection pressure.[14] Gene duplication generates functional redundancy and it is not normally advantageous to carry two identical genes. Mutations that disrupt either the structure or the function of any one of the two genes are not deleterious and will not be removed through the selection process. As a result, the gene that has been mutated gradually becomes a pseudogene and will be either unexpressed or functionless. This kind of evolutionary fate is shown by population genetic modeling[15][16] and also by genome analysis.[14][17] According to evolutionary context, these pseudogenes will either be deleted or become so distinct from the parental genes so that they will no longer be identifiable. Relatively young pseudogenes can be recognized due to their sequence similarity.[18]

43.2.3 Unitary pseudogenes

Various mutations can stop a gene from being successfully transcribed or translated, and a gene may become nonfunctional or deactivated if such a mutation becomes fixed in the population. This is the same mechanism by which non-processed genes become deactivated, but the difference in this case is that the gene was not duplicated before becoming disabled. Normally, such gene deactivation would be unlikely to become fixed in a population, but various population effects, such as genetic drift, a population bottleneck, or in some cases, natural selection, can lead to fixation. The classic example of a unitary pseudogene is the gene that presumably coded the enzyme L-gulono-γ-lactone oxidase (GULO) in primates. In all mammals studied besides primates (except guinea pigs), GULO aids in the biosynthesis of ascorbic acid (vitamin C), but it exists as a disabled gene (GULOP) in humans and other primates.[19][20] Another interesting and more recent example of a disabled gene links the deactivation of the caspase 12 gene (through a nonsense mutation) to positive selection in humans.[21]

Pseudogenes can complicate molecular genetic studies. For example, a researcher who wants to amplify a gene by PCR may simultaneously amplify a pseudogene that shares similar sequences. This is known as PCR bias or amplification bias. Similarly, pseudogenes are sometimes annotated as genes in genome sequences.

Processed pseudogenes often pose a problem for gene prediction programs, often being misidentified as real genes or exons. It has been proposed that identification of processed pseudogenes can help improve the accuracy of gene prediction methods.[22]

It has also been shown that the parent sequences that give rise to processed pseudogenes lose their coding potential faster than those giving rise to non-processed pseudogenes.[5]

43.3 Potential function

By definition, pseudogenes lack a functioning gene product. However, the classification of pseudogenes generally relies on computational analysis of genomic sequences using complex algorithms.[23] This has led to the incorrect identification of pseudogenes. Examples include

1. The *Drosophila jingwei* gene, a functional, chimeric gene which was once thought to be a processed pseudogene.[24]

2. *Makorin1* (*MKRN1*). In 2003, Hirotsune et al. identified a retrotransposed pseudogene whose transcript purportedly plays a *trans*-regulatory role in the expression of its homologous gene, *Makorin1* (MKRN1) (see also RING finger domain and ubiquitin ligases), and suggested this as a general model under which pseudogenes may play an important biological role.[25] Hirotsune's report prompted two molecular biologists to carefully review scientific literature on the subject of pseudogenes. To the surprise of many, they found a number of examples in which pseudogenes play a role in gene regulation and expression,[26] forcing Hirotsune's group to rescind their claim that they were the first to identify pseudogene function.[27] Furthermore, the original findings of Hirotsune et al. concerning *Makorin1* have recently been strongly contested;[28] thus, the possibility that some pseudogenes could have important biological functions was disputed.

3. Phosphoglycerate mutase 3 (*PGAM3P*). A processed pseudogene called phosphoglycerate mutase 3 (*PGAM3P*) actually produces a functional protein.[29]

4. siRNAs. Some endogenous siRNAs appear to be derived from pseudogenes, and thus some pseudogenes play a role in regulating protein-coding transcripts.[30][31]

5. piRNAs. Some Piwi-interacting RNAs (piRNAs) are derived from pseudogenes located in piRNA clusters. Those pseudogenes regulate their founding source genes via the piRNA pathway in mammalian testes.

6. *PTENP1* and *KRAS1P* (*KRASP1*). In June 2010, Nature published an article showing the mRNA levels of tumour suppressor *PTEN* and oncogenic*KRAS* is affected by their pseudogenes *PTENP1* and *KRASP1*. This discovery demonstrated an miRNA decoy function for pseudogenes and identified their transcripts as biologically active units in tumor biology; thus attributing a novel biological role to expressed pseudogenes, as they can regulate coding gene expression, and reveal a non-coding function for mRNAs in disease progression.[32]

43.3.1   Surveys

Svensson et al. have published a genome-wide survey of functional pseudogenes.[33]

A bioinformatics analysis has shown that processed pseudogenes can be inserted into introns of annotated genes and be incorporated into alternatively spliced transcripts.[11] This analysis showed strong evidence for transcription of 726 such retrogenes. However, their function was not studied experimentally.

43.3.2   Transcription

Quite a few pseudogenes can go through the process of transcription, either if their own promotor is still intact or in some cases using the promoter of a nearby gene; this expression of pseudogenes may be tissue-specific.[5] In the bacterium *Mycobacterium leprae*, 43% of its 1,133 pseudogenes are transcribed (as opposed to 49% overall and 57% of its ORFs[34]). However, that does not make them "functional" in the sense that these genes or proteins have an activity that benefits the organism.

43.4   Gene resurrection

For more details on this topic, see Ancestral gene resurrection.

The duplicated pseudogenic DNA can be resurrected to a functional protein in certain cases as a rare or occasional evolutionary event and may enable sampling of more sequence space for a protein or protein family.[18] The pseudogenes or parts of pseudogenes may be re-utilized once they have been drifted randomly without being subjected to selection pressure for certain period of evolution. Koch, for the first time, postulated an idea about such "untranslatable intermediates" in the evolution of protein.[35] Occasionally, this mechanism may yield a shorter evolutionary route to another desirable or favorable evolutionary energetic minimum although one would generally expect it to produce unviable or unfavorable leaps in sequence space. A longer time will be available to search sequence space by the pseudogene resurrection, but it is believed that it rarely brings into existence the proteins with new functions. The repair of lesions could be achieved by the reinsertion of a deleted segment, the removal (in frame) of an inserted segment, or other events that are likely to be improbable like gene conversion. Conversion of a pseudogene with a functional gene as a donor might improve the probability of pseudogene reactivation provided enough of the pseudogene sequence must be preserved throughout the course to maintain the benefits of expanding the sequence space explored after duplication.[36]

There are several examples that can be used to support such resurrection. The bovine seminal ribonuclease, which had lain dormant for about 20 million years as a pseudogene, appears to have been resurrected into a functional gene. It is believed that the event called gene conversion may be the cause of such resurrection.[37] The large group of pseudogenes for olfactory receptors (ORs) in metazoans, where 60% of the ORs in the human genome are pseudogenic, are resurrectable may be due to gene conversion events. In a cluster of ORs which contains 16 OR genes and 6 OR pseudogenes on chromosome 17, is appeared to be subjected to many number (20) of gene conversion events over the course of primate evolution.[38] These gene conversion events in OR gene clusters may aid to bring diversity in binding capability at the odorant binding site.[38] Finally, the resurrection of a pseudogene also led to the diversity of immunoglobulin heavy chain variable-region gene segments in the chicken which appears to be brought by the gene conversion event of a single functional gene with more than 80 pseudogenic gene segments.[39]

The era of molecular paleontology is just beginning. The surface of the pseudogene strata is barely studied, but if scientists conduct more research, they may be able to identify many more pseudogenes. The data mining process of large scale identification of pseudogenes is dynamic. The ancient and decayed pseudogenes are escaping from detection, although the recently generated pseudogenes are readily identified by the current techniques which are heavily based on the sequence comparison to well characterized genes. Characterization of pseudogenes will likely be improved as well since the sequence and annotation of the human genome itself are refined and updated. Modern clues may point to some possibilities of pseudogene resurrection- a dead gene become a living one and making a functional protein exist with the evidence.[40]

In addition to the seminal ribonuclease enzyme, the other incidents like slight differences in the pseudogene complements of individual people have also been found. For instance, in most people the olfactory receptor pseudogenes are dead but in few they are intact and functional genes. Some studies also suggested that however that in yeast, certain cell surface protein pseudogenes are resurrected due to stressful new environment challenged the organism. The two processed pseudogenes called the rat RC9 cytochrome c pseudogene[41] and the mouse L 32 ribosomal protein pseudogene rpL32-4A are implied to be potentially functional.[42] From the recent experiments, they found that in a bacterial genome a considerable segment of the intergenic regions are actively transcribed.[43] From the ENCODE project, scientists have found about 20% of the TARS were produced from previously unidentified 'potential unborn genes' which says that there are functional pseudogenes inside these regions.[44] To make sure that do the pseudogenes are transcribed in to RNA and to ascertain their functionality the studies on mouse oocyte are very useful where the small interfering RNAs (siRNAs) derived from pseudogene are found to be functional in regulating gene expression.[45] [31] Some pseudogenes are dead yet with some functions strengthen the fact that they are not 'junk DNA". With the embedded picture of genome annotation the real evolutionary history of pseudogenes will be revealed out in the near future of research.

43.5 See also

- Human disabled pseudogenes list

- Molecular evolution

- Retrotransposon

- Retroposon

- Molecular paleontology

43.6 References

[1] Vanin EF (1985). "Processed pseudogenes: characteristics and evolution". *Annu. Rev. Genet.* **19**: 253–72. doi:10.1146/annurev.ge.19.120185.0013 PMID 3909943.

[2]   Poliseno L (2010). "A coding-independent function of gene and pseudogene mRNAs regulates tumour biology". *Nature* **465**: 1033–1038. doi:10.1038/nature09144. PMC 3206313. PMID 20577206.

[3]   Herron, Jon C.; Freeman, Scott (2007). *Evolutionary analysis* (4th ed.). Upper Saddle River, NJ: Pearson Prentice Hall. ISBN 0-13-227584-8.

[4]   Jacq C, Miller JR, Brownlee GG (September 1977). "A pseudogene structure in 5S DNA of Xenopus laevis". *Cell* **12** (1): 109–20. doi:10.1016/0092-8674(77)90189-1. PMID 561661.

[5]   Zheng D, Frankish A, Baertsch R, Kapranov P, Reymond A, Choo SW, Lu Y, Denoeud F, Antonarakis SE, Snyder M, Ruan Y, Wei CL, Gingeras TR, Guigó R, Harrow J, Gerstein MB (June 2007). "Pseudogenes in the ENCODE regions: Consensus annotation, analysis of transcription, and evolution". *Genome Res.* **17** (6): 839–51. doi:10.1101/gr.5586307. PMC 1891343. PMID 17568002.

[6]   Mighell AJ, Smith NR, Robinson PA, Markham AF (February 2000). "Vertebrate pseudogenes". *FEBS Lett.* **468** (2–3): 109–14. doi:10.1016/S0014-5793(00)01199-6. PMID 10692568.

[7]   Jurka J (December 2004). "Evolutionary impact of human Alu repetitive elements". *Current Opinion in Genetics & Development* **14** (6): 603–8. doi:10.1016/j.gde.2004.08.008. PMID 15531153.

[8]   Dewannieux M, Heidmann T (2005). "LINEs, SINEs and processed pseudogenes: parasitic strategies for genome modeling". *Cytogenet. Genome Res.* **110** (1–4): 35–48. doi:10.1159/000084936. PMID 16093656.

[9]   Dewannieux M, Esnault C, Heidmann T (September 2003). "LINE-mediated retrotransposition of marked Alu sequences". *Nat. Genet.* **35** (1): 41–8. doi:10.1038/ng1223. PMID 12897783.

[10] Graur D, Shuali Y, Li WH (April 1989). "Deletions in processed pseudogenes accumulate faster in rodents than in humans". *J. Mol. Evol.* **28** (4): 279–85. doi:10.1007/BF02103423. PMID 2499684.

[11] Baertsch R, Diekhans M, Kent J, Haussler D, Brosius J (October 2008). "Retrocopy contributions to the evolution of the human genome". *BMC Genomics* **9**: 446–54. doi:10.1186/1471-2164-9-466. PMC 2584115. PMID 18842134.

[12] Pavlícek A, Paces J, Zíka R, Hejnar J (October 2002). "Length distribution of long interspersed nucleotide elements (LINEs) and processed pseudogenes of human endogenous retroviruses: implications for retrotransposition and pseudogene detection". *Gene* **300** (1–2): 189–94. doi:10.1016/S0378-1119(02)01047-8. PMID 12468100.

[13] Max EE (2003-05-05). "Plagiarized Errors and Molecular Genetics". TalkOrigins Archive. Retrieved 2008-07-22.

[14] Lynch M, Conery JS (November 2000). "The evolutionary fate and consequences of duplicate genes". *Science* **290** (5494): 1151–5. Bibcode:2000Sci...290.1151L. doi:10.1126/science.290.5494.1151. PMID 11073452.

[15] Walsh JB (January 1995). "How often do duplicated genes evolve new functions?". *Genetics* **139** (1): 421–8. PMC 1206338. PMID 7705642.

[16] Lynch M, O'Hely M, Walsh B, Force A (December 2001). "The probability of preservation of a newly arisen gene duplicate". *Genetics* **159** (4): 1789–804. PMC 1461922. PMID 11779815.

[17] Harrison PM, Hegyi H, Balasubramanian S, Luscombe NM, Bertone P, Echols N, Johnson T, Gerstein M (February 2002). "Molecular fossils in the human genome: identification and analysis of the pseudogenes in chromosomes 21 and 22". *Genome Res.* **12** (2): 272–80. doi:10.1101/gr.207102. PMC 155275. PMID 11827946.

[18] Zhang J (2003). "Evolution by gene duplication: an update.". *Trends in Ecology and Evolution* **18** (6): 292–298. doi:10.1016/S0169-5347(03)00033-8.

[19] Nishikimi M, Kawai T, Yagi K (October 1992). "Guinea pigs possess a highly mutated gene for L-gulono-gamma-lactone oxidase, the key enzyme for L-ascorbic acid biosynthesis missing in this species". *J. Biol. Chem.* **267** (30): 21967–72. PMID 1400507.

[20] Nishikimi M, Fukuyama R, Minoshima S, Shimizu N, Yagi K (May 1994). "Cloning and chromosomal mapping of the human nonfunctional gene for L-gulono-gamma-lactone oxidase, the enzyme for L-ascorbic acid biosynthesis missing in man". *J. Biol. Chem.* **269** (18): 13685–8. PMID 8175804.

[21] Xue Y, Daly A, Yngvadottir B, Liu M, Coop G, Kim Y, Sabeti P, Chen Y, Stalker J, Huckle E, Burton J, Leonard S, Rogers J, Tyler-Smith C (April 2006). "Spread of an Inactive Form of Caspase-12 in Humans Is Due to Recent Positive Selection". *American Journal of Human Genetics* **78** (4): 659–70. doi:10.1086/503116. PMC 1424700. PMID 16532395.

[22] van Baren MJ, Brent MR (May 2006). "Iterative gene prediction and pseudogene removal improves genome annotation". *Genome Res.* **16** (5): 678–85. doi:10.1101/gr.4766206. PMC 1457044. PMID 16651666.

[23] Harrison PM, Milburn D, Zhang Z, Bertone P, Gerstein M (February 2003). "Identification of pseudogenes in the Drosophila melanogaster genome". *Nucleic Acids Res.* **31** (3): 1033–7. doi:10.1093/nar/gkg169. PMC 149191. PMID 12560500.

[24] Long M, Langley CH (April 1993). "Natural selection and the origin of jingwei, a chimeric processed functional gene in Drosophila". *Science* **260** (5104): 91–5. Bibcode:1993Sci...260...91L. doi:10.1126/science.7682012. PMID 7682012.

[25] Hirotsune S, Yoshida N, Chen A, Garrett L, Sugiyama F, Takahashi S, Yagami K, Wynshaw-Boris A, Yoshiki A (May 2003). "An expressed pseudogene regulates the messenger-RNA stability of its homologous coding gene". *Nature* **423** (6935): 91–6. Bibcode:2003Natur.423...91H. doi:10.1038/nature01535. PMID 12721631.

[26] Balakirev ES, Ayala FJ (2003). "Pseudogenes: are they "junk" or functional DNA?". *Annu. Rev. Genet.* **37**: 123–51. doi:10.1146/annurev.genet.37.040103.103949. PMID 14616058.

[27] Hirotsune S, Yoshida N, Chen A, Garrett L, Sugiyama F, Takahashi S, Yagami K, Wynshaw-Boris A, Yoshiki A (November 2003). "Addendum: An Expressed Pseudogene Regulates the messenger-RNA Stability of Its Homologous Coding Gene". *Nature* **426** (6962): 100. Bibcode:2003Natur.426..100H. doi:10.1038/nature02094. PMID 14603326.

[28] Gray TA, Wilson A, Fortin PJ, Nicholls RD (August 2006). "The putatively functional Mkrn1-p1 pseudogene is neither expressed nor imprinted, nor does it regulate its source gene in trans". *Proc. Natl. Acad. Sci. U.S.A.* **103** (32): 12039–44. Bibcode:2006PNAS..10312039G. doi:10.1073/pnas.0602216103. PMC 1567693. PMID 16882727.

[29] Betrán E, Wang W, Jin L, Long M (May 2002). "Evolution of the phosphoglycerate mutase processed gene in human and chimpanzee revealing the origin of a new primate gene". *Mol. Biol. Evol.* **19** (5): 654–63. doi:10.1093/oxfordjournals.molbev.a004124. PMID 11961099.

[30] Tam OH, Aravin AA, Stein P, Girard A, Murchison EP, Cheloufi S, Hodges E, Anger M, Sachidanandam R, Schultz RM, Hannon GJ (May 2008). "Pseudogene-derived small interfering RNAs regulate gene expression in mouse oocytes". *Nature* **453** (7194): 534–8. Bibcode:2008Natur.453..534T. doi:10.1038/nature06904. PMC 2981145. PMID 18404147.

[31] Watanabe T, Totoki Y, Toyoda A, Kaneda M, Kuramochi-Miyagawa S, Obata Y, Chiba H, Kohara Y, Kono T, Nakano T, Surani MA, Sakaki Y, Sasaki H (May 2008). "Endogenous siRNAs from naturally formed dsRNAs regulate transcripts in mouse oocytes". *Nature* **453** (7194): 539–43. Bibcode:2008Natur.453..539W. doi:10.1038/nature06908. PMID 18404146.

[32] Poliseno, L, Salmena L, Zhang J, Carver B, Haveman WJ, Pandolfi PP (June 2010). "A coding-independent function of gene and pseudogene mRNAs regulates tumour biology". *Nature* **465** (7301): 1033–8. Bibcode:2010Natur.465.1033P. doi:10.1038/nature09144. PMC 3206313. PMID 20577206.

[33] Svensson O, Arvestad L, Lagergren J (May 2006). "Genome-Wide Survey for Biologically Functional Pseudogenes". *PLoS Comput. Biol.* **2** (5): e46. Bibcode:2006PLSCB...2...46S. doi:10.1371/journal.pcbi.0020046. PMC 1456316. PMID 16680195.

[34] Williams DL, Slayden RA, Amin A, et al. (2009). "Implications of high level pseudogene transcription in Mycobacterium leprae". *BMC Genomics* **10**: 397. doi:10.1186/1471-2164-10-397. PMC 2753549. PMID 19706172.

[35] Koch AL (October 1972). "Enzyme evolution. I. The importance of untranslatable intermediates". *Genetics* **72** (2): 297–316. PMC 1212829. PMID 4567287.

[36] Sassi SO, Braun EL, Benner SA (April 2007). "The evolution of seminal ribonuclease: pseudogene reactivation or multiple gene inactivation events?". *Mol. Biol. Evol.* **24** (4): 1012–24. doi:10.1093/molbev/msm020. PMID 17267422.

[37] Trabesinger-Ruef N, Jermann T, Zankel T, Durrant B, Frank G, Benner SA (March 1996). "Pseudogenes in ribonuclease evolution: a source of new biomacromolecular function?". *FEBS Lett.* **382** (3): 319–22. doi:10.1016/0014-5793(96)00191-3. PMID 8605993.

[38] Sharon D, Glusman G, Pilpel Y, Khen M, Gruetzner F, Haaf T, Lancet D (October 1999). "Primate evolution of an olfactory receptor cluster: diversification by gene conversion and recent emergence of pseudogenes". *Genomics* **61** (1): 24–36. doi:10.1006/geno.1999.5900. PMID 10512677.

[39] Pâques F, Haber JE (June 1999). "Multiple pathways of recombination induced by double-strand breaks in Saccharomyces cerevisiae". *Microbiol. Mol. Biol. Rev.* **63** (2): 349–404. PMC 98970. PMID 10357855.

[40] Gerstein M, Zheng D; Zheng (August 2006). "The real life of pseudogenes". *Sci. Am.* **295** (2): 48–55. Bibcode:2006SciAm.295b..48G. doi:10.1038/scientificamerican0806-48. PMID 16866288.

[41] Scarpulla RC (November 1984). "Processed pseudogenes for rat cytochrome c are preferentially derived from one of three alternate mRNAs". *Mol. Cell. Biol.* **4** (11): 2279–88. PMC 369056. PMID 6096691.

[42] Dudov KP, Perry RP (June 1984). "The gene family encoding the mouse ribosomal protein L32 contains a uniquely expressed intron-containing gene and an unmutated processed gene". *Cell* **37** (2): 457–68. doi:10.1016/0092-8674(84)90376-3. PMID 6327068.

[43] Fu LM, Shinnick TM (2007). "Genome-wide analysis of intergenic regions of Mycobacterium tuberculosis H37Rv using Affymetrix GeneChips". *EURASIP J Bioinform Syst Biol* **2007** (1): 23054. doi:10.1155/2007/23054. PMC 3171331. PMID 18253472.

[44] Rozowsky JS, Newburger D, Sayward F, Wu J, Jordan G, Korbel JO, Nagalakshmi U, Yang J, Zheng D, Guigó R, Gingeras TR, Weissman S, Miller P, Snyder M, Gerstein MB (June 2007). "The DART classification of unannotated transcription within the ENCODE regions: associating transcription with known and novel loci". *Genome Res.* **17** (6): 732–45. doi:10.1101/gr.5696007. PMC 1891334. PMID 17567993.

[45] Tam OH, Aravin AA, Stein P, Girard A, Murchison EP, Cheloufi S, Hodges E, Anger M, Sachidanandam R, Schultz RM, Hannon GJ (May 2008). "Pseudogene-derived small interfering RNAs regulate gene expression in mouse oocytes". *Nature* **453** (7194): 534–8. Bibcode:2008Natur.453..534T. doi:10.1038/nature06904. PMC 2981145. PMID 18404147.

43.7   Further reading

- Gerstein M, Zheng D (August 2006). "The real life of pseudogenes". *Sci. Am.* **295** (2): 48–55. doi:10.1038/scientificamerican08(48. PMID 16866288.

- Torrents D, Suyama M, Zdobnov E, Bork P (December 2003). "A Genome-Wide Survey of Human Pseudogenes". *Genome Res.* **13** (12): 2559–67. doi:10.1101/gr.1455503. PMC 403797. PMID 14656963.

- Bischof JM, Chiang AP, Scheetz TE, et al. (June 2006). "Genome-wide identification of pseudogenes capable of disease-causing gene conversion". *Hum. Mutat.* **27** (6): 545–52. doi:10.1002/humu.20335. PMID 16671097.

43.8   External links

- Pseudogene interaction database, miRNA-pseudogene and protein-pseudogene interaction maps database

- Yale University pseudogene database

- Hoppsigen database (homologous processed pseudogenes)

Chapter 44

Gene duplication

Gene duplication (or **chromosomal duplication** or **gene amplification**) is a major mechanism through which new genetic material is generated during molecular evolution. It can be defined as any duplication of a region of DNA that contains a gene. Gene duplications can arise as products of several types of errors in DNA replication and repair machinery as well as through fortuitous capture by selfish genetic elements. Common sources of gene duplications include ectopic homologous recombination, retrotransposition event, aneuploidy, polyploidy, and replication slippage.[1]

44.1 Mechanisms of duplication

44.1.1 Ectopic Recombination

Duplications arise from an event termed unequal crossing-over that occurs during meiosis between misaligned homologous chromosomes.The chance of this happening is a function of the degree of sharing of repetitive elements between two chromosomes. The products of this recombination are a duplication at the site of the exchange and a reciprocal deletion. Ectopic recombination is typically mediated by sequence similarity at the duplicate breakpoints, which form direct repeats. Repetitive genetic elements such as transposable elements offer one source of repetitive DNA that can facilitate recombination, and they are often found at duplication breakpoints in plants and mammals.[2]

44.1.2 Replication Slippage

Replication slippage is an error in DNA replication that can produce duplications of short genetic sequences. During replication DNA polymerase begins to copy the DNA. At some point during the replication process, the polymerase dissociates from the DNA and replication stalls. When the polymerase reattaches to the DNA strand, it aligns the replicating strand to an incorrect position and incidentally copies the same section more than once. Replication slippage is also often facilitated by repetitive sequences, but requires only a few bases of similarity.

44.1.3 Retrotransposition

During cellular invasion by a replicating retroelement or retrovirus, viral proteins copy their genome by reverse transcribing RNA to DNA. If viral proteins aberrantly attach to cellular mRNA, they can reverse transcribe copies of genes to create retrogenes. Retrogenes usually lack intronic sequences, and often contain poly A sequences that are also integrated into the genome. Many retrogenes display changes in gene regulation in comparison to their parental gene sequences, which sometimes results in novel functions.

44.1.4   Aneuploidy

Aneuploidy occurs when nondisjunction at a single chromosome results in an abnormal number of chromosomes. Aneuploidy is often harmful and in mammals regularly leads to spontaneous abortions (miscarriages). Some aneuploid individuals are viable, for example trisomy 21 in humans, which leads to Down syndrome. Aneuploidy often alters gene dosage in ways that are detrimental to the organism; therefore, it is unlikely to spread through populations.

44.1.5   Whole Genome Duplication

Whole genome duplication, or polyploidy, is a product of nondisjunction during meiosis which results in additional copies of the entire genome. Polyploidy is common in plants, but historically has also occurred in animals, with two rounds of whole genome duplication in the vertebrate lineage leading to humans.[3] After whole genome duplications many sets of additional genes are eventually lost, returning to singleton state. However, retention of many genes, most notably Hox genes, has led to adaptive innovation.

Polyploid is also a well known source of speciation, as offspring, which have different numbers of chromosomes compared to parent species, are often unable to interbreed with non-polyploid organisms. Whole genome duplications are thought to be less detrimental than aneuploidy as the relative dosage of individual genes should be the same.

44.2   Gene duplication as an evolutionary event

44.2.1   Neofunctionalization

Gene duplications are an essential source of genetic novelty that can lead to evolutionary innovation. Duplication creates genetic redundancy, where the second copy of the gene is often free from selective pressure — that is, mutations of it have no deleterious effects to its host organism. If one copy of a gene experiences a mutation that affects its original function, the second copy can serve as a 'spare part' and continue to function correctly. Thus, duplicate genes accumulate mutations faster than a functional single-copy gene, over generations of organisms, and it is possible for one of the two copies to develop a new and different function. Some examples of such neofunctionalization is the apparent mutation of a duplicated digestive gene in a family of ice fish into an antifreeze gene and duplication leading to a novel snake venom gene [4] and the synthesis of 1 beta-hydroxytestosterone.[5]

Gene duplication is believed to play a major role in evolution; this stance has been held by members of the scientific community for over 100 years.[6] Susumu Ohno was one of the most famous developers of this theory in his classic book *Evolution by gene duplication* (1970).[7] Ohno argued that gene duplication is the most important evolutionary force since the emergence of the universal common ancestor.[8] Major genome duplication events can be quite common. It is believed that the entire yeast genome underwent duplication about 100 million years ago.[9] Plants are the most prolific genome duplicators. For example, wheat is hexaploid (a kind of polyploid), meaning that it has six copies of its genome.

44.2.2   Subfunctionalization

Another possible fate for duplicate genes is that both copies are equally free to accumulate degenerative mutations, so long as any defects are complemented by the other copy. This leads to a neutral "subfunctionalization" or DDC (duplication-degeneration-complementation) model,[10][11] in which the functionality of the original gene is distributed among the two copies. Neither gene can be lost, as both now perform important non-redundant functions, but ultimately neither is able to achieve novel functionality.

Subfunctionalization can occur through neutral processes in which mutations accumulate with no detrimental or beneficial effects. However, in some cases subfunctionalization can occur with clear adaptive benefits. If an ancestral gene is pleiotropic and performs two functions, often neither one of these two functions can be changed without affecting the other function. In this way, partitioning the ancestral functions into two separate genes can allow for adaptive specialization of subfunctions, thereby providing an adaptive benefit.[12]

44.2.3 Loss

Often the resulting genomic variation leads to gene dosage dependent neurological disorders such as Rett-like syndrome and Pelizaeus-Merzbacher disease.[13] Such detrimental mutations are likely to be lost from the population and will not be preserved or develop novel functions. However, many duplications are, in fact, not detrimental or beneficial, and these neutral sequences may be lost or may spread through the population through random fluctuations via genetic drift.

44.3 Identifying duplications in sequenced genomes

44.3.1 Criteria and Single Genome Scans

The two genes that exist after a gene duplication event are called paralogs and usually code for proteins with a similar function and/or structure. By contrast, orthologous genes present in different species which are each originally derived from the same ancestral sequence. (See Homology of sequences in genetics).

It is important (but often difficult) to differentiate between paralogs and orthologs in biological research. Experiments on human gene function can often be carried out on other species if a homolog to a human gene can be found in the genome of that species, but only if the homolog is orthologous. If they are paralogs and resulted from a gene duplication event, their functions are likely to be too different. One or more copies of duplicated genes that constitute a gene family may be affected by insertion of transposable elements that causes significant variation between them in their sequence and finally may become responsible for divergent evolution. This may also render the chances and the rate of gene conversion between the homologs of gene duplicates due to less or no similarity in their sequences.

Paralogs can be identified in single genomes through a sequence comparison of all annotated gene models to one another. Such a comparison can be performed on translated amino acid sequences (e.g. BLASTp, tBLASTx) to identify ancient duplications or on DNA nucleotide sequences (e.g. BLASTn, megablast) to identify more recent duplications. Most studies to identify gene duplications require reciprocal-best-hits or fuzzy reciprocal-best-hits, where each paralog must be the other's single best match in a sequence comparison.[14]

Most gene duplications exist as low copy repeats (LCRs), rather highly repetitive sequences like transposable elements. They are mostly found in pericentronomic, subtelomeric and interstitial regions of a chromosome. Many LCRs, due to their size (>1Kb), similarity, and orientation, are highly susceptible to duplications and deletions.

44.3.2 Genomic microarrays detect duplications

Technologies such as genomic microarrays, also called array comparative genomic hybridization (array CGH), are used to detect chromosomal abnormalities, such as microduplications, in a high throughput fashion from genomic DNA samples. In particular, DNA microarray technology can simultaneously monitor the expression levels of thousands of genes across many treatments or experimental conditions, greatly facilitating the evolutionary studies of gene regulation after gene duplication or speciation.[15][16]

44.3.3 Next generation sequencing

Gene duplications can also be identified through the use of next-generation sequencing platforms. The simplest means to identify duplications in genomic resequencing data is through the use of paired-end sequencing reads. Tandem duplications are indicated by sequencing read pairs which map in abnormal orientations. Through a combination of increased sequence coverage and abnormal mapping orientation, it is possible to identify duplications in genomic sequencing data.

44.4   Gene duplication as amplification

Gene duplication does not necessarily constitute a lasting change in a species' genome. In fact, such changes often don't last past the initial host organism. From the perspective of molecular genetics, amplification is one of many ways in which a gene can be overexpressed. Genetic amplification can occur artificially, as with the use of the polymerase chain reaction technique to amplify short strands of DNA *in vitro* using enzymes, or it can occur naturally, as described above. If it's a natural duplication, it can still take place in a somatic cell, rather than a germline cell (which would be necessary for a lasting evolutionary change).

44.4.1   Role in cancer

Duplications of oncogenes are a common cause of many types of cancer. In such cases the genetic duplication occurs in a somatic cell and affects only the genome of the cancer cells themselves, not the entire organism, much less any subsequent offspring.

44.5   See also

- Pseudogenes

- Molecular evolution

- Unequal crossing over

- Human genome

- Comparative genomics

- Inparanoid

- Tandem exon duplication

44.6   References

[1] Zhang J (2003). "Evolution by gene duplication: an update". *Trends in Ecology & Evolution* **18** (6): 292–8. doi:10.1016/S0169-5347(03)00033-8.

[2] "Definition of Gene duplication". *medterms medical dictionary*. MedicineNet. 2012-03-19.

[3] Holland, Peter; Dehal, Paramvir; Boore, Jeffrey L (2005). "Two Rounds of Whole Genome Duplication in the Ancestral Vertebrate". *PLoS Biology* **3** (10): e314. doi:10.1371/journal.pbio.0030314. ISSN 1545-7885.

[4] Lynch VJ (2007). "Inventing an arsenal: adaptive evolution and neofunctionalization of snake venom phospholipase A2 genes". *BMC Evolutionary Biology* **7**: 2. doi:10.1186/1471-2148-7-2. PMC 1783844. PMID 17233905.

[5] Conant GC, Wolfe KH (2008). "Turning a hobby into a job: how duplicated genes find new functions". *Nature Reviews Genetics* **9** (12): 938–950. doi:10.1038/nrg2482. PMID 19015656.

[6] Taylor JS, Raes J (2004). "Duplication and divergence: the evolution of new genes and old ideas". *Annu. Rev. Genet.* **38**: 615–43. doi:10.1146/annurev.genet.38.072902.092831. PMID 15568988.

[7] Ohno, S. (1970). *Evolution by gene duplication*. Springer-Verlag. ISBN 0-04-575015-7.

[8] Ohno, S. (1967). *Sex Chromosomes and Sex-linked Genes*. Springer-Verlag. ISBN 91-554-5776-2.

[9] Kellis M, Birren BW, Lander ES (April 2004). "Proof and evolutionary analysis of ancient genome duplication in the yeast *Saccharomyces cerevisiae*". *Nature* **428** (6983): 617–24. doi:10.1038/nature02424. PMID 15004568.

[10] Force A., Lynch M., Pickett F.B., Amores A., Yan Y.L., Postlethwait J. (1999). "Preservation of duplicate genes by complementary, degenerative mutations.". *Genetics* **151** (4): 1531–45. PMC 1460548. PMID 10101175.

[11] Stoltzfus, A. (1999). "On the possibility of constructive neutral evolution". *J Mol Evol* **49** (2): 169–181. doi:10.1007/PL00006540. PMID 10441669.

[12] Des Marais DL, Rausher MD (2008). "Escape from adaptive conflict after duplication in an anthocyanin pathway gene". *Nature* **454** (7205): 762–5. doi:10.1038/nature07092. PMID 18594508.

[13] Lee JA, Lupski JR (October 2006). "Genomic rearrangements and gene copy-number alterations as a cause of nervous system disorders". *Neuron* **52** (1): 103–21. doi:10.1016/j.neuron.2006.09.027. PMID 17015230.

[14] Hahn MW, Han MV, Han S-G (2007). "Gene Family Evolution across 12 Drosophila Genomes". *PLoS Genet* **3** (11): e197. doi:10.1371/journal.pgen.0030197. PMC 2065885. PMID 17997610.

[15] Mao R, Pevsner J (2005). "The use of genomic microarrays to study chromosomal abnormalities in mental retardation". *Ment Retard Dev Disabil Res Rev* **11** (4): 279–85. doi:10.1002/mrdd.20082. PMID 16240409.

[16] Gu X, Zhang Z, Huang W (January 2005). "Rapid evolution of expression and regulatory divergences after yeast gene duplication". *Proc. Natl. Acad. Sci. U.S.A.* **102** (3): 707–12. doi:10.1073/pnas.0409186102. PMC 545572. PMID 15647348.

[17] Kinzler, Kenneth W.; Vogelstein, Bert (2002). *The genetic basis of human cancer*. McGraw-Hill. p. 116. ISBN 0-07-137050-1.

44.7 External links

- *A bibliography on gene and genome duplication*
- *A brief overview of mutation, gene duplication and translocation*

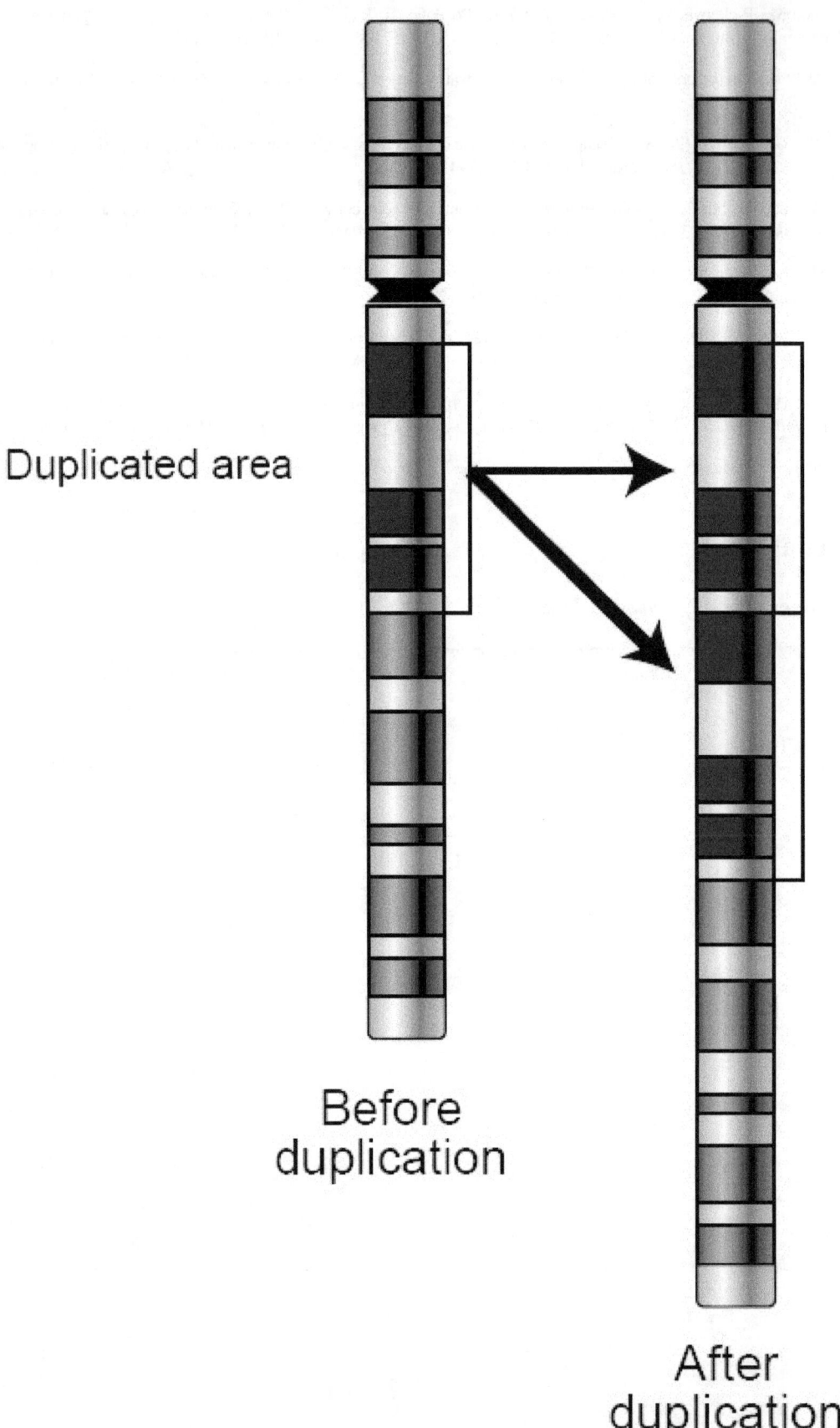

Duplicated area

Before
duplication

After
duplication

Schematic of a region of a chromosome before and after a duplication event

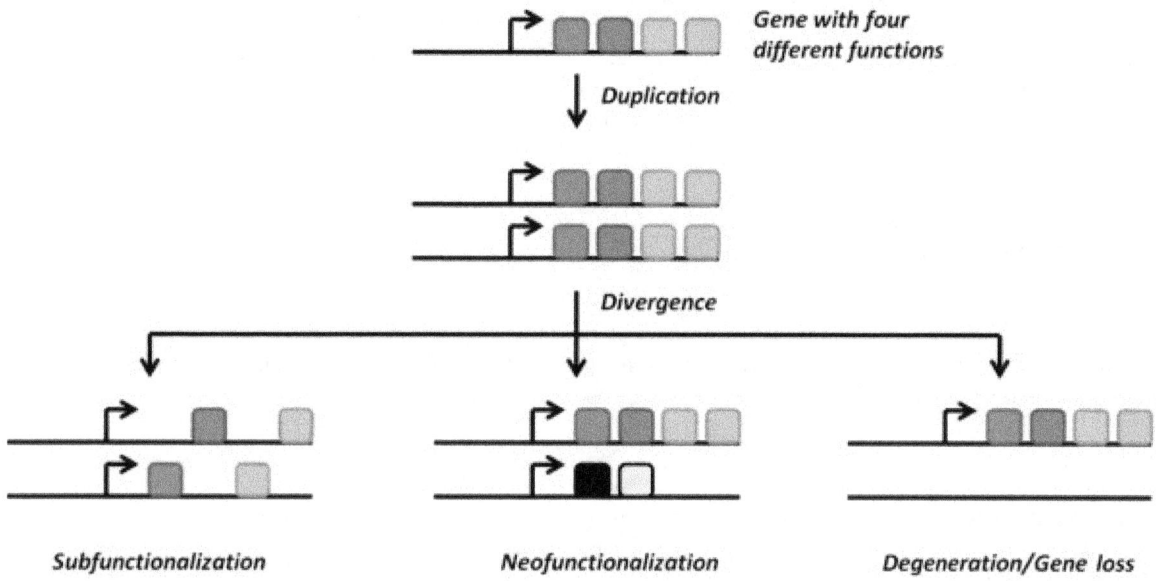

Evolutionary fate of duplicate genes

Chapter 45

Molecular evolution

Molecular evolution is a change in the sequence composition of cellular molecules such as DNA, RNA, and proteins across generations. The field of molecular evolution uses principles of evolutionary biology and population genetics to explain patterns in these changes. Major topics in molecular evolution concern the rates and impacts of single nucleotide changes, neutral evolution vs. natural selection, origins of new genes, the genetic nature of complex traits, the genetic basis of speciation, evolution of development, and ways that evolutionary forces influence genomic and phenotypic changes.

45.1 Forces in molecular evolution

The content and structure of a genome is the product of the molecular and population genetic forces which act upon that genome. Novel genetic variants will arise through mutation and will spread and be maintained in populations due to genetic drift or natural selection.

45.1.1 Mutation

Main article: Mutation

Mutations are permanent, transmissible changes to the genetic material (DNA or RNA) of a cell or virus. Mutations result from errors in DNA replication during cell division and by exposure to radiation, chemicals, and other environmental stressors, or viruses and transposable elements. Most mutations that occur are single nucleotide polymorphisms which modify single bases of the DNA sequence. Other types of mutations modify larger segments of DNA and can cause duplications, insertions, deletions, inversions, and translocations.

Most organisms display a strong bias in the types of mutations that occur with strong influence in GC-content. Transitions (A $\leftrightarrow$ G or C $\leftrightarrow$ T) are more common than transversions (purine $\leftrightarrow$ pyrimidine)[1] and are less likely to alter amino acid sequences of proteins.

Mutations are stochastic and typically occur randomly across genes. Mutation rates for single nucleotide sites for most organisms are very low, roughly 10^{-9} to 10^{-8} per site per generation, though some viruses have higher mutation rates on the order of 10^{-6} per site per generation. Among these mutations, some will be neutral or beneficial and will remain in the genome unless lost via Genetic drift, and others will be detrimental and will be eliminated from the genome by natural selection.

Because mutations are extremely rare, they accumulate very slowly across generations. While the number of mutations which appears in any single generation may vary, over very long time periods they will appear to accumulate at a regular pace. Using the mutation rate per generation and the number of nucleotide differences between two sequences, divergence times can be estimated effectively via the molecular clock.

This hedgehog has no pigmentation due to a genetic mutation.

45.1.2 Recombination

Further information: Genetic recombination
Recombination is a process that results in genetic exchange between chromosomes or chromosomal regions. Recombina-

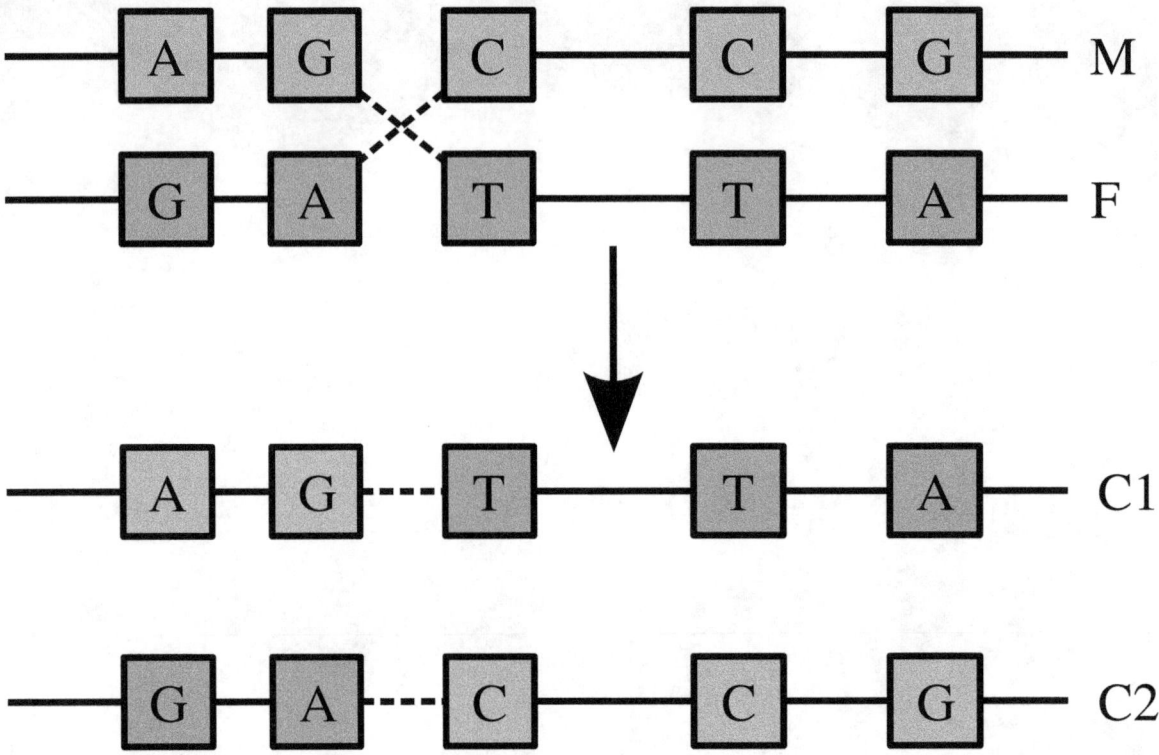

Recombination involves the breakage and rejoining of two chromosomes (M and F) to produce two re-arranged chromosomes (C1 and C2).

tion counteracts physical linkage between adjacent genes, thereby reducing genetic hitchhiking. The resulting independent inheritance of genes results in more efficient selection, meaning that regions with higher recombination will harbor fewer detrimental mutations, more selectively favored variants, and fewer errors in replication and repair. Recombination can also generate particular types of mutations if chromosomes are misaligned.

Gene conversion

Gene conversion is a type of recombination that is the product of DNA repair where nucleotide damage is corrected using orthologous genomic regions as a template. Damaged bases are first excised, the damaged strand is then aligned with an undamaged homolog, and DNA synthesis repairs the excised region using the undamaged strand as a guide. Gene conversion is often responsible for homogenizing sequences of duplicate genes over long time periods, reducing nucleotide divergence.

45.1.3 Genetic drift

Genetic drift is the change of allele frequencies from one generation to the next due to stochastic effects of random sampling in finite populations. Some existing variants have no effect on fitness and may increase or decrease in frequency simply due to chance. "Nearly neutral" variants whose selection coefficient is close to a threshold value of 1 / the effective population size will also be affected by chance as well as by selection and mutation. Many genomic features have been ascribed to accumulation of nearly neutral detrimental mutations as a result of small effective population sizes.[2] With a smaller effective population size, a larger variety of mutations will behave as if they are neutral due to inefficiency of selection.

45.1.4 Selection

Selection occurs when organisms with greater fitness, i.e. greater ability to survive or reproduce, are favored in subsequent generations, thereby increasing the instance of underlying genetic variants in a population. Selection can be the product of natural selection, artificial selection, or sexual selection. Natural selection is any selective process that occurs due to the fitness of an organism to its environment. In contrast sexual selection is a product of mate choice and can favor the spread of genetic variants which act counter to natural selection but increase desirability to the opposite sex or increase mating success. Artificial selection, also known as selective breeding, is imposed by an outside entity, typically humans, in order to increase the frequency of desired traits.

The principles of population genetics apply similarly to all types of selection, though in fact each may produce distinct effects due to clustering of genes with different functions in different parts of the genome, or due to different properties of genes in particular functional classes. For instance, sexual selection could be more likely to affect molecular evolution of the sex chromosomes due to clustering of sex specific genes on the X,Y,Z or W.

Selection can operate at the gene level at the expense of organismal fitness, resulting in a selective advantage for selfish genetic elements in spite of a host cost. Examples of such selfish elements include transposable elements, meiotic drivers, killer X chromosomes, selfish mitochondria, and self-propagating introns. (See Intragenomic conflict.)

45.2 Genome architecture

Main article: Genome evolution

45.2.1 Genome size

Genome size is influenced by the amount of repetitive DNA as well as number of genes in an organism. The C-value paradox refers to the lack of correlation between organism 'complexity' and genome size. Explanations for the so-called paradox are two-fold. First, repetitive genetic elements can comprise large portions of the genome for many organisms, thereby inflating DNA content of the haploid genome. Secondly, the number of genes is not necessarily indicative of the number of developmental stages or tissue types in an organism. An organism with few developmental stages or tissue types may have large numbers of genes that influence non-developmental phenotypes, inflating gene content relative to developmental gene families.

Neutral explanations for genome size suggest that when population sizes are small, many mutations become nearly neutral. Hence, in small populations repetitive content and other 'junk' DNA can accumulate without placing the organism at a competitive disadvantage. There is little evidence to suggest that genome size is under strong widespread selection in multicellular eukaryotes. Genome size, independent of gene content, correlates poorly with most physiological traits and many eukaryotes, including mammals, harbor very large amounts of repetitive DNA.

However, birds likely have experienced strong selection for reduced genome size, in response to changing energetic needs for flight. Birds, unlike humans, produce nucleated red blood cells, and larger nuclei lead to lower levels of oxygen transport. Bird metabolism is far higher than that of mammals, due largely to flight, and oxygen needs are high. Hence, most birds have small, compact genomes with few repetitive elements. Indirect evidence suggests that non-avian theropod dinosaur ancestors of modern birds [3] also had reduced genome sizes, consistent with endothermy and high energetic needs for running speed. Many bacteria have also experienced selection for small genome size, as time of replication and energy consumption are so tightly correlated with fitness.

45.2.2   Repetitive elements

Transposable elements are self-replicating, selfish genetic elements which are capable of proliferating within host genomes. Many transposable elements are related to viruses, and share several proteins in common.

DNA transposons are cut and paste transposable elements which excise DNA and move it to alternate sections of the genome.

non-LTR retrotransposons

LTR retrotransposons

Helitrons

Alu elements comprise over 10% of the human genome. They are short non-autonomous repeat sequences.

45.2.3   Chromosome number and organization

The number of chromosomes in an organism's genome also does not necessarily correlate with the amount of DNA in its genome. The ant *Myrmecia pilosula* has only a single pair of chromosomes[4] whereas the Adders-tongue fern *Ophioglossum reticulatum* has up to 1260 chromosomes.[5] Cilliate genomes house each gene in individual chromosomes, resulting in a genome which is not physically linked. Reduced linkage through creation of additional chromosomes should effectively increase the efficiency of selection.

Changes in chromosome number can play a key role in speciation, as differing chromosome numbers can serve as a barrier to reproduction in hybrids. Human chromosome 2 was created from a fusion of two chimpanzee chromosomes and still contains central telomeres as well as a vestigial second centromere. Polyploidy, especially allopolyploidy, which occurs often in plants, can also result in reproductive incompatibilities with parental species. *Agrodiatus* blue butterflies have diverse chromosome numbers ranging from n=10 to n=134 and additionally have one of the highest rates of speciation identified to date.[6]

45.2.4   Gene content and distribution

Different organisms house different numbers of genes within their genomes as well as different patterns in the distribution of genes throughout the genome. Some organisms, such as most bacteria, *Drosophila*, and *Arabidopsis* have particularly compact genomes with little repetitive content or non-coding DNA. Other organisms, like mammals or maize, have large amounts of repetitive DNA, long introns, and substantial spacing between different genes. The content and distribution of genes within the genome can influence the rate at which certain types of mutations occur and can influence the subsequent evolution of different species. Genes with longer introns are more likely to recombine due to increased physical distance over the coding sequence. As such, long introns may facilitate ectopic recombination, and result in higher rates of new gene formation.

45.2.5   Organelles

In addition to the nuclear genome, endosymbiont organelles contain their own genetic material typically as circular plasmids. Mitochondrial and chloroplast DNA varies across taxa, but membrane-bound proteins, especially electron transport chain constituents are most often encoded in the organelle. Chloroplasts and mitochondria are maternally inherited in most species, as the organelles must pass through the egg. In a rare departure, some species of mussels are known to inherit mitochondria from father to son.

45.3   Origins of new genes

New genes arise from several different genetic mechanisms including gene duplication, de novo origination, retrotransposition, chimeric gene formation, recruitment of non-coding sequence, and gene truncation.

In **gene duplication**, a gene sequence is copied to create redundancy. Duplicated gene sequences can then mutate to develop new functions or to specialize so that each new gene performs a subset of the original ancestral functions. In addition to duplicating whole genes, sometimes only a domain or part of a protein is duplicated so that the resulting gene is an elongated version of the parental gene.

Retrotransposition creates new genes by copying mRNA to DNA and inserting it into the genome. Retrogenes often insert into new genomic locations, and often develop new expression patterns and functions.

Chimeric genes form when duplication, deletion, or incomplete retrotransposition combine portions of two different coding sequences to produce a novel gene sequence. Chimeras often cause regulatory changes and can shuffle protein domains to produce novel adaptive functions.

Novel genes can also arise from previously non-coding DNA.[7] For instance, Levine and colleagues reported the origin of five new genes in the *D. melanogaster* genome from noncoding DNA.[8][9] Similar de novo origin of genes has been also shown in other organisms such as yeast,[10] rice[11] and humans.[12] De novo genes may evolve from transcripts that are already expressed at low levels.[13] Mutation of a stop codon to a regular codon or a frameshift may cause an extended protein that includes a previously non-coding sequence.

45.4 Molecular phylogenetics

Main articles: Molecular systematics and Phylogenetics

Molecular systematics is the product of the traditional fields of systematics and molecular genetics. It uses DNA, RNA, or protein sequences to resolve questions in systematics, i.e. about their correct scientific classification or taxonomy from the point of view of evolutionary biology.

Molecular systematics has been made possible by the availability of techniques for DNA sequencing, which allow the determination of the exact sequence of nucleotides or *bases* in either DNA or RNA. At present it is still a long and expensive process to sequence the entire genome of an organism, and this has been done for only a few species. However, it is quite feasible to determine the sequence of a defined area of a particular chromosome. Typical molecular systematic analyses require the sequencing of around 1000 base pairs.

45.5 The driving forces of evolution

Main articles: Neutral theory of molecular evolution, Modern evolutionary synthesis and Mutationism

Depending on the relative importance assigned to the various forces of evolution, three perspectives provide evolutionary explanations for molecular evolution.[14]

Selectionist hypotheses argue that selection is the driving force of molecular evolution. While acknowledging that many mutations are neutral, selectionists attribute changes in the frequencies of neutral alleles to linkage disequilibrium with other loci that are under selection, rather than to random genetic drift.[15] Biases in codon usage are usually explained with reference to the ability of even weak selection to shape molecular evolution.[16]

Neutralist hypotheses emphasize the importance of mutation, purifying selection, and random genetic drift.[17] The introduction of the neutral theory by Kimura,[18] quickly followed by King and Jukes' own findings,[19] led to a fierce debate about the relevance of neodarwinism at the molecular level. The Neutral theory of molecular evolution proposes that most mutations in DNA are at locations not important to function or fitness. These neutral changes drift towards fixation within a population. Positive changes will be very rare, and so will not greatly contribute to DNA polymorphisms.[20] Deleterious mutations will also not contribute very much to DNA diversity because they negatively affect fitness and so will not stay in the gene pool for long.[21] This theory provides a framework for the molecular clock.[20] The fate of neutral mutations are governed by genetic drift, and contribute to both nucleotide polymorphism and fixed differences between species.[22][23]

In the strictest sense, the neutral theory is not accurate.[24] Subtle changes in DNA very often have effects, but sometimes these effects are too small for natural selection to act on.[24] Even synonymous mutations are not necessarily neutral [24] because there is not a uniform amount of each codon. The nearly neutral theory expanded the neutralist perspective, suggesting that several mutations are nearly neutral, which means both random drift and natural selection is relevant to their dynamics.[24] The main difference between the the neutral theory and nearly neutral theory is that the latter focuses on weak selection, not strictly neutral.[21]

Mutationists hypotheses emphasize random drift and biases in mutation patterns.[25] Sueoka was the first to propose a modern mutationist view. He proposed that the variation in GC content was not the result of positive selection, but a consequence of the GC mutational pressure.[26]

45.6   Protein Evolution

This chart compares the sequence identity of different lipase proteins throughout the human body. It demonstrates how proteins evolve, keeping some regions conserved while others change dramatically.

Protein evolution describes the changes over time in protein shape, function, and composition. Through quantitative analysis and experimentation, scientists have strived to understand the rate and causes of protein evolution. Using the amino acid sequences of hemoglobin and cytochrome c from multiple species, scientists were able to derive estimations of protein evolution rates. What they found was that the rates were not the same among proteins.[21] Each protein has its own rate, and that rate is constant across phylogenies (i.e., hemoglobin does not evolve at the same rate as cytochrome c, but hemoglobins from humans, mice, etc. do have comparable rates of evolution.). Not all regions within a protein mutate at the same rate; functionally important areas mutate more slowly and amino acid substitutions involving similar amino acids occurs more often than dissimilar substitutions.[21] Overall, the level of polymorphisms in proteins seems to be fairly constant. Several species (including humans, fruit flies, and mice) have similar levels of protein polymorphism.[20]

45.6.1 Relation to Nucleic Acid Evolution

Protein evolution is inescapably tied to changes and selection of DNA polymorphisms and mutations because protein sequences change in response to alterations in the DNA sequence. Amino acid sequences and nucleic acid sequences do not mutate at the same rate. Due to the degenerate nature of DNA, bases can change without affecting the amino acid sequence. For example, there are six codons that code for leucine. Thus, despite the difference in mutation rates, it is essential to incorporate nucleic acid evolution into the discussion of protein evolution. At the end of the 1960s, two groups of scientists—Kimura (1968) and King and Jukes (1969)-- independently proposed that a majority of the evolutionary changes observed in proteins were neutral.[20][21] Since then, the neutral theory has been expanded upon and debated.[21]

45.7 Discordance with morphological evolution

There are sometimes discordances between molecular and morphological evolution, which are reflected in molecular and morphological systematic studies, especially of bacteria, archaea and eukaryotic microbes. These discordances can be categorized as two types: (i) one morphology, multiple lineages (e.g. morphological convergence, cryptic species) and (ii) one lineage, multiple morphologies (e.g. phenotypic plasticity, multiple life-cycle stages). Neutral evolution possibly could explain the incongruences in some cases.[27]

45.8 Journals and societies

The Society for Molecular Biology and Evolution publishes the journals "Molecular Biology and Evolution" and "Genome Biology and Evolution" and holds an annual international meeting. Other journals dedicated to molecular evolution include *Journal of Molecular Evolution* and *Molecular Phylogenetics and Evolution*. Research in molecular evolution is also published in journals of genetics, molecular biology, genomics, systematics, and evolutionary biology.

45.9 See also

- Abiogenesis

- Comparative phylogenetics

- Evolution

- *E. coli* long-term evolution experiment

- Evolutionary physiology

- Evolution of dietary antioxidants

- Genomic organization

- Genetic drift

- Genome evolution

- Heterotachy

- History of molecular evolution

- Horizontal gene transfer

- Human evolution

- Molecular clock

- Molecular paleontology

- Neutral theory of molecular evolution

- Nucleotide diversity

- Parsimony

- Population genetics

- Selection

45.10   References

[1] https://www.mun.ca/biology/scarr/Transitions_vs_Transversions.html

[2] Lynch, M. (2007). *The Origins of Genome Architecture*. Sinauer. ISBN 0-87893-484-7.

[3] Organ, C. L.; Shedlock, A. M.; Meade, A.; Pagel, M.; Edwards, S. V. (2007). "Origin of avian genome size and structure in nonavian dinosaurs". *Nature* **446**: 180–184. doi:10.1038/nature05621.

[4] Crosland, M.W.J., Crozier, R.H. (1986). "*Myrmecia pilosula*, an ant with only one pair of chromosomes". *Science* **231** (4743): 1278. Bibcode:1986Sci...231.1278C. doi:10.1126/science.231.4743.1278. PMID 17839565.

[5] Gerardus J. H. Grubben (2004). *Vegetables*. PROTA. p. 404. ISBN 978-90-5782-147-9. Retrieved 10 March 2013.

[6] Nikolai P. Kandul, Vladimir A. Lukhtanov, Naomi E. Pierce (2007), "KARYOTYPIC DIVERSITY AND SPECIATION IN AGRODIAETUS BUTTERFLIES", *The Society for the Study of Evolution*, 61(3):546-559: 546–559, doi:10.1111/j.1558-5646.2007.00046.x

[7] Tautz, Diethard and Domazet-Lošo, Tomislav (2011). "The evolutionary origin of orphan genes". *Nature Reviews Genetics* **12** (10): 692–702. doi:10.1038/nrg3053. PMID 21878963.

[8] Levine MT, Jones CD, Kern AD, et al. (2006). "Novel genes derived from noncoding DNA in Drosophila melanogaster are frequently X-linked and exhibit testis-biased expression". *Proc Natl Acad Sci USA* **103** (26): 9935–9939. Bibcode:2006PNAS..103.9935L. doi:10.1073/pnas.0509809103. PMC 1502557. PMID 16777968.

[9] Zhou Q, Zhang G, Zhang Y, et al. (2008). "On the origin of new genes in Drosophila". *Genome Res* **18** (9): 1446–1455. doi:10.1101/gr.076588.108. PMC 2527705. PMID 18550802.

[10] Cai J, Zhao R, Jiang H, et al. (2008). "De novo origination of a new protein-coding gene in *Saccharomyces cerevisiae*". *Genetics* **179** (1): 487–496. doi:10.1534/genetics.107.084491. PMC 2390625. PMID 18493065.

[11] Xiao W, Liu H, Li Y, et al. (2009). El-Shemy HA, ed. "A rice gene of de novo origin negatively regulates pathogen- induced defense response". *PLoS ONE* **4** (2): e4603. Bibcode:2009PLoSO...4.4603X. doi:10.1371/journal.pone.0004603. PMC 2643483. PMID 19240804.

[12] Knowles DG, McLysaght A (2009). "Recent de novo origin of human protein-coding genes". *Genome Res* **19** (10): 1752–1759. doi:10.1101/gr.095026.109. PMC 2765279. PMID 19726446.

[13] Wilson, Ben A.; Joanna Masel (2011). "Putatively Noncoding Transcripts Show Extensive Association with Ribosomes". *Genome Biology & Evolution* **3**: 1245–1252. doi:10.1093/gbe/evr099.

[14] Graur, D. and Li, W.-H. (2000). *Fundamentals of molecular evolution*. Sinauer. ISBN 0-87893-266-6.

[15] Hahn, Matthew W. (February 2008). "TOWARD A SELECTION THEORY OF MOLECULAR EVOLUTION". *Evolution* **62** (2): 255–265. doi:10.1111/j.1558-5646.2007.00308.x. PMID 18302709.

[16] Hershberg, Ruth; Petrov, Dmitri A. (December 2008). "Selection on Codon Bias". *Annual Review of Genetics* **42** (1): 287–299. doi:10.1146/annurev.genet.42.110807.091442. PMID 18983258.

[17] Kimura, M. (1983). *The Neutral Theory of Molecular Evolution*. Cambridge University Press, Cambridge. ISBN 0-521-23109-4.

[18] Kimura, Motoo (1968). "Evolutionary rate at the molecular level" (PDF). *Nature* **217** (5129): 624–626. Bibcode:1968Natur.217..624K. doi:10.1038/217624a0. PMID 5637732.

[19] King, J.L. and Jukes, T.H. (1969). "Non-Darwinian Evolution" (PDF). *Science* **164** (3881): 788–798. Bibcode:1969Sci...164..788L. doi:10.1126/science.164.3881.788. PMID 5767777.

[20] Akashi, H. "Weak Selection and Protein Evolution". *Genetics* **192** (1): 15–31. doi:10.1534/genetics.112.140178.

[21] Fay, JC, Wu, CI (2003). "Sequence divergence, functional constraint, and selection in protein evolution". *Annual Rev. Genomics Human Genetics* **4**: 213–35.

[22] Nachman M. (2006). C.W. Fox and J.B. Wolf, ed. ""Detecting selection at the molecular level" in: Evolutionary Genetics: concepts and case studies": 103–118.

[23] The nearly neutral theory expanded the neutralist perspective, suggesting that several mutations are nearly neutral, which means both random drift and natural selection is relevant to their dynamics.

[24] Ohta, T (1992). Missing or empty |title= (help)

[25] Nei, M. (2005). "Selectionism and Neutralism in Molecular Evolution". *Molecular Biology and Evolution* **22** (12): 2318–2342. doi:10.1093/molbev/msi242. PMC 1513187. PMID 16120807.

[26] Sueoka, N. (1964). "On the evolution of informational macromolecules". In In: Bryson, V. and Vogel, H.J. *Evolving genes and proteins*. Academic Press, New-York. pp. 479–496.

[27] Lahr, D. J.; Laughinghouse, H. D.; Oliverio, A. M.; Gao, F.; Katz, L. A. (2014). "How discordant morphological and molecular evolution among microorganisms can revise our notions of biodiversity on Earth". *BioEssays* **36** (10): 950–959. doi:10.1002/bies.201400056. PMID 25156897.

45.11 Further reading

- Li, W.-H. (2006). *Molecular Evolution*. Sinauer. ISBN 0-87893-480-4.

- Lynch, M. (2007). *The Origins of Genome Architecture*. Sinauer. ISBN 0-87893-484-7.

- A. Meyer (Editor), Y. van de Peer, "Genome Evolution: Gene and Genome Duplications and the Origin of Novel Gene Functions", 2003, ISBN 978-1-4020-1021-7

- T. Ryan Gregory, "The Evolution of the Genome", 2004, YSBN 978-0123014634

Chapter 46

Post-transcriptional modification

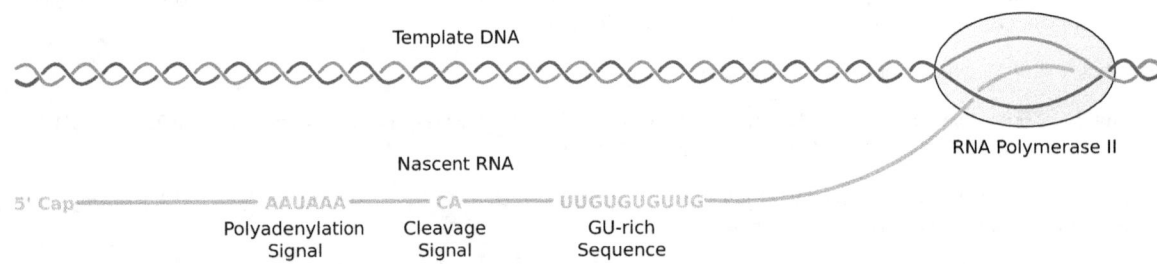

Example of a signal that directs post-transcriptional processing: the conserved eukaryotic polyadenylation signal directs cleavage at the cleavage signal and addition of a poly-A tail to the mRNA transcript

Post-transcriptional modification or **Co-transcriptional modification** is a process in cell biology by which, in eukaryotic cells, primary transcript RNA is converted into mature RNA. A notable example is the conversion of precursor messenger RNA into mature messenger RNA (mRNA), which includes splicing and occurs prior to protein synthesis. This process is vital for the correct translation of the genomes of eukaryotes because the human primary RNA transcript that is produced, as a result of transcription, contains both exons, which are coding sections of the primary RNA transcript and introns, which are the non-coding sections of the primary RNA transcript.[1]

46.1 mRNA processing

The pre-mRNA molecule undergoes three main modifications. These modifications are 5' capping, 3' polyadenylation, and RNA splicing, which occur in the cell nucleus before the RNA is translated.[2]

46.1.1 5' Processing

Main article: 5' cap

Capping

Capping of the pre-mRNA involves the addition of 7-methylguanosine (m^7G) to the 5' end. To achieve this, the terminal 5' phosphate requires removal, which is done with the aid of a phosphatase enzyme. The enzyme guanosyl transferase then catalyses the reaction, which produces the diphosphate 5' end. The diphosphate 5' end then attacks the alpha phosphorus atom of a GTP molecule in order to add the guanine residue in a 5'5' triphosphate link. The enzyme (guanine-N^7-)-methyltransferase ("cap MTase") transfers a methyl group from S-adenosyl methionine to the guanine ring.[3] This type of cap, with just the (m^7G) in position is called a cap 0 structure. The ribose of the adjacent nucleotide may also be methylated to give a cap 1. Methylation of nucleotides downstream of the RNA molecule produce cap 2, cap 3 structures and so on. In these cases the methyl groups are added to the 2' OH groups of the ribose sugar. The cap protects the 5' end of the primary RNA transcript from attack by ribonucleases that have specificity to the 3'5' phosphodiester bonds.[4]

46.1.2 3' Processing

Cleavage and polyadenylation

Main article: Polyadenylation

The pre-mRNA processing at the 3' end of the RNA molecule involves cleavage of its 3' end and then the addition of about 250 adenine residues to form a poly(A) tail. The cleavage and adenylation reactions occur if a polyadenylation signal sequence (5'- AAUAAA-3') is located near the 3' end of the pre-mRNA molecule, which is followed by another sequence, which is usually **(5'-CA-3')** and is the site of cleavage. A **GU-rich sequence** is also usually present further downstream on the pre-mRNA molecule. After the synthesis of the sequence elements, two multisubunit proteins called cleavage and polyadenylation specificity factor (CPSF) and cleavage stimulation factor (CStF) are transferred from RNA Polymerase II to the RNA molecule. The two factors bind to the sequence elements. A protein complex forms that contains additional cleavage factors and the enzyme Polyadenylate Polymerase (PAP). This complex cleaves the RNA between the polyadenylation sequence and the GU-rich sequence at the cleavage site marked by the (5'-CA-3') sequences. Poly(A) polymerase then adds about 200 adenine units to the new 3' end of the RNA molecule using ATP as a precursor. As the poly(A) tail is synthesised, it binds multiple copies of poly(A) binding protein, which protects the 3'end from ribonuclease digestion.[4]

46.1.3 Splicing

Main article: RNA splicing

RNA splicing is the process by which introns, regions of RNA that do not code for protein, are removed from the pre-mRNA and the remaining exons connected to re-form a single continuous molecule. Although most RNA splicing occurs after the complete synthesis and end-capping of the pre-mRNA, transcripts with many exons can be spliced co-transcriptionally.[5] The splicing reaction is catalyzed by a large protein complex called the spliceosome assembled from proteins and small nuclear RNA molecules that recognize splice sites in the pre-mRNA sequence. Many pre-mRNAs, including those encoding antibodies, can be spliced in multiple ways to produce different mature mRNAs that encode different protein sequences. This process is known as alternative splicing, and allows production of a large variety of proteins from a limited amount of DNA.

46.1.4 Histone mRNA processing

Main article: Histone

Histones H2A, H2B, H3 and H4 form the core of a nucleosome and thus are called core histones. Processing of core histones is done differently because typical histone mRNA lacks several features of other eucaryotic mRNAs, such as poly(A) tail and introns. Thus such mRNAs do not undergo splicing and their 3'-prime processing is done independent of most cleavage and polyadenylation factors. Core histone mRNAs have a special stem-loop structure at 3-prime end that is recognized by a stem–loop binding protein and a downstream sequence, called histone downstream element (HDE) that recruits U7 snRNA. Cleavage and polyadenylation specificity factor 73 cuts mRNA between stem-loop and HDE[6]

Histone variants, such as H2A.Z or H3.3, however, have introns and are processed as normal mRNAs including splicing and polyadenylation.[6]

46.2   Post-transcriptional Modification Resources

46.3   Citations

[1] Berg, Tymoczko & Stryer 2007, p. 836

[2] Berg, Tymoczko & Stryer 2007, p. 841

[3] Yamada-Okabe, Toshiko; Mio, Toshiyuki; Matsui, Mitsuaki; Arisawa, Mikio; Yamada-Okabe, Hisafumi (November 1999). "The Candida albicans gene for mRNA 5'-cap methyltransferase: identification of additional residues essential for catalysis". *Microbiology* **145** (11): 3023–3033. doi:10.1099/00221287-145-11-3023. ISSN 1350-0872. PMID 10589710. Retrieved January 7, 2011.

[4] Hames & Hooper 2006, p. 221

[5] Lodish HF, Berk A, Kaiser C, Krieger M, Scott MP, Bretscher A, Ploegh H, Matsudaira PT (2007). "Chapter 8: Post-transcriptional Gene Control". *Molecular Cell .Biology*. San Francisco: WH Freeman. ISBN 0-7167-7601-4.

[6] William F. Marzluff, Eric J. Wagner & Robert J. Duronio (November 2008). "Metabolism and regulation of canonical histone mRNAs: life without a poly(A) tail". *Nature Reviews Genetics* **9** (11): 843–854. doi:10.1038/nrg2438. PMID 18927579.

[7] Sun, WJ; Li, JH; Liu, S; Wu, J; Zhou, H; Qu, LH; Yang, JH (11 October 2015). "RMBase: a resource for decoding the landscape of RNA modifications from high-throughput sequencing data.". *Nucleic Acids Research*: gkv1036. doi:10.1093/nar/gkv1036. PMID 26464443.

[8] Machnicka, MA; Milanowska, K; Osman Oglou, O; Purta, E; Kurkowska, M; Olchowik, A; Januszewski, W; Kalinowski, S; Dunin-Horkawicz, S; Rother, KM; Helm, M; Bujnicki, JM; Grosjean, H (December 2012). "MODOMICS: a database of RNA modification pathways--2013 update.". *Nucleic Acids Research* **41** (Database issue): D262–7. doi:10.1093/nar/gks1007. PMID 23118484.

[9] Cantara, WA; Crain, PF; Rozenski, J; McCloskey, JA; Harris, KA; Zhang, X; Vendeix, FA; Fabris, D; Agris, PF (December 2010). "The RNA Modification Database, RNAMDB: 2011 update.". *Nucleic Acids Research* **39** (Database issue): D195–201. doi:10.1093/nar/gkq1028. PMC 3013656. PMID 21071406.

46.4   External links

RMBase RNA Modification Base is designed for decoding **the landscape of RNA modifications** identified from high-throughput sequencing data (**Pseudo-seq, Ψ-seq, CeU-seq, Aza-IP, MeRIP-seq, m6A-seq, miCLIP, m6A-CLIP, RiboMeth-seq**).

- Post-Transcriptional RNA Modification at the US National Library of Medicine Medical Subject Headings (MeSH)

46.5   See also

- messenger RNA
- translation
- RNA editing
- RNA-Seq

46.6 References

1. Berg, Jeremy M.; Tymoczko, John L.; Stryer, Lubert (2007), *Biochemistry* (6 ed.), New York: WH Freeman & Co., ISBN 0-7167-6766-X

2. Hames, David; Hooper, Nigel (2006), *Instant Notes Biochemistry* (3 ed.), Leeds: Taylor and Francis, ISBN 0-415-36778-6

Chapter 47

Transposable element

Bacterial composite transposon

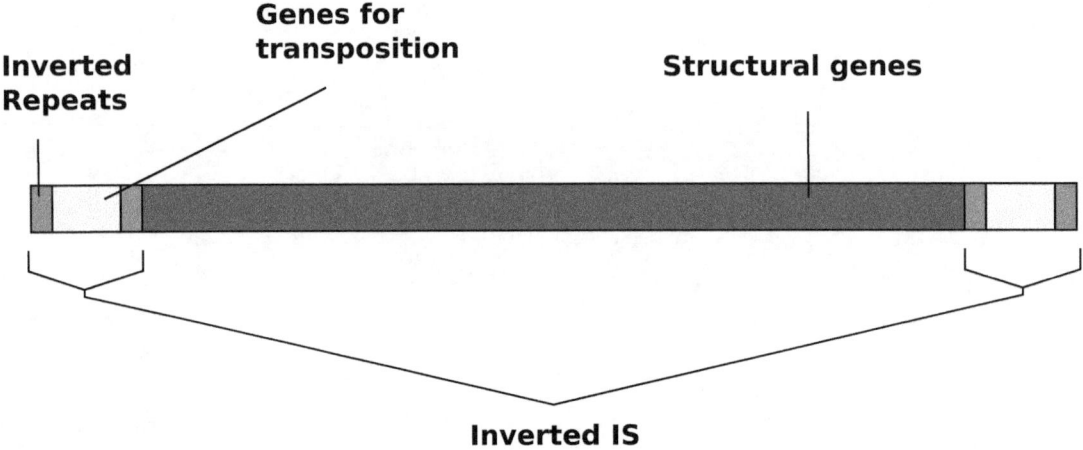

A bacterial DNA transposon

A **transposable element** (**TE** or **transposon**) is a DNA sequence that can change its position within the genome, sometimes creating or reversing mutations and altering the cell's genome size. Transposition often results in duplication of the TE. Barbara McClintock's discovery of these **jumping genes** earned her a Nobel prize in 1983.[1]

TEs make up a large fraction of the C-value of eukaryotic cells. There are at least two classes of TEs: class I TEs generally function via reverse transcription, while class II TEs encode the protein transposase, which they require for insertion and excision, and some of these TEs also encode other proteins.[2] It has been shown that TEs are important in genome function and evolution.[3] In *Oxytricha*, which has a unique genetic system, they play a critical role in development.[4] They are also very useful to researchers as a means to alter DNA inside a living organism.

47.1 Discovery

Barbara McClintock discovered the first TEs in maize, *Zea mays*, at the Cold Spring Harbor Laboratory. McClintock was experimenting with maize plants that had broken chromosomes.[5]

In the winter of 1944–1945 McClintock planted corn kernels that were self-pollinated, meaning that the flowers were pollinated by the silk of their own plant.[5] These kernels came from a long line of plants that had been self-pollinated, causing broken arms on the end of their ninth chromosome.[5] As the maize plants began to grow, McClintock noted unusual color patterns on the leaves.[5] For example, one leaf had two albino patches of almost identical size, located side by side on the leaf.[5] McClintock hypothesized that during cell division certain cells lost genetic material, while others gained what they had lost.[6] However, when comparing the chromosomes of the current generation of plants and their parent generation, she found certain parts of the chromosomes had switched positions on the chromosome.[6] She disproved the popular genetic theory of the time that genes were fixed in their position on a chromosome. McClintock found that genes could not only move, but they could also be turned on or off due to certain environmental conditions or during different stages of cell development.[6]

McClintock also showed that gene mutations could be reversed.[7] McClintock presented her report on her findings in 1951, and published an article on her discoveries in *Genetics* in November 1953 entitled, "Induction of Instability at Selected Loci in Maize."[8]

Her work would be largely dismissed and ignored until the late 1960s-1970s when it would be rediscovered after TEs were found in bacteria.[9] She was awarded a Nobel Prize in Medicine or Physiology in 1983 for her discovery of TEs, more than thirty years after her research and initial discovery.[10]

Approximately 90% of maize genome is made up of TEs, and 50% in the human genome.[11]

47.2 Classification

Transposable elements (TEs) represent one of several types of mobile genetic elements. TEs are assigned to one of two classes according to their mechanism of transposition, which can be described as either *copy and paste* (class I TEs) or *cut and paste* (class II TEs).[12]

47.2.1 *Class I* (retrotransposons)

Main article: retrotransposon

Class I TEs are copied in two stages: first they are transcribed from DNA to RNA, and the RNA produced is then reverse transcribed to DNA. This copied DNA is then inserted at a new position into the genome. The reverse transcription step is catalyzed by a reverse transcriptase, which is often encoded by the TE itself. The characteristics of retrotransposons are similar to retroviruses, such as HIV.

Retrotransposons are commonly grouped into three main orders:

- TEs with long terminal repeats (LTRs): encode reverse transcriptase, similar to retroviruses

- LINEs (LINE-1s or L1s): encode reverse transcriptase, lack LTRs, and are transcribed by RNA polymerase II

- SINEs: do not encode reverse transcriptase and are transcribed by RNA polymerase III.

Retroviruses can also be considered TEs. For example, after entering a host cell and conversion of the retroviral RNA into DNA, the newly produced retroviral DNA is integrated into the genome of the host cell. These integrated DNAs represent a provirus of the retrovirus. The provirus is a specialized form of eukaryotic retrotransposon, which can produce RNA intermediates that may leave the host cell and infect other cells. The transposition cycle of retroviruses has similarities to that of prokaryotic TEs, suggesting a distant relationship between these two TEs types.

47.2.2 *Class II* (DNA transposons)

The cut-and-paste transposition mechanism of class II TEs does not involve an RNA intermediate. The transpositions are catalyzed by several transposase enzymes. Some transposases non-specifically bind to any target site in DNA, whereas others bind to specific DNA sequence targets. The transposase makes a staggered cut at the target site resulting in single-strand 5' or 3' DNA overhangs (sticky ends). This step cuts out the DNA transposon, which is then ligated into a new target site; this process involves activity of a DNA polymerase that fills in gaps and of a DNA ligase that closes the sugar-phosphate backbone. This results in duplication of the target site. The insertion sites of DNA transposons may be identified by short direct repeats (created by the staggered cut in the target DNA and filling in by DNA polymerase) followed by a series of inverted repeats important for the TE excision by transposase. Cut-and-paste TEs may be duplicated if their transposition takes place during S phase of the cell cycle when a donor site has already been replicated, but a target site has not yet been replicated. Such duplications at the target site can result in gene duplication, which plays an important role in evolution.[13]:284 Not all DNA transposons transpose through the cut-and-paste mechanism. In some cases, a replicative transposition is observed in which a transposon replicates itself to a new target site (e.g. Helitron (biolog)).

Class II TEs make less than 2% of the human genome, making the rest Class I.[14]

47.2.3 Autonomous and non-autonomous TEs

Transposition can be classified as either "autonomous" or "non-autonomous" in both Class I and Class II TEs. Autonomous TEs can move by themselves while non-autonomous TEs require the presence of another TE to move. This is often because non-autonomous TEs lack transposase (for class II) or reverse transcriptase (for class I).

Activator element (*Ac*) is an example of an autonomous TE, and dissociation element (*Ds*) is an example of non-autonomous TE. Without *Ac, Ds* is not able to transpose.

47.3 Examples

- The first TEs were discovered in maize (*Zea mays*), by Barbara McClintock in 1948, for which she was awarded a Nobel Prize in 1983. She noticed insertions, deletions, and translocations, caused by these elements. These changes in the genome could, for example, lead to a change in the color of corn kernels. About 85% of the genome of maize consists in TEs.[15] The Ac/Ds system described by McClintock are class II TEs. Transposition of Ac in tobacco has been demonstrated by B. Baker (Plant Transposable Elements, pp 161–174, 1988, Plenum Publishing Corp., ed. Nelson).

- One family of TEs in the fruit fly *Drosophila melanogaster* are called *P elements*. They seem to have first appeared in the species only in the middle of the twentieth century. Within 50 years, they have spread through every population of the species. Gerald M. Rubin and Allan C. Spradling pioneered technology to use artificial P elements to insert genes into Drosophila by injecting the embryo.[16][17][18]

- Transposons in bacteria usually carry an additional gene for function other than transposition---often for antibiotic resistance. In bacteria, transposons can jump from chromosomal DNA to plasmid DNA and back, allowing for the transfer and permanent addition of genes such as those encoding antibiotic resistance (multi-antibiotic resistant bacterial strains can be generated in this way). Bacterial transposons of this type belong to the Tn family. When the transposable elements lack additional genes, they are known as insertion sequences.

- The most common form of transposable element in humans is the Alu sequence. It is approximately 300 bases long and can be found between 300,000 and a million times in the human genome. *Alu* is estimated to make up 15–17% of the human genome.[19]

- Mariner-like elements are another prominent class of transposons found in multiple species including humans. The Mariner transposon was first discovered by Jacobson and Hartl in *Drosophila*.[20] This Class II transposable element is known for its uncanny ability to be transmitted horizontally in many species.[21][22] There are an estimated 14 thousand copies of Mariner in the human genome comprising 2.6 million base pairs.[23] The first mariner-element transposons outside of animals were found in *Trichomonas vaginalis*.[24] These characteristics of the Mariner transposon have inspired the science fiction novel titled, "The Mariner Project".

- Mu phage transposition is the best known example of replicative transposition. Its transposition mechanism is somewhat similar to a homologous recombination.

- The five distinct yeast (*Saccharomyces cerevisiae*) retrotransposon families: Ty1, Ty2, Ty3, Ty4 and Ty5 [25]

- A helitron is a TE found in eukaryotes that are thought to replicate by a rolling-circle mechanism.

47.4   In disease

TEs are mutagens. They can damage the genome of their host cell in different ways:[26]

- a transposon or a retroposon that inserts itself into a functional gene will most likely disable that gene;

- after a DNA transposon leaves a gene, the resulting gap will probably not be repaired correctly;

- multiple copies of the same sequence, such as Alu sequences can hinder precise chromosomal pairing during mitosis and meiosis, resulting in unequal crossovers, one of the main reasons for chromosome duplication.

Diseases that are often caused by TEs include hemophilia A and B, severe combined immunodeficiency, porphyria, predisposition to cancer, and Duchenne muscular dystrophy.[27][28] *LINE1* (*L1*) TEs that land on the human Factor VIII caused haemophilia[29] and insertion of *L1* into *APC* gene caused colon cancer and this confirms that the TEs play an important role for disease development.[30]

Additionally, many TEs contain promoters which drive transcription of their own transposase. These promoters can cause aberrant expression of linked genes, causing disease or mutant phenotypes.

47.5   Silencing of TEs

If an organism is composed of mostly TEs, doesn't it affect their genetics? Surprisingly, in most cases TEs are silenced through epigenetics mechanisms like methylation, chromatin remodeling and piRNAs. So no phenotypic effects nor the movement of TEs occur as in some wild type plant TEs. Certain mutated plants were found to have defects in methylation related enzymes (methyl transferase) that cause the transcription of TEs, thus affecting the phenotype.[2][31]

A hypothesis suggests that only approximately 100 *LINE1* related sequences are active, despite their sequences making up 17% of the human genome. In human cells, silencing of *LINE1* sequences is triggered by an RNAi mechanism. Surprisingly, the RNAi sequences are derived from the 5' untranslated region (UTR) of the *LINE1*, a long terminal which repeats itself. Supposedly, the 5' *LINE1* UTR that codes for the sense promoter for *LINE1* transcription also encodes the antisense promoter for the miRNA that becomes the substrate for siRNA production. Inhibition of the RNAi silencing mechanism in this region showed an increase in *LINE1* transcription.[2][32]

47.6   Rate of transposition, induction and defense

One study estimated the rate of transposition of a particular retrotransposon, the Ty1 element in *Saccharomyces cerevisiae*. Using several assumptions, the rate of successful transposition event per single Ty1 element came out to be about once every few months to once every few years.[33]

Cells defend against the proliferation of TEs in a number of ways. These include piRNAs and siRNAs,[34] which silence TEs after they have been transcribed.

Some TEs contain heat-shock like promoters and their rate of transposition increases if the cell is subjected to stress,[35] thus increasing the mutation rate under these conditions, which might be beneficial to the cell.

47.7 Evolution

The scientific community is still exploring the evolution of TEs and their effect on genome evolution.

TEs are found in most life forms. They may have originated in the last universal common ancestor, arisen independently multiple times, or arisen once and then spread to other kingdoms by horizontal gene transfer.[36] While some TEs confer benefits on their hosts, most are regarded as selfish DNA parasites. In this way, they are similar to viruses. Various viruses and TEs also share features in their genome structures and biochemical abilities, leading to speculation that they share a common ancestor.

Because excessive TE activity can damage exons, many organisms have developed mechanisms to inhibit their activity. Bacteria may undergo high rates of gene deletion as part of a mechanism to remove TEs and viruses from their genomes while eukaryotic organisms use RNA interference (RNAi) to inhibit TE activity. Nevertheless, some TEs generated large families often associated with speciation events.

Evolution often deactivates DNA transposons, leaving them as introns (inactive gene sequences). In vertebrate animal cells nearly all >100,000 DNA transposons per genome have genes that encode inactive transposase polypeptides.[37] In humans, all Tc1-like transposons are inactive. The first synthetic transposon designed for use in vertebrate cells, the Sleeping Beauty transposon system, is a Tc1/mariner-like transposon. It exists in the human genome as an intron and was activated through reconstruction.[38]

Interspersed Repeats within genomes are created by transposition events accumulating over evolutionary time. Because interspersed repeats block gene conversion, they protect novel gene sequences from being overwritten by similar gene sequences and thereby facilitate the development of new genes.

TEs may have been co-opted by the vertebrate immune system as a means of producing antibody diversity. The V(D)J recombination system operates by a mechanism similar to that of some TEs.

TEs contain many type of genes, including those conferring antibiotic resistance and ability to transpose to conjugative plasmids. Some TEs also contain integrons (genetic elements that can capture and express genes from other sources). These contain integrase, which can integrate gene cassettes. There are over 40 antibiotic resistance genes identified on cassettes, as well as virulence genes.

Transposons do not always excise their elements precisely, sometimes removing the adjacent base pairs. This phenomenon is called exon shuffling. Shuffling two unrelated exons can create a novel gene product or, more likely, an intron.[39]

47.8 Applications

Main article: Transposons as a genetic tool

The first TE was discovered in the plant maize (*Zea mays*, corn species), and is named dissociator (Ds). Likewise, the first TE to be molecularly isolated was from a plant (Snapdragon). Appropriately, TEs have been an especially useful tool in plant molecular biology. Researchers use them as a means of mutagenesis. In this context, a TE jumps into a gene and produces a mutation. The presence of such a TE provides a straightforward means of identifying the mutant allele, relative to chemical mutagenesis methods.

Sometimes the insertion of a TE into a gene can disrupt that gene's function in a reversible manner, in a process called insertional mutagenesis; transposase-mediated excision of the DNA transposon restores gene function. This produces plants in which neighboring cells have different genotypes. This feature allows researchers to distinguish between genes that must be present inside of a cell in order to function (cell-autonomous) and genes that produce observable effects in cells other than those where the gene is expressed.

TEs are also a widely used tool for mutagenesis of most experimentally tractable organisms. The Sleeping Beauty transposon system has been used extensively as an insertional tag for identifying cancer genes [40]

The Tc1/mariner-class of TEs Sleeping Beauty transposon system, awarded as the Molecule of the Year 2009,[41] is active in mammalian cells and is being investigated for use in human gene therapy.[42][43][44]

47.9 *De novo* repeat identification

De novo repeat identification is an initial scan of sequence data that seeks to find the repetitive regions of the genome, and to classify these repeats. Many computer programs exist to perform *de novo* repeat identification, all operating under the same general principles.[45] As short tandem repeats are generally 1–6 base pairs in length and are often consecutive, their identification is relatively simple.[46] Dispersed repetitive elements, on the other hand, are more challenging to identify, due to the fact that they are longer and have often acquired mutations. However, it is important to identify these repeats as they are often found to be transposable elements (TEs).[45]

De novo identification of transposons involves three steps: 1) find all repeats within the genome, 2) build a consensus of each family of sequences, and 3) classify these repeats. There are three groups of algorithms for the first step. One group is referred to as the k-mer approach, where a k-mer is a sequence of length k. In this approach, the genome is scanned for overrepresented k-mers; that is, k-mers that occur more often than is likely based on probability alone. The length k is determined by the type of transposon being searched for. The k-mer approach also allows mismatches, the number of which is determined by the analyst. Some k-mer approach programs use the k-mer as a base, and extend both ends of each repeated k-mer until there is no more similarity between them, indicating the ends of the repeats.[45] Another group of algorithms employs a method called sequence self-comparison. Sequence self-comparison programs use databases such as AB-BLAST to conduct an initial sequence alignment. As these programs find groups of elements that partially overlap, they are useful for finding highly diverged transposons, or transposons with only a small region copied into other parts of the genome.[47] Another group of algorithms follows the periodicity approach. These algorithms perform a Fourier transformation on the sequence data, identifying periodicities, regions that are repeated periodically, and are able to use peaks in the resultant spectrum to find candidate repetitive elements. This method works best for tandem repeats, but can be used for dispersed repeats as well. However, it is a slow process, making it an unlikely choice for genome scale analysis.[45]

The second step of *de novo* repeat identification involves building a consensus of each family of sequences. A consensus sequence is a sequence that is created based on the repeats that comprise a TE family. A base pair in a consensus is the one that occurred most often in the sequences being compared to make the consensus. For example, in a family of 50 repeats where 42 have a T base pair in the same position, the consensus sequence would have a T at this position as well, as the base pair is representative of the family as a whole at that particular position, and is most likely the base pair found in the family's ancestor at that position.[45] Once a consensus sequence has been made for each family, it is then possible to move on to further analysis, such as TE classification and genome masking in order to quantify the overall TE content of the genome.

47.10 See also

- Insertion sequence
- Intragenomic conflict
- P element
- Tn3 Transposon
- Tn10
- Transposon tagging
- Signature tagged mutagenesis
- Evolution of sexual reproduction

47.11 Notes

- Kidwell, M.G. (2005). "Transposable elements". In ed. T.R. Gregory. *The Evolution of the Genome.* San Diego: Elsevier. pp. 165–221. ISBN 0-123-01463-8.

- Craig NL, Craigie R, Gellert M, and Lambowitz AM (ed.) (2002). *Mobile DNA II*. Washington, DC: ASM Press. ISBN 978-1-555-81209-6.

- Lewin B (2000). *Genes VII*. Oxford University Press. ISBN 978-0-198-79276-5.

47.11.1   References

[1] McClintock, Barbara (June 1950). "The origin and behavior of mutable loci in maize". *Proc Natl Acad Sci U S A*. **36** (6): 344–55. Bibcode:1950PNAS...36..344M. doi:10.1073/pnas.36.6.344. PMC 1063197. PMID 15430309.

[2] Pray, Leslie A. (2008). "Transposons: The jumping genes". *Nature Education* **1** (1): 204.

[3] Bucher E, Reinders J, Mirouze M. (Nov 2012). "Epigenetic control of transposon transcription and mobility in *Arabidopsis*". *Current Opinion in Plant Biology* **15** (5): 503–10. doi:10.1016/j.pbi.2012.08.006. PMID 22940592.

[4] "'Junk' DNA Has Important Role, Researchers Find". *Science Daily*. 21 May 2009.

[5] McGrayne, Sharon Bertsch (1998). *Nobel Prize Women in Science: Their Lives, Struggles, and Momentous Discoveries* (2nd ed.). Carol Publishing. p. 165. ISBN 978-0-9702256-0-3.

[6] McGrayne 1998, p. 166

[7] McGrayne 1998, p. 167

[8] McClintock B (November 1953). "Induction of Instability at Selected Loci in Maize". *Genetics* **38** (6): 579–99. PMC 1209627. PMID 17247459.

[9] Des Jardins, Julie (2010). *The Madame Curie Complex: The Hidden History of Women in Science*. Feminist Press at CUNY. p. 246. ISBN 978-1-55861-655-4.

[10] Fedoroff, Nina; Botstein, David, eds. (1 January 1992). *The Dynamic Genome: Barbara McClintock's Ideas in the Century of Genetics*. Cold Spring Harbor Laboratory Press. p. 2. ISBN 978-0-87969-422-7.

[11] SanMiguel P, Tikhonov A, Jin YK, et al. (November 1996). "Nested retrotransposons in the intergenic regions of the maize genome". *Science* **274** (5288): 765–8. doi:10.1126/science.274.5288.765. PMID 8864112.

[12] Wicker, T; et al. (December 2007). "A unified classification system for eukaryotic transposable elements". *Nature Reviews Genetics* **8** (12): 973–82. doi:10.1038/nrg2165. PMID 17984973.

[13] Madigan M, Martinko J (editors) (2006). *Brock Biolog of Microorganisms* (11th ed.). Prentice Hall. ISBN 0-13-144329-1.

[14] Kazazian HH Jr, Moran JV (May 1998). "The impact of L1 retrotransposons on the human genome". *Nature Genetics* **19** (1): 19–24. doi:10.1038/ng0598-19. PMID 9590283.

[15] Schnable; et al. (November 2009). "The B73 maize genome: complexity, diversity, and dynamics". *Science* **326** (5956): 1112–1115. Bibcode:2009Sci...326.1112S. doi:10.1126/science.1178534. PMID 19965430.

[16] Spradling AC, Rubin GM (October 1982). "Transposition of cloned P elements into Drosophila germ line chromosomes". *Science* **218** (4570): 341–347. Bibcode:1982Sci...218..341S. doi:10.1126/science.6289435. PMID 6289435.

[17] Rubin GM, Spradling AC (October 1982). "Genetic transformation of Drosophila with transposable element vectors". *Science* **218** (4570): 348–353. Bibcode:1982Sci...218..348R. doi:10.1126/science.6289436. PMID 6289436.

[18] Cesari F (15 October 2007). "Milestones in Nature: Milestone 9: Transformers, Elements in Disguise". *Nature*. doi:10.1038/nrg2254.

[19] Kazazian HH, Moran JV (May 1998). "The impact of L1 retrotransposons on the human genome". *Nat. Genet.* **19** (1): 19–24. doi:10.1038/ng0598-19. PMID 9590283.

[20] Jacobson JW, Medhora MM, Hartl DL (November 1986). "Molecular structure of a somatically unstable transposable element in Drosophila". *Proc. Natl. Acad. Sci. U.S.A.* **83** (22): 8684–8. doi:10.1073/pnas.83.22.8684. PMC 386995. PMID 3022302.

[21] Lohe AR, Moriyama EN, Lidholm DA, Hartl DL (January 1995). "Horizontal transmission, vertical inactivation, and stochastic loss of mariner-like transposable elements". *Mol. Biol. Evol.* **12** (1): 62–72. doi:10.1093/oxfordjournals.molbev.a040191. PMID 7877497.

[22] Lampe DJ, Witherspoon DJ, Soto-Adames FN, Robertson HM (April 2003). "Recent horizontal transfer of mellifera subfamily mariner transposons into insect lineages representing four different orders shows that selection acts only during horizontal transfer". *Mol. Biol. Evol.* **20** (4): 554–62. doi:10.1093/molbev/msg069. PMID 12654937.

[23] Mandal PK, Kazazian HH (October 2008). "SnapShot: Vertebrate transposons". *Cell* **135** (1): 192–192.e1. doi:10.1016/j.cell.2008.09.028. PMID 18854165.

[24] Carlton JM, Hirt RP, Silva JC, et al. (January 2007). "Draft genome sequence of the sexually transmitted pathogen *Trichomonas vaginalis*". *Science* **315** (5809): 207–12. doi:10.1126/science.1132894. PMC 2080659. PMID 17218520.

[25] Kim JM, Vanguri S, Boeke JD, Gabriel A, Voytas DF (May 1998). "Transposable elements and genome organization: a comprehensive survey of retrotransposons revealed by the complete Saccharomyces cerevisiae genome sequence". *Genome Res.* **8** (5): 464–78. doi:10.1101/gr.8.5.464. PMID 9582191.

[26] Belancio VP, Hedges DJ, Deininger P (March 2008). "Mammalian non-LTR retrotransposons: for better or worse, in sickness and in health". *Genome Res.* **18** (3): 343–58. doi:10.1101/gr.5558208. PMID 18256243.

[27] Kazazian H.H., Goodier J.L. (2002). "LINE drive: retrotransposition and genome instability". *Cell* **110** (3): 277–80. doi:10.1016/S0092-8674(02)00868-1. PMID 12176313.

[28] Kapitonov V.V., Pavlicek, A., Jurka, J. (2006). "Anthology of Human Repetitive DNA". *Encyclopedia of Molecular Cell Biology and Molecular Medicine.* doi:10.1002/3527600906.mcb.200300166.

[29] Kazazian, HH Jr; et al. (1988). "Haemophilia A resulting from de novo insertion of L1 sequences represents a novel mechanism for mutation in man". *Nature* **332** (6160): 164–6. doi:10.1038/332164a0. PMID 2831458.

[30] Miki, Y.; et al. (Feb 1992). "Disruption of the APC gene by a retrotransposal insertion of L1 sequence in colon cancer". *Cancer Research* **52** (3): 643–5. PMID 1310068.

[31] Miura A, Yonebayashi S, Watanabe K, Toyama T, Shimada H, Kakutani T (May 2001). "Mobilization of transposons by a mutation abolishing full DNA methylation in Arabidopsis". *Nature* **411** (6834): 212–4. doi:10.1038/35075612. PMID 11346800.

[32] Yang N, Kazazian HH (September 2006). "L1 retrotransposition is suppressed by endogenously encoded small interfering RNAs in human cultured cells". *Nat. Struct. Mol. Biol.* **13** (9): 763–71. doi:10.1038/nsmb1141. PMID 16936727.

[33] Paquin CE, Williamson VM (5 October 1984). "Temperature Effects on the Rate of Ty Transposition". *Science* **226** (4670): 53–55. Bibcode:1984Sci...226...53P. doi:10.1126/science.226.4670.53. PMID 17815421.

[34] Wei-Jen Chung,Katsutomo Okamura,Raquel Martin, Eric C. Lai (3 June 2008). "Endogenous RNA Interference Provides a Somatic Defense against Drosophila Transposons". *Current Biology* **18** (11): 795–802. doi:10.1016/j.cub.2008.05.006. PMC 2812477. PMID 18501606.

[35] Dennis J. Strand, John F. McDonald (1985). "Copia is transcriptionally responsive to environmental stress". *Nucleic Acids Research* **13** (12): 4401–4410. doi:10.1093/nar/13.12.4401. PMC 321795. PMID 2409535.

[36] Kidwell, M.G. (1992). "Horizontal transfer of P elements and other short inverted repeat transposons". *Genetica* **86** (1): 275–286. doi:10.1007/BF00133726. PMID 1334912.

[37] Plasterk R.H.A., Izsvák Z., Ivics Z. (1999). "Resident aliens: the Tc1/mariner superfamily of transposable elements". *Trends Genet* **15** (8): 326–32. doi:10.1016/S0168-9525(99)01777-1. PMID 10431195.

[38] Ivics Z., Hackett P.B., Plasterk R.H., Izsvak Z. (1997). "Molecular reconstruction of *Sleeping Beauty*, a Tc1-like transposon from fish, and its transposition in human cells". *Cell* **91** (4): 501–10. doi:10.1016/S0092-8674(00)80436-5. PMID 9390559.

[39] Moran JV, DeBerardinis RJ, Kazazian HH (March 1999). "Exon shuffling by L1 retrotransposition". *Science* **283** (5407): 1530–4. doi:10.1126/science.283.5407.1530. PMID 10066175.

[40] Carlson C.M., Largaespada D.A. (2005). "Insertional mutagenesis in mice: new perspectives and tools". *Nature Reviews Genetics* **6** (7): 568–80. doi:10.1038/nrg1638. PMID 15995698.

[41] Luft FC (May 2010). "*Sleeping Beauty* jumps to new heights". *Mol. Med* **88** (7): 641–643. doi:10.1007/s00109-010-0626-1. PMID 20467721.

[42] Ivics Z, Izsvák Z (October 2006). "Transposons for gene therapy!". *Curr Gene Ther* **6** (5): 593–607. doi:10.2174/156652306778520647. PMID 17073604.

[43] Wilson MH, Coates CJ, George AL (January 2007). "PiggyBac transposon-mediated gene transfer in human cells". *Mol. Ther.* **15** (1): 139–145. doi:10.1038/sj.mt.6300028. PMID 17164785.

[44] Hackett P.B., Largaespada D.A., Cooper L.J.N. (2010). "A transposon and transposase system for human application". *Mol. Ther.* **18** (4): 674–83. doi:10.1038/mt.2010.2. PMC 2862530. PMID 20104209.

[45] Makałowski W, Pande A, Gotea V, Makałowska I (2012). "Transposable elements and their identification". *Methods Mol. Biol.* **855**: 337–59. doi:10.1007/978-1-61779-582-4_12. PMID 22407715.

[46] Saha S, Bridges S, Magbanua ZV, Peterson DG (2008). "Computational Approaches and Tools Used in Identification of Dispersed Repetitive DNA Sequences". *Tropical Plant Biol.* **1**: 85–96. doi:10.1007/s12042-007-9007-5.

[47] Saha S, Bridges S, Magbanua ZV, Peterson DG (April 2008). "Empirical comparison of ab initio repeat finding programs". *Nucleic Acids Res.* **36** (7): 2284–94. doi:10.1093/nar/gkn064. PMC 2367713. PMID 18287116.

47.12   External links

- Kimball's Biology Pages: Transposons

- "An immune system so versatile it might kill you". *New Scientist* (2556). 21 June 2006. — A possible connection between aberrant reinsertions and lymphoma.

- GMO Safety: Transposons - research projects and basic infos

- A wiki specially dedicated to transposable elements and their classification

- Repbase — a database of transposable element sequences

- RepeatMasker — a computer program used by computational biologists to annotate transposons in DNA sequences

- Use of the Sleeping Beauty Transposon System for Stable Gene Expression in Mouse Embryonic Stem Cells

Chapter 48

Human genetic variation

Human genetic variation is the genetic differences both within and among populations. There may be multiple variants of any given gene in the human population (genes), leading to polymorphism. Many genes are not polymorphic, meaning that only a single allele is present in the population: the gene is then said to be fixed.[1] On average, in terms of DNA sequence all humans are 99.5% similar to any other humans.[2]

No two humans are genetically identical. Even monozygotic twins, who develop from one zygote, have infrequent genetic differences due to mutations occurring during development and gene copy-number variation.[3] Differences between individuals, even closely related individuals, are the key to techniques such as genetic fingerprinting. Alleles occur at different frequencies in different human populations, with populations that are more geographically and ancestrally remote tending to differ more.

Causes of differences between individuals include the exchange of genes during meiosis and various mutational events. There are at least two reasons why genetic variation exists between populations. Natural selection may confer an adaptive advantage to individuals in a specific environment if an allele provides a competitive advantage. Alleles under selection are likely to occur only in those geographic regions where they confer an advantage. The second main cause of genetic variation is due to the high degree of neutrality of most mutations. Most mutations do not appear to have any selective effect one way or the other on the organism. The main cause is genetic drift, this is the effect of random changes in the gene pool. In humans, founder effect and past small population size (increasing the likelihood of genetic drift) may have had an important influence in neutral differences between populations. The theory that humans recently migrated out of Africa supports this.

The study of human genetic variation has both evolutionary significance and medical applications. It can help scientists understand ancient human population migrations as well as how different human groups are biologically related to one another. For medicine, study of human genetic variation may be important because some disease-causing alleles occur more often in people from specific geographic regions. New findings show that each human has on average 60 new mutations compared to their parents.[4][5] Apart from mutations, many genes that may have aided humans in ancient times plague humans today. For example, it is suspected that genes that allow humans to more efficiently process food are those that make people susceptible to obesity and diabetes today.[6]

48.1 Measures of variation

Genetic variation among humans occurs on many scales, from gross alterations in the human karyotype to single nucleotide changes.[7]

Nucleotide diversity is the average proportion of nucleotides that differ between two individuals. The human nucleotide diversity is estimated to be 0.1%[8] to 0.4% of base pairs.[9] A difference of 1 in 1,000 amounts to approximately 3 million nucleotide differences, because the human genome has about 3 billion nucleotides.

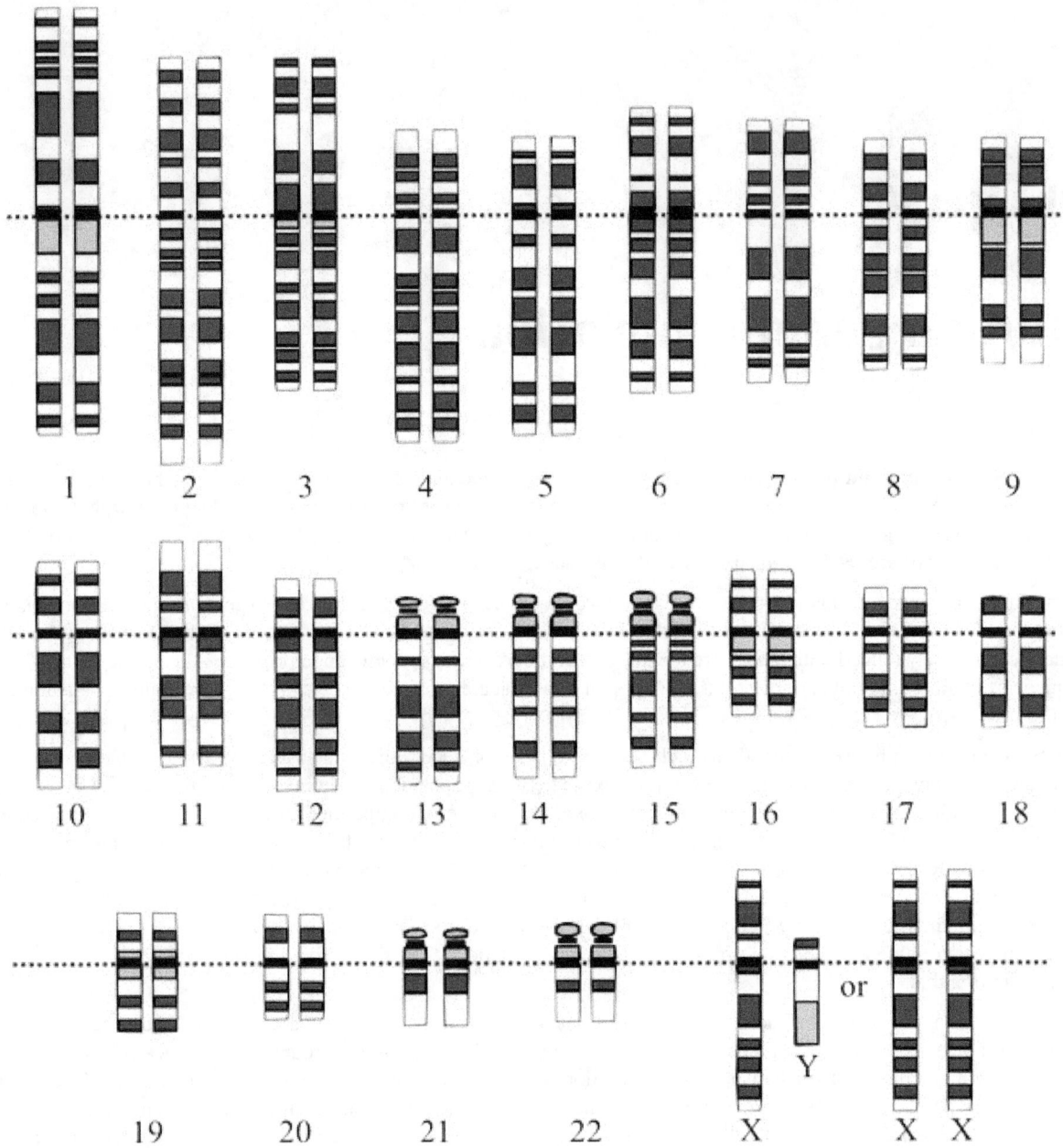

A graphical representation of the typical human karyotype

48.1.1   Single nucleotide polymorphisms

Main article: Single nucleotide polymorphism

A single nucleotide polymorphism (SNP) is difference in a single nucleotide between members of one species that occurs in at least 1% of the population. It is estimated that there are 10 to 30 million SNPs in humans.

SNPs are the most common type of sequence variation, estimated to account for 90% of all sequence variation.[10] Other sequence variations are single base exchanges, deletions and insertions.[11] SNPs occur on average about every 100 to 300 bases [12] and so are the major source of heterogeneity.

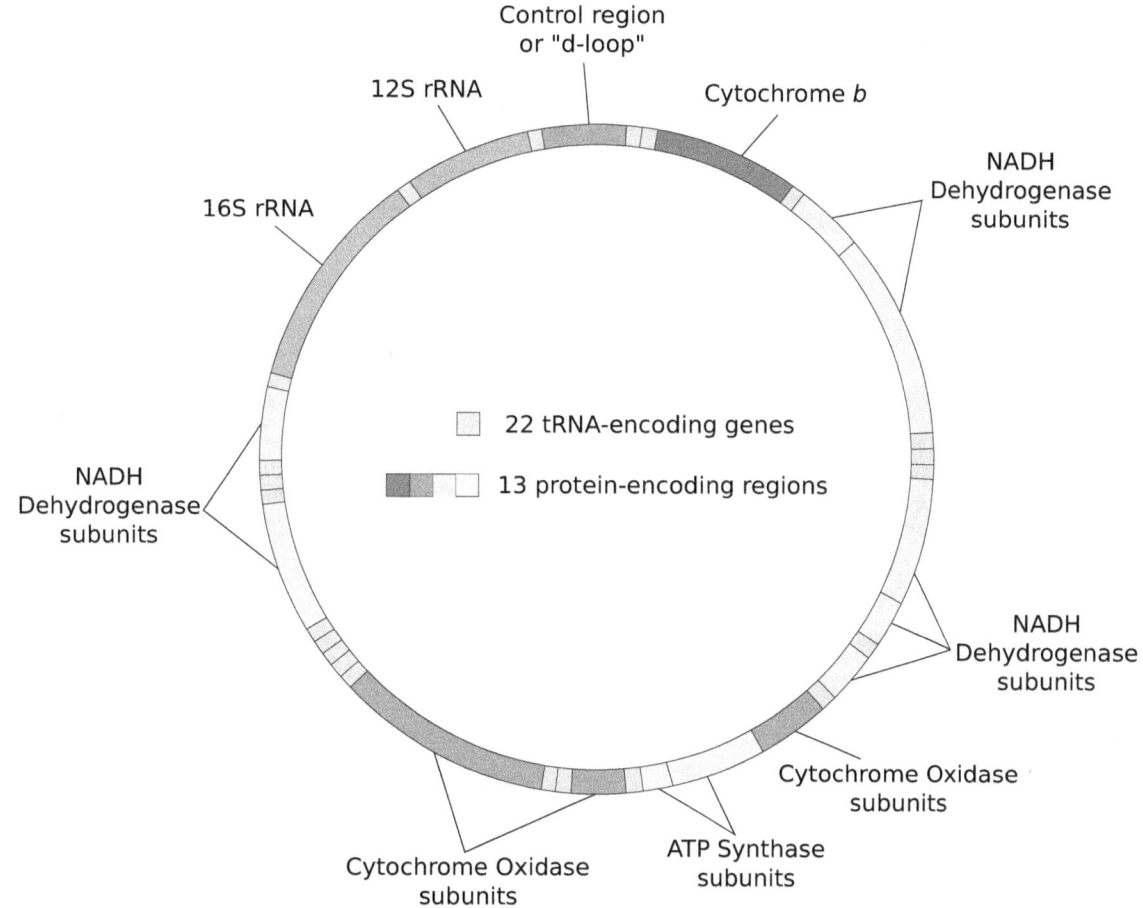

Mitochondrial DNA.

A functional, or non-synonymous, SNP is one that affects some factor such as gene splicing or messenger RNA, and so causes a phenotypic difference between members of the species. About 3% to 5% of human SNPs are functional (see International HapMap Project). Neutral, or synonymous SNPs are still useful as genetic markers in genome-wide association studies, because of their sheer number and the stable inheritance over generations.[10]

A coding SNP is one that occurs inside a gene. There are 105 Human Reference SNPs that result in premature stop codons in 103 genes. This corresponds to 0.5% of coding SNPs. They occur due to segmental duplication in the genome. These SNPs result in loss of protein, yet all these SNP alleles are common and are not purified in negative selection.[13]

48.1.2 Structural variation

Main article: Structural variation

Structural variation is the variation in structure of an organism's chromosome. Structural variations, such as copy-number variation and deletions, inversions, insertions and duplications, account for much more human genetic variation than single nucleotide diversity. This was concluded in 2007 from analysis of the diploid full sequences of the genomes of two humans: Craig Venter and James D. Watson. This added to the two haploid sequences which were amalgamations of sequences from many individuals, published by the Human Genome Project and Celera Genomics respectively.[14]

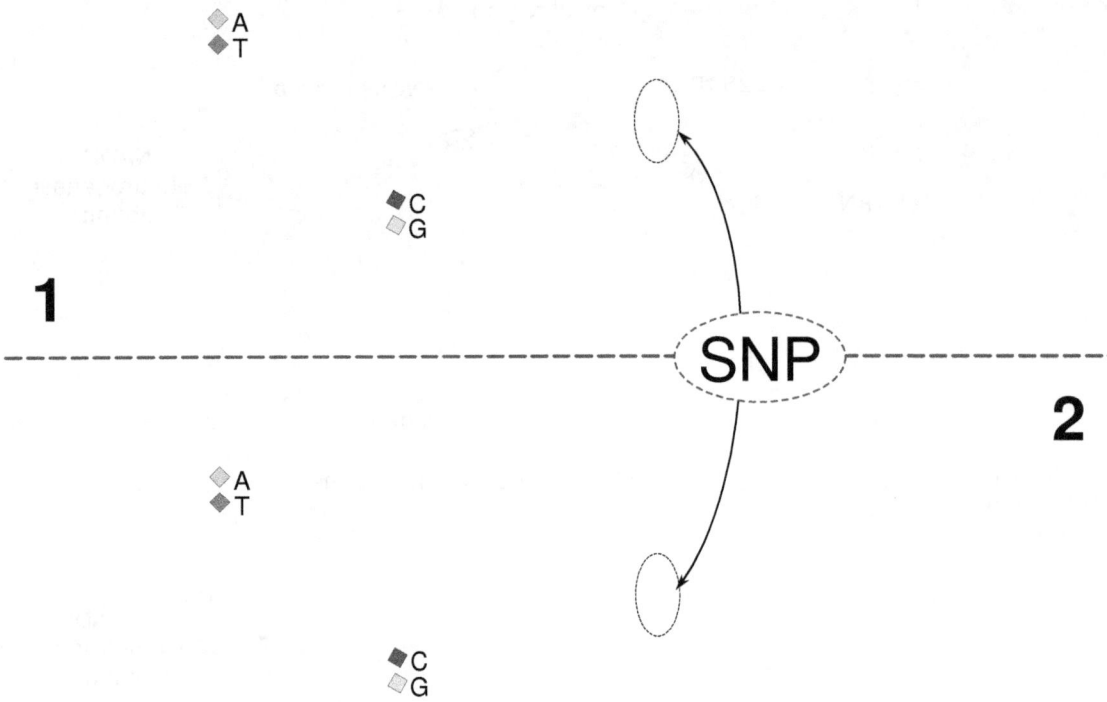

DNA molecule 1 differs from DNA molecule 2 at a single base-pair location (a C/T polymorphism).

Copy number variation

Main article: Copy number variation

A copy-number variation (CNV) is a difference in the genome due to deleting or duplicating large regions of DNA on some chromosome. It is estimated that 0.4% of the genomes of unrelated humans differ with respect to copy number. When copy number variation is included, human-to-human genetic variation is estimated to be at least 0.5% (99.5% similarity).[15][16][17][18] Copy number variations are inherited but can also arise during development.[19][20][21][22]

48.1.3   Epigenetics

Epigenetic variation is variation in the chemical tags that attach to DNA and affect how genes get read. The tags, "called epigenetic markings, act as switches that control how genes can be read."[23] At some alleles, the epigenetic state of the DNA, and associated phenotype, can be inherited across generations of individuals.[24]

48.1.4   Genetic variability

Main article: Genetic variability

Genetic variability is a measure of the tendency of individual genotypes in a population to vary (become different) from one another. Variability is different from genetic diversity, which is the amount of variation seen in a particular population. The variability of a trait is how much that trait tends to vary in response to environmental and genetic influences.

48.1.5 Clines

Main article: Cline (biology)

In biology, a cline is a continuum of species, populations, races, varieties, or forms of organisms that exhibit gradual phenotypic and/or genetic differences over a geographical area, typically as a result of environmental heterogeneity.[25][26][27] In the scientific study of human genetic variation, a gene cline can be rigorously defined and subjected to quantitative metrics.

48.1.6 Haplogroups

Main article: Haplogroup

In the study of molecular evolution, a haplogroup is a group of similar haplotypes that share a common ancestor with a single nucleotide polymorphism (SNP) mutation. Haplogroups pertain to deep ancestral origins dating back thousands of years.[28]

The most commonly studied human haplogroups are Y-chromosome (Y-DNA) haplogroups and mitochondrial DNA (mtDNA) haplogroups, both of which can be used to define genetic populations. Y-DNA is passed solely along the patrilineal line, from father to son, while mtDNA is passed down the matrilineal line, from mother to both daughter and son. The Y-DNA and mtDNA may change by chance mutation at each generation.

48.1.7 Variable number tandem repeats

Main article: Variable number tandem repeat

A variable number tandem repeat (VNTR) is the variation of length of a tandem repeat. A tandem repeat is the adjacent repetition of a short nucleotide sequence. Tandem repeats exist on many chromosomes, and their length varies between individuals. Each variant acts as an inherited allele, so they are used for personal or parental identification. Their analysis is useful in genetics and biology research, forensics, and DNA fingerprinting.

Short tandem repeats (about 5 base pairs) are called microsatellites, while longer ones are called minisatellites.

48.2 History and geographic distribution

See also: Human evolutionary genetics § Modern humans

The Out of Africa theory (more precisely called "*recent African origin of modern humans*") is the most widely accepted explanation of the origin and early dispersal of anatomically modern humans, *Homo sapiens sapiens*. The theory states that archaic *Homo sapiens* evolved into modern humans solely in Africa, 200,000 to 100,000 years ago; around that time, one African subpopulation of hominins among several was the subpopulation ancestral to all human beings today. Some members of that subpopulation left Africa by 60,000 years ago and over time replaced earlier hominin populations such as *Homo erectus* and Neanderthals on Earth. Alternative theories include the multiregional origin of modern humans hypothesis.

The theory is supported by both genetic and fossil evidence. The hypothesis originated in the 19th century, with Darwin's *Descent of Man*, but remained speculative until the 1980s when it was supported by study of present-day mitochondrial DNA, combined with evidence from physical anthropology of archaic specimens. A large study published in 2009 found that modern humans probably originated near the border of Namibia and South Africa (reported as Namibia and Angola by BBC[29]), and some left Africa through East Africa. Observations consistent with this are that Africa contains the most human genetic diversity anywhere on Earth, and the genetic structure of Africans traces to 14 ancestral population

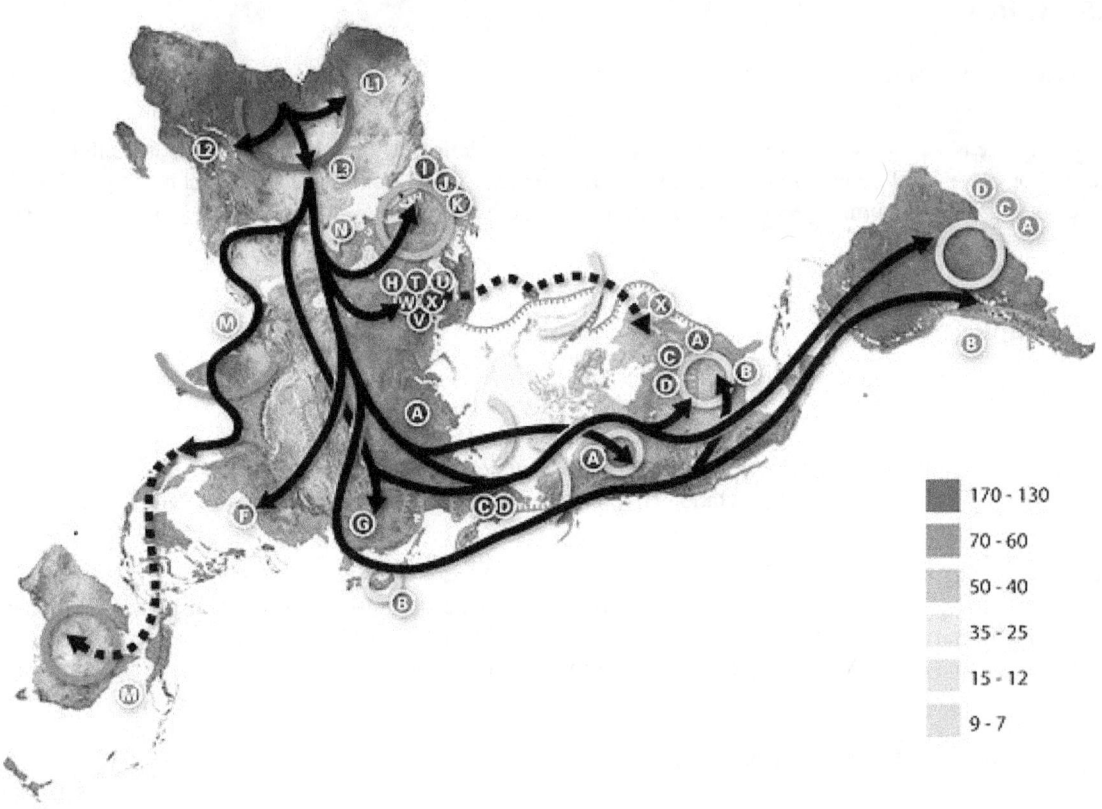

Map of the migration of modern humans out of Africa, based on mitochondrial DNA. Colored rings indicate thousand years before present.

clusters that correlate with ethnicity and culture or language. The study lasted ten years and analyzed variations at 1,327 DNA markers of 121 African populations, 4 African American populations, and 60 non-African populations.[30][31]

According to a 2000 study of Y-chromosome sequence variation,[32] human Y-chromosomes trace ancestry to Africa, and the descendants of the derived lineage left Africa and eventually were replaced by archaic human Y-chromosomes in Eurasia. The study also shows that a minority of contemporary East Africans and Khoisan are the descendants of the most ancestral patrilineages of anatomically modern humans that left Africa 35,000 to 89,000 years ago.[32] Other evidence supporting the theory is that variations in skull measurements decrease with distance from Africa at the same rate as the decrease in genetic diversity. Human genetic diversity decreases in native populations with migratory distance from Africa, and this is thought to be due to bottlenecks during human migration, which are events that temporarily reduce population size.[33][34]

48.2.1 Population genetics

In the field of population genetics, it is believed that the distribution of neutral polymorphisms among contemporary humans reflects human demographic history. It has been theorized that humans passed through a population bottleneck before a rapid expansion coinciding with migrations out of Africa leading to an African-Eurasian divergence around 100,000 years ago (ca. 5,000 generations), followed by a European-Asian divergence about 40,000 years ago (ca. 2,000 generations). Richard G. Klein, Nicholas Wade and Spencer Wells, among others, have postulated that modern humans did not leave Africa and successfully colonize the rest of the world until as recently as 60,000 - 50,000 years B.P., pushing back the dates for subsequent population splits as well.

The rapid expansion of a previously small population has two important effects on the distribution of genetic variation.

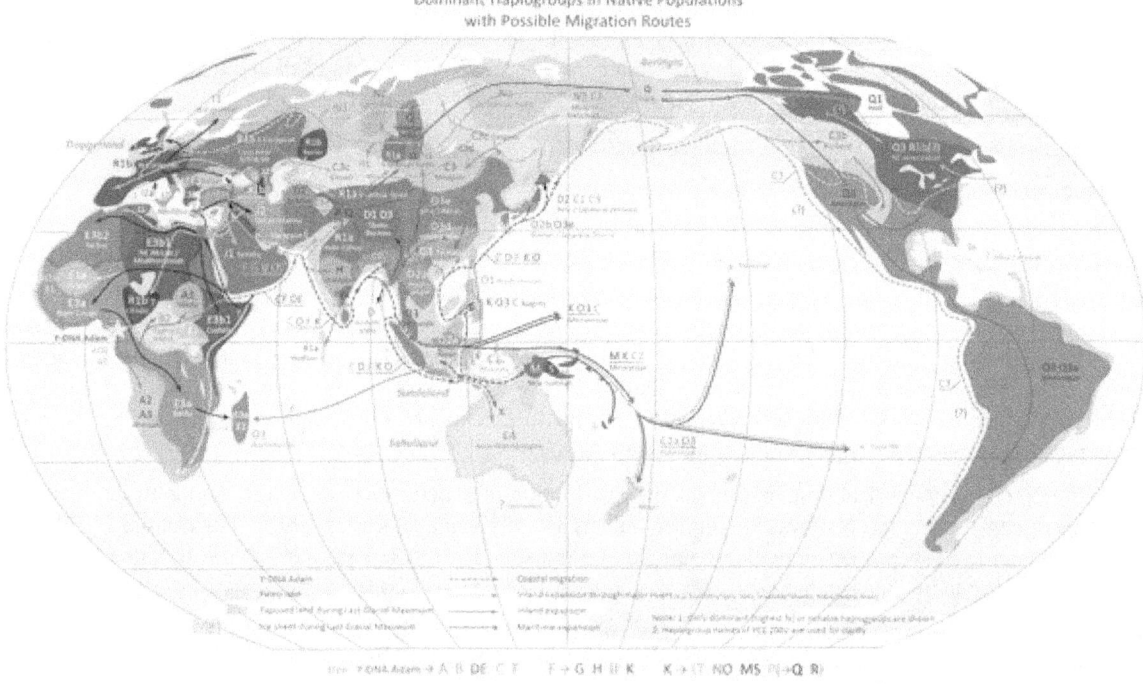

Dominant Y-chromosome haplogroups in pre-colonial world populations, with possible migrations routes according to the Coastal Migration Model.

First, the so-called founder effect occurs when founder populations bring only a subset of the genetic variation from their ancestral population. Second, as founders become more geographically separated, the probability that two individuals from different founder populations will mate becomes smaller. The effect of this assortative mating is to reduce gene flow between geographical groups, and to increase the genetic distance between groups. The expansion of humans from Africa affected the distribution of genetic variation in two other ways. First, smaller (founder) populations experience greater genetic drift because of increased fluctuations in neutral polymorphisms. Second, new polymorphisms that arose in one group were less likely to be transmitted to other groups as gene flow was restricted.

Our history as a species also has left genetic signals in regional populations. For example, in addition to having higher levels of genetic diversity, populations in Africa tend to have lower amounts of linkage disequilibrium than do populations outside Africa, partly because of the larger size of human populations in Africa over the course of human history and partly because the number of modern humans who left Africa to colonize the rest of the world appears to have been relatively low (Gabriel *et al.* 2002). In contrast, populations that have undergone dramatic size reductions or rapid expansions in the past and populations formed by the mixture of previously separate ancestral groups can have unusually high levels of linkage disequilibrium (Nordborg and Tavare 2002).

Many other geographic, climatic, and historical factors have contributed to the patterns of human genetic variation seen in the world today. For example, population processes associated with colonization, periods of geographic isolation, socially reinforced endogamy, and natural selection all have affected allele frequencies in certain populations (Jorde *et al.* 2000b; Bamshad and Wooding 2003). In general, however, the recency of our common ancestry and continual gene flow among human groups have limited genetic differentiation in our species.

48.2.2 Distribution of variation

The distribution of genetic variants within and among human populations are impossible to describe succinctly because of the difficulty of defining a "population," the clinal nature of variation, and heterogeneity across the genome (Long and Kittles 2003). In general, however, an average of 85% of genetic variation exists within local populations, ~7% is between local populations within the same continent, and ~8% of variation occurs between large groups living on different continents,. (Lewontin 1972; Jorde *et al.* 2000a). The recent African origin theory for humans would predict that in Africa there exists a great deal more diversity than elsewhere, and that diversity should decrease the further from Africa a population is sampled.

Phenotypic variation

For more details on this topic, see Phenotype § Phenotypic_variation.

Sub-Saharan Africa has the most human genetic diversity and the same has been shown to hold true for phenotypic diversity.[33] Phenotype is connected to genotype through gene expression. Genetic diversity decreases smoothly with migratory distance from that region, which many scientists believe to be the origin of modern humans, and that decrease is mirrored by a decrease in phenotypic variation. Skull measurements are an example of a physical attribute whose within-population variation decreases with distance from Africa.

The distribution of many physical traits resembles the distribution of genetic variation within and between human populations (American Association of Physical Anthropologists 1996; Keita and Kittles 1997). For example, ~90% of the variation in human head shapes occurs within continental groups, and ~10% separates groups, with a greater variability of head shape among individuals with recent African ancestors (Relethford 2002).

A prominent exception to the common distribution of physical characteristics within and among groups is skin color. Approximately 10% of the variance in skin color occurs within groups, and ~90% occurs between groups (Relethford 2002). This distribution of skin color and its geographic patterning — with people whose ancestors lived predominantly near the equator having darker skin than those with ancestors who lived predominantly in higher latitudes — indicate that this attribute has been under strong selective pressure. Darker skin appears to be strongly selected for in equatorial regions to prevent sunburn, skin cancer, the photolysis of folate, and damage to sweat glands.[35]

Understanding how genetic diversity in the human population impacts various levels of gene expression is an active area of research. While earlier studies focused on the relationship between DNA variation and RNA expression, more recent efforts are characterizing the genetic control of various aspects of gene expression including chromatin states,[36] translation,[37] and protein levels.[38] A study published in 2007 found that 25% of genes showed different levels of gene expression between populations of European and Asian descent.[39][40][41][42][43] The primary cause of this difference in gene expression was thought to be SNPs in gene regulatory regions of DNA. Another study published in 2007 found that approximately 83% of genes were expressed at different levels among individuals and about 17% between populations of European and African descent.[44][45]

48.2.3   Archaic admixture

Main article: Archaic human admixture with modern humans

There is a hypothesis that anatomically modern humans interbred with Neanderthals during the Middle Paleolithic. In May 2010, the Neanderthal Genome Project presented genetic evidence that interbreeding did likely take place and that a small but significant portion of Neanderthal admixture is present in the DNA of modern Eurasians and Oceanians, and nearly absent in sub-Saharan African populations.

Between 4% and 6% of the genome of Melanesians (represented by the Papua New Guinean and Bougainville Islander) are thought to derive from Denisova hominins - a previously unknown species which shares a common origin with Neanderthals. It was possibly introduced during the early migration of the ancestors of Melanesians into Southeast Asia. This history of interaction suggests that Denisovans once ranged widely over eastern Asia.[46]

Thus, Melanesians emerge as the most archaic-admixed population, having Denisovan/Neanderthal-related admixture of ~8%.

In a study published in 2013, Jeffrey Wall from University of California studied whole sequence-genome data and found higher rates of introgression in Asians compared to Europeans.[47] Hammer et al. tested the hypothesis that contemporary African genomes have signatures of gene flow with archaic human ancestors and found evidence of archaic admixture in African genomes, suggesting that modest amounts of gene flow were widespread throughout time and space during the evolution of anatomically modern humans.[48]

48.3   Categorization of the world population

See also: Race (human classification) and Race and genetics

New data on human genetic variation has reignited the debate about a possible biological basis for categorization of humans into races. Most of the controversy surrounds the question of how to interpret the genetic data and whether conclusions based on it are sound. Some researchers argue that self-identified race can be used as an indicator of geographic ancestry for certain health risks and medications.

Although the genetic differences among human groups are relatively small, these differences in certain genes such as duffy, ABCC11, SLC24A5, called ancestry-informative markers (AIMs) nevertheless can be used to reliably situate many individuals within broad, geographically based groupings. For example, computer analyses of hundreds of polymorphic loci sampled in globally distributed populations have revealed the existence of genetic clustering that roughly is associated with groups that historically have occupied large continental and subcontinental regions (Rosenberg *et al.* 2002; Bamshad *et al.* 2003).

Some commentators have argued that these patterns of variation provide a biological justification for the use of traditional racial categories. They argue that the continental clusterings correspond roughly with the division of human beings into sub-Saharan Africans; Europeans, Western Asians, Central Asians, Southern Asians and Northern Africans; Eastern Asians, Southeast Asians, Polynesians and Native Americans; and other inhabitants of Oceania (Melanesians, Micronesians & Australian Aborigines) (Risch *et al.* 2002). Other observers disagree, saying that the same data undercut traditional notions of racial groups (King and Motulsky 2002; Calafell 2003; Tishkoff and Kidd 2004[9]). They point out, for example, that major populations considered races or subgroups within races do not necessarily form their own clusters.

Furthermore, because human genetic variation is clinal, many individuals affiliate with two or more continental groups. Thus, the genetically based "biogeographical ancestry" assigned to any given person generally will be broadly distributed and will be accompanied by sizable uncertainties (Pfaff *et al.* 2004).

In many parts of the world, groups have mixed in such a way that many individuals have relatively recent ancestors from widely separated regions. Although genetic analyses of large numbers of loci can produce estimates of the percentage of a person's ancestors coming from various continental populations (Shriver *et al.* 2003; Bamshad *et al.* 2004), these estimates may assume a false distinctiveness of the parental populations, since human groups have exchanged mates from local to continental scales throughout history (Cavalli-Sforza *et al.* 1994; Hoerder 2002). Even with large numbers of markers, information for estimating admixture proportions of individuals or groups is limited, and estimates typically will have wide confidence intervals (Pfaff *et al.* 2004).

48.3.1   Genetic clustering

Main article: Human genetic clustering

Genetic data can be used to infer population structure and assign individuals to groups that often correspond with their self-identified geographical ancestry. Recently, Lynn Jorde and Steven Wooding argued that "Analysis of many loci now yields reasonably accurate estimates of genetic similarity among individuals, rather than populations. Clustering of individuals is correlated with geographic origin or ancestry."[8]

48.3.2   Forensic anthropology

Forensic anthropologists can determine aspects of geographic ancestry (i.e. Asian, African, or European) from skeletal remains with a high degree of accuracy by analyzing skeletal measurements.[49] According to some studies, individual test methods such as mid-facial measurements and femur traits can identify the geographic ancestry and by extension the racial category to which an individual would have been assigned during their lifetime, with over 80% accuracy, and in combination can be even more accurate. However, the skeletons of persons who have recent ancestry in different geographical regions, can exhibit characteristics of more than one ancestral group, and hence cannot be identified as belonging to any single ancestral group.

48.3.3   Gene flow and admixture

Main article: Gene flow

Gene flow between two populations reduces the average genetic distance between the populations, only totally isolated human populations experience no gene flow and most populations have continuous gene flow with other neighboring populations which create the clinal distribution observed for moth genetic variation. When gene flow takes place between well-differentiated genetic populations the result is referred to as "genetic admixture".

Admixture mapping is a technique used to study how genetic variants cause differences in disease rates between population.[50] Recent admixture populations that trace their ancestry to multiple continents are well suited for identifying genes for traits and diseases that differ in prevalence between parental populations. African-American populations have been the focus of numerous population genetic and admixture mapping studies, including studies of complex genetic traits such as white cell count, body-mass index, prostate cancer and renal disease.[51]

An analysis of phenotypic and genetic variation including skin color and socio-economic status was carried out in the population of Cape Verde which has a well documented history of contact between Europeans and Africans. The studies showed that pattern of admixture in this population has been sex-biased and there is a significant interactions between socio economic status and skin color independent of the skin color and ancestry.[52] Another study shows an increased risk of graft-versus-host disease complications after transplantation due to genetic variants in human leukocyte antigen (HLA) and non-HLA proteins.[53]

48.4   Health

See also: Race and health

Differences in allele frequencies contribute to group differences in the incidence of some monogenic diseases, and they may contribute to differences in the incidence of some common diseases (Risch *et al.* 2002; Burchard *et al.* 2003; Tate and Goldstein 2004). For the monogenic diseases, the frequency of causative alleles usually correlates best with ancestry, whether familial (for example, Ellis-van Creveld syndrome among the Pennsylvania Amish), ethnic (Tay-Sachs disease among Ashkenazi Jewish populations), or geographical (hemoglobinopathies among people with ancestors who lived in malarial regions). To the extent that ancestry corresponds with racial or ethnic groups or subgroups, the incidence of monogenic diseases can differ between groups categorized by race or ethnicity, and health-care professionals typically take these patterns into account in making diagnoses.[54]

Even with common diseases involving numerous genetic variants and environmental factors, investigators point to evidence suggesting the involvement of differentially distributed alleles with small to moderate effects. Frequently cited examples include hypertension (Douglas *et al.* 1996), diabetes (Gower *et al.* 2003), obesity (Fernandez *et al.* 2003), and prostate cancer (Platz *et al.* 2000). However, in none of these cases has allelic variation in a susceptibility gene been shown to account for a significant fraction of the difference in disease prevalence among groups, and the role of genetic factors in generating these differences remains uncertain (Mountain and Risch 2004).

Neil Risch of Stanford University has proposed that self-identified race/ethnic group could be a valid means of categorization in the USA for public health and policy considerations.[55][56] While a 2002 paper by Noah Rosenberg's group makes a similar claim "The structure of human populations is relevant in various epidemiological contexts. As a result of variation in frequencies of both genetic and nongenetic risk factors, rates of disease and of such phenotypes as adverse drug response vary across populations. Further, information about a patient's population of origin might provide health care practitioners with information about risk when direct causes of disease are unknown."[57]

48.5 Genome projects

For more details on this topic, see Category:Human genome projects.

Human genome projects are scientific endeavors that determine or study the structure of the human genome. The Human Genome Project was a landmark genome project.

48.6 See also

- Race and genetics

- Archaeogenetics

- Human evolutionary genetics

- Multiregional hypothesis

- Recent single origin hypothesis

- Isolation by distance

- Genealogical DNA test

- Y-chromosome haplogroups by populations

- Human genetic clustering

Regional:

- Genetic history of Europe

- Genetic history of South Asia

- African admixture in Europe

- Genetic history of indigenous peoples of the Americas

- Genetic history of the British Isles

Projects:

- 1000 Genomes Project

- Human Variome Project

48.7   References

[1] When all genes are fixed within a population, so every member of the population is genetically identical, the population is said to be clonal. This occurs in species that reproduce asexually.

[2] Dr.Craig Venter, Aaron. "In the Genome Race, the Sequel Is Personal".

[3] Bruder, CEG; et al. (2008). "Phenotypically Concordant and Discordant Monozygotic Twins Display Different DNA Copy-Number-Variation Profiles". *The American Journal of Human Genetics* **82** (3): 763–771. doi:10.1016/j.ajhg.2007.12.011.

[4] "We are all mutants: First direct whole-genome measure of human mutation predicts 60 new mutations in each of us". Science Daily. 13 June 2011. Retrieved 2011-09-05.

[5] Conrad, DF; et al. (2011). "Variation in genome-wide mutation rates within and between human families". *Nature Genetics* **43** (7): 712–4. doi:10.1038/ng.862. PMC 3322360. PMID 21666693.

[6] Tishkoff, S. A., & Verrelli, B. C. (2003). "PATTERNS OF HUMAN GENETIC DIVERSITY: Implications for human evolutionary history and disease". *Annual Review of Genomics and Human Genetics, 4* **4** (1).

[7] Kidd, JM; et al. (2008). "Mapping and sequencing of structural variation from eight human genomes". *Nature* **453** (7191): 56–64. Bibcode:2008Natur.453...56K. doi:10.1038/nature06862. PMC 2424287. PMID 18451855.

[8] Jorde, LB; Wooding, SP (2004). "Genetic variation, classification and 'race'". *Nature Genetics* **36** (11s): S28–33. doi:10.1038/ng1435. PMID 15508000.

[9] Tishkoff, SA; Kidd, KK (2004). "Implications of biogeography of human populations for 'race' and medicine". *Nature Genetics* **36** (11s): S21–7. doi:10.1038/ng1438. PMID 15507999.

[10] Collins, F. S.; Brooks, L. D.; Chakravarti, A. (1998). "A DNA polymorphism discovery resource for research on human genetic variation". *Genome Research* **8** (12): 1229–1231. PMID 9872978.

[11] Thomas, P. E.; Klinger, R.; Furlong, L. I.; Hofmann-Apitius, M.; Friedrich, C. M. (2011). "Challenges in the association of human single nucleotide polymorphism mentions with unique database identifiers". *BMC Bioinformatics* **12**: S4. doi:10.1186/1471-2105-12-S4-S4. PMC 3194196. PMID 21992066.

[12] Ke, X; Taylor, M. S.; Cardon, L. R. (2008). "Singleton SNPs in the human genome and implications for genome-wide association studies". *European Journal of Human Genetics* **16** (4): 506–15. doi:10.1038/sj.ejhg.5201987. PMID 18197193.

[13] Ng, P. C.; Levy, S.; Huang, J.; Stockwell, T. B.; Walenz, B. P.; Li, K.; Axelrod, N.; Busam, D. A.; Strausberg, R. L.; Venter, J. C. (2008). Schork, Nicholas J, ed. "Genetic Variation in an Individual Human Exome". *PLoS Genetics* **4** (8): e1000160. doi:10.1371/journal.pgen.1000160. PMC 2493042. PMID 18704161.

[14] Gross, L (2007). "A New Human Genome Sequence Paves the Way for Individualized Genomics". *PLoS Biology* **5** (10): e266. doi:10.1371/journal.pbio.0050266. PMC 1964778. PMID 20076646.

[15] "First Individual Diploid Human Genome Published By Researchers at J. Craig Venter Institute". J. Craig Venter Institute. 3 September 2007. Retrieved 2011-09-05.

[16] Levy, S; et al. (2007). "The Diploid Genome Sequence of an Individual Human". *PLoS Biology* **5** (10): e254. doi:10.1371/journal.pbio.005025 PMC 1964779. PMID 17803354.

[17] "Understanding Genetics: Human Health and the Genome". The Tech Museum of Innovation. 24 January 2008. Retrieved 2011-09-05.

[18] "First Diploid Human Genome Sequence Shows We're Surprisingly Different". Science Daily. 4 September 2007. Retrieved 2011-09-05.

[19] "Copy number variation may stem from replication misstep". EurekAlert!. 27 December 2007. Retrieved 2011-09-05.

[20] Lee, JA; Carvalho, CMB; Lupski, JR (2007). "A DNA Replication Mechanism for Generating Nonrecurrent Rearrangements Associated with Genomic Disorders". *Cell* **131** (7): 1235–47. doi:10.1016/j.cell.2007.11.037. PMID 18160035.

[21] Redon, R; et al. (2006). "Global variation in copy number in the human genome". *Nature* **444** (7118): 444–54. Bibcode:2006Natur.444..444R. doi:10.1038/nature05329. PMC 2669898. PMID 17122850.

[22] Dumas, L; et al. (2007). "Gene copy number variation spanning 60 million years of human and primate evolution". *Genome Research* **17** (9): 1266–77. doi:10.1101/gr.6557307. PMC 1950895. PMID 17666543.

[23] "Human Genetic Variation Fact Sheet". National Institute of General Medical Sciences. 19 August 2011. Retrieved 2011-09-05.

[24] Rakyan, V; Whitelaw, E (2003). "Transgenerational epigenetic inheritance". *Current Biology* **13** (1): R6. doi:10.1016/S0960-9822(02)01377-5. PMID 12526754.

[25] "Cline". *Microsoft Encarta Premium.* 2009.

[26] King, RC; Stansfield, WD; Mulligan, PK (2006). "Cline". *A dictionary of genetics* (7th ed.). Oxford University Press. ISBN 978-0195307610.

[27] Begon, M; Townsend, CR; Harper, JL (2006). *Ecology: From individuals to ecosystems* (4th ed.). Wiley-Blackwell. p. 10. ISBN 978-1405111171.

[28] "Haplogroup". *DNA-Newbie Glossary.* International Society of Genetic Genealogy. Retrieved 2012-09-05.

[29] Gill, V (1 May 2009). "Africa's genetic secrets unlocked". BBC World News. Retrieved 2012-09-05.

[30] "African Genetics Study Revealing Origins, Migration And 'Startling Diversity' Of African Peoples". Science Daily. 2 May 2009. Retrieved 2011-09-05.

[31] Tishkoff, SA; et al. (2009). "The Genetic Structure and History of Africans and African Americans". *Science* **324** (5930): 1035–44. Bibcode:2009Sci...324.1035T. doi:10.1126/science.1172257. PMC 2947357. PMID 19407144.

[32] Underhill, P. A.; Shen, P.; Lin, A. A.; Jin, L.; Passarino, G.; Yang, W. H.; Kauffman, E.; Bonné-Tamir, B.; Bertranpetit, J.; Francalacci, P.; Ibrahim, M.; Jenkins, T.; Kidd, J. R.; Mehdi, S. Q.; Seielstad, M. T.; Wells, R. S.; Piazza, A.; Davis, R. W.; Feldman, M. W.; Cavalli-Sforza, L. L.; Oefner, P. J. (2000). "Y chromosome sequence variation and the history of human populations". *Nature Genetics* **26** (3): 358–361. doi:10.1038/81685. PMID 11062480.

[33] "New Research Proves Single Origin Of Humans In Africa". Science Daily. 19 July 2007. Retrieved 2011-09-05.

[34] Manica, A; Amos, W; Balloux, F; Hanihara, T (2007). "The effect of ancient population bottlenecks on human phenotypic variation". *Nature* **448** (7151): 346–8. Bibcode:2007Natur.448..346M. doi:10.1038/nature05951. PMC 1978547. PMID 17637668.

[35] Jablonski, Nina G. (10 January 2014). *Living Color: The Biological and Social Meaning of Skin Color.* University of California Press. ISBN 978-0-520-28386-2. JSTOR 10.1525/j.ctt1pn64b. Lay summary (12 July 2015).

[36] Grubert F, Zaugg JB, Kasowski M, Ursu O, Spacek DV, Martin AR, Greenside P, Srivas R, Phanstiel DH, Pekowska A, Heidari N, Euskirchen G, Huber W, Pritchard JK, Bustamante CD, Steinmetz LM, Kundaje A, Snyder M (2015). "Genetic Control of Chromatin States in Humans Involves Local and Distal Chromosomal Interactions". *Cell* **162** (5): 1051–65. doi:10.1016/j.cell.2015.07.048. PMID 26300125.

[37] Cenik C, Cenik ES, Byeon GW, Grubert F, Candille SI, Spacek D, Alsallakh B, Tilgner H, Araya CL, Tang H, Ricci E, Snyder MP (2015). "Integrative analysis of RNA, translation, and protein levels reveals distinct regulatory variation across humans". *Genome Res.* **25**: 1610–21. doi:10.1101/gr.193342.115. PMID 26297486.

[38] Wu L, Candille SI, Choi Y, Xie D, Jiang L, Li-Pook-Than J, Tang H, Snyder M (2013). "Variation and genetic control of protein abundance in humans". *Nature* **499** (7456): 79–82. doi:10.1038/nature12223. PMC 3789121. PMID 23676674.

[39] Phillips, ML (9 January 2007). "Ethnicity tied to gene expression". *The Scientist.* Retrieved 2011-09-05.

[40] Spielman, RS; et al. (2007). "Common genetic variants account for differences in gene expression among ethnic groups". *Nature Genetics* **39** (2): 226–31. doi:10.1038/ng1955. PMC 3005333. PMID 17206142.

[41] Swaminathan, N (9 January 2007). "Ethnic Differences Traced to Variable Gene Expression". *Scientific American.* Retrieved 2011-09-05.

[42] Check, E (2007). "Genetic expression speaks as loudly as gene type". *Nature News.* doi:10.1038/news070101-8.

[43] Bell, L (15 January 2007). "Variable gene expression seen in different ethnic groups". BioNews.org. Retrieved 2011-09-05.

[44] Kamrani, K (28 February 2008). "Differences of gene expression between human populations". Anthropology.net. Retrieved 2011-09-05.

[45] Storey, JD; et al. (2007). "Gene-Expression Variation Within and Among Human Populations". *The American Journal of Human Genetics* **80** (3): 502–509. doi:10.1086/512017.

[46] Reich, D; et al. (2010). "Genetic history of an archaic hominin group from Denisova Cave in Siberia". *Nature* **468** (7327): 1053–60. Bibcode:2010Natur.468.1053R. doi:10.1038/nature09710. PMID 21179161.

[47] Wall, Jeffrey D.; et al. (2013). "Higher Levels of Neanderthal Ancestry in East Asians Than in Europeans". *Genetics* **194**: 199–209. doi:10.1534/genetics.112.148213.

[48] Hammer, Michael F.; et al. (2011). "Genetic evidence for archaic admixture in Africa". *Proceedings of the National Academy of Sciences* **108** (37): 15123–15128. Bibcode:2011PNAS..10815123H. doi:10.1073/pnas.1109300108. PMC 3174671. PMID 21896735.

[49] "Does Race Exist?". *NOVA*. PBS. 15 February 2000. Retrieved 2011-09-05.

[50] Winkler, C. A.; Nelson, G. W.; Smith, M. W. (2010). "Admixture mapping comes of age". *Annu Rev Genomics Hum Genet.* **11**: 65–89. doi:10.1146/annurev-genom-082509-141523. PMID 20594047.

[51] Bryc, K.; Auton, A.; Nelson, M. R.; Oksenberg, J. R.; Hauser, S. L.; Williams, S.; Froment, A.; Bodo, J. -M.; Wambebe, C.; Tishkoff, S. A.; Bustamante, C. D. (2009). "Genome-wide patterns of population structure and admixture in West Africans and African Americans". *Proceedings of the National Academy of Sciences* **107** (2): 786–791. Bibcode:2010PNAS..107..786B. doi:10.1073/pnas.0909559107. PMC 2818934. PMID 20080753.

[52] Beleza, S; Campos, J; Lopes, J; Araújo, I. I.; Hoppfer Almada, A; Correia e Silva, A; Parra, E. J.; Rocha, J (2012). "The admixture structure and genetic variation of the archipelago of Cape Verde and its implications for admixture mapping studies". *PLoS ONE* **7** (11): e51103. doi:10.1371/journal.pone.0051103. PMC 3511383. PMID 23226471.

[53] Arrieta-Bolaños, E; Madrigal, J. A.; Shaw, B. E. (2012). "Human leukocyte antigen profiles of latin american populations: Differential admixture and its potential impact on hematopoietic stem cell transplantation". *Bone Marrow Research* **2012**: 136087. doi:10.1155/2012/136087. PMC 3506882. PMID 23213535.

[54] Lu, YF; Goldstein, DB; Angrist, M; Cavalleri, G (24 July 2014). "Personalized medicine and human genetic diversity.". *Cold Spring Harbor perspectives in medicine* **4** (9): a008581. doi:10.1101/cshperspect.a008581. PMID 25059740.

[55] Tang, H; et al. (2005). "Genetic Structure, Self-Identified Race/Ethnicity, and Confounding in Case-Control Association Studies". *The American Journal of Human Genetics* **76** (2): 268–75. doi:10.1086/427888. PMC 1196372. PMID 15625622.

[56] Risch, N; Burchard, E; Ziv, E; Tang, H (2002). "Categorization of humans in biomedical research: genes, race and disease". *Genome Biology* **3** (7): 1–12. doi:10.1186/gb-2002-3-7-comment2007. PMC 139378. PMID 12184798.

[57] Rosenberg, NA; et al. (2002). "Genetic Structure of Human Populations". *Science* **298** (5602): 2381–5. Bibcode:2002Sci...298.2381R. doi:10.1126/science.1078311. PMID 12493913.

Bibliography

- Race, Ethnicity, and Genetics Working Group (2005). "The Use of Racial, Ethnic, and Ancestral Categories in Human Genetics Research". *The American Journal of Human Genetics* **77** (4): 519–32. doi:10.1086/491747. PMC 1275602. PMID 16175499.

- Altmüller, J; Palmer, L; Fischer, G; Scherb, H; Wjst, M (2001). "Genomewide Scans of Complex Human Diseases: True Linkage is Hard to Find". *The American Journal of Human Genetics* **69** (5): 936–50. doi:10.1086/324069. PMC 1274370. PMID 11565063.

- Aoki, K (2002). "Sexual selection as a cause of human skin colour variation: Darwin's hypothesis revisited". *Annals of Human Biology* **29** (6): 589–608. doi:10.1080/0301446021000019144. PMID 12573076.

- Bamshad, M; Wooding, S; Salisbury, BA; Stephens, JC (2004). "Deconstructing the relationship between genetics and race". *Nature Reviews Genetics* **5** (8): 598–609. doi:10.1038/nrg1401. PMID 15266342. reprint-zip

- Bamshad, M; Wooding, SP (2003). "Signatures of natural selection in the human genome". *Nature Reviews Genetics* **4** (2): 99–111. doi:10.1038/nrg999. PMID 12560807.

- Bamshad, MJ; et al. (2003). "Human Population Genetic Structure and Inference of Group Membership". *The American Journal of Human Genetics* **72** (3): 578–89. doi:10.1086/368061. PMC 1180234. PMID 12557124.

- Cann, RL; Stoneking, M; Wilson, AC (1987). "Mitochondrial DNA and human evolution". *Nature* **325** (6099): 31–6. Bibcode:1987Natur.325...31C. doi:10.1038/325031a0. PMID 3025745.

- Cardon, LR; Abecasis, GR (2003). "Using haplotype blocks to map human complex trait loci". *Trends in Genetics* **19** (3): 135–40. doi:10.1016/S0168-9525(03)00022-2. PMID 12615007.

- Cavalli-Sforza, LL; Feldman, MW (2003). "The application of molecular genetic approaches to the study of human evolution". *Nature Genetics* **33** (3s): 266–75. doi:10.1038/ng1113. PMID 12610536.

- Collins, FS (2004). "What we do and don't know about 'race', 'ethnicity', genetics and health at the dawn of the genome era". *Nature Genetics* **36** (11s): S13–5. doi:10.1038/ng1436. PMID 15507997.

- Collins, FS; Green, ED; Guttmacher, AE; Guyer, MS (2003). "A vision for the future of genomics research". *Nature* **422** (6934): 835–47. Bibcode:2003Natur.422..835C. doi:10.1038/nature01626. PMID 12695777.

- Ebersberger, I; Metzler, D; Schwarz, C; Pääbo, S (2002). "Genomewide Comparison of DNA Sequences between Humans and Chimpanzees". *The American Journal of Human Genetics* **70** (6): 1490–7. doi:10.1086/340787. PMC 379137. PMID 11992255.

- Edwards, AWF (2003). "Human genetic diversity: Lewontin's fallacy". *BioEssays* **25** (8): 798–801. doi:10.1002/bies.10315. PMID 12879450.

- Foster, MW; Sharp, RR (2004). "Opinion: Beyond race: Towards a whole-genome perspective on human populations and genetic variation". *Nature Reviews Genetics* **5** (10): 790–6. doi:10.1038/nrg1452. PMID 15510170.

- Foster, M; et al. (1999). "The Role of Community Review in Evaluating the Risks of Human Genetic Variation Research". *The American Journal of Human Genetics* **64** (6): 1719–27. doi:10.1086/302415. PMC 1377916. PMID 10330360.

- Gabriel, SB; et al. (2002). "The Structure of Haplotype Blocks in the Human Genome". *Science* **296** (5576): 2225–9. Bibcode:2002Sci...296.2225G. doi:10.1126/science.1069424. PMID 12029063.

- Harding, RM; et al. (2000). "Evidence for Variable Selective Pressures at MC1R". *The American Journal of Human Genetics* **66** (4): 1351–61. doi:10.1086/302863. PMC 1288200. PMID 10733465.

- Gyllensten, U; Ingman, M; Kaessmann, H; Pääbo, S (2000). "Mitochondrial genome variation and the origin of modern humans". *Nature* **408** (6813): 708–13. doi:10.1038/35047064. PMID 11130070.

- The International Hapmap Consortium (2003). "The International HapMap Project". *Nature* **426** (6968): 789–96. doi:10.1038/nature02168. PMID 14685227.

- The International Hapmap Consortium (2004). "Opinion: Integrating ethics and science in the International HapMap Project". *Nature Reviews Genetics* **5** (6): 467–75. doi:10.1038/nrg1351. PMC 2271136. PMID 15153999.

- The International Human Genome Sequencing Consortium (2001). "Initial sequencing and analysis of the human genome". *Nature* **409** (6822): 860–921. doi:10.1038/35057062. PMID 11237011.

- Jorde, LB; Bamshad, M; Rogers, AR (1998). "Using mitochondrial and nuclear DNA markers to reconstruct human evolution" (PDF). *BioEssays* **20** (2): 126–36. doi:10.1002/(SICI)1521-1878(199802)20:2<126::AID-BIES5>3.0.CO;2-R. PMID 9631658.

- Jorde, LB; et al. (2000). "The Distribution of Human Genetic Diversity: A Comparison of Mitochondrial, Autosomal, and Y-Chromosome Data". *The American Journal of Human Genetics* **66** (3): 979–88. doi:10.1086/302825. PMC 1288178. PMID 10712212.

• Jorde, LB; Watkins, WW; Kere, J; Nyman, D; Eriksson, AW (2000). "Gene Mapping in Isolated Populations: New Roles for Old Friends?". *Human Heredity* **50** (1): 57–65. doi:10.1159/000022891. PMID 10545758.

• Kaessmann, H; Heißig, D; von Haeseler, A; Pääbo, S (1999). "DNA sequence variation in a non-coding region of low recombination on the human X chromosome". *Nature Genetics* **22** (1): 78–81. doi:10.1038/8785. PMID 10319866.

• Kaessmann, H; Wiebe, V; Weiss, G; Pääbo, S (2001). "Great ape DNA sequences reveal a reduced diversity and an expansion in humans". *Nature Genetics* **27** (2): 155–6. doi:10.1038/84773. PMID 11175781.

• Keita, SOY; Kittles, RA (1997). "The Persistence of Racial Thinking and the Myth of Racial Divergence". *American Anthropologist* **99** (3): 534–544. doi:10.1525/aa.1997.99.3.534.

• Lewontin, RC (1972). "The apportionment of human diversity". *Evolutionary Biology* **6**: 381–398. doi:10.1007/978-1-4684-9063-3_14. ISBN 978-1-4684-9065-7.

• Marks, J (1995). *Human Biodiversity: Genes, Race, and History*. Aldine Transaction. ISBN 978-0-202-02033-4.

• Mountain, JL; Risch, N (2004). "Assessing genetic contributions to phenotypic differences among 'racial' and 'ethnic' groups". *Nature Genetics* **36** (11s): S48. doi:10.1038/ng1456. PMID 15508003.

• Pääbo, S (2003). "The mosaic that is our genome". *Nature* **421** (6921): 409–12. Bibcode:2003Natur.421..409P. doi:10.1038/nature01400. PMID 12540910.

• Ramachandran, S; et al. (2005). "Support from the relationship of genetic and geographic distance in human populations for a serial founder effect originating in Africa". *Proceedings of the National Academy of Sciences* **102** (44): 15942–7. Bibcode:2005PNAS..10215942R. doi:10.1073/pnas.0507611102. PMC 1276087. PMID 16243969.

• Relethford, JH (2002). "Apportionment of global human genetic diversity based on craniometrics and skin color". *American Journal of Physical Anthropology* **118** (4): 393–8. doi:10.1002/ajpa.10079. PMID 12124919.

• Sankar, P; Cho, MK (2002). "Toward a New Vocabulary of Human Genetic Variation". *Science* **298** (5597): 1337–8. doi:10.1126/science.1074447. PMC 2271140. PMID 12434037.

• Sankar, P; et al. (2004). "Genetic Research and Health Disparities". *JAMA: the Journal of the American Medical Association* **291** (24): 2985–9. doi:10.1001/jama.291.24.2985. PMC 2271142. PMID 15213210.

• Serre, D; Pääbo, S (2004). "Evidence for Gradients of Human Genetic Diversity Within and Among Continents". *Genome Research* **14** (9): 1679–85. doi:10.1101/gr.2529604. PMC 515312. PMID 15342553.

• Templeton, AR (1998). "Human Races: A Genetic and Evolutionary Perspective". *American Anthropologist* **100** (3): 632–650. doi:10.1525/aa.1998.100.3.632.

• Weiss, KM (1998). "Coming to Terms with Human Variation". *Annual Review of Anthropology* **27**: 273–300. doi:10.1146/annurev.anthro.27.1.273.

• Weiss, KM; Terwilliger, JD (2000). "How many diseases does it take to map a gene with SNPs?". *Nature Genetics* **26** (2): 151–7. doi:10.1038/79866. PMID 11017069.

• Yu, N; et al. (2003). "Low nucleotide diversity in chimpanzees and bonobos". *Genetics* **164** (4): 1511–8. PMC 1462640. PMID 12930756.

• Ziętkiewicz, E; et al. (2003). "Haplotypes in the Dystrophin DNA Segment Point to a Mosaic Origin of Modern Human Diversity". *The American Journal of Human Genetics* **73** (5): 994–1015. doi:10.1086/378777. PMC 1180505. PMID 14513410.

48.8 Further reading

- Pennisi, E (2007). "Breakthrough of the Year: Human Genetic Variation". *Science* **318** (5858): 1842–1843. doi:10.1126/science.318.5858.1842. PMID 18096770.

- Ramachandran, S; Tang, H; Gutenkunst, RN; Bustamante, CD (2010). "Genetics and Genomics of Human Population Structure". In Speicher, MR; Antonarakis, SE; Motulsky, AG. *Vogel and Motulsky's Human Genetics: Problems and Approaches* (4th ed.). Springer. ISBN 3-540-37653-4.

48.9 External links

- Human Genome Variation Society

Bantu (S. Africa)
Bantu (Kenya)

Mandenka

Yoruba

San
Mbuti Pygmy

Biaka Pygmy

Orcadian

Adygei

Russian

Basque

French

Italian

Sardinian

Tuscan

Mozabite

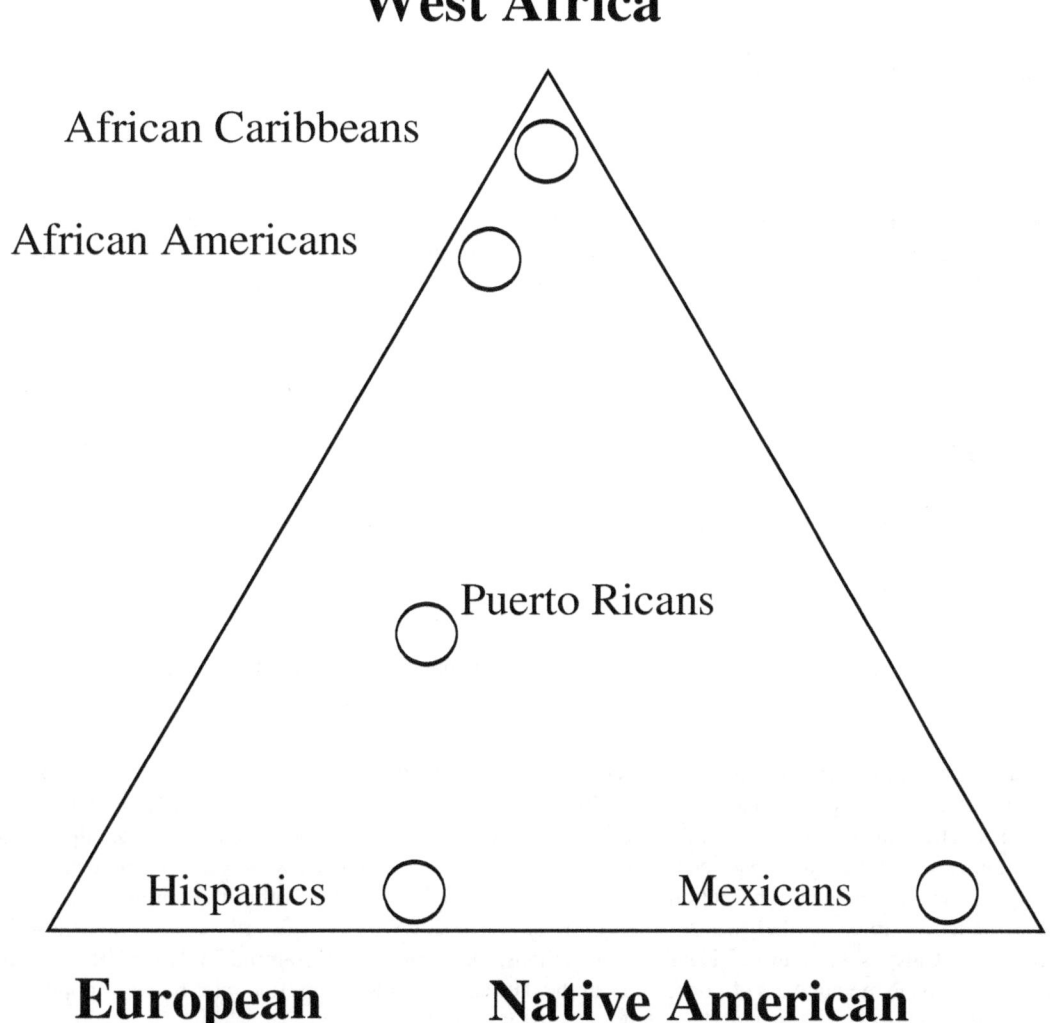

Triangle plot shows average admixture of five North American ethnic groups. Individuals that self-identify with each group can be found at many locations on the map, but on average groups tend to cluster differently.

Chapter 49

Human genetic clustering

Human genetic clustering analysis uses mathematical cluster analysis of the degree of similarity of genetic data between individuals and groups in order to infer population structures and assign individuals to groups. These groupings in turn often, but not always, correspond with the individuals' self-identified geographical ancestry. A similar analysis can be done using principal components analysis, which in earlier research was a popular method.[1] Many studies in the past few years have continued using principal components analysis.

49.1 Studies

49.1.1 Clusters by Rosenberg et al. (2006)

Main article: Race and genetics

In 2004, Lynn Jorde and Steven Wooding argued that "Analysis of many loci now yields reasonably accurate estimates of genetic similarity among individuals, rather than populations. Clustering of individuals is correlated with geographic origin or ancestry."[2]

Studies such as those by Risch and Rosenberg use a computer program called STRUCTURE to find human populations (gene clusters). It is a statistical program that works by placing individuals into one of an arbitrary number of clusters based on their overall genetic similarity, many possible pairs of clusters are tested per individual to generate multiple clusters.[3] These populations are based on multiple genetic markers that are often shared between different human populations even over large geographic ranges. The notion of a genetic cluster is that people within the cluster share on average similar allele frequencies to each other than to those in other clusters. (A. W. F. Edwards, 2003 but see also infobox "Multi Locus Allele Clusters") In a test of idealised populations, the computer programme STRUCTURE was found to consistently underestimate the numbers of populations in the data set when high migration rates between populations and slow mutation rates (such as single-nucleotide polymorphisms) were considered.[4]

Nevertheless the Rosenberg *et al.* (2002) paper shows that individuals can be assigned to specific clusters to a high degree of accuracy. One of the underlying questions regarding the distribution of human genetic diversity is related to the degree to which genes are shared between the observed clusters. It has been observed repeatedly that the majority of variation observed in the global human population is found within populations. This variation is usually calculated using Sewall Wright's Fixation index (FST), which is an estimate of between to within group variation. The degree of human genetic variation is a little different depending upon the gene type studied, but in general it is common to claim that ~85% of genetic variation is found within groups, ~6–10% between groups within the same continent and ~6–10% is found between continental groups. For example The Human Genome Project states "two random individuals from any one group are almost as different [genetically] as any two random individuals from the entire world."[5] Sarich and Miele, however, have argued that estimates of genetic difference between individuals of different populations fail to take into account human diploidy.

The point is that we are diploid organisms, getting one set of chromosomes from one parent and a second from the other. To the extent that your mother and father are not especially closely related, then, those two sets of chromosomes will come close to being a random sample of the chromosomes in your population. And the sets present in some randomly chosen member of yours will also be about as different from your two sets as they are from one another. So how much of the variability will be distributed where?

First is the 15 percent that is interpopulational. The other 85 percent will then split half and half (42.5 percent) between the intra- and interindividual within-population comparisons. The increase in variability in between-population comparisons is thus 15 percent against the 42.5 percent that is between-individual within-population. Thus, 15/42.5 is 32.5 percent, a much more impressive and, more important, more legitimate value than 15 percent.[6]

Additionally, Edwards (2003) claims in his essay "Lewontin's Fallacy" that: "It is not true, as *Nature* claimed, that 'two random individuals from any one group are almost as different as any two random individuals from the entire world'" and Risch *et al.* (2002) state "Two Caucasians are more similar to each other genetically than a Caucasian and an Asian." It should be noted that these statements are not the same. Risch *et al.* simply state that two indigenous individuals from the same geographical region are more similar to each other than either is to an indigenous individual from a different geographical region, a claim few would argue with. Jorde et al. put it like this:

> The picture that begins to emerge from this and other analyses of human genetic variation is that variation tends to be geographically structured, such that most individuals from the same geographic region will be more similar to one another than to individuals from a distant region.[2]

Whereas Edwards claims that it is not true that the differences between individuals from different geographical regions represent only a small proportion of the variation within the human population (he claims that within group differences between individuals are not almost as large as between group differences). Bamshad *et al.* (2004) used the data from Rosenberg *et al.* (2002) to investigate the extent of genetic differences between individuals within continental groups relative to genetic differences between individuals between continental groups. They found that though these individuals could be classified very accurately to continental clusters, there was a significant degree of genetic overlap on the individual level, to the extent that, using 377 loci, individual Europeans were about 38% of the time more genetically similar to East Asians than to other Europeans.

A study by the HUGO Pan-Asian SNP Consortium in 2009 using the similar principal components analysis found that East Asian and South-East Asian populations clustered together, and suggested a common origin for these populations. At the same time they observed a broad discontinuity between this cluster and South Asia, commenting "most of the Indian populations showed evidence of shared ancestry with European populations". It was noted that "genetic ancestry is strongly correlated with linguistic affiliations as well as geography".[7]

Criticism

The Rosenberg study has been criticised on several grounds.

The existence of allelic clines and the observation that the bulk of human variation is continuously distributed, has led some scientists to conclude that any categorization schema attempting to partition that variation meaningfully will necessarily create artificial truncations. (Kittles & Weiss 2003). It is for this reason, Reanne Frank argues, that attempts to allocate individuals into ancestry groupings based on genetic information have yielded varying results that are highly dependent on methodological design.[8] Serre and Pääbo (2004) make a similar claim:

> The absence of strong continental clustering in the human gene pool is of practical importance. It has recently been claimed that "the greatest genetic structure that exists in the human population occurs at the racial level" (Risch et al. 2002). Our results show that this is not the case, and we see no reason to assume that "races" represent any units of relevance for understanding human genetic history.

In a response to Serre and Pääbo (2004), Rosenberg *et al.* (2005) make three relevant observations. Firstly they maintain that their clustering analysis is robust. Secondly they agree with Serre and Pääbo that membership of multiple clusters can be interpreted as evidence for clinality (isolation by distance), though they also comment that this may also be due to admixture between neighbouring groups (small island model). Thirdly they comment that evidence of clusterdness is not evidence for any concepts of "biological race".[9]

Risch *et al.* (2002) state that "two Caucasians are more similar to each other genetically than a Caucasian and an Asian", but Bamshad *et al.* (2004)[10] used the same data set as Rosenberg *et al.* (2002) to show that Europeans are more similar to Asians 38% of the time than they are to other Europeans when only 377 microsatellite markers are analysed.

In agreement with the observation of Bamshad *et al.* (2004), Witherspoon *et al.* (2007) have shown that many more than 326 or 377 microsatellite loci are required in order to show that individuals are always more similar to individuals in their own population group than to individuals in different population groups, even for three distinct populations.[5]

Witherspoon et al. (2007) have argued that even when individuals can be reliably assigned to specific population groups, it may still be possible for two randomly chosen individuals from different populations/clusters to be more similar to each other than to a randomly chosen member of their own cluster. They found that many thousands of genetic markers had to be used in order for the answer to the question "How often is a pair of individuals from one population genetically more dissimilar than two individuals chosen from two different populations?" to be "never". This assumed three population groups separated by large geographic ranges (European, African and East Asian). The entire world population is much more complex and studying an increasing number of groups would require an increasing number of markers for the same answer. Witherspoon et al. conclude that "caution should be used when using geographic or genetic ancestry to make inferences about individual phenotypes."

Clustering does not particularly correspond to continental divisions. Depending on the parameters given to their analytical program, Rosenberg and Pritchard were able to construct between divisions of between 4 and 20 clusters of the genomes studied, although they excluded analysis with more than 6 clusters from their published article. Probability values for various cluster configurations varied widely, with the single most likely configuration coming with 16 clusters although other 16-cluster configurations had low probabilities. Overall, "there is no clear evidence that K=6 was the best estimate" according to geneticist Deborah Bolnick (2008:76-77).[12] The number of genetic clusters used in the study was arbitrarily chosen. Although the original research used different number of clusters, the published study emphasized six genetic clusters. The number of genetic clusters is determined by the user of the computer software conducting the study. Rosenberg later revealed that his team used pre-conceived numbers of genetic clusters from six to twenty "but did not publish those results because Structure [the computer program used] identified multiple ways to divide the sampled individuals". Dorothy Roberts, a law professor, asserts that "there is nothing in the team's findings that suggests that six clusters represent human population structure better than ten, or fifteen, or twenty."[13] When instructed to find two clusters, the program identified two populations anchored around by Africa and by the Americas. In the case of six clusters, the entirety of Kalesh people, an ethnic group living in Northern Pakistan, was added to the previous five.[14][15]

The law professor, Dorothy Roberts asserts that "the study actually showed that there are many ways to slice the expansive range of human genetic variation. In a 2005 paper, Rosenberg and his team acknowledged that findings of a study on human population structure are highly influenced by the way the study is designed.[15][16]

They reported that the number of loci, the sample size, the geographic dispersion of the samples and assumptions about allele-frequency correlation all have an effect on the outcome of the study. Rosenberg stated that their findings "should not be taken as evidence of our support of any particular concept of biological race (...). Genetic differences among human populations derive mainly from gradations in allele frequencies rather than from distinctive 'diagnostic' genotypes."[17] The study's overall results confirmed that genetic difference within populations is between 93 and 95%. Only 5% of genetic variation is found between groups.[15]

49.2   Controversy of genetic clustering and associations with "race"

In the late 1990s Harvard evolutionary geneticist Richard Lewontin stated that "no justification can be offered for continuing the biological concept of race. (...) Genetic data shows that no matter how racial groups are defined, two people from the same racial group are about as different from each other as two people from any two different racial groups.[18]

Lewontin's statement came under attack when new genomic technologies permitted the analysis of gene clusters. In 2003,

British statistician and evolutionary biologist A. W. F. Edwards faulted Lewontin's statement for basing his conclusions on simple comparison of genes and rather on a more complex structure of gene frequencies. Edwards charged Lewontin that he made an "unjustified assault on human classification, which he deplored for social reasons."[19]

According to Roberts, "Edwards did not refute Lewontin's claim: that there is more genetic variation within populations than between them, especially when it comes to races. (...) Lewontin did not ignore biology to support his social ideology (...). To the contrary, he argued that there is no biological support for the ideological project of race." "The genetic differences that exist among populations are characterized by gradual changes across geographic regions, not sharp, categorical distinctions. Groups of people across the globe have varying frequencies of polymorphic genes, which are genes with any of several differing nucleotide sequences. There is no such thing as a set of genes that belongs exclusively to one group and not to another. The clinal, gradually changing nature of geographic genetic difference is complicated further by the migration and mixing that human groups have engaged in since prehistoric times. Race [however defined] collapses infinite diversity into a few discrete categories that in reality cannot be demarcated genetically."[15]

Genetic clustering was also criticized by Penn State anthropologists Kenneth Weiss and Brian Lambert. They asserted that understanding human population structure in terms of discrete genetic clusters misrepresents the path that produced diverse human populations that diverged from shared ancestors in Africa. Ironically, by ignoring the way population history actually works as one process from a common origin rather than as a string of creation events, structure analysis that seems to present variation in Darwinian evolutionary terms is fundamentally non-Darwinian."[20]

In 2006, Lewontin wrote that any genetic study requires some priori concept of race or ethnicity in order to package human genetic diversity into defined, limited number of biological groupings. Informed by geneticist, zoologists have long discarded the concept of race for dividing up groups of non-human animal populations within a species. Defined on varying criteria, in the same species widely varying number of races could be distinguished. Lewontin notes that genetic testing revealed that "because so many of these races turned out to be based on only one or two genes, two animals born in the same litter could belong to different 'races'".[21]

Studies that seek to find genetic clusters are only as informative as the populations they sample. For example Risch and Burchard relied on two or three local populations from five continents, which together were supposed to represent the entire human race.[15] Another genetic clustering study used three sub-Saharan population groups to represent Africa; Chinese, Japanese, and Cambodian samples for East Asia; Northern European and Northern Italian samples to represent "Caucasians". Entire regions, subcontinents, and landmasses are left out of many studies. Furthermore, social geographical categories such "East Asia" and "Caucasians" were not defined. "A handful of ethnic groups to symbolize an entire continent mimic a basic tenet of racial thinking: that because races are composed of uniform individuals, anyone can represent the whole group" notes Roberts.[15][22][23]

The model of Big Few fails when including overlooked geographical regions such as India. The 2003 study which examined fifty-eight genetic markers found that Indian populations had their ancestral lineages to Africa, Central Asia, Europe, and southern China.[24][25] Reardon, from Princeton University, asserts that flawed sampling methods are built into many genetic research projects. The Human Genome Diversity Project (HGDP) relied on samples which were assumed to be geographically separate and isolated.[26] The relatively small sample sizes of indigenous populations for the HGDP do not represent the human species' genetic diversity, nor do they portray migrations and mixing population groups which has been happening since prehistoric times. Geographic areas such as the Balkans, the Middle East, North and East Africa, and Spain are seldom included in genetic studies.[15][27] East and North African indigenous populations, for example, are never selected to represent Africa because they do not fit the profile of "black" Africa. The sampled indigenous populations of the HGDP are assumed to be "pure"; the law professor Roberts claims that "their unusual purity is all the more reason they cannot stand in for all the other populations of the world that marked by intermixture from migration, commerce, and conquest."[15]

King and Motulsky, in a 2002 Science article, states that "While the computer-generated findings from all of these studies offer greater insight into the genetic unity and diversity of the human species, as well as its ancient migratory history, none support dividing the species into discrete, genetically determined racial categories".[28] Cavalli-Sforza asserts that classifying clusters as races would be a "futile exercise" because "every level of clustering would determine a different population and there is no biological reason to prefer a particular one." Bamshad, in 2004 paper published in Nature, asserts that a more accurate study of human genetic variation would use an objective sampling method. An objective sampling method would chose populations randomly and systematically across the world, including those populations which are characterized by historical intermingling, instead of cherry-picking population samples which fit a priori concept of racial classification. Roberts states that "if research collected DNA samples continuously from region to region throughout the world, they would find it impossible to infer neat boundaries between large geographical groups."[10][15][29][30]

Anthropologists such as C. Loring Brace,[31] philosophers Jonathan Kaplan and Rasmus Winther,[32][32][33][34] and geneticist Joseph Graves,[35] have argued that while there it is certainly possible to find biological and genetic variation that corresponds roughly to the groupings normally defined as "continental races", this is true for almost all geographically distinct populations. The cluster structure of the genetic data is therefore dependent on the initial hypotheses of the researcher and the populations sampled. When one samples continental groups the clusters become continental, if one had chosen other sampling patterns the clustering would be different. Weiss and Fullerton have noted that if one sampled only Icelanders, Mayans and Maoris, three distinct clusters would form and all other populations could be described as being clinally composed of admixtures of Maori, Icelandic and Mayan genetic materials.[36] Kaplan and Winther therefore argue that seen in this way both Lewontin and Edwards are right in their arguments. They conclude that while racial groups are characterized by different allele frequencies, this does not mean that racial classification is a natural taxonomy of the human species, because multiple other genetic patterns can be found in human populations that crosscut racial distinctions. Moreover, the genomic data underdetermines whether one wishes to see subdivisions (i.e., splitters) or a continuum (i.e., lumpers). Under Kaplan and Winther's view, racial groupings are objective social constructions (see Mills 1998 [37]) that have conventional biological reality only insofar as the categories are chosen and constructed for pragmatic scientific reasons.

49.2.1   Commercial ancestry testing and individual ancestry

Commercial ancestry testing companies, who use genetic clustering data, have been also heavily criticized. Limitations of genetic clustering are intensified when inferred population structure is applied to individual ancestry. The type of statistical analysis conducted by scientists translates poorly into individual ancestry because they are looking at difference in frequencies, not absolute differences between groups. Commercial genetic genealogy companies are guilty of what Pillar Ossorio calls the "tendency to transform statistical claims into categorical ones".[38] Not just individuals of the same local ethnic group, but two siblings may end up beings as members of different continental groups or "races" depending on the alleles they inherit.[15]

Many commercial companies use data from HapMap's initial phrase, where population samples were collected from four ethnic groups in the world: Han Chinese, Japanese, Yoruba Nigerian, and Utah residents of Northern European ancestry. If a person has ancestry from a region where the computer program does not have samples, it will compensate with the closest sample that may have nothing to do with the customer's actual ancestry: "Consider a genetic ancestry testing performed on an individual we will call Joe, whose eight great-grandparents were from southern Europe. The HapMap populations are used as references for testing Joe's genetic ancestry. The HapMap's European samples consist of "northern" Europeans. In regions of Joe's genome that vary between northern and southern Europe (such regions might include the lactase gene), the genetic ancestry test is using the HapMap reference population is likely to incorrectly assign the ancestry of that portion of the genome to a non-European population because that genomic region will appear to be more similar to the HapMap's Yoruba or Han Chinese samples than to Northern European samples.[39] Likewise, a person with East African ancestors may be classified as someone having part North European and part Western African ancestry.[40] "Telling customers that they are a composite of several anthropological groupings reinforces three central myths about race: that there are pure races, that each race contains people who are fundamentally the same and fundamentally different from people in other races, and that races can be biologically demarcated." Many companies base their findings on inadequate and unscientific sampling methods. Researchers have never sampled the world's populations in a systematic and random fashion.[15]

49.2.2   Geographical and continental groupings

Roberts argues against the use of broad geographical or continental groupings: "molecular geneticists routinely refer to African ancestry as if everyone on the continent is more similar to each other than they are to people of other continents, who may be closer both geographically and genetically.[15] Ethiopians have closer genetic affinity with Armenians and Norwegians than with Bantu populations.[41] Similarly, Somalis are genetically more similar to Gulf Arab populations than to other populations in Africa.[42] Braun and Hammonds (2008) asserts that the misperception of continents as natural population groupings is rooted in the assumption that populations are natural, isolated, and static. Populations came to be seen as "bounded units amenable to scientific sampling, analysis, and classification".[43] Human beings are not naturally organized into definable, genetically cohesive populations.

49.2.3 Usage in scientific journals

Some scientific journals have addressed previous methodological errors by requiring more rigorous scrutiny of population variables. Since 2000, Nature Genetics requires its authors to "explain why they make use of particular ethnic groups or populations, and how classification was achieved." Editors of Nature Genetics say that "[they] hope that this will raise awareness and inspire more rigorous design of genetic and epidemiological studies."[44]

49.3 See also

- Haplogroup

- Human genetic variation

- Gene cluster

- Genetic admixture

- Population groups in biomedicine

- Y-chromosome haplogroups by populations

49.4 References

[1] Patterson, Nick; Price, Alkes L.; Reich, David. "Population Structure and Eigenanalysis". *PLoS Genet* **2** (12): e190. doi:10.1371/journal.pgen.0020'' PMC 1713260. PMID 17194218.

[2] Lynn B Jorde & Stephen P Wooding, 2004, "Genetic variation, classification and 'race'" in *Nature Genetics* 36, S28–S33 Genetic variation, classification and 'race'

[3] "Genetic Similarities Within and Between Human Populations" (2007) by D.J. Witherspoon, S. Wooding, A.R. Rogers, E.E. Marchani, W.S. Watkins, M.A. Batzer and L.B. Jorde. *Genetics.* **176**(1) 351–359.

[4] Wapples, R., S. and Gaggiotti, O. *What is a population? An empirical evaluation of some genetic methods for identifying the number of gene pools and their degree of connectivity Molecular Ecology* (2006) **15:** 1419–1439. doi:10.1111/j.1365-294X.2006.02890.x PMID 16629801

[5] *Genetic Similarities Within and Between Human Populations* by D. J. Witherspoon, S. Wooding, A. R. Rogers, E. E. Marchani, W. S. Watkins, M. A. Batzer, and L. B. Jorde Genetics. 2007 May; 176(1): 351–359.

[6] Sarich VM, Miele F. Race: The Reality of Human Differences. Westview Press (2004). ISBN 0-8133-4086-1

[7] Mapping Human Genetic Diversity in Asia, The HUGO Pan-Asian SNP Consortium, 2009

[8] Back with a Vengeance: the Reemergence of a Biological Conceptualization of Race in Research on Race/Ethnic Disparities in Health Reanne Frank

[9] *Rosenberg NA, Mahajan S, Ramachandran S, Zhao C, Pritchard JK, et al. (2005)* Clines, Clusters, and the Effect of Study Design on the Inference of Human Population Structure. PLoS Genet *1(6): e70 doi:10.1371/journal.pgen.0010070 PMID 16355252*

[10] Michael Bamshad; et al. (2004). "Deconstructing the Relationship Between Genetics and Race". *Nature Reviews Genetics* **5** (598): 598–609. doi:10.1038/nrg1401. PMID 15266342.

[11] The table gives the percentage likelihood that two individuals from different clusters are genetically more similar to each other than to someone from their own population when 377 microsatellite markers are considered from Michael Bamshad; et al. (2004). "Deconstructing the Relationship Between Genetics and Race". *Nature Reviews Genetics* **5** (598): 598–609. doi:10.1038/nrg1401. PMID 15266342., original data from Rosenberg (2002).

[12] Bolnick, Deborah A. (2008). "Individual Ancestry Inference and the Reification of Race as a Biological Phenomenon". In Koenig, Barbara A.; Richardson, Sarah S.; Lee, Sandra Soo-Jin. *Revisiting race in a genomic age*. Rutgers University Press. ISBN 978-0-8135-4324-6.

[13] Kalinowski. "The Computer Program STRUCTURE Does Not Reliably Identify Main Genetic Clusters Within Species" **4**: 67–77.

[14] Sadaf Firasat, Shagufta Khalig, Aisha Mohyuddin, Myrto papioannou, Chris Tyler-Smith, Peter A. Underhill, and Qasim Ayub (2007). "Y-Chromosomal Evidence for a Limited Greek Contribution to the Pathan Population of Pakistan". *European Journal of Human Genetics* **15**: 121–6. doi:10.1038/sj.ejhg.5201726. PMC 2588664. PMID 17047675.

[15] Roberts, Dorothy (2011). *Fatal Invention*. London, New York: The New Press.

[16] Noah A. Rosenberg, Saurabh Mahajan, Sohini Ramachandran, Chengfeng Zhao, Jonathan K. Pritchard, and Marcus Feldman (2005). "Clines, Clusters, and the Effects of Study Design on the Inference of Human Population Science". *PloS Genetics* **1**: 660, 668. doi:10.1371/journal.pgen.0010070. PMC 1310579. PMID 16355252.

[17] Rosenberg; et al. "Genetic Structure of Human Populations": 2384.

[18] "Response to OMB Directive 15". American Anthropological Association. 1997.

[19] A.W.F. Edwards (2003). "Human Genetic Diversity: Lewontin's Fallacy". *BioEssays* **25** (8): 798–801. doi:10.1002/bies.10315. PMID 12879450.

[20] Kenneth M. Weiss and Brian W. Lambert (2010). "Does History Matter? Do the Facts of Human Variation Package Our Views or Do Our Views Package the Facts?". *Evolutionary Anthropology* **19**: 92, 97. doi:10.1002/evan.20261.

[21] "Confusion About Human Races". Social Science Research Council. 26 July 2006.

[22] Charles N. Rotini and Lynn B. Jorde (2010). "Ancestry and Disease in the Age of Genomic Medicine". *New England Journal of Medicine* **363**: 1551–1552. doi:10.1056/nejmra0911564.

[23] S.O.Y. Keita and Rick A. Kittles (1997). "The Persistence of Racial Thinking and the Myth of Racial Divergence". *American Anthropologist* **99**: 534–544. doi:10.1525/aa.1997.99.3.534.

[24] Rick A. Kittles and Kenneth M. Wells (2003). "Race, Ancestry, and Genes: Implications for Defining Disease Risk". *Annual Review of Genomics and Human Genetics* **4**: 33, 38. doi:10.1146/annurev.genom.4.070802.110356. PMID 14527296.

[25] Analabha Basul; et al. (2003). "Ethnic India: A Genomic View with Special Reference to Peopling and Structure". *Genome Research* **13** (10): 2277–90. doi:10.1101/gr.1413403. PMC 403703. PMID 14525929.

[26] Reardon, Jenny (2005). *Race to the Finish: Identity and Governance in the Age of Genomics*. Princeton, NJ: Princeton University Press.

[27] Graves, Joseph (2004). *The Race Myth*. New York: Dutton. p. 113.

[28] Mary-Claire King and Arno G. Motulsky (2002). "Mapping Human History". *Science* **298** (5602): 2342–2343. doi:10.1126/science.1080373.

[29] John H. Fujimura, Ramya Rajagopalan, Pilar N. Ossorio, and Kjell A. Doksum (2010). "Race and Ancestry: Operationalizing Populations in Human Genetic Variation Studies". *What's the Use of Race? Modern Governance and the Biology of Difference* (Cambridge MIT Press).

[30] L. Luca Cavalli-Sforza, Paolo Menozzi and Alberto Piazza (1994). *The History and Geography of Human Genes*. Princeton, NJ: Princeton University Press.

[31] Loring Brace, C. 2005. Race is a four letter word. Oxford University Press.

[32] Kaplan, Jonathan Michael (January 2011) 'Race': What Biology Can Tell Us about a Social Construct. In: Encyclopedia of Life Sciences (ELS). John Wiley & Sons, Ltd: Chichester

[33] Winther, Rasmus Grønfeldt (2011) ¿La cosificación genética de la 'raza'? Un análisis crítico in C López-Beltrán (ed.) *Genes (&) Mestizos. Genómica y raza en la biomedicina mexicana*. Ficticia editorial http://philpapers.org/archive/WINLCG.1.pdf

[34] Kaplan, Jonathan Michael, Winther, Rasmus Grønfeldt (2012). Prisoners of Abstraction? The Theory and Measure of Genetic Variation, and the Very Concept of 'Race' *Biological Theory* 7 http://philpapers.org/archive/KAPPOA.14.pdf

[35] Graves, Joseph. 2001. The Emperor's New Clothes. Rutgers University Press

[36] Weiss KM and Fullerton SM (2005) Racing around, getting nowhere. Evolutionary Anthropology 14: 165–169

[37] Mills CW (1988) "But What Are You Really? The Metaphysics of Race" in *Blackness visible: essays on philosophy and race*, pp. 41-66. Cornell University Press, Ithaca, NY

[38] Pillar Ossorio (2005). "Race, Genetic Variation, and the Haplotype Mapping Project". *Louisiana Law Review* **66** (131, 141).

[39] Royal, Novembre, Fullerton; et al. "Inferring Genetic Ancestry" (667-68).

[40] Mark D., Shriver; Rick A. Kittles (2004). "Genetic Ancestry and the Search for Personalized Genetic Histories". *Nature Reviews Genetics* **5**: 611–8. doi:10.1038/nrg1405. PMID 15266343.

[41] Wilson, James F.; Weale, Michael E.; Smith, Alice C.; Gratrix, Fiona; Fletcher, Benjamin; Thomas, Mark G.; Bradman, Neil; Goldstein, David B. (2001). "Population genetic structure of variable drug response". *Nature Genetics* **29** (3): 265–9. doi:10.1038/ng761. PMID 11685208.

[42] Mohamoud, A. M. (October 2006). "P52 Characteristics of HLA Class I and Class II Antigens of the Somali Population". *Transfusion Medicine* **16** (Supplement s1): 47. doi:10.1111/j.1365-3148.2006.00694_52.x.

[43] Braun, Lundy; Evelynn Hammonds (2008). "Race, Populations, and Genomics: Africa as Laboratory". *Social Science & Medicine* **67**: 1580–8. doi:10.1016/j.socscimed.2008.07.018.

[44] "Census, Race and Science". *Nature Genetics* **24** (2): 97–98. 2000. doi:10.1038/72884.

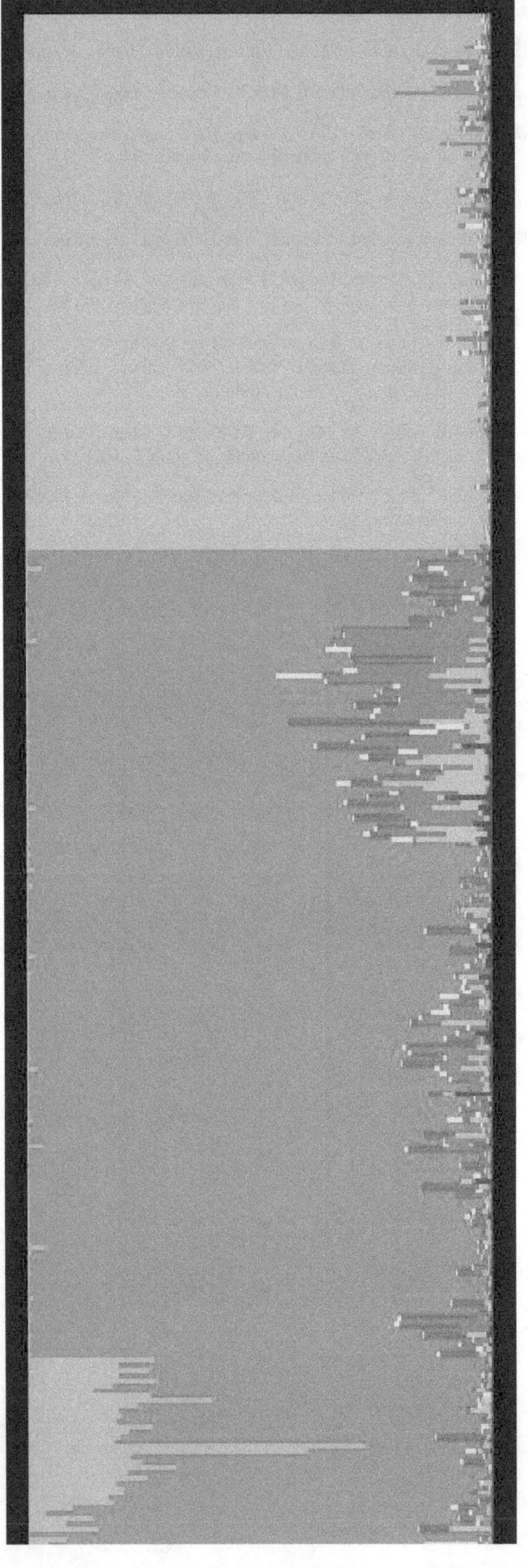

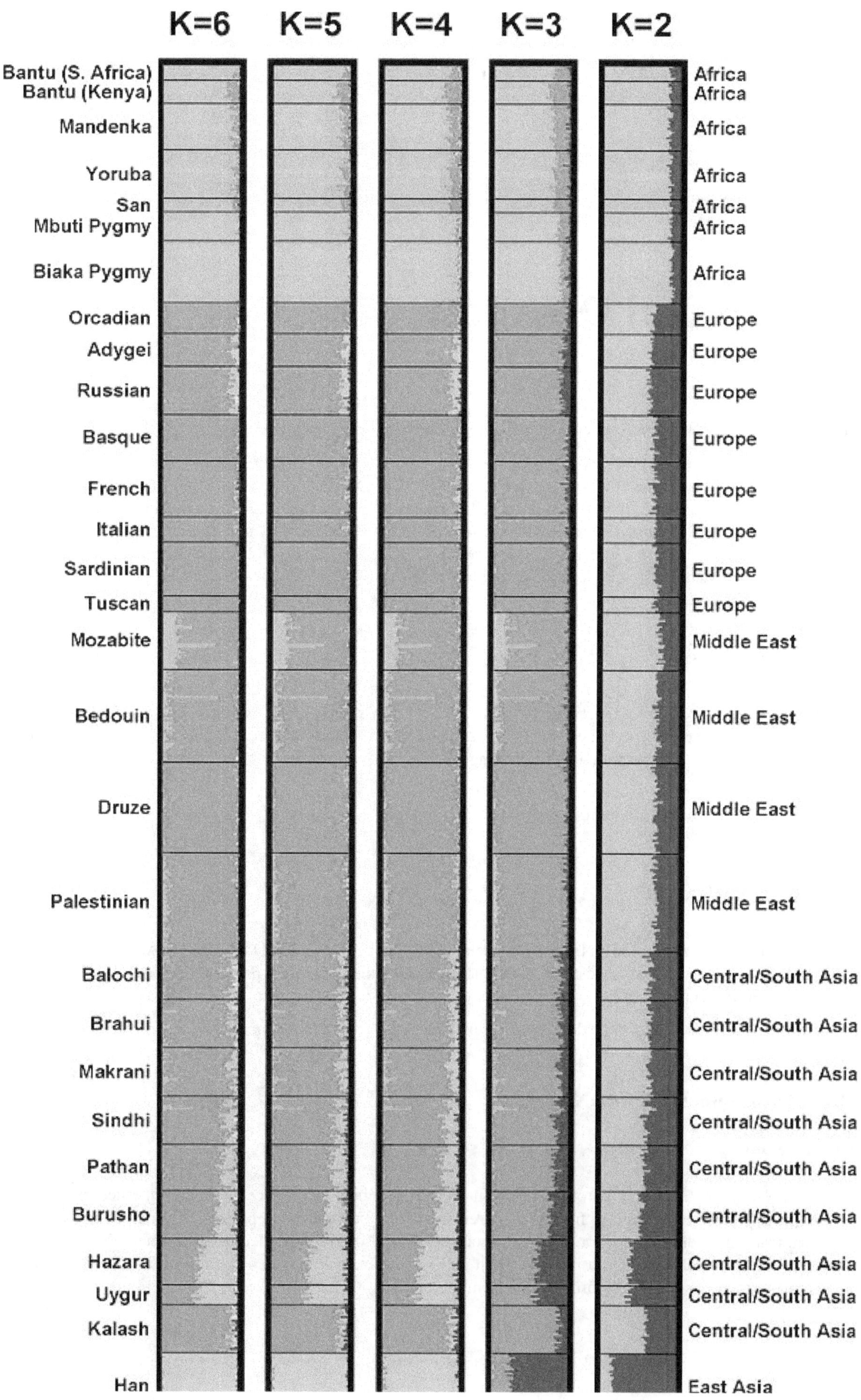

Chapter 50

Genealogical DNA test

For a non-technical introduction to genetics in general, see Introduction to genetics.

A **genealogical DNA test** looks at a person's genome at specific locations. Results give information about genealogy or personal ancestry. In general, these tests compare the results of an individual to others from the same lineage or to current and historic ethnic groups. The test results are not meant for medical use, where different types of genetic testing are needed. They do not determine specific genetic diseases or disorders (see possible exceptions in *Medical information* below). They are intended only to give genealogical information.

50.1 Procedure

Taking a genealogical DNA test requires the submission of a DNA sample. The most common way to collect a DNA sample is by a cheek-scraping (also known as a buccal swab). Other methods include spit-cups, mouthwash, and chewing gum. After collection, the sample is mailed to a testing lab.

Some laboratories, such as the Human Origins Genotyping Laboratory (HOGL) at the University of Arizona, offer to store DNA samples for ease of future testing.

50.2 Types of tests

There are three types of genealogical DNA tests, autosomal (atDNA), mitochondrial DNA (mtDNA), and Y-Chromosome (Y-DNA). Autosomal tests for all ancestry. Y-DNA tests a male along his direct paternal line. mtDNA tests a man or woman along their direct maternal line.

Any of these tests can be used to some degree for recent genealogy or for ethnic ancestry.

50.2.1 Autosomal DNA (atDNA)

What gets tested

Autosomal DNA is the 22 pairs of chromosomes that do not contribute to gender. These are inherited exactly equally from both parents and roughly equally from grandparents to about 3x great-grand parents. Inheritance is more random and unequal from more distant ancestors. The X-chromosome is also often included in Autosomal DNA tests. The X-chromosome has a special path of transmission. Both males and females receive an X-chromosome from their mother, but only females receive a second X-chromosome from their father.

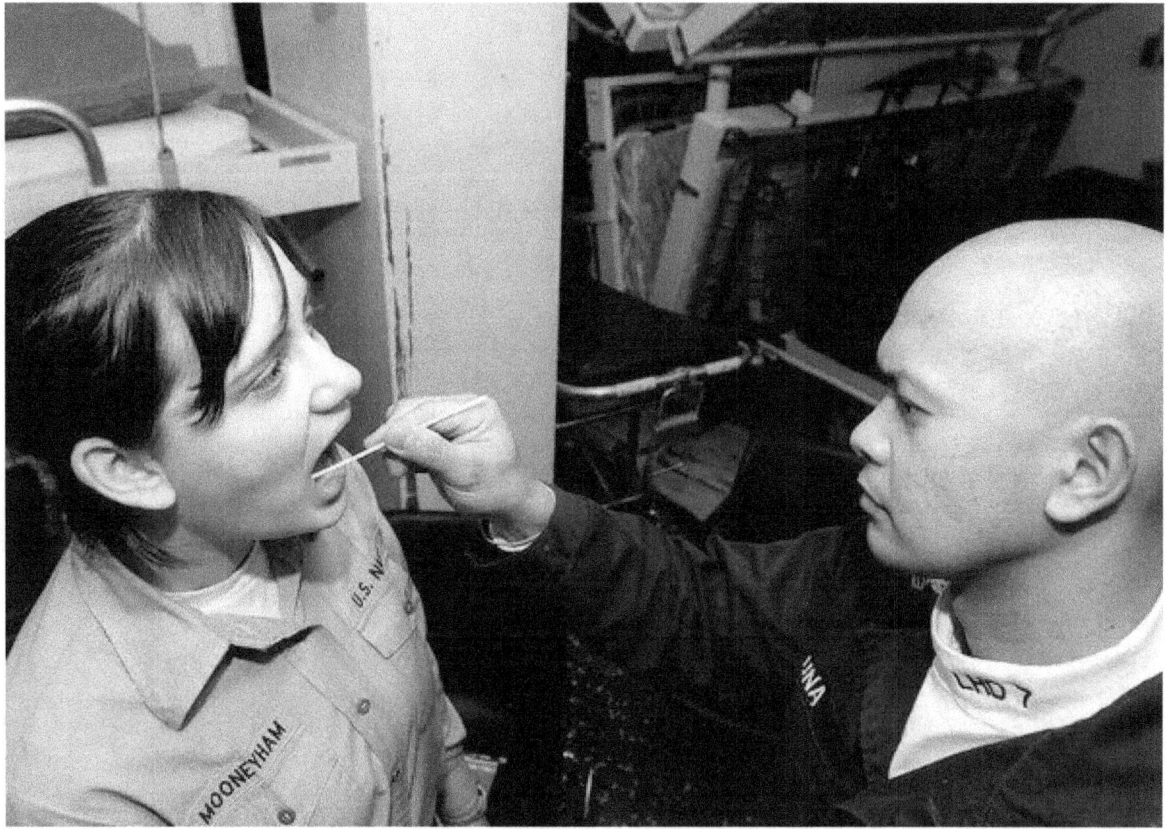

Hospital Corpsman 1st Class uses a swab to take a DNA sample from a Fireman aboard USS Iwo Jima (LHD 7)

Generally, a genealogical DNA test might test about 700,000 SNPs.(single-nucleotide polymorphisms).Like mtDNA and Y-DNA SNPs, autosomal SNPs are changes at a single point in genetic code. Autosomal DNA recombines each generation. Therefore, the number of markers shared with a specific ancestor decreases by half each generation. Some type of microarray chip is used by the laboratory. Different chips test different SNPs.

STRs Some genealogical companies offer autosomal STRs. These are similar to Y-DNA STRs. The number of STRs offered is limited, and not genealogically useful.

Matching process

The major component of an autosomal DNA test is matching other individuals. If two individuals share more than a certain threshold of DNA, they are considered a match by the testing company. Based on the amount of shared DNA, usually expressed in centiMorgens(cM), their relationship may be predicted. Due to the random nature of DNA inheritance, the exact relationship cannot be exactly predicted. Depending on the threshold, all 5th cousins and closer should be a match. Whether 6th cousins and further match depends on how DNA has been inherited.

More detailed analysis of matches including the method of triangulation can reveal more precisely how two matches are related genealogically. GedMatch offers free tools for these purposes. This area of DNA testing can be the most complex and difficult to understand. Many popular blogs and websites explain these tools for beginners.

Bio-geographical ancestry

Most companies offer a percentage breakdown by ethnicity or region. Generally the world is specified into about 20-25 regions, and the approximate percentage of DNA inherited from each is stated.This is usually done by comparing the frequency of each Autosomal DNA marker tested to many population groups. The reliability of this type of test is dependent on comparative population size, the number of markers tested, the ancestry informative value of the SNPs tested, and the degree of admixture in the person tested. Earlier ethnicity estimates were often wildly inaccurate, but their accuracies have since improved greatly.

50.2.2   Mitochondrial DNA (mtDNA) testing

The Mitochondrion is a component of a human cell, and is technically not DNA. It is very small, at just 16,569 base pairs (by comparison the human genome is 3.2 Billion base pairs). It is transmitted from mother to child, thus a direct maternal ancestor can be traced using mtDNA. The transmission occurs with very little recombination or mutation. A perfect match found to another person's mtDNA test results indicates shared recent ancestry. More distant matching to a specific haplogroup or subclade may be linked to a common geographic origin.

Some people cite paternal mtDNA transmission as invalidating mtDNA testing,[1] but this has not been found problematic in genealogical DNA testing, nor in scholarly population genetics studies.

What gets tested

mtDNA, by current conventions, is divided into three regions. They are the coding region (00577-16023) and two Hyper Variable Regions (HVR1 [16024-16569], and HVR2 [00001-00576]).[2] All test results are compared to the mtDNA of a European in Haplogroup H2a2a.

The two most common mtDNA tests are a sequence of HVR1 and HVR2 and a full sequence of the mitochondria. Some mtDNA tests may only analyze a partial range in these regions.Generally, testing only the HVRs has limited genealogical use so it is increasingly popular and accessible to have a full sequence. The full sequence is still somewhat controversial because it may reveal medical information.

Understanding test results

It is not normal for test results to give a base-by base list of results. Instead results are compared to the Cambridge Reference Sample (CRS), which is the mitochondria of a European women from Haplogroup H. Differences between the CRS and the tester are usually very few, thus it is more convenient than listing ones raw results for each base pair.

Examples

Note that in HVR1, instead of reporting the base pair exactly, for example 16,111, the 16 is often removed to give in this example 111. The Letters refer to one of the 4 bases (A,T,G,C) that make up human DNA.

Haplogroups

All humans descend in the direct female line from Mitochondrial Eve, a female who lived probably around 200,000 years ago in Africa. Different branches of her descendants are different haplogroups. Most mtDNA results include a prediction or exact assertion of ones mtDNA Haplogroup. Mitochrondial haplogroups were greatly popularized by the popular book *The Seven Daughters of Eve*, which explores mitochondrial DNA from an European perspective.

mtDNA in the news

mtDNA testing was used by University of Leicester archaeologists to verify the skeletal remains of King Richard III, found in September 2012.[3]

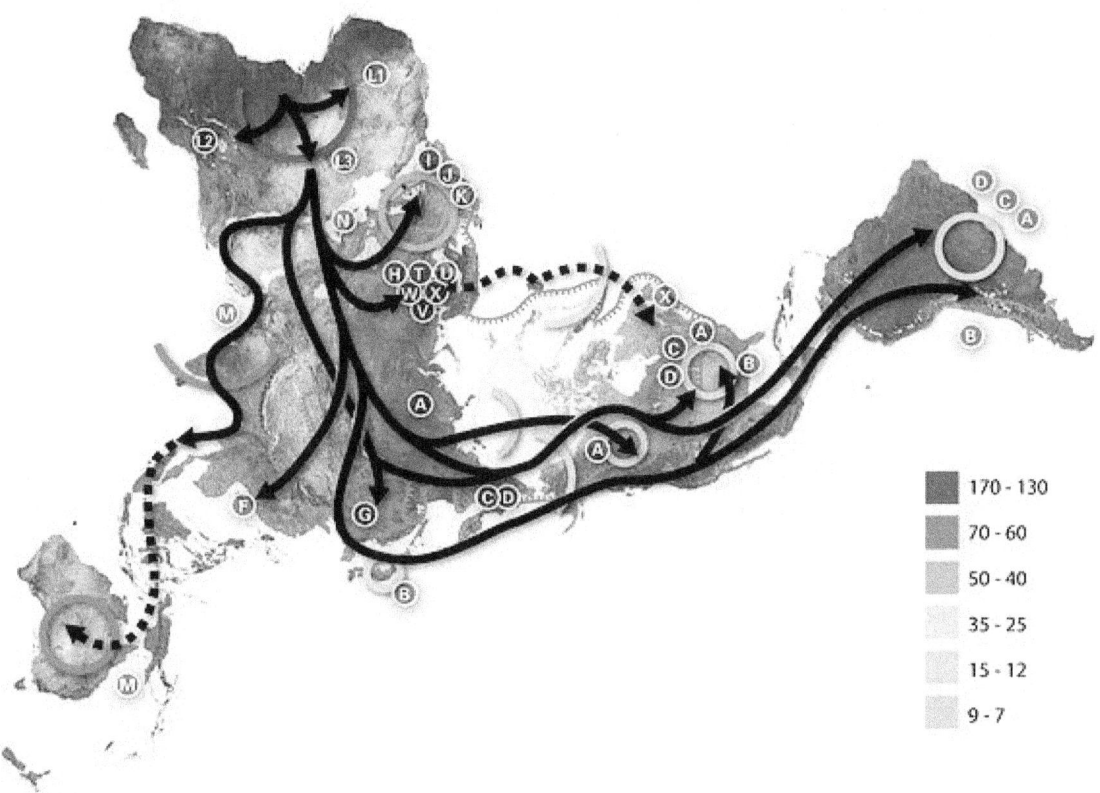

Map of human migration out of Africa, according to Mitochondrial DNA. The numbers represent thousands of years before present time. The blue line represents the area covered in ice or tundra during the last great ice age. The North Pole is at the center. Africa, the center of the start of the migration, is at the top left and South America is at the far right.

50.2.3 Y chromosome (Y-DNA) testing

The Y-Chromosome is part of the 23rd pair of human chromosomes. Only males have a Y-chromosome, because women have 2 X chromosomes in their 23rd pair. A man's patrilineal ancestry, or male-line ancestry, can be traced using the DNA on his Y chromosome (Y-DNA), because the majority of the Y-chromosome is transmitted father to son nearly unchanged. A man's test results are compared to another man's results to determine the time frame in which the two individuals shared a most recent common ancestor, or MRCA, in their direct patrilineal lines. If their test results are a perfect, or nearly perfect match, they are related within genealogy's time frame.[4] A surname project is where many individuals whose Y-chromosomes match collaborate to find their common ancestry.

Women who wish to determine their direct paternal DNA ancestry can ask their father, brother, paternal uncle, paternal grandfather, or a paternal uncle's son(their cousin) to take a test for them.

What gets tested

There are two types of DNA testing-STRs and SNPs.

STR markers   Most common is STRs (short tandem repeat). A certain section of DNA is examined for a pattern that repeats (e.g. ATCG). The number of times it repeats is the value of the marker. Typical tests test between 30 and 120 STR markers. STRs mutate fairly frequently. The results of two individuals are then compared to see if there is a match. Close matches may often join a surname project. DNA companies will usually provide information about how closely related two matches are, based on the difference between their results. One's haplogroup can be predicted but not confirmed by a STR test. Confirmation requires a SNP test.

Haplotype   Further information: Y-chromosome haplogroups by populations

A Y-DNA haplotype is the numbered results of a genealogical Y-DNA STR test. Each allele value has a distinctive frequency within a population. For example, at DYS455, the results will show 8, 9, 10, 11 or 12 repeats, with 11 being most common.[5] For high marker tests the allele frequencies provide a signature for a surname lineage.

The test results are then compared to another project member's results to determine the time frame in which the two people shared a most recent common ancestor (MRCA). Testing companies usually provide information on this.

Haplogroup prediction

A person's haplogroup can often be inferred from their haplotype, but can be proven only with a Y-chromosome SNP tests (Y-SNP test). In addition, some companies offer sub-clade tests, such as for Haplogroup G. For example, Haplogroup G has a known modal haplotype:

Few haplotypes will exactly match the modal values for Haplogroup G. One can consult an allele frequency table to determine the likelihood of remaining in Haplogroup G based on the variations observed.

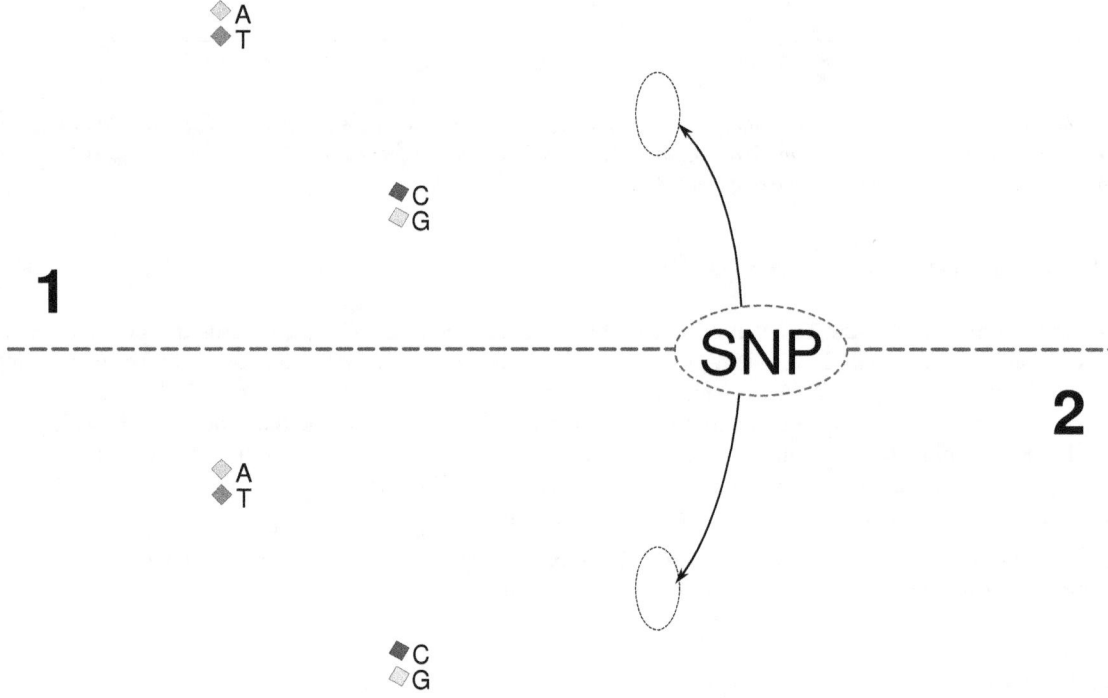

Strand 1 differs from strand 2 at a single base pair location (a C → T polymorphism).

SNP markers and Haplogroups   A single-nucleotide polymorphism (SNP) is a change to a single nucleotide in a DNA sequence. Typical Y-DNA SNP tests test about 20,000 SNPs. Getting a SNP test allows a much higher resolution than STRs. It can be used to provide additional information about the relationship between two individuals and to confirm haplogroups.

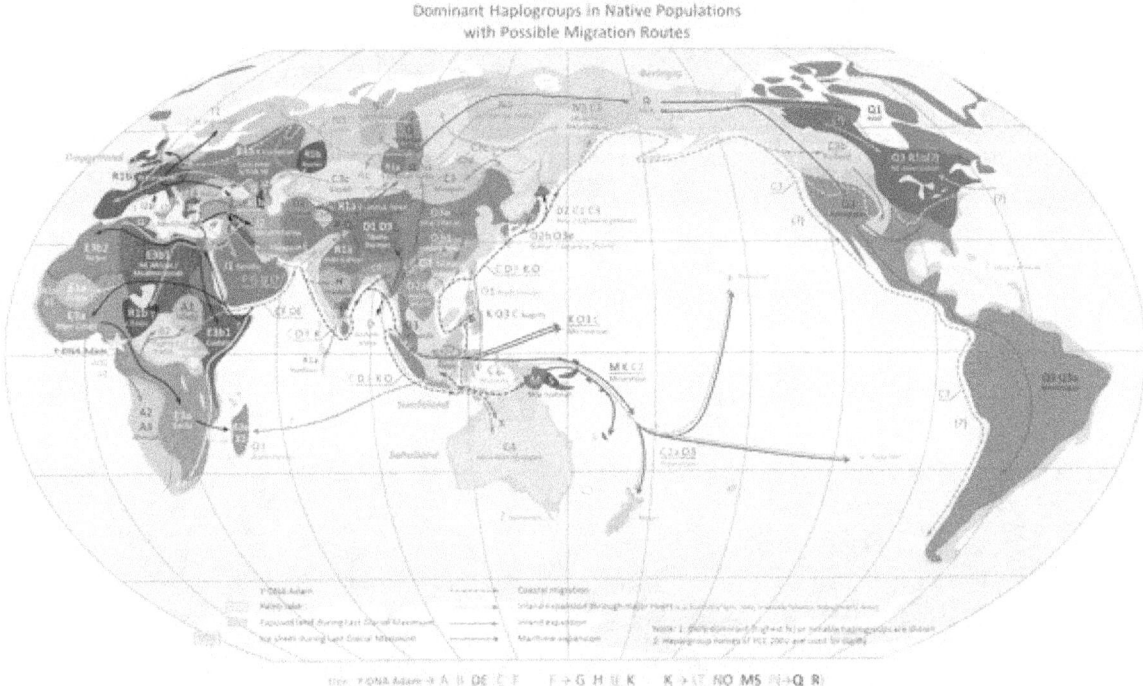

Dominant Y-chromosome haplogroups in pre-colonial world populations, with possible migrations routes according to the Coastal Migration Model.

All human men descend in the paternal line from a single man dubbed Y-chromosomal Adam, who lived probably between 200,000 and 400,000 years ago. A 'family tree' can be drawn showing how men today descend from him. Different branches of this tree are different haplogroups. Most haplogroups can be further subdivided multiple times into sub-clades. Some known sub-clades were founded in the last 1000 years, meaning their timeframe approaches the genealogical era(c.1500 onwards).

New sub-clades of haplogroups may be discovered if an individual tests, especially if they are non-European. Most significant of these new discoveries was in 2013 when the haplogroup A00 was discovered, which required theories about Y-chromosomal Adam to be significantly revised. The haplogroup was discovered when an African-American man tested STRs at FamilyTreeDNA and his results were found to be unusual. SNP testing confirmed that he does not descend patrilineally from the "old" Y-chromosomal Adam and so a much older man became Y-Chromosomal Adam. If enough individuals belong to a newly discovered subclade, the subclade is added to the ISOGG Y-DNA haplogroup Tree, which is the most up-to-date and respected tree of Y-DNA haplogroups.

50.3   Audience

The interest in genealogical DNA tests has been linked to both an increase in curiosity about traditional genealogy and to more general personal origins. Those who test for traditional genealogy often utilize a combination of autosomal, mitochondrial, and Y-Chromosome tests. Those with an interest in personal ethnic origins are more likely to use an autosomal test. However, answering specific questions about the ethnic origins of a particular lineage may be best suited to an mtDNA test or a Y-DNA test.

Y-DNA-haplogroups in Europe

50.3.1   Maternal origin tests

For recent genealogy, exact matching on the mtDNA full sequence is used to confirm a common ancestor on the direct maternal line between two suspected relatives. Because mtDNA mutations are very rare, a *nearly* perfect match is not usually considered relevant to the most recent 1 to 16 generations.[6] In cultures lacking matrilineal surnames to pass down, neither relative above is likely to have as many generations of ancestors in their matrilineal information table as in the above patrilineal or Y-DNA case: for further information on this difficulty in *traditional genealogy*, due to lack of *matrilineal* surnames (or matrinames), see Matriname.[7] However, the foundation of testing is still two suspected descendants of one person. This hypothesize and test DNA pattern is the same one used for autosomal DNA and Y-DNA.

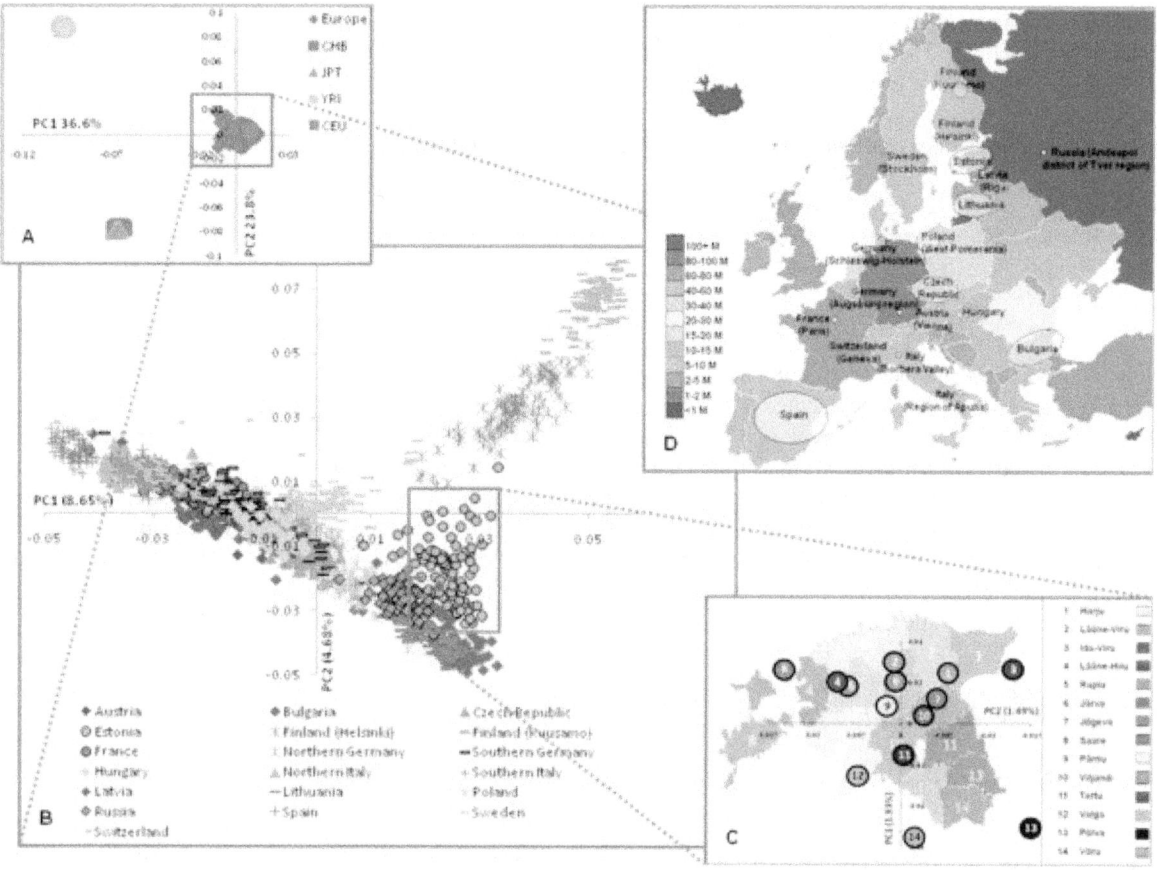

European genetic structure (based on Autosomal SNPs) by PCA

50.3.2   Geographic origin tests

As discussed above, autosomal tests usually report the ethnic proportions of the individual. These attempt to measure an individual's mixed geographic heritage by identifying particular markers, called ancestry informative markers or AIM, that are associated with populations of specific geographical areas. Earlier tests were often imprecise and inaccurate, but more recent tests have been dramatically improved, though different companies results may still much discrepancy for the same individual. Results may still be inaccurate for individuals with ancestry from Scandinavia and the British Isles. Anomalous findings most often result from databases too small to associate markers with all the areas where they occur in indigenous populations.

African ancestry

Y-DNA and mtDNA testing may be able to determine with which peoples in present-day Africa a person shares a direct line of part of his or her ancestry, but patterns of historic migration and historical events cloud the tracing of ancestral groups. Testing company African Ancestry[8] maintains an "African Lineage Database" of African lineages from 30 countries and over 160 ethnic groups. Due to joint long histories in the US, approximately 30% of African American males have a European Y-Chromosome haplogroup[9] Approximately 58% of African Americans have the equivalent of one great-grandparent (12.5 percent) of European ancestry. Only about 5% have the equivalent of one great-grandparent of Native American ancestry. By the early 19th century, substantial families of Free Persons of Color had been established in the Chesapeake Bay area who were descended from people free during the colonial period; most of those have been documented as descended from white men and African women (servant, slave or free). Over time various groups married more within mixed-race, black or white communities.[10]

According to authorities like Salas, nearly three-quarters of the ancestors of African Americans taken in slavery came from regions of West Africa. The African-American movement to discover and identify with ancestral tribes has burgeoned since DNA testing became available. Often members of African-American churches take the test as groups. African Americans usually cannot easily trace their ancestry during the years of slavery through surname research, census and property records, and other traditional means. Genealogical DNA testing may provide a tie to regional African heritage.

United States - Melungeon testing

Main article: Melungeon § DNA testing

Melungeons are one of numerous multiracial groups in the United States with origins wrapped in myth. The historical research of Paul Heinegg has documented that many of the Melungeon groups in the Upper South were descended from mixed-race people who were free in colonial Virginia and the result of unions between the Europeans and Africans. They moved to the frontiers of Virginia, North Carolina, Kentucky and Tennessee to gain some freedom from the racial barriers of the plantation areas.[11] Several efforts, including a number of ongoing studies, have examined the genetic makeup of families historically identified as Melungeon. Most results point primarily to a mixture of European and African, which is supported by historical documentation. Some may have Native American heritage as well. Though some companies provide additional Melungeon research materials with Y-DNA and mtDNA tests, any test will allow comparisons with the results of current and past Melungeon DNA studies

50.3.3   Native American ancestry

Further information: Genetic history of indigenous peoples of the Americas

The pre-columbian indigenous people of the United States are called "Native Americans" in American English.[12] Autosomal testing, Y-DNA, and mtDNA testing can be conducted to determine the ancestry of Native Americans. A mitochondrial Haplogroup determination test based on mutations in Hypervariable Region 1 and 2 may establish whether a person's direct female line belongs to one of the canonical Native American Haplogroups, A, B, C, D or X. If one's DNA belonged to one of those groups, the implication would be that he or she is, in whole or part, Native American.

As political entities, tribes have established their own requirements for membership, often based on at least one of a person's ancestors having been included on tribal-specific Native American censuses (or final rolls) prepared during treaty-making, relocation to reservations or apportionment of land in the late 19th century and early 20th century. One example is the Dawes Rolls. In addition, the U.S. government does not consider DNA as admissible evidence for enrollment in any federally recognized tribe or reception of benefits. Tribes are political constructs, not genetic populations.

The vast majority of Native American individuals belong to one of the five identified mtDNA Haplogroups. Many Americans are now just discovering they have some percentage of Native ancestry. Some attempt to validate their heritage with the goal of gaining admittance into a tribe, but most tribes do not use DNA results in that way.

50.3.4   Cohanim ancestry

Main article: Y-chromosomal Aaron

The Cohanim (or Kohanim) is a patrilineal priestly line of descent in Judaism. According to the Bible, the ancestor of the Cohanim is Aaron, brother of Moses. Many believe that descent from Aaron is verifiable with a Y-DNA test: the first published study in genealogical Y-Chromosome DNA testing found that a significant percentage of Cohens had distinctively similar DNA, rather more so than general Jewish or Middle Eastern populations. These Cohens tended to belong to Haplogroup J, with Y-STR values clustered unusually closely around a haplotype known as the Cohen Modal Haplotype (CMH). This could be consistent with a shared common ancestor, or with the hereditary priesthood having originally been founded from members of a single closely related clan.

Nevertheless, the original studies tested only six Y-STR markers, which is considered a low-resolution test. In response to the low resolution of the original 6-marker CMH, the testing company FTDNA released a 12-marker CMH signature that was more specific to the large closely related group of Cohens in Haplogroup J1.

A further academic study published in 2009 examined more STR markers and identified a more sharply defined SNP haplogroup, J1e* (now J1c3, also called J-P58*) for the J1 lineage. The research found "that 46.1% of Kohanim carry Y chromosomes belonging to a single paternal lineage (J-P58*) that likely originated in the Near East well before the dispersal of Jewish groups in the Diaspora. Support for a Near Eastern origin of this lineage comes from its high frequency in our sample of Bedouins, Yemenis (67%), and Jordanians (55%) and its precipitous drop in frequency as one moves away from Saudi Arabia and the Near East (Fig. 4). Moreover, there is a striking contrast between the relatively high frequency of J-58* in Jewish populations (»20%) and Kohanim (»46%) and its vanishingly low frequency in our sample of non- Jewish populations that hosted Jewish diaspora communities outside of the Near East."[13]

Recent phylogenetic research for haplogroup J-M267 placed the "Extended Cohen Modal Haplotype" in a subhaplogroup of J-L862,L147.1 (age estimate 5631-6778yBP yBP): YSC235>PF4847/CTS11741>YSC234>ZS241>ZS227>Z18271 (age estimate 2731yBP).[14]

50.3.5 European testing

Further information: Genetic history of Europe

For people with European maternal ancestry, mtDNA tests are offered to determine which of eight European maternal "clans" the direct-line maternal ancestor belonged to. This mtDNA haplotype test was popularized in the book *The Seven Daughters of Eve*.

50.3.6 Hindu testing

The 49 established *gotras* are clans or families whose members trace their descent to a common ancestor, usually a sage of ancient times. The gotra proclaims a person's identity and a "gotra-pravara" is required to be presented at Hindu ceremonies. People of the same gotra are not allowed to marry.

One company says it can use a 37-marker Y-DNA test to "verify genetic relatedness and historical gotra genealogies for Hindu and Buddhist engagements, marriages and business partnerships." This has not been supported by independent research. Any Y-DNA test can be used to compare results with another person whose gotra is known.

50.4 Popularity

The number of testers has dramatically increased with time. In 2014 and 2015 all three major DNA testing companies(23andMe, Ancestry and FamilytreeDNA) reported to have more than 1 million customers.[15]

According to the ISOGG, the following atDNA SNP testing customers had these numbers of testers in the database as of 2015. These lists are not exhaustive, but all companies not mentioned on these lists are very small.

23andMe : ~ 1,000,000
Ancestry.com : ~1,000,000
National Geographic : ~200,000
FamilyTreeDNA : ~150,000
Chromo2 (Britains DNA+subsidaries) : ~10,000

The following companies offered Y-DNA STR tests

FamilytreeDNA : ~539,000
Genebase Systems
Oxford Ancestors

YSEQ
The following companies offered Y-DNA SNP tests

23andMe: included in atDNA test
National Geographic : included in atDNA test
FamilytreeDNA
Chromo2
Full Genomes Corporation
The following companies offered mtDNA tests [16]

FamilytreeDNA : ~200,000
23andMe : included in atDNA test
National Geographic : included in atDNA test
Chromo2
Oxford Ancestors
Full Genomes Corporation
Genebase systems

50.5   Benefits

Main article: Genetic genealogy

Genealogical DNA tests have become popular due to the ease of testing at home and their usefulness in supplementing genealogical research. Genealogical DNA tests allow for an individual to determine with high accuracy whether he or she is related to another person within a certain time frame, or with certainty that he or she is not related. DNA tests are perceived as more scientific, conclusive and expeditious than searching the civil records. However, they are limited by restrictions on lines that may be studied. The civil records are always only as accurate as the individuals having provided or written the information.

Y-DNA testing results are normally stated as probabilities: For example, with the same surname a perfect 37/37 marker test match gives a 95% likelihood of the most recent common ancestor (MRCA) being within 8 generations,[17] while a 111 of 111 marker match gives the same 95% likelihood of the MRCA being within only 5 generations back.[18]

As presented above in mtDNA testing, if a perfect match is found, the mtDNA test results can be helpful. In some cases, research according to traditional genealogy methods encounters difficulties due to the lack of regularly recorded matrilineal surname information in many cultures (see Matrilineal surname).[7]

50.6   Drawbacks

Common concerns about genealogical DNA testing are cost and privacy issues (some testing companies retain samples and results for their own use without a privacy agreement with subjects). The most common complaint from DNA test customers is the failure of the company to make results understandable to them.

DNA tests can do some things well, but there are constraints. Testing of the Y-DNA lineage from father to son may reveal complications, due to unusual mutations, secret adoptions, and false paternity (i.e., the father in one generation is not the father in birth records). According to some genomics experts, autosomal tests may have a margin of error up to 15% and blind spots.

Some users have recommended that there be government or other regulation of ancestry testing to ensure more standardization.[19]

50.7 Medical information

Though genealogical DNA test results in general have no informative medical value and are not intended to determine genetic diseases or disorders, a correlation exists between a lack of DYS464 markers and infertility, and between mtDNA haplogroup H and protection from sepsis. Certain haplogroups have been linked to longevity.[20]

The testing of full mtDNA sequences is still somewhat controversial as it may reveal medical information. The field of linkage disequilibrium, unequal association of genetic disorders with a certain mitochondrial lineage, is in its infancy, but those mitochondrial mutations that have been linked are searchable in the genome database Mitomap.[21] The National Human Genome Research Institute operates the Genetic And Rare Disease Information Center[22] that can assist consumers in identifying an appropriate screening test and help locate a nearby medical center that offers such a test.

50.8 DNA in genealogy software

Some genealogy software programs now allow recording DNA marker test results, allowing for tracking of both Y-chromosome and mtDNA tests, and recording results for relatives. DNA-family tree wall charts are available.

50.9 See also

Main article: List of genetic genealogy topics

- 23andMe

- Archaeogenetics

- Electropherogram

- Family Tree DNA

- Family name (Patrilineal surname)

- Genetic fingerprinting

- Genetic genealogy

- Genetic Information Nondiscrimination Act

- Genetic testing

- Human genetic clustering

- International HapMap Project

- List of DNA tested mummies

- Matriname (Matrilineal surname)

- National Geographic Geno 2.0

- DNA paternity testing

50.10   References

[1] for example: M. Pickford, "Paradise lost: Mitochondrial eve refuted", SpringerLink, July 2006

[2] "mtDNA regions". *Phylotree.org*. Retrieved 2011-06-15.

[3] "DNA Tests prove that the body found under a parking lot belongs to King Richard III; but was he truly a "hunchback?"".

[4] "Y-DNA matches". *Smgf.org*. Retrieved 2011-06-15.

[5] "Ybase statistics". *Ybase.org*. 19 April 2011. Retrieved 2011-06-15.

[6] "mtDNA matches". *Smgf.org*. Retrieved 2011-06-15.

[7] Sykes, Bryan (2001). *The Seven Daughters of Eve*. W. W. Norton. ISBN 0-393-02018-5, pp. 291-92. Sykes discusses the difficulty in genealogically tracing a maternal lineage, due to the lack of matrilineal surnames (or matrinames).

[8] "African Ancestry". African Ancestry. Retrieved 2011-06-15.

[9] "Patriclan: Trace Your Paternal Ancestry". African Ancestry. Retrieved 2011-06-15.

[10] Paul Heinegg, *Free African Americans of Virginia, North Carolina, South Carolina, Maryland and Delaware*, accessed 15 February 2008

[11] Paul Heinegg, *Free African Americans of Virginia, North Carolina, South Carolina, Maryland and Delaware*, accessed 15 February 2008

[12] http://www.merriam-webster.com/dictionary/native%20american

[13] Hammer, Michael F; Doron M Behar and 7 others (8 August 2009). "Extended Y chromosome haplotypes resolve multiple and unique lineages of the Jewish priesthood". *Hum Genet* **126** (5): 707–717. doi:10.1007/s00439-009-0727-5. PMC 2771134. PMID 19669163. Cite uses deprecated parameter |coauthors= (help)

[14] Mas, V. (2013). "Y-DNA Haplogroup J1 phylogenetic tree". doi:10.6084/m9.figshare.741212.

[15] http://www.isogg.org/wiki/Timeline:History_of_genetic_genealogy ISOGG Genetic Genealogy timeline

[16] http://www.isogg.org/wiki/mtDNA_testing_comparison_chart ISOGG mtDNA testing Comparison chart

[17] ftdna.com (kept uptodate). http://www.familytreedna.com/faq/answers/default.aspx?faqid=9#922 "FAQ: ...how should the genetic distance at 37 Y-chromosome STR markers be interpreted?" Retrieved 2012-01-13.

[18] ftdna.com (kept uptodate). http://www.familytreedna.com/faq/answers/default.aspx?faqid=9#925 "FAQ: ...how should the genetic distance at 111 Y-chromosome STR markers be interpreted?" Retrieved 2012-01-13.

[19] Lee et al., "The Illusive Gold Standard in Genetic Ancestry Testing", *Science* 3, July 2009: 38-39 http://www.sciencemag.org/content/325/5936/38.full

[20] De Benedictis, G; Rose, G; Carrieri, G; De Luca, M; Falcone, E; Passarino, G; Bonafe, M; Monti, D; Baggio, G; Bertolini, S; Mari, D; Mattace, R; Franceschi, C (September 1999). "Mitochondrial DNA inherited variants are associated with successful aging and longevity in humans". *FASEB J.* **13** (12): 1532–6. PMID 10463944.

[21] "Mitomap". Mitomap. Retrieved 2011-06-15.

[22] "Genetic And Rare Disease Information Center (GARD)". *Genome.gov*. 22 March 2011. Retrieved 2011-06-15.

50.11   Further reading

- Anne Hart; Anne Hart M a (April 2004). *How to Interpret Family History and Ancestry DNA Test Results for Beginners*. iUniverse. ISBN 978-0-595-31684-7.

- Megan Smolenyak; Ann Turner (12 October 2004). *Trace your roots with DNA: using genetic tests to explore your family tree*. Rodale. ISBN 978-1-59486-006-5.

- Chris Pomery; Steve Jones (1 October 2004). *DNA and family history: how genetic testing can advance your genealogical research*. Dundurn Press Ltd. ISBN 978-1-55002-536-1.

50.12 External links

Tutorials

- Tutorial for Y-DNA STR tests (phrased as FAQs)
- Tutorial for Y-DNA SNP tests (also as FAQs)
- Tutorial for mtDNA tests (as FAQs)

Societies

- International Society for Genetic Genealogy

Foundations and research projects

- The National Geographic Genographic Project
- Sorenson Molecular Genealogy Foundation A wholly owned subsidiary of Sorenson

Information and Maps on Y-DNA haplogroups

- Y-haplogroups World Map
- Y-Haplogroups brief descriptions and regional origins
- Y-DNA Ethnographic and Genographic Atlas and Open-Source Data Compilation

External

- A list of DNA testing companies

Chapter 51

European Bioinformatics Institute

Warning: Page using Template:Infobox organization with unknown parameter "fgcolor" (this message is shown only in preview).
Warning: Page using Template:Infobox organization with unknown parameter "websitetwitter" (this message is shown only in preview).
Warning: Page using Template:Infobox organization with unknown parameter "bgcolor" (this message is shown only in preview).

The **European Bioinformatics Institute** (**EMBL-EBI**) is a centre for research and services in bioinformatics, and is part of European Molecular Biology Laboratory (EMBL).
EMBL-EBI is listed in the Registry of Research Data Repositories re3data.org.[3]

51.1 About

The roots of the EMBL-EBI lie in the EMBL Nucleotide Sequence Data Library[4][5] (now known as EMBL-Bank), which was established in 1980 at the EMBL laboratories in Heidelberg, Germany and was the world's first nucleotide sequence database.[6] The original goal was to establish a central computer database of DNA sequences, to supplement sequences submitted to journals. What began as a modest task of abstracting information from literature soon became a major database activity with direct electronic submissions of data and the need for highly skilled informatics staff. The task grew in scale with the start of the genome projects, and grew in visibility as the data became relevant to research in the commercial sector. It soon became apparent that the EMBL Nucleotide Sequence Data Library needed better financial security to ensure its long-term viability and to cope with the sheer scale of the task.

There was also a need for research and development to provide services, to collaborate with global partners to support the project, and to provide assistance to industry. To this end, in 1992, the EMBL Council voted to establish the European Bioinformatics Institute and to locate it at the Wellcome Trust Genome Campus in the United Kingdom where it would be in close proximity to the major sequencing efforts at the Wellcome Trust Sanger Institute. From 1992 through to 1994, a gradual transition of the activities in Heidelberg took place, until in September 1994 the EMBL-EBI occupied its current location on the Wellcome Trust Genome Campus.

When the EMBL-EBI moved to Hinxton it hosted two databases, one for nucleotide sequences (the EMBL Data Library, which was renamed EMBL-Bank and eventually became part of the European Nucleotide Archive) and one for protein sequences (Swiss-Prot–TrEMBL, now known as UniProt). Since then, the EMBL-EBI has diversified to provide data resources in all the major molecular domains and expanded to include a broad research base. It provides user support and offers advanced training in bioinformatics.[7]

550

51.2   Funding

EMBL-EBI logo

As part of EMBL, the largest part of EMBL-EBI's funding comes from the governments of EMBL's 21 member states. Other major funders include the European Commission, Wellcome Trust, US National Institutes of Health, UK Research Councils, EMBL-EBI's industry partners and the UK Department of Trade and Industry. In addition, the Wellcome Trust provides the facilities for the EMBL-EBI on its Genome Campus at Hinxton, and the UK Research Councils have provided funds for EMBL-EBI's facilities in Hinxton.[8]Coordinates: 52°4′47.76″N 0°11′12.25″E / 52.0799333°N 0.1867361°E

51.3   Resources at the EMBL-EBI

The EMBL-EBI hosts a number of publicly open, free to use life science resources, including biomedical databases, analysis tools and bio-ontologies. These include:

- ArrayExpress - archive of gene expression experiments

- BioModels Database - a database of computational models relevant to the life sciences

- Chemical Entities of Biological Interest (ChEBI) - database and ontology of molecular entities

- European Nucleotide Archive (ENA) - resource of nucleotide sequencing information

- Ensembl project - genome databases for vertebrates and other eukaryotic species (joint with Wellcome Trust Sanger Institute)

- Europe PubMed Central - database offering free access to collection of biomedical research literature

- Experimental Factor Ontology (EFO) - ontology of experimental variables for biomedical data

- Expression Atlas - database of summary information on which genes are expressed under which conditions

- Gene ontology - ontology of gene functions and processes

- InterPro - database of protein functional domains and families

- Protein Data Bank in Europe - European resource for the collection, organisation and dissemination of data on biological macromolecular structures

- Proteomics Identifications Database (PRIDE) - repository of mass spectrometry (MS) based proteomics

- UniProt - database of protein sequence and functional information (joint with Swiss Institute of Bioinformatics and Protein Information Resource)

51.4   Other bioinformatics organisations

- National Center for Biotechnology Information United States National Library of Medicine

- DNA Data Bank of Japan

- Expasy Swiss Institute of Bioinformatics

- BRAEMBL Bioinformatics Resource Australia - EMBL

51.5   References

[1] http://www.ebi.ac.uk/about/background Background

[2] http://www.ebi.ac.uk/about/jobs

[3] "EBI Entry in re3data.org". *www.re3data.org*. Retrieved 28 August 2014.

[4] Attwood T.K., Gisel A., Eriksson N-E. and Bongcam-Rudloff E. (2011). "Concepts, Historical Milestones and the Central Place of Bioinformatics in Modern Biology: A European Perspective". *Bioinformatics - Trends and Methodologies*. InTech. Retrieved 8 Jan 2012.

[5] Stoesser, G.; Sterk, P.; Tuli, M.; Stoehr, P.; Cameron, G. (1997). "The EMBL Nucleotide Sequence Database". *Nucleic Acids Research* **25** (1): 7–14. doi:10.1093/nar/25.1.7. PMC 146376. PMID 9016493.

[6] Kneale, G.; Kennard, O. (1984). "The EMBL nucleotide sequence data library". *Biochemical Society transactions* **12** (6): 1011–1014. doi:10.1042/bst0121011. PMID 6530028.

[7] Wright, V. A.; Vaughan, B. W.; Laurent, T.; Lopez, R.; Brooksbank, C.; Schneider, M. V. (2010). "Bioinformatics training: Selecting an appropriate learning content management system--an example from the European Bioinformatics Institute". *Briefings in Bioinformatics* **11** (6): 552–562. doi:10.1093/bib/bbq023. PMID 20601435.

[8] "ELIXIR – European life science infrastructure for biological information".

Chapter 52

Wellcome Trust Sanger Institute

The **Wellcome Trust Sanger Institute** (previously known as '**The Sanger Centre'**) is a non-profit, British genomics and genetics research institute, primarily funded by the Wellcome Trust.[1]

It is located on the Wellcome Trust Genome Campus by the village of Hinxton, outside Cambridge. It shares this location with the European Bioinformatics Institute. It was established in 1992 as **The Sanger Centre**, named after double Nobel Laureate, Frederick Sanger.[2][3] It was conceived as a large scale DNA sequencing centre to participate in the Human Genome Project, and went on to make the largest single contribution to the gold standard sequence of the human genome. From its inception the Institute established and has maintained a policy of data sharing, and does much of its research in collaboration.

Since 2000, the Institute expanded its mission to understand "the role of genetics in health and disease".[4] The Institute now employs around 900 people and engages in four main areas of research: Human genetics, pathogen genetics, mouse and zebrafish genetics and bioinformatics.

52.1 Facilities and resources

52.1.1 Campus

The Wellcome Trust Genome Campus

In 1993 the then 17 Sanger Institute staff moved into temporary laboratory space at Hinxton Hall[5] in Cambridgeshire.

Commemorative stain window located in the Sulston building of the Wellcome Trust Sanger Institute, to mark the opening of the Genome Campus.

This 55-acre (220,000 m^2) site was to become the Wellcome Trust Genome Campus, which has a growing population of around 1300 staff, approximately 900 of whom work at the Sanger Institute.[6] The Genome Campus also includes the Wellcome Trust Conference Centre[7] and the European Bioinformatics Institute. A major extension of the campus was officially opened in 2005;[8] the buildings accommodate new laboratories, a data centre and staff amenities.[9] In discussing the name of the centre, Sanger (still alive when the centre was opened) told John Sulston, the founding director, that the centre 'had better be good.' Sulston commented, "I rather wished I hadn't asked."[10]

52.1.2   Sequencing

The Sanger Institute's sequencing staff handle millions of DNA samples each week. The Institute "capitalises on leading-edge technologies to answer questions unanswerable only a few years ago".[11] The advances in technology allow the Sanger Institute to carry out sequencing of the genomes of individual humans, vertebrate species and pathogens, at an ever increasing pace and reducing cost. The Institute has more than 100 ongoing pathogen sequencing projects.[12] The output of the Sanger Institute is around 10 billion bases of raw sequence data per day.[13]

52.1.3   Scientific Resources

Bioinformatic databases resources are one of the outcomes of research programmes that the Sanger Institute is involved in. Those hosted by the Sanger Institute include:

- COSMIC,[14] a catalogue of somatic mutations in cancer

- DECIPHER, a database of chromosomal imbalance and phenotype in humans, using Ensembl resources[15]

- Ensembl,[16] a genome browser co-hosted by the European Bioinformatics Institute.

- GeneDB,[17] a pathogen sequence database

- MEROPS,[18] a peptidase database

- Mouse Genetics Project, including a database of standardised phenotypic analysis for many hundreds of mutant mice.

- Pfam,[19] a protein family database

- Rfam,[20] an RNA family database

- TreeFam,[21] a database of phylogenetic trees for animal genes

- Vega,[22] a vertebrate genome annotation resource

- WormBase,[23] a database on the biology and sequence of the model organism *C. elegans* and other related Nematodes.

- WormBase ParaSite, a database for the genomics for parasitic helminths (both Nematodes and Platyhelminthes).

52.2 Research

The Morgan Building (right), part of the Sanger Institute

Since 2000, the Sanger Institute has built on its sequencing skills to develop new programmes in postgenomic biology - understanding the messages in genes. The Institute engages in several areas of research:

52.2.1 Human genetics

The hall Institute's research in human genetics focuses on the characterisation of human genetic variation in health and disease. Aside from the Institute's contribution to the Human Genome Project, researchers at the Sanger Institute have made contributions in various research areas relating to disease, population comparative and evolutionary genetics. In January 2008, the launch of the 1000 Genomes Project, a collaboration with scientists around the globe, signalled an effort to sequence the genomes of 1000 individuals in order to create the "most detailed map of human genetic variation to support disease studies".[24] The data from the pilot projects was made freely available in public databases in June 2010.[25] In 2010, the Sanger Institute announced its participation in the UK10K project,[26][27][28] which will sequence the genomes of 10,000 individuals to identify rare genetic variants and their effects on human health. The Sanger Institute is also part of the International Cancer Genome Consortium, an international effort to describe different cancer tumour types.[29] It is also part of the GENCODE and ENCODE research programmes[30] to create an encyclopaedia of DNA elements.

52.2.2 Model organisms

The mouse and zebrafish genetics programme at the Sanger Institute uses genome sequence of these model organisms to understand basic biological mechanisms, and gene function in human health and disease. Projects include the study of development, cancer, hearing and behaviour.

52.2.3 Pathogen genetics

The Institute carries out research in pathogen genetics to bolster understanding of the effects of genome variation on the biology of host-pathogen interactions. Research is underway at the Sanger Institute into the genomes of pathogens including many bacteria, viruses and parasites. The Pathogen Sequencing Advisory Group (PSAG) considers proposals or suggestions for sequencing of any genome of particular importance. All the genomes after sequencing are made available at the web-based onsite-maintained database, GeneDB.

52.2.4 Bioinformatics

The Sanger Institute's bioinformatics teams have developed IT systems for sequencing and postgenomic research. The Institute houses genome resources, RNA, protein and other family resources and functional annotation databases and resources. Researchers worldwide are able to use these resources to make inferences of genomic knowledge through computational analysis and integration of data.

52.2.5 Collaborations

Much of the Sanger Institute's research is carried out in partnership with the wider scientific community; over 90 percent of the Institute's research papers involve collaborations with other organisations.[31] Significant collaborations include:

- 1000 Genomes Project

- GENCODE and ENCODE (ENCyclopedia Of DNA Elements)

- International Cancer Genome Consortium

- International HapMap Project[32]

- International Knockout Mouse Consortium

- International Mouse Phenotyping Consortium

- SNP (Single nucleotide polymorphism) Consortium

- The Copy Number Variation Project[33]

- The genome sequencing of *S. pombe*, *C. elegans*, mouse and the Malaria parasite.

- The Human Genome Project

- The UK 10,000 Genomes Project (UK10K)[34]

- Wellcome Trust - Department of Biotechnology, India Alliance[35]

52.2.6 Public engagement

Children at a public outreach event thread bracelets in four colours to spell out a DNA sequence.

The Sanger Institute has a programme of public engagement activity. The programme aims to make complex biomedical research accessible to a range of audiences including school students and their teachers, and local community members.

The Communication and Public Engagement programme aims to "encourage informed discussion about issues relevant to Sanger Institute research"[36] and "foster a community of researchers who can engage effectively with different audiences".[36] The Institute hosts visits for more than 1,500 students, teachers and community groups per year. Visitors may meet scientific staff, tour the Institute and its facilities, and participate in ethical debates and activities. The programme also offers professional development sessions for teachers of GCSE and post 16 science through the national network of Science Learning Centres, and by hosting visits for groups interested in updating their knowledge in contemporary genetics. Videoconferencing into the Sanger Institute is also offered for Science Learning Centres, Science Centres and schools.

The programme maintains a dedicated public website, yourgenome.org,[37] that is intended to help people understand genetics and genomics science and its implications for society. The website includes teaching resources for secondary school science teachers that have been developed with Institute researchers.

Scientific and public engagement staff also collaborate on and contribute to national projects such as the UK's *InsidedeDNA*[38] traveling exhibition and the *Who am I?* gallery at The Science Museum.[39] They also participate in public events such as the Cambridge Science Festival.

52.2.7   Graduate training

The Institute operates two PhD training programmes: a four year course for basic science graduates, and a thee year course for clinicians. The four year course permits students to rotate around three different laboratories in order to broaden their scientific horizons before choosing a PhD project. Each student is required to choose at least one experimental one informatics-based rotation project.[40] Institute houses approximately 50 pre-doctoral students, all of whom are registered at the University of Cambridge.[41]

52.3   Staff

As of 2015 the Sanger employs around 900 people, and is led by Michael Stratton. Notable scientific staff, faculty and alumni are listed below:

52.3.1   Academic faculty

As of 2015 a faculty of 42 scientists lead hypothesis-driven research, seeking answers to biomedical questions.[42] These include:

- Allan Bradley FRS- Mouse genomics

- Peter Campbell - Cancer Genome Project

- Gordon Dougan[43] FRS - Microbial pathogenesis

- Richard Durbin[44] FRS - Genome informatics

- Ultan McDermott - Cancer Genome Project

- Julian Parkhill FRS - Pathogen genomics

- Julian Rayner - Malaria

- Michael Stratton - Cancer Genome Project

- Sarah Teichmann[45] - Gene expression genomics

- Eleftheria Zeggini[46] human genetics

52.3.2   Associate faculty and international fellows

Associate Faculty members work part-time at the Sanger usually jointly with another organisation.[42] As of 2015 there are 16 associate faculty including:

- Adrian Bird FRS - Epigenetic mechanisms in health and disease, Professor at the University of Edinburgh

- Ewan Birney FRS,[47] based at the European Bioinformatics Institute

- John Danesh, Professor in the Department of Public Health and Primary Care at the University of Cambridge

- Chris Ponting - Computational genome biology, based at the University of Oxford

- Fiona Powrie, FRS, University of Oxford[48]

- Stephen O'Rahilly, Professor at the University of Cambridge

- Wolf Reik - Epigenetic reprogramming, based at the Babraham Institute

- Toni Vidal-Puig, Professor of Molecular Nutrition and Metabolism at the Institute of Metabolic Science, University of Cambridge

52.3.3 Scientific Advisory Board

The Sanger Board of Management are guided by the scientific advisory board[49] whose members include Professor David Altshuler, Professor Anton Berns, Professor David J. Lipman, Professor Kevin Marsh and Professor Sir Paul Nurse.

52.3.4 Honorary faculty

The Sanger has an Honorary Faculty of researchers based at other research centres[50] This includes:

- Martin Bobrow

- Gad Frankel - Cell and molecular biology of bacterial infection, Professor at Imperial College London

- Andy Futreal - Cancer Genome Project, based at the University of Texas MD Anderson Cancer Center

- John Sulston, University of Manchester - Science ethics, former Director

52.3.5 Alumni

As of 2015 previous faculty members at the Sanger include:[51]

- Alex Bateman, EMBL-EBI[52]

- Tim Hubbard, Professor at King's College London

- Leena Peltonen-Palotie (1952 – 2010)[53]

52.4 History

52.4.1 Management

John E. Sulston was the founding Director of the Sanger Institute. Sulston was instrumental in the choice of the Hinxton site for the Institute and remained there as Director until the announcement of the completion of the draft human genome in 2000.[54] Sulston graduated from the University of Cambridge in 1963 and completed his PhD on the chemical synthesis of DNA in 1966.[55] He shared the 2002 Nobel Prize in Physiology or Medicine with Robert Horvitz and Sydney Brenner,[56] two years after standing down as Director of the Institute.

In 2000, Allan Bradley left his appointment as Professor at the Baylor College of Medicine, in the USA, to take up the position as Director of the Sanger Institute. Bradley wanted to build on the achievements made by the Sanger Institute in the Human Genome Project by "concentrating on gene function, cancer genomics, and the genomes of model organisms such as the mouse and the zebrafish".[57] Bradley received his BA, MA and PhD in Genetics from the University of Cambridge.[58][59][60][61]

In 2010, Bradley stepped down from his leadership role to form a startup company, but remains on the faculty of the Institute as Director Emeritus. Mike Stratton, who is a leader of the Cancer Genome Project and the International Cancer Genome Consortium, was appointed Director of the Sanger Institute in May of that year.[62]

Wellcome Trust Sanger Institute

The Wellcome Trust Sanger Institute was established in 1992, funded by the Wellcome Trust and the UK's Medical Research Council. One of the primary goals of the Institute on its creation, was to "play a role in mapping, sequencing and decoding the human genome and the genomes of other organisms".[5] The Sanger Institute now hosts several research programmes aiming to elucidate the associations between genes and biological traits - most often disease susceptibilities. The Sanger Institute has, since inception, maintained a policy that "aims to provide rapid access to data sets of use to the research community".[63]

52.4.2   Human Genome Project

The Sanger Institute was opened in 1993, three years after the inception of the Human Genome Project, and went on to make the largest single contribution to the gold standard sequence of the human genome, published in 2004.[64] The Institute was engaged in collaborations to sequence 8 of the 23 human pairs of chromosomes (1, 6, 9, 10, 13, 20, 22, and X).[65] Since the publishing of the human genome, research carried out at the Institute has developed beyond sequencing of organisms into various biomedical research areas, including studies into diseases such as cancer, malaria and diabetes.

52.5   References

[1] "MRC Centre United Kingdom: Wellcome Trust Sanger Institute". Medical Research Council. Retrieved 2008-12-22.

[2] Walker, John (2014). "Frederick Sanger (1918–2013) Double Nobel-prizewinning genomics pioneer". *Nature* **505** (7481): 27. doi:10.1038/505027a. PMID 24380948.

[3] Sanger, F. (1988). "Sequences, Sequences, and Sequences". *Annual Review of Biochemistry* **57**: 1–29. doi:10.1146/annurev.bi.57.070188.000 PMID 2460023.

[4] "Wellcome Trust Sanger Institute - About us". Wellcome Trust Sanger Institute. Retrieved 2010-06-28.

[5] "Wellcome Trust Sanger Institute - History". Wellcome Trust Sanger Institute. Retrieved 2010-06-28.

[6] "Wellcome Trust Sanger Institute - Work and study". Wellcome Trust Sanger Institute. Retrieved 2010-06-28.

[7] Wellcome Trust Conference Centre

[8] "Wellcome Trust Genome Campus Extension Opened: Visit by Her Royal Highness, The Princess Royal". Wellcome Trust Sanger Institute. Retrieved 2009-01-07.

[9] Doctorow, C. (2008). "Big data: Welcome to the petacentre". *Nature* **455** (7209): 16–21. doi:10.1038/455016a. PMID 18769411.

[10] Sulston, John. "Interview with John Sulston". Cold Spring Harbor Laboratories digital archive. Retrieved 13 November 2015.

[11] "Wellcome Trust Sanger Institute - Sequencing". Wellcome Trust Sanger Institute. Retrieved 2010-06-28.

[12] "Wellcome Trust Sanger Institute - Pathogen genomics". Wellcome Trust Sanger Institute. Retrieved 2010-06-28.

[13] Jones, I. "Feature: Highly cited". Wellcome Trust. Retrieved 2009-01-22.

[14] Forbes, S. A.; Beare, D; Gunasekaran, P; Leung, K; Bindal, N; Boutselakis, H; Ding, M; Bamford, S; Cole, C; Ward, S; Kok, C. Y.; Jia, M; De, T; Teague, J. W.; Stratton, M. R.; McDermott, U; Campbell, P. J. (2015). "COSMIC: Exploring the world's knowledge of somatic mutations in human cancer". *Nucleic Acids Research* **43** (Database issue): D805–11. doi:10.1093/nar/gku1075. PMC 4383913. PMID 25355519.

[15] Bragin, E; Chatzimichali, E. A.; Wright, C. F.; Hurles, M. E.; Firth, H. V.; Bevan, A. P.; Swaminathan, G. J. (2014). "DECIPHER: Database for the interpretation of phenotype-linked plausibly pathogenic sequence and copy-number variation". *Nucleic Acids Research* **42** (Database issue): D993–D1000. doi:10.1093/nar/gkt937. PMC 3965078. PMID 24150940.

[16] Cunningham, F; Amode, M. R.; Barrell, D; Beal, K; Billis, K; Brent, S; Carvalho-Silva, D; Clapham, P; Coates, G; Fitzgerald, S; Gil, L; Girón, C. G.; Gordon, L; Hourlier, T; Hunt, S. E.; Janacek, S. H.; Johnson, N; Juettemann, T; Kähäri, A. K.; Keenan, S; Martin, F. J.; Maurel, T; McLaren, W; Murphy, D. N.; Nag, R; Overduin, B; Parker, A; Patricio, M; Perry, E; et al. (2015). "Ensembl 2015". *Nucleic Acids Research* **43** (Database issue): D662–9. doi:10.1093/nar/gku1010. PMID 25352552.

[17] Logan-Klumpler, F. J.; De Silva, N; Boehme, U; Rogers, M. B.; Velarde, G; McQuillan, J. A.; Carver, T; Aslett, M; Olsen, C; Subramanian, S; Phan, I; Farris, C; Mitra, S; Ramasamy, G; Wang, H; Tivey, A; Jackson, A; Houston, R; Parkhill, J; Holden, M; Harb, O. S.; Brunk, B. P.; Myler, P. J.; Roos, D; Carrington, M; Smith, D. F.; Hertz-Fowler, C; Berriman, M (2012). "GeneDB-- an annotation database for pathogens". *Nucleic Acids Research* **40** (Database issue): D98–108. doi:10.1093/nar/gkr1032. PMC 3245030. PMID 22116062.

[18] Rawlings, N. D.; Waller, M; Barrett, A. J.; Bateman, A (2014). "MEROPS: The database of proteolytic enzymes, their substrates and inhibitors". *Nucleic Acids Research* **42** (Database issue): D503–9. doi:10.1093/nar/gkt953. PMC 3964991. PMID 24157837.

[19] Finn, R. D.; Bateman, A; Clements, J; Coggill, P; Eberhardt, R. Y.; Eddy, S. R.; Heger, A; Hetherington, K; Holm, L; Mistry, J; Sonnhammer, E. L.; Tate, J; Punta, M (2014). "Pfam: The protein families database". *Nucleic Acids Research* **42** (Database issue): D222–30. doi:10.1093/nar/gkt1223. PMC 3965110. PMID 24288371.

[20] Nawrocki, E. P.; Burge, S. W.; Bateman, A; Daub, J; Eberhardt, R. Y.; Eddy, S. R.; Floden, E. W.; Gardner, P. P.; Jones, T. A.; Tate, J; Finn, R. D. (2015). "Rfam 12.0: Updates to the RNA families database". *Nucleic Acids Research* **43** (Database issue): D130–7. doi:10.1093/nar/gku1063. PMC 4383904. PMID 25392425.

[21] Schreiber, F; Patricio, M; Muffato, M; Pignatelli, M; Bateman, A (2014). "Tree *Fam* v9: A new website, more species and orthology-on-the-fly". *Nucleic Acids Research* **42** (Database issue): D922–5. doi:10.1093/nar/gkt1055. PMC 3965059. PMID 24194607.

[22] Wilming, L. G.; Gilbert, J. G.; Howe, K; Trevanion, S; Hubbard, T; Harrow, J. L. (2008). "The vertebrate genome annotation (Vega) database". *Nucleic Acids Research* **36** (Database issue): D753–60. doi:10.1093/nar/gkm987. PMC 2238886. PMID 18003653.

[23] Harris, T. W.; Baran, J; Bieri, T; Cabunoc, A; Chan, J; Chen, W. J.; Davis, P; Done, J; Grove, C; Howe, K; Kishore, R; Lee, R; Li, Y; Muller, H. M.; Nakamura, C; Ozersky, P; Paulini, M; Raciti, D; Schindelman, G; Tuli, M. A.; Van Auken, K; Wang, D; Wang, X; Williams, G; Wong, J. D.; Yook, K; Schedl, T; Hodgkin, J; Berriman, M; et al. (2014). "Worm *Base* 2014: New views of curated biology". *Nucleic Acids Research* **42** (Database issue): D789–93. doi:10.1093/nar/gkt1063. PMC 3965043. PMID 24194605.

[24] "1000 Genomes: A Deep Catalog of Human Genetic Variation". 1000 Genomes Project. Retrieved 2008-12-22.

[25] "1000 Genomes Project releases data from pilot projects on path to providing database for 2,500 human genomes". Wellcome Trust Sanger Institute. Retrieved 2010-06-28.

[26] Kaye, J; Hurles, M; Griffin, H; Grewal, J; Bobrow, M; Timpson, N; Smee, C; Bolton, P; Durbin, R; Dyke, S; Fitzpatrick, D; Kennedy, K; Kent, A; Muddyman, D; Muntoni, F; Raymond, L. F.; Semple, R; Spector, T; Uk, 10K (2014). "Managing clinically significant findings in research: The UK10K example". *European Journal of Human Genetics* **22** (9): 1100–4. doi:10.1038/ejhg.2013.290. PMC 4026295. PMID 24424120.

[27] Muddyman, D; Smee, C; Griffin, H; Kaye, J (2013). "Implementing a successful data-management framework: The UK10K managed access model". *Genome Medicine* **5** (11): 100. doi:10.1186/gm504. PMC 3978569. PMID 24229443.

[28] "Wellcome Trust launches study of 10,000 human genomes in UK". Wellcome Trust Sanger Institute. Retrieved 2010-06-28.

[29] "International Cancer Genome Consortium Homepage". International Cancer Genome Consortium. Retrieved 2009-01-23.

[30] "Wellcome Trust Sanger Institute - ENCODE and GENCODE". Retrieved 2010-06-28.

[31] Figure based on data for 2008 retrieved from SCOPUS website

[32] -, -; Gibbs, R. A.; Belmont, J. W.; Hardenbol, P.; Willis, T. D.; Yu, F.; Yang, H.; Ch'Ang, L. Y.; Huang, W.; Liu, B.; Shen, Y.; Tam, P. K. H.; Tsui, L. C.; Waye, M. M. Y.; Wong, J. T. F.; Zeng, C.; Zhang, Q.; Chee, M. S.; Galver, L. M.; Kruglyak, S.; Murray, S. S.; Oliphant, A. R.; Montpetit, A.; Hudson, T. J.; Chagnon, F.; Ferretti, V.; Leboeuf, M.; Phillips, M. S.; Verner, A.; Kwok, P. Y. (2003). "The International HapMap Project". *Nature* **426** (6968): 789–796. doi:10.1038/nature02168. PMID 14685227.

[33] Copy Number Variation Project

[34] UK10K (the UK 10,000 Genomes Project)

[35] http://www.wellcomedbt.org/Mini_symposium_final_poster_schedule.pdf

[36] "Team 104: Communication and Public Engagement Programme". Wellcome Trust Sanger Institute. Retrieved 2009-01-12.

[37] yourgenome.org

[38] InsideDNA

[39] Who am I? gallery

[40] "Wellcome Trust Sanger Institute PhD Programmes". Wellcome Trust Sanger Institute. Retrieved 2009-11-13.

[41] "Sanger PhD and MPhil theses". Archived from the original on 2015-04-07.

[42] "Sanger Faculty". Archived from the original on 2015-03-16.

[43] *DOUGAN, Prof. Gordon*. Who's Who **2015** (online Oxford University Press ed.). A & C Black, an imprint of Bloomsbury Publishing plc. (subscription required)

[44] *DURBIN, Richard Michael*. Who's Who **2015** (online edition via Oxford University Press ed.). A & C Black, an imprint of Bloomsbury Publishing plc. (subscription required)

[45] Sarah Teichmann's publications indexed by the Scopus bibliographic database, a service provided by Elsevier.

[46] "Professor Eleftheria Zeggini, Group Leader". Hinxton: Wellcome Trust Sanger Institute. Archived from the original on 2015-11-11.

[47] Pennisi, Elizabeth (2012). "Profile of Ewan Birney: Genomics' Big Talker". *Science* **337** (6099): 1167–1169. doi:10.1126/science.337.6099.1 PMID 22955814.

[48] *POWRIE, Prof. Fiona Margaret*. Who's Who **2015** (online Oxford University Press ed.). A & C Black, an imprint of Bloomsbury Publishing plc. (subscription required)

[49] "Sanger Advisory Board". Wellcome Trust Sanger Institute. Archived from the original on 2015-04-07.

[50] "Honorary Faculty". Welcome Trust Sanger Institute. Archived from the original on 2015-04-07.

[51] "Previous faculty at the Sanger". Archived from the original on 2015-04-07.

[52] Logan, D. W.; Sandal, M.; Gardner, P. P.; Manske, M.; Bateman, A. (2010). "Ten Simple Rules for Editing Wikipedia". *PLOS Computational Biology* **6** (9): e1000941. doi:10.1371/journal.pcbi.1000941. PMC 2947980. PMID 20941386.

[53] van Ommen, Gertjan (2010). "Obituary: Leena Peltonen-Palotie (1952–2010) A visionary in medical genetics.". *Nature* **464** (7291): 992–992. doi:10.1038/464992a. PMID 20393553.

[54] "International Human Genome Sequencing Consortium Announces "Working Draft" of Human Genome". National Human Genome Research Institute. Retrieved 2008-12-09.

[55] Sulston J, Ferry G (2002). *The Common Thread: A story of Science, Politics, Ethics, and the Human Genome*. The Joseph Henry Press. p. 18.

[56] "The Nobel Prize in Physiology or Medicine 2002". Nobel Prize. Retrieved 2009-01-07.

[57] "Sanger Institute looks to the future". Genome Biology. Retrieved 2008-12-09.

[58] "Professor Allan Bradley". Wellcome Trust Sanger Institute. Archived from the original on 2015-04-07.

[59] Gura, T (2000). "Changing of the guard: Allan Bradley is to head the Sanger Centre, a powerhouse of the Human Genome Project". *Nature* **405** (6785): 389. doi:10.1038/35013238. PMID 10839511.

[60] Adam, D (2001). "Sanger Centre welcomes gene funds with a new name". *Nature* **413** (6857): 660. doi:10.1038/35099707. PMID 11606985.

[61] Dickson, D (2000). "Geneticist from Baylor named as new head of UK's Sanger Centre". *Nature* **405** (6784): 264. doi:10.1038/35012766. PMID 10830929.

[62] "Professor Mike Stratton appointed new Director". Wellcome Trust Sanger Institute. Retrieved 2010-05-17.

[63] "Wellcome Trust Sanger Institute - Data conditions". Wellcome Trust Sanger Institute. Retrieved 2010-06-28.

[64] Human Genome Sequencing Consortium (2004). "Finishing the euchromatic sequence of the human genome". *Nature* **431** (7011): 931–945. doi:10.1038/nature03001. PMID 15496913.

[65] Pennisi E (2003). "Reaching Their Goal Early, Sequencing Labs Celebrate". *Science* **300** (5618): 409. doi:10.1126/science.300.5618.409. PMID 12702850.

Coordinates: 52°05′N 0°11′E / 52.083°N 0.183°E

Chapter 53

International HapMap Project

The **International HapMap Project** was an organization that aimed to develop a **hap**lotype **map** (**HapMap**) of the human genome, to describe the common patterns of human genetic variation. HapMap is used to find genetic variants affecting health, disease and responses to drugs and environmental factors. The information produced by the project is made freely available for research.

The International HapMap Project is a collaboration among researchers at academic centers, non-profit biomedical research groups and private companies in Canada, China, Japan, Nigeria, the United Kingdom, and the United States. It officially started with a meeting on October 27 to 29, 2002, and was expected to take about three years. It comprises two phases; the complete data obtained in Phase I were published on 27 October 2005. The analysis of the Phase II dataset was published in October 2007. The Phase III dataset was released in spring 2009.

53.1 Background

Unlike with the rarer Mendelian diseases, combinations of different genes and the environment play a role in the development and progression of common diseases (such as diabetes, cancer, heart disease, stroke, depression and asthma), or in the individual response to pharmacological agents. To find the genetic factors involved in these diseases, one could in principle obtain the complete genetic sequence of several individuals, some with the disease and some without, and then search for differences between the two sets of genomes. At the time, this approach was not feasible because of the cost of full genome sequencing. The HapMap project proposed a shortcut.

Although any two unrelated people share about 99.5% of their DNA sequence, their genomes differ at specific nucleotide locations. Such sites are known as single nucleotide polymorphisms (SNPs), and each of the possible resulting gene forms is called an allele. The HapMap project focuses only on common SNPs, those where each allele occurs in at least 1% of the population.

Each person has two copies of all chromosomes, except the sex chromosomes in males. For each SNP, the combination of alleles a person has is called a genotype. Genotyping refers to uncovering what genotype a person has at a particular site. The HapMap project chose a sample of 269 individuals and selected several million well-defined SNPs, genotyped the individuals for these SNPs, and published the results.

The alleles of nearby SNPs on a single chromosome are correlated. Specifically, if the allele of one SNP for a given individual is known, the alleles of nearby SNPs can often be predicted. This is because each SNP arose in evolutionary history as a single point mutation, and was then passed down on the chromosome surrounded by other, earlier, point mutations. SNPs that are separated by a large distance on the chromosome are typically not very well correlated, because recombination occurs in each generation and mixes the allele sequences of the two chromosomes. A sequence of consecutive alleles on a particular chromosome is known as a haplotype.

To find the genetic factors involved in a particular disease, one can proceed as follows. First a certain region of interest in the genome is identified, possibly from earlier inheritance studies. In this region one locates a set of tag SNPs from the

HapMap data; these are SNPs that are very well correlated with all the other SNPs in the region. Thus, learning the alleles of the tag SNPs in an individual will determine the individual's haplotype with high probability. Next, one determines the genotype for these tag SNPs in several individuals, some with the disease and some without. By comparing the two groups, one determines the likely locations and haplotypes that are involved in the disease.

53.2   Samples used

Haplotypes are generally shared between populations, but their frequency can differ widely. Four populations were selected for inclusion in the HapMap: 30 adult-and-both-parents Yoruba trios from Ibadan, Nigeria (YRI), 30 trios of Utah residents of northern and western European ancestry (CEU), 44 unrelated Japanese individuals from Tokyo, Japan (JPT) and 45 unrelated Han Chinese individuals from Beijing, China (CHB). Although the haplotypes revealed from these populations should be useful for studying many other populations, parallel studies are currently examining the usefulness of including additional populations in the project.

All samples were collected through a community engagement process with appropriate informed consent. The community engagement process was designed to identify and attempt to respond to culturally specific concerns and give participating communities input into the informed consent and sample collection processes.

In phase III, 11 global ancestry groups have been assembled: ASW (African ancestry in Southwest USA); CEU (Utah residents with Northern and Western European ancestry from the CEPH collection); CHD (Chinese in Metropolitan Denver, Colorado); GIH (Gujarati Indians in Houston, Texas); LWK (Luhya in Webuye, Kenya); MEX (Mexican ancestry in Los Angeles, California); MKK (Maasai in Kinyawa, Kenya); TSI (Tuscans in Italy); YRI (Yoruba in Ibadan, Nigeria). [1]

Three combined panels have also been created, which allow better identification of SNPs in groups outside the nine homogenous samples: CEU+TSI (Combined panel of Utah residents with Northern and Western European ancestry from the CEPH collection and Tuscans in Italy); JPT+CHB (Combined panel of Japanese in Tokyo, Japan and Han Chinese in Beijing, China) and JPT+CHB+CHD (Combined panel of Japanese in Tokyo, Japan, Han Chinese in Beijing, China and Chinese in Metropolitan Denver, Colorado). CEU+TSI, for instance, is a better model of UK British individuals than is CEU alone. [1]

53.3   Scientific strategy

For the Phase I, one common SNP was genotyped every 5,000 bases. Overall, more than one million SNPs were genotyped. The genotyping was carried out by 10 centres using five different genotyping technologies. Genotyping quality was assessed by using duplicate or related samples and by having periodic quality checks where centres had to genotype common sets of SNPs.

The Canadian team was led by Thomas J. Hudson at McGill University in Montreal and focused on chromosomes 2 and 4p. The Chinese team was led by Huanming Yang with centres in Beijing, Shanghai and Hong Kong and focused on chromosomes 3, 8p and 21. The Japanese team was led by Yusuke Nakamura at the University of Tokyo and focused on chromosomes 5, 11, 14, 15, 16, 17 and 19. The British team was led by David R. Bentley at the Sanger Institute and focused on chromosomes 1, 6, 10, 13 and 20. There were four United States' genotyping centres: a team led by Mark Chee and Arnold Oliphant at Illumina Inc. in San Diego (studying chromosomes 8q, 9, 18q, 22 and X), a team led by David Altshuler at the Broad Institute in Cambridge, USA (chromosomes 4q, 7q, 18p, Y and mitochondrion), a team led by Richard A. Gibbs at the Baylor College of Medicine in Houston (chromosome 12), and a team led by Pui-Yan Kwok at the University of California, San Francisco (chromosome 7p).

To obtain enough SNPs to create the Map, the Consortium had to fund a large re-sequencing project to discover millions of additional SNPs. These were submitted to the public dbSNP database. As a result, by August 2006, the database included more than ten million SNPs, and more than 40% of them were known to be polymorphic. By comparison, at the start of the project, fewer than 3 million SNPs were identified, and no more than 10% of them were known to be polymorphic.

During Phase II, more than two million additional SNPs have been genotyped throughout the genome by the company Perlegen Sciences and 500,000 by the company Affymetrix.

53.4   Data access

All of the data generated by the project, including SNP frequencies, genotypes and haplotypes, were placed in the public domain and are available for download. This website also contains a genome browser which allows to find SNPs in any region of interest, their allele frequencies and their association to nearby SNPs. A tool that can determine tag SNPs for a given region of interest is also provided. These data can also be directly accessed from the widely used Haploview program.

53.5   Criticisms

It has been argued that the HapMap project has broadly misrepresented itself as a tool for uncovering causal agents of common diseases in a bid to maintain its funding. Increasing evidence suggests that HapMap data is far more useful for studies of population structure than it is for its alleged purpose of controlling for population structure in genome wide association studies.[2]

53.6   Publications

- International HapMap Consortium. (2003) The International HapMap Project. *Nature* **426**(6968):789-96.

- International HapMap Consortium. (2004) Integrating ethics and science in the International HapMap Project. *Nat Rev Genet.* **5**(6):467-75.

- International HapMap Consortium. (2005) A haplotype map of the human genome. *Nature* **437**(7063):1299-320.

- International HapMap Consortium. (2007) A second generation human haplotype map of over 3.1 million SNPs. *Nature* **449**(7164):851-861.

- International HapMap Consortium. (2010) Integrating common and rare genetic variation in diverse human populations. *Nature* **467**(7311):52-58.

- Deloukas P, Bentley D. (2004) The HapMap project and its application to genetic studies of drug response. *Pharmacogenomics J.* **4**(2):88-90.

- Secko, David Phase I of the HapMap Complete The Scientist (October, 2005)

- Thorisson GA, Smith AV, Krishnan L, Stein LD. (2005) The International HapMap Project Web site. *Genome Res.* **15**(11):1592-3.

- Terwilliger JD and Hiekkalinna T (2006). An utter refutation of the 'Fundamental Theorem of the HapMap' *European Journal of Human Genetics 14, 426–437*

53.7   See also

- Genealogical DNA test

- The 1000 Genomes Project

- Population groups in biomedicine

- Human Variome Project

- Human genetic variation

53.8 References

[1] International HapMap consortium et al. (2010). Integrating common and rare genetic variation in diverse human populations. *Nature*, **467**, 52-8. doi

[2] Terwilliger JD and Hiekkalinna T (2006). An utter refutation of the 'Fundamental Theorem of the HapMap' *European Journal of Human Genetics 14, 426–437*.

53.9 External links

- International HapMap Project (HapMap Homepage)

- National Human Genome Research Institute (NHGRI) HapMap Page

- Browsing HapMap Data Using the Genome Browser

- What is the HapMap Project? - An Introduction to HapMap

- The Mexican Genome Diversity Project

Chapter 54

Genetics

This article is about the general scientific term. For the scientific journal, see Genetics (journal).
For a more accessible and less technical introduction to this topic, see Introduction to genetics.

Genetics is the study of genes, heredity, and genetic variation in living organisms.[1][2] It is generally considered a field of biology, but it intersects frequently with many of the life sciences and is strongly linked with the study of information systems.

The father of genetics is Gregor Mendel, a late 19th-century scientist and Augustinian friar. Mendel studied 'trait inheritance', patterns in the way traits were handed down from parents to offspring. He observed that organisms (pea plants) inherit traits by way of discrete "units of inheritance". This term, still used today, is a somewhat ambiguous definition of what is referred to as a gene.

Trait inheritance and molecular inheritance mechanisms of genes are still a primary principle of genetics in the 21st century, but modern genetics has expanded beyond inheritance to studying the function and behavior of genes. Gene structure and function, variation, and distribution are studied within the context of the cell, the organism (e.g. dominance) and within the context of a population. Genetics has given rise to a number of sub-fields including epigenetics and population genetics. Organisms studied within the broad field span the domain of life, including bacteria, plants, animals, and humans.

Genetic processes work in combination with an organism's environment and experiences to influence development and behavior, often referred to as nature versus nurture. The intra- or extra-cellular environment of a cell or organism may switch gene transcription on or off. A classic example is two seeds of genetically identical corn, one placed in a temperate climate and one in an arid climate. While the average height of the two corn stalks may be genetically determined to be equal, the one in the arid climate only grows to half the height of the one in the temperate climate, due to lack of water and nutrients in its environment.

54.1 Etymology

The word genetics stems from the Ancient Greek γενετικός *genetikos* meaning "genitive"/"generative", which in turn derives from γένεσις *genesis* meaning "origin".[3][4][5]

54.2 The gene

The modern working definition of a gene is a portion (or sequence) of DNA that codes for a known cellular function or process (e.g. the function "make melanin molecules"). A single 'gene' is most similar to a single 'word' in the English language. The nucleotides (molecules) that make up genes can be seen as 'letters' in the English language. Nucleotides are named according to which of the four nitrogenous bases they contain. The four bases are cytosine, guanine, adenine, and thymine. A single gene may have a small number of nucleotides or a large number of nucleotides, in the same way that a word may be small or large (e.g. 'cell' vs. 'electrophysiology'). A single gene often interacts with neighboring genes to produce a cellular function and can even be ineffectual without those neighboring genes. This can be seen in the same way that a 'word' may have meaning only in the context of a 'sentence.' A series of nucleotides can be put together without forming a gene (non coding regions of DNA), like a string of letters can be put together without forming a word (e.g. udkslk). Nonetheless, all words have letters, like all genes must have nucleotides.

A quick heuristic that is often used (but not always true) is "one gene, one protein" meaning a singular gene codes for a singular protein type in a cell (enzyme, transcription factor, etc.)

The sequence of nucleotides in a gene is read and translated by a cell to produce a chain of amino acids which in turn folds into a protein. The order of amino acids in a protein corresponds to the order of nucleotides in the gene. This relationship between nucleotide sequence and amino acid sequence is known as the genetic code. The amino acids in a protein determine how it folds into its unique three-dimensional shape, a structure that is ultimately responsible for the protein's function. Proteins carry out many of the functions needed for cells to live. A change to the DNA in a gene can alter a protein's amino acid sequence, thereby changing its shape and function and rendering the protein ineffective or even malignant (e.g. sickle cell anemia). Changes to genes are called mutations.

54.3 History

Main article: History of genetics

The observation that living things inherit traits from their parents has been used since prehistoric times to improve crop plants and animals through selective breeding.[6] The modern science of genetics, seeking to understand this process, began with the work of Gregor Mendel in the mid-19th century.[7]

Although the science of genetics began with the applied and theoretical work of Mendel, other theories of inheritance preceded his work. A popular theory during Mendel's time was the concept of blending inheritance: the idea that individuals inherit a smooth blend of traits from their parents.[8] Mendel's work provided examples where traits were definitely not blended after hybridization, showing that traits are produced by combinations of distinct genes rather than a continuous blend. Blending of traits in the progeny is now explained by the action of multiple genes with quantitative effects. Another theory that had some support at that time was the inheritance of acquired characteristics: the belief that individuals inherit traits strengthened by their parents. This theory (commonly associated with Jean-Baptiste Lamarck) is now known to be wrong—the experiences of individuals do not affect the genes they pass to their children,[9] although evidence in the field of epigenetics has revived some aspects of Lamarck's theory.[10] Other theories included the pangenesis of Charles Darwin (which had both acquired and inherited aspects) and Francis Galton's reformulation of pangenesis as both particulate and inherited.[11]

54.3.1 Mendelian and classical genetics

Modern genetics started with Gregor Johann Mendel, a scientist and Augustinian friar who studied the nature of inheritance in plants. In his paper "*Versuche über Pflanzenhybriden*" ("Experiments on Plant Hybridization"), presented in 1865 to the *Naturforschender Verein* (Society for Research in Nature) in Brünn, Mendel traced the inheritance patterns of certain traits in pea plants and described them mathematically.[12] Although this pattern of inheritance could only be observed for a few traits, Mendel's work suggested that heredity was particulate, not acquired, and that the inheritance patterns of many traits could be explained through simple rules and ratios.

The importance of Mendel's work did not gain wide understanding until the 1890s, after his death, when other scientists working on similar problems re-discovered his research. William Bateson, a proponent of Mendel's work, coined the word *genetics* in 1905.[13][14] (The adjective *genetic*, derived from the Greek word *genesis*—γένεσις, "origin", predates the noun and was first used in a biological sense in 1860.)[15] Bateson both acted as a mentor and was aided significantly by the work of female scientists from Newnham College at Cambridge, specifically the work of Becky Saunders, Nora Darwin Barlow, and Muriel Wheldale Onslow.[16] Bateson popularized the usage of the word *genetics* to describe the study of inheritance in his inaugural address to the Third International Conference on Plant Hybridization in London, England, in 1906.[17]

After the rediscovery of Mendel's work, scientists tried to determine which molecules in the cell were responsible for inheritance. In 1911, Thomas Hunt Morgan argued that genes are on chromosomes, based on observations of a sex-linked white eye mutation in fruit flies.[18] In 1913, his student Alfred Sturtevant used the phenomenon of genetic linkage to show that genes are arranged linearly on the chromosome.[19]

54.3.2 Molecular genetics

Although genes were known to exist on chromosomes, chromosomes are composed of both protein and DNA, and scientists did not know which of the two is responsible for inheritance. In 1928, Frederick Griffith discovered the phenomenon of transformation (see Griffith's experiment): dead bacteria could transfer genetic material to "transform" other still-living bacteria. Sixteen years later, in 1944, the Avery–MacLeod–McCarty experiment identified DNA as the molecule responsible for transformation.[20] The role of the nucleus as the repository of genetic information in eukaryotes had been established by Hämmerling in 1943 in his work on the single celled alga *Acetabularia*.[21] The Hershey–Chase experiment in 1952 confirmed that DNA (rather than protein) is the genetic material of the viruses that infect bacteria, providing further evidence that DNA is the molecule responsible for inheritance.[22]

James Watson and Francis Crick determined the structure of DNA in 1953, using the X-ray crystallography work of Rosalind Franklin and Maurice Wilkins that indicated DNA had a helical structure (i.e., shaped like a corkscrew).[23][24] Their double-helix model had two strands of DNA with the nucleotides pointing inward, each matching a complementary nucleotide on the other strand to form what looks like rungs on a twisted ladder.[25] This structure showed that genetic information exists in the sequence of nucleotides on each strand of DNA. The structure also suggested a simple method for replication: if the strands are separated, new partner strands can be reconstructed for each based on the sequence of the old strand. This property is what gives DNA its semi-conservative nature where one strand of new DNA is from an original parent strand.[26]

Although the structure of DNA showed how inheritance works, it was still not known how DNA influences the behavior of cells. In the following years, scientists tried to understand how DNA controls the process of protein production.[27] It was discovered that the cell uses DNA as a template to create matching messenger RNA, molecules with nucleotides very similar to DNA. The nucleotide sequence of a messenger RNA is used to create an amino acid sequence in protein; this translation between nucleotide sequences and amino acid sequences is known as the genetic code.[28]

With the newfound molecular understanding of inheritance came an explosion of research.[29] A notable theory arose from Tomoko Ohta in 1973 with her amendment to the neutral theory of molecular evolution through publishing the nearly neutral theory of molecular evolution. In this theory, Ohta stressed the importance of natural selection and the environment to the rate in which genetic evolution occurs.[30] One important development was chain-termination DNA sequencing in 1977 by Frederick Sanger. This technology allows scientists to read the nucleotide sequence of a DNA molecule.[31] In 1983, Kary Banks Mullis developed the polymerase chain reaction, providing a quick way to isolate and amplify a specific section of DNA from a mixture.[32] The efforts of the Human Genome Project, Department of Energy, NIH, and parallel private effort by Celera Genomics led to the sequencing of the human genome in 2003.[33]

54.4   Features of inheritance

54.4.1   Discrete inheritance and Mendel's laws

Main article: Mendelian inheritance

At its most fundamental level, inheritance in organisms occurs by passing discrete heritable units, called genes, from parents to progeny.[34] This property was first observed by Gregor Mendel, who studied the segregation of heritable traits in pea plants.[12][35] In his experiments studying the trait for flower color, Mendel observed that the flowers of each pea plant were either purple or white—but never an intermediate between the two colors. These different, discrete versions of the same gene are called alleles.

In the case of pea, which is a diploid species, each individual plant has two copies of each gene, one copy inherited from each parent.[36] Many species, including humans, have this pattern of inheritance. Diploid organisms with two copies of the same allele of a given gene are called homozygous at that gene locus, while organisms with two different alleles of a given gene are called heterozygous.

The set of alleles for a given organism is called its genotype, while the observable traits of the organism are called its phenotype. When organisms are heterozygous at a gene, often one allele is called dominant as its qualities dominate the phenotype of the organism, while the other allele is called recessive as its qualities recede and are not observed. Some alleles do not have complete dominance and instead have incomplete dominance by expressing an intermediate phenotype, or codominance by expressing both alleles at once.[37]

When a pair of organisms reproduce sexually, their offspring randomly inherit one of the two alleles from each parent. These observations of discrete inheritance and the segregation of alleles are collectively known as Mendel's first law or the Law of Segregation.

54.4.2 Notation and diagrams

Geneticists use diagrams and symbols to describe inheritance. A gene is represented by one or a few letters. Often a "+" symbol is used to mark the usual, non-mutant allele for a gene.[38]

In fertilization and breeding experiments (and especially when discussing Mendel's laws) the parents are referred to as the "P" generation and the offspring as the "F1" (first filial) generation. When the F1 offspring mate with each other, the offspring are called the "F2" (second filial) generation. One of the common diagrams used to predict the result of cross-breeding is the Punnett square.

When studying human genetic diseases, geneticists often use pedigree charts to represent the inheritance of traits.[39] These charts map the inheritance of a trait in a family tree.

54.4.3 Multiple gene interactions

Organisms have thousands of genes, and in sexually reproducing organisms these genes generally assort independently of each other. This means that the inheritance of an allele for yellow or green pea color is unrelated to the inheritance of alleles for white or purple flowers. This phenomenon, known as "Mendel's second law" or the "Law of independent assortment", means that the alleles of different genes get shuffled between parents to form offspring with many different combinations. (Some genes do not assort independently, demonstrating genetic linkage, a topic discussed later in this article.)

Often different genes can interact in a way that influences the same trait. In the Blue-eyed Mary (*Omphalodes verna*), for example, there exists a gene with alleles that determine the color of flowers: blue or magenta. Another gene, however, controls whether the flowers have color at all or are white. When a plant has two copies of this white allele, its flowers are white—regardless of whether the first gene has blue or magenta alleles. This interaction between genes is called epistasis, with the second gene epistatic to the first.[40]

Many traits are not discrete features (e.g. purple or white flowers) but are instead continuous features (e.g. human height and skin color). These complex traits are products of many genes.[41] The influence of these genes is mediated, to varying degrees, by the environment an organism has experienced. The degree to which an organism's genes contribute to a complex trait is called heritability.[42] Measurement of the heritability of a trait is relative—in a more variable environment, the environment has a bigger influence on the total variation of the trait. For example, human height is a trait with complex causes. It has a heritability of 89% in the United States. In Nigeria, however, where people experience a more variable access to good nutrition and health care, height has a heritability of only 62%.[43]

54.5 Molecular basis for inheritance

54.5.1 DNA and chromosomes

Main articles: DNA and Chromosome

The molecular basis for genes is deoxyribonucleic acid (DNA). DNA is composed of a chain of nucleotides, of which there are four types: adenine (A), cytosine (C), guanine (G), and thymine (T). Genetic information exists in the sequence of these nucleotides, and genes exist as stretches of sequence along the DNA chain.[44] Viruses are the only exception to this rule—sometimes viruses use the very similar molecule RNA instead of DNA as their genetic material.[45] Viruses cannot reproduce without a host and are unaffected by many genetic processes, so tend not to be considered living organisms.

DNA normally exists as a double-stranded molecule, coiled into the shape of a double helix. Each nucleotide in DNA preferentially pairs with its partner nucleotide on the opposite strand: A pairs with T, and C pairs with G. Thus, in its two-stranded form, each strand effectively contains all necessary information, redundant with its partner strand. This structure of DNA is the physical basis for inheritance: DNA replication duplicates the genetic information by splitting the strands and using each strand as a template for synthesis of a new partner strand.[46]

Genes are arranged linearly along long chains of DNA base-pair sequences. In bacteria, each cell usually contains a single circular genophore, while eukaryotic organisms (such as plants and animals) have their DNA arranged in multiple linear chromosomes. These DNA strands are often extremely long; the largest human chromosome, for example, is about 247 million base pairs in length.[47] The DNA of a chromosome is associated with structural proteins that organize, compact and control access to the DNA, forming a material called chromatin; in eukaryotes, chromatin is usually composed of nucleosomes, segments of DNA wound around cores of histone proteins.[48] The full set of hereditary material in an organism (usually the combined DNA sequences of all chromosomes) is called the genome.

While haploid organisms have only one copy of each chromosome, most animals and many plants are diploid, containing two of each chromosome and thus two copies of every gene.[36] The two alleles for a gene are located on identical loci of the two homologous chromosomes, each allele inherited from a different parent.

Many species have so-called sex chromosomes that determine the gender of each organism.[49] In humans and many other animals, the Y chromosome contains the gene that triggers the development of the specifically male characteristics. In evolution, this chromosome has lost most of its content and also most of its genes, while the X chromosome is similar to the other chromosomes and contains many genes. The X and Y chromosomes form a strongly heterogeneous pair.

54.5.2 Reproduction

Main articles: Asexual reproduction and Sexual reproduction

When cells divide, their full genome is copied and each daughter cell inherits one copy. This process, called mitosis, is the simplest form of reproduction and is the basis for asexual reproduction. Asexual reproduction can also occur in multicellular organisms, producing offspring that inherit their genome from a single parent. Offspring that are genetically identical to their parents are called clones.

Eukaryotic organisms often use sexual reproduction to generate offspring that contain a mixture of genetic material inherited from two different parents. The process of sexual reproduction alternates between forms that contain single copies of the genome (haploid) and double copies (diploid).[36] Haploid cells fuse and combine genetic material to create a diploid cell with paired chromosomes. Diploid organisms form haploids by dividing, without replicating their DNA, to create daughter cells that randomly inherit one of each pair of chromosomes. Most animals and many plants are diploid for most of their lifespan, with the haploid form reduced to single cell gametes such as sperm or eggs.

Although they do not use the haploid/diploid method of sexual reproduction, bacteria have many methods of acquiring new genetic information. Some bacteria can undergo conjugation, transferring a small circular piece of DNA to another bacterium.[50] Bacteria can also take up raw DNA fragments found in the environment and integrate them into their genomes, a phenomenon known as transformation.[51] These processes result in horizontal gene transfer, transmitting fragments of genetic information between organisms that would be otherwise unrelated.

54.5.3 Recombination and genetic linkage

Main articles: Chromosomal crossover and Genetic linkage
 The diploid nature of chromosomes allows for genes on different chromosomes to assort independently or be separated from their homologous pair during sexual reproduction wherein haploid gametes are formed. In this way new combinations of genes can occur in the offspring of a mating pair. Genes on the same chromosome would theoretically never recombine. However, they do via the cellular process of chromosomal crossover. During crossover, chromosomes exchange stretches of DNA, effectively shuffling the gene alleles between the chromosomes.[52] This process of chromosomal crossover generally occurs during meiosis, a series of cell divisions that creates haploid cells.

The first cytological demonstration of crossing over was performed by Harriet Creighton and Barbara McClintock in 1931. Their research and experiments on corn provided cytological evidence for the genetic theory that linked genes on paired chromosomes do in fact exchange places from one homolog to the other.

The probability of chromosomal crossover occurring between two given points on the chromosome is related to the distance between the points. For an arbitrarily long distance, the probability of crossover is high enough that the inheritance of the genes is effectively uncorrelated.[53] For genes that are closer together, however, the lower probability of crossover means that the genes demonstrate genetic linkage; alleles for the two genes tend to be inherited together. The amounts of linkage between a series of genes can be combined to form a linear linkage map that roughly describes the arrangement of the genes along the chromosome.[54]

54.6 Gene expression

54.6.1 Genetic code

Main article: Genetic code
 Genes generally express their functional effect through the production of proteins, which are complex molecules responsible for most functions in the cell. Proteins are made up of one or more polypeptide chains, each of which is composed of a sequence of amino acids, and the DNA sequence of a gene (through an RNA intermediate) is used to produce a specific amino acid sequence. This process begins with the production of an RNA molecule with a sequence matching the gene's DNA sequence, a process called transcription.

This messenger RNA molecule is then used to produce a corresponding amino acid sequence through a process called translation. Each group of three nucleotides in the sequence, called a codon, corresponds either to one of the twenty possible amino acids in a protein or an instruction to end the amino acid sequence; this correspondence is called the genetic code.[55] The flow of information is unidirectional: information is transferred from nucleotide sequences into the amino acid sequence of proteins, but it never transfers from protein back into the sequence of DNA—a phenomenon Francis Crick called the central dogma of molecular biology.[56]

The specific sequence of amino acids results in a unique three-dimensional structure for that protein, and the three-dimensional structures of proteins are related to their functions.[57][58] Some are simple structural molecules, like the fibers formed by the protein collagen. Proteins can bind to other proteins and simple molecules, sometimes acting as enzymes by facilitating chemical reactions within the bound molecules (without changing the structure of the protein itself). Protein structure is dynamic; the protein hemoglobin bends into slightly different forms as it facilitates the capture, transport, and release of oxygen molecules within mammalian blood.

A single nucleotide difference within DNA can cause a change in the amino acid sequence of a protein. Because protein structures are the result of their amino acid sequences, some changes can dramatically change the properties of a protein by destabilizing the structure or changing the surface of the protein in a way that changes its interaction with other proteins and molecules. For example, sickle-cell anemia is a human genetic disease that results from a single base difference within the coding region for the β-globin section of hemoglobin, causing a single amino acid change that changes hemoglobin's physical properties.[59] Sickle-cell versions of hemoglobin stick to themselves, stacking to form fibers that distort the shape of red blood cells carrying the protein. These sickle-shaped cells no longer flow smoothly through blood vessels, having a tendency to clog or degrade, causing the medical problems associated with this disease.

Some DNA sequences are transcribed into RNA but are not translated into protein products—such RNA molecules are called non-coding RNA. In some cases, these products fold into structures which are involved in critical cell functions (e.g. ribosomal RNA and transfer RNA). RNA can also have regulatory effect through hybridization interactions with other RNA molecules (e.g. microRNA).

54.6.2 Nature and nurture

Main article: Nature and nurture

Although genes contain all the information an organism uses to function, the environment plays an important role in determining the ultimate phenotypes an organism displays. This is the complementary relationship often referred to as "nature and nurture". The phenotype of an organism depends on the interaction of genes and the environment. An interesting example is the coat coloration of the Siamese cat. In this case, the body temperature of the cat plays the role of the environment. The cat's genes code for dark hair, thus the hair producing cells in the cat make cellular proteins resulting in dark hair. But these dark hair-producing proteins are sensitive to temperature (i.e. have a mutation causing temperature-sensitivity) and denature in higher-temperature environments, failing to produce dark-hair pigment in areas where the cat has a higher body temperature. In a low-temperature environment, however, the protein's structure is stable and produces dark-hair pigment normally. The protein remains functional in areas of skin that are colder – such as its legs, ears, tail and face – so the cat has dark-hair at its extremities.[60]

Environment plays a major role in effects of the human genetic disease phenylketonuria.[61] The mutation that causes phenylketonuria disrupts the ability of the body to break down the amino acid phenylalanine, causing a toxic build-up of an intermediate molecule that, in turn, causes severe symptoms of progressive mental retardation and seizures. However, if someone with the phenylketonuria mutation follows a strict diet that avoids this amino acid, they remain normal and healthy.

A popular method in determining how genes and environment ("nature and nurture") contribute to a phenotype is by studying identical and fraternal twins or siblings of multiple births.[62] Because identical siblings come from the same zygote, they are genetically the same. Fraternal siblings are as genetically different from one another as normal siblings. By analyzing statistics on how often a twin of a set has a certain disorder compared to other sets of twins, scientists can determine whether that disorder is caused by genetic or environmental factors (i.e. whether it has 'nature' or 'nurture' causes). One famous example is the multiple birth study of the Genain quadruplets, who were identical quadruplets all diagnosed with schizophrenia.[63]

54.6.3 Gene regulation

Main article: Regulation of gene expression

The genome of a given organism contains thousands of genes, but not all these genes need to be active at any given moment. A gene is expressed when it is being transcribed into mRNA and there exist many cellular methods of controlling the expression of genes such that proteins are produced only when needed by the cell. Transcription factors are regulatory proteins that bind to DNA, either promoting or inhibiting the transcription of a gene.[64] Within the genome of *Escherichia coli* bacteria, for example, there exists a series of genes necessary for the synthesis of the amino acid tryptophan. However, when tryptophan is already available to the cell, these genes for tryptophan synthesis are no longer needed. The presence of tryptophan directly affects the activity of the genes—tryptophan molecules bind to the tryptophan repressor (a transcription factor), changing the repressor's structure such that the repressor binds to the genes. The tryptophan repressor blocks the transcription and expression of the genes, thereby creating negative feedback regulation of the tryptophan synthesis process.[65]

Differences in gene expression are especially clear within multicellular organisms, where cells all contain the same genome but have very different structures and behaviors due to the expression of different sets of genes. All the cells in a multicellular organism derive from a single cell, differentiating into variant cell types in response to external and intercellular signals and gradually establishing different patterns of gene expression to create different behaviors. As no single gene is responsible for the development of structures within multicellular organisms, these patterns arise from the complex interactions between many cells.

Within eukaryotes, there exist structural features of chromatin that influence the transcription of genes, often in the form of modifications to DNA and chromatin that are stably inherited by daughter cells.[66] These features are called "epigenetic" because they exist "on top" of the DNA sequence and retain inheritance from one cell generation to the next. Because of epigenetic features, different cell types grown within the same medium can retain very different properties. Although epigenetic features are generally dynamic over the course of development, some, like the phenomenon of paramutation, have multigenerational inheritance and exist as rare exceptions to the general rule of DNA as the basis for inheritance.[67]

54.7 Genetic change

54.7.1 Mutations

Main article: Mutation

During the process of DNA replication, errors occasionally occur in the polymerization of the second strand. These errors, called mutations, can have an impact on the phenotype of an organism, especially if they occur within the protein coding sequence of a gene. Error rates are usually very low—1 error in every 10–100 million bases—due to the "proofreading" ability of DNA polymerases.[68][69] Processes that increase the rate of changes in DNA are called mutagenic: mutagenic chemicals promote errors in DNA replication, often by interfering with the structure of base-pairing, while UV radiation induces mutations by causing damage to the DNA structure.[70] Chemical damage to DNA occurs naturally as well and cells use DNA repair mechanisms to repair mismatches and breaks. The repair does not, however, always restore the original sequence.

In organisms that use chromosomal crossover to exchange DNA and recombine genes, errors in alignment during meiosis can also cause mutations.[71] Errors in crossover are especially likely when similar sequences cause partner chromosomes to adopt a mistaken alignment; this makes some regions in genomes more prone to mutating in this way. These errors create large structural changes in DNA sequence – duplications, inversions, deletions of entire regions – or the accidental exchange of whole parts of sequences between different chromosomes (chromosomal translocation).

54.7.2 Natural selection and evolution

Main article: Evolution
Further information: Natural selection

Mutations alter an organism's genotype and occasionally this causes different phenotypes to appear. Most mutations have little effect on an organism's phenotype, health, or reproductive fitness.[72] Mutations that do have an effect are usually deleterious, but occasionally some can be beneficial.[73] Studies in the fly *Drosophila melanogaster* suggest that if a mutation changes a protein produced by a gene, about 70 percent of these mutations will be harmful with the remainder being either neutral or weakly beneficial.[74]

Population genetics studies the distribution of genetic differences within populations and how these distributions change over time.[75] Changes in the frequency of an allele in a population are mainly influenced by natural selection, where a given allele provides a selective or reproductive advantage to the organism,[76] as well as other factors such as mutation, genetic drift, genetic draft,[77] artificial selection and migration.[78]

Over many generations, the genomes of organisms can change significantly, resulting in evolution. In the process called adaptation, selection for beneficial mutations can cause a species to evolve into forms better able to survive in their environment.[79] New species are formed through the process of speciation, often caused by geographical separations that prevent populations from exchanging genes with each other.[80] The application of genetic principles to the study of population biology and evolution is known as the "modern synthesis".

By comparing the homology between different species' genomes, it is possible to calculate the evolutionary distance between them and when they may have diverged. Genetic comparisons are generally considered a more accurate method of characterizing the relatedness between species than the comparison of phenotypic characteristics. The evolutionary distances between species can be used to form evolutionary trees; these trees represent the common descent and divergence of species over time, although they do not show the transfer of genetic material between unrelated species (known as horizontal gene transfer and most common in bacteria).[81]

54.7.3   Model organisms

Although geneticists originally studied inheritance in a wide range of organisms, researchers began to specialize in studying the genetics of a particular subset of organisms. The fact that significant research already existed for a given organism would encourage new researchers to choose it for further study, and so eventually a few model organisms became the basis for most genetics research.[82] Common research topics in model organism genetics include the study of gene regulation and the involvement of genes in development and cancer.

Organisms were chosen, in part, for convenience—short generation times and easy genetic manipulation made some organisms popular genetics research tools. Widely used model organisms include the gut bacterium *Escherichia coli*, the plant *Arabidopsis thaliana*, baker's yeast (*Saccharomyces cerevisiae*), the nematode *Caenorhabditis elegans*, the common fruit fly (*Drosophila melanogaster*), and the common house mouse (*Mus musculus*).

54.7.4   Medicine

Medical genetics seeks to understand how genetic variation relates to human health and disease.[83] When searching for an unknown gene that may be involved in a disease, researchers commonly use genetic linkage and genetic pedigree charts to find the location on the genome associated with the disease. At the population level, researchers take advantage of Mendelian randomization to look for locations in the genome that are associated with diseases, a method especially useful for multigenic traits not clearly defined by a single gene.[84] Once a candidate gene is found, further research is often done on the corresponding gene – the orthologous gene – in model organisms. In addition to studying genetic diseases, the increased availability of genotyping methods has led to the field of pharmacogenetics: the study of how genotype can affect drug responses.[85]

Individuals differ in their inherited tendency to develop cancer,[86] and cancer is a genetic disease.[87] The process of cancer development in the body is a combination of events. Mutations occasionally occur within cells in the body as they divide. Although these mutations will not be inherited by any offspring, they can affect the behavior of cells, sometimes causing them to grow and divide more frequently. There are biological mechanisms that attempt to stop this process; signals are given to inappropriately dividing cells that should trigger cell death, but sometimes additional mutations occur that cause cells to ignore these messages. An internal process of natural selection occurs within the body and eventually mutations accumulate within cells to promote their own growth, creating a cancerous tumor that grows and invades various tissues of the body.

Normally, a cell divides only in response to signals called growth factors and stops growing once in contact with surrounding cells and in response to growth-inhibitory signals. It usually then divides a limited number of times and dies, staying within the epithelium where it is unable to migrate to other organs. To become a cancer cell, a cell has to accumulate mutations in a number of genes (3–7) that allow it to bypass this regulation: it no longer needs growth factors to divide, it continues growing when making contact to neighbor cells, and ignores inhibitory signals, it will keep growing indefinitely and is immortal, it will escape from the epithelium and ultimately may be able to escape from the primary tumor, cross the endothelium of a blood vessel, be transported by the bloodstream and will colonize a new organ, forming deadly metastasis. Although there are some genetic predispositions in a small fraction of cancers, the major fraction is due to a set of new genetic mutations that originally appear and accumulate in one or a small number of cells that will divide to form the tumor and are not transmitted to the progeny (somatic mutations). The most frequent mutations are a loss of function of p53 protein, a tumor suppressor, or in the p53 pathway, and gain of function mutations in the ras proteins, or in other oncogenes.

54.7.5   Research methods

DNA can be manipulated in the laboratory. Restriction enzymes are commonly used enzymes that cut DNA at specific sequences, producing predictable fragments of DNA.[88] DNA fragments can be visualized through use of gel electrophoresis, which separates fragments according to their length.

The use of ligation enzymes allows DNA fragments to be connected. By binding ("ligating") fragments of DNA together from different sources, researchers can create recombinant DNA, the DNA often associated with genetically modified organisms. Recombinant DNA is commonly used in the context of plasmids: short circular DNA molecules with a few genes on them. In the process known as molecular cloning, researchers can amplify the DNA fragments by inserting plasmids into bacteria and then culturing them on plates of agar (to isolate clones of bacteria cells). ("Cloning" can also refer to the various means of creating cloned ("clonal") organisms.)

DNA can also be amplified using a procedure called the polymerase chain reaction (PCR).[89] By using specific short sequences of DNA, PCR can isolate and exponentially amplify a targeted region of DNA. Because it can amplify from extremely small amounts of DNA, PCR is also often used to detect the presence of specific DNA sequences.

54.7.6 DNA sequencing and genomics

DNA sequencing, one of the most fundamental technologies developed to study genetics, allows researchers to determine the sequence of nucleotides in DNA fragments. The technique of chain-termination sequencing, developed in 1977 by a team led by Frederick Sanger, is still routinely used to sequence DNA fragments.[90] Using this technology, researchers have been able to study the molecular sequences associated with many human diseases.

As sequencing has become less expensive, researchers have sequenced the genomes of many organisms, using a process called genome assembly, which utilizes computational tools to stitch together sequences from many different fragments.[91] These technologies were used to sequence the human genome in the Human Genome Project completed in 2003.[33] New high-throughput sequencing technologies are dramatically lowering the cost of DNA sequencing, with many researchers hoping to bring the cost of resequencing a human genome down to a thousand dollars.[92]

Next generation sequencing (or high-throughput sequencing) came about due to the ever-increasing demand for low-cost sequencing. These sequencing technologies allow the production of potentially millions of sequences concurrently.[93][94] The large amount of sequence data available has created the field of genomics, research that uses computational tools to search for and analyze patterns in the full genomes of organisms. Genomics can also be considered a subfield of bioinformatics, which uses computational approaches to analyze large sets of biological data. A common problem to these fields of research is how to manage and share data that deals with human subject and personally identifiable information. See also genomics data sharing.

54.8   Society and culture

On 19 March 2015, a leading group of biologists urged a worldwide ban on clinical use of methods, particularly the use of CRISPR and zinc finger, to edit the human genome in a way that can be inherited.[95][96][97][98] In April 2015, Chinese researchers reported results of basic research to edit the DNA of non-viable human embryos using CRISPR.[99][100]

54.9   See also

- Bacterial genome size
- Eugenics
- Embryology
- Evolution
- Genetic disorder
- Genetic engineering
- Genetic enhancement
- Index of genetics articles

- Medical genetics

- Molecular tools for gene study

- Mutation

- Outline of genetics

- Timeline of the history of genetics

54.10   References

[1] Griffiths, Anthony J. F.; Miller, Jeffrey H.; Suzuki, David T.; Lewontin, Richard C.; Gelbart, eds. (2000). "Genetics and the Organism: Introduction". *An Introduction to Genetic Analysis* (7th ed.). New York: W. H. Freeman. ISBN 0-7167-3520-2.

[2] Hartl D, Jones E (2005)

[3] "Genetikos (γενετ-ικός)". *Henry George Liddell, Robert Scott, A Greek-English Lexicon.* Perseus Digital Library, Tufts University. Retrieved 20 February 2012.

[4] "Genesis (γένεσις)". *Henry George Liddell, Robert Scott, A Greek-English Lexicon.* Perseus Digital Library, Tufts University. Retrieved 20 February 2012.

[5] "Genetic". Online Etymology Dictionary. Retrieved 20 February 2012.

[6] DK Publishing (2009). *Science: The Definitive Visual Guide.* Penguin. p. 362. ISBN 978-0-7566-6490-9.

[7] Weiling, F (1991). "Historical study: Johann Gregor Mendel 1822–1884.". *American journal of medical genetics* **40** (1): 1–25; discussion 26. doi:10.1002/ajmg.1320400103. PMID 1887835.

[8] Matthew Hamilton (2011). *Population Genetics.* Georgetown University. p. 26. ISBN 978-1-4443-6245-9.

[9] Lamarck, J-B (2008). In Encyclopædia Britannica. Retrieved from Encyclopædia Britannica Online on 16 March 2008.

[10] Singer, Emily (4 February 2009). "A Comeback for Lamarckian Evolution?". *Technology Review.* Retrieved 14 March 2013.

[11] Peter J. Bowler, *The Mendelian Revolution: The Emergency of Hereditarian Concepts in Modern Science and Society* (Baltimore: Johns Hopkins University Press, 1989): chapters 2 & 3.

[12] Blumberg, Roger B. "Mendel's Paper in English".

[13] genetics, *n.*, Oxford English Dictionary, 3rd ed.

[14] Bateson W. "Letter from William Bateson to Alan Sedgwick in 1905". The John Innes Centre. Retrieved 15 March 2008. Note that the letter was to an Adam Sedgwick, a zoologist and "Reader in Animal Morphology" at Trinity College, Cambridge

[15] genetic, *adj.*, Oxford English Dictionary, 3rd ed.

[16] Richmond, Marsha L. (November 2007). "Opportunities for women in early genetics". *Nature Review Genetics* **8**: 897–902. doi:10.1038/nrg2200. Retrieved April 23, 2015.

[17] Bateson, W (1907). "The Progress of Genetic Research". In Wilks, W. *Report of the Third 1906 International Conference on Genetics: Hybridization (the cross-breeding of genera or species), the cross-breeding of varieties, and general plant breeding.* London: Royal Horticultural Society.

> Initially titled the "International Conference on Hybridisation and Plant Breeding", the title was changed as a result of Bateson's speech. See: Cock AG, Forsdyke DR (2008). *Treasure your exceptions: the science and life of William Bateson.* Springer. p. 248. ISBN 978-0-387-75687-5.

[18] Moore, John A. (1983). "Thomas Hunt Morgan—The Geneticist". *Integrative and Comparative Biology* **23** (4): 855–865. doi:10.1093/icb/23.4.855.

[19] Sturtevant AH (1913). "The linear arrangement of six sex-linked factors in Drosophila, as shown by their mode of association" (PDF). *Journal of Experimental Biology* **14**: 43–59. doi:10.1002/jez.1400140104.

[20] Avery, OT; MacLeod, CM; McCarty, M (1944). "Studies on the Chemical Nature of the Substance Inducing Transformation of Pneumococcal Types: Induction of Transformation by a Desoxyribonucleic Acid Fraction Isolated from Pneumococcus Type III". *The Journal of experimental medicine* **79** (2): 137–58. doi:10.1084/jem.79.2.137. PMC 2135445. PMID 19871359. Reprint: Avery, OT; MacLeod, CM; McCarty, M (1979). "Studies on the chemical nature of the substance inducing transformation of pneumococcal types. Inductions of transformation by a desoxyribonucleic acid fraction isolated from pneumococcus type III". *The Journal of experimental medicine* **149** (2): 297–326. doi:10.1084/jem.149.2.297. PMC 2184805. PMID 33226.

[21] Cell and Molecular Biology", Pragya Khanna. I. K. International Pvt Ltd, 2008. p. 221. ISBN 81-89866-59-1, ISBN 978-81-89866-59-4

[22] Hershey, AD; Chase, M (1952). "Independent functions of viral protein and nucleic acid in growth of bacteriophage". *The Journal of General Physiology* **36** (1): 39–56. doi:10.1085/jgp.36.1.39. PMC 2147348. PMID 12981234.

[23] Judson, Horace (1979). *The Eighth Day of Creation: Makers of the Revolution in Biology*. Cold Spring Harbor Laboratory Press. pp. 51–169. ISBN 0-87969-477-7.

[24] Watson, J. D.; Crick, FH (1953). "Molecular Structure of Nucleic Acids: A Structure for Deoxyribose Nucleic Acid" (PDF). *Nature* **171** (4356): 737–8. Bibcode:1953Natur.171..737W. doi:10.1038/171737a0. PMID 13054692.

[25] Watson, J. D.; Crick, FH (1953). "Genetical Implications of the Structure of Deoxyribonucleic Acid" (PDF). *Nature* **171** (4361): 964–7. Bibcode:1953Natur.171..964W. doi:10.1038/171964b0. PMID 13063483.

[26] Stratmann, S. A. (1 Nov 2013). "DNA replication at the single molecule level". *Chemical Society Reviews* **43** (4): 1201–20. doi:10.1039/c3cs60391a. PMID 24395040.

[27] Frederick Betz (2010). *Managing Science: Methodology and Organization of Research*. Springer. p. 76. ISBN 978-1-4419-7488-4.

[28] Stanley A. Rice (2009). *Encyclopedia of Evolution*. Infobase Publishing. p. 134. ISBN 978-1-4381-1005-9.

[29] Sahotra Sarkar (1998). *Genetics and Reductionism*. Cambridge University Press. p. 140. ISBN 978-0-521-63713-8.

[30] Ohta, Tomoko (1973). "Slightly Deleterious Mutant Substitutions in Evolution". *Nature* **246** (5428): 96–98. Bibcode:1973Natur.246...96O. doi:10.1038/246096a0. PMID 4585855.

[31] Sanger, F; Nicklen, S; Coulson, AR (1977). "DNA sequencing with chain-terminating inhibitors". *Proceedings of the National Academy of Sciences of the United States of America* **74** (12): 5463–7. Bibcode:1977PNAS...74.5463S. doi:10.1073/pnas.74.12.5463. PMC 431765. PMID 271968.

[32] Saiki, RK; Scharf, S; Faloona, F; Mullis, KB; Horn, GT; Erlich, HA; Arnheim, N (1985). "Enzymatic amplification of beta-globin genomic sequences and restriction site analysis for diagnosis of sickle cell anemia". *Science* **230** (4732): 1350–4. Bibcode:1985Sci...230.1350S. doi:10.1126/science.2999980. PMID 2999980.

[33] "Human Genome Project Information". Human Genome Project. Retrieved 15 March 2008.

[34] Griffiths, Anthony J. F.; Miller, Jeffrey H.; Suzuki, David T.; Lewontin, Richard C.; Gelbart, eds. (2000). "Patterns of Inheritance: Introduction". *An Introduction to Genetic Analysis* (7th ed.). New York: W. H. Freeman. ISBN 0-7167-3520-2.

[35] Griffiths, Anthony J. F.; Miller, Jeffrey H.; Suzuki, David T.; Lewontin, Richard C.; Gelbart, eds. (2000). "Mendel's experiments". *An Introduction to Genetic Analysis* (7th ed.). New York: W. H. Freeman. ISBN 0-7167-3520-2.

[36] Griffiths, Anthony J. F.; Miller, Jeffrey H.; Suzuki, David T.; Lewontin, Richard C.; Gelbart, eds. (2000). "Mendelian genetics in eukaryotic life cycles". *An Introduction to Genetic Analysis* (7th ed.). New York: W. H. Freeman. ISBN 0-7167-3520-2.

[37] Griffiths, Anthony J. F.; Miller, Jeffrey H.; Suzuki, David T.; Lewontin, Richard C.; Gelbart, eds. (2000). "Interactions between the alleles of one gene". *An Introduction to Genetic Analysis* (7th ed.). New York: W. H. Freeman. ISBN 0-7167-3520-2.

[38] Cheney, Richard W. "Genetic Notation". Archived from the original on 3 January 2008. Retrieved 18 March 2008.

[39] Griffiths, Anthony J. F.; Miller, Jeffrey H.; Suzuki, David T.; Lewontin, Richard C.; Gelbart, eds. (2000). "Human Genetics". *An Introduction to Genetic Analysis* (7th ed.). New York: W. H. Freeman. ISBN 0-7167-3520-2.

[40] Griffiths, Anthony J. F.; Miller, Jeffrey H.; Suzuki, David T.; Lewontin, Richard C.; Gelbart, eds. (2000). "Gene interaction and modified dihybrid ratios". *An Introduction to Genetic Analysis* (7th ed.). New York: W. H. Freeman. ISBN 0-7167-3520-2.

[41] Mayeux, R (2005). "Mapping the new frontier: complex genetic disorders". *The Journal of Clinical Investigation* **115** (6): 1404–7. doi:10.1172/JCI25421. PMC 1137013. PMID 15931374.

[42] Griffiths, Anthony J. F.; Miller, Jeffrey H.; Suzuki, David T.; Lewontin, Richard C.; Gelbart, eds. (2000). "Quantifying heritability". *An Introduction to Genetic Analysis* (7th ed.). New York: W. H. Freeman. ISBN 0-7167-3520-2.

[43] Luke, A; Guo, X; Adeyemo, AA; Wilks, R; Forrester, T; Lowe Jr, W; Comuzzie, AG; Martin, LJ; Zhu, X; Rotimi, CN; Cooper, RS (2001). "Heritability of obesity-related traits among Nigerians, Jamaicans and US black people". *International journal of obesity and related metabolic disorders* **25** (7): 1034–41. doi:10.1038/sj.ijo.0801650. PMID 11443503.

[44] Pearson, H (2006). "Genetics: what is a gene?". *Nature* **441** (7092): 398–401. Bibcode:2006Natur.441..398P. doi:10.1038/441398a. PMID 16724031.

[45] Prescott, L (1993). *Microbiology*. Wm. C. Brown Publishers. ISBN 0-697-01372-3.

[46] Griffiths, Anthony J. F.; Miller, Jeffrey H.; Suzuki, David T.; Lewontin, Richard C.; Gelbart, eds. (2000). "Mechanism of DNA Replication". *An Introduction to Genetic Analysis* (7th ed.). New York: W. H. Freeman. ISBN 0-7167-3520-2.

[47] Gregory, SG; Barlow, KF; McLay, KE; Kaul, R; Swarbreck, D; Dunham, A; Scott, CE; Howe, KL; et al. (2006). "The DNA sequence and biological annotation of human chromosome 1". *Nature* **441** (7091): 315–21. Bibcode:2006Natur.441..315G. doi:10.1038/nature04727. PMID 16710414.

[48] Alberts et al. (2002), II.4. DNA and chromosomes: Chromosomal DNA and Its Packaging in the Chromatin Fiber

[49] Griffiths, Anthony J. F.; Miller, Jeffrey H.; Suzuki, David T.; Lewontin, Richard C.; Gelbart, eds. (2000). "Sex chromosomes and sex-linked inheritance". *An Introduction to Genetic Analysis* (7th ed.). New York: W. H. Freeman. ISBN 0-7167-3520-2.

[50] Griffiths, Anthony J. F.; Miller, Jeffrey H.; Suzuki, David T.; Lewontin, Richard C.; Gelbart, eds. (2000). "Bacterial conjugation". *An Introduction to Genetic Analysis* (7th ed.). New York: W. H. Freeman. ISBN 0-7167-3520-2.

[51] Griffiths, Anthony J. F.; Miller, Jeffrey H.; Suzuki, David T.; Lewontin, Richard C.; Gelbart, eds. (2000). "Bacterial transformation". *An Introduction to Genetic Analysis* (7th ed.). New York: W. H. Freeman. ISBN 0-7167-3520-2.

[52] Griffiths, Anthony J. F.; Miller, Jeffrey H.; Suzuki, David T.; Lewontin, Richard C.; Gelbart, eds. (2000). "Nature of crossing-over". *An Introduction to Genetic Analysis* (7th ed.). New York: W. H. Freeman. ISBN 0-7167-3520-2.

[53] Jack E. Staub (1994). *Crossover: Concepts and Applications in Genetics, Evolution, and Breeding*. University of Wisconsin Press. p. 55. ISBN 978-0-299-13564-5.

[54] Griffiths, Anthony J. F.; Miller, Jeffrey H.; Suzuki, David T.; Lewontin, Richard C.; Gelbart, eds. (2000). "Linkage maps". *An Introduction to Genetic Analysis* (7th ed.). New York: W. H. Freeman. ISBN 0-7167-3520-2.

[55] Berg JM, Tymoczko JL, Stryer L, Clarke ND (2002). "I. 5. DNA, RNA, and the Flow of Genetic Information: Amino Acids Are Encoded by Groups of Three Bases Starting from a Fixed Point". *Biochemistry* (5th ed.). New York: W. H. Freeman and Company.

[56] Crick, F (1970). "Central dogma of molecular biology" (PDF). *Nature* **227** (5258): 561–3. Bibcode:1970Natur.227..561C. doi:10.1038/227561a0. PMID 4913914.

[57] Alberts et al. (2002), I.3. Proteins: The Shape and Structure of Proteins

[58] Alberts et al. (2002), I.3. Proteins: Protein Function

[59] "How Does Sickle Cell Cause Disease?". Brigham and Women's Hospital: Information Center for Sickle Cell and Thalassemic Disorders. 11 April 2002. Retrieved 23 July 2007.

[60] Imes, DL; Geary, LA; Grahn, RA; Lyons, LA (2006). "Albinism in the domestic cat (*Felis catus*) is associated with a tyrosinase (TYR) mutation". *Animal genetics* **37** (2): 175–8. doi:10.1111/j.1365-2052.2005.01409.x. PMC 1464423. PMID 16573534.

[61] "MedlinePlus: Phenylketonuria". NIH: National Library of Medicine. Retrieved 15 March 2008.

[62] For example, Ridley M (2003). *Nature via nurture: genes, experience and what makes us human*. Fourth Estate. p. 73. ISBN 978-1-84115-745-0.

[63] Rosenthal, David (1964). *The Genain quadruplets: a case study and theoretical analysis of heredity and environment in schizophrenia*. New York: Basic Books. doi:10.1002/bs.3830090407.

[64] Brivanlou, AH; Darnell Jr, JE (2002). "Signal transduction and the control of gene expression". *Science* **295** (5556): 813–8. Bibcode:2002Sci...295..813B. doi:10.1126/science.1066355. PMID 11823631.

[65] Alberts et al. (2002), II.3. Control of Gene Expression – The Tryptophan Repressor Is a Simple Switch That Turns Genes On and Off in Bacteria

[66] Jaenisch, R; Bird, A (2003). "Epigenetic regulation of gene expression: how the genome integrates intrinsic and environmental signals". *Nature Genetics*. 33 Suppl (3s): 245–54. doi:10.1038/ng1089. PMID 12610534.

[67] Chandler, VL (2007). "Paramutation: from maize to mice". *Cell* **128** (4): 641–5. doi:10.1016/j.cell.2007.02.007. PMID 17320501.

[68] Griffiths, Anthony J. F.; Miller, Jeffrey H.; Suzuki, David T.; Lewontin, Richard C.; Gelbart, eds. (2000). "Spontaneous mutations". *An Introduction to Genetic Analysis* (7th ed.). New York: W. H. Freeman. ISBN 0-7167-3520-2.

[69] Freisinger, E; Grollman, AP; Miller, H; Kisker, C (2004). "Lesion (in)tolerance reveals insights into DNA replication fidelity". *The EMBO Journal* **23** (7): 1494–505. doi:10.1038/sj.emboj.7600158. PMC 391067. PMID 15057282.

[70] Griffiths, Anthony J. F.; Miller, Jeffrey H.; Suzuki, David T.; Lewontin, Richard C.; Gelbart, eds. (2000). "Induced mutations". *An Introduction to Genetic Analysis* (7th ed.). New York: W. H. Freeman. ISBN 0-7167-3520-2.

[71] Griffiths, Anthony J. F.; Miller, Jeffrey H.; Suzuki, David T.; Lewontin, Richard C.; Gelbart, eds. (2000). "Chromosome Mutation I: Changes in Chromosome Structure: Introduction". *An Introduction to Genetic Analysis* (7th ed.). New York: W. H. Freeman. ISBN 0-7167-3520-2.

[72] Moselio Schaechter (2009). *Encyclopedia of Microbiology*. Academic Press. p. 551. ISBN 978-0-12-373944-5.

[73] Mike Calver; Alan Lymbery; Jennifer McComb; Mike Bamford (2009). *Environmental Biology*. Cambridge University Press. p. 118. ISBN 978-0-521-67982-4.

[74] Sawyer, SA; Parsch, J; Zhang, Z; Hartl, DL (2007). "Prevalence of positive selection among nearly neutral amino acid replacements in Drosophila". *Proceedings of the National Academy of Sciences of the United States of America* **104** (16): 6504–10. Bibcode:2007PNAS..104.6504S. doi:10.1073/pnas.0701572104. PMC 1871816. PMID 17409186.

[75] Griffiths, Anthony J. F.; Miller, Jeffrey H.; Suzuki, David T.; Lewontin, Richard C.; Gelbart, eds. (2000). "Variation and its modulation". *An Introduction to Genetic Analysis* (7th ed.). New York: W. H. Freeman. ISBN 0-7167-3520-2.

[76] Griffiths, Anthony J. F.; Miller, Jeffrey H.; Suzuki, David T.; Lewontin, Richard C.; Gelbart, eds. (2000). "Selection". *An Introduction to Genetic Analysis* (7th ed.). New York: W. H. Freeman. ISBN 0-7167-3520-2.

[77] Gillespie, John H. (2001). "Is the population size of a species relevant to its evolution?". *Evolution* **55** (11): 2161–2169. doi:10.1111/j.0014-3820.2001.tb00732.x. PMID 11794777.

[78] Griffiths, Anthony J. F.; Miller, Jeffrey H.; Suzuki, David T.; Lewontin, Richard C.; Gelbart, eds. (2000). "Random events". *An Introduction to Genetic Analysis* (7th ed.). New York: W. H. Freeman. ISBN 0-7167-3520-2.

[79] Darwin, Charles (1859). *On the Origin of Species* (1st ed.). London: John Murray. p. 1. ISBN 0-8014-1319-2.
Earlier related ideas were acknowledged in Darwin, Charles (1861). *On the Origin of Species* (3rd ed.). London: John Murray. xiii. ISBN 0-8014-1319-2.

[80] Gavrilets, S (2003). "Perspective: models of speciation: what have we learned in 40 years?". *Evolution; international journal of organic evolution* **57** (10): 2197–215. doi:10.1554/02-727. PMID 14628909.

[81] Wolf, YI; Rogozin, IB; Grishin, NV; Koonin, EV (2002). "Genome trees and the tree of life". *Trends in genetics* **18** (9): 472–9. doi:10.1016/S0168-9525(02)02744-0. PMID 12175808.

[82] "The Use of Model Organisms in Instruction". University of Wisconsin: Wisconsin Outreach Research Modules. Retrieved 15 March 2008.

[83] "NCBI: Genes and Disease". NIH: National Center for Biotechnology Information. Retrieved 15 March 2008.

[84] Davey Smith, G; Ebrahim, S (2003). "'Mendelian randomization': can genetic epidemiology contribute to understanding environmental determinants of disease?". *International Journal of Epidemiology* **32** (1): 1–22. doi:10.1093/ije/dyg070. PMID 12689998.

[85] "Pharmacogenetics Fact Sheet". NIH: National Institute of General Medical Sciences. Retrieved 15 March 2008.

[86] Frank, SA (2004). "Genetic predisposition to cancer – insights from population genetics". *Nature reviews. Genetics* **5** (10): 764–72. doi:10.1038/nrg1450. PMID 15510167.

[87] Strachan T, Read AP (1999). *Human Molecular Genetics 2* (second ed.). John Wiley & Sons Inc. Chapter 18: Cancer Genetics

[88] Lodish et al. (2000), Chapter 7: 7.1. DNA Cloning with Plasmid Vectors

[89] Lodish et al. (2000), Chapter 7: 7.7. Polymerase Chain Reaction: An Alternative to Cloning

[90] Brown TA (2002). "Section 2, Chapter 6: 6.1. The Methodology for DNA Sequencing". *Genomes 2* (2nd ed.). Oxford: Bios. ISBN 1-85996-228-9.

[91] Brown (2002), Section 2, Chapter 6: 6.2. Assembly of a Contiguous DNA Sequence

[92] Service, RF (2006). "Gene sequencing. The race for the $1000 genome". *Science* **311** (5767): 1544–6. doi:10.1126/science.311.5767.1544. PMID 16543431.

[93] Hall, Nell (May 2007). "Advanced sequencing technologies and their wider impact in microbiology". *J. Exp. Biol.* **209** (Pt 9): 1518–1525. doi:10.1242/jeb.001370. PMID 17449817.

[94] Church, George M. (January 2006). "Genomes for all". *Sci. Am.* **294** (1): 46–54. doi:10.1038/scientificamerican0106-46. PMID 16468433.(subscription required)

[95] Wade, Nicholas (19 March 2015). "Scientists Seek Ban on Method of Editing the Human Genome". *New York Times*. Retrieved 20 March 2015.

[96] Pollack, Andrew (3 March 2015). "A Powerful New Way to Edit DNA". *New York Times*. Retrieved 20 March 2015.

[97] Baltimore, David; Berg, Paul; Botchan, Dana; Charo, R. Alta; Church, George; Corn, Jacob E.; Daley, George Q.; Doudna, Jennifer A.; Fenner, Marsha; Greely, Henry T.; Jinek, Martin; Martin, G. Steven; Penhoet, Edward; Puck, Jennifer; Sternberg, Samuel H.; Weissman, Jonathan S.; Yamamoto, Keith R. (19 March 2015). "A prudent path forward for genomic engineering and germline gene modification". *Science* **348**: 36–8. Bibcode:2015Sci...348...36B. doi:10.1126/science.aab1028. PMID 25791083. Retrieved 20 March 2015.

[98] Lanphier, Edward; Urnov, Fyodor; Haecker, Sarah Ehlen; Werner, Michael; Smolenski, Joanna (26 March 2015). "Don't edit the human germ line". *Nature (journal)* **519**: 410–411. Bibcode:2015Natur.519..410L. doi:10.1038/519410a. PMID 25810189. Retrieved 20 March 2015.

[99] Kolata, Gina (23 April 2015). "Chinese Scientists Edit Genes of Human Embryos, Raising Concerns". *New York Times*. Retrieved 24 April 2015.

[100] Liang, Puping; et al. (18 April 2015). "CRISPR/Cas9-mediated gene editing in human tripronuclear zygotes". *Protein & Cell* **6**: 363–72. doi:10.1007/s13238-015-0153-5. PMC 4417674. PMID 25894090. Retrieved 24 April 2015.

54.11 Further reading

See also: Bibliography of biology § Genetics

- Bruce Alberts; Dennis Bray; Karen Hopkin; Alexander Johnson; Julian Lewis; Martin Raff; Keith Roberts; Peter Walter (2013). *Essential Cell Biology, 4th Edition*. Garland Science. ISBN 978-1-317-80627-1.

- Griffiths, Anthony J. F.; Miller, Jeffrey H.; Suzuki, David T.; Lewontin, Richard C.; Gelbart, eds. (2000). *An Introduction to Genetic Analysis* (7th ed.). New York: W. H. Freeman. ISBN 0-7167-3520-2.

- Hartl D, Jones E (2005). *Genetics: Analysis of Genes and Genomes* (6th ed.). Jones & Bartlett. ISBN 0-7637-1511-5.

- King, Robert C; Mulligan, Pamela K; Stansfield, William D (2013). *A Dictionary of Genetics* (8th ed.). New York: Oxford University Press. ISBN 0-1997-6644-4.

- Lodish H, Berk A, Zipursky LS, Matsudaira P, Baltimore D, and Darnell J (2000). *Molecular Cell Biology* (4th ed.). New York: Scientific American Books. ISBN 0-7167-3136-3.

54.12 External links

-

- Genetics on *In Our Time* at the BBC. (listen now)

- Genetics at DMOZ

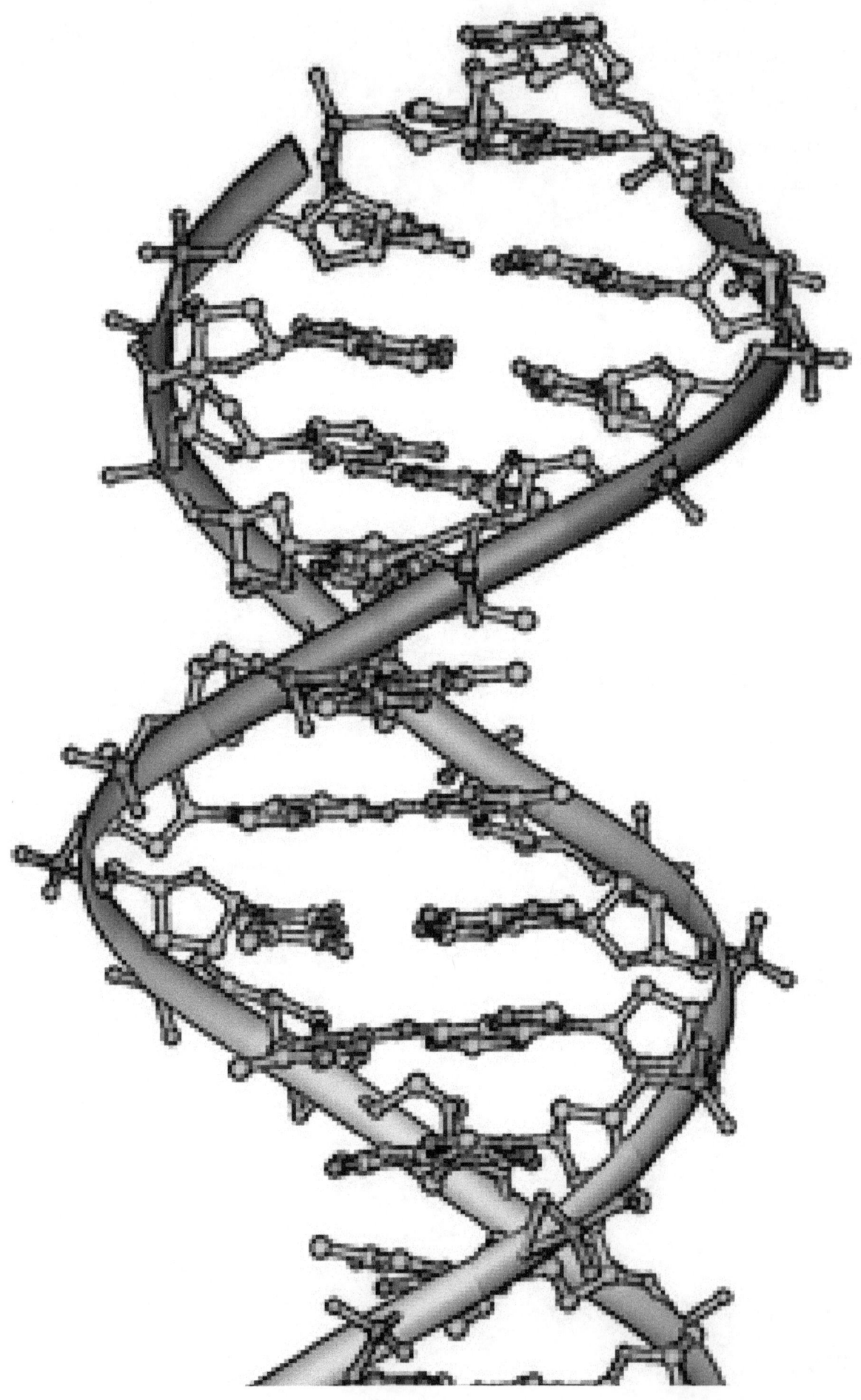

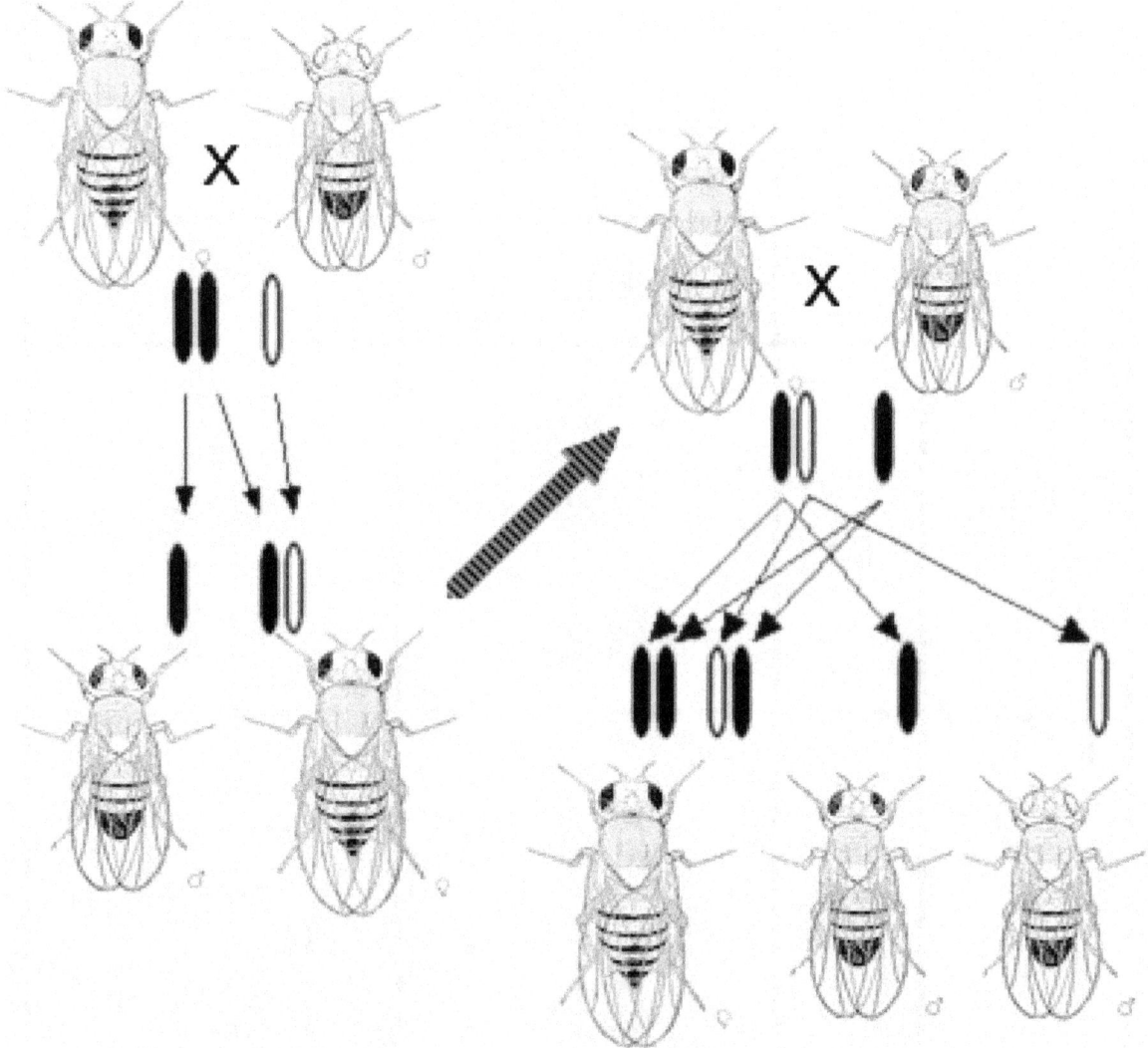

Morgan's observation of sex-linked inheritance of a mutation causing white eyes in Drosophila *led him to the hypothesis that genes are located upon chromosomes.*

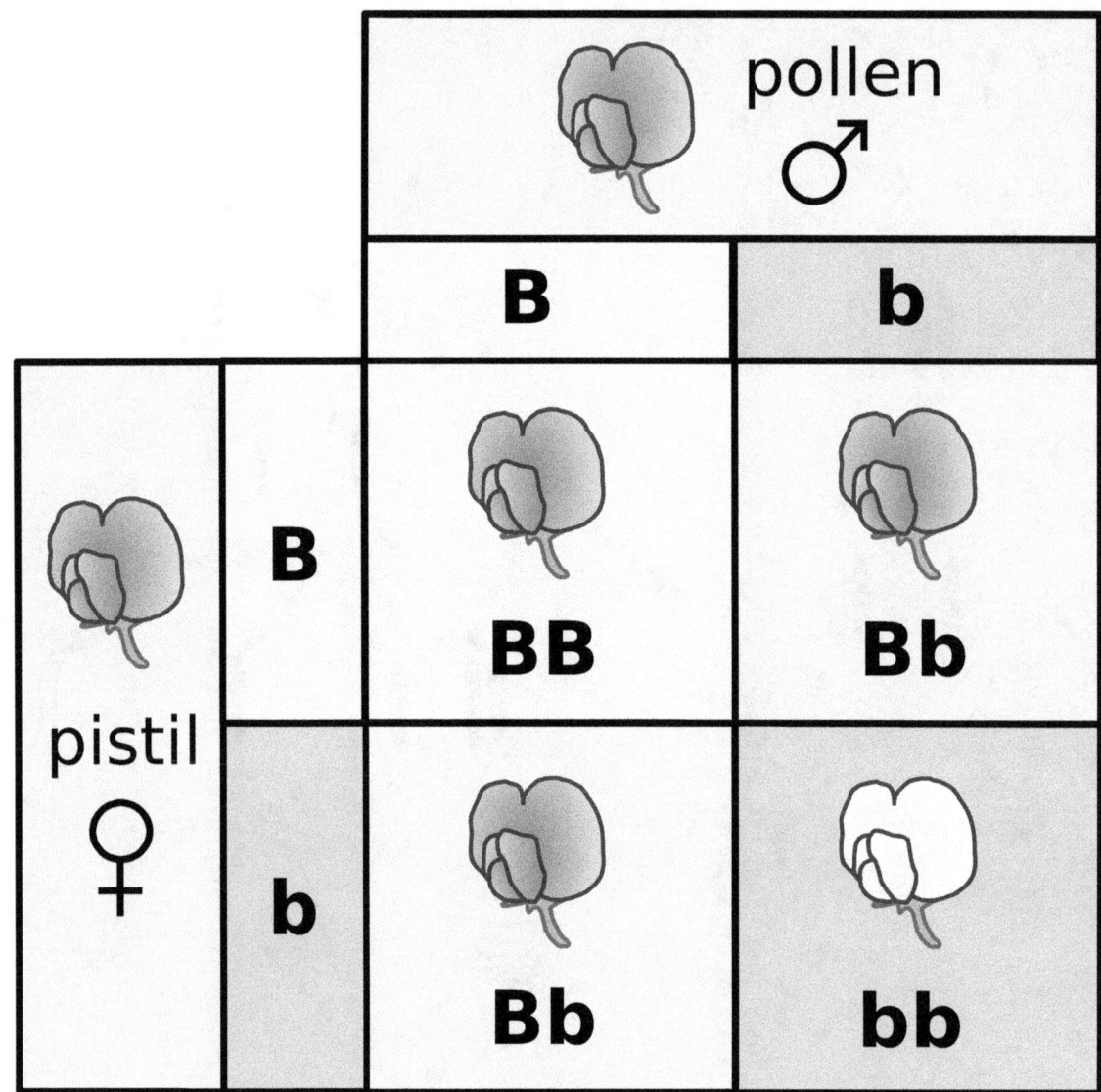

A Punnett square depicting a cross between two pea plants heterozygous for purple (B) and white (b) blossoms.

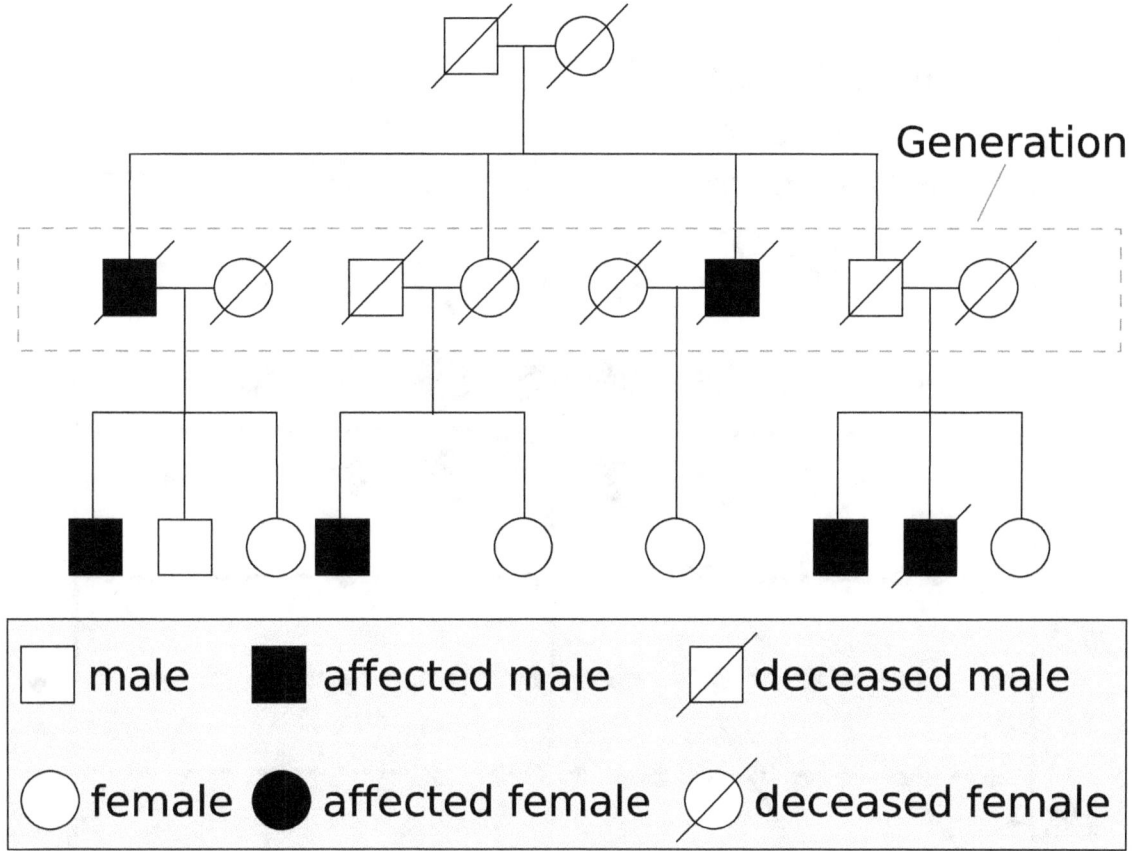

Genetic pedigree charts help track the inheritance patterns of traits.

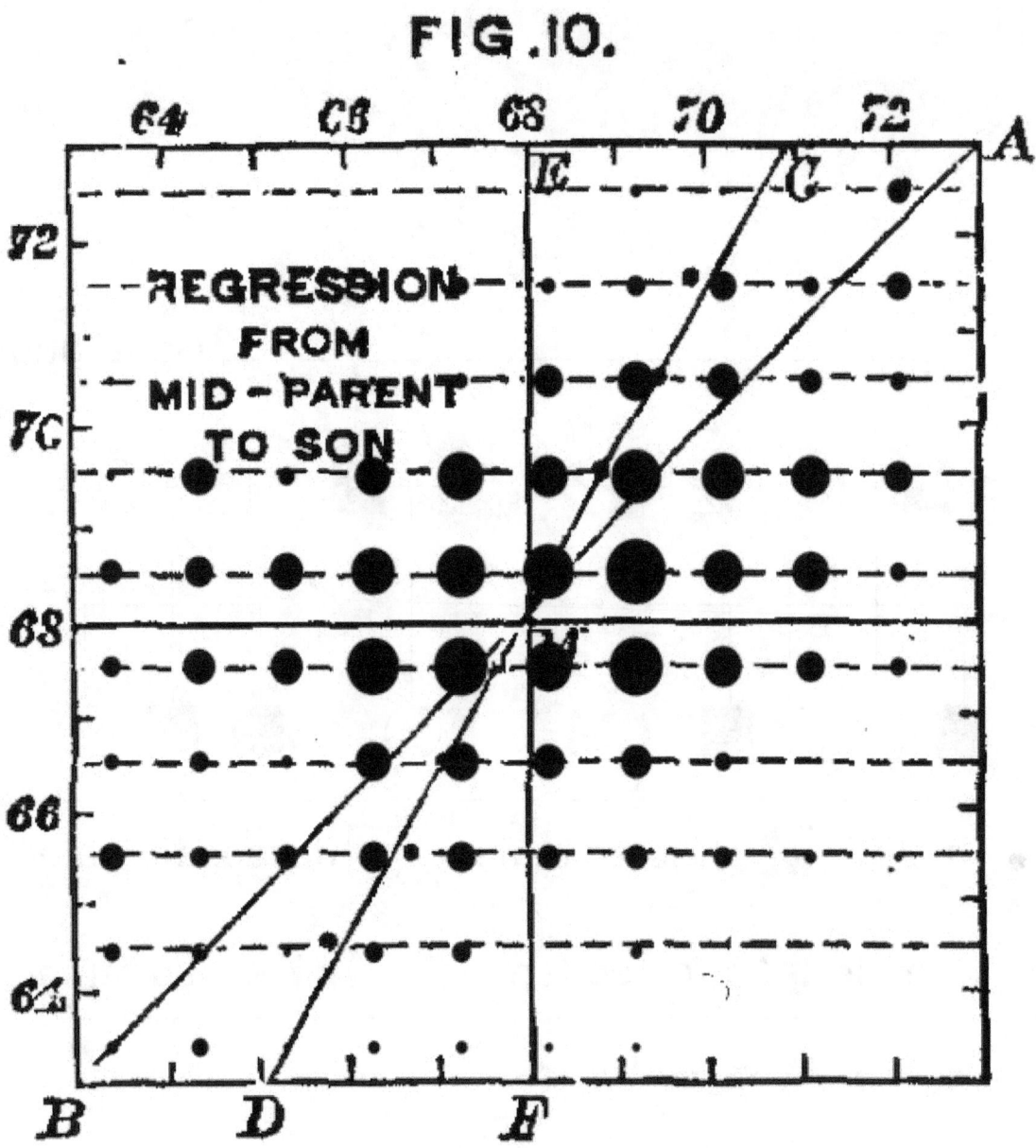

Human height is a trait with complex genetic causes. Francis Galton's data from 1889 shows the relationship between offspring height as a function of mean parent height. While correlated, remaining variation in offspring heights indicates environment is also an important factor in this trait.

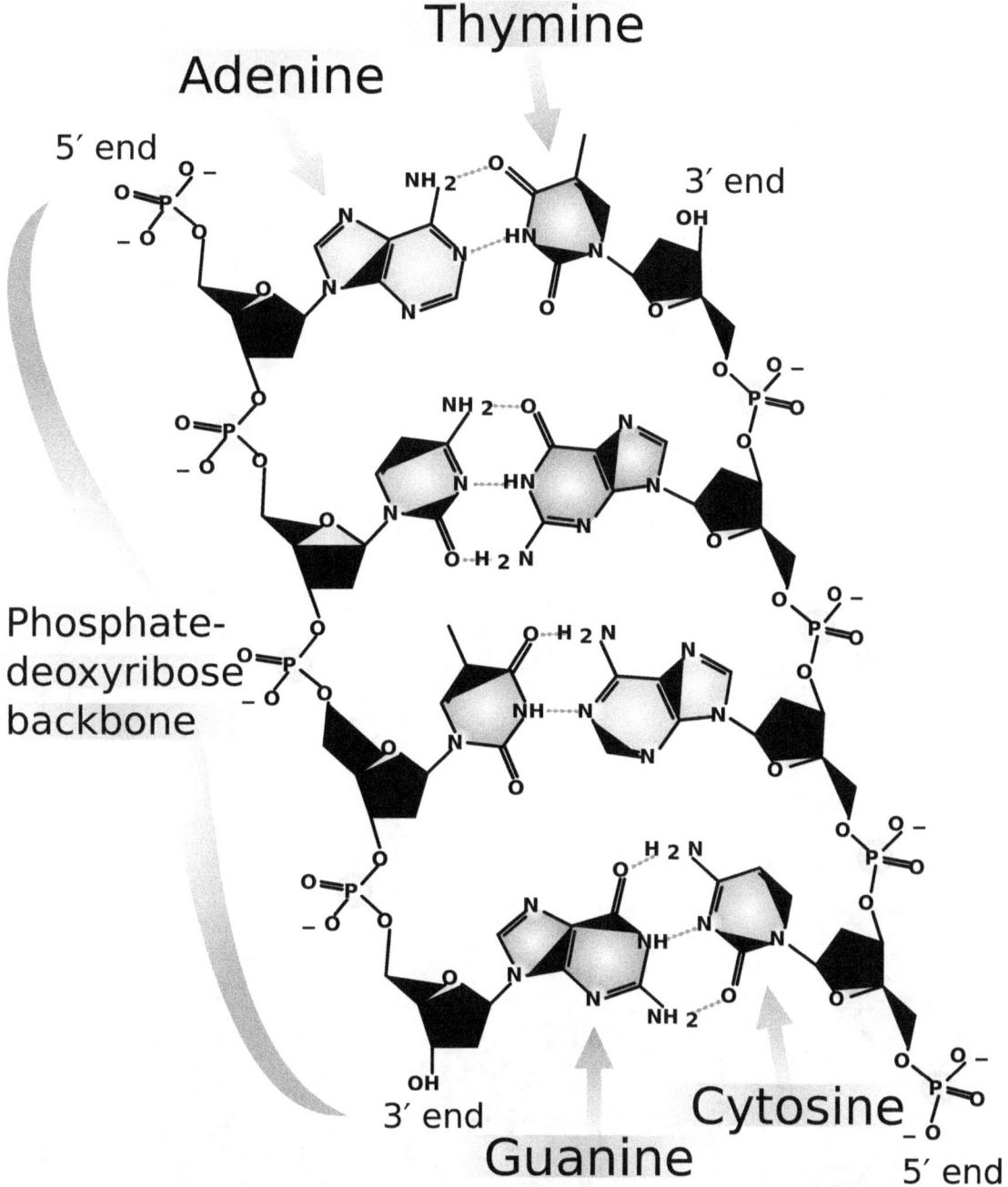

The molecular structure of DNA. Bases pair through the arrangement of hydrogen bonding between the strands.

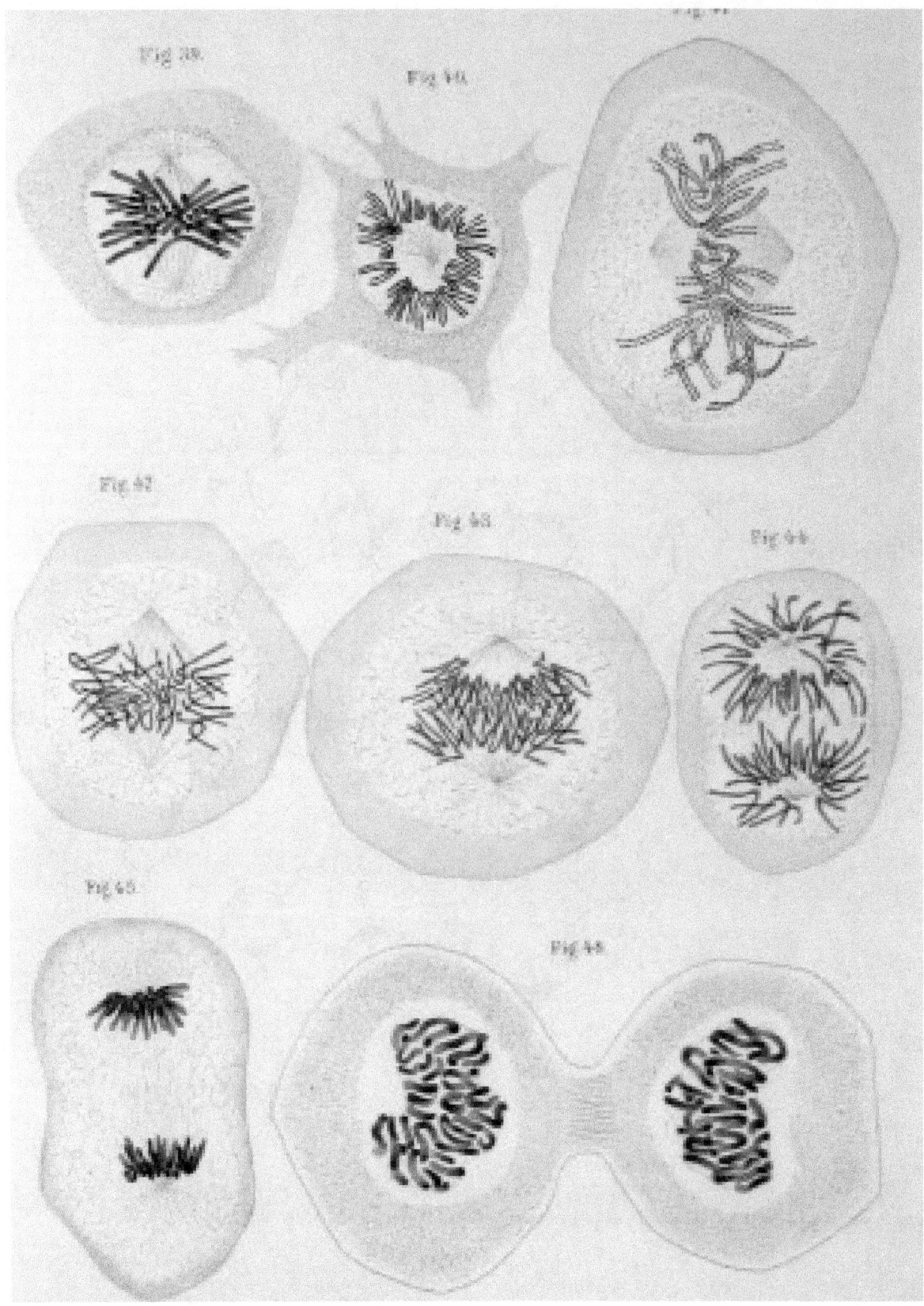

Walther Flemming's 1882 diagram of eukaryotic cell division. Chromosomes are copied, condensed, and organized. Then, as the cell divides, chromosome copies separate into the daughter cells.